A DICTIONARY OF
Environmental
& Civil Engineering

This one's for Engel

A DICTIONARY OF
Environmental
& Civil Engineering

Len Webster

The Parthenon Publishing Group
International Publishers in Medicine, Science & Technology

NEW YORK   LONDON

Published in the UK by
The Parthenon Publishing Group Limited
Casterton Hall
Carnforth, Lancs. LA6 2LA, UK

Published in the USA by
The Parthenon Publishing Group Inc.
One Blue Hill Plaza
PO Box 1564, Pearl River
New York 10965, USA

Copyright © 2000 Len Webster

British Library Cataloguing in Publication Data

Webster, L. F. (Len F.), 1929–
Dictionary of civil and environmental engineering
1. Environmental engineering – Dictionaries
2. Environmental sciences – Dictionaries
I. Title
628′.03

ISBN 1-85070-075-3

Library of Congress Cataloging-in-Publication Data

Dictionary of civil and environmental engineering /edited by L. F.
    Webster
        p.   cm.
        ISBN 1-85070-075-3
        1. Civil engineering Dictionaries.   2. Environmental engineering
    Dictionaries.   3. Ecology Dictionaries.   I. Webster, L. F.
    (Len F.), 1929–
    TA9.D46 1999
    624'.03—dc21

                                            99–16958
                                                CIP

Design and phototypesetting: Len Webster.
Printed and bound by Bookcraft (Bath) Ltd., Midsomer Norton, UK

PREFACE

A mere three decades ago, this dictionary would have been half its present size and covered a significantly less range of subjects under the umbrella term 'the environment'.

Global warming, resource conservation, urban pollution, solid waste disposal and recycling — these and a dozen more subjects are the daily diet of news reporters and commentators. Simultaneously, the proliferation of new terms by which to describe the constantly emerging environmental technology continues unabated.

This dictionary is the result of almost forty years accumulation of references ranging from scraps of paper to full-blown glossaries. Beginning as editor of a now-defunct magazine called *Canadian Municipal Utilities* and on through a range of other publications — a number of which have been dedicated to water supply, pollution control, environmental protection, etc. — the author has accumulated terms and definitions whenever they became available. Added to those are the thousands contributed by companies, associations, government agencies, professionals and, by no means last, friends who have taken an interest in the project.

The major focus of this work are the mechanical aspects of environmental engineering, and in particular the machinery and equipment that is used. All those years ago, the limits of this interest embraced the municipal supply of potable water and the treatment of solid and liquid wastes. Industry did whatever it felt like doing, or not doing. In the forests, nobody was hugging trees, and most people regarded mining as a staple of the economy — and left it at that. Smog was a slang term only just invented (and it was limited to the Los Angeles basin, wasn't it?). Remember, nuclear power stations were only then being imagined, and lead-free gasoline had still to be invented.

In all, this dictionary contains more than 15,000 terms. That number could be a lot bigger if some of the more esoteric aspects of industrial pollution control were to be included. But maybe that is for another work, or another time.

AASHTO (soil compaction) TEST - T-99
See **Soil compaction test.**

AASHTO (soil compaction) TEST - T-180
See **Soil compaction test.**

AASHTO SOIL CLASSIFICATION SYS-TEM *See* **Soil classification systems.**

AB Prefix used in the centimeter-gram-sec-ond (CGS) system of measurement and de-noting a practical quantification of a theo-retical electromagnetic unit. It links the CGS system with the SI unitary system, making conversion from one to the other unneces-sary.

ABAMPERE Unit of electric current in the centimeter-gram-second (CGS) system of measurement equal to the current, flowing through two infinitely long parallel conduc-tors, 1 cm apart, that produces a force of 2 dynes per cm between them. 1 abampere is equal to 10 amperes of the SI system.

ABANDONED AREA Deserted area in which work has ceased and in which further work is not intended.

ABANDONED MINE Mine where opera-tions have occurred in the past and where either the applicable reclamation bond or financial assurance has been released or for-feited or, if no reclamation bond or other financial assurance has been posted, no min-ing operations have occurred for five years or more.

ABANDONED WELL Well whose use has

been permanently discontinued or which is in a state of disrepair such that it cannot be used for its intended purpose or for observa-tion purposes.

ABANDONMENT 1. Relinquishing of the public interest in the right to use water for any purpose, or the method of using such water, or of a right of way or activity thereon with no intention to reclaim or use again for highway (or transportation) purposes. *Also called* **Vacation**; **2.** Cessation of work by a contractor and removal of all equipment and uninstalled materials from site.

ABATE Reduce in amount, degree, or in-tensity.

ABATEMENT Reduction in environmen-tal release of pollutants through improved waste management practices, including source reduction and recycling.

ABATSON Baffle that deflects airborne sound waves downward.

ABATVENT Device that breaks the force of, or prevents the entrance of wind to a duct, passage, chimney, etc.

ABATVOIX Reflector of airborne sound waves.

ABAXIAL Surface of a leaf that faces away from the stem.

ABC EXTINGUISHER Fire extinguisher that can be used on fires involving ordinary combustibles, flammable liquids, and ener-gized electrical equipment. *Also called* **Mul-tipurpose extinguisher.**

ABFARAD Unit of electric capacitance in the centimeter-gram-second (CGS) system of measurement, equal to the capacitance of a capacitor carrying a charge of 1 abcou-lombe with a potential difference of 1 abvolt between its plates. An abfarad is equal to 10^9 farads.

ABHENRY Unit of electrical inductance in the centimeter-gram-second system of mea-surement, equal to the inductance produced when a rate of change of current of 1 ampere per second generates an induced electromo-tive force of 1 abvolt. 1 abhenry is equal to 10^{-9} henry.

ABIOCEN Nonliving components of the environment.

ABIOSESTON Nonliving matter floating in water.

ABIOTIC Nonbiological.

ABLATION Formation of residual deposits by washing away of loose or soluble materials.

ABLUTION FITTING Sanitary washing fitting designed for communal use, often with water sprays controlled by touchless or foot valves.

ABNORMAL EROSION Erosion rates at variance with the geological norm, due to human activity.

ABOHM Unit of electric resistance in the centimeter-gram-second system of measurement. 1 abohm equals 10^{-9} ohm.

ABOVEGROUND BIOMASS Aboveground portion of a plant, excluding the root system.

ABOVEGROUND RELEASE Release to the surface of the land, or to surface water.

ABOVEGROUND STORAGE FACILITY Tank or other container, the bottom of which is on a plane not more than 150 mm (6 in.) below the surrounding surface.

ABOVEGROUND TANK Tank situated in such a way that its entire surface area is completely above the plane of the adjacent surrounding surface.

ABRUPT WAVE Transitionary wave or increase in depth of water in an open channel due to sudden change in the conditions of flow.

ABS (alkyl benzene Sulfonate) Type of surface active agent which, in waste treatment processes, tends to cause foaming and resist biological degradation. ABS has generally been replaced in detergents by linear alkyl sulfonate (LAS) which is biodegradable.

ABSOLUTE 1. Chemical substance free, for all practical purposes, of impurities; **2.** Independent of any arbitrary standard; non-relative.

ABSOLUTE FILTRATION RATING Diameter of the largest spherical particle that will pass through a filter under specified test conditions. This is an indication of the largest opening in the filter element. It does not indicate the largest particle that will pass through the element, since particles of greater length than diameter may pass.

ABSOLUTE HUMIDITY *See* **Humidity.**

ABSOLUTE MANOMETER Instrument whose calibration can be determined from its measurable physical constants and which is the same for all vapors and gases.

ABSOLUTE PRESSURE *See* **Pressure.**

ABSOLUTE SPECIFIC GRAVITY Ratio of the mass (referred to a vacuum) of a given volume of a solid or liquid at a stated temperature (in the U.S., 15.5°C (50°F)) to the mass (referred to a vacuum) of an equal volume of gas-free distilled water at a stated temperature.

ABSOLUTE TEMPERATURE Temperature measured or calculated on the absolute scale where absolute zero is the minimum and scale units equal in magnitude to centigrade degrees.

ABSOLUTE VELOCITY Velocity of a body in relation to a fixed point on the earth's surface.

ABSOLUTE VISCOSITY *See* **Viscosity.**

ABSOLUTE VOLUME In the case of solids, the displacement volume of the particles themselves, including their permeable and impermeable voids, but excluding any space between them; in the case of fluids, their volume.

ABSOLUTE ZERO Hypothetical temperature characterized by complete absence of heat, equivalent to approximately -273°C (-460°F).

absorb. *Abbreviation for:* absorbed; absorbing.

ABSORBED MOISTURE Moisture that has entered a solid by absorption and has physical properties not substantially different from ordinary water at the same temperature and pressure. *See also* **Absorption, Capillary water,** and **Gravitational water.**

ABSORBENCY Property of a substance to imbibe liquids.

ABSORBENT Material capable of taking in water, moisture, or other fluid.

ABSORBENT SEPARATOR *See* **Separator.**

ABSORBER Material or device for removing a specific fluid or gas from a substance to a designed degree, either chemically or mechanically.

ABSORBER PLATE In a solar energy system, that part of the collector system that transfers solar heat to a working medium.

ABSORBING DRAIN *See* **Well.**

ABSORBING WELL Well dug or drilled through an impermeable strata into a dry strata and used to drain away water.

ABSORPTION 1. Weight of fluid taken in by a material when immersed. *See also* **Absorbed moisture; 2.** Process in which a substance combines with or extracts one or more substances from a mixture of gases or liquids; **3.** Property of a porous or fibrous substance to attract and hold liquids and gases; **4.** Taking up of radiant energy by a material.

ABSORPTION BED Relatively large pit filled with coarse aggregate, used to receive the effluent from a waste treatment system.

ABSORPTION CHILLER Cooler that operates by the evaporation of a fluid in the low pressure side of a closed system; the refrigerant being recovered by the absorbing action of a second fluid. In the high pressure side of the system the absorbent containing the dissolved refrigerant is boiled, the refrigerant vapor and the absorbent being returned to the low pressure side for reuse.

ABSORPTION COEFFICIENT Volume of a gas that is dissolved (referred to 0°C and 760 mm Hg pressure).

ABSORPTION FIELD Effluent disposal system, particularly from septic tanks, consisting of a network of perforated pipes laid in coarse gravel in trenches that slope away from the tank outlet. The trenches are backfilled with progressively finer material, finishing with top soil.

ABSORPTION HYGROMETER Instrument used to measure the amount of water vapor in air by measuring the increase in weight of a drying agent over which a sample of air is drawn.

ABSORPTION LOSS Water lost to the surrounding soil from unlined reservoirs, irrigation ditches, sloughs, etc.

ABSORPTION RATE Weight of fluid taken in by a material following immersion for a specific period in a liquid of known properties and temperature.

ABSORPTION TEST Test made to determine the ability of materials to absorb vapors or fluids under specified conditions.

ABSORPTION TRENCH Trench containing a perforated pipe, or distribution tile surrounded by coarse aggregate, into which the effluent from a wastewater treatment process is discharged.

ABSORPTIVE CAPACITY Volume of waste materials that can be deposited in a particular environment without ecological damage or aesthetic impact.

ABSORPTIVE TERRACE Terrace of varying width formed on either side of a ridge so as to cause moisture to spread and be retained over the widest possible area.

ABSORPTIVITY 1. Capacity of a material to absorb (light, heat, moisture, gases); **2.** Fraction of a radiant energy absorbed by the surface that it strikes.

ABS PIPE *See* **Pipe.**

ABSTRACT Brief background of all information and a history of transactions, typically of a plot from the date of the original deed.

ABSTRACTING Process of assembling data of a similar character under an item ready for billing; the first stage of working up a bill of materials. *See also* **Bill of materials.**

ABSTRACTION Temporary or permanent withdrawal of water from a source.

ABSTRACTIVE USE Use of a resource that is temporarily removed from its natural environment.

ABSTRACT OF BIDS Summary of unit prices supplied by the owner, used as the basis for selecting the low bid on a tendered contract.

ABSTRACT OF TITLE Document showing the condensed history of the title to property, containing portions of all conveyances or other pertinent instruments relating to the estate or interest in the property, and all liens, charges, encumbrances, and releases.

ABUTMENT 1. Solid wall placed to counteract the lateral thrust of a bridge deck, vault or arch; **2.** In bridges, the end structure that supports the beams, girders, and deck of a bridge and sometimes retains the earthen bank, or supports the end of the approach pavement slab; **3.** In dams, the side of the gorge or bank of the stream against which a dam abuts. *See also* **Arch.**

ABUTMENT FLASHING Flashing that protects the head and sides of an abutment — level at the head; stepped at slopes.

ABUTMENT PIECE Lowest structural member in framing that receives and distributes loads from an upright; a soleplate.

ABUTTAL Parts in which land abuts on other land; boundaries.

ABVOLT Unit of potential difference or electromotive force in the centimeter-gram-second (CGS) system of measurement, equal to the potential that exists between two points when 1 erg of work must be done to transfer 1 abcoulomb of charge from one of the points to the other. 1 abvolt is equal to 10^{-8} volt.

ABWATT Unit of power in the centimeter-gram-second (CGS) system of measurement, equal to the power dissipated when a current of 1 abampere flows across a potential difference of 1 abvolt. 1 abwatt is equal to 10^{-7} watt.

ABYSSAL Depth in a water body below that to which light will penetrate.

ABYSSINIAN WELL Perforated tube driven into the ground through which groundwater is pumped.

ABYSSOBENTHOS Ocean floor at depths below that to which light will penetrate.

ABYSSOPELAGIC Organisms that exist in water at depths greater than 3000 m.

ACCELERATED ABSORPTION TEST Test in which the end result is hastened by testing at conditions more severe than those anticipated in typical service.

ACCELERATED COMPLETION Requirement for a contractor to complete the work at an earlier date than was scheduled in the contract documents.

ACCELERATED DELIVERY *See* **Accelerated design and construction.**

ACCELERATED DESIGN AND CONSTRUCTION Technique whereby construction is started before completion of final detail drawings. *Also called* **Accelerated delivery, Fast-track construction, Fast tracking, Phased construction,** and **Phased design and construction.**

ACCELERATED LIFE TEST Method of designing to approximate in a short time the deteriorating effects obtained under normal service conditions.

ACCELERATED SERVICE TEST Service or bench test in which some service condition, such as speed, or temperature, or continuity of operation, is exaggerated in order to obtain a result in a shorter time.

ACCELERATED WEATHERING Testing of material samples to cycles of excess dryness, frost, heat, wetness, etc. *Also called* **Artificial weathering.**

ACCELERATION Rate of change of velocity with time.

ACCEPTABLE 1. Acceptable to the authority having jurisdiction; **2.** An installation or equipment is acceptable if it meets any of the following conditions:

(a) If it is accepted, or certified, or listed, or labelled, or otherwise determined to be safe by a qualified testing laboratory capable of determining the suitability of materials and equipment for installation;

(b) If no qualified testing laboratory accepts, certifies, lists, or determines the installation or equipment to be safe, if it is inspected or tested by a regulatory responsible authority and found to be in compliance with its requirements; or

(c) With respect to custom-made equipment or related installations that are designed, fabricated for, and intended for use by a particular customer, if it is determined to be safe for its intended use by its manufacturer on the basis of test data that the employer keeps and makes available for inspection.

ACCEPTABLE DAILY INTAKE Daily amount of a substance which, over an entire lifetime, may be ingested without apparent risk based on all known facts.

ACCEPTANCE 1. Written indication that goods and/or materials supplied and/or installed are as specified; **2.** Consignee's receipt of a shipment that terminates a common carrier contract for transportation.

ACCEPTANCE INSPECTION Inspection by the owner's representative of work scheduled for full or partial payment.

ACCEPTANCE OF A BATCH SEQUENCE Condition where the number of rejected batches in a sequence is less than or equal to the sequence acceptable number as determined by the appropriate sampling plan.

ACCEPTANCE OF WORK Handing over of a completed construction project, excepting possibly any agreed-upon deficiencies or make-goods, with an obligation for payment less any holdback. *See also* **Interim acceptance, Final acceptance,** and **Partial acceptance.**

ACCEPTANCE PLAN Prescribed method of sampling, measuring and testing, together with criteria for the acceptability of a lot of material or construction.

ACCEPTANCE TEST Inspection that affirms if a material meets acceptance criteria.

ACCEPTED An installation or equipment item is 'accepted' if it has been inspected and found to be safe by a person qualified to make such an inspection, or by a qualified laboratory.

ACCEPTED BID Bid accepted by the owner or his representative and that forms the basis of a contract.

ACCEPTED ENGINEERING PRACTICES Practices that are compatible with standards of practice required by a registered professional engineer.

ACCESS Way or means of approach.

ACCESS CHAMBER Space large enough for a worker to enter.

ACCESS CLEANOUT Pipe fitting with a removable plate.

ACCESS DOOR Door giving access to machine rooms, machine and equipment spaces, or other normally closed areas.

ACCESS EYE *See* **Cleanout.**

ACCESS FLOOR Floor, usually modular and demountable, above a structural floor, beneath which cables and other services are located.

ACCESS FLOORING Modular floor components that can be easily removed to give access to underfloor ducting.

ACCESS FOR THE PHYSICALLY CHALLENGED Special measures taken in design and construction to accommodate the special needs of the physically challenged.

ACCESS HOLE Opening large enough for the intended use: a worker, a tool, etc.

ACCESSIBLE 1. Electrical and/or communications wiring capable of being removed or exposed without damaging the building structure or finish, or not permanently closed in by the structure or finish of the building. *See also* **Concealed,** and **Exposed**; **2.** Equipment admitting close approach; not guarded by locked doors, elevation, or other effective means. *See also* **Readily accessible.**

ACCESSIBLE ENVIRONMENT Atmosphere, land surfaces, surface waters, oceans, and all of the lithosphere that is beyond the controlled area.

ACCESSION Additions to property resulting from the annexation of fixtures or of alluvial deposits.

ACCESSORY 1. Any supplemental device that cannot be classified as an attachment; a secondary part or assembly of parts that contributes to the overall function and usefulness of a machine; **2.** Components used to join, attach or connect structural elements or services.

ACCESSORY BOX Weatherproof rectangular box, mounted on or within a wall, used to house electrical junctions and components.

ACCESSORY SEAL Seal that is used for sealing accessory equipment. On various engines, 'accessory seal' pertains to a seal that is employed for sealing an accessory shaft in the gear box, such as a shaft for operating an oil pump, fuel pump, generator, starter, or de-oiler.

ACCESSORY SPECIES Species of tree that represents between 25% and 50% of a stand.

ACCESS RIGHT *See* **Right of access.**

ACCESS ROAD Paved or unpaved road, whose use requires authorization, either of public or private ownership, that provides a connection between a public road and a utility.

ACCESS SWITCH *See* **Switch.**

ACCIDENT 1. Unexpected, undesirable event that adversely affects man or the environment; **2.** Occurring by chance; a sudden, unexpected event identifiable as to time and place.

ACCIDENTAL OCCURRENCE An accident, including continuous or repeated exposure to conditions, which results in bodily injury or property damage.

ACCIDENTAL SPECIES Species of tree

that represents less than 25% of a stand.

ACCIDENTAL RELEASE Sudden or non-sudden release of petroleum from an underground storage tank that results in a need for corrective action and/or compensation for bodily injury or property damage.

ACCIDENT INSURANCE Insurance indemnity for loss of time, loss of facility, medical expenses, and other defined perils and expenses due to accident.

ACCIDENT PREVENTION Planning and conscious acts taken to reduce the likelihood and opportunity for the occurrence of accidents.

ACCLIMATION Physiological or behavioral adaptation of organisms to one or more environmental conditions associated with a test method (e.g., temperature, hardness, pH).

ACCLIMATIZATION Process of adapting to changing environmental conditions.

ACCLIVITY Upward slope or steepness of a line or plane relative to horizontal.

ACCOMMODATION BRIDGE Temporary bridge.

ACCOUNTABILITY Being answerable for the exercise of one's authority and the performance of one's duties.

ACCOUNTABILITY MATRIX *See* **Responsibility assignment matrix.**

ACCOUNT CODE STRUCTURE Assignment of alphanumeric identification to work components within individual cost packages.

ACCOUNT NUMBER Identification of a work package so that all costs incurred for its completion can be properly assigned and allocated.

ACCOUNT PAYABLE Sum owing to a creditor, usually resulting from the purchase of goods or services.

ACCOUNT RECEIVABLE Sum claimed against a debtor, usually resulting from the sale of goods or services.

ACCREDIT Authorize; give credentials to;

give authority to; certify.

ACCREDITED Person or laboratory recognized by a competent authority as qualified and capable to perform a specified duty or task.

ACCRETION Addition to land from natural causes, such as alluvial deposits.

ACCRUAL DATE Date of an incident causing loss or damage, or the date on which loss or damage should have been discovered through the exercise of reasonable care.

ACCRUAL METHOD Accounting procedure that recognizes revenues and expenses when they occur regardless of when cash is received or paid. *See also* **Cash method.**

ACCRUED ASSET Accumulating but not yet enforceable claim resulting from the provision of services or the expenditure of money (as during construction or development) that has been only partly performed, is not yet billable, and has not been paid for.

ACCRUED DEPRECIATION Difference between reproduction cost new, or replacement cost new, and the present worth of those improvements, plant or utilities, as measured on the date of appraisal.

ACCRUED EXPENSE Incurred expense for which there currently is no enforceable claim due to the project or contract not having been completed to a billable stage.

ACCRUED INTEREST Interest that has been earned but is not due or payable.

ACCRUED LIABILITY Accumulating but not yet enforceable claim by a consultant or contractor resulting from partially completed work which is not yet billable.

ACCRUED REVENUE Earned revenue resulting from partially completed work that cannot yet be billed.

ACCUMULATED DEFICIENCY Total deficiency from an established normal condition of temperature, precipitation, etc., from a stated time, to date. *See also* **Accumulated excess.**

ACCUMULATED DEPRECIATION To-tal by which an asset has been depreciated since being acquired or placed in service.

ACCUMULATED EXCESS Total excess from an established normal condition of temperature, precipitation, etc., from a stated time, to date. *See also* **Accumulated deficiency.**

ACCUMULATED SPECULATIVELY Material accumulated before being recycled.

ACCURACY Comparison of an indicated value to a known reference; the quality of accuracy is often expressed by stating the difference of two values as a percentage of the known reference value.

ACID Compound having a pH of less than 7.0; hydrogen-containing compound that reacts with metals to form salts, and with metal oxides and bases to form a salt and water. The strength of an acid depends on the extent to which its molecules ionize, or dissociate, in water, and on the resulting concentration of hydrogen ions (H+) in solution.

ACID ATTACK *See* **Aggressive solution.**

ACID COPPER CHROMATE Waterborne wood preservative consisting of copper sulfate, sodium dichromate, and chromic acid.

ACID DEWPOINT Temperature at which dilute sulfuric acid appears as a condensate when a flue gas containing sulfur trioxide and water vapor is cooled below saturation temperature.

ACID GAS 1. Gas stream of hydrogen sulfide (H_2S) and carbon dioxide (CO_2) that has been separated from sour natural gas by a sweetening unit. **2.** By-product of incomplete combustion having a pH value of less than 6.5.

ACIDIFICATION Forced or accelerated leaching by exposing a material to a higher concentration of solvents (e.g., acids) than would normally be encountered.

ACIDITY Quantitative capacity of an aqueous solution to react with hydroxyl ions. *See also* **pH.**

ACID LEAD Fully refined, 99.9% pure lead containing a small amount of copper.

ACID MINE DRAINAGE Drainage from certain types of mines that contains acidic salts in solution, formed primarily by oxidation of iron pyrites in the presence of moisture.

ACID MINE WATER Water present in, or discharging from a mine that contains minerals in solution and having a low pH, commonly caused by oxidation of iron sulfide to sulfuric acid.

ACID MIST Sulfuric acid mist.

ACID NEUTRALIZER Device within a drainage system into which an acid discharge is probable, consisting of a chamber filled with limestone chips through which the flow is conducted to produce a neutral effluent.

ACID OR FERRUGINOUS MINE DRAINAGE Mine drainage which, before any treatment, either has a pH of less than 6.0 or a total iron concentration equal to or greater than 10 mg/I.

ACID RAIN Form of air and water pollution caused by the release of acid gases from the combustion of fossil fuels in internal combustion engines, combustors, factories, mine processing plants, or coal and oil burning electrical generators.

ACID RECOVERY Sulfuric acid pickling operations that include processes for recovering the unreacted acid from spent pickling acid solutions.

ACID REGENERATION those hydrochloric acid pickling operations that include processes for regenerating acid from spent pickling acid solutions.

ACID RESISTANCE Extent to which a surface will resist attack by acids.

ACID SOOT Carbon particles, 1 - 3 mm in diameter, resulting from incomplete combustion, held together by water made acid due to combination with sulfur trioxide.

ACOUSTIC Relating to sound and hearing.

ACOUSTIC ABSORPTION COEFFICIENT Acoustic energy absorbed by a surface, divided by the total energy incident on it, subject to the frequency of the particular sound.

ACOUSTIC ABSORPTIVITY *See* **Sound absorption coefficient.**

ACOUSTICAL ASSURANCE PERIOD Specified period of time during which an exhaust system, properly used and maintained, must continue in compliance with a prescribed standard.

ACOUSTICAL BARRIER Baffle or barrier designed and placed to reduce or prevent the passage of sound waves.

ACOUSTICAL BOARD Any type of special material, such as insulating boards, used in the control of sound or to prevent the passage of sound from one room to another.

ACOUSTICAL CEILING Ceiling system designed to reduce the transmission of sound: may consist of a layer of absorbent material, or a suspended system supporting acoustical board.

ACOUSTICAL CORRECTION Measures taken to improve the acoustical performance of a completed room or auditorium.

ACOUSTICAL IMPEDANCE Mathematical expression for characterizing a material as to its energy transfer properties (the product of its unit density and its sound velocity pV).

ACOUSTICAL MATERIALS Sound absorbing or attenuating materials.

ACOUSTICAL REDUCTION FACTOR *See* **Sound reduction factor.**

ACOUSTICAL TILE Wall and ceiling tiles composed of materials, and of a design intended to attenuate sound waves.

ACOUSTICAL TRANSMISSION FACTOR Ratio of original sound volume to that which is transmitted through and beyond any obstruction to the sound waves, such as a wall, panel, or baffle.

ACOUSTIC CLIP Floor clip with a flexible rubber pad, used to reduce the transmission of sound and vibrations.

ACOUSTIC CONSTRUCTION Construction techniques aimed at measurably reducing or attenuating sound transmission.

ACOUSTIC DESCRIPTOR Numeric, symbolic, or narrative information describing a product's acoustic properties as they are determined according to a prescribed test methodology.

ACOUSTIC FINISHINGS Materials, usually porous, inside which sound energy is adsorbed.

ACOUSTIC PLASTER Gypsum plaster that, when applied to a wall or ceiling, has a high sound absorbency.

ACOUSTIC REFLEX Muscular action of the mechanical mechanism of the human ear to protect against loud and sudden sound.

ACOUSTICS Study of sound; quality of sound.

ACOUSTIC TILE Rectangle of acoustic board, commonly interlocking for direct attachment to a subbase, or square edged and mounted in a suspended ceiling system.

ACOUSTIC TRACE Line on a vibration record that records a sound level.

ACQUIESCENCE Concurrence by owners of adjoining property that resolves a boundary or rights dispute.

ACQUIRED CHARACTER Change in an organism in response to environmental change.

ACQUIRED IMMUNE DEFICIENCY SYNDROME Condition in humans, AIDS, in which the immune system suffers progressive failure.

ACQUISITION Process of obtaining a property or right of way. *Also called* **Taking.** *See also* **Conveyance, Dedication, Eminent domain, Expropriation, Negotiation, Option, Remainder,** and **Severalty.**

ACRE Nonmetric unit of area, equal to 43,560 square feet or 4840 square yards. Symbol: ac. Multiply by 0.404 to obtain hectares, symbol: ha; by 404.856 to obtain square meters, symbol: m^2. *See also the appendix:* **Metric and nonmetric measurement.**

ACRE-FOOT Nonmetric unit of volume, equal to the amount of water, gravel, coal, or other minerals required to cover 1 acre to a depth of 1 ft, equal to 1613 yd^3. Symbol: ac/ft. Multiply by 1233.482 to obtain cubic meters, symbol: m^3. *See also the appendix:* **Metric and nonmetric measurement.**

ac/ft *Abbreviation for:* acre-foot.

ACRE-INCH Nonmetric unit of volume, equal to the amount of water, gravel, coal, or other minerals required to cover 1 acre to a depth of 1 in., equal to 3630 ft^3. Symbol: ac/in. Multiply by 102.815 to obtain cubic meters, symbol: m^3. *See also the appendix:* **Metric and nonmetric measurement.**

ACRE-INCH-DAY Nonmetric unit of volume equal to the flow required to cover 1 acre to a depth of 1 in. in a 24-hour period, or 0.042 cfs.

ACRYLIC FIBER Manufactured synthetic fiber in which the fiber-forming substance is any long-chain synthetic polymer composed of at least 85% by weight of acrylonitrile units.

ACT Legislation, or amendment to legislation, authorized into law by a governing body.

ACTINIDES Group of elements, having atomic numbers ranging from 89 to 103, produced in nuclear reactions or from radioactive decay.

ACTINOMETER Instrument used to measure the heat absorbed by a blackened disk or chamber.

ACTINOMYCETES Large group of moldlike microorganisms that give off an odor characteristic of rich earth and that are significant in the stabilization of solid waste by composting.

ACTION LEVEL 1. Concentration of lead or copper in water which determines, in some cases, the treatment requirements that a water system is required to complete; **2.** Airborne concentration of asbestos of 0.1 fibers per cubic centimeter (f/cc) of air calculated as an 8-hour time-weighted average; **3.** Exposure level (concentration of the material in the air) at which certain OSHA regulations to protect employees take effect.

ACTIVATED ALUMINA Granular, porous form of aluminum trioxide capable of absorbing water and other substances, used in pollution control.

ACTIVATED CARBON Porous carbon, produced by heating carbon to 900°C with steam or carbon dioxide, used to absorb odors.

ACTIVATED CARBON PROCESS Technique for removing sulfur dioxide from flue gases.

ACTIVATED CHARCOAL Charcoal produced by carbonizing organic material, usually in the absence of oxygen, used in granular or powder form to absorb odors.

ACTIVATED MANGANESE DIOXIDE PROCESS Technique for removing sulfur dioxide from flue gases by dry absorption to produce ammonium sulfate.

ACTIVATED SILICA Negatively charged colloidal particle formed by the reaction of a dilute sodium silicate solution with a dilute solution of an acidic material or other activant; used primarily as a coagulant aid.

ACTIVATED SLUDGE Sludge floc developed in raw or settled wastewater by the growth of zoogleal bacteria and other organisms in the presence of dissolved oxygen and accumulated in sufficient concentration by returning floc previously formed.

ACTIVATED SLUDGE LOADING Weight of biochemical oxygen demand (BOD) in the applied liquor per unit volume of aeration capacity, or per weight of activated sludge per day.

ACTIVATED SLUDGE PROCESS Aerobic sewage treatment method in which air is introduced to raw sewage to promote mixing and bacterial digestion.

ACTIVATED SOLIDS Combination of organisms in wastewater solids produced in the presence of dissolved oxygen.

ACTIVATION 1. Generation, under aerobic conditions, or organisms capable of absorbing organic material from water in the activated sludge process; **2.** Process of making a material radioactive by bombardment with neutrons, protons, or other nuclear particles; **2.** Process of moving plant or equipment from a dormant state to a working state.

ACTIVATION PRODUCT Material made radioactive as a result of irradiation, particularly by neutrons in a nuclear reactor.

ACTIVATOR Compound ingredient used to increase the effectiveness of an accelerator.

ACTIVE EARTH PRESSURE *See* **Pressure.**

ACTIVE FACTORS That which supplies energy and nutrients for natural processes in plants.

ACTIVE FIRE PROTECTION System that includes electrical and mechanical sensors and equipment to warn of and/or extinguish fires, usually within enclosed spaces and structures. *See also* **Passive fire protection.**

ACTIVE INGREDIENT 1. Any substance or group of structurally similar substances that will prevent, destroy, repel or mitigate any pest, or that functions as a plant regulator, desiccant, or defoliant; **2.** Pesticide which will prevent, destroy, repel, or mitigate any pest; **3.** Plant regulator which, through physiological, biochemical action, will accelerate or retard the rate of growth or rate of maturation or otherwise alter the behavior of ornamental or crop plants or their product; **4.** Defoliant which will cause the leaves or foliage to drop from a plant; **5.** Desiccant which will artificially accelerate the drying of plant tissue.

ACTIVE INSTITUTIONAL CONTROL 1. Controlling access to a disposal site by any means other than passive institutional controls; **2.** Performing maintenance operations or remedial actions at a site; **3.** Controlling or cleaning up releases from a site; **4.** Monitoring parameters related to disposal system performance.

ACTIVE LAYER 1. Ground surface layer that moves seasonally owing to volume changes occasioned by temperature differentials; **2.** Area above permafrost that thaws each, or most years.

ACTIVE LIFE OF A HAZARDOUS

WASTE FACILITY Period from the initial receipt of hazardous waste at the facility until receipt of certification by a competent authority of final closure.

ACTIVE LIFE OF A SOLID WASTE FACILITY Period of operation beginning with the initial receipt of solid waste and ending at completion of closure activities.

ACTIVE MINE Surface or underground workings from which minerals, ore, aggregate, etc. is being obtained.

ACTIVE MINING AREA Area, on and beneath land, used or disturbed in activity related to the extraction, removal, or recovery of coal, ore, minerals or aggregate from its natural deposits.

ACTIVE PORTION Portion of a facility where treatment, storage, or disposal operations are being or have been conducted.

ACTIVE REPAIR TIME *See* **Machine time.**

ACTIVE SERVICE Drain that is receiving wastewater from a process unit that will continuously maintain a water seal.

ACTIVE SOLAR SYSTEM Solar system that uses mechanical devices such as collectors, thermal storage devices, transfer fluid, fans and pumps to collect, store, and distribute useful energy. *See also* **Hybrid solar system.**

ACTIVE SOUND ATTENUATOR Sound generator that produces waves asynchronous to many produced by the fans in a heating/ventilating/air-conditioning system.

ACTIVE STORAGE Volume of a storage facility available for use.

ACTIVE TRANSPORT Movement of a substance from a region of low concentration to one of higher concentration.

ACTIVE WATER Water having corrosive qualities.

ACTIVITY 1. Task performed over a period of time; **2.** Element of work performed during the course of a project. An activity normally has an expected duration, an ex-

pected cost and expected resource requirements. *Also called* **Tasks; 3.** Total flow of energy through a system in a unit of time; **4.** Amount of a radioactive material; a measure of the transformation rate of radioactive nuclei at a given time.

ACTIVITY DEFINITION Identifying the specific activities that must be performed in order to produce the various project deliverables.

ACTIVITY DESCRIPTION (AD) Short phrase or label used in a project network diagram which normally describes the scope of the work of the activity.

ACTIVITY DURATION Actual time or estimated time required to accomplish something.

ACTIVITY DURATION ESTIMATING Estimation of the number of work periods that will be needed to complete individual activities.

ACTIVITY OF CLAY Ratio of plasticity index to percent by weight of the total sample of a clay that is smaller than 0.002 mm (0.00008 in.) in grain size.

ACTIVITY-ON-ARROW *See* **Arrow diagramming method.**

ACTIVITY-ON-NODE *See* **Precedence diagramming method.**

ACTIVITY ORIENTED SYSTEMS Systems that present information in terms of activities (i.e., periods of time).

ACT OF GOD Circumstance assumed to be beyond human control.

ACTUAL *See* **Cost types.**

ACTUAL AGE (OF A PROPERTY) Actual number of years that have passed since a structure was built. *Also called* **Chronological age,** and **Historical age.**

ACTUAL CASH VALUE Actual market value of something insured at the time of its loss or damage.

ACTUAL COST OF WORK PERFORMED Total costs incurred (direct and indirect) in accomplishing work during a given time period. *See also* **Earned value.**

ACTUAL CREDITS Emission credits based on actual production volumes as contained in the end-of-year reports submitted to a regulatory authority.

ACTUAL DIMENSION Dimension as measured or calculated.

ACTUAL EMISSIONS Actual rate of emissions of a pollutant from an emissions unit.

ACTUAL FINISH DATE Point in time that work actually ended on an activity. (Note: in some application areas, the activity is considered 'finished' when work is 'substantially complete.')

ACTUAL GROUNDWATER VELOCITY Effective rate of movement of groundwater percolating through water-bearing material, as measured by the volume of groundwater passing through a unit cross-sectional area in a unit of time divided by the effective porosity. *Also called* **Effective groundwater velocity, Field groundwater velocity,** and **True groundwater velocity.**

ACTUAL PRODUCTIVE TIME *See* **Machine time.**

ACTUAL SLOPE Slope to which an excavation face is excavated.

ACTUAL START DATE Point in time that work actually started on an activity.

ACTUATE Set in operation.

ACTUATOR 1. Linear or rotary device that converts hydraulic or fluid energy into mechanical energy; **2.** Mechanical linkage that transmits movement.

ACULTURAL VIBRATION Vibration that is strange and unfamiliar to the observer.

ACUTE Disease or medical condition at or approaching crisis.

ACUTE ANGLE Angle between 0° and 90°.

ACUTE DERMAL LD$_{50}$ Statistically derived estimate of the single dermal dose of a substance that would cause 50% mortality to a test population under specified conditions.

ACUTE EFFECT Toxic effect in mammals and aquatic life that rapidly follows exposure to a toxic substance. *Also called* **Acute toxicity.**

ACUTE INHALATION LC$_{50}$ Statistically derived estimate of the concentration of a substance that would cause 50% mortality to a test population under specified conditions.

ACUTE INHALATION TOXICITY Adverse effects caused by a substance following a single uninterrupted exposure by inhalation over a short period of time (24 hours or less).

ACUTE LC$_{50}$ Concentration of a substance, expressed as parts per million of a medium, that is lethal to 50% of a test population of animals under test conditions.

ACUTE LETHAL TOXICITY Lethal effect produced on an organism within a short period of time of exposure to a chemical.

ACUTE ORAL LD$_{50}$ Statistically derived estimate of the single oral dose of a substance that would cause 50% mortality to a test population under specified conditions.

ACUTE ORAL TOXICITY Adverse effects occurring within a short time of oral administration of a single dose of a substance, or multiple doses given within 24 hours.

ACUTE TOXICITY 1. Property of a substance or mixture of substances that causes adverse effects on an organism through a single short-term exposure; **2.** Discernible adverse effects induced in an organism within a short period of time (days) of exposure to a chemical. For aquatic animals this usually refers to continuous exposure to the chemical in water for a period of up to four days. The effects (lethal or sublethal) occurring may usually be observed within the period of exposure with aquatic organisms. Shell deposition is sometimes used as the measure of toxicity.

ACUTE TOXICITY TEST 1. Method used to determine the concentration of a substance that produces a toxic effect (death) on a specified percentage of test organisms in a short period of time (e.g., 96 hours); **2.** Chemical substance that produces acutely toxic effects if it kills within a short time period (usually 14 days): (a) At least 50% of the exposed mammalian test animals following oral administration of a single dose of the test substance at 25 milligrams or less per kilogram of body weight (LD_{50}), (b) at least 50% of the exposed mammalian test animals following dermal administration of a single dose of the test substance at 50 milligrams or less per kilogram of body weight (LD_{50}), (c) at least 50% of the exposed mammalian test animals following administration of the test substance for 8 hours or less by continuous inhalation at a steady concentration in air at 0.5 milligrams or less per litre of air (LC_{50}).

ADAPT Make suitable for a particular purpose by modification or change.

ADAPTATION Change in structure or function to better deal with environmental or other change.

ADAPTER Fitting that joins pipes and other devices and shapes of different materials or different sizes, or different threads, etc.

ADAXIAL Surface of a leaf that faces toward the stem.

ADDENDA Additions or supplements issued before bid opening that clarifies, corrects, or changes the bidding documents or requirements of the contract documents.

ADDITIONAL WORK Increase in the scope of work originally contracted for. This results in a variation, which may result in an extra to the contract sum.

ADDITIVE Chemical substance that is intentionally added to food or drink to improve its stability or impart some other desirable quality.

ADDITIVE ALTERNATE *See* **Alternate bid.**

ADDRESSABLE SYSTEM SMOKE DE-TECTOR Smoke detection system in which individual detectors are programmed to signal their location when activated.

ADDS Summated quantities that bring to-

gether all work and materials of a similar nature.

ADEQUATE SO₂ EMISSION LIMITA-TION Emission limitation approved or promulgated as adequate to attain and maintain the atmospheric quality standard in the areas affected by the stack emissions without the use of any unauthorized dispersion technique.

ADEQUATELY WET Sufficiently mixed or penetrated with liquid to prevent the release of particulates.

ADEQUATELY WETTED Sufficiently mixed or coated with water or an aqueous solution to prevent dust emissions.

ADEQUATE STORAGE Placing of pesticides in proper containers and in safe areas so as to minimize the possibility of escape which could result in unreasonable adverse effects on the environment.

ADHESION 1. State in which two surfaces are held together by interfacial effects, which may consist of molecular forces, interlocking action, or both; **2.** Soil quality of sticking to metal parts of mechanical equipment: buckets, blades, tracks, etc.

ADHESIVE WATER Film of water surrounding a grain of water-bearing material after gravity water has been drained off. *Also called* **Water of adhesion.** *See also* **Pellicular water.**

ADIABATIC Condition in which heat neither enters nor leaves a system.

ADIT Horizontal or near horizontal shaft driven from the surface to reach an ore body, and which may be used to drain the resulting mine.

ADJACENT Bordering, contiguous, or neighboring. Typically, wetlands separated from other waters by man-made dikes or barriers, natural river berms, beach dunes, and the like.

ADJUSTABLE PARALLELOGRAM RIPPER *See* **Ripper.**

ADJUSTABLE RADIAL RIPPER *See* **Ripper.**

ADJUSTABLE, TIMBER, SINGLE-POST SHORE Individual timber used with a fabricated clamp to obtain adjustment. Not normally manufactured as a complete unit. *See also* **Fabricated single-post shore, Post shore,** and **Timber single-post shore.**

ADJUSTED BASE COST Base construction cost adjusted for alternate components, plus or minus.

ADJUSTING NUT Threaded nut that, when turned, will vary some already established tolerance.

ADJUSTING SCREW 1. Threaded screw that, when turned, will cause two connected faces to move further apart or closer together; **2.** A leveling device or jack composed of a threaded screw and an adjusting handle, typically used for vertical adjustment of shoring and formwork.

ADJUSTMENT Behavioral response to environmental conditions.

ADJUTAGE TUBE Tube inserted into an orifice; an efflux tube.

ADMINISTRATIVE AUTHORITY Individual, group, agency, etc., authorized under law to enforce the provisions of a code, regulation, etc.

ADMINISTRATIVE CHARGES Overhead costs resulting from administration.

ADMINISTRATIVE CLOSURE Generating, gathering, and disseminating information to formalize project completion.

ADOBE Fine clay or loess deposits resulting from ice abrasion during glacation.

ADOLESCENT RIVER River in the second stage of a new drainage system; one that has largely obliterated the lakes and waterfalls of its youthful stage, and which possesses a well-cut channel that may reach base level at its mouth.

ADSORBED WATER Water held on the surfaces of material by electrochemical forces and having physical properties substantially different from those of absorbed water or chemically combined water at the same temperature and pressure. *See also* **Adsorption.**

ADSORBENT Material having the ability to contain another material within its structure without changing the character of the original material.

ADSORBER Material or device used to adsorb another material.

ADSORPTION Development (at the surface of either a liquid or solid) of a higher concentration of a substance than exists in the bulk of the medium; especially formation of one or more layers or molecules of gases, or dissolved substances, or of liquids at the surface of a solid (such as cement, cement paste, or aggregates), or of air-entraining agents at the air-water interfaces; also the process by which a substance is adsorbed. *See also* **Adsorbed water.**

ADSORPTION RATIO, K_d Amount of a test chemical adsorbed by a sediment or soil (i.e., the solid phase) divided by the amount of test chemical in the solution phase, which is in equilibrium with the solid phase, at a fixed solid/solution ratio.

ADSORPTION WATER Water retained on the surface of solid particles by molecular forces with emission of heat.

ADVANCE 1. Payment made on account of, but before completion of, a contract, or before receipt of goods or services. *See also* **Deposit; 2.** Set something to occur ahead of a set mark or time. *See also* **Retard.**

ADVANCED AIR EMISSION CONTROL DEVICES Air pollution control equipment, such as electrostatic precipitators and high energy scrubbers, that are used to treat an air discharge which has been treated initially by equipment including knockout chambers and low energy scrubbers.

ADVANCED GAS-COOLED REACTOR Nuclear reactor operating at 675°C and higher that is fuelled with enriched uranium dioxide, cooled by gaseous carbon dioxide and which uses graphite as a moderator.

ADVANCED SEWAGE TREATMENT Treatment necessary to remove suspended and dissolved substances not normally removed during secondary treatment including organic matter, suspended solids, inorganic ions such as calcium, phosphate, nitrate and potassium, as well as synthetic organic compounds.

ADVECTION Transfer of matter or energy in a horizontal stream of air.

ADVENTIVE PLANT Introduced or alien plant that grows without aid but which is not permanently established.

ADVERSE POSSESSION Right of an occupant to acquire title in defiance of someone else's legal title following occupancy for a required statutory period in actual, open, notorious, exclusive use.

AEOLIAN DEPOSIT Carried or blown by the wind; deposited as loess.

AERATE Act of mixing with air.

AERATED CONTACT BED *See* **Contact bed.**

AERATED POND Natural or artificial wastewater treatment pond in which mechanical or diffused-air aeration supplements the oxygen supply.

AERATION 1. Injection of air into a body of water.

AERATION PERIOD 1. Time during which mixed liquor is subjected to mechanical aeration while undergoing activated sludge treatment; **2.** Time during which water is subjected to mechanical aeration.

AERATION TANK Large, open-topped chamber where air is introduced under pressure to a fluid, commonly sewage and wastewater, to oxidize the organic fraction. The air is introduced in a variety of ways, from large, to very fine bubbles, to surface agitators that each lift and throw water from the surface, entraining air and mixing the tank contents by agitation.

AERATION ZONE Depth of soil below ground to the top of groundwater level: divided into three layers, an upper soil belt, intermediate belt, and lower capillary fringe.

AERATOR Device that adds air to water.

AERIAL ATTACK Use of aircraft in fight-

ing forest fires, mainly to drop fire retarding or extinguishing solutions on, or in the path of, the conflagration.

AERIAL PHOTOGRAMMETRY Interpreting information from aerial photographs.

AERIAL PHOTOGRAPHY Photos taken from the air at regular intervals and used in photo interpretation to provide information about land development, vegetation, and land forms.

AERIAL PHOTOMAP Aerial photograph or photomosaic to which basic mapping data has been added.

AERIAL PHOTOMOSAIC Composite of aerial photographs depicting a portion of the earth's surface.

AERIAL PLANKTON Spores, bacteria and other microorganisms floating in the air.

AERIAL SURVEYING Surveying done through photo-interpretation of images taken from aircraft flying a grid pattern at a fixed altitude over the target area.

AEROALLERGEN Airborne pollen and organic dust that is an irritant to some people, resulting in hay fever and other allergenic conditions.

AEROBE Organism that uses oxygen to sustain life, used in water purification and, more particularly, sewage treatment.

AEROBIC Environment rich in free or uncombined oxygen.

AEROBIC BACTERIA *See* **Bacteria.**

AEROBIC DECOMPOSITION Decay or breaking down of organic material in the presence of free or dissolved oxygen.

AEROBIC DIGESTION Breakdown of organic components by microbial action in the presence of oxygen.

AEROBIC PROCESS Process that relies on the presence of dissolved oxygen.

AEROBIC RESPIRATION Oxidation of organic compounds by oxygen.

AEROBIC WASTE TREATMENT PRO-CESS Breakdown of wastes by microorganisms in the presence of free or dissolved oxygen.

AEROBIOSIS Biological processes that require oxygen.

AEROCHLORINATION Injection of a mixture of chlorine gas and compressed air into a wastewater for the removal of grease.

AERODYNAMIC DIAMETER Diameter of a sphere of unit density which behaves aerodynamically like the particles of a test substance. It is used to compare particles of different sizes, shapes, and densities and to predict where in the respiratory tract such particles may be deposited.

AERODYNAMIC DRAG Resistance of a solid body to move through air.

AERODYNAMIC ROUGHNESS Turbulence in moving air caused by the unevenness of the surface over which it passes.

AERODYNAMICS Branch of physics that deals with the various forces associated with air or other gases in motion.

AEROFILTER Layer of coarse aggregate, used for the rapid filtering of sewage effluent.

AEROGENERATOR Device that relies on the movement of air to turn rotor blades to generate electricity.

AEROHYDROUS Of both water and air — commonly applied to minerals containing water in pores or cavities.

AEROSOL Small, solid or liquid particles, from 0.01 to 100 microns in diameter, suspended in air.

AEROSOL PROPELLANT Inert liquid with a low boiling point which vaporizes instantaneously at room temperature on release of pressure; used to carry and disperse a compound.

AEROSOL SPRAY Container in which a propellant is mixed with a substance held under pressure and which releases its contents as a fine mist.

AEROSPHERE Lower portion of the atmosphere surrounding earth, consisting of the troposphere and stratosphere..

AESTIDURILIGNOSA Mixed evergreen and deciduous forest.

AESTILIGNOSA Deciduous forest and brush.

AESTHETIC OBJECTIVE Objectives applied to certain substances or characteristics of drinking water that can affect its acceptance by consumers or interfere with practices for supplying good-quality water.

AFFLUENT STREAM Stream or river that flows into another or other body of water; a tributary.

AFFLUX 1. Difference in elevation between the high flood levels downstream and upstream of a weir; **2.** Upstream increase of water level above normal caused by obstruction in a channel or the constriction of a bridge or weir, or by regulation.

AFFORESTATION Establishment of forest crops by artificial methods, such as planting or sowing on land where trees have never before grown.

AFTERBAY 1. Initial section of a stream or conduit forming the discharge from a storage pond or reservoir; **2.** Tailrace of a hydroelectric generating station, with restricted outflow at the discharge of the turbines.

AFTERBURNER Device used to burn or oxidize the combustible constituents remaining in effluent gases to destroy smoke and odors.

AFTERFILTER High-efficiency filter located near a terminal unit in an air-conditioning system.

AFTER-FLUSH Water bled from the stream refilling a cistern following flushing of a toilet, and diverted to the bowl to ensure integrity of the seal.

AFTERPRECIPITATION Deposition of colloidal calcium carbonate on the individual grains of a sand filter and/or on the walls of pipes of the associated distribution system following treatment of the water with lime.

AFTERSHOCK Small seismic tremor following an earthquake of significant magnitude and occurring close to the focus of the main shock.

AGENCY 1. Relationship between principal and agent whereby the agent is employed to undertake certain acts on the principal's behalf when dealing with a third party; **2.** Administrative branch of government.

AGENT 1. Person authorized to act for and/or represent another; **2.** General term for a material that may be used either as an addition or additive to another. *See also* **Additive.**

AGEOSTROPHIC Air movement due to an imbalance between the horizontal pressure gradient force and a deviating force due to wind velocity and which is associated with vertical motion, cloud formation and weather change.

AGE/STRENGTH RELATIONSHIP Increase in strength of a material due to a maturing process and/or chemical reaction. *Also called* **Aging process.**

AGE TANK Tank in which a known concentration of a chemical solution is stored prior to being delivered to a chemical feeder.

AGGLOMERATE 1. Coarse-grained pyroclastic rock; **2.** Gathering together into a floc of suspended colloidal particles.

AGGLOMERATION Particle accumulation.

AGGRADATION Building up of a portion of land toward a uniformity of grade or slope.

AGGRADING RIVER River in process of building up its valley bottom through the deposition of materials carried forward by the flow.

AGGREGATE Collection of granulated rock particles of different substances into a compound or agglomerated mass. There are many classifications, including:

> **Angular:** Aggregate particles that possess well-defined edges formed at the intersection of roughly planar faces.

Artificial: Most lightweight aggregates (except pumice) and blastfurnace slag and clinker.

As-dug: *See* **Pit-run aggregate** (below).

Coarse: Aggregate predominantly retained on the No. 4 (4.75 mm, 0.187 in.) sieve.

Crushed: Product resulting from the artificial crushing of gravel with a specified minimum percentage of fragments having one or more faces resulting from fracture.

Crushed gravel: Product produced by mechanically crushing large-size gravel with all resulting fragments having at least one face resulting from fracture.

Crushed stone: Product resulting from the artificial crushing of rocks, boulders, or large cobblestones, substantially all faces of which possess well-defined edges and having resulted from the crushing operation.

Crusher-run: Aggregate that has been mechanically broken and that has not been subjected to subsequent screening.

Dense-graded: Aggregates graded to produce low void content and maximum weight when compacted.

Fine: Aggregate passing the 9.5 mm (0.375 in.) sieve and almost entirely passing the No. 4 (4.75 mm, 0.187 in.) sieve and predominantly retained on the No. 200 (0.074 mm, 0.0029 in.) sieve.

Gap-graded: Aggregate so graded that certain intermediate sizes are substantially absent.

Heavyweight: Aggregate of high density, such as barite, magnetite, hematite, limonite, ilmenite, iron, or steel, used in heavyweight concrete.

Lightweight: Aggregate of low density, such as (a) expanded or sintered clay, shale, slate, diatomaceous shale, perlite, vermiculite, or slag, (b) natural pumice, scoria, volcanic cinders, tuff, and diatomite, (c) sintered fly ash or industrial cinders, used in lightweight concrete.

Normal weight: Aggregate that is neither heavyweight nor lightweight.

Open-graded: Aggregate in which the voids are relatively large when the aggregate is compacted.

Pit-run: Quarry material that has not been crushed or screened but loaded direct for delivery. *Also called* **As-dug aggregate.**

Reactive: Aggregate containing substances capable of reacting chemically with the products of solution or hydration of portland cement in concrete or mortar under ordinary conditions of exposure, resulting in some cases in harmful expansion, cracking, or staining.

Refractory: Aggregate having refractory properties such that, when bound together into a conglomerate mass by a matrix, it forms a refractory body.

Single-sized: Aggregate in which a major portion of the particles are in a narrow size range.

Ultralightweight: Lightweight aggregate such as expanded perlite and exfoliated perlite and exfoliated vermiculite, used principally for thermal insulation.

Well-graded: Aggregate having a particle size distribution that produces maximum density, i.e., minimum void space.

AGGREGATE BLENDING Process of intermixing two or more aggregates to produce a different set of properties, generally, but not exclusively, to improve grading.

AGGREGATE FACILITY Individual drain system together with ancillary downstream sewer lines and oil-water separators, down to and including any secondary oil-water separator.

AGGRESSIVE CONDITIONS Environmental conditions such as heat, cold, chemical fumes, etc., that require special consideration in the selection of materials and/or construction.

AGGRESSIVE SOLUTION Any chemical that attacks the material on which it is spread: in concrete, a chemical that attacks the cement matrix, and to a lesser extent, the aggregate or hardened concrete. *Also called* **Acid attack.**

AGING Process in which a material or complex changes over time: in some cases improving as it gets older, in others deteriorating.

AGING PROCESS *See* **Age/strength relationship.**

AGITATION Shaking or stirring so as to mix various constituents.

AGITATOR Mechanical equipment for mixing and/or aerating.

AGONISTIC BEHAVIOR Competitive or aggressive behavior displayed by an animal toward another.

AGREEMENT Written agreement between two parties, i.e., between the owner and contractor pertaining to the work to be performed by the contractor, and the obligations for payment by the owner, and that is part of the contract documents.

AGRIC HORIZON Depositional soil horizon resulting from cultivation.

AGRICULTURAL DRAIN Unsocketed, unglazed, earthenware, plastic, or concrete tiles, usually about 100 mm (4 in.) in diameter, laid end-to-end without closing the joints so as to drain the subsoil.

AGRICULTURAL LIME Hydrated lime, used to condition soil.

AGRICULTURAL SOLID WASTE Solid waste that is generated by the rearing of animals, and the producing and harvesting of crops or trees.

AGROECOSYSTEM Community of microorganisms, plants and animals, with their abiotic environment, that exists on farmed land.

AGROFORESTRY Practice of raising trees and agricultural products such as forage and/or livestock on the same area at the same time.

AGRONOMIC RATES Digested sewage sludge application rates that provide the amount of nitrogen needed by a specific crop or vegetation being grown while minimizing the amount that passes below the root zone.

AGRONOMY Study and application of soil and plant sciences to agriculture.

A-HORIZON Uppermost of the three layers (A-, B-, C-) of soil, up to 2 m (6 ft 6 in.) deep, which accumulates organic matter from plant roots and leaches its fine grains and soluble constituents into the B-horizon, below. *See also* **B-horizon, C-horizon,** and **Podzol.**

AIMLESS DRAINAGE Flow or stream pattern common to low, swampy ground.

AIR Mixture of gases that constitute the earth's atmosphere which, by volume at sea level, comprise 78% nitrogen, 20.95% oxygen, 0.93% argon, and 0.03% carbon dioxide, plus trace amounts of other constituents. *See* also **Standard air.**

AIR-ADMITTANCE VALVE *See* **Valve.**

AIR BALANCING Series of adjustments made to an air-conditioning system to ensure that conditioned air is evenly distributed to all rooms and areas.

AIR BARRIER Material used in the house envelope to retard the passage of air.

AIR BINDING 1. Decrease in the rate of infiltration into soil resulting from an increase in the pressure of the air trapped in the interstices between soil grains; **2.** Reduction in the efficiency of a filter, conduit, or pump due to entrained air.

AIRBLAST Sound pressure wave from a blast travelling through the atmosphere.

AIRBLAST FOCUSING Concentration of sound energy in a small region at ground

level due to refraction of the sound waves back to earth from the atmosphere.

AIR BLOWING *See* **Air sparging.**

AIRBORNE PARTICULATE MATTER Minute particles of solid material that remain in suspension under the prevailing local conditions of temperature, humidity and air movement.

AIRBORNE SOUND Noise or sound passing through the air only.

AIR BOTTLE Thick-walled, steel or composite metal cylinder equipped with a control valve, pressure gauge, and nipple outlet for a high-pressure hose connection, used to contain compressed air or oxygen as an emergency breathing supply. *Also called* **Air tank.**

AIR-BOUND Condition where air trapped in the high point of a pipeline or pump obstructs the free flow of water.

AIR BREAK Piping arrangement whereby the waste from an appliance, fixture, etc. discharges into the open, and then into another fixture so as to prevent backflow or siphonage.

AIR-BUBBLE CURTAIN *See* **Bubble barrier.**

AIR CHAMBER Closed pipe chamber on the discharge of a reciprocating pump that absorbs irregularities in hydraulic conditions and helps induce a uniform flow in suction and discharge lines and protect the pump against shocks caused by pulsating flow.

AIR-CHAMBER PUMP *See* **Pump.**

AIR CHANGE Replacement of one complete volume of air (house, rented/leased area, plant, etc.) by natural or mechanical means, measured in air changes per hour.

AIR CIRCULATION Movement of air from one area to another, either through convection or under pressure in a circulation system.

AIR CLASSIFICATION Process employing an air stream to separate materials by difference in density and aerodynamic properties.

AIR CLASSIFIER Mechanical device using air currents to separate solid components into 'light' or 'heavy' fractions.

AIR CLEANER FILTER ELEMENT Device to remove particulates from the primary air that enters the air induction system of an engine.

AIR COLLECTOR Solar collector that uses air as the heat transfer medium.

AIR COMPRESSOR Machine used to take in air at ambient pressure and compress it to pressures up to approximately 8.5 atmospheres (861 kPa, 125 psi). Compressed air is usually stored in a reservoir before being piped to operate equipment or appliances.

AIR CONDITIONER Mechanical device capable of adjusting the temperature and humidity of collected air and delivering it back to the rooms and areas contained within the system.

AIR CONDITIONING Process of heating or cooling, cleaning, humidifying or dehumidifying, and circulating air.

AIR CONDITIONING SYSTEM Assembly of equipment, sensor and controls necessary for the conditioning and circulation of air to defined standards.

AIR CONDITIONING UNIT Compact, independent, self-contained device that conditions and circulates air.

AIR CONTAMINANTS Airborne substances, solid, liquid, or gaseous, that are considered deleterious to life forms, or to the operation of machinery.

AIR CONTAMINANT CONTROL Process that neutralizes air contaminants and odors. *Also called* **Odor control.**

AIR-COOLED FURNACE WALL *See* **Wall.**

AIR COOLER Device that lowers the temperature of air between its inlet and outlet stages.

AIR CURTAIN Use of bubbles rising from a perforated pipe placed across the bottom of a tank to prevent horizontal mixing of the tank contents.

AIR DEFICIENCY Lack of sufficient air in an air-fuel mixture, to supply the quantity of oxygen required to completely oxidize the fuel.

AIR DIFFUSER Device that converts a stream of air into a diffuse and irregular flow to promote the mixing of primary air with secondary air, typically within a room or space.

AIR-DISPLACEMENT PUMP *See* **Pump.**

AIR-DISTRIBUTING ACOUSTICAL CEILING Suspended acoustical ceiling that permits the passage of a measured volume of air from a pressurized plenum above.

AIR DISTRIBUTION Supply of conditioned air from a central plant under pressure through ductwork to exit louvres and vents.

AIR DUCT Pipe, tube, or passageway for conveying air, normally associated with heating, ventilating, and air conditioning.

AIR ELIMINATOR Plumbing fitting for the automatic release of trapped air and gas from a supply system.

AIR EMISSIONS Airborne solid particulates (such as unburned carbon), gaseous pollutants (such as oxides of nitrogen or sulfur), or odors emanating from any of a broad variety of sources.

AIR ENTRAINING Process of injecting air into material in a manner such that it forms minute bubbles of consistent size and dispersion. *See also* **Air entrainment.**

AIR ESCAPE Mechanism that permits excess air to discharge from a water pipe. *Also called* **Air valve.**

AIR EXCAVATION EQUIPMENT Equipment designed to force compressed air through a wand so as to remove soil from around pipe, cable or other buried services by means of jetting.

AIR-EXHAUST VENTILATOR Fan-equipped hood, ducted to a grill on an external wall, that collects and removes odors and fumes from an area.

AIRFIELD SOIL CLASSIFICATION System based on sieve analyses and consistency limits of soils in which cohesive soils are divided into those with a liquid limit below or above 50%, the former being generally silts, the latter clays.

AIR FILTER Device that traps solid and/or gaseous pollutants from air.

AIR GAP Vertical distance from the top of the flood rim (highest point water can reach in a fixture) to the faucet or spout that supplies water to the fixture.

AIR HANDLING EQUIPMENT Any of a variety of devices designed to move air under controlled conditions. This may include devices to heat, cool, humidify, dehumidify, filter, or other wise condition and/or change air.

AIR-HANDLING SYSTEM System incorporating fans, ducts, controls, dampers, filters, intakes, exhausts, heating and cooling equipment, etc., installed in a building to circulate and recondition air.

AIR HEATER 1. Heat exchanger through which air passes and is heated by a medium of a higher temperature, such as hot combustion gases or steam; **2.** Device for warming the charge of intake air to a diesel's cylinders, especially during cold weather.

AIR INLET Port, duct, grille, etc. through which air is admitted to a device, or vented from a system to a room..

AIR INTAKE Device through which air is drawn.

AIR LIFT Means of raising fluids by forcing air into the bottom of an open discharge pipe submerged in a well or tank. *See also* **Pump.**

AIRLINE MASK Breathing apparatus connected to an air supply outside a contaminated area.

AIRLINE PRESSURE REGULATOR Regulator that transforms a fluctuating air pressure supply to provide a constant lower pressure output.

AIR LOCK 1. Chamber that can be made airtight, separating two other rooms, chambers, or areas; **2.** Bubble of air, located in a pipe, or in a pump that has not been properly primed, that prohibits the movement of liquid.

AIR MAINTENANCE DEVICE Device that automatically maintains the required air pressure in a sprinkler system.

AIR MASK *See* **Air pack.**

AIR MASS Body of air covering a large area, i.e., 1000 km or more, and which possesses distinct characteristics: temperature, humidity, pressure, etc.

AIR MIXING PLENUM Chamber in an air conditioning system where recirculating air is mixed with fresh air.

AIR MONITORING Means taken to measure the quality of air and its constituents.

AIR OUTLET Device that permits air to pass from within a conditioning or ducting system.

AIR OXIDATION REACTOR Device or process vessel in which one or more organic reactants are combined with air, or a combination of air and oxygen, to produce one or more organic compounds, including ammoxidation and oxychlorination reactions.

AIR PACK Self-contained breathing equipment that provides an air supply from tanks through a regulator. *Also called* **Air mask.**

AIR PADDING Process of pumping dry air (dewpoint -40°C (-40°F)) into a container to assist with the withdrawal of a liquid or to force a liquefied gas such as chlorine or sulfur dioxide out of a container.

AIR PARCEL Theory involving a body of air whose properties and transformations may be determined from the laws of physics and mechanics.

AIR PERMEABILITY Property that allows the passage of air through a mass.

AIR POLLUTION Presence of unwanted material in the atmosphere, which includes any material present in sufficient concentrations for a sufficient time, and under circumstances to interfere significantly with the comfort, health, or welfare of persons, or with the full use and enjoyment of property.

AIR POLLUTION CONTROL AGENCY Any agency which is charged with responsibility and has authority to monitor air quality and enforce means by which to maintain a defined quality.

AIR POLLUTION INDEX Arbitrary expression of the concentration of one or more pollutants present in an air sample, used to scale the severity of air pollution at the place of sampling.

AIR PRESSURE SPRINKLER SYSTEM Sprinkler system in which air pressure is used to force water from a storage tank into the system.

AIR PUMP *See* **Pump.**

AIR PURGE VALVE *See* **Valve.**

AIR QUALITY 1. An arbitrarily-established criterion that establishes the constituents, and the quantities of those constituents in a measured sample, that comprises a standard or measure of air; **2.** Concentration in a measured sample of air of one or more pollutants.

AIR QUALITY ACT Legislation that establishes a minimum acceptable air quality: for a region; or under defined circumstances.

AIR QUALITY CRITERIA Levels of pollution and lengths of exposure to pollution above which negative effects on human health and/or welfare may occur.

AIR QUALITY MANAGEMENT Principles and strategy by which an Air Quality Act or similar legislation shall be administered.

AIR QUALITY RESTRICTED OPERATION OF A SPRAY TOWER Operation utilizing formulations (e.g., those with high nonionic content) which require a very high rate of wet scrubbing to maintain desirable quality of stack gases, and thus generate much greater quantities of waste water than can be recycled to process.

AIR QUALITY STANDARDS Level of pollutants prescribed by law that cannot be exceeded during a specified time in a defined area.

AIR STRIPPING Desorption operation employed to transfer one or more volatile components from a liquid mixture into a gas (air) either with or without the application of heat to the liquid.

AIR REHEATER Device that adds heat to air being circulated through a system.

AIR RELEASE VALVE *See* **Valve.**

AIR RELIEF VALVE *See* **Valve.**

AIR RIGHTS Property rights for the control or specific use of a designated airspace.

AIR SCOURING Technique for cleaning water mains up to approximately 200 mm (8 in.) in diameter by introducing short, high-pressure air blasts which remove most loose deposits from the inside walls.

AIR-SLACK Condition where a soft-body clay, after absorbing moisture and being exposed to the atmosphere, will spall a piece of clay and/or glaze.

AIR SLAKED Wetted by exposure to moisture in the atmosphere.

AIR SPACE Landfill's remaining volume or capacity for waste disposal at a given time, not including volume occupied by cap, cover, and liner.

AIR SPARGING Mixing process employing compressed air as the mixer where air is injected into the bottom of a tank through multiple ports, and its upward movement through the liquid provides the mixing.

AIR STRIPPING Process for the removal of ammonia or volatile organic compounds.

AIR SUPPLY UNIT Machine to refill, on-site, exhausted compressed self-contained breathing apparatus air cylinders.

AIR TANK *See* **Air bottle.**

AIR TEST Test of the integrity of a drain in which the section of the system is sealed and subjected to a positive air pressure, which must not drop below a specified percent of the original pressure over a defined period of time.

AIRTIGHT Not capable of permitting the passage of air at designed pressures.

AIRTIGHT INSPECTION COVER Sealable, nonventilating drain cover.

AIRTIGHTNESS Ability of the building envelope to resist infiltration and exfiltration of air.

AIRTIGHT WOOD STOVE Closed and sealed metal container, often decoratively finished, having one or more means of controlling air input and fume exhaust. Heat distribution is by radiation from the outer casing.

AIR-TO-AIR RESISTANCE Resistance to the flow of air, or the mixing of air, due to such factors as laminar flow, pressure differential, temperature differential, or other factors.

AIR TREATMENT Any process that changes the nature of air: temperature, humidity, dust content, purity, etc.

AIR VALVE *See* **Valve.**

AIR VENT Means of releasing trapped air from within a water distribution system, usually located at its highest point.

AIR VENTILATION Quantity of fresh air necessary to maintain the desired quality of air within a space.

AIR VESSEL Tank containing air under pressure.

AIR WASH Process in which compressed air and water are used to wash the filtering medium of a rapid sand filter.

AIR WASHER System for cleaning air, usually incorporating a water spray system.

AIR-WATER STORAGE TANK Water storage vessel in which the air above the water level is under pressure.

AIR WAVES Airborne vibrations or pulses, usually (but not always) accompanied by sound waves.

A_k Area of each natural draft opening (k) in a total enclosure, in square meters.

ALANINE Essential amino acid, $CH_3CH(NH_2)COOH$, having a molecular weight of 89.1.

ALARM Any signal indicating an out-of-normal condition.

ALARM BELL Bell that is rung automatically by a device set to detect an out-of-limit condition, or that can be manually activated to alert to a situation.

ALARM CIRCUIT Electrical circuit connecting two points in an alarm system.

ALARM CONTACT Switch that operates when some preset low, high, or abnormal condition exists.

ALARM SIGNAL Signal indicating an emergency requiring immediate action, such as an alarm for fire from a manual box, a water flow alarm, an alarm from an automatic fire alarm, etc.

ALARM SWITCH *See* **Switch.**

ALARM VERIFICATION Automatic fire detection and alarm system that must report or confirm alarm conditions for a minimum period after being reset to be accepted as a valid alarm initiation signal.

ALBEDO Ratio of light reflected from an object to that falling on it.

ALBIC Soil horizon from which clay and iron oxides have been removed.

ALCOHOLIC FERMENTATION Production of ethanol from sugar by fermentation.

ALDEHYDES Air pollutant consisting of organic compounds containing the group -CHO attached to a hydrocarbon, characterized by an unpleasant smell and which can be an irritant to nose and eyes.

ALDRIN + DIELDRIN Organochlorine insecticides. Aldrin is rapidly converted to dieldrin under most environmental and biological conditions. Dieldrin is a highly persistent compound, with low mobility in soil. Both insecticides have been detected in drinking water samples and surface waters.

ALGAE Range of aquatic plants that in sunlight convert carbon dioxide into oxygen by photosynthesis.

ALGAL BLOOM Population explosion of micro-algae in surface waters.

ALGICIDAL Having the property of killing algae.

ALGICIDE Substance or chemical capable of killing or controlling the growth of algae.

ALGISTATIC Having the property of inhibiting algal growth.

ALIPHATIC Organic compounds having open chains of carbon atoms: paraffins, olefins and acetylenes and their derivatives.

ALIQUOT Portion of a sample; an equally divided portion of a sample.

ALKALI A base that dissolves in water producing hydroxide ions. Soluble salt, commonly sodium, potassium, magnesium or calcium, having the property of combining with acid to form neutral salts that may be used in chemical processes such as water conditioning and wastewater treatment.

ALKALI ACCUMULATION Concentration over time of carbonates and hydroxides of alkali metals in the upper layers of a soil due to evaporation of water containing carbonate and bicarbonate or hydroxide.

ALKALI FLAT Flat area, usually the bottom of an undrained basin in an arid region, containing an excess of alkali.

ALKALINE 1. Water, wastewater, or soil containing sufficient amount of alkali substances to raise the pH above 7.0; **2.** Igneous rocks containing an abundance of alkaline metals: sodium-rich pyroxenes and amphiboles; and sodium- and/or potassium-rich feldspar.

ALKALINE CLEANING Use of a solution (bath), usually detergent, to remove lard, oil, and other such compounds from a metal surface. Alkaline cleaning is usually followed by a water rinse. The rinse may consist of single or multiple stage rinsing.

ALKALINE CLEANING BATH Bath consisting of an alkaline cleaning solution through which a workpiece is processed.

ALKALINE CLEANING RINSE Rinse following an alkaline cleaning bath through which a workpiece is processed.

ALKALINE MINE DRAINAGE Mine drainage which, before any treatment, has a pH equal to or greater than 6.0 and total iron concentration of less than 10 mg/L.

ALKALINE SOIL Soil containing soluble salts of magnesium, sodium, or the like, and having a pH of between 7.0 and 8.5.

ALKALINITY Alkali concentration or alkaline quality of an alkali-containing substance; the capacity of water to neutralize an acid based on its content of carbonate, bicarbonate, hydroxide, and occasionally borate, silicate, and phosphate.

ALKALI RESISTANCE Extent to which a material resists reaction with alkali.

ALKALINE WATER 1. Water having a pH greater than 7.0; **2.** Water having a high sodium content, but relatively low dissolved solids.

ALKALI SOIL Soil in which the growth of most plant crops is reduced due to having a high degree of alkalinity (pH 8.5 or higher), and/or a 15% or higher exchangeable sodium content.

ALKYL BENZONE SULFONATE (ABS) Type of surface active agent which, in waste treatment processes, tends to cause foaming and resist biological degradation. ABS has generally been replaced in detergents by linear alkyl sulfonate (LAS) which is biodegradable.

ALKYLI MERCURY Highly toxic mercurial compound used as a fungicide and seed dressing.

ALKYL SULFONATES Surface-active agents used in synthetic detergents.

ALL-AIR Air conditioning system in which

the air is exhausted to atmosphere following one complete circulation cycle.

ALLELOCHEMISTRY Chemistry of substances released by one population that affect another population.

ALLELOPATHY Influence that the chemical exudates of one living plant have upon another.

ALLEY COLLECTION Collection of solid waste from containers placed adjacent to or in an alley.

ALLOCHTHONOUS Material that originated elsewhere.

ALLOWABLE COSTS Project costs that are: eligible, reasonable, necessary, and allocable to the project.

ALLOWABLE-CUT EFFECT Allocation of anticipated future forest timber yields to the present allowable cut. This is employed to increase current harvest levels (especially when constrained by even flow) by spreading anticipated future growth over all the years in rotation.

ALLOWABLE EMISSIONS An emissions rate of a stationary source calculated using the maximum rated capacity of the source.

ALLOY PIPE *See* **Pipe.**

ALLOY STEEL *See* **Steel.**

ALLUVIAL CLAY Clayey material transported by streams from the place of origin.

ALLUVIAL CONE Alluvial material deposited where a stream debouches from a range of hills or mountains undergoing erosion, shaped like a segment of a cone.

ALLUVIAL DEPOSIT Sediment deposited along the course of a stream.

ALLUVIAL FAN Sloped, spreading deposit of boulders, gravel, and sand left by a stream where it emptied from a more confined channel.

ALLUVIAL PLAIN Area of relatively flat land formed by alluvial material eroded from areas of higher elevation.

ALLUVIAL RIVER River that has formed its channel through the process of aggradation.

ALLUVIAL-SLOPE SPRING Spring occurring on the lower slope of an alluvial cone at the point where the water-table slope and surface gradient are equal.

ALLUVIAL SOIL Soil, sand, or gravel deposited by flowing water.

ALLUVIATION Process of accumulating deposits of gravel, sand, silt, clay and other materials at places in streams, rivers, lakes and estuaries where the velocity of flow is sufficiently reduced to cause their settling out.

ALLUVIUM Sediments deposited by running or flowing water.

ALPHA DECAY Radioactive process in which an alpha particle is emitted from the nucleus of an atom, decreasing the atom's atomic number by two.

ALPHANUMERIC DISPLAY Type of digital readout that displays both letters and numbers, separately or mixed.

ALPHA PARTICLE Positively-charged particle emitted by certain radioactive materials and consisting of two neutrons and two protons.

ALPINE Land elevations above the tree line, but below the permanent snow line.

ALPINE OROGENY Mountain building resulting from the movement of tectonic plates and which culminated in the Miocene Period, forming the Alpine-Himalayan belt.

ALTERED DISCHARGE Any discharge other than a current discharge or improved discharge

ALTERNATE BID Amount stated in a bid to be added to or subtracted from the amount of the base bid under specified conditions, such as substitution of materials or change in scope of the project or method of construction. *Also called* **Additive alternate.**

ALTERNATE PROPOSAL Submission that varies from what is described in the bid documents.

ALTERNATING DEVICE Device, part of a larger complex, to or through which a stream may be diverted or directed automatically or manually in accordance with a predetermined sequence.

ALTERNATING SPRINKLER SYSTEM Fire protection sprinkler system that can be changed from a wet-pipe system in summer to a dry-pipe system in winter.

ALTERNATIVE EFFLUENT LIMITATIONS All effluent limitations or standards of performance for the control of the thermal component of any discharge.

ALTERNATIVE ENERGY Energy derived from the exploitation of a renewable resource: hydroelectric power, wind power, biogas, solar energy, etc.

ALTERNATIVE METHOD Any method of sampling and analyzing for an air pollutant which is not a reference or equivalent method but which has been demonstrated to produce adequate results.

ALTERNATIVE TECHNOLOGY Proven wastewater treatment processes and techniques which provide for the reclaiming and reuse of water, productively recycle wastewater constituents or otherwise eliminate the discharge of pollutants, or recover energy. Specifically, alternative technology includes land application of effluent and sludge; aquifer recharge; aquaculture; direct reuse (nonpotable); horticulture; revegetation of disturbed land; containment ponds; sludge composting and drying prior to land application; self-sustaining incineration; and methane recovery.

ALTERNATIVE TO CONVENTIONAL TREATMENT WORKS FOR A SMALL COMMUNITY Alternative technology used by treatment works in small communities include specific defined alternative technologies as well as individual and onsite systems; small diameter gravity, pressure or vacuum sewers conveying treated or partially treated wastewater. Such systems may also include small diameter gravity sewers carrying raw wastewater to cluster systems.

ALTERNATIVE WATER SUPPLIES Includes, but is not limited to, drinking water and household water supplies.

ALTIMETER Aneroid barometer calibrated to read height above a reference level, whose barometric pressure is set on the instrument's dial.

ALTITUDE Height above some reference level.

ALTITUDE-CONTROL VALVE *See* **Valve.**

ALTOCUMULUS Clouds generated by up-currents not having their origin at ground or sea level.

ALTOSTRATUS High-level layer of cloud of uniform appearance, but lacking dappled or fibrous features.

ALUM Aluminum sulfate $(Al_2(SO_4)_3.18H_2O)$, used as a coagulant in potable water treatment and in wastewater treatment.

ALUMINA Aluminium trioxide (Al_2O_3), used as a desiccant.

ALUMINIZED STEEL PIPE *See* **Pipe.**

ALUMINUM Very light, ductile metallic element (Al) produced from bauxite.

ALUMINUM HYDRAULIC SHORING Pre-engineered shoring system comprised of aluminum hydraulic cylinders (cross braces) used in conjunction with vertical rails (uprights) or horizontal rails (walers), designed specifically to support the sidewalls of an excavation and prevent cave-ins.

ALUMINUM PIPE Pipe fabricated of an aluminum alloy, restricted for use to service where no chemical reaction can occur, usually irrigation and potable water supply.

ALUMINUM SULFATE Chemical used in potable water and wastewater treatment prepared by combining bauxite with sulfuric acid.

ALUMINOUS CEMENT Portland cement containing 30% to 50% alumina, 30% to 50% lime (calcium hydroxide), and not more than 30% to 50% iron oxide, silica, and other ingredients.

AMBIENT 1. Surrounding, encircling; **2.** Average background noise at a given location at a given time.

AMBIENT AIR 1. The surrounding air; **2.** That portion of the atmosphere, external to buildings, to which the general public has access.

AMBIENT CONDITIONS Quality of the surrounding environment.

AMBIENT MOISTURE Amount of moisture permissible in the air surrounding a shipment of secondary material.

AMBIENT TEMPERATURE Surrounding air temperature.

AMBIENT WATER CRITERION Concentration of a toxic pollutant in a navigable water that, based upon available data, will not result in adverse impact on important aquatic life, or on consumers of such aquatic life, after exposure of that aquatic life for periods of time exceeding 96 hours and continuing at least through one reproductive cycle; and will not result in a significant risk of adverse health effects in a large human population based on available information such as mammalian laboratory toxicity data, epidemiological studies of human occupational exposures, or human exposure data, or other relevant data.

AMEBA Type of small, one-celled animal or protozoan.

AMENDED WATER Water to which a surfactant has been added.

AMENDMENT Substance added to a soil to improve its physical properties such as texture, as opposed to fertilizer, that is added to improve chemical properties.

AMENITY AREA Area or areas within the boundaries of a project intended for recreation purposes and that may include landscaped site areas, patios, common areas, communal lounges, swimming pools, and areas for similar purposes.

AMERICAN STANDARD FITTINGS Standardized types and dimensions of various pipe fittings, as established in standards published by the American Standards Institute.

AMERICAN WHEEL Water turbine with the flow first inward, then axial.

AMINO ACID Any organic compound containing both an amino group (NH_2) and a carboxylic acid group (COOH); an essential component of the protein molecule.

AMMONIA Chemical combination of hydrogen (H) and nitrogen (N); used in potable water and wastewater engineering and expressed as NH_3.

AMMONIACAL NITROGEN Nitrogen present as ammonia and ammonium ion in liquid effluents.

AMMONIA-N (or ammonia-nitrogen) Value obtained by manual distillation (at pH 9.5) followed by the Nesslerization method.

AMMONIA SLIP Excess ammonia discharged in stack gases from selective noncatalytic reduction systems used for nitrogen-oxide control using ammonia or urea injection.

AMMONIATOR Apparatus used to apply ammonia, or ammonium compounds, in potable water or wastewater treatment.

AMMONIFICATION Bacterial decomposition of organic nitrogen to ammonia.

AMORTIZATION 1. Process of writing off a charge, as for an item of equipment or machinery, by prorating its capital cost over time; **2.** Repayment of debt, principal and interest, in (usually) equal installments at regular periods over a given period; **3.** Procedure by which the capital cost of forestry-related projects, such as roads or bridges, is written off over a longer period as the timber volumes developed by the projects are harvested and extracted.

AMOUNT OF PESTICIDE OR ACTIVE INGREDIENT Weight or volume of the pesticide or active ingredient used in producing a pesticide, expressed as weight for solid or semisolid products and as weight or volume of liquid products.

AMPERAGE Measurement of the amount of electricity flowing past a given point in a conductor per second. *Also called* **Current.**

AMPERE One of the seven base units of the SI system of measurement: a unit of electrical current equal to that current which, if maintained in two straight parallel conductors of infinite length, or negligible circular cross-section, and placed 1 m apart in vacuum, would produce between these conductors a force equal to 0.2 µN/m of length. Symbol: A. *See also the appendix:* **Metric and nonmetric measurement.**

AMPERE-HOUR Derived unit of electric charge over a specific length of time at 26.6°C (80°F). Symbol: A/h. Multiply by 3.6 to obtain kilocoulomb, symbol: kC. *See also the appendix:* **Metric and nonmetric measurement.**

AMPERE PER METER A derived unit of magnetic field strength with a compound name of the SI system of measurement. Symbol: A/m. *See also the appendix:* **Metric and nonmetric measurement.**

AMPERE TURN SI unit of magnetomotive force equal to one ampere flowing through one turn of a magnetizing coil.

AMPEROMETRIC Measurement of electric current flowing or generated, rather than by voltage.

AMPEROMETRIC TITRATION Electrometric detection of the equivalence point in a titration by observation of the change in diffusion current at a suitable applied voltage as a function of the volume of titrating solution.

AMPHIBIAN AND REPTILE POISONS AND REPELLENTS Includes all substances or mixtures intended for preventing, destroying, repelling, or mitigating amphibians and reptiles declared to be pests. Amphibian and reptile poisons and repellents include, but are not limited to: **(a)** substances or mixtures intended for use in baits or sprays for killing or repelling snakes, frogs, or lizards; and **(b)** reproductive inhibitors intended to reduce or alter the reproductive capacity or potential of amphibian or reptile pests.

AMPHOTERIC Having the capacity to behave as either an acid or a base; e.g., aluminum hydroxide, $Al(OH)_3$, which neutralizes acids to form aluminum salts and reacts with strong bases to form aluminates.

AMPLIFICATION Process of increasing the strength of a signal.

AMPLITUDE Peak or maximum value in a wave-type transmission or motion.

AMPLITUDE OF TIDE Semi-range of a specific tidal occurrence.

ANABAENA Genus of blue-green algae able to fix atmospheric nitrogen and which form symbiotic relationship with some aquatic ferns.

ANABATIC WIND Wind generated as a slope is warmed, causing an upward movement of air.

ANABRANCH Effluent of a stream that rejoins the main stream, forming an island between the two watercourses.

ANACLINAL STREAM Stream that descends in a direction opposite to the dip of the underlying rock layers.

ANAEROBE Microorganism, as a bacterium, able to live in the absence of oxygen.

ANAEROBIC Without oxygen; activity occurring in the absence of oxygen.

ANAEROBIC BACTERIA *See* **Bacteria.**

ANAEROBIC CONTACT PROCESS Wastewater treatment process in which the microorganisms responsible for waste stabilization are removed from the treated effluent and returned upstream to enhance the rate of treatment.

ANAEROBIC DECOMPOSITION Decay or breaking down of organic material in an environment containing no free or dissolved oxygen.

ANAEROBIC DIGESTION Breakdown of organic components by microbial components in the absence of oxygen.

ANAEROBIC RESPIRATION Process by which organisms release energy by the chemical breakdown of food substances without consuming gaseous or dissolved oxygen.

ANAEROBIC WASTE TREATMENT
Wastewater treatment accomplished through
the action of microorganisms in the absence
of air or elemental oxygen, commonly employing methane fermentation.

ANA-FRONT Atmospheric front at which
the air on one side, commonly cold air, is
ascending.

ANALOG Of or pertaining to the general
class of devices whose output varies as a
continuous function of its input.

ANALOG DETECTOR Fire detector that
generates electronic signals to represent what
it senses.

ANALOG DISPLAY Display that provides
a continuous indication relative to reference
points.

ANALOGOUS ARTICLES Articles not
detailed in a classification but having characteristics similar to those classified.

ANALOG PROCESSING Method of processing information using the representative
value of a physical variable. (Digital processing, that uses direct numerical values, is
many times faster and more reliable.)

ANALOGY Correspondence in some respects between objects which are otherwise
dissimilar.

ANALYSIS 1. Record of an examination of
potable water or wastewater; **2.** Investigation of a situation through detailed consideration of its essential elements. There are many
approaches, including:

Benefit/cost: Evaluation of the profits,
income, output, or other benefits
against the cost of obtaining them.

Correlation: Measurement of the degree
of relationship, if any, between variables.

Cost/benefit: *See* **Benefit/cost** above.

Cost/volume/profit: Study of the effects
of changes in fixed costs, variable
costs, sales quantities, sales prices,
and/or sales mix.

Incremental: Evaluation of the small increase, usually in revenue and expenses or cash flow, that would result from a course of action.

Input/output: Summary of the transactions between all economic units involved in a project, showing those
consumed and the resulting product.

Network: Method of planning and scheduling a project, commonly displayed
in diagrammatic form, so as to identify the interrelated sequences that
must be completed to finish a project.

Sensitivity: Measurement of the degree
of change in one variable as a response to a measured change in another variable.

ANALYZER Device that conducts periodic
or continuous measurement of some factor.

ANCHORAGE 1. Block of material installed
along a pipeline, large, and heavy enough to
resist forces resulting from fluids passing
through the pipe, or due to changes in temperature of the pipe; **2.** Place where ships
may be anchored.

ANCHOR BLOCK Concrete block cast
around or adjacent to a pipeline, particularly
where it changes direction, of sufficient mass
to resist forces generated by fluids passing
through the pipe.

ANCHOR GATE Heavy gate supported at
its upper bearing, used to close off a canal or
channel.

ANCHOR ICE Ice formation that extends
from the surface to the bottom of a stream or
pond and is thus not free to move.

ANCIENT LIGHTS Windows in an external wall that have 'enjoyed' daylight for at
least twenty years and which are protected
against being closed or built over without
the owner's consent.

ANCIENT MONUMENT Building, structure or site listed and protected under regulation.

ANCIENT SOIL Older, weathered subsurface soil zone, the product of past climates

and processes.

ANCILLARY EQUIPMENT 1. Devices including, but not limited to, piping, fittings, flanges, valves, and pumps, that are used to distribute, meter, or control the flow of a hazardous or nonhazardous waste from its point of generation to a storage or treatment tank(s), between waste storage and treatment tanks to a point of disposal onsite, or to a point of shipment for disposal off-site; **2.** Devices including, but not limited to piping, fittings, flanges, valves, and pumps used to distribute, meter, or control the flow of regulated substances to and from an underground storage tank.

'AND' DEVICE Control device that has its output in the logical '1' state if all the control signals assume the logical '1' state.

ANDESITE Fine-grained, intermediate volcanic rock associated with continental crust.

ANECHOIC ROOM Room whose internal boundaries almost completely absorb any sound generated within it.

ANEMOMETER Instrument for measuring wind speed.

ANEMOPHILY Wind pollination.

ANEROID BAROMETER Portable instrument that incorporates a sealed box that expands and contracts with the rise and fall of air pressure and which is coupled to a needle capable of registering such movement on a scale.

ANESTHETIC EFFECT Loss of sensation with or without the loss of consciousness. It can be caused by the inhalation of volatile hydrocarbons.

ANGLE DOZER Tractor dozer whose blade can be pivoted about a vertical center pin so as to cast its load to one side or the other.

ANGLED TEE Pipe fitting with the junction or inlet leg at a shallow angle.

ANGLE GATE VALVE *See* **Valve.**

ANGLE GLOBE VALVE *See* **Valve.**

ANGLE OF CURRENT In stream gauging, the angular difference between 90° and the angle made by the current with a measuring section.

ANGLE OF INCIDENCE Angle formed by the path of a body or radiation incident on a surface and a perpendicular to the surface at the point of impact.

ANGLE OF INDRAFT Angle of inclination of a steady wind to the isobars (called positive when the direction is toward low pressure).

ANGLE OF NATURAL REPOSE Angle to the horizontal at which a material will no longer be affected by gravity and slide downward of its own accord.

ANGLE OF THREAD Angle included between the sides of a thread measured in a plane through the centerline.

ANGLE STOP VALVE *See* **Valve.**

ANGLE VALVE *See* **Valve.**

ANGSTROM Obsolete unit of length, equal to 0.1 nanometer, that should not be used with the SI system of measurement. Symbol: Å. *See also the appendix:* **Metric and nonmetric measurement.**

ANGULAR AGGREGATE *See* **Aggregate.**

ANGULAR FREQUENCY Frequency of sound, expressed in radians per second.

ANGULARITY 1. State, condition or quality of being angular; **2.** Angle between the centerline of a stream or conduit and a line normal to the centerline of a structure crossing it.

ANGULARITY CORRECTION Correction made to an observed velocity when the direction of the current of a flow stream is not exactly at right angles to the discharge section line.

ANGULAR MISALIGNMENT Minor angle formed between the intersecting axes of two pipes.

ANGULAR PARTICLE Angular shape of some coarse aggregate particles.

ANGULAR INCONFORMITY Angular divergence between rocks of different age.

ANHYDRITE Evaporite mineral, calcium sulfate ($CaSO_4$), found in sedimentary rocks.

ANHYDROUS Without water, especially water of crystallization.

ANHYDROUS PRODUCT Theoretical product that would result if all water were removed from the actual product.

ANICUT Dam constructed across a stream to regulate irrigation.

ANIMAL FEED Any crop grown for consumption by animals, such as pasture crops, forage, and grain.

ANIMAL FEEDING OPERATION Lot or facility (other than an aquatic animal production facility) where the following conditions are met: (a) animals (other than aquatic animals) have been, are, or will be stabled or confined and fed or maintained for a total of 45 days or more in any 12-month period, and (b) crops, vegetation forage growth, or postharvest residues are not sustained in the normal growing season over any portion of the lot or facility.

ANIMALS Organism having the characteristics of a fixed structure and limited growth, specialized sense organs and the power of locomotion.

ANIMAL UNIT Management unit of livestock related to food needs.

ANION Negatively charged ion.

ANIONIC SURFACTANT Ionic surface-active substance, the hydrophilic group of which carries a negative charge in a washing solution.

ANISOTROPIC Unequal, uneven, irregular: not having the same physical properties in all directions.

ANISOTROPIC SOIL Soil mass having different properties in different directions at any given point, referring primarily to stress/strain or permeability characteristics.

ANKERITE Carbonate mineral (Ca,Mg,Fe)

$(CO_3)_2$.

ANNEALING WITH OIL Use of oil to quench a workpiece as it passes from an annealing furnace.

ANNEALING WITH WATER Use of a water spray or bath, of which water is the major constituent, to quench a workpiece as it passes from an annealing furnace.

ANNEXATION Legal process by which an incorporated municipality expands its boundaries.

ANNUAL 1. Occurring once each year; **2.** Corporate fiscal year.

ANNUAL ALLOWABLE HARVEST Quantity of timber scheduled to be removed from a particular management unit in one year.

ANNUAL AVERAGE Maximum allowable discharge of BODS or TSS, as calculated by multiplying the total mass (kkg or 1000 lb) of each final product produced for the entire processing season or calendar year by the applicable annual average limitation.

ANNUAL CAPACITY FACTOR Ratio between the actual heat input to a steam generating unit from the listed fuels during a calendar year and the potential heat input to the steam generating unit had it been operated for 8,760 hours during a calendar year at the maximum steady state design heat input capacity.

ANNUAL COKE PRODUCTION Coke produced in the batteries connected to the coke by-product recovery plant over a 12-month period.

ANNUAL CYCLE One of the two cycles (the other being diurnal) dominating the terrestrial environment and observed as a result of the inclination of the earth's axis.

ANNUAL DOCUMENT LOG Detailed information maintained at a PCB waste handling facility.

ANNUAL FLOOD Maximum 24-hour average rate of flow occurring in a stream during any consecutive 12-month period.

ANNUAL GROWTH Average annual increase in the biomass of plants of a specified area.

ANNUAL GROWTH RING Growth layer put on by a tree in a single growth year, including springwood and summerwood. *Also called* **Growth ring.**

ANNUALIZED COST Stream of annual payments for a determined time period, equal in value to a onetime payment based on a selected rate of interest.

ANNUAL LOAD FACTOR Design calculation based on the actual load measured over an entire year, with allowance for any design consideration which may increase or decrease that figure.

ANNUAL VARIATION General pattern of an element throughout the year, obtained by plotting its normal values at regular intervals and connecting the plotted values by a smooth curve.

ANNULAR Ring-shaped.

ANNULUS 1. Circular space or groove; the space between two concentric circles; **2.** Space between a drill string, pump column, or casing and the wall of a borehole or outer casing.

ANNULUS GROUTING Grouting the ring-shaped gap between two circular objects, one inside the other, as between two pipes.

ANNUNCIATOR Device, usually electronic or electromagnetic, but also mechanical, that indicates the origin of a signal or message.

ANODE 1. Positive terminal of an electrical circuit; **2.** Opposite pole from the cathode in electrolysis.

ANODIZE Electrolytically form a thin film of alumina on the surface of aluminum to improve resistance to corrosion.

ANOMALY Deviation from the norm or common order for which an explanation is not apparent on the basis of available data.

ANORTHOSITE Coarse-grained, igneous rock, usually Precambrian, composed almost entirely of plagioclase feldspar.

ANOXIC Absence of oxygen.

ANOXIC DENITRIFICATION Biological process in which nitrate nitrogen is converted by microorganisms to nitrogen gas in the absence of oxygen.

ANTAGONISM State in which the presence of two or more substances diminishes or decreases the toxic effects of the substances acting independently.

ANTECEDENT Occurrence or event prior to another.

ANTECEDENT DRAINAGE Drainage system that maintains its original direction across a line of localized uplift that occurred after the system was established.

ANTECEDENT MOISTURE Amount or degree of wetness of a soil at the beginning of a runoff period.

ANTECEDENT PRECIPITATION Rainfall that occurred prior to the event under consideration.

ANTECEDENT-PRECIPITATION INDEX Index of moisture stored in a drainage basin prior to onset of a storm.

ANTECEDENT STREAM Stream that maintained and developed its course during and after a period of disturbance which altered the surrounding drainage area.

ANTHRACITE FILTER Type of filter used in large water treatment plants and consisting of multiple layers of anthracite arranged in diminishing size to trap particulate matter from water flowing through the bed. The filter is cleaned by reversing the flow and flushing out the trapped material.

ANTHROPIC Of or pertaining to humans.

ANTHROPOGENIC Resulting from human activity.

ANTIBIOSIS Antagonistic action of one or more microorganisms to the detriment of some or all.

ANTIBIOTIC Chemical substance, produced by microorganisms, which has the capacity to inhibit the growth of, or destroy, other microorganisms.

ANTICHLORS Reagents, such as sulfur dioxide, sodium bisulfite, and sodium thiosulfate, used to remove excess chlorine residuals from potable water or treated wastewater effluent.

ANTICLINE Geological fold that is convex upward with the oldest rocks occupying the inner core.

ANTICLINAL SPRING Contact spring occurring along the surface outcrop of an anticline, from a pervious stratum overlying one less pervious.

ANTICORROSION TREATMENT Mechanical or chemical treatment aimed at eliminating corrosion-producing qualities of a substance.

ANTICYCLONE Extensive system of winds spiralling outward from a high-pressure center, circling clockwise in the Northern Hemisphere, and counterclockwise in the Southern Hemisphere.

ANTIDUNE Sandhill with a steep upstream face.

ANTIDUNE MOVEMENT Transport phenomenon in water flowing on a sand bar in which a sand dune moves against the current.

ANTI-EXTRUSION RING Rigid or semi-rigid ring employed at one or both ends of a packing set, primarily to prevent extrusion into clearances. *Also called* a **Bull ring.**

ANTIFER Type of armored revetment consisting of precast concrete blocks shaped as modified cubes having a central hole and indentations on four of its six faces.

ANTIFLOOD VALVE *See* **Valve.**

ANTIFREEZE SPRINKLER SYSTEM System, usually 20 heads or less, connected to a wet pipe sprinkler system to protect small unheated areas of heated occupancies.

ANTIFRICTION BEARING *See* **Bearing.**

ANTIGRAFFITI TREATMENT Textured or applied finish of a type likely to discour-

age marking and/or similar abuse.

ANTIHAMMER DEVICE Air chamber, such as a closed length of pipe or coil, designed to absorb the shock caused by a rapidly closed valve.

ANTIHEAVE MEASURE Precaution taken to prevent soil heave.

ANTI-INTRUDER FENCING Woven wire fencing product.

ANTIMICROBIAL AGENTS Substances or mixtures, except those defined as fungicides and slimicides, intended for inhibiting the growth of, or destroying any bacteria, fungi pathogenic to man and other animals, or viruses declared to be pests, including disinfectants intended to destroy or irreversibly inactivate infectious or other undesirable bacteria, pathogenic fungi, or viruses on surfaces or inanimate objects; sanitizers intended to reduce the number of living bacteria or viable virus particles on inanimate surfaces, in water, or in air; bacteriostats intended to inhibit the growth of bacteria in the presence of moisture; sterilizers intended to destroy viruses and all living bacteria, fungi and their spores, on inanimate surfaces; fungicides and fungistats intended to inhibit the growth of, or destroy fungi (including yeasts), pathogenic to man or other animals on inanimate surfaces; and commodity preservatives and protectants intended to inhibit the growth of, or destroy bacteria in or on raw materials (such as adhesives and plastics) used in manufacturing

ANTINODE Point, line or surface where the amplitude of sound is at its maximum.

ANTINOISE *See* **White sound.**

ANTIOZONANT Compound ingredient used to retard deterioration caused by ozone.

ANTISIPHON PIPE Branch vent.

ANTISIPHON TRAP Trap in a drainage system designed to preserve a water seal by preventing siphonage.

ANTISIPHON VALVE *See* **Valve.**

ANTISTATIC AGENT Compound(s) that prevent the accumulation of static electricity by increasing surface conductivity.

ANTIVIBRATION MOUNTING Flexible pad or system of springs for mounting mechanical equipment so as to prevent transmission of vibration and equipment noise to the structure.

AOX Organic halogens subject to absorption: a measure of the amount of chlorine (and other halogens) combined with organic compounds.

APERIODIC Taking place at unequal intervals; not occurring regularly.

APHOTIC ZONE Depth in water below which light does not penetrate.

APHYTIC ZONE Ground level below water that, due to its depth below the surface, lacks plant life.

API SCALE Arbitrary gravity scale adopted by the American Petroleum Institute for use with crude oil and equal to: degrees API = (141.5-131.5)/specific gravity at 60°F.

APOGEAN RANGE Range of tide at the occurrence of apogean tides.

APOGEAN TIDES Tides with decreased range occurring at the time when the moon is in apogee.

APOGEE Point in the orbit of the moon or of an artificial satellite most distant from earth.

APPARATUS 1. Equipment designed to perform a specific function; **2.** Collective term for equipment specially designed to fight fires.

APPARENT COHESION Cohesion of moist soils due to surface tension in capillary interstices that disappears on immersion of the soil.

APPARENT DENSITY Mass per unit volume, including pore space, of oven-dried soil.

APPARENT DIRT CAPACITY In fluid filter evaluation, the amount of dirt that can be added to the filter test system before the terminal differential pressure is reached.

APPARENT GROUNDWATER VELOCITY Apparent distance covered by groundwater over a unit of time.

APPARENT VISCOSITY *See* **Viscosity.**

APPLETON LAYER Atmospheric region approximately 150-1000 km above sea level containing the highest concentration of free electrons and useful for radio transmissions.

APPLICABLE REQUIREMENTS Cleanup standards, standards of control, and other substantive requirements, criteria, or limitations that specifically address a hazardous substance, pollutant, contaminant, remedial action, location, or other circumstance.

APPLICABLE STANDARD Any requirement relating to the quality of water containing or potentially containing pollutants that may affect groundwater, the discharge of a sludge or disposal practice, including standards for sewage sludge use or disposal, effluent limitations, water quality standards, standards of performance, toxic effluent standards or prohibitions, best management practices, and pretreatment standards.

APPLICABLE WATER QUALITY STANDARDS Water quality standards adopted by a competent and authorized organization.

APPLICATION FACTOR Ratio between that concentration of a substance which produces a particular chronic response in an organism and that which causes death in 50% of the population within a given time.

APPLICATION FOR PAYMENT Written request by a contractor for partial payment under the terms of the contract; may be for work done, materials on site, etc.

APPLICATOR Fire hose pipe or nozzle for applying foam or water fog to fires. *Also called* **Foam applicator,** and **Fog applicator.**

APPLIED CLIMATOLOGY Study of weather and its effects on man and his activities.

APPLIED COST Cost that has been assigned to a product or activity.

APPRAISAL Written statement independently and impartially prepared by a qualified appraiser setting forth an opinion of defined value of an adequately described property as of a specific date.

APPRAISAL APPROACH Methods for estimating the value of real property. There are several methods, including:

  Capitalization: Estimation of value according to capitalization of productivity and income.

  Cost: Appraisal based on the depreciated new replacement cost of improvements, plus the current market value of the site.

  Income: Appraisal based on a property's anticipated future income.

  Market comparison: Appraisal based on analysis of the recent sales prices of similar properties.

  Summation: Adding together the parts of a property, each part being considered separately, e.g., value of the land considered as vacant, plus the cost of reproduction of structures, less depreciation.

APPRAISAL INCREASE Credit resulting from an increase in the recorded value of a fixed asset arising from a reappraisal.

APPRAISAL REPORT Report presenting a value estimate of real property. It should contain the data on which the appraisal is made together with any analysis of the data, plus conclusions leading to the estimate.

APPRAISED INVENTORY Detailed list of items constituting an assembled property, with or without unit costs or prices appended.

APPRAISED PRICE Price of a particular item based on the estimate of its actual market value; the minimum accepted price for the item at sale.

APPRAISED VALUE Market value: the amount a qualified assessor calculates a property will fetch on the open market.

APPRAISER One who is experienced and/ or qualified in the procedures necessary to complete an evaluation of real property.

APPRECIATION Increase in the value of property and/or goods.

APPROPRIATE SENSITIVE BENTHIC MARINE ORGANISMS At least one species each representing filter-feeding, deposit-feeding, and burrowing species chosen from among the most sensitive species as being reliable test organisms to determine the anticipated impact on a site.

APPROPRIATE SENSITIVE MARINE ORGANISM At least one species each representative of phytoplankton or zooplankton, crustacean or mollusc, and fish species chosen from among the most sensitive species documented in the scientific literature as being reliable test organisms to determine the anticipated impact of waste disposal or deposition on the ecosystem at a disposal site.

APPROPRIATE TECHNOLOGY Technology that is suitable for the end to be achieved, at the time, and at the location of application.

APPROPRIATION 1. Taking possession of or setting aside for a specific purpose; **2.** Act of a legislative body that makes funds available for expenditures with specific limitations as to amount, purpose, and period. *See also* **Allotment,** and **Apportionment**; **3.** Allocation of funds to a project. *Also called* **Feasibility budget.** *See also* **Project budget**; **4.** Setting aside private land for public use.

APPROPRIATIVE RIGHT Right of beneficial use.

APPROPRIATOR OF WATER RIGHTS One who diverts and puts to beneficial use that water of a stream or other body of water under a water right obtained through appropriation.

APPROVAL Indication or certification that work has reached a specific stage, that something is now ready for further development, or that a product or commodity complies with established standards.

APPROVED Sanctioned, endorsed, accredited, certified, or accepted as satisfactory by a duly constituted and nationally recognized authority or agency.

APPROVED EQUAL Material, equipment, or method approved by an architect or engineer as being an acceptable substitute for its specified equivalent. *Also called* **Or equal.**

APPROVING AUTHORITY Agency, board, department, etc., authorized under statute to enforce specified requirements.

APPROXIMATE QUANTITIES Preliminary estimate of the amounts of materials and labor needed to complete a project.

APPURTENANCE Subordinate but necessary accessory; in plumbing, for instance, the fittings, valves, traps, etc., that are necessary to complete a house drain.

APPURTENANT WORKS All labor and materials necessary for satisfactory completion of a project.

APRON Hard surface created on the sea bed or to the bed or banks of a stream or waterway to prevent scour.

APRON CONVEYER *See* **Conveyer.**

AQUACULTURE Breeding and rearing of fish in captivity.

AQUAFALFA Ground having a high water table.

AQUATIC ANIMALS Animals that live wholly or mostly in water.

AQUATIC ENVIRONMENT AND AQUATIC ECOSYSTEM Waters that include wetlands that serve as habitat for interrelated and interacting communities and populations of plants and animals.

AQUATIC FLORA Plant life associated with the aquatic ecosystem including algae and higher plants.

AQUATIC GROWTH Live floating, drifting or attached organisms in a body of water.

AQUEDUCT 1. Conduit designed to transport water from a remote source, usually by gravity; **2.** Elevated structure supporting a

conduit or canal passing over a river or low ground.

AQUEOUS METAMORPHISM Change in the texture and composition of rocks by solution and redeposition of mineral matter through the influence of water which they contain.

AQUEOUS VAPOR Gaseous form of water.

AQUICLUDE Geologic formation which, although porous and capable of absorbing water slowly, will not transmit it rapidly enough to furnish an appreciable supply for a well or spring.

AQUIFER Underground formation of sands, gravel, or fractured porous rock, that is saturated with water, and that supplies water for wells and springs.

AQUIFER SERVICE AREA Area above an aquifer including the area where the entire population served by the aquifer lives.

AQUIFER TEST Procedure in which measured quantities of water are removed or added to a well and the resulting changes in head are measured during and/or after the period of discharge or recharge.

AQUIFUGE Ground that neither contains nor transmits water in useful quantities.

AQUIHERBOSA Submerged aquatic vegetation.

AQUITARD Layer of soil or rock that is relatively impermeable and restricts the flow of water from one aquifer to another. *Also called* **Aquiclude.**

ARABLE AREA Land suitable for cultivation.

ARBORETUM Place where trees and shrubs are grown for educational, scientific and other purposes.

ARBORICULTURE Cultivation of trees for ornament or timber..

ARC FURNACE *See* **Furnace.**

ARCH CLOUDS Stationary wave clouds

extending a considerable distance along a mountain range.

ARCH DAM Dam held in position by the restraint of the walls it joins, i.e., the rock faces of the valley it closes. Resistance to the weight of water impounded upstream of the dam is provided by the dam also being arched in that direction.

ARCHEAN Oldest known Precambrian rocks.

ARCHIMEDEAN SCREW Large, slow-pitch spiral-feed screw, sometimes contained within a tube, set at an incline, that, when turned, lifts and carries water and other material to a higher elevation.

ARCHIMEDES PRINCIPLE Principle of buoyancy, stating that the resultant force on a wholly or partly submerged body acts vertically upward through the center of gravity of the displaced fluid and is equal to the weight of the fluid displaced.

ARCTIC SMOKE Steaming fog produced when very cold air passes over warm water.

ARE Obsolete metric unit of area, equal to 100 m^2, that should not be used with the SI system of measurement. Symbol: a. *See also the appendix:* **Metric and nonmetric measurement.**

AREA 1. Square measurement; **2.** Amount of surface enclosed by defined boundaries, expressed as a square measurement. *See also* **Area of building, Floor area, Gross area, Net area,** and **Net room area.**

AREA-CAPACITY CURVE Graph representing the relationship between the surface area of the water in a reservoir and the corresponding volume.

AREA CURVE Curve expressing the relationship between area and some other variable.

AREA DIFFERENTIAL SYSTEM Valve operated by means of opposing pistons of different displacement.

AREA DRAIN Drain set in a floor that cannot be otherwise drained.

AREA ESTIMATING METHOD Method of estimating costs by multiplying the adjusted gross floor area by a unit cost.

AREA OF DIVERSION Portion of an adjacent area beyond the normal groundwater or watershed divide that contributes water to the groundwater basin or watershed under consideration.

AREA OF INFLUENCE Land area having the same horizontal extension as the part of the watertable or other piezometric surface that is perceptibly lowered by withdrawal of water through a well at a given rate.

AREA TECHNIQUE *See* **Sanitary landfilling.**

ARENACEOUS Composed primarily of sand; sandy.

ARGILLACEOUS Composed primarily of clay or shale; clayey.

ARGILLITE Rock containing chiefly clayey materials.

ARGON Chemical element (Ar) that is a colorless, odorless, inert gas forming approximately 1% of the earth's atmosphere.

ARID 1. Region where precipitation is so deficient in quantity, or occurs at such intervals, that agriculture is impractical without artificial irrigation; **2.** Climates with rainfall insufficient to support vegetation.

ARITHMETIC MEAN Sum of a set of variables divided by the number of variables in the set.

ARMOR Revetment of large, heavy rocks or precast concrete shapes to protect surfaces and faces from scour.

ARMORED PIPE *See* **Pipe.**

AROMATIC Organic compounds derived from benzene and having closed rings of carbon atoms.

ARRAY Series of related elements or components placed in relationship to each other.

ARROW 1. Representation of an activity with an indication adjacent to the arrow

37

(above or below) of its duration, used in both CPM and PERT systems. *See also* **Critical Path Method (CPM),** and **Program Evaluation and Review Technique (PERT);** **2.** Large skewer made of stiff wire with a ring at one end (to which colored rag or tape can be tied) and a point at the other, used for marking precisely on the ground the point to which measurements are made with a tape or chain.

ARROW DIAGRAMMING METHOD Network diagramming technique in which activities are represented by arrows. The tail of the arrow represents the start and the head represents the finish of the activity (the length of the arrow does not represent the exact duration of the activity). Activities are connected at points called nodes (usually drawn as small circles) to illustrate the sequence in which activities are expected to be performed.. Used in CPM or PERT. *Also called* **Activity-on-arrow.** *See also* **Precedence diagramming method.**

ARROYO Stream channel or gully, usually small, walled with steep banks, and dry much of the time.

ARSENIC Metalloid with four oxidation states and which is widely distributed throughout the earth's crust and present in trace amounts in all living matter.

ARTERIAL DRAINAGE Drainage system where the flow from a number of branch drains is fed into one main channel.

ARTERIAL HIGHWAY *See* **Highway.**

ARTERIAL VENT Vent serving a main drain or sewer, not necessarily via a manhole.

ARTESIAN Groundwater, or things connected with groundwater where the water is under hydrostatic pressure.

ARTESIAN AQUIFER Aquifer confined above and below by material of lesser permeability and in which the water level stands above the top of the water body it draws upon.

ARTESIAN BASIN Subsurface bowl filled with water under sufficient pressure that it will rise above the bottom of the overlying impervious strata if given the opportunity.

ARTESIAN CAPACITY Rate at which a well will yield water at the surface as a result of artesian pressure.

ARTESIAN DISCHARGE Rate of flow from an artesian well.

ARTESIAN FLOW Flow from an underground water-bearing stratum under sufficient hydrostatic pressure to force water above ground level.

ARTESIAN-FLOW AREA Surface area within which the water in an underlying aquifer is under sufficient hydrostatic pressure to rise above the ground surface.

ARTESIAN HEAD Distance above (or below) ground level to which water in an artesian aquifer would rise if free to do so.

ARTESIAN PRESSURE Pressure exerted by groundwater against an overlying impermeable, or less permeable formation, the bottom of which is at a lower elevation that that of the free water surface of the groundwater.

ARTESIAN SPRING Spring issuing under pressure through a fissure or other opening in the confining formation above the aquifer.

ARTESIAN WATER Subsurface water under sufficient head of pressure to raise the water in wells above the groundwater table.

ARTESIAN WATER POWER Energy developed at the mouth of an artesian well by the pressure of the water discharged from the well.

ARTESIAN WELL Drilled well that penetrates a strata containing water under pressure.

ARTICLE Manufactured item that (a) is formed to a specific shape or design during manufacture, (b) has end-use function(s) dependent in whole or in part upon its shape and design during end use, and (c) does not release, or otherwise result in exposure to a hazardous chemical or promote a hazardous condition under normal conditions of use.

ARTICULATED DUMP TRUCK On- and off-highway vehicle having the load-carrying dump body and its associated frame, suspension, and drive wheels connected to the operator's compartment, engine compartment, front suspension, and steering wheels through an articulated joint that gives a limited range of vertical and horizontal movement.

ARTIFICIAL Man-made; not occurring naturally.

ARTIFICIAL CONTAMINANT Contaminant of known composition and particle size distribution that is introduced into a fluid system or fluid system components for test purposes. *See also* **Built-in contaminant, Contaminated,** and **General contaminant.**

ARTIFICIAL FUEL Man-made fuel, including all manufactured and by-product fuels. Examples are water gas, blast furnace gas, gasahol, coke, etc.

ARTIFICIAL HARBOR Protected area created by building one or more breakwaters or by placing artificial barriers around a body of water.

ARTIFICIAL ISLAND Fill placed, or caused to be deposited on the bottom of a lake or in a stream of river in sufficient quantities, and in such a manner that it permanently breaks the surface.

ARTIFICIAL LAKE Man-made basin (other than a swimming pool), 23 m² (250 ft²) or greater, permanently containing water to a minimum depth of 600 mm (2 ft).

ARTIFICIAL NAVIGABLE WATER-WAY Stream, channel or canal made navigable by dredging and/or construction of controlling works.

ARTIFICIAL OBSTRUCTION Any barrier other than a natural obstruction that impedes or prevents passage or flow.

ARTIFICIAL RAIN Rain induced by an artificial stimulus, commonly called 'cloud seeding' which involves the distribution of common salt or silver iodide by aircraft in the atmosphere at elevations and where conditions are such that the formation of supercooled droplets is promoted with the possibility of subsequent condensation of moisture into rain.

ARTIFICIAL RECHARGE Recharge of an aquifer at a rate greater than that due to natural circumstances, as a result of human activity, for instance.

ARTIFICIAL REEF Use of man-made materials, solid wastes, rubble, scrap, etc., to provide cover for marine organisms.

ARTIFICIAL REGENERATION Establishing a new forest by planting seedlings or by direct seeding (as opposed to natural regeneration).

ARTIFICIAL RESPIRATION *See* **Artificial resuscitation.**

ARTIFICIAL RESUSCITATION Breathing maintained by an artificial means. *Also called* **Artificial respiration,** and **Artificial ventilation.**

ARTIFICIAL SLUDGE Substances such as iron, aluminum and manganese hydroxides, and silica gels, used as substitutes for return sludge in the activated sludge process.

ARTIFICIAL VENTILATION *See* **Artificial resuscitation.**

ASBESTOS 1. Asbestiform varieties of chrysolite (serpentine); crocidolite (riebeckite); amosite (cummingtonite-grunerite); tremolite; anthophyllite; and actinolite. **2.** Fibrous mineral used as a heat and fire barrier.

ASBESTOS ABATEMENT Processes and techniques employed in the removal and/or containment of asbestos materials used in structures.

ASBESTOS BLANKET Small asbestos-filled foil blanket wrapped around pipes being welded to reduce heat loss.

ASBESTOS CEMENT Fire-resisting weatherproof building material made from portland cement and asbestos fibers in forms such as plain and corrugated sheets, shingles, pipes, etc.

ASBESTOS-CEMENT BOARD Fire-resistant sheet made from asbestos cement.

ASBESTOS-CEMENT (A/C) PIPE Asbestos-containing product made of cement and intended for use as pipe or fittings for joining pipe. Major applications of this product include: pipe used for transmitting water or sewage; conduit pipe for the protection of utility or telephone cable; and pipes used for air ducts.

ASBESTOS-CEMENT PRODUCTS Products manufactured in rigid material composed essentially of asbestos fiber and portland cement.

ASBESTOS-CONTAINING WASTE MATERIALS Mill tailings or any waste that contains commercial asbestos. This term includes filters from control devices, friable asbestos waste material, and bags or other similar packaging contaminated with commercial asbestos. As applied to demolition and renovation operations, it also includes regulated asbestos-containing material waste and materials contaminated with asbestos, including disposable equipment and clothing.

ASBESTOS DEBRIS Pieces of asbestos-cement building materials that can be identified by color, texture, or composition.

ASBESTOS DIAPHRAGM Asbestos-containing product that is made of paper and intended for use as a filter in the production of chlorine and other chemicals, and which acts as a mechanical barrier between the cathodic and anodic chambers of an electrolytic cell.

ASBESTOS-DIATOMITE Mixture of asbestos fibers, diatomaceous earth, and portland cement used to make insulating, fire-resistant building products.

ASBESTOS ENCAPSULATION In-place sealing of sprayed asbestos with a durable and impact-resistant coating.

ASBESTOS FELT Sheet of matted asbestos fibers, the basis for saturated asbestos felt and asbestos asphalt-coated roofing products.

ASBESTOS FIBER Naturally occurring fibrous rock used in the production of asbestos-cement goods.

ASBESTOS-FREE PRODUCTS Range of products that perform the same functions as those that have historically contained asbestos fibers but which are manufactured with alternative materials.

ASBESTOSIS Disease manifested by scarring of the lungs caused by inhalation of asbestos fibers.

ASBESTOS JOINT RUNNER Asbestos rope that is wrapped around a pipe above a bell-and-spigot joint, clamped, and used as a dam while molten lead is poured in a caulked joint.

ASBESTOS MIXTURE Mixture which contains bulk asbestos or another asbestos mixture as an intentional component. It may be either amorphous or a sheet, cloth fabric, or other structure. This term does not include mixtures which contain asbestos as a contaminant or impurity.

ASBESTOS PLASTER Nonflammable pipe insulation containing asbestos fibers, diatomaceous earth, and gypsum plaster.

ASBESTOS REMOVAL Specialist work for removal of asbestos from building and structures.

ASBESTOS-CEMENT SHEETS Plain or corrugated sheets made of asbestos cement, used as roof or wall cladding.

ASBESTOS SHINGLES Type of shingle made for fireproofing.

ASBESTOS TAILINGS Any solid waste that contains asbestos and is a product of asbestos mining or milling operations.

AS-BUILT DRAWING Scale drawing that shows a project as it was built and that incorporates details of any additions, deletions, or variations from the drawings used at the time of accepting a bid to construct the project and/or as approved by the client.

A-SCALE Sound level measurement scale that discriminates against low frequencies and approximates the human ear.

AS-DUG AGGREGATE *See* **Aggregate.**

ASEISMIC DESIGN Structures designed to withstand earthquakes.

ASEISMIC REGION Region not liable to earth tremors.

ASEPTIC Free of pathogenic organisms.

ASEXUAL REPRODUCTION Reproduction in which new individuals are produced from a single parent without formation of gametes: spore formation and propagation by means of bulbs, cutting, etc.

ASH 1. Mineral content of a substance that remains following complete combustion; **2.** Pyroclastic fragments, less than 4 mm long, ejected by a volcano.

ASH CONTENT Measure of the residue which remains after combustion under specified conditions.

ASH DUMP Flat, metal door with a flange, built flush with the inner hearth of a fireplace, used for dumping ashes into an ash pit.

ASHLESS DISPERSANT *See* **Dispersant.**

ASH SLUICE Trench or channel in which water flushes residue from an ash pit to a storage or disposal point.

ASIATIC CLOSET Squatting-type latrine.

ASKAREL Generic term for a group of non-flammable synthetic chlorinated hydrocarbons used as electrical insulating media.

ASME-TYPE LP GAS CYLINDER Fuel container for liquefied petroleum gas made and inspected under the American Society of Mechanical Engineers (ASME) Boiler and Pressure Vessel Code, Section V-iii, for Unfired Pressure Vessels. Where used for storage of an industrial truck's fuel supply, it may be permanently or removably affixed to the vehicle.

ASPECT Compass direction to which a slope, or a building, faces. *Also called* **Exposure.**

ASPHALT Brown to black, solid or semi-solid bituminous substance that occurs naturally or which can also be obtained as a residue in refining some petroleum products.

ASPHALT PROCESSING Storage and blowing of asphalt.

ASPHALT PROCESSING PLANT Plant which blows asphalt for use in the manufacture of asphalt products.

ASPHALT SOIL STABILIZATION Treatment of naturally occurring, nonplastic or moderately plastic soil with cutback or emulsified asphalt at ambient temperatures. Aeration and compaction of the asphalt-soil mixture to produce water-resistant base and sub-base courses of improved load-bearing qualities.

ASPHALT STABILIZATION Mixing of asphalt with sand or soil to improve granular cohesion, waterproofing, or other factor.

ASPHALT STORAGE TANK Tank used to store asphalt at asphalt roofing plants, petroleum refineries, and asphalt processing plants.

ASPHYXIATING POLLUTANTS Airborne pollutants capable of reducing the ability of blood to absorb and carry oxygen, either through obstruction of the air passages or by an effect on the body processes.

ASPHYXIATION Condition that causes death because of a deficient amount of oxygen and an excessive amount of carbon monoxide and/or other gases in the blood.

ASPIRATE To draw by suction.

ASPIRATION Drawing of air at atmospheric pressure into a combustion chamber.

ASPIRATOR Suction device for removing undesirable material from the throat of a victim.

ASPIRATOR MIXER Gas-air proportioning device that uses the venturi principle to cause the flow of combustion air to induce the proper amount of gas into the air stream, used with low-pressure and zero-pressure gas. *Also called* a **Suction-type mixer.**

ASSEMBLAGE Combination of two or more parcels of land.

ASSEMBLY 1. Group of materials or parts, designed to act as a coordinated whole; **2.** Smallest community unit of plants or animals.

ASSEMBLY BOLT Threaded bolt used to hold together temporarily several parts during installation and/or construction.

ASSEMBLY DRAWING Drawing that shows the composite parts of a unit and how, and in what order they should be combined and fastened.

ASSEMBLY LANGUAGE Language used in computer programming that uses mnemonics for the binary words of machine language.

ASSEMBLY ROD External threaded rod or bolt used in machine assembly.

ASSEMBLY SEQUENCE Designated sequence in which parts or components must be assembled and/or fastened.

ASSEMBLY TIME Time necessary to assemble a group of materials and/or parts designed to form a complete unit.

ASSEMBLY TORQUE Designed torque applied at final assembly.

ASSESSED VALUE Value of property, established for the purpose of levying a tax.

ASSESSMENT Valuation and listing of property for the purpose of levying a tax upon it.

ASSESSMENT RATIO Ratio of assessed value of property to its established or estimated market value.

ASSESSMENT ROLL Listing of all taxable property within a property taxing jurisdiction.

ASSESSOR One who determines the value of property for tax purposes.

ASSETS Property of value owned by an individual or enterprise.

ASSIMILATIVE CAPACITY Capacity of a natural body of water to receive wastewater without deleterious effects, and/or toxic materials without damage to aquatic life or humans who consume the water.

ASSOCIATION 1. Ecological unit containing two or more species in closer proximity than would occur by chance; **2.** Degree of dependence or independence between two or more variables.

ASTHENOSPHERE Lower part of the earth's crust in which seismic waves travel at much reduced speeds.

ASTRONOMICAL TWILIGHT Part of evening during sunset when the Sun's center is optically 18° below below the horizon.

AT-GRADE INTERSECTION *See* **Intersection types.**

ATMOMETER Device that measures the rate of evaporation of water to the atmosphere.

ATMOSPHERE 1. A unit of pressure; a standard atmosphere for purposes of direct comparison is considered to be sea level atmosphere of $1 \text{ kg}/645 \text{ mm}^2$ (14.7 lb/in.^2) at 15°C (59°F), or measured as 760 mm (29.95 in.) of mercury on a barometer; **2.** The gaseous envelope surrounding a celestial body, e.g. Earth.

ATMOSPHERE FOR TESTING Air at ambient humidity and temperature.

ATMOSPHERIC CEILING Level in the atmosphere at which a heated column ceases to rise.

ATMOSPHERIC CORRECTION Correction applied to a measured distance to allow for the effects of air temperature and atmospheric pressure on the speed of light.

ATMOSPHERIC DISPERSION Dilution of gaseous or smoke pollution in which the concentration is progressively decreased.

ATMOSPHERIC DISPLACEMENT System or method of applying water fog in a superheated area causing the water to be converted into steam that expands and displaces the atmosphere in a burning room or building.

ATMOSPHERIC INVERSION Condition in which the air temperature increases with height, preventing the normal tendency of heat to rise from the ground.

ATMOSPHERIC MOISTURE Water as it occurs in various forms in the atmosphere.

ATMOSPHERIC PRESSURE *See* **Pressure.**

ATMOSPHERIC STORAGE TANK Storage tank designed to operate at pressures from atmospheric to 3.5 kPa (0.51 psi) gauge.

ATMOSPHERIC-TYPE VACUUM BREAKER Backflow preventer in which a float, kept elevated by normal water flow, falls to a check seat if the flow is reversed.

ATOMIC ENERGY Energy, as heat and radiation, derived from controlled fission or fusion in a nuclear reactor.

ATOMIC NUMBER Number (Z) of electrons surrounding the nucleus of the neutral atom of an element, or the number of protons in the nucleus.

ATOMIC SPECTRUM Spectrum emitted by an excited atom due to changes within it.

ATOMIC WEIGHT Relative atomic mass (A_r): being the ratio of the average mass per atom of a specific isotope composition of an element 10 1/12 of the mass of an atom of $^{12}_{6}C$.

ATOMIZATION Process in which a stream of water or gas impinges upon a molten metal stream, breaking it into droplets which solidify as powder particles.

ATOMIZING HUMIDIFIER Device that introduces tiny particles of water into a stream of air.

ATRAZINE + METABOLITES Atrazine is a chloro-N-dialkyl-substituted triazine herbicide used extensively as a pre- and post-emergent weed control agent in many agricultural sectors, and for total vegetation control in non-cropland and industrial areas. It has low soil adsorption and soil-water distribution coefficients and is ranked high for potential groundwater contamination, being one of the most frequently detected pesticides in surface and well water.

AT REST EARTH PRESSURE *See* **Pressure.**

ATTACHED GROWTH PROCESSES Wastewater treatment processes in which the microorganisms and bacteria used to condition the organic fraction are attached to the media in the reactor.

ATTACHMENT Device specifically designed to perform a function or purpose and for a particular vehicle, machine, or item of equipment, that does not form part of the basic vehicle, machine, or equipment.

ATTENUATOR Sound deadening unit incorporated in air conditioning and air handling equipment to lessen mechanical equipment noise.

ATTERBERG LIMITS Arbitrary water contents (shrinkage limit, plastic limit, liquid limit) determined by a standard test that define the boundaries between the different states of consistency of plastic soils. *Also called* **Consistency limits.** *See also* **Plasticity index,** and **Plastic limit.**

ATTERBERG TEST Method for determining the plasticity of soils.

AT-THE-SOURCE At or before the commingling of delacquering scrubber liquor blowdown with other process or nonprocess wastewaters.

ATTO Prefix representing 10^{-18}. Symbol: a. Used in the SI system of measurement. *See also the appendix:* **Metric and nonmetric measurement.**

ATTRACTANTS Substances or mixtures which, through their property of attracting certain animals, are used as pesticides to mitigate or destroy vertebrate or invertebrate animals.

ATTRACTING GROIN Groin that attracts the current in moving water toward itself, fixing the deep channel close by.

ATTRITION Removal of material from a surface by friction wear.

AUDIBLE ALARM Horn, siren, bell, or buzzer that is used to attract the attention of a machine operator when a fault occurs.

AUDIBLE SIGNAGE Mechanical annunciation of information or directions. *See also*

43

Signage.

AUDIBLE SOUND Sound frequencies between approximately 15 and 20,000 Hz with sufficient sound pressure to be detectable.

AUDIO FREQUENCY Any frequency of oscillation of a sound wave that is audible; usually between 15 and 20,000 Hz.

AUDIOGRAPH Printout from a machine that determines the rate at which sound levels diminish over distance; when used in an enclosed area this is an indication of the sound absorption quality of the materials used in construction.

AUDIOMETRIC ROOM Room constructed and finished of materials that, in combination, will absorb over 99% of sound generated within the room, isolated from the remainder of the structure by springs to eliminate the transfer of vibrations.

AUDIOVISUAL WARNING SYSTEM Alarm device that signals the operator of equipment of out-of-tolerance conditions or potentially hazardous situations. These include such circumstances as low engine oil pressure, high engine coolant temperature, high hydraulic and/or transmission oil temperature, two-block condition, potential overload, etc.

AUDIT Examination to determine the reliability of a record or claim, or to evaluate compliance with rules or policies or with conditions of an agreement.

AUFEIS Layer upon layer of frozen water, discharged sequentially, as a spring.

AUGER Tool used to drill and collect cores of soil and surface deposits.

AURORA Streams or bands of light appearing in the sky at night.

AURORA AUSTRALIS Streamers or bands of light appearing in the southern sky at night in high latitudes.

AURORA BOREALIS Streamers of bands of light appearing in the northern sky at night.

AUTHORITY Sum of the rights and powers assigned to a position or vested in a person.

AUTHORITY HAVING JURISDICTION Entity or person having statutory authority.

AUTHORIZED APPROPRIATION *See* **Cost types.**

AUTHORIZED PERSON Person approved or assigned by the employer to perform a specific type of duty or duties or to be at a specific location or locations at the jobsite. *See also* **Competent person,** and **Designated person.**

AUTOCHTHONOUS Indegenous; native to a particular place.

AUTOCLAVE Strong vessel for effecting chemical reactions or sterilization under high pressure and/or high temperature.

AUTOGAMY Self-fertilization in plants.

AUTOGENETIC DRAINAGE Drainage due to erosion caused by the flow of constituent streams.

AUTOGENIC SUCCESSION Succession resulting from changes by the organisms themselves.

AUTOIMMUNITY Condition in which the immune responses of an animal are directed against its own tissue.

AUTOLYSIS Self-destruction of animal and plant tissue carried out by enzymes.

AUTOMATED METHOD Method for measuring concentrations of an ambient air pollutant in which sample collection, analysis, and measurement are performed automatically.

AUTOMATIC ALARM Alarm actuated when a preset or predetermined condition is exceeded or voided.

AUTOMATIC CONTROL VALVE *See* **Valve.**

AUTOMATIC FIRE ALARM SYSTEM Fire alarm system that detects the presence of combustion and automatically initiates a sequence of events.

AUTOMATIC FIRE PUMP *See* **Pump.**

AUTOMATIC FIRE VENT Roof-mounted vent in a single-story structure that operates automatically in the event of fire, providing for the removal of smoke.

AUTOMATIC GATE Gate that operates without human assistance when prescribed conditions are reached.

AUTOMATIC OPERATION Mode of operation of a system wherein it performs its specific function without operator adjustments after the initial setup.

AUTOMATIC RECORDING GAUGE Instrument that automatically measures and graphically records conditions.

AUTOMATIC SEAL Seal that is activated by the pressure of the fluid that it seals.

AUTOMATIC SMOKE ALARM Device that triggers an alarm upon the detection of the products of combustion.

AUTOMATIC SPILLWAY Device for discharging water into drainage courses built along canal sides. It may be an uncontrolled overflow crest, a crest equipped with automatic gates, or an enclosed crest designed for siphon operation.

AUTOMATIC SPRINKLER Nozzle, attached to a water supply system, fitted with an orifice with a flow control device and a deflector that will distribute water over a given area at a prescribed rate and in a designed manner. There are several types, including:

Frangible-bulb: Fire protection sprinkler head incorporating a bulb filled with a liquid and a bubble of air. When the liquid is heated it absorbs the air causing the bulb to break, allowing water to flow from the wet-type distribution system through the sprinkler head.

Frangible-pellet: Fire protection sprinkler head incorporating a pellet of solder that melts at a given temperature, releasing the control device of the sprinkler and allowing water from the distribution system to flow

through the sprinkler head.

Solder-link: Sprinkler head connected to a water supply system by a link constructed of an alloy of tin, lead, cadmium, and bismuth in proportions that will melt at a determined temperature. When connected to a wet-type fire-protection sprinkler system, such a link will melt if the ambient temperature at its location rises to the preset limit, allowing water to pass from the distribution system and through the sprinkler head.

AUTOMATIC SPRINKLER KIT Kit containing the tools and equipment necessary to close and service an open sprinkler head.

AUTOMATIC SPRINKLER SYSTEM An integrated fire protection system of underground and overhead piping designed in accordance with fire protection engineering standards and including a water supply, such as a gravity tank, fire pump, reservoir, or pressure tank and/or connection by underground piping to a municipal supply. The system is usually activated by heat from a fire and discharges water over the fire area. There are several types of system:

Combination system: One that combines two or more defined systems.

Deluge system: Used where simultaneous operation of all sprinklers is required for complete discharge over a total area.

Dry system: Used in unheated areas subject to freezing.

Preaction system: Used where accidental discharge or where water leakage in the absence of fire can cause damage to contents, usually of high or critical value.

Wet system: Used only in heated areas not subject to freezing.

AUTOMATIC TEMPERATURE COMPENSATOR Device that continuously senses the temperature of fluid flowing through a metering device and automatically adjusts the registration of the measured volume to the equivalent volume at a base

45

temperature.

AUTOMATIC VALVE *See* **Valve.**

AUTO-SUPPRESSION SYSTEM Automatic fire suppression system that is activated by a fire detector.

AUTOTROPHY Organisms that thrive by using inorganic materials for energy and growt; e.g. all plants and bacteria.

AUXILIARY Anything additional or supplementary: a helper, standby generator, etc.

AVAILABILITY FACTOR Measure of the time that a machine is available to be used against the time it is unavailable due to servicing or mechanical failure: actual working time divided by available working time, in hours.

AVAILABLE CHLORINE Measure of the total oxidizing power of chlorinated lime and hypochlorite.

AVAILABLE DILUTION Ratio of the quantity of untreated wastewater, or partly or completed effluent, to the average quantity of diluting flow available, effective at the point of discharge.

AVAILABLE MOISTURE Moisture content of soil in excess of the permanent wilting point.

AVAILABLE MOISTURE CAPACITY Difference between the percentage of water at field capacity and the percentage of water at the wilting point.

AVAILABLE OXYGEN Quantity of dissolved oxygen available for oxidation of organic matter in a body of water.

AVAILABLE WATER Moisture content of soil in excess of the permanent wilting point at any time, expressed as a percentage by weight of the dry soil or as the measure of water per measure of depth of soil. This is the moisture that can be extracted by plants.

AVAILABLE WATER SUPPLY Quantity of water in a source such as a stream or groundwater basin in excess of the quantity necessary to supply valid prior rights and demands.

AVALANCHE Moving mass of debris, snow, and ice sliding rapidly down a slope.

AVERAGE Arithmetic mean obtained by adding quantities and dividing the sum by the number of quantities.

AVERAGE ANNUAL FLOOD Flood discharge equal to the mean of the discharged of all of the maximum annual floods during the period of record.

AVERAGE CONCENTRATION Average of analyses made over a single period.

AVERAGE DAILY FLOW Total quantity of liquid in a tributary to a point, divided by the number of days of flow measurement.

AVERAGE DAILY TRAFFIC *See* **Volume.**

AVERAGE DIMENSION Average of corresponding dimensions of random sample units.

AVERAGE EFFICIENCY Efficiency of a machine or mechanical device over the range of load through which the machine operates.

AVERAGE END AREA Method of calculating the volume of earthwork between two cross sections where the volume equals the average cross-sectional areas multiplied by the distance between the two cross sections.

AVERAGE FLOW Arithmetic average of flows measured at a given point.

AVERAGE FREQUENCY OF OCCURRENCE Average number of years between events of a similar nature or magnitude: storms, floods, rainfall above a defined magnitude, wind storms, etc.

AVERAGE FROST PENETRATION Numerical average based on historical data of the depth of frost penetration for a given area or locality.

AVERAGE GRADE Average of the elevations of various ground positions throughout a construction site.

AVERAGE GROUNDWATER VELOCITY Average distance travelled per unit of time by a body of groundwater, equal to the total volume of groundwater passing through a unit cross-sectional area divided by the porosity of the water-transmitting material.

AVERAGE HAUL Average distance material is hauled from the cut to the fill (sometimes calculated from a mass diagram). *See also* **Free haul, Haul, Overhaul,** and **Station yards.**

AVERAGE MONTHLY DISCHARGE LIMITATION Highest allowable average of daily discharges over a calendar month, calculated as the sum of all daily discharges measured during a calendar month divided by the number of daily discharges measured during that month.

AVERAGE RELATIVE HUMIDITY Average of the maximum and minimum relative humidities.

AVERAGE SEA LEVEL Mean average elevation of sea water at a designated location, used as a reference for other points measured above or below seal level.

AVERAGE STREAMFLOW Average rate of discharge of a stream during a specified period.

AVERAGE STREAMFLOW VELOCITY Average velocity of a stream flowing in a channel or conduit at a given cross section or in a given reach.

AVERAGE YEAR Year for which the observed quantities of hydrologic phenomena, such as precipitation, evaporation, temperature, and streamflow, approximate the mathematical mean of such observed quantities for a considerably longer period.

AVOIDED COST Tipping fees that are avoided by reducing the volume of solid waste destined for a landfill or combustor, using waste reduction, recycling, and/or composting.

AVULSION Natural transfer of land from one parcel to another, as when a stream or river changes its channel or floods..

AVULSIVE CUTOFF 1. River's action when an avulsion occurs; 2. Area of land circumscribed by the old and the new channels when a river changes its course.

AWARD OF CONTRACT Formal process, in writing, that accepts a tender or bid and gives the contractor/supplier authority to proceed.

AWASH Nearly flush with the water level.

AXENIC Uncontaminated.

AXIAL FLOW Movement in the direction of the axis.

AXIAL-FLOW COMPRESSOR *See* **Compressor.**

AXIAL-FLOW FAN Fan that draws in air and expels air in the direction of the shaft on which its propeller is mounted.

AXIAL-FLOW PUMP *See* **Pump.**

AXIAL FORCE DIAGRAM In statics, a graphical representation of the axial load acting at each section of a structural member.

AXIAL LOADING Force applied along the length of a member.

AXIAL MOTION Motion occurring parallel to the centerline of the object; in the case of an expansion joint, parallel to the centerline of the bellows (can be either expansion or compression).

AXIAL-PISTON HYDRAULIC PUMP *See* **Pump.**

AXIAL SEAL Shaft seal in which the packing member approaches the surface to be sealed in an axial direction. The primary seal function is at right angles to the axis of rotation.

AXIAL TO IMPELLER Direction in which material being pumped flows around the impeller, or flows paralleled to the impeller shaft.

AXIS 1. Real or imaginary line on which an object supposedly (as in a pictorial projection) or actually (as with a shaft) rotates; 2. Centerline of a tunnel or shaft. *See also* **Arch.**

AXIS OF IMPELLER Theoretical line along the center of a shaft, such as an impeller shaft.

AXIS OF ROTATION Vertical line representing the axis around which an object rotates.

AXIS OF SYMMETRY Line about which something geometrically balanced is developed, and on which the center of gravity is located.

AXIS OF WELD Imaginary line along the center of gravity, and perpendicular to a cross-section of the weld metal.

AZINPHOS-METHYL An organophosphorus insecticide and acaricide used for the control of a variety of pests for many fruit, vegetable, grain and forage crops.

AZODRIN Organophosphorus compounds.

AZOLLA Genus of aquatic ferns that form a symbiotic relationship with the blue-green algae *Anabaena* which can fix nitrogen.

AZOTOBACTER Genus comprising the most important of the aerobic, nitrogen-fixing bacteria.

B

Bacillariophyta Diatoms. Green or brown unicellular algae containing chlorophyll and having siliceous walls; the origin of diatomaceous earth.

Bacillus Genus of rod-shaped aerobic bacteria.

BACK 1. Roof or top of an underground opening; **2.** Ore located between a given level and the surface, or between two levels; **3.** Top surface of a horizontal or inclined timber.

BACK BLADING Pushing soil in a windrow back into the position from which it came, usually with a grader.

BACKBLOW Reversal of the normal flow of water, under pressure, in a well to free the screen or strainer and the aquifer of clogging material.

BACKBOARD Temporary board on the outside of a scaffold.

BACKCROSS Offspring of a cross-mating between a hybrid with one of its parents or with an organism genetically similar to one of its parents.

BACK DRAIN Perforated drain pipe, usually of 100 mm (4 in.) internal diameter, placed longitudinally behind the bottom of a retaining wall and surrounded by a stone filter, used to reduce water pressure acting against the rear of the wall.

BACK DROP *See* **Drop manhole.**

BACK-END MATERIALS RECOVERY Engineered system that provides for collection of discrete reusable materials from mixed wastes that have been burned or treated.

BACK-END SYSTEM Combination of system components that changes the chemical properties of a waste or converts its components into energy or compost. *See also* **Front-end process.**

BACKFALL Slope opposite to an intended direction of flow that will result in backflow or ponding.

BACKFILL 1. Material used to replace earth, etc., previously excavated, commonly into a trench or pier excavation, around and against a basement foundation. *Also called* **Soil fill.** *See also* **Fill,** and **Structural fill.**

BACKFILL BLADE Small dozing blade fitted to a machine having a prime function other than dozing (trenching, excavating, etc.) so as to permit an operator to replace material excavated from a trench or hole.

BACKFILL DENSITY Percent compaction required or expected for backfill.

BACKFILL GROUT *See* **Low-pressure grout.**

BACKFIRE 1. Fire started upwind of an existing brush fire, to burn against and cut off its spread. *See also* **Area ignition, Controlled burning,** and **Prescribed burning; 2.** Explosion in the intake or exhaust passages of an engine.

BACKFLOW Flow of water or other liquids, mixtures, or substances into the distributing pipes of a potable supply of water from any source other than that intended, produced by the differential pressure existing between two systems, either or both of which are at pressure greater than atmospheric.

BACKFLOW CONNECTION Any arrangement whereby backflow can occur. *Also called* **Interconnection,** and **Cross connection.**

BACKFLOW PREVENTER Device in a potable water supply system that prevents the backflow of water from the connections on its outlet end.

BACKFLOW PREVENTER VALVE *See* **Valve.**

BACKGROUND CONCENTRATION OF POLLUTANTS Concentration of undesirable substances normally or naturally present at a location, independent of any local or particular generator or source of a pollutant.

BACKGROUND CONTAMINATION Total of the extraneous particles that are introduced in the process of obtaining, storing, moving, transferring, and analyzing a fluid sample.

BACKGROUND RADIATION Radiation naturally occurring in the Earth and its atmosphere and waters, the main sources of which are potassium-40, thorium, uranium, and their decay products including radium, and cosmic rays.

BACKGROUND SOIL pH pH of the soil prior to the addition of substances that alter the hydrogen ion concentration.

BACKHOE Bucket, boom, and stick assembly that digs by pulling the bucket toward itself. Used as an attachment to a self-propelled machine such as a tractor-dozer; as a companion tool on a loader-backhoe; as an attachment on a tool-carrier; as an attachment to mount on the bed of a pickup truck, etc. *Also called* **Hoe,** and **Pullshovel.**

BACKHOE/LOADER Multipurpose machine equipped at its front end with a bucket for loading, and at its rear with a backhoe. The bucket can often be interchanged with a range of types and sizes, and with other attachments such as snowplows, blades, forks, etc. The backhoe is usually a permanent fixture and also can be equipped with a range of attachments and buckets.

BACKHOE TAMPING Processing step, often used in direct-dump transfer systems, in which a conventional backhoe is used to compact waste contained in an open-top transfer trailer.

BACKING WIND Wind showing a turn of direction.

BACK-INLET GULLEY Trapped branch entry to a drain into which rainwater or waste pipes discharge below a surface grating but

above the water seal.

BACK OF LEVEE Face of a levee away from the river it contains, facing the protected area. *Also called* **Inside of levee,** and **Land side of levee.**

BACKPACK Personal equipment used to fight a brush or forest fire, consisting of a 19 L (5 gal) pump-tank extinguisher carried on a person's back and supported by straps, and having a hand-operated pump and spray nozzle.

BACK PRESSURE *See* **Pressure.**

BACK-PRESSURE VALVE *See* **Valve.**

BACK PROP Raking strut used to transfer the mass of timber to the ground in deep trenches; usually placed below every second or third frame.

BACK RIPPER Short spikes pivoted on the back of the blade of an angledozer, bulldozer, or other machine and mounted to face toward the rear. They bite into the surface when the machine moves backwards and float on the surface when the machine moves forward. *See also* **Ripper.**

BACKRUSH *See* **Backwash.**

BACKSCATTER Dispersion of radiation in a direction opposite to that of its origin.

BACK SHORE 1. That part of the shore covered by water during exceptional storms only, especially those combined with exceptionally high tides; **2.** Outer support, under the rider shore, in raking shores.

BACKSIPHONAGE Backflow of water from a plumbing fixture, vessel or other sources into a water supply pipe due to a negative pressure in the pipe.

BACKSIPHON PREVENTER Plumbing fixture designed to prevent the backflow of water into a potable supply.

BACKSLOPE 1. Less-sloping face of a ridge; **2.** Portion beyond the ditch sides of an earth embankment for a highway cut that rejoins the original ground.

BACKUP 1. Second unit of equipment installed or held ready in case of failure of the prime unit; **2.** Overflow, or reverse of flow from or in a piping system caused by blockage or excess volume.

BACK VENT Branch vent connected to the main vent stack of a plumbing system and extending to a location near a fixture trap to prevent the trap from siphoning. *Also called* **Individual vent.**

BACKWARD PASS Calculation of the latest finish and latest start dates within the **Critical Path Method (CPM).**

BACKWASH 1. Seaward return of water following the uprush of waves across a beach or shore. *Also called* **Backrush; 2.** Water, spray or waves thrown back by an obstruction such as a cliff, breakwater, ship, etc.; **3.** Reversal of flow through a rapid sand filter to free material clogging the filtering media, flush it to a sewer, and thus reduce conditions causing loss of head across the filter. *Also called* **Filter wash.**

BACKWASHING Cleaning a sand or similar filter by reversing the flow through it and piping the washwater to waste.

BACKWATER 1. Water held back from its natural flow by a natural or artificial obstruction; **2.** Body of relatively tranquil water held in coves or bays, or covering low-lying areas, connected to the main body but less influenced by currents; **3.** Water reserve gained at high tide and held behind a barrier for later discharge at low tide.

BACKWATER CURVE Concave curve of the surface of a stream flowing in an open channel immediately upstream of a weir or dam.

BACKWATER EFFECT Effect that a dam or other obstruction has in raising the upstream elevation of the water surface.

BACKWATER FUNCTION Quantity occurring in the formula for determining the shape of the backwater curve, or the increased depth, of water in a waterway at various distances upstream of a dam or other obstruction that is raising the upstream elevation of the water surface.

BACKWATER GATE Gate installed at the end of a drain, culvert or outlet pipe to prevent the reverse flow of water or wastewater for any reason. *Also called* **Tide gate.**

BACKWATER VALVE *See* **Valve.**

BACTERIA Large, diverse group of one-celled microorganisms found wherever there is life; any of numerous unicellular microorganisms, of the class Schizomycetes, occurring in a wide variety of forms. They exist as free-living organisms or as parasites, usually appearing as spheroid, rodlike, or curved entities, but occasionally as sheets, chains, or branched filaments having a wide range of biochemical, often pathogenic, properties. There are many types, including:

Aerobic: Bacteria that require free elemental oxygen for growth.

Anaerobic: Bacteria that live and are active in the absence of free oxygen; the working bacteria in an anaerobic waste disposal system such as a septic tank.

Chemosynthetic: Bacteria that synthesize organic compounds by using energy derived from the oxidation of organic and inorganic materials without the aid of light.

Coli-aerogenes: Bacteria predominantly inhabiting the intestines of man or animals, but also occasionally found elsewhere.

Coliform-group: Group of bacteria predominantly inhabiting the intestines of man or animals. Includes all aerobic and facultative anaerobic types. Gram-negative, non-spore-forming bacilli that ferment lactose with the production of gas. Also included are all bacteria that produce a dark, purplish-green colony with metallic sheen by the membrane-filter technique used for coliform identification.

Facultative: Bacteria that can adapt to growth in the presence, as well as the absence, of oxygen.

Iron: Bacteria that assimilate iron and excrete its compounds in their life processes.

Manganese: Bacteria capable of using dissolved manganese and depositing it as hydrated manganic hydroxide.

Nitrobacteria : Bacteria that oxidize nitrite to nitrate.

Parasitic: Bacteria that thrive on other living organisms.

Pathogenic: Bacteria that may cause disease in the host organisms.

Photosynthetic: Bacteria that obtain their energy for growth from light by photosynthesis. *See also* **Blue-green algae.**

Saprophytic: Bacteria that thrive on dead organic matter.

Sulfate-reducing: Bacteria capable of assimilating oxygen from sulfate compounds, reducing them to sulfides.

Sulfur: Bacteria capable of using dissolved sulfur compounds in their growth.

BACTERIA BED *See* **Trickling filter.**

BACTERIAL ANALYSIS Examination of potable water, raw water and wastewater to determine the presence, number and identification of bacteria.

BACTERIAL CORROSION Destruction of a material by chemical processes brought about by the activity of certain bacteria that produce substances such as hydrogen sulfide, ammonia, and sulfuric acid.

BACTERIAL TREATMENT Treatment of wastewater involving bacteria. *See also* **Biological wastewater treatment.**

BACTERIOCHLOROPHYLLS Bacterial pigments that trap light energy.

BACTERIOLOGICAL AFTER-GROWTH Development of bacterial growth in the storage vessels, distribution pipes and appurtenances following treatment of the water supply.

BACTERIOLOGICAL COUNT Means for quantifying numbers of organisms from a sample. *See also* **Indicated number, Most probable number,** and **Plate count.**

BACTERIOPHAGES Viruses that infect bacteria.

BACTERIOSTAT Substance that retards or inhibits the growth of bacteria.

BAD AIR Contaminated air, or gas, or air containing insufficient oxygen that constitutes a hazard.

BADGER 1. Tool used to remove excess mortar from the inside joints of drain tile as it is laid; **2.** *See* **Plane.**

BADLANDS Area of barren land characterized by roughly eroded ridges, peaks, and mesas.

BAFFLE 1. Vane set at an angle to the main direction of the flow of vapor or water to impede the flow or influence its direction; **2.** Vane dividing a fluid storage vessel wholly or partially into compartments, primarily to reduce rapid movement of the contents when the vessel is being transported.

BAFFLE AERATOR Aerator in which baffles cause turbulence and minimize short-circuiting.

BAFFLE CHAMBER Chamber following a furnace combustion chamber, in which baffles change the direction of and/or reduce the velocity of the combustion gases in order to promote the settling of fly ash or coarse particulate matter.

BAFFLE PIER Obstruction set in the path of fast-flowing water to break up the stream and to dissipate energy.

BAG DAM Low-rise, temporary dam constructed of bags filled with a mixture of soil or sand and cement.

BAG FILTER Type of filter used in air handling systems in which return air is passed through a bag of material fine enough to trap particulate matter.

BAGGING OPERATION Mechanical process by which bags are filled with nonmetallic minerals.

BAG HOUSE Chamber, part of a flue or exhaust system, containing fabric bags through which the gas streams are vented and which serves to trap particulate matter.

BAG PLUG Inflatable device used to block and seal the outflow of a drain when testing.

BAG TRAP Plumbing fitting: an S-trap in which the inlet and outlet pipes are in vertical alignment.

BAGWORK Revetment or dike made of bags filled with a mixture of soil, sand, gravel, or sand and cement.

BAILER 1. Hollow cylinder used to remove rock chips and water from holes made while churn drilling; **2.** Length of pipe with a foot valve which, when lowered down a well casing, can raise oil or water mixed with sand that cannot otherwise be pumped.

BAILING BUCKET Bucket-like tool for removing water from the hole during drilling or in preparation for concrete placement. *See also* **Bucket.**

BAILEY BRIDGE Demountable temporary bridge constructed of modular components. The principal components are a series of panels that are linked to each other by steel pins and which can be stacked horizontally and vertically to increase the allowable span and carrying capacity. No lifting tackle is needed to assemble a bridge, and no component is heavier than a two-person load. The bridge, which is assembled on rollers on one bank of a crossing and has a temporary extension to its length that acts as a counterweight, is then launched over the gap from one abutment to the other, or to an intermediate pier.

BAKE OVEN Device which uses heat to dry or cure coatings.

BALANCE 1. Excess of debits over credits, or credits over debits in an account; **2.** Instrument for measuring weight.

BALANCE BAR Large, heavy beam extending along the upper edge of a lock gate and extending back a considerable distance beyond the hinge, used to open and close the gate when the water level is the same on each side of the gate.

BALANCED BID Bid where all the details, individual prices, and total are appropriate for the work to be done. *See also* **Bid,** and **Unbalanced bid.**

BALANCED CUT-AND-FILL Earthworks in which the volume of material to be excavated equals the volume of material needed as fill. *Also called* **Balanced earthworks.**

BALANCED EARTHWORKS *See* **Balanced cut-and-fill.**

BALANCED GATE Gate that operates automatically according to the upstream head to release water.

BALANCED, INDIGENOUS COMMUNITY Biotic community typically characterized by diversity, the capacity to sustain itself through cyclic seasonal changes, and the presence of necessary food chain species. Such a community may include species introduced in connection with a program of wildlife management and species whose presence or abundance results from substantial, irreversible environmental modifications.

BALANCED, INDIGENOUS POPULATION Ecological community which: **(a)** Exhibits characteristics similar to those of nearby, healthy communities existing under comparable but unpolluted environmental conditions; or **(b)** May reasonably be expected to become reestablished in a polluted water body segment from adjacent waters if the source(s) of pollution was removed.

BALANCED VALVE *See* **Valve.**

BALANCE PIPE Pipe connection that equalizes the pressure at two points in a system.

BALANCE POINT Point where all excavated material has been used as fill and all additional fill must be imported or additional excavation exported.

BALANCING 1. Adjustment of air or water through a system; **2.** Use of permanent or temporary storage areas to attenuate stormwater runoff.

BALANCING RESERVOIR Surface or elevated water storage facility such as a reservoir or tower that acts to smooth the flow within the system.

BALANCING VALVE See **Valve.**

BALER Machine that compacts materials, such as solid waste into high density packages.

BALLAST 1. Any heavy material placed in the hold of a vessel to increase stability; **2.** Crushed rock providing a foundation for road or rail tracks.

BALLAST GRATING See **Gravel guard.**

BALLAST VOYAGE Journey of a ship during which its holds or other storage spaces are filled with heavy material to ensure stability of the vessel.

BALL CLAY Secondary clay, commonly characterized by the presence of organic matter, high plasticity, high dry strength, long vitrification range, and a light color when fired.

BALL COCK Valve or faucet controlled by a change in water level and generally consisting of a device floating near the surface of the water that operates the valve.

BALLED AND BURLAPPED Plant material that has been dug with a solid ball of earth around the roots, that is then wrapped and tied in burlap for shipment.

BALL JOINT Flexible pipe joint shaped like a ball or sphere.

BALL LIGHTNING Form of atmospheric electric discharge, usually in the form of an incandescent sphere moving at a rate similar to ambient air movement.

BALL NOZZLE Sprinkler nozzle having a cup at its end and a ball within the cup. In operation, the ball is held in proximity to the tip of the nozzle by atmospheric pressure; the negative-pressure head being caused by the high-velocity sheet of water surround the ball.

BALL TEST Drain test in which a ball of approximately 12 mm (0.5 in.) less diameter

than the drain, is flushed through the drain from one manhole to the next to detect any partial blockage.

BALL VALVE See **Valve.**

BANDED STEEL PIPE See **Pipe.**

BAND SCREEN Endless moving belt of wire mesh that traps large solids at the intake to a power station, water or wastewater treatment plant. *Also called* **Belt screen.**

BANK 1. Steep rise or slope; **2.** In excavation, a mass of soil rising above digging or trucking level; **3.** Soil to be excavated from its natural position; **4.** Grouping of common devices mounted in a common frame or in a common panel; **5.** Continuous margin along a river or stream where all upland vegetation ceases; **6.** Elevation of land that confines the waters of a stream or river to their channel; **7.** Subaqueous area, detached from the shore, over which the water is relatively shallow.

BANK CUBIC METER/CUBIC YARD One cubic meter (yard) of material as it lies naturally. *See also* **Loose cubic meter/cubic yard.**

BANK GRAVEL Natural mixture of fines, sand, gravel, and cobbles.

BANK MATERIAL Soil, gravel, rock, etc., in place before excavation or blasting.

BANK MEASURE Measurement of material in its original place in the ground.

BANK PROTECTION Measures taken to prevent or reduce erosion of an earthen sloping face.

BANK REVETMENT Type of bank protection that covers the entire wetted face, from the deepest part of the river bed to above the elevation of the highest anticipated flood and/or wave.

BANK-RUN GRAVEL Material 150 mm (6 in.) maximum to less than 6 mm (0.25 in.) minimum, excavated from a naturally occurring deposit (finer material being screened out).

BANK SAND See **Soil types.**

BANK SLOPING Working a bank, usually with a grader or blade-equipped dozer, to form a uniform slope to the required profile.

BANK STORAGE Water absorbed by the banks of a waterway or stream that is returned to the stream as the water level falls.

BANK YARD Cubic yard of soil or rock measured in situ, before excavation.

BANNER CLOUD Wave cloud fixed to a mountain peak and extending downwind.

BANQUETTE Bench or berm constructed on the land side of a levee to increase the width of its lower potion.

BAR 1. Silt, sand and/or gravel deposited at the mouth of an estuary of a river; **2.** Non-SI unit of pressure, equal to 100 kilopascals (14.504 psi), permitted for use with the SI system of measurement for a limited time. Symbol: bar. *See also the appendix:* **Metric and nonmetric measurement.**

BARCHAN Crescent-shaped sand dune with its horns pointing downwind and the steeper slope to leeward.

BAR CHART 1. Graphical representation of frequencies or magnitudes with vertical or horizontal bands drawn in proportion of the frequencies or magnitudes involved; **2.** Graphic display of schedule-related information in which activities or other project elements are listed down the left side of the chart, dates are shown across the top, and activity durations are shown as date-placed horizontal bars. *Also called* **Gantt chart.**

BARE SITE Designated area of land, cleared of any existing structures, debris, and unwanted vegetation.

BARE SOIL Exposed soil, free of vegetation.

BARGE BED Natural or made mud bottom near a river bank where barges can rest at low tide while being loaded or unloaded.

BARITE Mineral, barium sulfate ($BaSO_4$), used in either pure or impure form as concrete aggregate, primarily for the construction of high-density radiation shielding.

BARIUM Silvery-white, soft metallic element (Ba) that tarnishes readily in air, occurring as baryte and as barium carbonate. It occurs in a number of compounds, most commonly barite (barium sulfate) and, to a lesser extent, witherite (barium carbonate).

BARIUM SULFATE TEST Quick chemical color test to determine approximately the pH value of soil.

BARK Tough outer covering of the stems and roots of woody plants, including an outside layer of dead cells, a cortex (which in twigs and small branches may contain chlorophyll), and an inner layer of phloem and xylem containing the tubes which carry sugar and water along the stem and root.

BARMINUTOR Combined bar screen and comminutor, used to trap and shred solids in the primary influent to sewage treatment plants.

BARNES' FORMULA Means for calculating flow in slimy sewers that states: velocity in m/s and which is equal to $46.5\mu m^{0.7}\ \sqrt{\iota}$ where μ is the hydraulic mean depth in meters and μ is the slope.

BARN WASTEWATER Wastewater from animal enclosures containing considerable quantities of animal manure, straw, bedding, etc.

BAROCLINIC Having a different distribution of pressure at different heights due to horizontal variations in air density.

BAROGRAPH Recording barometer.

BAROMETER Instrument for measuring atmospheric pressure, usually in bars, pascals, millimeters, or inches of mercury column.

BAROTHERMOGRAPH Instrument that simultaneously records barometric pressure and temperature on the same chart.

BAROMETRIC DAMPER *See* **Damper.**

BAROMETRIC PRESSURE Atmospheric pressure at a specific place according to the current reading of a barometer.

BAROTROPIC Having the same pressure

field at all heights and no horizontal density gradients.

BAR RACK Screen made up of vertical, horizontal, or inclined parallel bars positioned in a channel or waterway to catch floating and partially-submerged debris.

BARRAGE Dam to divert a flow of water or to increase its depth.

BARRANCA Steep bluff.

BARREL 1. Cylindrical container with sides that bulge outward and having flat ends. *See also* **Drum**; **2.** Water passage in a culvert; **3.** That part of a pipe throughout which the bore and wall thickness remain uniform; **4.** Water cylinder of a pump. **5.** Non-SI measure of volume, commonly of crude oil, equivalent to 42 US gallons, 35 Imperial gallons, or 159 liters.

BARREL NIPPLE Short length of pipe, outside threaded on each end, bare in the middle.

BARRIER 1. Dam or sill across a small stream, usually on steep slopes, built to retard flow and reduce the cutting action of the stream; **2.** Any material or structure that prevents or substantially delays movement of water or radionuclides toward the accessible environment.

BARRIER BAR Ridge of sand and/or gravel more-or-less parallel to the shore, separating a bay or large river from the body of water into which it discharges; usually developed by the breaking of storm waves on shallow, shelving shores.

BARRIER BEACH Beach, essentially parallel to the shoreline, formed some distance out and forming a lagoon.

BARRIER EFFECT Formation of a surface film that protects a material from further corrosion.

BARRIER FILTER Filter that removes particles from a fluid stream with a porous structure in the flow path.

BARRIER FOREST Treed slopes that hold back snow and so prevent avalanches.

BARRIER ISLAND Unstable rows of parallel ridges of sand lying close to shore and parallel to it that form an offshore island.

BARRIER PLANTING Strategic use of trees, shrubs, and other vegetation to form a physical, visual, acoustic, or other type of screen.

BARRIER SPIT Beach barrier connected at one end to the mainland.

BARRIER SPRING Spring rising through a raised bedrock fault covered by a thick deposit of alluvium.

BARRING AND WEDGING Removal, by prying with bars and the insertion of wedges, typically of the final portion of unsuitable rock from the foundation surface of a dam.

BARROW PIT *See* **Borrow pit.**

BAR SCREEN Frame containing a series of similar-size bars or rods placed parallel with each other.

BASCULE GATE Crest gate rotating about a horizontal torque tube and raised by the action of a hydraulic piston on the tube.

BASE Any substance that reacts with acid (H^+ ions) to form salts, plus water.

BASE BID Amount for which the bidder offers to do the work or supply the goods, not including work for which alternate bids are also submitted.

BASED FLOOD Flood that has a 1% or greater chance of recurring in any year, or a flood of a magnitude equalled or exceeded once in 100 years on average over a significantly long period.

BASE ELBOW Pipe elbow with a baseplate by which it is supported.

BASE EXCHANGE Water softening process in which water is passed through a pressurized tank containing the salt-absorbing mineral zeolite.

BASE-EXCHANGE PROCESS Process permitting the exchange of positive ions on a prepared medium.

BASE FLOOD Flood having a 1% chance of being equaled or exceeded in a given year.

BASE FLOOD ELEVATION Maximum elevation that a base flood would reach.

BASE FLOW That part of the flow of a stream that originates in groundwater, long-term lake storage, or snowmelt.

BASE-FLOW DEPRECIATION CURVE Characteristic rate, expressed as a curve, at which the base flow of a stream at a given point diminishes until supplemented by precipitation or runoff.

BASE LEVEL 1. Lowest elevation to which running water will erode a land surface; **2.** Profile at which a stream neither degrades nor aggrades its course.

BASELINE Tolerance level of an organism to a particular concentration of a substance.

BASELINE FINISH DATE *See* **Scheduled finish date.**

BASELINE START DATE *See* **Scheduled start date.**

BASE LOAD Minimum load over a given period at a stated rate.

BASE LOG First log in a cribbing face (face log) resting directly on the ground or on the mud sills.

BASE PERIOD Time during which the flow in a stream exceeds the base flow as a result of an isolated storm. *See also* **Hydrograph.**

BASE PROCESS Manufacturing process in which raw materials are introduced to a plant in batches, as against continuously, and which tends also to result in the generation of wastes in slugs, rather than a continuous stream.

BASE RUNOFF Sustained, or dry-weather flow.

BASE SATURATION Soil condition where the cation exchange capacity is saturated with exchangeable bases of Ca^{2+}, K^+ or Na^+.

BASE SUBSISTENCE DENSITY Human population density above which the contin-

ued survival of the population is impossible.

BASE TEMPERATURE Arbitrary reference temperature for determining liquid densities or adjusting the measured volume of a liquid quantity.

BASIC DATA Records of facts, occurrences, and conditions, as they have occurred, excluding anything developed by means of computation or estimate.

BASIC HYDROLOGIC DATA Records of measurements and observations of the quantity per unit of time of precipitation (including snowfall), streamflow, evaporation from water, and water elevations (both surface and underground, etc.).

BASIC ROCK Dark colored igneous rock containing little or no silica; basalt, lava, gabbro, etc.

BASIC-STAGE FLOOD Arbitrary selected rate of flow of a stream, used as the lower limit in selecting floods to be analyzed; usually taken as the minimum annual flood.

BASIDIOMYCETES Group of fungi having external spores, usually in fours, on mother cells, i.e. mushrooms, puffballs, bracket fungus, etc.

BASIN 1. Ground depression with sloping sides; **2.** Surface area within a given drainage system; **3.** Shallow tank through which liquids may be passed or in which they are detained for treatment or storage; **4.** Receptacle for water; in plumbing, a lavatory.

BASIN MIXER Mixing faucet for a wash basin in which hot and cold water and rate of flow can be adjusted with a single control.

BATCH-FED INCINERATOR *See* **Incinerator.**

BATCH LOADER *See* **Refuse truck.**

BATCH PROCESS Treatment process in which a tank or reactor is filled, the contents are treated or a chemical solution is prepared, and the tank is emptied.

BATCH-TYPE FURNACE Furnace that is shut down periodically to remove old charge and add a new charge, as opposed to a continuous-type furnace.

BATCH WATER HEATER In a domestic solar heating system, a water storage tank in an insulated container fitted with a glazed cover, used to heat and store water.

BATESIAN MIMICRY Defensive ability of an otherwise harmless animal to mimic the resemblance of a poisonous, dangerous or distasteful one.

BATH MIXER Blending faucet for a bathtub in which hot and cold water, plus rate of flow, and sometimes the shower control, are combined in one control.

BATHOLITH Large igneous intrusion, mostly granitic, with steeply dipping contacts and no apparent floor.

BATHYAL Sediments and the sea bed between the edge of the continental shelf and the start of the abyssal zone.

BATHYMETER Instrument used to determine ocean depths.

BATHYPELAGIC Marine organisms living at depths between 1000 and 3000 m (3280 and 9842 ft).

BATTERY OF WELLS Group of interconnected wells from which water is drawn by a single pump.

BATTURE Elevated river bed.

BAUMÉ SCALE One of several useful measurements of specific gravity of a liquid. *See* **Specific gravity.**

BAY Curved shoreline enclosed between headlands or promontories; larger than a cove, smaller than a gulf.

BAY AND PROMONTORY Bay, shaped by a ridge line, and a promontory that is the dominant upland feature that shapes the bay. *Also called* a **Headland.**

BAYHEAD BARRIER Barrier enclosing a lagoon near the inland head of a bay.

BAYMOUTH BARRIER Barrier across the mouth of a bay.

BEACH 1. Gently sloping, sand and pebble edge to a body of water; **2.** Face of a hydraulically filled earth dam.

BEACH DRIFT Zigzag movement of material along a beach caused by waves that wash obliquely on to the shore.

BEACH EROSION Retrogression of the shoreline of large lakes and coastal water caused by wave action and shore currents.

BEACHING Loose-graded stones, from 0.07 to 0.2 m (2.75 to 8 in.) in diameter, used in a layer 0.3 to 0.6 m (12 to 24 in.) thick as a revetment.

BEACH PIPE Hydraulic fill pipe used to carry a slurry of earth and water to construct an earth dam.

BEARER Horizontal scaffold member upon which the platform rests and which may be supported by ledgers or runners; the transverse member that joins scaffold uprights, posts, poles, and similar members.

BEAT Periodic increase and decrease in amplitude.

BEAUFORT SCALE Scale of wind strength between calm and hurricane that classifies conditions between 1 and 12:

Beaufort #	Description	Wind speed (km/h)
0	Calm	0-1
1	Light air	1-5
2	Light breeze	6-11
3	Gentle breeze	12-18
4	Moderate breeze	19-26
5	Fresh breeze	27-34
6	Strong breeze	35-43
7	Moderate gale	44-53
8	Fresh gale	54-64
9	Strong gale	65-77

10	Whole gale	78-90
11	Storm	91-104
12	Hurricane	105+

In the US and the UK, an earlier table of wind speeds is ordinarily in use, and in 1955 added Beaufort scale numbers 13 to 17 to be able to announce projected wind speed in greater detail in hurricane forecasts:

Description	mph	km/h
Light	0-1	0-2
Light	1-2	2-5
Light	4-7	6-11
Gentle	8-12	12-19
Moderate	13-18	20-29
Fresh	19-24	30-38
Strong	25-31	39-50
Strong	32-38	51-61
Gale	39-46	62-74
Gale	47-54	75-86
Whole gale	55-63	87-101
Whole gale	64-72	102-115
Hurricane	73-82	116-131
Hurricane	83-92	132-147
Hurricane	93-103	148-165
Hurricane	104-114	166-182
Hurricane	115-125	183-200
Hurricane	126-136	201-218

BECQUEREL One of 17 derived units of the SI system of measurement: a unit of radioactivity identified as one disintegration per second (1 Bq=2.7x 10^{-11} curies). Symbol: Bq. *See also the appendix:* **Metric and non-**metric measurement.

BED 1. Bottom of a watercourse or body of water; **2.** Seam or deposit earlier in origin than the material overlying it.

BED AND HAUNCHING Weak concrete, not less than 150 mm (6 in.) thick, partly surrounding a pipe laid in a trench; the concrete reaches the sides of the trench and extends up to the horizontal diameter of the pipe.

BEDDING 1. Supports or prepared ground on which buried pipe is laid; **2.** Thick layer of material laid over a surface to fill any irregularities between it and something placed on top of it; **3.** Raised mound on which seedlings are planted (technique common to the southeastern US).

BEDDING PLANE Plane or surface separating the individual layers, beds, or strata in sedimentary or stratified rocks.

BED LOAD Volume of sand and gravel or debris moved along the bed of a stream by the flow of water.

BEDROCK Continuous solid rock that everywhere underlies the regolith (noncemented rock fragments and deposited materials).

BEEHIVE BURNER *See* **Burner.**

BEGIN ACTUAL CONSTRUCTION Initiation of physical onsite construction activities on an emissions unit which are of a permanent nature. Such activities include, but are not limited to, installation of building supports and foundations, laying of underground pipework, and construction of permanent storage structures.

BEGINNING EVENT Start of the first activity in a critical path program.

BEHAVIOR Manner in which an organism responds to a stimulus.

BELANGER'S CRITICAL VELOCITY Velocity in an open channel for which the velocity head is equal to one-half the mean depth.

BELL Expanded termination at one end of a pipe section into which the regular diameter end of the next pipe or pipe fitting fits.

BELL AND SPIGOT Joint system used on large-diameter cast-iron, plastic, or clay pipe, one end of each length being swelled to form a bell into which the narrow end (spigot) of another length will fit, the resulting joint (in cast iron work) being caulked with oakum and lead. *Also called* a **Bell joint.**

BELL DOLPHIN Large bell-shaped steel or concrete fender supported on a cluster of piles in open water, used to moor vessels.

BELLMOUTH Flared end of a tube.

BELLMOUTHED ORIFICE Orifice with a short tube on the outside.

BELLMOUTH INLET *See* **Flaring inlet.**

BELLMOUTH OVERFLOW Progressively larger overflow structure from a reservoir.

BELLOWS SEAL Metal bellows used in a shaft seal, in place of packing for valves, and in place of gaskets in long pipelines, to allow for expansion and contraction of the line.

BELL TILE Clay pipe sections with one end enlarged to loosely join with the next pipe section.

BELLY DUMP Material hauler, prime mover or trailer, that discharges its load from the bottom.

BELLY PAN Steel pan-shaped sheet mounted on the underside of heavy earthmoving equipment to protect the upper structure, engine, and hydraulic components.

BELOWGROUND RELEASE Any release to the subsurface of the land and to groundwater. This includes, but is not limited to, releases from the belowground portions of an underground storage tank system and belowground releases associated with overfills and transfer operations as the regulated substance moves to or from an underground storage tank.

BELOWGROUND STORAGE FACILITY Tank or other container located other than as defined as above ground.

BELT CONVEYOR Conveying device that transports material from one location to another by means of an endless belt that is carried on a series of idlers and routed around a pulley at each end.

BELT OF CEMENTATION Lower portion of the katamorphic zone, extending from the watertable to the lower limits of the rock fracture zone, in which the physical and chemical processes are constructive and result in the deposition, increase in volume, and induration of material.

BELT OF PHREATIC FLUCTUATION Belt of watertable fluctuation.

BELT OF SOIL WATER That part of the lithosphere, immediately below the surface, from which water is discharged into the atmosphere by action of plants and/or by evaporation of soil water. *Also called* **Soil-water belt.**

BELT OF WATERTABLE FLUCTUATION That part of the lithosphere lying within the saturation zone part of the time, and in in the overlying aeration zone for the remainder.

BELT OF WEATHERING Top layer of the lithosphere in the katamorphic zone, extending from the ground surface to the groundwater table, in which the physical and chemical processes are destructive and result in the breaking down of the material of which the belt is composed.

BELT SCREEN *See* **Band screen.**

BENCH FLUME Flume constructed on a bench cut into a sloping hillside and resting on the ground.

BENCHING 1. Process of excavating whereby terraces or ledges are worked in a stepped shape; **2.** Concrete around a half-round drainage channel in a manhole, sloping up from the tile edge to the manhole wall; **3.** Berm above a ditch.

BENCH SCALE ANALYSIS Small-scale laboratory analysis.

BENCH TERRACE More-or-less level step in a slope of cut.

BEND Short piece of curved pipe.

BENDERGAIN Allowance incorporated when measuring the length of sheet metal, conduit, or other material that must be of accurate length when bent.

BENEDICTION PROCESS Dressing or processing of gold-bearing ores for the purpose of: **(a)** regulating the size of, or recovering, the ore or product, **(b)** removing unwanted constituents from the ore, and **(c)** improving the quality, purity, or assay grade of a desired product.

BENEFICATION Improvement of the chemical or physical properties of a raw material or intermediate product by the removal or modification of undesirable components or impurities.

BENEFICATION SYSTEM In recycling, the mechanical process of removing contaminants and cleaning scrap; the treatment of a material to improve its form or properties.

BENEFICIAL ORGANISM Any pollinating insect, or any pest, predator, parasite, pathogen or other biological control agent which functions naturally or as part of an integrated pest management program to control another pest.

BENEFICIAL USE Use of some aspect of the natural environment to human benefit.

BENEFICIATION Separation of a valuable mineral from its host rock.

BENEFIT–COST ANALYSIS *See* **Analysis.**

BENEFIT–COST RATIO Ratio obtained by dividing the anticipated benefits of a project by its anticipated costs. Either gross or net benefits may be used as the numerator.

BENEFITS (MANDATORY AND CUSTOMARY) Individual benefits, in addition to salary or wages, required by law (such as social security, worker's compensation, unemployment insurance, disability insurance, statutory holidays, etc.), and by custom (such as sick leave, holidays, etc.), and optional within the organization or firm (life insurance, supplementary health insurance, pension plans, etc.).

BENTGRASS *See* **Grass.**

BENTHAL DEMAND Dissolved oxygen demand of water overlying benthal deposits resulting from the upward diffusion of the products of decomposition.

BENTHAL DEPOSITS Mud, sand or gravel deposited by flowing water.

BENTHIC Organisms living close to the bottom of a lake, river or sea.

BENTHOS 1. Sea or lake bottom; **2.** The aggregate of organisms living from high water mark to the bottom of a body of water.

BENTONITE Clay composed principally of minerals of the montmorillonoid group, characterized by high adsorption and very large volume change with wetting or drying; the basis of driller's mud, used to prevent collapse of holes augered into cohesionless soils.

BENTONITE SLURRY TRENCH Trench excavated through common material and filled with bentonite slurry to maintain the dug sides.

BENZENE Aromatic hydrocarbon (C_6H_6) which is a colorless, volatile, and flammable liquid, used mainly in the manufacture of other organic chemicals. It is present in gasoline at concentrations of approximately 1% to 2%, and vehicular emissions constitute the main source of benzene in the environment. It is introduced into water from industrial effluents and atmospheric pollution. It is a documented human carcinogen.

BERM 1. Shoulder of a paved road or ditch; **2.** Artificial ridge of earth; **3.** Earth embankment, often combined with fencing or planting to create a visual and/or sound barrier.

BERNOULLI'S THEORY If no work is done on or by a flowing, frictionless liquid, its energy, due to pressure and velocity, remains constant at all points along the streamline.

BERTH Docking place for a ship.

BERTHING IMPACT Live forces imposed on piers during vessel docking.

BEST AVAILABLE CONTROL TECH-NOLOGY Emissions limitation (including a visible emission standard) based on the maximum degree of reduction for each pollutant subject to a regulation or Act which would be emitted from any proposed major stationary source.

BEST BID/TENDER Bid to complete work or supply services/materials that most closely meets the needs of the owner. This may not necessarily be the lowest price submitted.

BEST MANAGEMENT PRACTICES Measures or practices used to reduce the amount of pollution entering a sanitary sewer system, surface water, air, land, or groundwater.

BEST PRACTICABLE WASTE TREAT-MENT TECHNOLOGY Best and most cost-effective technology that can treat wastewater, combined sewer overflows and non-excessive infiltration and inflow in publicly owned or individual wastewater treatment works.

BETA PARTICLE Elementary particle emitted from a nucleus during radioactive decay and having a single electrical charge, and a mass equal to 1/1837 that of a proton.

BETA RADIATION Stream of beta particles, i.e., electrons or positrons emitted from radioactive nuclei.

BETA X ($ß_x$) Filtration ratio obtained during a multi-pass test. Under specified test conditions, it is the number of particles in an influent fluid greater than C (in μm) per unit volume, divided by the number of particles in the effluent fluid in the same size range and unit volume.

BETA (ß) RATIO In fluid filtration, the ratio of a number of particles greater than a given size (z μm) in a given volume of influent fluid to the number of particles greater than the same size (x μm) in the same volume of effluent fluid.

BETHELL PROCESS Method of preserving wood by impregnating the cells under pressure with creosote.

B-HORIZON Lowest part of topsoil, between the A-horizon and C-horizon, that contains metal oxides and other soluble materials leached from the A-horizon. *See also* **A-horizon,** and **C-horizon.**

BIAS Systematic statistical distortion which balances out overall.

BIAT Timber bearer giving support to guard rails, decking, walkways, etc.

BIB Tap or faucet threaded for connection of a hose. *Also called* a **Hose bib,** and **Sill cock.**

BICARBONATE ALKALINITY Alkalinity caused by bicarbonate ions.

BID Submitted tender or bidding price for the purchase of something, completion of specified work and/or supply of specified materials. *See also* **Balanced bid,** and **Unbalanced bid.**

BID ABSTRACT List of bidders to a project with details of bid items per item. *Also called* **Bid tabulation.**

BID BOND *See* **Bond.**

BID CALL Published invitation to submit bids for specified work or materials at a designated place and time.

BID DATE Date, and time at which bids must be filed. *Also called* **Bid time.**

BIDDER One who is qualified, and who has taken out a bid package, in order to prepare a price for the supply of goods and/or materials or to complete a job. *Also called* **Tenderer.**

BIDDER'S LIST List of those who have taken documents in order to prepare a bid. *Also called* **Tenderer's list.**

BIDDING DOCUMENTS Range of documents, including an invitation to bid, instructions to bidders, bid form, proposed contract documents, and drawings.

BIDDING OR NEGOTIATION PHASE Fourth phase of an architect's basic services, when competitive bids or negotiated proposals are sought as the basis for awarding a contract.

BIDDING PERIOD Elapsed time between the final day for taking bid documents and the prescribed time to submit a completed bid.

BIDDING REQUIREMENTS Specific details establishing procedures and conditions for the submission of bids, contained in the bidding documents.

BID DOCUMENTS Documents, including working drawings, specifications, contract details, etc., that are available to all contractors bidding the same job. *Also called* **Tender documents.**

BIDET Plumbing fixture designed to facilitate washing the perineal area of the body.

BID FORM Form supplied to a bidder that must be completed and that becomes part of the bidding documents. *Also called* **Tender form.**

BID GUARANTEE Security against completion of the contract and securing of any required bonds, required to be submitted with the bid documents.

BID OPENING Process of opening sealed bids and evaluating the submissions according to the terms indicated in the bid notice. *Also called* **Opening of tenders.**

BID PACKAGE Documents and other submissions that make up a bid to complete work and that may include offers from more than one company for parts of the work, and/or details of materials or construction techniques that vary from those in the bidding document. *Also called* **Tender package.**

BID PRICE Sum stated in a bid for which the bidder offers to complete the work.

BID QUANTITIES Quantities in a bid schedule for unit-price items; extended by unit prices to arrive at a total bid price for each unit-price bid item.

BID SECURITY Deposit of a bid bond, bank draft, cash, cashier's check, certified check, or other specified instrument required to be submitted with a bid to serve as a guarantee that the bidder, if awarded the contract, will complete the contract in accordance with the bidding and contract docu-

ments.

BID SHOPPING Practice of some contractors on being awarded a contract of soliciting subcontract bids from other than those who submitted prices during the bid compilation stage.

BID TABULATION *See* **Bid abstract.**

BID/TENDER BOND *See* **Bond.**

BID/TENDER NOTICE Advertisement calling for bids.

BID TIME *See* **Bid date.**

BIENNIAL PLANT Plant with a life span of two years.

BIFURCATE Divide into two branches.

BIFURCATION Division of one penstock into two or more smaller units, each of which is separately connected to a turbine scroll case.

BIFURCATION GATE Gate positioned where a conduit is divided into two branches to divert the flow into, or out of, one or the other.

BIG-CONE DOUGLAS FIR TREE *Pseudotsuga menziesii.* Commonly planted timber tree originating from the Western Rocky Mountains.

BIGHT Slight bend in a coast forming an open, crescent-shaped bay.

BIG LINE Fire hose at least 65 mm (2.5 in.) in diameter.

BILL OF MATERIALS List detailing the materials and products to be used in a project, usually forming part of the bidding document. *Also called* **Bill of quantities.**

BILL OF QUANTITIES *See* **Bill of materials.**

BILLOW CLOUD Clouds arranged in several parallel bars in close proximity in a layer.

BILTMORE STICK Stick graduated in such a way that the diameter of a standing tree may be estimated when the stick is held across the main axis of the tree at the distance from the eye for which the stick is graduated (usually 60 cm (24 in.)). *See also* **Cruiser's stick.**

BIMETALLIC CORROSION Corrosion that develops from the reaction between dissimilar metals in contact.

BINAURAL Able to hear with both ears.

BINDER SOIL Material consisting primarily of fine soil particles (fine sand, silt, true clays and colloids) and having good binding properties. *Also called* **Clay binder.**

BINDING ARBITRATION Settlement of a dispute between parties to an agreement or contract using the experience and knowledge of an independent expert or panel who, after hearing submissions by the various parties, will decide upon the form of settlement, which will be binding upon all parties to the dispute. *See also* **Arbitration.**

BINOCULAR VISION Ability of both eyes to focus simultaneously on an object, thus seeing the same scene from a slightly different angle to give stereoscopic vision.

BIOACCUMULATION The increasing concentration of a compound in the bodies of living organisms..

BIOAERATION Technique and associated equipment by which oxygen is introduced to a biological waste by a device that promotes mixing and development of aerobic bacteria within the mass.

BIOASSAY 1. Evaluation method using a change in biological activity to analyze a response to treatment; **2.** Use of viable organisms, e.g. live fish, to determine the toxic effects of wastewater.

BIOAVAILABILITY Relative amount of administered test substance which reaches the systemic circulation.

BIOCHEMICAL Substance relating to the chemistry of plant and animal life.

BIOCHEMICAL ACTION Chemical change resulting from the metabolism of living organisms.

BIOCHEMICAL OXIDATION Oxidation resulting from biological activity; the chemical combination of oxygen and organic matter.

BIOCHEMICAL OXYGEN DEMAND (BOD) Measure of the amount of oxygen used by microorganisms to break down organic waste materials in water. This is commonly determined under standard laboratory conditions in tests lasting 5 days conducted at a temperature of 20°C (68°F), expressed in mm/L. There are several stages, including:

> **First stage:** That part associated with biochemical oxidation of carbonaceous (as distinct from nitrogenous) material.
>
> **Immediate:** Initial quantity of oxygen used by polluted liquid immediately upon being introduced into water containing dissolved oxygen.
>
> **Second stage:** Oxygen demand associated with the biochemical oxidation of nitrogenous materials.
>
> **Standard:** Biochemical oxygen demand as determined under standard laboratory procedure for 5 days at 20°C, expressed in mg/L.
>
> **Ultimate:** Total quantity of oxygen required to satisfy the biochemical oxygen demand of a sample.

BIOCHEMICAL OXYGEN DEMAND TEST Procedure that measures the rate of oxygen use under controlled conditions of time and temperature. Standard test conditions include dark incubation at 20°C (68°F) for a specified time (usually 5 days).

BIOCHEMICAL PROCESS Chemicl process in living systems whereby organic substances are transformed through the action of enzymes.

BIOCIDES Agents that kill living organisms.

BIOCLASTIC Made up of fragmented organic remains.

BIOCLIMATOLOGY Study of the relationship of climate to living organisms.

BIOCOEN All living components of the environment.

BIOCONCENTRATION 1. Net accumulation of a substance directly from water into and onto aquatic organisms; **2.** Increase in concentration of test material in or on test organisms relative to the concentration of test material in the ambient water or soil.

BIOCONCENTRATION FACTOR 1. Quotient of the concentration of a test chemical in tissues of aquatic organisms at or over a discrete time period of exposure divided by the concentration of test chemical in the test water at or during the same time period; **2.** Ratio of a test substance concentration in a test fish (C_t) to the concentration in the test water (C_w) at steady state.

BIOCONVERSION Conversion of organic waste via biological decomposition to produce usable gas, liquid fuels, or compost products.

BIODEGRADABLE Substance susceptible to break down by bacterial attack.

BIODEGRADATION Chemical breakdown of materials by living organisms in the environment. The process depends on certain microorganisms, such as bacteria, yeast, and fungus, which break down molecules for sustenance.

BIODIESEL Diesel fuel substitute manufactured from rape seed oil (*Brassica napus*) or sugar cane.

BIOENGINEERING In the context of mechanical engineering and construction, the use of natural materials and living organisms, i.e. the use of sedges and other plants as revetment or for waste water cleaning..

BIOFILTRATION Use of natural materials and organisms to effect change and/or degrade a component, i.e. the use of compost or peat to remove odors from air passed through it; or the use of microorganisms to break down the organic fraction of sewage.

BIOFLOCCULATION Clumping together of fine, dispersed organic particles by the action of bacteria and algae.

BIOFUEL Fuel that has a biological origin, i.e. biodiesel, biogas, or methane.

BIOGAS Gas generated by anaerobic decomposition of organic materials and typically composed of 62% methane and 38% carbon dioxide.

BIOGASIFICATION Resource recovery process for the extraction of methane resulting from anaerobic decomposition of organic matter.

BIOGEOCLIMATIC CLASSIFICATION Delineation of biotic regions or zones on the basis of vegetation, soils, topography, and climate.

BIOGEOSPHERE Outer part of the Earth's crust, as far below the surface as life exists.

BIOLOGICAL ADDITIVES Microbiological cultures, enzymes, or nutrient additives that are deliberately introduced into an oil discharge for the specific purpose of encouraging biodegradation to mitigate the effects of the discharge.

BIOLOGICAL AMPLIFICATION Accumulative concentration of a persistent substance, in or by an organism, or plant.

BIOLOGICAL BENCHMARK Use of a plant or animal species to establish a level of something, the absence or presence of which can be determined by subsequent measurement.

BIOLOGICAL CONCENTRATOR Class of filter feeders, including limpets, oysters and other shellfish, plant or bacterium that concentrate heavy metals and other stable compounds present in dilute concentrations in sea or freshwater or soil.

BIOLOGICAL CONTROL AGENT Any living organism applied to or introduced into the environment that is intended to function as a pesticide against another organism.

BIOLOGICAL FILTER *See* **Trickling filter.**

BIOLOGICAL FILTRATION Process in which a liquid is passed through the media

65

of a biological filter, bringing it into contact with attached zoogleal films that adsorb and absorb fine suspended, colloidal, and dissolved solids, break down macromolecules, releasing the end products of biochemical action.

BIOLOGICAL INDICATOR Species of organisms, the presence of which, or changes in which, are a mark of some environmental quality.

BIOLOGICALLY-ACTIVE FLOC Floc resulting from the action of biological activities; activated sludge.

BIOLOGICAL MONITORING Direct measurement of change to habitat, based on evaluations of the number and distribution of individuals or species before and after a change.

BIOLOGICAL NUTRIENT REMOVAL Process by which the nitrogen content of sewage treatment plant effluent is decreased in order to reduce the growth of algae, bacteria and aquatic plants in receiving waters, simultaneously lowering the ammonia and nitrate content.

BIOLOGICAL OXIDATION Process whereby living organisms in the presence of oxygen convert organic matter into a more stable form.

BIOLOGICALS Preparations made from living organisms and their products, including vaccines, cultures, etc., intended for use in diagnosing, immunizing or treating humans or animals or in research.

BIOLOGICAL SHIELDING Shielding provided to attenuate or absorb nuclear radiation, such as neutron, proton, alpha, and beta particles, and gamma radiation. The shielding is provided mainly by the density of the concrete or lead, except that in the case of neutrons, the attenuation is achieved by compounds of some lighter elements, e.g. hydrogen and boron. *See also* **Boron-loaded concrete,** and **Shielding concrete.**

BIOLOGICAL SLIME *See* **Microbial film.**

BIOLOGICAL TREATMENT Reduction

of waterborne organic wastes by living organisms.

BIOLUMINESCENCE Production of light of various colors by living organisms.

BIOMASS 1. Mass, volume or energy of organic material consisting of living organisms feeding on the wastes in wastewater, dead organisms, and other debris. **2.** Total woody material in a forest. Refers to both merchantable material and material left following a conventional logging operation. In the broad sense, all of the organic material in a given area; in the narrow sense, burnable vegetation to be used for fuel in a combustion system.

BIOMASS FUEL Carbonaceous fuel derived from dead plant material.

BIOMASS HARVESTING Harvesting of all material including limbs, tops, unmerchantable stems and stumps, usually for wood energy.

BIOME Extensive climatically controlled ecological community, distributed over a wide area.

BIOMETALLURGY Technique of employing microbiological processes to effect mineral treatment or extraction, typically the extraction of nonferrous metals from sulfide ores.

BIOMETEOROLOGY Study of the relationship between weather and living organisms.

BIOMONITORING Evaluation or measurement of the effects of toxic substances in effluents on aquatic or other organisms in receiving waters or soils.

BIOPHOTOLYSIS Storage and use of electrons produced in the initial stages of photosynthesis through breakdown of water into oxygen and hydrogen.

BIOPLEX System in which the wastes produced at each stage of a sequence of stages are used as raw materials for a succeeding stage.

BIOREACTOR Waste reduction or manufacturing device in which a gas–liquid stream is brought into intimate contact with oxygen.

BIOREMEDIATION Use of biological agents to mitigate and/or restore contaminated soil.

BIOSCRUBBER Odor removal filter consisting of a tower filled with plastic screens through which sewage effluent is passed. The plastic medium promotes the growth of bacteria which remove the hydrogen sulfide from the flow, converting it to dilute sulfuric acid.

BIOSESTON Total of the living microplankton floating or swimming in water.

BIOSOLIDS Primarily organic solid product, a by-product of wastewater treatment processes, that can be beneficially recycled.

BIOSPHERE Portion of the Earth and its atmosphere capable of sustaining life.

BIOSPHERIC CYCLES Natural series of phases in which materials are recycled so that after use they are returned in a reusable form. The major natural cycles are the carbon cycle, oxygen cycle, nitrogen cycle, and hydrological cycle — all of which operate simultaneously and maintain all life on Earth.

BIOSTIMULANT Substance that stimulates plant or animal growth.

BIOSURVEY Survey of the types and numbers of organisms naturally present in an area.

BIOSYSTEMATICS Study of the variation and evolution of organisms.

BIOTA Total animal and plant life of a region.

BIOTECHNOLOGY Use of industrial techniques to simulate, enhance or change a biological process.

BIOTIC Of or having to do with life or living things.

BIOTIC FACTORS Influences on the environment resulting from the activities of living organisms.

BIOTIC INDEX Rating used in assessing the quality of the environment in ecological terms.

BIOTIC INFLUENCE Biological influence on vegetable life, in contrast to climatic influence.

BIOTIC POTENTIAL Estimate of the maximum rate of increase of a species in the absence of competition from predators.

BIOTIC PROVINCE Major ecological region of a continent.

BIOTOPE Habitat that is uniform in its main climate, soil and biotic conditions.

BITTERNS Saturated brine solution remaining after precipitation of sodium chloride in the solar evaporation process.

BLACK ALUM Aluminum sulfate containing a small percentage of activated carbon, used in water treatment.

BLACK CHERRY (TREE) *Prunus serotina* An aromatic tree with a tall trunk. Height to 24 m (80 ft); diameter to 0.6 m (2 ft), the largest and most important native cherry for timber and fruit. *Also called* **Cherry (tree).**

BLACK ICE Clear ice that can form on a wet road at night.

BLACK IRON PIPE *See* **Pipe.**

BLACK LIQUOR OXIDATION SYSTEM Vessels used to oxidize, with air or oxygen, the black liquor, and associated material in storage tanks.

BLACK LIQUOR SOLIDS Dry weight of the solids which enter the recovery furnace in the black liquor.

BLACK SMOKE Smoke consisting of carbon particles; an indicator of incomplete combustion.

BLACK SPRUCE TREE *Picea mariana (nigra)* Northern coniferous tree used in plantations for timber.

BLADE Equipment part or accessory that digs and pushes broken rock or earth but that does not carry it; typically fitted to a tractor dozer, trencher, or excavator.

BLADE MIXING Mixing of materials, on site, with the blade of a motor grader.

BLANK Bottle containing only dilution water or distilled water, used as a standard.

BLANKET 'Fertile' material surrounding a fast reactor core to trap and retain neutrons and create more fissile material.

BLANKET BOG Area, often extensive, of acid peatland, characteristic of broad, flat, upland watersheds in constantly wet climates.

BLANKET INSULATION Relatively flat and flexible insulation in coherent sheet form, furnished in units of substantial area, including batt insulation.

BLANK FLANGE Undrilled flange.

BLEACH Dry or liquid compound containing chlorine, sometimes combined with calcium or sodium.

BLEACHING POWDER Slaked lime and chlorine in which calcium oxychloride is the active ingredient.

BLEED 1. Exudation, percolation or seepage of a liquid through a surface; **2.** Removal of unwanted air or fluid from passages, pipes, etc. *See also* **Purge.**

BLEEDER 1. Small valve, used to drain small quantities of fluid from a system; **2.** Intentional leak, usually used to reduce pressure in an impulse line.

BLEEDER PIPE Pipe through a retaining wall to drain water and relieve pressure.

BLEEDER TILE Drainage pipe connecting an external foundation drain to an internal drain, passing through or under the foundation of the building.

BLEEDING Surface exudation.

BLEED OFF To pass by or circumvent.

BLEED VALVE *See* **Valve.**

BLENDED WATER Mixture of hot and cold water, usually established and controlled by a mixing valve.

BLENDING VALVE *See* **Valve.**

BLIND DRAIN 1. Stone-filled covered ditch; **2.** Ground drain without an apparent outlet.

BLIND FLANGE Flange that closes the end of the pipe to which it is attached.

BLINDING 1. Filling or plugging of the openings in a screen or sieve by the material being separated; **2.** Clogging of filter media; **3.** Compaction of soil immediately over a tile drain so as to reduce the tendency of fines to wash into the tile.

BLIND NIPPLE Pipe nipple, capped at one end.

BLIZZARD Violent, long-lasting, blinding snowstorm usually accompanied by temperatures of -10°C (12°F) or colder and high winds.

BLOATED CLAY Clay swollen from absorbed moisture.

BLOCK Small dam or obstruction sufficient to prevent flow in an earthen channel.

BLOCKING Condition where an anticyclone remains stationary and cyclones move around it.

BLOCKOUT Means to physically prevent the operation of equipment.

BLOOM Mass of microscopic and macroscopic plant life occurring in a water body, typically green algae.

BLOWBY Substance that moves under pressure past a seal or gasket.

BLOWDOWN 1. Stand of trees uprooted or broken by wind. *Also called* **Wind throw; 2.** Removal of water from a system or vessel by the injection of air under pressure; **3.** Minimum discharge of recirculating water for the purpose of discharging materials contained in the water, the further buildup of which would cause concentration in amounts exceeding limits established by best engineering practice.

BLOWING Upward movement of soil material in the base of a cofferdam or excavation due to groundwater pressure; normally associated with insufficient toe penetration of the sheeting. *See also* **Blowout, Boiling,** and **Piping.**

BLOWING WELL Well from which a current of air issues.

BLOWOFF 1. Vent to atmosphere of a fluid or vapor held under pressure; **2.** Outlet for the discharge of sediment or water from a sewer.

BLOWOFF VALVE *See* **Valve.**

BLOWOUT 1. Release of water from a confined area due to rupture; **2.** Upward flow of water in a sandy formation due to unbalanced hydrostatic pressure in the surrounding groundwater; **3.** Small basin or bowl from which loose surface material has been removed by wind erosion.

BLOW SAND Wind-borne sand.

BLUE COPPERAS *See* **Copper sulfate.**

BLUEGRASS *See* **Grass.**

BLUE-GREEN ALGAE OR BACTERIA Algae containing red and blue pigments in addition to chlorophylls, some of which are capable of fixing nitrogen directly; others convert nitrates to nitrites; many are capable of producing harmful toxins.

BLUE STAIN A discoloration of lumber due to a fungus growth in unseasoned wood. Although it mars the appearance, the stain has no serious affect on the lumber's strength. *See also* **Brown stain,** and **Chemical brown stain.**

BLUE STAIN FUNGUS Most common form of fungal stain occurring in sapwood. Conifers are most susceptible but it may also occur in light-colored heartwood of perishable timbers. Commonly develops in dead trees, logs, lumber, and other wood products until the wood is dry. Reduces the grade of wood, but does not significantly reduce the strength. Some blue-stain lumber is highly valued for speciality products.

BLUE STONE *See* **Copper sulfate.**

BLUE VITRIOL Waterworks name for copper sulfate.

BLUFF Abrupt rise of rock on the terrain.

BOARD MEASURE Non-SI system of measurement for lumber, the unit of measure being one board foot, that is represented by a piece of lumber 300 mm (1 ft) square and approximately 25 mm (1 in.) thick.

BOARD METER Non-SI measure of timber representing a piece 1 m^2 (10.76 ft^2) and 25 mm (1.0 in.) thick.

BOD Five-day biochemical oxygen demand.

BOD$_5$ Biochemical oxygen demand of the materials entered into process and measured over 5 days. It can be calculated by multiplying the fats, proteins and carbohydrates by factors of 0.890, 1.031 and 0.691 respectively. Organic acids (e.g. lactic acids) should be included as carbohydrates. Composition of input materials may be based on either direct analyses or generally accepted published values.

BOD$_7$ Biochemical oxygen demand of a liquid as determined by incubation at 20°C for a period of 7 days. Agitation employing a magnetic stirrer set at 200 to 500 rpm may be used.

BODILY INJURY Has the meaning given to this term by applicable law; however, it does not include those liabilities which, consistent with standard insurance industry practices, are excluded from coverage in liability insurance policies for bodily injury.

BODILY INJURY AND PROPERTY DAMAGE In liability insurance requirements, the terms bodily injury and property damage have the meanings given these terms by applicable law. However, these terms do not include those liabilities which, consistent with standard industry practices, are excluded from coverage in liability policies for bodily injury and property damage.

BODY Main part of a valve or similar plumbing fitting.

BODY BURDEN Amount of radioactive or

toxic material in a human or animal body at a point in time.

BODY FLUIDS Liquid emanating or derived from humans and limited to blood; amniotic, cerebrospinal, synovial, pleural, peritoneal and pericardial fluids; and semen and vaginal secretions.

BOG Soft, spongy ground, usually wet and composed of decomposing vegetable matter.

BOIL 1. Uplift of a portion of the water surface caused by turbulent upward movement of water; **2.** Vertical flow of water and soil into the bottom of an excavation.

BOILER 1. Approved vessel in which water can be heated or steam generated; **2.** Appliance intended to supply hot water or steam for space heating, processing, or power purposes.

BOILER FEEDWATER Water forced into a boiler to replace that evaporated in the generation of steam.

BOILER HORSEPOWER *See* **Horsepower.**

BOILER OPERATING DAY 24-hour period during the whole of which fossil fuel is combusted in a steam generating unit.

BOILER SCALE Encrustation on boiler heating surfaces caused by precipitation of minerals from the water used.

BOILING POINT Temperature at which a specified liquid boils; 100°C (212°F) for water at sea level.

BOILING RANGE Temperature spread between the initial boiling point and final boiling point.

BOLSON Geologic basin with, or originally having, an inward-flowing drainage system.

BOND Promissory note. An instrument of prequalification, representing that the principal has been examined by the surety and found to be qualified to complete the obligation or undertaking in question. There are several types, including:

Bid/tender: Sum of money required to

be put in escrow at the time of preparing and submitting a bid/tender for work.

Blanket: Bond covering a group of persons, articles, or properties.

Blanket fidelity: Bond that covers an employer's losses due to dishonest acts of employees.

Completion: Form of surety that contains the promise of a third party that a project will be completed or, in the case of the construction company defaulting, the cost of completion will be covered.

Contract: Approved form of security, executed by the contractor and his surety or sureties, guaranteeing complete execution of a contract and all supplemental agreements and the payment of all legal debts pertaining to the construction of the project.

Contract performance: Security furnished to the contracting agency to guarantee completion of the work in accordance with the contract.

Contract surety: Three-party instrument by which one party (surety) guarantees or promises a second party (obligee) the successful performance by a third party (principal).

Fidelity: Surety bond that reimburses an employer for loss sustained through dishonest acts of his employees.

Guaranty: Assurance that the contractor will either complete the work as specified or pay all obligations, or both.

Labor and material payment: Surety to provide for the payment of labor done and materials supplied to a project.

Lease: Bond guaranteeing that a lessee will erect a structure as described in the lease.

Liability: Protection for the assured against court-directed damages.

License: Surety against loss or damages resulting from operations permitted and licensed by law.

Lien: Assurance given by a contractor protecting an owner against liens against his property resulting from work done there; surety that mechanic's or material liens will be filed against a bond and not against the project itself.

Maintenance: Bond in favor of an owner guaranteeing for a specified period, rectification by the contractor of defects in workmanship of materials.

No lien: Bond that denies the right to file a lien against the protected contract or resulting works.

Performance: Bond required by a client to insure that a contractor fulfills contractual obligations.

Release of lien: Mechanism whereby a surety bond is issued to a lienholder to permit the clearance of title to a project.

Release of retained percentage: Owner's protection against loss resulting from premature release of the retained percentage of the contract sum following completion of the work.

Subcontract: Performance and payment bond given by a subcontractor to the general contractor.

Subdivision: Surety provided to a public authority by a subdivider guaranteeing construction of all required public facilities and improvements.

Supply: Bond furnished by a manufacturer and/or supplier/distributor guaranteeing that materials or equipment will be delivered as contracted for.

Surety: Promise to be liable for the debt, default, or failure of another.

Termite: Bond protecting against damage to a structure by termites.

Union wage: Bond provided by a contractor guaranteeing a union that he will pay scale wages and remit any benefit funds withheld.

BOND RELEASE Time at which the appropriate regulatory authority returns a reclamation or performance bond based upon its determination that reclamation work (including, in the case of underground mines, mine sealing and abandonment procedures) has been satisfactorily completed.

BONNET 1. Upper portion of a gate valve into which the disk rises when the valve is opened; **2.** Wire mesh, usually in the shape of a sphere, used to cover the top of vent pipes.

BONNET PACKING Pliable material around a valve stem that prevents water from leaking.

BONUS-AND-PENALTY CLAUSE Provision in a construction contract for payment of a bonus for early completion of the project, and for a financial penalty for failure to complete on time.

BONUS FOR EARLY COMPLETION Additional money to be paid a contractor for completion of the works ahead of the date stipulated for completion by the owner at the time the contract was awarded.

BOOK INVENTORY Inventory shown in the records and presumed to be on hand. *See also* **Physical inventory.**

BOOK OF SPECIFICATIONS *See* **Specification of work.**

BOOK VALUE Capital amount at which property or equipment is shown in the accounts. Normally it is the original cost less depreciation plus additions (not including maintenance or repairs).

BOOM Floating structure, often flexible or comprising hinged sections, used to protect against wave action, or to exclude or contain floating debris.

BOOMING Sudden discharge of a large quantity of accumulated water in placer mining.

BOOSTER COMPRESSOR Compressor that discharges into the suction line of another compressor.

BOOSTER CYCLE Period during which additional hydraulic pressure is exerted to push the last charge of solid waste into a transfer trailer or a container attached to a stationary compactor.

BOOSTER FAN Fan installed in heating ducts to increase air movement.

BOOSTER PUMP *See* **Pump.**

BOOSTER RELAY VALVE Relay-type valve used to accelerate the application and release of pressurized air to towed vehicles or relay emergency valves.

BOOSTER STATION Pumping station in a water distribution system, used to increase system pressure on the discharge side of the pump(s).

BORDER Earth berm built to contain irrigation water in a field.

BORDER IRRIGATION Method of irrigation by flooding the area with the water retained and confined by parallel ridges or borders.

BORE 1. Internal diameter of a cylinder, hose, or pipe; 2. Borehole drilled or excavated into the ground; 3. Wave advancing with a near vertical front upstream during the flowing tide in an estuary.

BOREAL Northern biogeographic zone with short, warm summers and snowy, very cold winters.

BORED WELL Well constructed by boring a hole in the ground with an auger and installing a casing.

BOREHOLE 1. Hole made by drilling, or boring: a well; 2. Interior of a well.

BOREHOLE LOG Record of the findings at a borehole; a detailed account of the operation and results.

BOREHOLE PUMP *See* **Pump.**

BOREHOLE SAMPLE Material obtained by borehole drilling, kept in as undisturbed state as possible and in the sequence removed from the drill. Used for analysis of the material through which drilling is taking place.

BORON Ubiquitous in the environment, boron occurs naturally in over 80 minerals and in the Earth's crust. The predominant form of boron in water is boric acid with levels in well water more variable and often higher than those in surface waters.

BORROW Suitable material from off-site sources, used primarily for fill.

BORROW PIT Excavation from which material is taken for use in another location. *Also called* **Barrow pit.**

BOSCH NUMBER Measure of diesel smoke determined by passing the exhaust gas through a white filter paper. The darkening of the paper is determined using a reflecto-meter, and Bosch numbers are reported on a scale of 1 (clear) to 10 (black).

BOSS Cone for opening pipes.

BOTTOM ASH 1. Solid material that remains on a hearth or falls off the grate after thermal processing is complete; 2. Ash that drops out of the gas stream in the furnace and in the economizer sections. Economizer ash is included when it is collected with bottom ash.

BOTTOM CONTRACTION Reduction in the area of overflowing water caused by the crest of a weir contracting the nappe.

BOTTOM DUMP *See* **Hopper body.**

BOTTOM-DUMP SCRAPER *See* **Scraper.**

BOTTOM-DUMP TRUCK *See* **Truck.**

BOTTOM SAMPLER Sounding lead (leed) attached to a coring device for obtaining samples of the ground below water.

BOTTOM SETS Gently sloping layers of fine-grained sediment at the seaward end of a delta.

BOTTOM SHORE Member nearest the wall

in a series of raking shores supporting a wall.

BOTTOM VENTILATION Air movement through the media of a wastewater filter, facilitated by vent stacks or means for the admission or exit of air at the filter's base.

BOULDER Rock, usually rounded by weathering and abrasion, greater than 200 mm (8 in.) in size.

BOULDER CLAY Stiff, hard, usually unstratified clay that contains boulders.

BOUNDARY Perimeter of something; the edge of a construction site.

BOUNDARY LAYER Layer of fluid or air which because of its proximity to a boundary, or a layer having significantly different characteristics or properties, moves more slowly than the main stream.

BOUNDARY PIEZOMETER Disk piezometer used to measure pore water pressure at the interface between a structure and the adjacent soil.

BOUND WATER 1. Water held on the surface of colloidal particles; **2.** Water contained in a substance present in hygroscopic materials in the lowest concentration that is in equilibrium with saturated air.

BOWL 1. Body or bucket of a scraper; **2.** Moldboard or blade of a dozer; **3.** Exterior shell of an expansion ring-type coupling; **3.** That part of a sanitary fitting made to contain water and having a waste outlet; **4.** Case that is closed at one end and mates with a filter head. *Also called* **Shell.**

BOWL URINAL Semi-spherical vitreous china bowl that mounts on the wall, and which is fitted with a flush pipe and waste outlet.

BOX CULVERT Tunnellike reinforced concrete structure consisting of single or multiple openings, usually square or rectangular in cross section.

BOX DAM Cofferdam completely surrounding an area.

BOX DRAIN Small, rectangular, brick or concrete drain.

BOX GAUGE *See* **Gauge.**

BRACKISH WATER Water having a mineral content between that of freshwater and seawater; water containing from 1000 to 10 000 mg/L of dissolved solids.

BRAIDED CHANNEL Large flood channel of a fast-flowing stream made up of many small stream beds crossing each other. These stream channels are associated with low stream banks. They are also associated with frequent runoffs and/or ice flows. *See also* **Channel.**

BRAKE HORSEPOWER *See* **Horsepower.**

BRANCH 1. Specially shaped section of pipe used to form a connection for a sewer or water main, available as T, Y, T-Y, double Y, and V shapes; **2.** Soil-or-waste pipe that (a) is in one story, (b) is connected at its upstream end to the junction of two or more soil-or-waste pipes, or to a soil-or-waste stack, and (c) is connected at its downstream end to another branch, a soil-or-waste stack, or a building drain.

BRANCH DRAIN Pipe connecting a building's soil line or plumbing fixtures to the main sewer.

BRANCH FITTING Special pipe shape that permits the junction of one or more branch pipes to a main, typically a Tee, Y, Tee-Y, double-Y, or V. *Also called* **Pipe branch.**

BRANCH INTERVAL Vertical distance between the connection of pipes to a main drain or waste-and-vent stack, usually one-story, but not less than 2.4 m (8 ft).

BRANCH LINE Piping into which fire protection sprinklers are fitted and fed by a cross main.

BRANCH MANHOLE Manhole in which a branch connection is made.

BRANCH SEWER Arbitrary term for a sewer that receives sewage from more than one pubic sewer from a relatively small area.

BRANCH VENT Vent that connects with one or more vents from fixtures and which leads to a vent stack.

BRASS PIPE *See* **Pipe.**

BREACH OF CONTRACT Actions by one who has signed a contract in violation of the terms and/or conditions set out in the document.

BREAKAWAY POINT Least rate of flow that a water meter is capable of registering.

BREAKER LINE Line offshore where waves commence to break — a variable distance from the shoreline and dependant upon many variables, including wave height, wind direction and velocity, height of tide, etc.

BREAKERS Waves whose crests are tumbling.

BREAKING THE SEAL Breaching the airtightness of the water seal in the trap of a plumbing fitting.

BREAKOUT OF CHLORINE Point at which chlorine leaves solution as a gas because the chlorine feed rate is too high.

BREAKPOINT CHLORINATION Addition of sufficient chlorine to sewage effluent to ensure elimination of microbiological hazard without threat to downstream flora or fauna.

BREAK-PRESSURE TANK Open pressure-relief vessel sited on a gravity main at a location calculated to reduce the head before the possibility of it becoming a threat to the system and its appurtenances.

BREAKTHROUGH CAPACITY Capacity of an ion-exchange column at a fixed regeneration level.

BREAKUP Period of time in the spring when melting snow creates soft soil conditions and high water in streams. Logging must usually be curtailed during this time. *See also* **Freeze-up.**

BREAKWATER Structure designed and placed so as to break the force of waves, tide, or flood. *Also called* **Mole.**

BREAST WHEEL Wheel that operates partly by impulse of water striking its vanes, and partly by the weight of the water.

BREATHING WELL Well in which air is alternately drawn in and blown out.

BRECCIA Sedimentary rock consisting of angular fragments of older rocks naturally cemented together in a matrix.

BREEDER REACTOR Nuclear power generator in which additional fissionable fuels are produced from the neutrons provided by an initial charge of uranium-235.

BREEZE Light wind from 6.5–50 km/h, force 2–6 on the Beaufort scale.

BRIDGE In a drainage structure, a span of 6 m (20 ft) or more, having a designed clearance over the highest expected floods.

BRIDGING Condition of a filter element loading in which contaminant spans the space between adjacent sections of a filter element, thus blocking a portion of the useful filtration area.

BRINE 1. Concentrated salt solution remaining after removal of distilled product; **2.** Concentrated brackish saline or seawater containing more than 36 000 mg/L of total dissolved solids (NaCl).

BRITISH IMPERIAL GALLON *See* **Imperial gallon.**

BRITISH THERMAL UNIT Nonmetric unit of energy, equal to the quantity of energy required to raise 1 lb of water through 1°F. Symbol: Btu. Multiply by 1055.06 to obtain joules, symbol: J. *See also the appendix:* **Metric and nonmetric measurement.**

BRITISH THERMAL UNIT PER GALLON (IMP) Nonmetric unit of heat. Symbol: Btu/gal. Multiply by 238.08 to obtain kilojoules per cubic meter, symbol: kJ/m^3. *See also the appendix:* **Metric and nonmetric measurement.**

BRITISH THERMAL UNIT PER GALLON (US) Nonmetric unit of heat. Symbol: Btu/gal (US). Multiply by 278.717 to obtain kilojoules per cubic meter, symbol: kJ/m^3. *See also the appendix:* **Metric and nonmetric measurement.**

BRITISH THERMAL UNIT PER HOUR Nonmetric unit of heat. Symbol: Btu/h. Multiply by 0.292 072 to obtain watts, symbol: W. *See also the appendix:* **Metric and nonmetric measurement.**

BRITISH THERMAL UNIT PER POUND Nonmetric unit of heat. Symbol: Btu/lb. Multiply by 2.326 to obtain kilojoule per kilogram, symbol: kJ/kg. *See also the appendix:* **Metric and nonmetric measurement.**

BRITISH THERMAL UNIT PER POUND °F Nonmetric unit of heat. Symbol: Btu/lb°F. Multiply by 4.1868 to obtain kilojoules per kilogram °C, symbol: kJ/(kg°C). *See also the appendix:* **Metric and nonmetric measurement.**

BRITISH THERMAL UNIT PER SQUARE FOOT PER HOUR Nonmetric unit of work. Symbol: $Btu/(ft^2/h)$. Multiply by 3.154 60 to obtain watts per square meter, symbol: W/m^2. *See also the appendix:* **Metric and nonmetric measurement.**

BRITISH THERMAL UNIT PER SQUARE FOOT PER HOUR PER DEGREE FAHRENHEIT Nonmetric unit of work. Symbol: $Btu/(ft^2/h)°F$. Multiply by 5.678 29 to obtain watts per square meter per degree Centigrade, symbol: $W/(m^2°C)$. *See also the appendix:* **Metric and nonmetric measurement.**

BRITISH THERMAL UNIT PER (2000-lb) TON Nonmetric unit of heat. Symbol: Btu/ton. Multiply by 1.163 to obtain kilojoules per tonne, symbol: kJ/t. *See also the appendix:* **Metric and nonmetric measurement.**

BROAD-BASE TERRACE Long ridge of earth, 4.5 to 9 m (15 to 30 ft) wide and 254 to 760 mm (10 to 30 in.) high with sloping sides, a rounded crown, and a broad shallow channel along the upper side, designed to control erosion by diverting surface water along the contour at low velocity.

BROADCAST To toss granular material, such as sand, over a horizontal surface so that a thin, uniform layer is obtained.

BROADCAST BURNING Controlled forest burn, where the fire is intentionally ignited and allowed to proceed over a designated area within well-defined boundaries, for the reduction of fuel hazard or for site preparation. *Also called* **Slash burning.**

BROAD-CRESTED WEIR Weir having a substantial width of crest parallel to the direction of the flow passing over it, and which produces no bottom contraction of the nappe. *Also called* **Wide-crested weir.**

BROAD IRRIGATION Irrigation of crops as a means of wastewater effluent disposal. *See also* **Wastewater irrigation.**

BROMINE Heavy, volatile, corrosive, reddish-brown, nonmetallic liquid element (Br) which fumes readily and mixes easily with water. Used as a water and soil disinfectant.

BROMOXYNIL Bromoxynil and its octanoate ester are phenolic benzonitril-based herbicides employed for the control of broadleaved weeds in grain crops and are ranked high with respect to potential for groundwater contamination, with traces found in both municipal and private water supplies.

BROOK Small, shallow stream, usually in continuous, turbulent flow.

BROWNIAN MOVEMENT Random movement of microscopic particles suspended in a liquid or gas medium.

BROWN ORES Soils leached of the more soluble minerals, but retaining rich amounts of iron oxides and hydroxides.

BROWN PODZOLIC SOIL Acid forest soil comprising a surface layer of litter over a dark greyish-brown, organic and mineral soil, with a pale leached layer beneath.

BROWN SMOKE Smoke produced by volatile tarry substances when coal is burned at a low temperature.

BROWN STOCK WASHER Washer and associated knotters, vacuum pumps, and filtrate used to wash wood pulp following the digestion system.

BROWSE Buds, shoots, and leaves of woody plants that can be eaten by livestock or wild animals.

BRUSH Trees and shrubs less than 100 mm (4 in.) in diameter.

BRUSH OUT To clear an area of tree limbs, saplings, and debris. *Also called* **Swamp out.**

BRUSH RAKE Lightly constructed rake blade with a high top; an attachment for a crawler-tractor, used in mechanical site preparation to penetrate and mix soil and tear roots.

BUBBLE Notional dome-shaped boundary around a defined entity or area, within which an upper limit for pollutants is established by which to regulate the generation and treatment of emissions. The concept accepts the possibility of a low-emitting entity trading emission credits to a high-emitting entity.

BUBBLE BARRIER Air bubbles released from a perforated tube secured to the bottom of a storage or treatment tank across its width or length. The rising bubbles prevent the fluid on either side from intermixing, thus preventing cross-contamination. However, mixing of the fluid within the compartments created by the bubble barrier is promoted by the rising action of the bubbles. An additional effect is the entrainment of oxygen to the liquid as it is stirred by the air bubbles rising within it.

BUBBLE BURSTING Chief mechanism for the projection of minute water droplets from the sea surface.

BUBBLE POINT Pressure at which the first steady stream of gas bubbles is emitted from a wetted filter element under specified test conditions.

BUBBLING FLUIDIZED-BED COM-BUSTOR Combustor in which the majority of the bed material remains in a fluidized state in the primary combustion zone.

BUBBLING SPRING Spring originating in a pool where ascending water currents and/or escaping gases cause the water to bubble on the surface.

BUCKET 1. Cup on the perimeter of a Pelton wheel; **2.** Inverted curve at the bottom of a spillway that deflects the water horizontally from the steep overflow face to the apron; **3.** Transition curve between the face of an overflow dam and its apron; **4.** Receptacle fixed to a manhole frame to retain grit carried by surface washings; **5.** Fitting or attachment to an item of mechanical equipment, such as an excavator, backhoe, or dragline, that digs, lifts, loads, and carries material. *See also* **Bailing bucket, Belling bucket,** and **Slat bucket.** There are many types of bucket attachment, including:

Claw: Bucket with positive clamping jaws for garbage pickup, snow removal, debris cleanup or any loose bundled loads.

Extreme service: Designed for tougher trenching applications such as fragmented rock, frozen ground, caliche, etc. It is equipped with pockets on the rear to accept optional ripper shanks.

Four-in-one: *See* **Multipurpose,** below.

General-purpose: Designed for general excavation digging dirt and mass excavation.

Grading: Bucket with a long, flat floor and straight lip, used for finish work in housing developments, concrete pours, landscaping, and light dozing.

Heavy-duty: Designed for rougher conditions dense clay and light rock.

High dump: Gives extended dump height of light materials.

Light material: For excavating, loading, and easy-digging in light material. Can also be used as a finishing/cleanup bucket. Can be equipped with a bottom cutting edge.

Loose material: Designed for snow, woodchips, hay, coal, etc. Can be fitted with an independently controlled top clamp for such material as brush, silage, or compost.

Mass excavation: Designed for volume truck loading with a shorter tip radii and greater bite width.

Multipurpose: Loader or backhoe clam bucket whose floor is hinged to the top of the bucket back, and is under separate control. It enables the operator to perform multiple tasks with one bucket, such as dig, load, doze, grade, backfill, grab and, with flip-over pallet forks, lift. *Also called* **Six-in-one bucket.**

Rock-ripping: Specially designed for extreme digging and rock conditions, with a staggered tooth design that allows the center tooth to enter the ground first at a 45° angle, allowing it to use all the machine force to rip. The two teeth on either side enter the ground next. The outer teeth, which project outward, enter the ground last and slice the trench wall, leaving it clean and straight.

Severe-duty: Designed for highly abrasive applications: shot rock and demolition.

Side-dump: Able to dump forward or to the left; particularly useful in close quarters or to reduce turning time.

Slat: Openwork bucket made of bars instead of solid plates, used for digging sticky soil.

Trenching: A narrow but deep bucket whose bite width is usually dictated by the pipe diameter.

BUCKET AUGER Cylindrical rotary drilling tool with a hinged bottom containing a soil cutting blade. Spoil enters the bucket which is then lifted out of the hole and swung aside where the contents are dumped by releasing the latch on the hinged bottom. *Also called* **Bucket loader,** and **Drilling bucket.**

BUCKET CAPACITY (*See also* **Bucket rating.**) Actual volume of material that can be carried in an equipment bucket. It is rated in two ways:

Heaped: Volume in the bucket under the strike-off plane plus the volume of the heaped material above the strike-off plane, having an angle of repose of 1:1, without consideration for any material supported or carried by the spillgate or bucket teeth.

Struck: The volume actually enclosed inside the outline of the sideplates and rear and front bucket enclosures

without consideration for any material supported or carried by the spillgate or bucket teeth.

BUCKET ELEVATOR Conveying device consisting of a head and foot assembly that supports and drives an endless single or double strand chain or belt to which buckets are attached.

BUCKET-LADDER DREDGE Dredge equipped with a bucket elevator reaching below its keel into the mud to be dredged.

BUCKET-LADDER EXCAVATOR *See* **Trencher.**

BUCKET LOADER Usually a chain-bucket loader, but also a tractor-loader. *See also* **Loader.**

BUCKET PUMP 1. Large-capacity reservoir-and-pump for lubricants; **2.** *See* **Pump.**

BUCKET RATING (*See also* **Bucket Capacity.**) Volume of material that a bucket is designed to hold. There are two ratings:

Heaped: Bucket struck capacity plus the additional material needed to create a 2:1 angle of repose (1:1 for hydraulic excavators) with a struck line parallel to the ground.

Struck: Volume of material in a bucket after a load is leveled by a straight edge resting on the cutting edge and the back of the bucket (not including see-through rock guards).

BUCKET-WHEEL EXCAVATOR Excavating machine consisting of a rotating wheel fitted with tooth-edged buckets, used to excavate narrow trenches.

BUDGET Plan or schedule adjusting the management of finances or activities for a given period. Budgets can be designed to achieve a range of objectives, including:

Capital: Budget for proposed additions to capital assets and their financing.

Cash: Budget of cash receipts, payments, and periodic balances.

Fixed: Budget prepared for a single level

of activity.

Flexible: Budget prepared for a range of levels of activity.

Project: Budget for authorized appropriation of funds, generally based on an estimate of the scope of work in the construction phase of a project, and including the cost of the feasibility phase.

BUDGETARY CONTROL Process of planning, executing, and evaluating a project or program through the use of a budget.

BUDGETED COST OF WORK PERFORMED Sum of the approved cost estimates (including any overhead allocation) for activities (or portions of activities) completed during a given period (usually project-to-date). *See also* **Earned value.**

BUDGETED COST OF WORK SCHEDULED Sum of the approved cost estimates (including any overhead allocation) for activities (or portions of activities) scheduled to be performed during a given period (usually project-to-date). *See also* **Earned value.**

BUDGET ESTIMATES Estimates of anticipated costs prepared in order to establish a budget.

BUDGET VARIANCE *See* **Variance.**

BUFFER 1. Screen of vegetation, a berm or other physical barrier erected between the public and a disposal, transfer, or recycling function; **2.** Any of certain combinations of chemicals used to stabilize the pH values of solutions.

BUFFER ACTION Action of certain ions in solution in opposing a change in hydrogen-ion concentration.

BUFFER CAPACITY Measure of the capacity of a solution to neutralize acids or bases.

BUFFER SOLUTION Solution containing two or more compounds that, when combined, resist marked changes in pH after moderate quantities of either strong acid or base have been added.

BUFFER STRIP Strip of land (often including undisturbed vegetation) where disturbance is not allowed or is closely monitored to preserve or enhance aesthetic or other qualities along or adjacent to roads, trails, watercourses and recreation sites. *Also called* **Green strip, Leave strip,** and **Streamside management zone.**

BUFFER ZONE Transitional area between two areas of significantly different land use.

BUILDER Alkaline phosphate added to a detergent to adjust its alkalinity or add to its soil-suspending ability.

BUILDING (POLLUTING) ACTIVITIES Pollutant emitting activities that belong to the same industrial grouping, are located on one or more contiguous or adjacent properties, and under common control.

BUILDING CODE Regulations governing building design and construction, established by regulatory authorities. Building codes are promulgated and administered by various levels of Government, and originate also from a number of quasi-governmental and private organizations. They include:

Building bylaw: Building regulation enacted by a local authority for application within its jurisdiction.

City building code: Regulations enacted by a municipal authority governing the design and construction of all types of buildings and structures. In Canada, these regulations supplement standards established in the National Building code.

County and township building code: In the US, some counties and townships have enacted ordinances (regulations) governing construction work done outside incorporated cities.

Housing code (Canada) *See* **National building code,** below.

Housing code (US) National set of standards that all housing must meet on final construction. May be augmented by state or local requirements. *Also called* **Minimum standards bylaw.**

Minimum standards bylaw *See* **Housing code (US),** above, and **National building code,** below.

Model building code: In the US, three organizations have sponsored model building codes. These are:

BOCA National Building Code Sponsored by the Building Officials and Code Administrators International.

Southern Standard Building Code Sponsored by the Southern Building Code International Congress.

Uniform Building Code Sponsored by the International Conference of Building Officials.

National Building Code: In Canada, a publication of the National Research Council of Canada that sets forth the minimum national standards for construction of buildings. These standards may be altered by provincial or local authorities. The National Building Code forms the basis for all building regulations in Canada. In the US, the BOCA (Building Officials and Code Administrators International) National Building Code forms the basis for all building regulations. Buildings erected on federal property are exempt from the requirements of local codes. *Also called* **Housing code,** and **Minimum standards bylaw.**

Provincial building codes: Building regulations enacted by the provincial authorities in Canada that may supplement the National Building Code, and local building codes. Typically they include an electrical code, plumbing and gas installation codes.

State building codes: In the US, most states have passed ordinances that apply to buildings of a particular type of construction or use; these are not uniform across the country.

Township building code: Construction regulations that supplement national

or model building codes, applicable to specific jurisdictions.

BUILDING COMMITTEE Persons appointed or assembled to develop and manage a project.

BUILDING CONNECTION Extension from a sewer tap to the property line, or to the easement line of the property to be served.

BUILDING CONTRACTOR Person knowledgeable of, and skilled in building construction and who contracts to build.

BUILDING DEPARTMENT 1. Department of local government responsible for the regulation of construction within its jurisdiction; **2.** Department of local government responsible for municipal works, including maintenance; **3.** Department within an organization responsible for the maintenance of physical property.

BUILDING DRAIN That part of a plumbing drainage system that receives discharge from soil, waste, and other stacks inside a building. *Also called* **House drain.**

BUILDING DRAINAGE SYSTEM *See* **Drainage system.**

BUILDING INSPECTOR Person charged with administration and enforcement of the applicable building codes. Also *called* **Building official.**

BUILDING INSULATION Material, primarily designed to resist heat flow, installed between the conditioned volume of a building and adjacent unconditioned volumes or the outside including, but not limited to, insulation products such as blanket, board, spray-in-place, and loose-fill that are used as ceiling, floor, foundation, and wall insulation.

BUILDING MAIN Water supply piping beginning at the source of supply and ending at the first branch inside the building.

BUILDING OFFICIAL *See* **Building inspector.**

BUILDING PERMIT Permit issued for a fee by local authorities that authorize various stages of construction.

BUILDING REGULATIONS Regulatory requirements, adopted by a local authority, governing the design and construction of physical structures.

BUILDING RESTRICTIONS Provisions in a building code or other applicable ordinance or regulation that affect the siting, orientation, size, appearance, construction, or other aspect of a building.

BUILDING SEWER Pipe that is connected to a building drain 900 mm (36 in.) outside of a wall of a building to conduct sewage, clear water waste, or storm water to a public sewage disposal system.

BUILDING STORM DRAIN Drainage system used to receive rainwater, surface water, and groundwater, connected to the building sewer outside the building.

BUILDING STORM SEWER Piping that connects to the end of a building storm drain to receive and convey the contents to a public storm sewer or combined sewer.

BUILDING SUBDRAIN Portion of a building drainage system that can drain by gravity into the building sewer.

BUILDING TRAP Trap that is installed in a building drain or building sewer to prevent circulation of air between a drainage system and a public sewer.

BUILT-IN DIRT Material passed into the effluent stream composed of foreign materials incorporated into a filter medium.

BULK ASBESTOS Any quantity of asbestos fiber of any type or grade, or combination of types or grades, that is mined or milled with the purpose of obtaining asbestos.

BULK CONTAINER Large container that can either be pulled or lifted mechanically onto a service vehicle or emptied mechanically into a service vehicle.

BULKHEAD 1. Partition or wall securely dividing one part of a structure from another; **2.** Wall or partition erected to resist ground or water pressure; **3.** Structure erected along a shore to arrest wave action; **4.** Permanent or movable wall across a waterway; **5.** Air restraining barrier constructed for long-term control of radon-222 and radon-222 decay product levels in mine air.

BULKING SLUDGE Activated sludge that settles poorly because of a floc of low density.

BULK SPREADER Machine for carrying cement or other material and spreading it on a prepared surface for soil stabilization.

BULKY WASTE Large items of solid waste such as household appliances, furniture, large auto parts, trees, branches, stumps, and other oversize wastes whose size precludes or complicates their handling by normal solid wastes collection, processing, or disposal methods.

BULLDOZER Tractor fitted with a front pushing blade.

BULL'S LIVER Inorganic silt of slight plasticity; quakes like jelly from vibration.

BUND Continuous wall around a storage tank containing potentially hazard liquid, built to such a height that the volume it encloses is larger than that of the tank it surrounds.

BUOY Floating object, usually of recognizable shape and color, moored to the bottom to mark a channel, anchorage, shoal, rock, etc.

BUOYANCY Tendency or capacity to remain afloat in a liquid, or to rise in air or gas.

BUREAU OF PUBLIC ROADS SOIL CLASSIFICATION *See* **Soil classification systems.**

BURIED CHANNEL Former stream channel that has filled with alluvial or glacial deposits and which has been later covered by other material to a degree that makes it virtually indiscernible from surface observation.

BURIED RIVER River bed that has been buried beneath a stream of basalt or alluvial drift.

BURIED SERVICES Any of the conventional building services placed belowground.

BURNER Device that positions a flame in the desired location by delivering fuel (and sometimes air) to that location. Some burners also atomize the fuel, and some mix the fuel and the air. There are several basic types, including:

Conical: A vertical, hollow, cone-shaped combustion chamber that has an exhaust vent at its apex and a door in its side at its base through which waste materials, usually wood, are charged; air is drawn through louvers to the burning solid waste inside the cone. *Also called* **Beehive burner,** and **Teepee burner.**

Primary: Burner that dries out and ignites a fuel in the primary combustion chamber.

Refuse: Device for either central or onsite volume reduction of solid waste by burning.

Secondary: Burner installed in the secondary combustion chamber to maintain a minimum temperature to complete the combustion of incompletely burned gases.

BURNING AGENTS Additives that, through physical or chemical means, improve the combustibility of the materials to which they are applied.

BURNING CONDITIONS Environmental conditions that affect fire in a given fuel association.

BURNING INDEX Number related to the effort needed to contain a forest fire of a particular fuel type within a rating area.

BURNING OUT Setting fire inside a control line to consume fuel between the edge of a forest fire and the control line. *Also called* **Clean burning,** and **Firing out.**

BURNING RATE Quantity of fuel per unit of time that is charged to a furnace, or the amount of heat released during combustion.

BURN OUT 1. Failure of a nuclear reactor heat-exchanger surface that can result in the loss of heat-exchange fluids and may, ultimately, result in the release of radioactive materials; **2.** Measure of the efficiency of incineration that relies on determination of fixed carbon and putrescible matter in the residue following combustion.

BURST PRESSURE *See* **Pressure.**

BURY BARGE Submerged equipment that moves a submerged pipeline after it has been laid into the trench that it digs and then backfills.

BURY LENGTH Depth from the surface of the ground to the bottom of the pipe to which a hydrant is connected.

BUSHING Short tube threaded internally and/or externally, used to connect pipe fittings.

BUSHING FITTING *See* **Fitting.**

BUTTE Steep-sided formation with a nearly flat top (usually igneous intrusions that have been exposed to erosion).

BUTTERFLY GATE Gate that opens like a damper, turning on a shaft within the pipe to which it is attached.

BUTTERFLY VALVE *See* **Valve.**

BUTTRESS DAM Dam supported on its downstream side by a series of buttresses.

BUTTRESS DRAIN Stone- or rock-filled trench drain constructed in steps down the surface of a soil slope, used to assist in removing water from the soil slope and to provide weighty buttresses capable of resisting any local tendency to slip.

BUTT-WELD STEEL PIPE *See* **Pipe.**

BYATT Horizontal timber supporting decking and walkways in trench excavations.

BYE CHANNEL Spillway leading water around a reservoir when it is full.

BYLAW Local law, made by an organized community for application within its boundaries.

BY OTHERS Work done under subcontract or by another contractor.

BYPASS

BYPASS 1. Secondary passage of fluid, air or electrical flow, in addition to the main flow path; **2.** Intentional diversion of sewage wastes from any portion of a treatment facility; **3.** System that prevents all or a portion of kiln or clinker cooler exhaust gases from entering the main control device and ducts the gases through a separate control device.

BYPASS CHANNEL Man-made channel or conduit designed to carry excessive flows from a stream.

BYPASS FILTER 1. Filter that removes contaminants from a fluid by bypassing a percentage of the flow through its filter medium; **2.** Filter incorporating a ball valve which, when closed, isolates the filter from the circuit into which it is coupled, permitting media exchange or other service to be completed without closing down the circuit; **3.** Component that allows the proper hydraulic flow to continue when the system's filter becomes sufficiently clogged to inhibit the normal flow of filtered hydraulic oil.

BYPASS GATE Gate used to equalize pressure on both sides of a larger gate.

BYPASS STACK Stack that vents exhaust gases to the atmosphere from the bypass control device.

BYPASS VALVE *See* **Valve.**

BYPASS VENT Vent stack running parallel to a soil stack and connected to it at intervals.

BY-PRODUCT Material that is not one of the primary products of a production process.

BY-PRODUCT WASTE Liquid or gaseous substance produced at chemical manufacturing plants or petroleum refineries (except natural gas, distillate oil, or residual oil) and combusted in a steam generating unit for heat recovery or for disposal, excluding gaseous substances with carbon dioxide levels greater than 50% or carbon monoxide levels greater than 10%.

C Thermal conductance. C is similar to k, but applies to the actual thickness of a material; it is a measure of the rate of heat flow for the thickness of a material for a given area at one degree F difference between the inner and outer surfaces.

CABLE ENTRY Sealed enclosure through which a power cable passes from outside to inside a building.

CABLE GLAND Seal formed around electrical conductors where they enter and/or emerge from a buried duct.

CABLE GRIP Device temporarily attached to the end of a cable to assist in pulling the cable during installation.

CABLE JACKET Outside cover of an electrical cable.

CABLE LADDER Ladder-like metal frame used to support cable runs, which are secured to the frame at intervals with cleats or ties.

CABLE LOCATOR Portable, battery-operated instrument capable of locating buried cables and other services.

CABLE PULLING COMPOUND Substance that when applied to the exterior of a cable, assists when pulling wires.

CABLE PULLOUT UNLOADING METHOD Procedure in which a landfill tractor assists in unloading a transfer trailer by pulling a cable network especially placed within the transfer van for that purpose.

CABLE SHEATH Protective covering applied to cables.

CABLE TAIL Length of electrical conductor left coiled inside a connection box in readiness for connection.

CABLE-TOOL DRILLING *See* **Churn drilling.**

CABLE TRAY Length of preformed galvanized steel sheet with turned-up edges, used to support cable runs.

CADMIUM Soft, silvery heavy metal (Cd) having an atomic weight of 112.4. Cadmium is poisonous in minute quantities.

CAGE Circular frame that limits the sideways motion of balls or rollers in a bearing.

CAGE SCREEN Filtration screen in the form of a drum, hinged so that it may be raised or lowered for cleaning.

CAISSON Structural chamber used to keep soil and water from entering into a deep excavation or construction area.

C_{aj} Concentration of volatile organic compounds in each gas stream (j) exiting an emission control device, in parts per million by volume

CALCAREOUS SPRING Spring water containing a considerable quantity of calcium salts in solution.

CALCINE Solid materials produced by a roaster.

CALCINER Unit in which the moisture and organic matter of phosphate rock is reduced within a combustion chamber.

CALCIUM Silvery-white metallic element: symbol Ca. It is an abundant natural element, entering freshwater systems through the weathering of rocks, especially limestone, and from the soil through seepage, leaching and runoff. Surface water generally contains lower concentrations of calcium than groundwater. Raw water supplies that receive lime treatment show significant increases in the amount of calcium in the treated water. Calcium is one of the principal cations associated with hardness in drinking water, ranging from 75 mg/L (considered soft) to more than 300 mg/L (considered very hard).

CALCIUM CARBIDE Material containing 70% to 85% calcium carbide by weight.

CALCIUM DEFICIENCY Condition in birds caused by DDT and other chlorinated hydrocarbons such as polychlorinated biphenyls that interferes with their ability to metabolize calcium and resulting in the production of eggs having thin shells that can easily be crushed by the weight of the birds when nesting.

CALCIUM HYPOCHLORITE Dry powder consisting of lime and chlorine combined so that when dissolved in water it releases active chlorine.

CALCIUM OXIDE Quicklime (CaO), which becomes calcium hydroxide when water is added.

CALCIUM SULFATE STORAGE PILE RUNOFF calcium sulfate transports water runoff from or through a calcium sulfate pile, and the precipitation that falls directly on the storage pile and which may be collected in a seepage ditch at the base of the outer slopes of the storage pile.

CALCULATED CETANE Number representing the ignition properties of diesel fuel oils from API gravity and mid-boiling point as determined by ASTM standard method D 976-80, entitled 'Standard Methods for Calculated Cetane Index of Distillate Fuels'. ASTM test method D 976-80 is incorporated by reference.

CALENDAR UNIT Smallest unit of time used in scheduling a project — generally hours, days and weeks, but can also be shifts, or even minutes. Used primarily in relation to project management software.

CALIBRE Bore or internal diameter of a cylinder or pipe.

CALIBRATE To check and adjust the graduations of a measuring instrument.

CALIBRATED SAND METHOD Weight of a cubic measure of soil, in place.

CALIBRATING GAS Gas of known concentration used to establish the response curve of an analyzer.

CALIBRATION 1. Precise adjustment against, or to an established measure or standard; **2.** Set of specifications, including tolerances, unique to a particular design, version, or application of a component or components.

CALIBRATION OF EQUIPMENT Measurement of dispersal or output of application equipment and adjustment of such equipment to control the rate of dispersal, and droplet or particle size of a substance being dispersed.

CALICHE Naturally occuring fertilizer containing sodium nitrate.

CALIFORNIA BEARING RATIO Ratio of the force per unit area required to penetrate a soil mass with a 193.5 mm^2 (3 in.^{2}) circular piston at the rate of 1.27 mm (0.05 in.) per min to the force required for corresponding penetration of a standard material. The ratio is usually determined at 0.1 in. penetration.

CALL FOR BIDS Request to submit a price for the completion of specified work.

CALL FOR PROPOSALS Invitation to qualified bidders to complete a project in which the bidder may be requested to include proposals for structural systems, construction details, financing, project timing, etc. *Also called* **Invitation for proposal call,** and **Proposal call.**

CALL FOR TENDERS Invitation to qualified contractors and/or suppliers to meet the requirements of a proposal.

CALL LOAN Loan payable immediately upon the demand of the lender.

CALLOUT Note on a drawing with a leader to the feature.

CALORIE Unit of energy, equal to 4.19 joules, that should not be used with the SI system of measurement. Symbol: cal. *See* *also the appendix:* **Metric and nonmetric measurement.**

CALORIFIC VALUE Heat, measured in calories or Btu, released by combustion of a unit quantity of fuel.

CALORIFIER Sealed tank in which water is heated, usually by a heat exchanger in the form of a submerged coil carrying steam or hot water.

CALORIMETER Instrument for measuring heat exchange during a chemical reaction, typically as the quantities of heat liberated by the combustion of a fuel or hydration of a cement.

CAMBIUM MINER Insect (*Agromyza carbonaria*) that burrows in the cambium and causes pith flecks in numerous woods.

CAN Container formed from sheet metal and consisting of a body and two ends or a body and a top.

CANAL Channel built to carry water and, in some cases, vessels.

CANALIZATION Division of a river or waterway into reaches separated by locks and weirs, to control the rate of flow, and to make passage possible.

CANAL LIFT Large, water-filled tank, moved on wheels up and down an incline, or vertically, used in place of a lock for lifts greater than about 15 m (50 ft) to transport barges and other vessels.

CANAL LOCK Crest gate, the face of which is a section of a cylinder, that rotates about a horizontal axis downstream from a gate. *Also called* **Tainter gate.**

CANAL RAPIDS Inclined conduit or chute used to convey water at high velocity to a lower elevation.

CANAL SECTION Cross sectional profile of a canal, at right angles to its axis.

CANAL SEEPAGE Loss of water from a canal by capillary action and percolation, once the canal has stabilized.

CANAL SYSTEM All the interconnected canals constituting a complete irrigation or water transmission system.

CANAL TRIMMER Equipment that performs the final shaping of the bottom and sides of an earthen canal in preparation for placing a waterproof membrane.

CANCELLATION CLAUSE Provision in a contract, typically a lease, which confers upon one or more or all of the signatories the right to terminate their obligations under defined conditions and/or circumstances.

CANDELA One of the seven base units of the SI system of measurement: it is the luminous intensity, in the perpendicular direction, of a surface of 1/600 000 m^2 of a black body (full radiator) at the temperature of freezing platinum (2046°K) under a pressure of 101.325 kPa. Symbol: cd. Derived units are illumination lux (symbol: lx), and luminous flux lumen (symbol lm). *See also the appendix:* **Metric and nonmetric measurement.**

CANDELA PER SQUARE METER A derived unit of luminance with a compound name of the SI system of measurement. Symbol: cd/m^2. *See also the appendix:* **Metric and nonmetric measurement.**

CANDIDATE METHOD Method of sampling and analyzing the ambient air for an air pollutant.

CAP 1. Pipe plug with female threads; **2.** Layer of clay or other highly impermeable material installed over the top of a closed landfill to prevent entry of rainwater and minimize production of a leachate.

CAPACITANCE Property of a device which permits storage of electrically-separated charges when differences in electrical potential exist between the conductors and measured as the ratio of stored charge to the difference in electrical potential between conductors.

CAPACITOR Device for accumulating and holding a charge of electricity and consisting of conducting surfaces separated by a dielectric.

CAPACITOR/CONDENSER Device for the storage of electrical energy consisting of two oppositely charged conducting plates separated by a dielectric and which resists the flow of direct current.

CAPACITOR MOTOR Single-phase induction motor having its main winding connected to a source of power and an auxiliary winding connected in series with a capacitor to facilitate starting.

CAPACITY 1. Exact quantity; flow rate that can be carried exactly; **2.** Load for which a machine, apparatus, process, or system is rated and/or designed.

CAPACITY CURVE 1. Graphic, two-dimensional presentation of the data representing some aspect of performance; **2.** Graph showing the volume contained in a storage vessel at any given water level.

CAPACITY FACTOR Ratio of the average load on a machine or equipment for the period of time considered to its capacity rating.

CAPACITY REDUCER Device that permits an adjustment of the output capacity of an air compressor without alteration to any other operating condition.

CAP CLOUD Stationary cloud on or over a mountain peak.

CAPILLARITY Movement of a liquid in the interstices of a porous material due to surface tension. *See also* **Properties of soil.**

CAPILLARY BREAK Gap or space between two surfaces large enough to prevent capillary action of water.

CAPILLARY CAPACITY Approximate quantity of water that can be permanently retained in the soil in opposition to the downward pull of gravity.

CAPILLARY FLOW Flow of moisture through a capillary pore system.

CAPILLARY FRINGE Zone immediately above the watertable where some or all of the interstices are filled with water at a pressure less than atmosphere.

CAPILLARY GROOVE Space between two surfaces, large enough to prevent cap-

illary movement of water; a water break.

CAPILLARY HEAD Head in a capillary tube at any moment before the water has been raised to its ultimate position in the tube by capillary force, equal to the difference in elevation between the ultimate position of the meniscus and its position at a given time.

CAPILLARY INTERSTICE Interstice small enough for water to be held in it at an appreciable height in a capillary fringe above a watertable or hydrostatic pressure level, and large enough to preclude molecular attraction of its walls from spanning the interstice.

CAPILLARY LIFT Height to which water or other liquid rises in a capillary tube.

CAPILLARY MOISTURE Available soil moisture easily abstracted by roots of plants.

CAPILLARY MOVEMENT Movement of groundwater as a result of capillary attraction.

CAPILLARY OPENING Opening small enough in cross-sectional area to permit a condition of capillarity.

CAPILLARY PRESSURE Internal pressure that helps earth to stand in an excavation being drained from outside the excavated area. *Also called* **Seepage force.**

CAPILLARY SPACE Void resembling a microscopic channel small enough to draw liquid through it by the molecular attraction of the water adsorbed on its inner surface (capillarity).

CAPILLARY STRESS Pore water pressure less than atmospheric value produced by the surface tension of pore water acting on the meniscus formed in void spaces between soil particles.

CAPILLARY TUBE Tube with a very small bore that conducts a small quantity of liquid, under pressure, to a readout. Typical system components are: a capillary tube and dial gauge for an oil pressure readout.

CAPILLARY WATER Subsurface water held above the watertable by capillary attraction. *See also* **Absorbed water, Gravitational water,** and **Moisture content of soils.**

CAPILLARY ZONE Zone in which soil water is held by capillary forces.

CAPITAL ASSETS Assets of a relatively permanent nature.

CAPITAL BUDGET *See* **Budget.**

CAPITAL EXPENDITURE Improvement that is not a repair and that will have a life of more than one year; an expenditure for a physical or operational change to a stationary source.

CAPPED END Hose end covered to protect its internal elements.

CAPPING PIECE Horizontal timber over the ends of two walings butted together that takes the thrust of a strut, transferring it to the walings.

CAP ROCK Relatively impermeable upper seal of an underground reservoir.

CAPTURE Water that wholly or in part replaces that withdrawn from an aquifer.

CAPTURE SYSTEM Equipment (including ducts, hoods, fans, dampers, etc.) used to capture or transport particulate matter to an air pollution control device.

CARBOHYDRATES Group of organic compounds, $C_x(H_2O)_y$, composed of carbon, hydrogen and oxygen and containing sugars, i.e., monosaccharides and disaccharides and the polysaccharides starch and cellulose.

CARBON Nonmetallic element, C, having an atomic number of 6, relative atomic mass of 12.011 that makes up all living matter, and, subsequently, fuels such as coal, petroleum, and natural gas.

CARBONACEOUS Of or containing carbon.

CARBONATE ALKALINITY Alkalinity caused by carbonate ions.

CARBONATED SPRING Spring water containing carbon dioxide gas.

CARBONATE HARDNESS Hardness of water caused by the presence of carbonates and bicarbonates of calcium and magnesium.

CARBONATE MINERALS Minerals that contain the carbonate group (CO_3^{2+}).

CARBONATION Diffusion of carbon dioxide gas through a liquid to render it stable with respect to precipitation or dissolution of alkaline constituents.

CARBONATOR Apparatus for the carbonation or recarbonation of water.

CARBONACEOUS STAGE Stage of decomposition that occurs in biological treatment processes when aerobic bacteria, using dissolved oxygen, change carbon compounds to carbon dioxide.

CARBON BLACK Finely divided amorphous carbon characterized by a high oil absorption and low specific gravity, produced by burning natural gas in a supply of air insufficient for complete combustion.

CARBON CHLOROFORM EXTRACTION Method for assessing organic pollutants in water.

CARBON CYCLE Cycle of natural processes in which atmospheric carbon in the form of carbon dioxide is converted to carbohydrates by photosynthesis, metabolized by animals, and ultimately returned to the atmosphere as a carbon dioxide waste or decomposed product.

CARBON DIOXIDE One of the products (CO_2) of combustion; a heavier than air gas used to extinguish Class B fires my smothering or by displacing the oxygen necessary for combustion.

CARBON DIOXIDE FIRE EXTINGUISHER Extinguisher that emits carbon dioxide, which reduces the oxygen content of air from 21% to 15%, smothering most fires.

CARBONIZATION Treatment of fossil fuels by heat in a closed vessel to produce coke and/or gas.

CARBON MONOXIDE Colorless, odorless and poisonous gas; a by-product of the burning of carbon or carbon-based fuels, as in an internal combustion engine.

CARBON REGENERATION UNIT Enclosed thermal treatment device used to regenerate spent activated carbon.

CARCINOGEN Compound or element that can or will induce cancer in man and other animals.

CARNOT EFFICIENCY Direct outcome of the Second Law of Thermodynamics that places an upper limit on the maximum possible conversion of thermal energy (heat) to work in a heat engine.

CARRIER 1. Liquid, such as water, solvent, or oil, in which an active ingredient is dissolved or dispersed; **2.** Material, including but not limited to feed, water, soil, nutrient media, with which a test substance is combined for administration to a test system or test organisms.

CARRYING CAPACITY 1. Maximum rate of flow that a channel, conduit, or other hydraulic structure can carry; **2.** Electrical load that a cable or fuse can carry without overload; **3.** Maximum number of species an area can support during the harshest part of the year.

CARRYING CONTAINER *See* **Waste container.**

CARRYOUT COLLECTION *See* **Collection.**

CARRYOVER Liquid or solid particles entrained in the vapor evolved by a boiling liquid.

CARRYOVER SOIL MOISTURE Moisture stored in the root-zone soils during winter while vegetable material is dormant, or prior to it being planted.

CAR-SEALED Having a seal placed on a device used to change the position of a valve (e.g. from open to closed) such that the position of the valve cannot be changed without breaking the seal.

CARTRIDGE FILTER 1. Filter with a disposable element; **2.** Discrete filter unit containing both filter paper and activated carbon that traps and removes contaminants from, typically, petroleum solvent, together with its piping and ductwork.

CARTRIDGE FUSE Electrical fuse that is gripped between conductors and which, when subject to overload, fails by melting, thus creating a nonconducting gap.

CASCADE One, or a series of steps that interrupts an otherwise steady flow of water to dissipate energy and/or to produce agitation and aeration.

CASCADE AERATOR Series of steps over which water flows and which cause agitation and aeration.

CASCADING FAILURE Sequence triggered by the failure of one part of a system that causes an increasingly unsupportable load to be passed to the adjacent component or subsystem.

CASE DRAIN LINE Line connecting fluid from a component housing to the reservoir.

CASH ALLOWANCE Sum included in a contract to cover the cost of prescribed items not fully detailed, the amount subject to change orders based on actual expenditures.

CASING Pipe or tubing of appropriate material, of varying diameter and weight, lowered into a borehole during or after drilling in order to support the sides of the hole and thus prevent the walls from caving, to prevent loss of drilling mud into porous ground, or to prevent water, gas, or other fluid from entering or leaving the hole.

CASING BLOWS Blows, usually of a 136 kg (300 lb) hammer falling 450 mm (18 in.) on to a soil sampler casing while making a soil boring.

CASING HEAD Heavy metal cap on the top of a string or casing that absorbs the blows of a driver in well boring.

CASTELLATUS Turret-shaped cumulus clouds, often in the form of lines.

CAST IRON PIPE *See* **Pipe.**

CASUALTY INSURANCE Insurance against losses caused by injuries to persons or their property.

CATALYST Substance that initiates a chemical reaction and enables it to proceed under milder conditions than otherwise required and that does not, itself, alter or enter into the reaction.

CATALYTIC COMBUSTION SYSTEM Process in which a substance is introduced into an exhaust gas stream to oxidize vaporized hydrocarbons or odorous contaminants; the substance itself remains intact.

CATALYTIC CONVERTER Emission control device incorporated into an internal combustion engine's exhaust system and containing catalysts such as platinum, palladium, or rhodium that reduce the levels of hydrocarbons (HC), carbon monoxide (CO) and nitrogen oxides (NOx) emitted to the air.

CATALYTIC CRACKING Process of breaking of carbon–carbon bonds with the aid of a catalyst.

CATALYTIC REACTION Chemical reaction in which the amount of one of the substances involved (the catalyst) is not decreased, although its presence is necessary for the reaction to occur at all.

CATARACT Large waterfall.

CATARACT ACTION Excavating back-action of a rapid current of water on a river bed.

CATASTROPHE Sudden and severe calamity: for insurance purposes, an event that causes a loss of extraordinary size.

CATASTROPHIC COLLAPSE Sudden and utter failure of overlying strata caused by removal of underlying materials.

CATCH BASIN Receptacle connected with a sewer or drain tile into which water from a roof, floor, etc. will drain.

CATCH FEEDER Subsidiary waterway in a catchment area.

CATCHMENT Structure, such as a basin or reservoir, for collecting or draining water.

CATCHMENT AREA Watershed or drainage area.

CATCHMENT BASIN Intake area of an aquifer.

CATCHPIT Accessible receptacle in a drainage system in which grit and other debris is deposited.

CATCHWATER Collection channel along the lip of an embankment through which streams and surface water are diverted from flooding the low-lying land below.

CATEGORICAL EXCLUSION Category of actions that do not individually or cumulatively have a significant effect on the human environment.

CATEGORICAL STANDARD Standard promulgated by a competent authority.

CATENA Group of soils derived from uniform or similar parent material but exhibiting variations in type due to differences in topography or drainage, as between ridges and valleys.

CATENARY Curve made by a flexible line hung between two points at the same elevation.

CATENARY SCAFFOLD Suspension scaffold consisting of a platform fastened to two essentially horizontal and parallel wire ropes, that are secured to structural members.

CATERPILLAR GATE Spillway control gate mounted on crawler tracks having steel rollers that bear on steeply sloping rails on each side of the opening.

CATERPILLAR TRACK *See* **Crawler track.**

CATHETOMETER Instrument for measuring comparatively small heights, consisting of a telescope that can slide up and down a calibrated vertical pillar. Readings are taken with the telescope focused on the top and then on the bottom of the object whose height is being measured (and one reading is subtracted from the other).

CATHODE Electrical term for the negative terminal.

CATHODIC PROTECTION Technique to prevent corrosion of a metal surface by making that surface the cathode of an electrochemical cell. For example, a tank system can be cathodically protected through the application of either galvanic anodes or impressed current.

CATHODIC PROTECTION TESTER Person who can demonstrate an understanding of the principles and measurements of all common types of cathodic protection systems as applied to buried or submerged metal piping and tank systems. At a minimum, such persons would have education and experience in soil resistivity, stray current, structure-to-soil potential, and component electrical isolation measurements of buried metal piping and tank systems.

CAT ICE *See* **Shell ice.**

CATION Ion having a positive charge and, in electrolytes, characteristically moving toward a negative electrode.

CATION EXCHANGE CAPACITY Sum total of exchangeable cations that a sediment or soil can adsorb, expressed in milliequivalents of negative charge per 100 grams (meq/100g) or milliequivalents of negative charge per gram (meq/g) of soil or sediment.

CATION-EXCHANGE WATER SOFTENER Device that removes dissolved ionic contaminants in hard water, such as magnesium and calcium, and replaces them with sodium ions.

CATIONIC POLYMER Polymer that contains one or more covalently charged subunits that bear a net positive charge.

CATIONIC SURFACTANT Surfactant in which the hydrophilic group is positively charged and which exhibits excellent disinfectant properties.

CATWALK Narrow elevated walkway.

CAULK 1. Waterproof sealant used to fill joints or seams; available as putties, ropes, or compounds extruded from cartridges; **2.** To fill a joint with mastic, usually under pressure.

CAULKED JOINT Non-mechanical joint sealed with caulk injected between the faces.

CAULKING 1. Process of filling in cracks or cavities and expansion joints using an elastic material, usually from a caulking gun; **2.** Material used for joint sealing where minor or no elastomeric properties are required.

CAULKING CHISEL Plumber's tool for compacting the lead in a lead-caulked joint.

CAULKING COMPOUND Soft, plastic material consisting of pigment and a vehicle, used for sealing joints in buildings and other structures where normal structural movement may occur. Such compounds retain their plasticity for extended periods after application by gun or knife, or in the form of preformed extruded shapes.

CAULKING GUN Hand- or power-operated tool that holds prepackaged cartridges of caulk.

CAULKING HAMMER Medium-weight hammer used to strike a caulking chisel.

CAULKING RECESS Space between the bell of one piece of pipe and the spigot of the pipe length to which it is joined that is filled with caulking to seal the joint.

CAULKING TOOL Tool used for driving a caulking compound into seams and crevices to make joints air- and watertight.

CAUSEWAY Raised way or road, usually across wet or marshy ground, or through shallow water.

CAUSTIC Substance that can corrode, burn or dissolve through chemical action. Caustic liquids are often used to etch materials that are acid resistant. Caustics have a lesser effect on ferrous materials, and are often used in cleaning processes for engine castings, forgings, and formed parts.

CAUSTIC ALKALINITY Alkalinity caused by hydroxyl ions.

CAUSTIC EMBRITTLEMENT Intergranular failure of some types of steel, particularly those for hot water boilers, resulting from the combination of stress beyond the yield point of the steel and attack by a concentrated caustic solution.

CAUSTIC SCRUBBING Process for removing sulfur dioxide from flue gases by passing them through a solution of sodium hydroxide.

CAUSTIC SODA Sodium hydroxide, NaOH; a strong alkali.

CAVE-IN Sudden collapse of a trench or excavation wall.

CAVING SOIL Soil that tends to fall into an uncased or unshored hole or excavation.

CAVITATION 1. Starvation of pumps created by a blockage or restriction that prevents the supply from reaching the pumps. The extreme low-pressure vacuum developed can cause pieces of metal to be pulled from parts of the pump; **2.** Formation of a void due to reduced pressure in lubricating grease dispensing systems.

C_{bi} Concentration of volatile organic compounds in each gas stream (i) entering the emission control device, in parts per million by volume.

$CBOD_5$ Five-day measure of the pollutant parameter carbonaceous biochemical oxygen demand (CBOD).

CDD AND D INSURANCE Comprehensive dishonesty, disappearance, and destruction insurance coverage against employee dishonesty, loss of money and securities, and theft of materials and equipment.

CEILING (In meteorology) the distance between the surface of the Earth and the lowest clouds.

CEILING INSULATION Material, primarily designed to resist heat flow, which is installed between the conditioned area of a building and an unconditioned attic, as well as common ceiling–floor assemblies between separately conditioned units in multiunit structures. Where the conditioned area of a building extends to the roof, ceiling

insulation includes such a material used between the underside and upper side of the roof.

CELERITY Speed at which a wave develops through a fluid medium, relative to the undisturbed speed of the fluid through which the disturbance passes. *See also* **Wave velocity.**

CELL 1. Basic unit of living matter, bounded by a thin membrane and, in plants, surrounded by a cell wall usually made of cellulose; **2.** Compacted solid wastes that are enclosed by natural soil or cover material in a land disposal site.

CELL HEIGHT Vertical distance between the top and bottom of compacted solid waste enclosed by natural soil or cover material in a sanitary landfill.

CELL PAIR Two opposite-type membranes and two separators, used together to form one product cell and one concentrating cell.

CELL-PAIR RESISTANCE Electrical resistance of one cell pair.

CELL ROOM Structure(s) housing one or more mercury electrolytic chlor-alkali cells.

CELL-TYPE INCINERATOR *See* **Incinerator.**

CELLULAR COFFERDAM Cofferdam enclosed by a wall consisting of a series of filled cells. *See also* **Cofferdam,** and **Crib.**

CELLULOSE Carbohydrate polymer constructed from the glucose building blocks formed by the action of photosynthesis: the principal structural element and major constituent of the cell walls of trees and other higher plants where it is bound with lignin to stiffen plant stems.

CELLULOSE DEBRIS Fine wood debris that is often evidence of the presence of wood borers and termites.

CELLULOSE ECONOMY Use of the Sun's energy through photosynthesis to produce cellulose, which is then processed by enzymatic means or by chemical hydrolysis to yield glucose, which can be fermented to yield products of industrial value, protein or fuels.

CELLULAR POLYISOCYANURATE INSULATION Insulation produced principally by the polymerization of polymeric polyisocyanates, usually in the presence of polyhydroxl compounds with the addition of catalysts, cell stabilizers, and blowing agents.

CELLULAR POLYSTYRENE INSULATION Foam composed principally of polymerized styrene resin processed to form a homogenous rigid mass of cells.

CELLULAR POLYURETHANE INSULATION Insulation composed principally of the catalyzed reaction product of polyisocyanurates and polyhydroxi-compounds, processed usually with a blowing agent to form a rigid foam having a predominantly closed cell structure.

CELLULOSE FIBER FIBERBOARD Insulation composed principally of cellulose fibers usually derived from paper, paperboard stock, cane, or wood, with or without binders.

CELLULOSE FIBER LOOSE-FILL Basic material of recycled wood-based cellulosic fiber made from selected paper, paperboard stock, or ground wood stock, excluding contaminated materials which may reasonably be expected to be retained in the finished product, with suitable chemicals introduced to provide properties such as flame resistance, processing and handling characteristics. The basic cellulosic material may be processed into a form suitable for installation by pneumatic or pouring methods.

CELSIUS Modern name for the centigrade scale on which water boils at 100° and freezes at 0°. *See also* **Centigrade.**

CEMENT 1. Mineral matter that binds together the fragments in sedimentary rock; **2.** Fine grey powder made by burning clay and limestone that sets hard after having been mixed with water.

CEMENTED SOIL *See* **Soil types.**

CEMENTING Operation whereby a ce-

ment slurry is pumped into a drilled hole and/or forced behind a casing.

CEMENT-LINED PIPE *See* **Pipe.**

CENSUS Enumeration of a whole population with respect to specific variables.

CENTER OF BUOYANCY Center of gravity of the space occupied in a body of water by a floating object.

CENTER OF FLOTATION Centroid of the water plane area of a floating body.

CENTER OF GRAVITY Point in a body about which the weights of the various parts balance; point within a machine around which its weight is evenly distributed. The three measurements necessary to determine the center of gravity are:

> **Horizontal:** Measured fore-and-aft from a reference plane.

> **Lateral:** Measured from the center line of the body to the side.

> **Vertical:** Measured up and down from a reference plane.

CENTER OF MASS Cross-sectional line of a cut or fill that divides the volume into halves.

CENTER OF MOMENTS 1. Point at which a body tends to rotate; **2.** Point, arbitrarily selected, for determining the resultant moment of a series of forces.

CENTER OF PRESSURE Point, or an area, subjected to hydraulic or pneumatic pressure over which the whole force due to pressure is taken to act.

CENTER-TO-CENTER Spacing of elements or components, as in studs for frame construction being on 400-mm (16-in.) centers. *See also* **On center.**

CENTI Prefix representing 10^{-2}. Symbol: c. Used in the SI system of measurement. *See also the appendix:* **Metric and nonmetric measurement.**

CENTIGRADE Scale of temperature that features 0° and 100° as the freezing and boiling point of water, respectively. *See also* **Celsius.**

CENTIGRAM Unit of mass, equal to 1/100 gram. Symbol: cg. Used in the SI system of measurement. *See also the appendix:* **Metric and nonmetric measurement.**

CENTILITER Unit of volume, equal to 1/100 litre (0.6102 in.³) Symbol: cL. Used in the SI system of measurement. *See also the appendix:* **Metric and nonmetric measurement.**

CENTIMETER Unit of length, equal to 1/100 meter (0.3937 in.). Symbol: cm. Used in the SI system of measurement. *See also the appendix:* **Metric and nonmetric measurement.**

CENTIPOISE Unit of absolute (dynamic) viscosity.

CENTISTOKE Unit of kinematic viscosity.

CENTRAL COLLECTION POINT Location where a generator consolidates regulated (typically medical) waste brought together from original generation points prior to its transport off-site or its treatment on-site (e.g. incineration).

CENTRAL INCINERATOR *See* **Incinerator.**

CENTRAL LUBRICATION System whereby a number of remotely located points may be lubricated from one location, either manually as required, or automatically at preset intervals.

CENTRATE Fluid leaving a centrifuge after most of the solids have been removed.

CENTRIFUGAL DEWATERING Partial removal of water from a sludge or slurry by centrifugal action.

CENTRIFUGAL DRYING Partial drying of a sludge or slurry by centrifugal action.

CENTRIFUGAL FORCE 1. Outward force exerted by a body moving through a curve; **2.** Force generated by eccentric weight(s) rotating at a specific frequency inside the drum of a vibratory compactor.

CENTRIFUGAL PUMP *See* **Pump.**

CENTRIFUGAL PUMP CAPACITY Factor that varies directly as the speed or the diameter of the impeller.

CENTRIFUGAL PUMP HEAD Factor that varies as the square of the speed or as the square of the impeller diameter.

CENTRIFUGAL PUMP POWER Factor that varies as the cube of the speed or the cube of the impeller diameter.

CENTRIFUGAL SCREW PUMP *See* **Pump.**

CENTRIFUGAL SEPARATOR *See* **Separator.**

CENTRIFUGE 1. Machine with a compartment spun around a central axis; **2.** Apparatus for separating particles in suspension by rotating them at high speeds: the rate of sedimentation depends on the speed of rotation and the size of the particles; at any given speed, the rate is quicker for larger particles than for smaller ones; **3.** Equipment used to dewater a slurry.

CENTRIFUGE MOISTURE EQUIVALENT Water content retained by a presaturated soil that has then been subjected to a force equal to 1,000 times gravity for one hour.

CENTRIFUGE VOLUME Volume of a liquid or solid, or both, separated from a volume of liquid exposed to centrifugal force.

CENTRIPETAL DRAINAGE Drainage pattern generally toward a central point.

CERAMIC-DISC VALVE *See* **Valve.**

CERAMIC FILTER Filter made of an inert, inorganic material shaped and fired at a high temperature.

CERTIFICATE Document that attests to something; written authorization or approval.

CERTIFICATE FOR PAYMENT Certification by an authorized professional that work to a specific stage has been completed

CERTIFICATE OF ANALYSIS

and that specific expenditures have been made, qualifying a contractor for an interim or final payment under the contract for the works.

CERTIFICATE OF ANALYSIS List of laboratory test results that a supplier affirms to be representative of the quality of a product shipped to a particular customer.

CERTIFICATE OF COMPLIANCE Certification by an authorized authority that a completed project is in compliance with the appropriate regulations and requirements.

CERTIFICATE OF INSURANCE Document issued by an insurance company that verifies coverage.

CERTIFICATE OF OCCUPANCY Document issued by a competent authority certifying that a building complies with relevant public standards and permitting occupancy for its designated use.

CERTIFICATE OF PRACTICAL COMPLETION Certificate issued when a building has been substantially completed, but with minor defects and/or omissions.

CERTIFICATE OF TITLE Written legal opinion declaring that a title to land is vested as stated, following a review and examination of the abstracts or chains of title.

CERTIFICATION Written statement that an inspection has been carried out by an authorizing entity and that specific criteria has been met.

CERTIFICATION EXAMINATION Examination administered by a professional association or other competent organization to indicate a level of professional competence.

CERTIFIED APPLICATOR Individual certified to use or supervise the use of any restricted-use pesticides covered by his or her certification.

CERTIFYING AGENCY Person or agency designated by statute, or by other governmental act, to certify compliance with applicable (typically water) quality standards.

CERTIFYING OFFICIAL 1. For a corporation, a president, secretary, treasurer, or vice-president of the corporation in charge of a principal business function, or any other person who performs similar policy- or decision-making functions for the corporation; **2.** For a partnership or sole proprietorship, a general partner or the proprietor, respectively; **3.** For a government entity or other public agency, either a principal executive officer or ranking elected official.

CESIUM-137 Radioactive alkali metal with an ability to concentrate in some food chains and a half-life of 30.17 years. The maximum acceptable concentration in drinking water is 10 Bq/L.

CESSPOOL Chamber below grade for collecting and holding disposal from house drains.

CETANE Colorless liquid hydrocarbon, $C_{16}H_{34}$, used as a standard in determining diesel fuel ignition performance.

CETANE IMPROVER Additive for raising the cetane number of a diesel fuel.

CETANE INDEX Prediction of the cetane number of diesel fuel based on physical property measurements.

CETANE NUMBER Comparitive measurement of a fuel's volatility, the comparison being with the characteristics of cetane, one of the many fractions of crude petroleum. The higher the cetane value, the easier the fuel is to ignite. High cetane values are important in winter or cold climates. The cetane number is the volatility standard for diesel fuels, in the same way in which octane is the volatility standard (in this case, for antiknock qualities) for gasoline engine fuels.

C-FRAME C- or U-shaped frame that connects the blade to a dozer.

CGS SYSTEM System of units based on the centimeter (for length), gram (mass), and second (time). It was superceded first by the meter–kilogram–second (MKS) system, and then by the SI system.

CHAIN-BUCKET DREDGER Ladder-bucket dredging machine.

CHAIN GAUGE Gauge comprising a tagged or indexed chain, tape, or other type of line attached to a weight which is lowered to touch a water surface, with its elevation read from a graduated staff or index.

CHAIN OF LOCKS Series of interconnected hydraulic locks on a waterway.

CHAIN PIPE VICE Portable vice that uses a clamped length of chain to hold and restrain a length of pipe being worked on. *Also called* **Chain tongs.**

CHAIN PUMP *See* **Pump.**

CHAIN REACTION Situation in which one event causes a second and further events.

CHALYBEATE SPRING Spring containing a significant quantity of iron compounds, especially iron sulfate in solution.

CHAMBER Space enclosed by walls: a compartment, often prefixed by an adjective indicating its function.

CHANGED CONDITIONS Job conditions that differ substantially from those represented in the plans and specifications and/ or the contract documents.

CHANGED USE Significant change from a use pattern approved in connection with the registration of a pesticide product. Examples of significant changes include, but are not limited to, changes from nonfood to food use, outdoor to indoor use, ground to aerial application, terrestrial to aquatic use, and nondomestic to domestic use.

CHANGE IN SCOPE *See* **Scope of change.**

CHANGE IN THE WORK Modification to the work as described in the contract documents.

CHANGE OF AIR One complete change of all the air contained within a room or rooms and enclosed spaces.

CHANGE OF STATE Change from one phase to another: solid to liquid, liquid to gas, etc.

CHANGE ORDER Written order issued by a recipient, or its designated agent, to its contractor authorizing an addition to, deletion from, or revision of, a contract, usually initiated at the contractor's request.

CHANGES TO THE WORK Changes to work described or specified in a signed contract and authorized under the terms of a change order.

CHANNEL 1. Long groove or furrow; **2.** Tubelike passage for liquids. **3.** Deep portion of a river or waterway where the current flows; **4.** Bed of a stream or waterway. *See also* **Braided channel, Flood channel,** and **Meandering channel.**

CHANNEL ACCRETION Gradual buildup of a channel bottom or bank resulting from sediment deposition.

CHANNEL AXIS Line joining the middle points of the water surface of a channel in successive cross sections.

CHANNEL BEND Open U-shaped channel, curved on plan, that guides the flow from a branch pipe in a manhole into the principal sewer.

CHANNEL FLOW ACCRETION Gradual increase in a stream flow in dry weather resulting from influent seepage.

CHANNEL FLOW DEPLETION Gradual downstream decrease in a stream flow due to seepage from the stream to an adjacent or underlying groundwater body.

CHANNEL GULLY Trapped gully fitted with a channel through which sullage must pass before reaching the gully proper.

CHANNEL LINE Route of strongest flow of a river.

CHANNEL LOSS Loss of water from a channel by capillary action and percolation.

CHANNEL OF APPROACH Reach of a channel immediately upstream of a control structure.

CHANNEL-PHASE RUNOFF Phase of runoff that occurs in a channel and which is governed by the laws of channel hydraulics.

CHANNEL PIPE Fired clay pipe, semicircular or three-quarter-round in crosssection, used principally in manholes gathering branch drains.

CHANNEL ROUGHNESS Roughness of a channel; friction offered to flowing water by the surface of the submerged profile of a channel.

CHANNEL SPRING Spring on the bank of a stream that has cut a channel below the watertable.

CHANNEL STORAGE Volume of water stored in a channel at a point in time.

CHANNEL TERRACE Contour ridge built of soil moved from its uphill slope, used to divert surface water.

CHAPARRAL Vegetation dominated by shrubs with small, broad, hard, evergreen leaves.

CHAR Solid product of destructive distillation or carbonization of an organic material.

CHARACTERISTIC SPECIES Plant species localized within a given association.

CHARACTERISTIC SPEED Speed, expressed in rpm, at which the runner of a specific type of turbine would operate if it were reduced in size and proportion so that it would develop one horsepower under a 300 mm (1 ft) head.

CHARACTERISTIC STRESS Stress at the assumed yield point or limit of proportionality of a material.

CHARCOAL Form of carbon produced by the destructive distillation of wood.

CHARGE Quantity of fuel fed to a furnace.

CHARGED LINE Rigid or flexible service connection or hose under pressure.

CHARGE FOR CONDITIONAL WATER SERVICE Amount that a consumer is charged for readiness to supply up to a stated quantity of water from a supply that is additional to and separate from an existing source of water.

CHARGING CHUTE Overhead passage through which waste materials and solid fuels are introduced into a furnace.

CHARGING GATE Horizontal movable cover that closes the opening of a top-charging furnace.

CHARGING HOPPER Receptacle into which constituent materials are placed and which discharges to a mixer.

CHARGING PORT Port in the hydraulic head of an injection pump through which fuel passes to fill the pumping chamber.

CHARLES' LAW The volume of a fixed mass of gas varies directly with absolute temperature, provided the pressure remains constant.

CHAROPHYTA Class of algae found in still or slow-moving fresh or brackish water.

CHART OF ACCOUNTS 1. List of account numbers and designations; **2.** Any numbering system used to monitor project costs by category (e.g. labor, supplies, materials), usually based upon the corporate chart of accounts of the primary performing organization. *See also* **Code of accounts.**

CHASE Continuous recess built into a wall to receive pipes, ducts, etc.

CHECK Area of land between ridges that confines irrigation water.

CHECK CHAINS Heavy-duty chain and spring assembly designed to prevent dump body kickup at the end of the hoist stroke and to assist in returning the body to a level position. *Also called* **Restraining devices, Snubber chains,** and **Stop chains.**

CHECK DAM Dam that divides a drainage course into two or more sections with reduced slopes.

CHECKERPLATE Modular metal plates on which a nonslip pattern has been stamped.

CHECK GATE Gate set in the cutoff wall of an irrigation ditch to permit water delivery from the ditch to subsidiary ditches, or to adjacent land.

CHECK IRRIGATION Method of irrigation whereby water flows from one check to another along the slope of land.

CHECKLIST Detailed list of equipment, or tasks, or sequence of actions or events, all of which comprise what is necessary to complete something.

CHECK VALVE See **Valve.**

CHELATE Of or pertaining to a heterocyclic ring containing a metal ion attached by coordinate bonds to at least two nonmetal ions in the same molecule.

CHELATING AGENTS Group of organic compounds that can incorporate metal ions into their structure and so obtain a soluble, stable, and readily excretable substance known chemically as a chelate complex.

CHEMICAL Chemical substance or mixture.

CHEMICAL AGENTS Those elements, compounds, or mixtures that coagulate, disperse, dissolve, emulsify, foam, neutralize, precipitate, reduce, solubilize, oxidize, concentrate, congeal, entrap, fix, make the pollutant mass more rigid or viscous, or otherwise facilitate the mitigation of deleterious effects or the removal of the pollutant from the water.

CHEMICAL ANALYSIS Analysis by chemical methods to show the composition and concentration of substances.

CHEMICAL CLOSET See **Chemical toilet.**

CHEMICAL COAGULATION Use of a floc-forming chemical to destabilize colloidal and finely divided suspended matter and cause its initial aggregation. See also **Flocculation.**

CHEMICAL COMPOSITION Name and percentage by weight of each compound in an additive and the name and percentage by weight of each element in an additive.

CHEMICAL COMPOUND Substance made up of two or more elements combined together in definite proportions by weight.

CHEMICAL CONTAMINANT Foreign material in a fluid which is either in solution or in a gas or liquid bulk phase.

CHEMICAL CORROSION Direct chemical attack, typically by a liquid or a gas, usually upon something metallic. See also **Electrochemical corrosion.**

CHEMICAL DISSOLVING BOX See **Chemical solution tank.**

CHEMICAL DOSE Specific quantity of a chemical or chemical mixture to be added to a specific quantity of fluid for a specific purpose.

CHEMICAL ENERGY Energy liberated by a chemical reaction, usually mainly in the form of heat.

CHEMICAL ENGINEERING Branch of engineering concerned with the conversion of industrial raw material and the industrial manufacture of chemical products.

CHEMICAL EQUIVALENT Weight in grams of a substance that combines with or displaces one gram of hydrogen; the formula weight divided by its valence.

CHEMICAL FATE STUDIES Studies performed for the characterization of physical, chemical, and persistence factors that may be used to evaluate transport and transformation processes.

CHEMICAL FEEDER Mechanism for dispensing a chemical or mixture of chemicals at a predetermined rate. The feed rate may be changed manually, or automatically; models are designed to handle solids, liquids, or gases.

CHEMICAL GAUGING Estimating the quantity of water flow by determining the dilution of a chemical solution introduced upstream at a known rate and concentration. Also called **Chemihydrometry.**

CHEMICALLY PURE WATER Water that has no material in solution or suspension.

CHEMICAL MANUFACTURING PLANT Facility engaged in the production of chemicals by chemical, thermal, physical, or biological processes for use as a product, co-product, by-product, or intermediate including but not limited to industrial organic chemicals, organic pesticide products, pharmaceutical preparations, paint and allied products, fertilizers, and agricultural chemicals.

CHEMICAL METAL CLEANING WASTE Wastewater resulting from the cleaning of any metal process equipment with chemical compounds, including, but not limited to, boiler tube cleaning.

CHEMICAL NAME Scientific designation of a chemical substance in accordance with the nomenclature system developed by the International Union of Pure and Applied Chemistry or the Chemical Abstracts Service's rules of nomenclature, or a name which will clearly identify a chemical substance for the purpose of conducting a hazard evaluation.

CHEMICAL OXYGEN DEMAND Measure of the oxygen equivalent of the organic matter in a sample of sewage, liquid waste, leachate, or polluted water that is susceptible to oxidation by a strong chemical oxidant.

CHEMICAL PRECIPITATION Settling out of a solid fraction from a solution, caused by the addition of a chemical, hastened by floccing.

CHEMICAL PROTECTIVE CLOTHING Items of clothing that provide a protective barrier to prevent dermal contact with chemical substances of concern. Examples can include, but are not limited to: full body protective clothing, boots, coveralls, gloves, jackets, and pants.

CHEMICAL SLUDGE Sludge obtained by treatment of wastewater with chemicals.

CHEMICAL SOLUTION TANK Tank, basin, vessel, etc., in which chemicals are placed in solution prior to their introduction to raw water or wastewater.

CHEMICAL STRUCTURE Molecular structure of a compound.

CHEMICAL SUBSTANCE Organic or inorganic substance of a particular molecular identity, including any combination of such substances occurring in whole or in part as a result of a chemical reaction or occurring in nature, and any chemical element or uncombined radical; except: (a) any mixture, (b) any pesticide when manufactured, processed, or distributed in commerce for use as a pesticide, (c) tobacco or any tobacco product, but not including any derivative products, (d) any source material, special nuclear material, or by-product material, (e) any pistol, firearm, revolver, shells, and cartridges, and (f) any food, food additive, drug, cosmetic, or device, when manufactured, processed, or distributed in commerce for use as a food, food additive, drug, cosmetic, or device.

CHEMICAL SYMBOL Use of upper- and lower-case alphabetical letters alone or in combination used to represent an atom of a chemical element.

CHEMICAL THINNING Any forest thinning in which the unwanted trees are killed by chemical poisoning; band or frill girdling may be done at the same time.

CHEMICAL TOILET Toilet without water or sewer connections, in which human wastes are neutralized by chemicals or by biological action.

CHEMICAL TREATMENT Process in which the original material is changed from the addition of chemicals.

CHEMICAL WASTE LANDFILL Landfill at which protection against risk of injury to health or the environment from migration of PCBs to land, water, or the atmosphere is provided.

CHEMIHYDROMETRY *See* **Chemical gauging.**

CHEMAUTOTROPHIC Organisms that produce organic material from inorganic compounds using simple inorganic reactions as a source of energy.

CHEMOSPHERE Portion of the lower section of the Earth's atmosphere in which chemical processes such as molecular dissociation and recombination take place during the day and night, respectively, under the influence of ultraviolet radiation.

CHEMOSYNTHETIC BACTERIA *See* **Bacteria.**

CHEMOTROPHIC Organisms that obtain energy from any source other than light.

CHEVRON DRAIN *See* **Herringbone drain.**

CHIEF FACILITY OPERATOR Person responsible for daily onsite supervision, technical direction, management, and overall performance of the facility.

CHINOOK 1. Moist, warm wind blowing from the Pacific coast; **2.** Warm, dry wind that descends from the eastern slopes of the Rocky Mountains.

CHIP CONTROL FILTER Filter intended to prevent only large particles from entering a component immediately downstream. *Also called* **Grit control filter,** and **Last-chance filter.**

CHLORACNE Sublethal skin disorder resulting from exposure to dioxin and other chlorinated hydrocarbons.

CHLOR-ALKALI PROCESS Process involving the electrolysis of brine, used for the manufacture of chlorine.

CHLORAMINES Compounds of organic or inorganic nitrogen and chlorine. See also **Monochloramine.**

CHLORIDE Compound of chlorine with another element. Its presence in natural waters can be attributed to the dissolution of salt deposits, salting of roads for ice control, and industrial effluents. Chloride in water supplies is objectionable because it imparts undesirable tasts to water and beverages prepared with water.

CHLORIDE INDEX Quantity of chlorides in wastewater relative to the quantity of chlorides in the water supply.

CHLORINATED COPPERAS Solution of ferrous sulfate and ferric chloride produced by chlorinating a solution of ferrous sulfate.

CHLORINATED HYDROCARBONS One of three major groups of synthetic insecticides (the others being organophosphorus compounds and synthetic pyrethrins) that includes DDT, aldrin, endrin, benzene hexachloride, dieldrin, and others.

CHLORINATED ISOCYANURATES Chlorinated sodium or potassium salts of cyanuric acid, used as a source of chlorine for disinfection.

CHLORINATED LIME Combination of slaked lime and chlorine in which calcium oxychloride is the active ingredient.

CHLORINATED POLYVINYL CHLORIDE (CPVC) PIPE *See* **Pipe.**

CHLORINATION Application of chlorine to water or wastewater as a disinfectant.

CHLORINATION CHAMBER *See* **Chlorine contact chamber.**

CHLORINE Nonmetallic chemical element (Cl) of the halogen (salt-producing) family; a greenish-yellow poisonous gas about 2.5 times the weight of air. One of several chemicals used as an oxidant in water sterilization and purification.

CHLORINE–AMMONIA PROCESS Application of chlorine and ammonia to water, or of ammonia to water containing chlorine, in such ratios as to provide combined residual chlorination.

CHLORINE CONTACT CHAMBER Detention chamber in which chlorine is diffused through a liquid flow. *Also called* **Chlorination chamber.**

CHLORINE DEMAND 1. Amount of chlorine needed to kill all the pathogens in a sample of water; **2.** Difference between the amount of chlorine added to water or wastewater and the residual chlorine remaining after a specified contact period.

CHLORINE DIOXIDE Orange, water-soluble, unstable, explosive gas (ClO_2), used in water treatment primarily to remove tastes and odors.

CHLORINE HYDRATE Yellowish ice that forms in a chlorinator when chlorine gas comes in contact with water at 10°C (49°F) or lower.

CHLORINE REQUIREMENT Amount of chlorine needed for a particular purpose.

CHLORINE ROOM Room, or building, in which chlorination equipment is housed, with facilities for protecting personnel and equipment.

CHLOROFLUOROCARBONS Class of chemically inert compounds, CFCs, commonly used as solvents, aerosol propellants, refrigerants, and in the production of foams, including CFC-11 (cci$_3$F or trichlorofluoromethane) or CFC-12 (CCl$_2$F$_2$ or dichlorodifluoromethane).

CHLOROGANIC Organic compounds combined with chlorine.

CHLOROPHENOLS Products formed from the reaction of phenolic compounds with chlorine.

CHLOROPHYLL Green pigments, present in all 'green' plants, which capture light energy and enable the plants to form carbohydrates from carbon dioxide and water through the process of photosynthesis.

CHLOROPHYTA Green algae; the largest and most diverse division of algae, occurring in fresh and salt water and in damp places on land.

CHLOROPLAST Cytoplasmic body in plant cells that contain chlorophyll; the site of photosynthesis.

CHLOROSIS Condition marked by yellowing or whitening of green plants that is indicative of a state of reduced chlorophyll, possibly caused by acid rain.

CHOLERA Bacterial infection spread in humans by contamination of drinking water supplies by sewage effluent and infected food.

C-HORIZON Parent material, without humus, below the topsoil of the A-horizon and B-horizon. *See also* **A-horizon,** and **B-horizon.**

CHOROLOGY Study of areas and their floral and faunal development.

CHROMATOGRAPHY Analytical technique used principally for separating and identifying the components of a sample by distributing them between a stationary and a moving phase.

CHROMIUM Hard, white metal (Cr), used principally as a steel-alloying element and in plating. Trivalent chromium, the most common natural state, is essential in humans and animals for efficient lipid, glucose and protein metabolism, It is considered to be non-toxic; however, if present in raw water, it may be oxidzed to hexavalent chromium during chlorination.

CHRONIC Effects of a single or repeated exposures over a long period of time which eventually cause symptoms to continue for a long time.

CHRONIC EFFECT Cumulative physiological damage resulting from prolonged exposure or series of exposures to toxic substances.

CHRONIC TOXICITY Property of a substance or mixture of substances to cause adverse effects in an organism upon repeated or continuous exposure.

CHRONIC TOXICITY TEST Method used to determine the concentration of a substance in water that produces an adverse effect on a test organism over an extended period of time.

CHRONOLOGICAL AGE *See* **Actual age.**

CHRONON Very small unit of time, equal to the time a photon (at the speed of light) would take to cross the width of an electron; about 10^{-24} second.

CHRONOTHERM Thermostat and clock combination that can be set to activate a furnace or air-conditioner at predetermined times and/or temperatures.

CHUNK SAMPLE METHOD Determining the weight per cubic measure of soil in place.

CHUNK UP Clean up and pile debris after logging an area.

CHURN DRILL Machine that drills a hole in ground by raising and dropping a string of drilling tools suspended by a reciprocating cable. *Also called* **Cable drill.**

CHUTE 1. Sloping trough or tube for conveying fluids and free-flowing materials from a higher point to a lower point; **2.** Waterfall or rapid.

CHUTE-FED INCINERATOR *See* **Incinerator.**

CHUTE GATE Opening in the tailgate of a dump truck body to allow a limited flow of material during dumping, normally fitted with a sliding door, control handle, and lip for directing the flow. Chutes may also be located on the body side or in the floor.

CIENEGA Area where the water table is at or near the surface, producing standing water in depressions.

CILIATES Class of protozoans distinguished by short hairs on all or parts of their bodies.

CINDER 1. Solid particle remaining following the incomplete burning of carbon-containing material; **2.** Scora 4 - 32 mm in diameter.

CIPOLLETTI WEIR Trapezoidal-shaped measuring weir, widest at the top and with sides sloped 1 vertical to 4 horizontal.

CIRCADIAN Recurring every 24 hours, especially in relation to body rhythms.

CIRCADIAN RHYTHM Rhythmic changes that occur every 24 hours in plants and animals, i.e., cycles of sleep and waking and the leaf movements of plants.

CIRCANNUAL Recurring once every year, especially in relation to the human body and to the body rythms of animals that hibernate or aestivate on an annual basis.

CIRCLE Rotary table that supports the blade of a grader and regulates its angle.

CIRCLE OF INFLUENCE Horizontal perimeter of a depression, roughly conical in shape, produced in a watertable by extraction of water from a well at a given rate.

CIRCLE REVERSE Mechanism that changes the angle of a grader blade.

CIRCUIT BREAKER Automatic mechanical device that serves the same purpose as a fuse, i.e. to prevent overheating in a circuit through overloading.

CIRCUIT PROTECTIVE CONDUCTOR Earth, or ground wire.

CIRCUIT PROTECTOR Device designed to protect a single electric circuit, such as a circuit breaker or fuse.

CIRCUIT VENT Vent pipe that is connected at its lower end to a branch and at its upper end to a vent stack, or which is terminated in open air.

CIRCULAR-TYPE CELLULAR COFFERDAM Structure constructed of interlocking steel-sheet piling consisting of circular cells joined with connecting arcs. The arcs are installed after the cells are completed; the cells are then filled with granular soils. *See also* **Diaphragm-type cellular cofferdam.**

CIRCULATING PUMP *See* **Pump.**

CIRCULATION PIPE Flow or return pipe in a circuit.

CIRCUMFERENTIAL FLOW Flow parallel to the circumference of a storage tank that is circular in plan.

CIRCUMPOLAR VORTEX System of westerly winds in high latitudes, particularly at altitudes between 1000 and 12 000 m (3280 and 39,370 ft), that circulate air around the north and south poles.

CIRQUE Steep, bowl-shaped hollow in a mountainous region made by glacial ice and often forming the head of a valley.

CIRROCUMULUS Cloud made up of

rows or groups of small, fleecy clouds.

CIRROSTRATUS Thin, veil-like cloud high in the atmosphere.

CIRRUS Thin, curling, wispy cloud very high in the atmosphere.

CISTERN Reservoir or tank, often underground, for storing water.

CITY AND REGIONAL PLANNING Land and community planning that embraces areas larger than a single community, usually on a cooperative basis between adjacent communities and/or jurisdictions.

CITY BUILDING CODE *See* **Building codes.**

CITY GROWTH *See* **Urban growth.**

CITY PLANNER *See* **Urban planner.**

CIVIL AUTHORITY CLAUSE Provision in fire insurance that covers the policyholder against loss caused by civil authorities while attempting to prevent the spread of fire.

CIVIL ENGINEER Professional engineer specializing in public works and heavy engineering works.

CIVIL ENGINEERING CONTRACTOR Contractor who specializes in structures and plant.

CIVIL TWILIGHT Part of evening during sunset when the Sun's center is optically 6° below the horizon. *See also* **Astronomical twilight.**

CLACK VALVE *See* **Valve.**

CLADISTICS System of classifying living organisms according to their common ancestry.

CLADOGRAM Treelike diagram depicting similarities among groups of organisms, degrees of relationship and evolutionary descent.

CLAIM 1. Demand or assertion by one of the parties to an agreement seeking an increase, decrease, or other adjustment of the terms of the contract documents; **2.** Demand

for payment under an insurance policy.

CLAIMANT Individual or entity contracted to furnish labor, materials, or equipment to a contractor or subcontractor for use within a contract.

CLAMSHELL 1. Bucket with two jaws that are held open by the opening line, and that clamp together to load under their own weight when lifted by the closing line; **2.** Shovel equipped with a clamshell bucket.

CLAPOTIS Lapping of waves on a wall, the effect of which is to raise the water level and increase the effective pressure on a retaining wall.

CLARIFICATION 1. Process or processes designed to reduce the concentration of suspended matter in liquids; **2.** Additional information requested when the contract documents fail to show exactly how something is to be built or installed, or when documents of equal precedence disagree.

CLARIFICATION DRAWING Graphic interpretation of a detail of the contract drawings or specifications.

CLARIFIED SEWAGE Sewage from which all or part of the suspended matter has been removed.

CLARIFIER Mechanism that removes, by gravity (sometimes with chemical additives), the particles suspended in a liquid.

CLASS Type, set, configuration, collection, etc., having one or more attributes in common, or so grouped or defined within a classification or specification.

CLASS (OF FIRE)

> **Class A:** Fire involving combustible materials such as wood, cloth, and paper.

> **Class B:** Fire involving a flammable liquid, fat or grease.

> **Class C:** Fire involving energized electrical equipment.

> **Class D:** Fire involving a combustible metal.

CLASS (OF WILDFIRE) Size of a wildfire:

Class A: Fire of 0.1 ha (0.25 ac) or less.

Class B: Fire greater than 0.1 ha (0.25 ac) but less than 4.0 ha (10 ac).

Class C: Fire greater than 4.0 ha (10 ac) but less than 40 ha (100 ac).

Class D: Fire greater than 40 ha (100 ac) but less than 120 ha (300 ac).

Class E: Fire greater than 120 ha (300 ac) but less than 400 ha (1000 ac).

Class F: Fire greater than 400 ha (1000 ac) but less than 2000 ha (5000 ac).

Class G: Fire of 2000 ha (5000 ac) or more.

CLASS H WELLS Wells that inject fluids: (a) which are brought to the surface in connection with conventional oil or natural gas production and may be comingled with waste waters from gas plants which are an integral part of production operations, unless those waters would be classified as a hazardous waste at the time of injection; (b) for enhanced recovery of oil or natural gas; and (c) for storage of hydrocarbons which are liquid at standard temperature and pressure.

CLASSIFICATION Grouping of subjects or materials according to a description, size, etc.

CLASSIFIED PRODUCT Product listed and bearing the label of an approved laboratory.

CLASSIFIER Device capable of separating materials into defined groups, types, etc.

CLASSIFIER RAKE Machine for separating coarse and fine particles of granular materials temporarily suspended in water; the coarse particles settle to the bottom of a vessel and are scraped up an incline by a set of blades, while the fine particles remain in suspension to be carried over the edge of the classifier.

CLASTIC Sediments composed of fragments or parent rocks or minerals that have been transported and deposited mechanically.

CLAUS SULFUR RECOVERY PLANT Process unit which recovers sulfur from hydrogen sulfide by a vapor-phase catalytic reaction of sulfur dioxide and hydrogen sulfide.

CLAW BUCKET *See* **Bucket.**

CLAY *See* **Soil types,** and **Soil groups.**

CLAY BINDER *See* **Binder soil.**

CLAY CONTENT Percentage of clay by dry weight of a heterogeneous material.

CLAY CUTTER 1. Bit fitted to the suction pipe of a dredge; **2.** Section of steel pipe dropped into clay with a cable drill, the contained clay being ejected on the surface.

CLAY FILTRATION Refining process using fuller's earth or bauxite to adsorb minute solids from lubricating oil, as well as remove traces of water, acids, and polar compounds.

CLAY LOAM *See* **Soil types.**

CLAY MINERAL ANALYSIS Estimation or determination of the kinds of clay-size minerals and the amount present in a sediment or soil.

CLAY MINERALS Stable secondary minerals formed by the weathering of some primary minerals.

CLAYPAN Stratum or horizon of accumulated stiff, compact, relatively impervious clay which, while not cemented, may interfere with the movement of groundwater.

CLAY PIPE *See* **Pipe.**

CLAY SIZE FRACTION Portion of soil that is finer than 0.002 mm (0.00008 in.); not a positive measure of the material's plasticity or its clay characteristics.

CLAY SPADE Tool with a wide, flat chisel-like blade, used in a pneumatic or hydraulic hammer on cohesive soils.

CLEAN 1. Free of foreign matter; **2.** Sand or gravel lacking a binder; **3.** New or properly cleaned filter element.

CLEANABLE Filter element which, when loaded, can be restored by a suitable process to an acceptable percentage of its original dirt capacity.

CLEANABILITY Ability of a cleanable filter element to withstand repeated field cleanings and to retain adequate dirt capacity and service life.

CLEAN AGGREGATE Sand or gravel, free from clay or silt.

CLEAN AIR STANDARDS Enforceable rules, regulations, guidelines, standards, limitations, orders, etc., of a Clean Air Act.

CLEAN BURNING *See* **Burning out.**

CLEAN FILL Uncontaminated, inert solid material used to bring a site to grade.

CLEANING EYE Access eye, opening, or cleanout.

CLEANING OR ETCHING Process involving a chemical solution bath and a rinse or series of rinses designed to produce a desired surface finish on the workpiece. This term includes air pollution control scrubbers which are sometimes used to control fumes from chemical solution baths.

CLEANING WATER Process water used to clean the surface of an intermediate or final plastic product or to clean the surfaces of equipment used in plastics molding and forming that contact an intermediate or final plastic product. It includes water used in both the detergent wash and rinse cycles of a cleaning process.

CLEANLINESS LEVEL Analogue of contamination level.

CLEANOUT 1. Any opening or orifice left so as to permit the removal of debris from the space behind; **2.** Plumbing unit with a removable plate or plug affording access into the pipe for rodding, usually provided at pipe bends. *Also called* **Access eye, Cleaning eye,** and **Rodding eye.**

CLEANOUT DOOR Metal door placed below a flue at the chimney bottom, opening to the cleanout with access to the ash pit.

CLEAN RIVER River that gives no sensible evidence of pollution and from which drinking water may be obtained using water purification.

CLEAN ROOM Room in which the controlled atmosphere is filtered to remove 99.99% of all dust and contaminants and in which a positive atmospheric pressure is maintained to prevent the infiltration of unfiltered air.

CLEAN WATER STANDARD Enforceable limitation, control, condition, prohibition, standard, or other requirement.

CLEAR-AIR TURBULENCE Small-scale disturbance in cloudless air, often above cloud tops, although sometimes caused artificially by the prior passage of jet aircraft, that causes uneven flight for aircraft.

CLEAR AND GRUB Removal of all vegetation, trees, and rubble, or anything that will interfere with construction inside the limits of the project.

CLEARCUT When all trees and saplings of a logging area are felled, bucked, and removed.

CLEARING 1. Area from which the natural brush and vegetation have been removed; **2.** Cutting down and removing trees and brush.

CLEARING AWAY Removal of excess topsoil from a site.

CLEARING THE SITE *See* **Site clearing.**

CLEAR TITLE Title to something, clear of encumbrances or disputes.

CLEAR-WATER RESERVOIR Filtered water reservoir of sufficient capacity to buffer the variable demands made on a filtration system. *Also called* **Filtered-water reservoir.**

CLIENT Person or entity engaging another

to perform a duty or hold a responsibility; one to whom those engaged to carry out some act are responsible.

CLIMATE Meteorological conditions, including temperature, precipitation, and wind, that characterize a region or locality.

CLIMATIC CYCLE Cyclic recurrences of weather patterns and phenomena.

CLIMATIC PROVINCE Area characterized by a general similarity of climate throughout.

CLIMATIC VARIATION 1. Gradual change in the climate of a given locality; **2.** Extremes of climate of a given locality.

CLIMATIC YEAR Any continuous 12-month period during which a complete annual cycle occurs.

CLIMATOLOGY Study of regional climatic variations and changes; planning and siting of structures with regard to local climate.

CLIMAX 1. Point of greatest intensity in any series or progression of events; **2.** Stage in ecological development or evolution in which the community of organisms becomes stable.

CLIMAX FOREST Forest community that represents the final stage of natural forest succession for its locality, i.e. for its environment. Often identified as those forests that can reproduce indefinitely, i.e. in their own shade.

CLIMAX SPECIES Plant species that will remain essentially unchanged in terms of species composition for as long as the site remains undisturbed.

CLOGGING INDICATOR Indicator that is activated when a predetermined pressure differential across a filter is reached.

CLOSE-COUPLED PUMP *See* **Pump.**

CLOSED BASIN Depression or pond from which water is lost only by evaporation.

CLOSED BURNER Sealed-in burner that, in most cases, supplies all the air for combustion through the burner itself.

CLOSED CANOPY Forest stand where the crowns of the main level of trees forming the canopy are touching and intermingled so that sunlight cannot reach the forest floor.

CLOSED CENTRIFUGAL PUMP *See* **Pump.**

CLOSED-CIRCUIT BREATHING APPARATUS Breathing apparatus in which the wearer's exhalations are recycled. Carbon dioxide and moisture are removed from the exhalation and, after some oxygen is added, the wearer rebreathes the mixture.

CLOSED-CIRCUIT SURVEILLANCE SYSTEM Network of strategically-placed TV cameras linked to a central monitoring station by coaxial cable.

CLOSED COMMUNITY Community in which all niches are occupied, precluding colonization.

CLOSED CONDUIT Closed artificial or natural duct, used to convey liquids.

CLOSED CONTAINER Container so sealed with a lid or other seal that neither liquid nor vapor will escape at ordinary temperatures.

CLOSED COOLING WATER SYSTEM Configuration of equipment in which heat is transferred by circulating water that is contained within the equipment and not discharged to the air; chilled water loops are included.

CLOSED-END NUT Type of plumbing cap nut.

CLOSED IMPELLER Pump impeller consisting of a series of rotating blades or vanes, similar to the old-fashioned paddle wheel. In the case of an open impeller, the impeller blades rotate between the stationary walls of the blower housing. These walls tend to channel the air so that most of it flows out through the tips of the blades, but some air slips out sideways from between the blades and short-circuits back to the impeller inlet. A closed impeller has cover plate disks attached to the sides of the blades, and thus short-circuiting is minimized.

CLOSED LIST OF BIDDERS *See* **Invited bidders.**

CLOSED SPECIFICATIONS Specifications stipulating the use of specific products and/or processes.

CLOSED STACK Unvented plumbing system.

CLOSED-VENT SYSTEM System that is not open to the atmosphere, composed of piping, connections, and, if necessary, flow-inducing devices that transport gas or vapor from a piece or pieces of equipment to a control device.

CLOSE NIPPLE Pipe nipple so short that its two sets of threads meet in the middle. *See also* **Nipple.**

CLOSEOUT Actions taken to assure satisfactory completion of project work and to fulfil administrative requirements, including financial settlement, submission of acceptable required final reports, and resolution of any outstanding issues.

CLOSET BEND Elbow drainage fitting connecting a water closet to a branch drain.

CLOSE TIMBERING Planks placed adjacent to each other against the ground in an excavation.

CLOSET SPUD Connector between the base of a ball-cock assembly in a water closet tank and the water supply pipe.

CLOSING DIKE Structure across the branch channel of a river to stop or reduce the flow entering the channel.

CLOSING LINE Wire rope that performs two functions: (a) closes a clamshell or orange-peel bucket, and (b) operates as a hoisting rope. *See also* **Crowd, Digging line, Drag,** and **Dragline.**

CLOSING OF EXCAVATION Final backfill over services and around foundations to finished ground level.

Clostridium pasteurianum Species of anaerobic bacteria responsible, in part, for the gradual increase in the nitrogen content of unmanured grassland.

CLOSURE Act of securing a Hazardous Waste Management facility pursuant to regulatory requirements.

CLOSURE PERIOD Time beginning with the cessation, with respect to a waste impoundment, of uranium ore processing operations and ending with completion of requirements specified under a closure plan.

CLOUD Visible body of very fine droplets of water or particles of ice dispersed in the atmosphere above the Earth's surface.

CLOUD AMOUNT Proportion of the visible sky covered by cloud.

CLOUD BASE Elevation of the lowest part of a cloud, or layer of cloud, usually in reference to low cloud.

CLOUD BOW Arc in the sky caused by the refraction of light by spherical cloud droplets.

CLOUDBURST 1. Sudden rainstorm; **2.** Rainstorm in which the rate of fall is or exceeds 100 mm (3.94 in.)/h.

CLOUD CHAMBER Laboratory instrument in which air may be expanded adiabatically to produce a simulation of cloud.

CLOUD GENERATION Spontaneous production of clouds, usually from the adiabatic cooling due to the ascent of air.

CLOUD SEEDING Scattering of solid carbon dioxide (dry ice), sea salt or silver iodide crystals into cloud to promote rainfall.

CLOUD SHADOW Shadow cast on the ground by clouds.

COAGULANT Substance that, when added to water, causes suspended solids to bulk together or floc.

COAGULANT AID Substance such as activated silica or lime added to a coagulant to intensify settling of particulate matter and densify the floc.

COAGULANTS Chemicals that cause very fine particles in suspension or partial suspension in water to clump together into larger particles that will either sink or float depending on their specific gravity.

COAGULATION Process in which suspended solids in water and wastewater are induced to combine to form a floc of greater density that will settle.

COAGULATION BASIN Basin in which a liquid, with or without the addition of a coagulant, is mixed gently to induce agglomeration with a consequent increase in settling velocity of particulates.

COAL Black or brownish-black combustible substance containing varying amounts of carbon, used as a natural fuel and for the manufacture of coal gas, coal tar, etc. Coal is a kind of sedimentary rock formed over geologic time from the partial decomposition of vegetable matter away from air and under-varying degrees of pressure.

COAL EQUIVALENT Amount of coal that, on combustion, would produce the same quantity of heat as a given mass of a given fuel.

COALESCENCE Capacity of materials to grow together, fusing and binding.

COALESCENT DEBRIS CONE Debris cone composed of two or more debris cones formed by debris contributed by large streams.

COALESCING SEPARATOR See **Separator.**

COAL GAS Mixture of combustible gases made by distilling bituminous coal, used for heating and lighting.

COAL REFUSE Waste products of coal mining, cleaning, and coal preparation operations (e.g. culm, gob, etc.) containing coal, matrix material, clay, and other organic and inorganic material and/or with an ash content greater than 50%, by weight, and a heating value less than 13 900 kJ/kg (6 000 Btu/lb) on a dry basis.

COAL REFUSE DISPOSAL PILE Coal refuse deposited on the earth and intended as permanent disposal or long-term storage (greater than 180 days), but not including coal refuse deposited within the active mining area or coal refuse never removed from the active mining area.

COAL TAR Dark brown or black, heavy, sticky liquid obtained as a residue after the distillation of bituminous coal.

COANDA EFFECT Tendency of a stream of air to cling to a surface when blown parallel to it.

COARSE AGGREGATE See **Aggregate.**

COARSE-AGGREGATE FACTOR Ratio, expressed as a decimal, of the amount (mass or solid volume) of coarse aggregate in a unit volume of well-proportioned concrete to the amount of dry-rodded coarse aggregate compacted into the same volume. *Also called b/b°.*

COARSE GRADED AGGREGATE Sample of aggregate containing a high proportion of large particles.

COARSE-GRAINED FILTER Filter made up of gravel, stone, slag and similar relatively bulky material; a filter made of other than sand.

COARSE-GRAINED SOIL Soil in which the larger grain sizes, such as sand and gravel, predominate.

COARSE GRAVEL Rock fragments ranging in size from 20 to 60 mm (0.75 to 2.5 in.).

COARSE RACK Rack with bars spaced at distances of 25 mm (1 in.) or more.

COARSE SAND Rock fragments ranging in size from 0.6 to 2 mm (0.024 to 0.78 in.).

COASTAL Zone extending from the highwater mark on land to the edge of the continental shelf.

COASTAL EFFECTS Climatic differences experienced near a coast compared with the inland weather.

COASTAL PLAIN Plain bordering the coastline or seashore.

COASTLINE Line that separates the land surface and the surface of a sea or ocean;

the shape or boundary of a coast.

COATING 1. Material applied to another; **2.** Foreign or deleterious substances found adhering to another.

COBALT-60 Radioactive isotope of cobalt that emits high-energy gamma radiation and which has a half-life of more than five years.

COCCI Spherical bacteria.

COCK Control valve.

CO-COMPOSTING Composting of municipal solid waste and wastewater treatment plant sludge.

COD Chemical oxygen demand.

CODE Any systematic collection or set of rules pertaining to one main subject and drafted for the purpose of securing uniformity of standards of workmanship or for maintaining proper standards of procedure.

CODE OF PRACTICE Collection of rules pertaining to a specific subject setting out procedures, lines of authority, powers such as purchasing authority, etc.

CODISPOSAL Disposal of municipal solid waste and ash in a municipal waste landfill.

CODOMINANT 1. In forest stands with a closed canopy, those trees whose crowns form the general level of the canopy and that receive full light from above, but comparatively little from the sides; in young stands, those trees with above average height growth; **2.** Two species that are equally dominant in a climax community.

COEFFICIENT Factor that remains constant in a system within specified conditions.

COEFFICIENT OF AREA Ratio of the area of a trickling filter surface receiving a wastewater to the total area of the filter surface.

COEFFICIENT OF COMPRESSIBIL-ITY Change in the ratio of voids per unit increased from pressure.

COEFFICIENT OF CONSOLIDATION Value of the coefficient of permeability di-vided by the coefficient of compressibility.

COEFFICIENT OF CONTRACTION Ratio of the smallest cross-sectional area of a jet discharged under pressure from an orifice, to the area of the orifice.

COEFFICIENT OF DISCHARGE Factor used in calculating flow through an orifice. It takes into account the facts that a fluid flowing through an orifice will contract to a cross-sectional area that is even smaller than that of the orifice, and that there is some dissipation of energy due to turbulence.

COEFFICIENT OF EARTH PRESSURE Ratio of the active earth pressure, normal to a plane surface, to the corresponding pressure in a fluid of the same density.

COEFFICIENT OF EARTH PRESSURE AT REST Ratio of the earth pressure, normal to a plane surface, to the corresponding pressure in a fluid of the same density, when there is no movement.

COEFFICIENT OF EXPANSION Rate at which a material expands with a rise in temperature.

COEFFICIENT OF FINENESS Ratio of suspended solids to turbidity; a measure of the size of particles causing turbidity.

COEFFICIENT OF FRICTION Ratio of the forces required to move an object resting on a horizontal plane to the weight of the object. *See also* **Coefficient of kinetic friction.**

COEFFICIENT OF HEAT TRANSMIS-SION Constant that represents the ability of a certain material to transmit heat.

COEFFICIENT OF INTERNAL FRIC-TION Tangent of the angle ϕ, the angle of internal friction.

COEFFICIENT OF KINETIC FRIC-TION Ratio of the tangential force sustaining motion at constant velocity in a sliding system to the load perpendicular to the motion. *See also* **Coefficient of friction.**

COEFFICIENT OF LINEAR EXPAN-SION Fractional change in length of a material for a unit change in temperature.

COEFFICIENT OF PASSIVE EARTH PRESSURE Ratio of the passive earth pressure, normal to a plane surface, to the corresponding pressure in a fluid of the same density.

COEFFICIENT OF PERFORMANCE Measure of heat pump efficiency; the ratio of useful heat energy transferred, to the amount of energy put into the system.

COEFFICIENT OF PERMEABILITY Average velocity of water through the total area of soil under a hydraulic gradient of 1.

COEFFICIENT OF PERMEABILITY TO WATER Rate of discharge of water under lamina flow conditions through a unit cross-sectional area of a porous medium under a unit hydraulic gradient and standard temperature conditions, usually 20°C (68°F).

COEFFICIENT OF REGIME Ratio of a maximum annual flow at peak to the lowest annual flow.

COEFFICIENT OF RUNOFF Factor that accounts for stormwater losses due to evaporation, exfiltration, etc., in the design of stormwater drainage systems.

COEFFICIENT OF STORAGE Volume of water that an aquifer receives per unit surface area per unit change in head.

COEFFICIENT OF SUBGRADE FRICTION Coefficient of friction between a slab and its subgrade.

COEFFICIENT OF SUBGRADE REACTION Ratio of load per unit area on soil to the corresponding deformation. *Also called* **Modulus of subgrade reaction,** and **Subgrade modulus.**

COEFFICIENT OF THERMAL EXPANSION Change in linear dimension per unit length, or change in volume, per unit volume per degree of temperature change.

COEFFICIENT OF THERMAL TRANSMISSION Amount of heat transferred through a unit area of a material per hour, per degree temperature difference between the air on each side.

COEFFICIENT OF UNIFORMITY Ratio between the number of particles having a diameter that is larger than 60% by mass of the particles in a soil, or aggregate sample of that diameter, to the effective size which is larger by 10% by mass of the particles.

COEFFICIENT OF VARIATION The standard deviation expressed as a percentage of the average. *See also* **Standard deviation.**

COEFFICIENT OF VELOCITY Ratio of measured velocity to the theoretical discharge velocity.

COEFFICIENT OF VISCOSITY Measure of the internal resistance of a fluid to flow, equal to the shearing force in dynes per square centimeter from one fluid plane to another parallel plane one centimeter distant.

COENOCLINE Sequence of natural communities associated with an environmental gradient.

COFFERDAM Structure that enables construction work to be carried out below water level; a caisson. A cofferdam does not have to be entirely watertight. It may be cheaper or more practical to permit some flow into the working area with the excess water then being removed with pumps. *See also* **Cellular cofferdam,** and **Crib.**

CO-FIRED COMBUSTOR Unit combusting up to 30% municipal solid wastes with a non-municipal solid-waste fuel.

COGENERATION 1. Generation of heat and/or electricity using more than one type of fuel; **2.** Development of more than one type of usable energy from a fuel, i.e. electricity and heat from waste wood fiber.

COGENERATION STEAM GENERATING UNIT Steam generating unit that simultaneously produces both electrical (or mechanical) and thermal energy from the same primary energy source.

COGENERATION SYSTEM Power system which simultaneously produces both electrical (or mechanical) and thermal energy from the same energy source.

COHESION Mutual attraction by which the elements of a body are held together.

COHESIONLESS SOIL Soil or sand that, when unconfined, has little or no strength when air-dried and which has little or no cohesion when submerged.

COIL Helically coiled pipe through which hot, or cold water flows in a heat exchanger.

COKE Smokeless fuel made from coal by heating it in a closed oven until the gases have been removed.

COKE BURN-OFF Coke removed from the surface of the fluid catalytic cracking unit catalyst by combustion in the catalyst regenerator.

COKE BY-PRODUCT RECOVERY PLANT Facility designed and operated for the separation and recovery of coal tar derivatives (by-products) evolved from coal during the coking process of a coke oven battery.

COKE-TRAY AERATOR Aerator in which a flow is passed through or sprayed on coke-filled trays.

COL Saddleback-shaped pass between adjacent mountain peaks.

COLD FRONT Leading portion of a cold air mass moving against, and eventually replacing, a warm air mass.

COLD LOW Center of low atmospheric pressure caused by a strong vertical uplifting of air, in mid or low latitudes, accompanied by strong convection but little precipitation.

COLD SPRING Nonthermal spring in which the water has a temperature appreciably lower than ambient.

COLD TROUGH Projection of air at low pressure with cold air above it.

COLD-WORKED PIPE AND TUBE Cold forming operations that process unheated pipe and tube products using either water or oil solutions for cooling and lubri-cation.

COLI-AEROGENES BACTERIA *See* **Bacteria.**

COLIFORM GROUP BACTERIA *See* **Bacteria.**

COLIFORM COUNT Measure of the purity of water based on the number of presumptive coliform bacteria present in 100 milliliters of water.

COLLAPSIBLE SOIL Soil in which the voids ratio rapidly decreases on the introduction of water.

COLLAR Enlargement to the I.D. or O.D. of a pipe.

COLLAR BOSS Pipe fitting for a plumbing stack, with bosses that can be drilled out for future connections.

COLLECTING SYSTEM Drains and sewers between point(s) of influent and discharge to a main sewer.

COLLECTION Act of removing accumulated containerized solid waste from the generating source. There are several methods, including:

Alley: Picking up of solid waste from containers placed adjacent to an alley.

Carryout: Crew collection of solid waste from on on-premise storage area using a carrying container, carry-cloth, or a mechanical method.

Contract: Collection of solid waste carried out in accordance with a written agreement in which the rights and duties of the contractual parties are set forth.

Curb: Collection of solid waste from containers placed adjacent to a thoroughfare.

Franchise: Collection made by a private firm that is given exclusive right to collect for a fee paid by customers in a specific territory or from specific types of customers.

Municipal: Collection of solid waste by public employees and equipment under the supervision and direction of municipal authorities.

Private: Collection of solid waste by individuals or companies from residential, commercial, or industrial premises; the arrangements for the service are made directly between the owner or occupier of the premises and the collector.

Setout/setback: Removal of full and the return of empty containers between the one premise storage and the curb by a collection crew.

COLLECTION CENTER Site designed to accept secondary materials from homes, businesses, institutions, and industrial sites.

COLLECTION FREQUENCY Number of times refuse collection is provided in a given period of time.

COLLECTION LINE Plumbing drain.

COLLECTION METHOD Means by which refuse is collected from its point of origin. There are several methods, including:

Daily route: Each collection crew is assigned a weekly route that is divided into daily routes.

Definite working day: Collection proceeds along a route for a length of time adopted for a working day. The next day, collection begins where the crew stopped the day before. This procedure continues until the whole route is covered, whereupon the crew returns to the beginning of the route.

Group task: Responsibility for collecting on assigned routes is shared by more than one crew. Any crew that finishes a particular route works on another until all are completed.

Inter-route relief: Regular crews help collect on other routes when they finish their own.

Large route: Each crew is assigned a weekly route. The crew works each day without a fixed stopping point or work time, but it completes the route within the working week.

Reservoir route: Several crews are used to pick up on a centrally located route after having collected on peripheral routes.

Single load: Areas or routes are laid out that normally provide a full load of solid waste. Each crew usually has at least two such routes for a day's work. The crew quits for the day when the assigned number of routes is completed.

Swing crew: One or more reserve work crews go anywhere help is needed.

Variable-size crew: A variable number of collectors is provided for individual crews, depending on the amount and condition of work on particular routes.

COLLECTION STOP Stop made by a vehicle and crew to collect solid waste from one or more service sites.

COLLECTION SYSTEM Diverse system that collects and transmits all forms of sewage and wastewater, plus in some instances storm water, to a discharge point or treatment facility.

COLLECTOR Part of a solar system that intercepts the Sun's rays and converts them directly to a form of transportable energy.

COLLECTOR ANGLE Angle at which a solar collector must be tilted relative to the horizontal to maximize collection of solar radiation.

COLLECTOR EFFICIENCY Ratio of solar energy arriving at a collector to that collected by the device.

COLLECTOR SEWER Common lateral sewers within a publicly owned treatment system that are primarily installed to receive wastewaters directly from facilities which convey wastewater from individual systems, or from private property, and which include service 'Y' connections designed for connection with those facilities including: (a) crossover sewers connecting more than one property on one side of a major street, road, or highway to a lateral sewer on the other side when more cost effective than parallel sewers; and (b) pumping units and pressurized lines serving individual structures or groups of structures.

COLLECTOR SUBSYSTEM Assembly necessary for absorbing solar radiation, converting it into useful thermal energy, and transferring the thermal energy to a heat transfer fluid.

COLLECTOR TILT Angle of a solar collector assembly or the roof supporting it to the horizontal.

COLLOID Suspension of finely divided particles in a continuous medium producing a substance that is in a state of division, preventing passage through a semipermeable membrane.

COLLOIDAL DISPERSION Mixture resembling a true solution but containing one or more substances that are finely divided but large enough to prevent passage through a semipermeable membrane. It consists of particles which are larger than molecules, which settle out very slowly with time, which scatter a beam of light, and which are too small for resolution with an ordinary light microscope.

COLLOIDAL MATTER Finely divided solids that will not settle but which may be removed by coagulation or biochemical action, or membrane filtration.

COLLOIDAL PARTICLE Electrically charged particle dispersed in a second continuous medium.

COLLOIDORS Surfaces suspended vertically in a wastewater tank to aid in coagulating colloidal matter.

COLLUVIAL SOIL Soil deposited mainly through the action of landslides.

COLLUVIARIUM Maintenance and ventilation opening in an aqueduct.

COLOR Color in drinking water may be due to the presence of colored organic substances, metals such as iron, manganese and copper, or highly colored industrial wastes. Although the presence of color in drinking water is not directly linked to health, consumers may turn to alternative, possibly unsafe sources when their drinking water contains aesthetically displeasing levels of color. The aesthetic objective for color in drinking water is <15 true color units.

COLOR CODING Identification of similar parts of a system according to an applied color.

COLORIMETRY Spectroscopic measurement of color naturally present in samples, or developed by the addition of reagents.

COLOR TEMPERATURE For a particular source of light, the temperature in degrees Kelvin of a black body (an ideal radiator) that radiates the same wavelength(s). Typical 'standard' color temperatures are: incandescent (tungsten) lamp — 2854° K; direct noon sunlight — 4810° K; overcast sky — 6770° K.

COMBINATION FAUCET Device in which hot and cold water supplies are combined in varying proportions and discharged via a common spout.

COMBINATION MEDIUM Filter medium composed of two or more types, grades, or arrangements of filter media to provide properties that are not available in a single filter medium.

COMBINATION WASTE AND VENT Wastewater piping layout sized and configured so as to convey the flow of waste while simultaneously providing adequate protection of fixture traps against loss of seal.

COMBINATION WELL Well system of an open well connected to one or more wells or infiltration galleries.

COMBINED AVAILABLE CHLORINE Concentration of chlorine combined with ammonia as chloramine or other chloro derivatives, yet still available as an oxidant.

COMBINED AVAILABLE RESIDUAL CHLORINE That portion of the total residual chlorine remaining in water or wastewater following a specified contact period that will react chemically and biologically as chloramines or organic chloramines.

COMBINED CARBON Carbon that is chemically combined with oxygen.

COMBINED-CYCLE GAS TURBINE Stationary gas turbine which recovers heat from the gas turbine exhaust gases to heat water or generate steam.

COMBINED-CYCLE GENERATING STATION Electrical generating plant that produces power from both a gas turbine fuelled by natural gas or oil, plus a steam turbine supplied with steam generated by the exhaust gases from the gas turbine.

COMBINED-CYCLE SYSTEM System in which a separate source, such as a gas turbine, internal combustion engine, kiln, etc., provides exhaust gas to a heat recovery steam generating unit.

COMBINED DRAIN Drainage piping that carries both sanitary sewage and storm water. *See also* **Drain.**

COMBINED HEAT AND POWER Production of heat, often in the form of steam, and power, usually on the form of electricity.

COMBINED HYDROGEN Hydrogen that is chemically combined with oxygen.

COMBINED LOAD Sum of two or more different types of load (dead load, live load, wind load, etc.) occurring simultaneously.

COMBINED MOISTURE Moisture in combination with organic or inorganic matter.

COMBINED RESIDUAL CHLORINE Application of chlorine to water or wastewater in quantities sufficient to produce a combined available chlorine residual, which may consist of chlorine compounds formed by the reaction of chlorine with natural or added ammonia (NH_3) or with certain organic nitrogen compounds.

COMBINED SEWER Sewer intended to carry both sanitary sewage and stormwater, or industrial wastes and stormwater.

COMBINED STRESSES Stress more complex than simple tension, compression, or shear.

COMBINED SYSTEM Internal system of pipework in which soil water and waste water are carried in one soil pipe.

COMBINED WASTEWATER Mixture of domestic and industrial wastes, plus surface runoff.

COMBINED WATER Soil water held in chemical combination and which can only be driven off by heating, remaining after hygroscopic water evaporates.

COMBUSTIBLE GAS INDICATOR Device for measuring the concentration of potentially explosive fumes.

COMBUSTIBLES Materials that can be ignited at a specific temperature in the presence of air to release heat energy.

COMBUSTIBLE WASTE *See* **Solid waste.**

COMBUSTION 1. Burning of fuel in the presence of oxygen or other gas; **2.** Chemical change, especially oxidation, accompanied by the production of light and heat.

COMBUSTION AIR Air introduced into a combustion chamber to ensure complete combustion of a fuel.

COMFORT COOLING TOWERS Cooling towers that are dedicated exclusively to and are an integral part of heating, ventilation, and air conditioning or refrigeration systems.

COMMENCE As applied to construction of a major stationary source or major modification, means that the owner or operator has all necessary preconstruction approvals or permits and either has: (a) begun, or caused to begin, a continuous program of actual onsite construction of the source, to be completed within a reasonable time; or (b) entered into binding agreements or contractual obligations, which cannot be cancelled or modified without substantial loss to the owner or operator, to undertake a program of actual construction of the source to be completed within a reasonable time.

COMMENCEMENT OF WORK Date when work on a project begins on site and the contractor assumes responsibility.

COMMERCIAL AND INDUSTRIAL FRICTION MATERIAL Asbestos-containing product that is either molded or woven, intended for use as a friction material in braking and gear changing components in industrial and commercial machinery and consumer appliances.

COMMERCIAL APPLICATOR Certified applicator who uses or supervises the use of any pesticide which is classified for restricted use for any purpose or on any property other than as provided by the definition of private applicator.

COMMERCIALLY DRY SLUDGE Sludge containing a maximum 10% moisture by weight (5% if it is to be incorporated into a fertilizer product).

COMMERCIAL SOLID WASTE All types of solid wastes generated by stores, offices, restaurants, warehouses, and other non-manufacturing activities, excluding residential and industrial wastes and non-processing waste generated at industrial facilities such as office and packing wastes.

COMMERCIAL THINNING Silviculture treatment that thins out an overstocked stand by removing trees that are large enough to be sold as products, such as poles or fence posts (as opposed to juvenile spacing). It is carried out to improve the health and growth rate of the remaining crop trees.

COMMERCIAL USER Nonresidential user (of a sewer system) providing a service, or one connected with commerce, not otherwise classified as an industrial user.

COMMERCIAL WASTE *See* **Solid waste.**

COMMINUTED SOLIDS Solids that have been cut and sliced into fine particles.

COMMINUTING SCREEN Device for screening wastewater and cutting and slicing the screenings into particles sufficiently fine to pass through the screen openings.

COMMINUTION Process of cutting and slicing screened solids contained in a wastewater flow prior to its entering flow pumps or process units.

COMMINUTOR Device that reduces the size of solid material, typically installed at the influent of a waste treatment facility.

COMMISSIONING Procedures aimed at taking a completed but inanimate plant or project to the point where it can perform according to design and as specified.

COMMITMENTS Contractual agreement to buy or sell specified materials or services to named parties or their agents.

COMMON EXPOSURE ROUTE Likely way (oral, dermal, respiratory) by which a pesticide may reach and/or enter an organism.

COMMON NAME Designation or identification such as code name, code number, trade name, brand name, or generic chemical name used to identify a chemical substance other than by its chemical name.

COMMON SEWER Sewer to which subdrains from various properties are connected.

COMMON TRENCH Services such as gas, water, sewer, electrical, telephone, cable, etc., laid in a common trench and separated from each other by at least 150 mm (6 in.).

COMMON VENT Vertical vent pipe serving two fixture drains installed at the same level in a vertical stack.

COMMON-FACILITIES PLAN Written and graphic statement expressing the actual or desired pattern of public facilities within an area.

COMMUNITY Any naturally occurring group of organisms that occupy a common environment.

COMMUNITY WATER SYSTEM Public water system which serves at least 15 service connections used by year round residents or regularly serves at least 25 residents.

COMPACT Reduce in bulk; densify.

COMPACT COARSE SAND *See* **Soil types.**

COMPACTED Materials are considered compacted if their relative compaction exceeds 90%.

COMPACTED CUBIC YARD Measurement of material after it has been placed and compacted as fill, equivalent to 0.765 m^3.

COMPACTED EARTH Area of bare soil, made dense by artificial means or by pedestrian and vehicular traffic.

COMPACTED FILL Fill that is placed and compacted in layers under controlled conditions to achieve a uniform and dense soil mass capable of supporting a defined load. *See also* **Fill.**

COMPACTED VOLUME Volume of a material, such as soil, after it has been compacted a known percent.

COMPACT FINE SAND *See* **Soil types.**

COMPACTING FACTOR Ratio obtained by dividing the observed mass that fills a container of standard size and shape when allowed to fall into it under standard conditions of test, by the mass of fully compacted material that fills the same container.

COMPACTION 1. Increase in density from loose material to compacted material. *See also* **Consolidation**; **2.** Mechanically increasing the density of a material such as soil or waste material by reducing the voids between the particles of their composition. *Also called* **Forces of compaction.** There are several forces involved that can be applied singly or in combination, including:

Impact: Creation of a pressure wave from the force of a falling object hitting the surface to be compacted, such as from a tamper or impact hammer.

Manipulation: Process of densifying created by kneading, using sheepsfoot rollers and staggered-wheel rubber-tired rollers. This is especially effective at the surface of the lift of material to be compacted and helps to close the small, hairline cracks through which moisture can penetrate.

Static pressure: This force can be applied in two ways:

Preload - The application of a static load equal to a proportion of the load that will be applied following construction, created by the heaping of soil on the area of ground to be compacted prior to commencement of foundation work.

Rolling load - Application of weighted loads applied by rollers to produce shear stresses in soil or asphalt, causing individual particles to break their natural bonds and slide across each other to move into a more stable position within the material. *See also* **Nijboer quotient.**

Vibration: Increased compaction produced by a rapid succession of pressure waves which spread in all directions, breaking the bonds between the particles of the material which are reoriented into a denser state by the application of pressure.

COMPACTION EQUIPMENT Machinery and equipment used to densify soil and other materials, including steel-wheel static rollers, pneumatic-tire rollers, vibratory rollers, and combination rollers.

COMPACTION PIT TRANSFER SYSTEM Transfer system in which solid waste is compacted in a storage pit by a crawler tractor before being pushed into an open-top transfer trailer.

COMPACTION RATIO Initial volume of solid waste divided by the final volume attained after compaction.

COMPACTION SPECIFICATION Minimum performance standards for material placement, preparation and compaction established by an owner or sponsoring agency. There are several types, including:

Method only: Where the type of equip-

115

ment to be used, number of passes, roller speed, layer thickness and details such as moisture content are stated, but nothing about results.

Method and end result: Typically, this would call for achievement of 95% AASHTO with a minimum number of passes on a defined lift thickness using a named type of roller.

Suggested method and end result: This gives the experienced contractor the flexibility to employ initiative and experience, while providing the less experienced contractor with useful guidelines.

End result: This gives the contractor the choice to select the equipment and methods, which must produce results that meet the stated levels of named test procedures.

COMPACTION TARGETS Target densities established by a specifying agency, generally set on one of two basis:

Relative: Density measure usually based on a laboratory standard.

Absolute: Density measure based on the maximum theoretical measure of a voidless mix as determined by a recognized test method.

COMPACTIVE EFFORT Energy transferred into a material that presses the particles together, expels air and moisture from the mass, and fills the voids to make the material more dense.

COMPACT MATERIAL Granular soil with a relative compaction of 90% that can be dug or loosened with a pick.

COMPACTOR 1. Machine designed and used specifically to compact materials. It densifies material through the application of static force, or dynamic force combined with static force; **2.** Self-propelled or towed vehicle used to densify materials through the application of static force, or centrifugal force combined with static force; **3.** Equipment used to compact waste materials. There are several types, including:

Mobile: Vehicle with an enclosed body containing mechanical devices that convey and compress solid waste in the main compartment of the body.

Sanitary landfill: Vehicle equipped with a blade and wheels with load concentrators to provide compaction and a crushing effect.

Stationary: Machine that reduces the volume of solid waste by forcing and compressing it into a container.

COMPACTOR ATTACHMENT Attachment designed for backhoe/loaders and excavators that provides compaction and/or driving capability; may be attached directly to the stick or attached to the bucket.

COMPACTOR COLLECTION VEHICLE Vehicle with an enclosed body containing mechanical devices that convey solid waste into the main compartment of the body and compress it into a smaller volume of greater density.

COMPARTMENT 1. Forest management subdivision of a block of land, usually of continuous land ownership; **2.** Subdivision of an enclosed space; a chamber.

COMPATIBLE Ability of two or more substances to maintain their respective physical and chemical properties upon contact with one another.

COMPENSATION POINT 1. Light intensity, or concentration of carbon dioxide in the air surrounding an organic plant, at which the rates of respiration and photosynthesis in a plant are equal; **2.** Depth in a sea or lake below which plants use up more organic matter in respiration than they make during photosynthesis.

COMPENSATION WATER Water that must be allowed to pass a dam in satisfaction of water rights extant prior to construction of the dam.

COMPETENT Properly qualified to perform functions associated with the nature of the activity and the associated responsibility.

COMPETENT BED Layer of rock that is

relatively strong and which flexes rather than flows during folding.

COMPETENT PERSON 1. One who has been proven through examination or demonstrated experience and ability to have a broad knowledge of a subject or skill; **2.** One who is capable of identifying existing and predictable hazards in the surroundings or working conditions that are unsanitary, hazardous, or dangerous to employees, and who has authorization to take prompt corrective measures to eliminate them. *See also* **Authorized person** and **Designated person.**

COMPETITION Conflict that exists when two or more species have requirements that exceed the available supply.

COMPETITIVE BID CONTRACT Contract awarded as a result of competitive bidding.

COMPETITIVE BIDDING/TENDERING Process in which several independent, qualified contractors or suppliers are invited to bid on a contract.

COMPLETE DIVERSION Taking water from one location in a natural drainage area and discharging it into another drainage area.

COMPLETED OPERATIONS INSURANCE Indemnity against accidents that occur following completion of a project that has been turned over to the owner, or abandoned.

COMPLETED VALUE INSURANCE Policy written at the start of a project in an amount, usually derived from the contract sum, less specified exclusions, and adjusted to the final insurable cost on completion.

COMPLETELY-CLOSED DRAIN SYSTEM Individual drain system that is not open to the atmosphere and is equipped and operated with a closed vent system and control device.

COMPLETE TREATMENT Processing of defined wastewater to the maximum state of purification required by applicable legislation.

COMPLETE TREE Every component of a tree from leaves or needles to root hairs.

COMPLETE TREE HARVESTING Harvesting of a complete tree, including the roots.

COMPLETE WASTE TREATMENT SYSTEM All the treatment works involving: (a) the transport of wastewater from individual homes or buildings to a plant or facility where treatment of the wastewater is accomplished; (b) the treatment of the wastewater to remove pollutants; and (c) the ultimate disposal, including recycling or reuse, of the treated wastewater and residues which result from the treatment process.

COMPLETION End of work on a contract or subcontract, or for a trade or activity.

COMPLETION BOND *See* **Bond.**

COMPLETION DATE Date upon which the work described in the contract documents must be completed.

COMPLETION OF PROJECT *See* **Project completion.**

COMPLEXATION Inactivation of an ion by the addition of a reagent that combines with it, having the effect of preventing it from participating in other reactions.

COMPLEX COVER Specified combination of soil, crops or vegetation, and tillage.

COMPLEXES Compounds formed by the union of two or more simple salts.

COMPLIANCE Compliance with clean air standards or clean water standards, or with a schedule or plan ordered or approved by a court of competent jurisdiction, or an air or water pollution control agency.

COMPLIANCE CYCLE Nine-year calendar year cycle during which public water systems must be monitored. Each compliance cycle consists of three three-year compliance periods. The first calendar year cycle begins January 1, 1993 and ends December 31, 2001; the second begins January 1, 2002 and ends December 31, 2010; the third begins January 1, 2011 and ends December 31, 2019.

COMPLIANCE SCHEDULE Date or dates by which a source or category of sources is required to comply with specific emission limitations contained in an implementation plan and with any increments of progress toward such compliance.

COMPONENT Manufactured product that serves a specific function.

COMPONENT TIDE Each of the simple tides into which the tide of nature is resolved: principal lunar, principal solar, N_2, K, and O.

COMPOSITE LINER Landfill liner that includes both a synthetic liner and compacted clay components.

COMPOSITE ROCK-FILL DAM Dam consisting of rock fill on the downstream side and an earth fill on the upstream side.

COMPOSITE SAMPLE 1. Sample obtained by blending two or more individual samples of a material; **2.** Combination of individual samples obtained at regular intervals over a specified period. *See also* **Composite sample quality.** *Also called* **Composite wastewater sample.**

COMPOSITE SAMPLE QUALITY Concentration of some parameter obtained and/or tested in a composite sample. *See also* **Composite sample.**

COMPOSITE UNIT HYDROGRAPH Presentation of unit hydrographs for the significant subdivisions of a large area, with the times of beginning of rise appropriately lagged by the times of travel from the subarea outlets to the gauging station.

COMPOSITE WASTEWATER SAMPLE *See* **Composite sample.**

COMPOST Mixture of decayed vegetable and animal matter, such as leaves or manure, used to fertilize and condition soil.

COMPOSTING Process of controlled decay which enables aerobic bacteria and other microorganisms to decompose organic mat-

ter to produce a stable end product.

COMPOUND Mixture of materials that are combined to give desired properties when used in the manufacture of a product. *See also* **Mixture.**

COMPOUND ALLUVIAL FAN *See* **Piedmont alluvial plain.**

COMPOUND DREDGER Dredge equipped with a bucket ladder and a means to scour or excavate bottom material.

COMPOUND HYDROGRAPH Hydrograph of an intermittent storm.

COMPOUND INGREDIENT Material added to another to form a mix.

COMPOUND MATERIAL Substance composed of two or more separately identifiable materials.

COMPOUND METER Type of water meter consisting of two meters of different capacities and a regulating valve to automatically divert all or part of the flow from one meter to the other, depending on the flow rate.

COMPOUND PIPE Pipe made up of lengths of different diameter pipe in series.

COMPREHENSIVE GENERAL LIABILITY Insurance policy that automatically includes all forms of general (as against specific) liability.

COMPREHENSIVE MATERIAL DAMAGE INSURANCE Indemnity against loss or damage from such causes as fire, flood, theft, vandalism, windstorm, etc.

COMPRESSED AIR Air compressed to a pressure greater than one atmosphere. *See also* **Free air,** and **Standard air.**

COMPRESSED GAS Any contained mixture or material with either an absolute pressure exceeding 275 kPa (40 psi) at 21°C (69.8°F) or an absolute pressure exceeding 717 kPa (104 psi) at 54°C (130°F), or both.

COMPRESSIBILITY 1. Ratio of the percentage change in the volume of a material or fluid to a percentage change in pressure;

2. Soil property that allows it to deform under load.

COMPRESSIBILITY EFFECT Change of density of a gas under conditions of compressible flow.

COMPRESSION 1. Force that tends to densify a granular material and shorten a structural member; **2.** Compacting effect of the weight of steel-wheel rollers, measured at the bottom of the roll in kg per mm (lbs per linear inch) of roll width.

COMPRESSIONAL WAVE Seismic wave whose motion is compression–dilation, or push–pull, generated by rock's resistance to compression.

COMPRESSION COUPLING Pipe fitting used to connect sections of hubless pipe and which relies for its integrity upon the compression of a gasket.

COMPRESSION FITTING See **Fitting.**

COMPRESSION FORCE See **Force.**

COMPRESSION ROLL Drive wheel of a steel-wheel roller. *Also called* **Drive roll.**

COMPRESSIVE STRESS Stress that resists a force attempting to crush a body.

COMPRESSION-TYPE HYDRANT Hydrant that opens against the flow of water by movement of the operating stem and in which the main valve is kept closed by water pressure.

COMPRESSOR Machine for densifying air or gas from an initial intake pressure to a higher storage or discharge pressure. There are several types, including:

> **Axial flow:** Compressor that uses the rotation of a propeller-type blade to move the gas in the axial direction of the shaft.

> **Centrifugal:** Compressor that uses the rotation of an impeller and a shaft to push the gas outward.

> **Dynamic:** See **Rotary compressor** (below).

> **Kinetic:** See **Rotary compressor** (below).

> **Multistage:** Compressor that raises a gas to the desired pressure in several steps through a sequence of chambers.

> **Positive displacement:** There are two types:

>> **Reciprocating:** Compressor that employs a piston and cylinder to compress a gas.

>> **Rotary:** Compressor that uses a rotating vane or screw to move and compress a gas. *Also called* **Dynamic compressor, Kinetic compressor,** and **Screw compressor.**

> **Screw:** See **Rotary compressor** (above).

COMPUTER-AIDED DESIGN Use of computer-generated solutions to assist in the determination of design problems.

COMPUTER-AIDED DESIGN AND DRAFTING Use of a computer to generate solutions to assist in the determination of design problems and to drive a drafting machine.

CONCAVE BANK Stream bank having its center of curvature toward the channel.

CONCENTRATE AND CONTAIN Waste management technique that relies on the wastes being concentrated and contained and kept separate from other environmental media.

CONCENTRATED FALL Fall used in a hydroelectric project which is concentrated at one point on the stream.

CONCENTRATE STREAM Stream increasing in salt concentration in an electrodialysis process.

CONCENTRATION 1. Amount of a given substance dissolved in a unit volume of solution; **2.** Process of increasing the dissolved solids per unit volume of a solution.

CONCENTRATION FACTOR

CONCENTRATION FACTOR Ratio of
the average flow of a stream during that
part of the day when pondage or temporarily stored streamflow is being drawn upon
to the 24-hour average flow.

CONCENTRATION OF A SOLUTION
Amount of solute in a given amount of solvent or solution expressed as a
weight:weight or weight:volume relationship. The conversion from a weight relationship to one of volume incorporates density as a factor. For dilute aqueous solutions, the density of the solvent is approximately equal to the density of the solution;
thus, concentrations in mg/L are approximately equal to $10^{-6}g/10^3g$ or parts per million (ppm); ones g/L are approximately
equal to $10^{-6}g/10^3g$ or parts per billion (ppb).
In addition, concentration can be expressed
in terms of molarity, normality, molality,
and mole fraction. For example, to convert
from weight/volume to molarity one incorporates molecular mass as a factor.

CONCENTRATION POLARIZATION
1. Buildup of retained particles on a membrane surface due to dewatering of the feed
closest to the membrane; the thickness of
the concentration polarization layer is controlled by the flow velocity across the membrane; **2.** Technique used in corrosion studies to indicate a depletion of ions near an
electrode; **3.** The basis for chemical analysis by a polarograph.

CONCENTRATION TANK Short detention settling tank in which sludge is concentrated by sedimentation or flotation prior
to treatment, dewatering, or disposal.

CONCENTRATION TIME 1. Time necessary for storm runoff to flow from the
most remote point of a catchment or drainage area to the outlet or a specific point; **2.**
Point in time when the rate of runoff is equal
to the rate of rainfall of uniform intensity.

CONCENTRATION *vs* TIME STUDY
Results in a graph which plots the measured
concentration of a given compound in a
solution as a function of elapsed time. Usually, it provides a more reliable determination of equilibrium water solubility of hy-
drophobic compounds than can be obtained
by single measurements of separate samples.

CONCENTRATOR Solids contact unit
that decreases the water content of a sludge
or slurry.

CONCRETE Mixture of crushed stone or
gravel, sand, cement and water that hardens
as it cures and dries.

CONCRETION Solid mass of matter
formed by the cohesion or coalescence of
its constituent particles.

CONCURRENT Period in which construction of an emission control device serving
an affected facility is commenced or completed, beginning 6 months prior to and ending 2 years after the facility is completed.

CONCURRENT ENGINEERING Ap-
proach to project staffing that, in its most
general form, calls for implementors to be
involved in the design phase, and which is
sometimes confused with fast tracking.

CONDENSATE 1. Hydrocarbon liquid
separated from natural gas which condenses
due to changes in the temperature and/or
pressure and remains liquid at standard conditions; **2.** Condensed steam.

CONDENSATION Physical process by
which a liquid is removed from a vapor or
vapor mixture.

CONDENSATION LEVEL Altitude to
which a mass of unsaturated air must as-
cend before it becomes saturated.

CONDENSATION NUCLEUS Solid particle in the atmosphere on to which water
vapor condenses when the air is saturated,
so forming a liquid droplet.

CONDENSATION SAMPLING Gas-sam-
pling technique in which a gas is passed
through tubes immersed in a refrigerant, thus
trapping and condensing various fractions
of the gas mixture.

CONDENSATION TRAIL Visible trail of
cloud formed by the mixing of hot aircraft
engine exhaust with the surrounding cold
air.

CONDENSER 1. Machine that compresses
air; **2.** Apparatus for storing or intensifying
an electric charge; **3.** Heat-transfer device

that reduces a thermodynamic fluid from its vapor phase to its liquid phase.

CONDITIONAL INSTABILITY Instability of cold air that comes into contact with a warmer land mass, causing the warmed air to rise convectively.

CONDITIONAL SUPPLY Type of water supply agreement that is conditional upon some limitation, circumstance, etc.

CONDITIONS OF CONTRACT Fundamental terms that collectively describe the rights, obligations, procedures, and recourses of contracting parties.

CONDUCTANCE Measure of the conducting power of a solution, equal to the reciprocal of its resistance.

CONDUCTIVITY Measure of the ability of a material to conduct an electrical charge.

CONDUCTIVITY BRIDGE Means of measuring conductivity whereby a conductivity cell forms one arm of a Wheatstone bridge, a standard fixed resistance forms another arm, and a calibrated slide wire resistance with end coils provides the remaining two arms.

CONDUCTOR PIPE Round, square, or rectangular pipe used to lead water from a roof to the sewer.

CONDUIT Channel or pipe for conveying water or other fluids.

CONE Surface generated by a straight line passing through a fixed point and moving along the intersection with a fixed curve.

CONE OF DEPRESSION Area of soil around an underground suction point depressed below normal elevation and dried out.

CONE OF INFLUENCE Roughly conical depression produced in a water table or other piezometric surface by the extraction of water from a well at a given rate.

CONE-SHEET Funnel-shaped dike, usually surrounding an igneous intrusion.

CONE VALVE *See* **Valve.**

CONFIDENCE LIMITS Limits within which, at some specified level of probability, the true value of a result lies.

CONFINED AQUIFER Aquifer in which groundwater is confined under pressure that is significantly greater than atmospheric pressure.

CONFINED GROUNDWATER *See* **Groundwater.**

CONFINED SPACE Space which, by design, has limited openings for entry and exit and unfavorable natural ventilation which could contain or produce dangerous air contaminants, and which is not intended for continuous employee occupancy.

CONFINED-WATER WELL Well having confined groundwater as its sole source of supply.

CONFINING BED Geologic formation or stratum beneath, above, and surrounding an aquifer, which causes a hydrostatic head to be created or retained in the basin.

CONFINING STRATUM Impervious stratum or confining layer directly above or below one bearing water.

CONFINING ZONE Geological formation, group of formations, or part of a formation that is capable of limiting fluid movement above an injection zone.

CONFLUENCE Place where two or more rivers, streams, etc., come together.

CONFLUENT Flowing together; blending into one.

CONFLUENT GROWTH Continuous bacterial growth covering the entire filtration area of a membrane filter, or a portion thereof, in which bacterial colonies are not discrete.

CONFLUENT STREAM Stream that melds or unites with another.

CONFORMITY MARK Tag, stamp, label or other mark put on a product by its manufacturer to indicate its certification status.

CONGENER Any one particular member of a class of chemical substances. A specific congener is denoted by unique chemical structure, for example 2,3,7,8-tetrachlorodibenzofuran.

CONING 1. Behavior of a plume when it expands to form a roughly conical shape; **2.** Development of a cone-shaped depression in the surface, or upper layer of a fluid from which a continuous volume is being withdrawn at a lower level.

CONJUGATE DEPTHS Depths preceding and following a hydraulic jump.

CONNATE WATER Water entrapped in the interstices of rock when the material was deposited; water that neither escapes or is renewed.

CONNECTED PIPING All piping including valves, elbows, joints, flanges, and flexible connectors attached to a system through which substances flow.

CONNECTION BAND Coupling that fits over adjacent pipe ends and which, when drawn tight, holds the pipe together either by friction or mechanical bond.

CONNECTOR Flanged, screwed, welded, or other joined fittings used to connect two pipelines or a pipeline and a piece of equipment.

CONSECUTIVE DIGESTION Thermophilic digestion followed immediately by digestion and concentration under mesophilic conditions.

CONSENT OF SURETY Written consent of the surety on a performance bond and/or labor and material payment bond to contract changes such as change orders, reduction of retainage, etc.

CONSEQUENT Applied to rivers whose course is determined by the existing slope and pattern of the landscape.

CONSEQUENTIAL DAMAGES Loss of value of a parcel, no portion of which is acquired, resulting from highway improvement.

CONSEQUENTIAL LOSS Loss not directly caused, but that may arise from prior damage.

CONSEQUENT STREAM Stream course that has been controlled, in direction and location, by surface topography.

CONSERVANCY SYSTEM Disposal of human wastes using earth closets and privies.

CONSERVATION Protection, improvement, and wise use of natural resources according to principles that will assure utilization of the resources to obtain the highest economic and/or social benefits.

CONSERVATION OF ENERGY *See* **Energy conservation.**

CONSERVATION OF LAND Dedication of areas of land for the preservation of defined human activity and/or species of animals.

CONSERVATION STORAGE Storage of water for some future useful purpose.

CONSIGNEE Ultimate treatment, storage or disposal facility to which the hazardous waste will be sent.

CONSOCIATION Climax community of plants dominated by one species.

CONSOLE Any frame containing an array of control devices .

CONSOLIDATION Natural processes that result in the gradual compression of cohesive soils of low permeability, such as clay, due to the imposition of weight, which drives the water and air out of the voids in the soil. *See also* **Compaction.**

CONSOLIDATION SEDIMENTATION Process in which loose, soft, or liquid earth materials become firm and coherent.

CONSOLIDATION SETTLEMENT Settlement of loaded soil that occurs over a period of years due to dissipation of excess pore pressure as water, or air and water are expelled from the voids in the soil.

CONSOLIDATION TEST Method of determining the coefficient of permeability of soils.

CONSTANT SPRING Spring in which variation in discharge from maximum to minimum does not exceed one third of its average discharge.

CONSTANCY Degree of frequency with which a particular species occurs in different stands or samples of the same association.

CONSTRICTION Obstruction that confines a flow.

CONSTRUCTION 1. Placement, assembly, or installation of facilities or equipment (including contractual obligations to purchase such facilities or equipment) at the premises where such equipment will be used, including preparation work at such premises; **2.** Erection, building, acquisition, alteration, remodelling, modification, improvement, or extension of any facility; **3.** Erection or building of new structures, or the replacement, expansion, remodelling, alteration, modernization, or extension of existing structures including engineering and architectural surveys, designs, plans, working drawings, specifications, and other actions necessary to complete the project; **4.** Remedial actions in response to a release, or a threat of a release, of a hazardous substance into the environment; **5.** Preliminary planning to determine the feasibility of treatment works, engineering, architectural, legal, fiscal, or economic investigations or studies, surveys, designs, plans, working drawings, specifications, procedures, field testing of innovative or alternative wastewater treatment processes and techniques or other necessary actions, erection, building, acquisition, alteration, remodelling, improvement, or extension of treatment works.

CONSTRUCTION AND DEMOLITION WASTES Waste building materials and rubble resulting from construction, remodelling, repair, and demolition operations on houses, commercial buildings, pavements, and other structures.

CONSTRUCTION BUDGET 1. Stipulated highest acceptable bid for a specified construction; **2.** Sum established by the owner as available for construction of the described project.

CONSTRUCTION CONTRACT Written agreement between an owner and a contractor, referencing all contract documents, calling for construction of the described and illustrated project.

CONSTRUCTION COST Any of the cost types (appropriation, commitment, expenditure, or estimate to completion) associated with the scope of the work.

CONSTRUCTION CONTRACTOR Corporation or individual who has entered into a contract to perform construction work.

CONSTRUCTION DEFECT Anything about a construction project that is at variance to the contract documents.

CONSTRUCTION DOCUMENTS *See* **Contract document.** *See also* **Working drawings.**

CONSTRUCTION DOCUMENTS PHASE Third phase of the architects basic services in which working drawings and specifications are prepared from the approved design development documents.

CONSTRUCTION ESTIMATOR Person who estimates the costs of materials and labor necessary to complete a project.

CONSTRUCTION INSPECTION *See* **Site inspection.**

CONSTRUCTION JOINT a. Rigid, immovable joint creating a single structural unit from two or more individual parts or materials, or between stages of construction, not necessarily intended to accommodate movement; **b.** Surface where two successive placements of concrete meet; **3.** Temporary joint employed when the placing of concrete must be interrupted for any reason.

CONSTRUCTION LOAD Load to which a permanent or temporary structure is subjected during construction.

CONSTRUCTION LOAN Loan to the builder sufficient to cover construction prior to either permanent financing or staged payment by the owner under the terms of the contract.

CONSTRUCTION MANAGEMENT Project delivery system directed by a construction manager, a member of the owner–architect/engineer–construction manager team, whose responsibility is to clarify time and cost consequences of design decisions and their construction practicability and to manage the bidding, award, and construction phases of the project.

CONSTRUCTION MANAGEMENT FIRM Corporation that performs the duties of the construction manager and his staff. Especially on large construction projects, such firms take over the role of a general contractor by providing day-to-day supervision and direction of activities by general and subtrade construction contractors.

CONSTRUCTION MANAGER Title of a position on the project team of a person responsible to the project manager for directing the construction of a project within authority and responsibility limits established by the project manager. On smaller projects these duties may be the responsibility of the project manager.

CONSTRUCTION PHASE That part of a project life during which working drawings are prepared, contracts are tendered and awarded, and the construction work is carried out. *Also called* **Construction process.** *See also* **Feasibility phase.**

CONSTRUCTION PROCESS *See* **Construction phase.**

CONSTRUCTION PROGRAM Plan that sets out the sequence in which construction operations will be carried out, materials delivered to site, strength of labor force at any point in time, expenditures, cash flow, etc.

CONSTRUCTION SCHEDULE Sequence of events that will see the completion of a construction program. *Also called* **Schedule of work,** and **Work schedule.**

CONSTRUCTION SITE *See* **Job site.**

CONSTRUCTION STAGE PLANNING Process of determining the principal stages necessary to complete a construction project, sequencing them, and ensuring that there is adequate provision for materials and labor

and that there is no conflict in operations between one stage and another.

CONSTRUCTION STAKEOUT *See* **Stakeout.**

CONSTRUCTION START Point in a project when new construction, alterations, renovations, or demolition begin, following such work as site surveying.

CONSTRUCTION STATUS REPORT *See* **Progress report.**

CONSTRUCTION TIME Time between commencement of site work and handing over of a completed project.

CONSTRUCTION TIME OVERRUN *See* **Time overrun,** and **Overrun.**

CONSTRUCTION WARRANTY *See* **Warranty.**

CONSTRUCTION WORK Work that contributes to a physical structure (as distinct to decoration, that contributes to the appearance of a structure).

CONSULTANT One who offers expert and specialized knowledge and opinion.

CONSULTING ENGINEER Professional engineer who offers expert opinion but who has subordinate responsibility to the prime consultant or contractor.

CONSUMER WASTE Materials that have been used and discarded by the buyer, or consumer, as opposed to waste discarded in-house during the manufacturing process.

CONSUMPTION RESIDUES Wastes that result from the final consumption of goods or services.

CONSUMPTIVE USE 1. Use of water that depletes an available supply; **2.** Volume of water lost by transpiration and evaporation from fields under irrigation.

CONSUMPTIVE WASTE Water that returns to the atmosphere without benefiting man.

CONTACT AERATOR Biological waste treatment process in which air, usually in the form of bubbles, is introduced under pressure into the liquid waste in a specially shaped tank.

CONTACT BED In wastewater treatment, a bed of broken rock over which the waste flow is distributed so as to come in intimate contact with aerobic bacteria.

CONTACT FILTER Filter for the partial removal of turbidity prior to final filtration.

CONTACT STABILIZATION Modification of the activated sludge process in which raw wastewater is mixed with a high concentration of activated sludge and aerated for a short period to obtain BOD removal by absorption. The solids are subsequently removed by sedimentation and transferred to a stabilization tank where further aeration oxidizes and conditions them prior to their reintroduction to the incoming wastewater flow.

CONTACT TANK Tank in which contact between chemicals and other materials and the body of liquid to be treated is promoted.

CONTAINER Nonspecific term for a receptacle capable of closure.

CONTAMINANT Material that, by reason of its action upon, within, or to a person, is likely to cause physical harm. *See also* **Artificial contaminant, Built-in contaminant,** and **Generated contaminant.**

CONTAMINATE Introduction of a substance that would cause: (a) the concentration of that substance in the groundwater to exceed the maximum allowed contaminant level; or (b) an increase in the concentration of that substance in the ground water where the existing concentration of that substance exceeds the allowable maximum contaminant level.

CONTAMINATED LAND Land contaminated through uses during which little or no means was in effect to contain pollutants from entering the soil.

CONTAMINATED NONPROCESS WASTEWATER Water, including precipitation runoff which, during manufacturing or processing, comes into incidental contact with any raw material, intermediate

product, finished product, by-product or waste product by means of: (a) precipitation runoff; (b) accidental spills; (c) accidental leaks caused by the failure of process equipment; and (d) discharges from safety showers and related personal safety equipment, and from equipment washings for the purpose of safe entry, inspection and maintenance.

CONTAMINATED RUNOFF Runoff that comes into contact with any raw material, intermediate product, finished product, by-product or waste product.

CONTAMINATION Addition of something that diminishes the original quality of a resource.

CONTAMINATION LEVEL Quantitative term specifying the degree of contamination.

CONTEMPLATED CHANGE NOTICE (CCN) Form issued to a construction contractor as a means for obtaining his quotation on the price change associated with a contemplated change in the work included in his contract.

CONTENTS HAZARD CLASSIFICATION Classification of the potential danger of building contents as ordinary, high, or low.

CONTINENTAL CLIMATE Climate characteristic that can be linked to one or more features associated with a large land mass.

CONTINENTAL DRIFT Phenomenon of the movement of continental land masses as part of seafloor spreading and plate tectonics.

CONTINENTAL RISE Gentle slope connecting the bottom of the continental slope with the abyssal plains.

CONTINENTAL SHELF Comparatively shallow sea area surrounding a continental mass, extending seaward to the continental slope.

CONTINENTAL SLOPE Portion of the continental margin that begins at the outer edge of the continental shelf and descends into the ocean deeps.

CONTINGENCY Component of the authorized appropriation of estimated final cost types for the scope of work associated with a cost class. The contingency is an estimated allowance for the cost of unknowns or changes. The anticipated award price of a cost class contains contingencies for escalation and estimating error. It is added to the estimated award price for further contingencies for design changes and contractor claims. NOTE: The reserve for scope changes is not a contingency in this same sense; rather it is an allowance that is transferred into specific cost classes when the scope of work in the class is amended. The authorization appropriation for that class is correspondingly amended following the transfer. *Also called* **Contingency allowance.** *See also* **Reserve,** and **Contingency planning.**

CONTINGENCY ALLOWANCE *See* **Contingency,** and **Reserve.**

CONTINGENCY FUND Cash or investments set aside or reserved for unforeseen expenditures.

CONTINGENCY PLAN Document setting out an organized, planned, and coordinated course of action to be followed in case of a fire, explosion, or release of hazardous waste or hazardous waste constituents which could threaten human health or the environment.

CONTINGENCY PLANNING Development of a management plan that identifies alternative strategies to be used to ensure project success if specified risk events occur.

CONTINGENCY RESERVE Separately planned quantity used to allow for future situations which may be planned for only in part (sometimes called 'known unknowns') — i.e. rework is certain, the amount of rework is not.

CONTINGENT AGREEMENT Agreement that relies upon some future condition or circumstance in order to remain valid, i.e. an agreement to complete work being contingent on the owner successfully securing financing.

CONTIGUOUS ZONE Zone of the high seas, established under Article 24 of the Convention on the Territorial Sea and Contiguous Zone, which is contiguous to the territorial sea and which extends a specified distance seaward from the outer limit of the territorial sea.

CONTINUING ENVIRONMENTAL PROGRAMS Pollution control programs which will not be completed within a definable time period.

CONTINUITY EQUATION Statement that the mass rate of fluid flow into a fixed space is equal to the mass flow rate out. Hence, the mass flow rate of fluid past all cross sections of a conduit is equal.

CONTINUOUS DISCHARGE Discharge that occurs without interruption throughout the operating hours of a facility, except for infrequent shutdowns for maintenance, process changes, or other similar activities.

CONTINUOUS DISPOSAL Method of tailings management and disposal in which tailings are dewatered by mechanical methods immediately after generation. The dried tailings are then placed in trenches or other disposal areas and immediately covered to limit emissions.

CONTINUOUS EMISSION MONITORING SYSTEM Monitoring system for continuously measuring the emissions of a pollutant from an affected facility.

CONTINUOUS-FEED INCINERATOR *See* **Incinerator.**

CONTINUOUS IRRIGATION System whereby each irrigator receives the allotted quantity of water at a continuous rate during the irrigation season.

CONTINUOUS-FLOW PUMP *See* **Pump.**

CONTINUOUS-FLOW TANK Tank, through which liquid flows continuously at its normal flow rate (as distinct from a fill-and-draw or batch tank).

CONTINUOUS FOREST INVENTORY Timber sampling system that provides for periodic remeasurement of specific stands or plots of individual trees to show status and periodic change over time for the forest as a whole and major subdivisions therein.

CONTINUOUS FURNACE Furnace operated on an uninterrupted cycle, in which the charge is being constantly added to, moved through, and removed from the furnace, as opposed to a batch-type furnace.

CONTINUOUS HORSEPOWER *See* **Horsepower.**

CONTINUOUS INTERSTICE Interstice in granular material that is connected with other interstices.

CONTINUOUSLY REGENERATING TRAP Device that removes particulates from the exhaust of diesel engines and which uses nitrogen dioxide (NO_2) to oxidize the particulates.

CONTINUOUS MECHANICAL VENTILATION Mechanical ventilation with continuously-operating, slow-running fans: a ducted extract system.

CONTINUOUS OPERATIONS Where an industrial user introduces regulated wastewaters to a treatment facility throughout its operating hours, except for infrequent shutdowns for maintenance, process changes, or other similar activities.

CONTINUOUS PROCESS Treatment process in which the material to be treated flows into the treatment tank at essentially the same rate as it overflows or is withdrawn.

CONTINUOUS RECORDER Data-recording device recording an instantaneous data value at least once every 15 minutes.

CONTINUOUS SAMPLING Sampling without interruptions throughout an operation or for a predetermined time.

CONTINUOUS SLUDGE-REMOVAL TANK Sedimentation tank equipped for the continuous removal of sludge at a controlled rate.

CONTINUOUS STREAM Stream that is uninterrupted in its course.

CONTINUOUS VENT Upward continuation of a drain to produce a vent.

CONTINUOUS WASTE Two or more fixtures connected to the same trap.

CONTINUOUS WASTE AND VENT Open vertical continuation to atmosphere of a waste collection stack connected to a drain.

CONTOUR 1. Line that defines or bounds something; **2.** Line on a topographical map that connects all points of identical elevation.

CONTOUR BASIN Basin formed by borders or levees that follow a contour.

CONTOUR CHECK Areas in an irrigated section of land separated by borders that are sited along contours.

CONTOUR FARMING Cultivating the surface of the land along lines parallel to contours, rather than directly or randomly across them, thus increasing the capacity of the soil to retain water.

CONTOUR FLYING Technique of flying so that the aircraft, and the geophysical instruments it is carrying, is at a constant height above the ground.

CONTOUR GAUGE Series of slender metal or plastic rods about 100 mm (4 in.) long, held parallel in, and free to slide through a narrow metal frame about 150 mm (6 in.) wide. Used to create a template of an irregular surface such as a moulding.

CONTOUR GRADIENT Zero slope line.

CONTOUR INTERVAL Vertical distance represented by two consecutive contour lines. This interval is normally constant throughout a contour map.

CONTOUR IRRIGATION Irrigation through channels formed by ridges that follow contours.

CONTOUR LINE 1. Imaginary line on the surface of the ground, every part of which is at the same elevation; **2.** Line on a drawing that identifies points of common elevation.

CONTOUR MAP Topographic map that portrays relief by means of contour lines.

CONTOUR SCALING Crust forming across the surface of sandstones and limestones that follows the contour of the surface rather than the bedding planes of the stones. The result of direct pollution, the pores of the stone are blocked by formations of recrystalized calcium sulfates.

CONTRACT Agreement in law; a commitment document. The term contract is used to refer to the document itself. Contracts are often classified and described by the terms of payment they contain: stipulated price contract, lump sum contract, unit price contract, cost-plus contract, etc. *See also* **Contract item,** and **Contract time.**

CONTRACT ADMINISTRATION Responsibilities of the owner's designated agent, architect, and engineer during the construction phase of a project.

CONTRACT AMENDMENT *See* **Contract change.**

CONTRACT BOND *See* **Bond.**

CONTRACT CHANGE Alteration to a completed and signed contract that must be agreed to and acknowledged in writing by all parties represented in the original contract. *Also called* **Contract amendment.**

CONTRACT CLOSEOUT Completion and settlement of the contract, including resolution of all outstanding items.

CONTRACT COLLECTION *See* **Collection.**

CONTRACT DATE Date that a legally binding contract between two or more parties is signed by all parties, not withstanding any other date contained within the text of the contract. *Also called* **Date of agreement.**

CONTRACT DOCUMENT(S) 1. Document forming part of a contract; **2.** Documents that define the responsibilities of the parties involved in bidding, purchasing, supplying, and installing and/or erecting components and materials necessary for completion of a project. *Also called* **Construction documents.**

CONTRACT DOCUMENT(S) STAGE Stage in a project where planning has developed to the point where documents (working drawings, specifications, etc.) have been completed and bids for completion of the work can be called.

CONTRACTED-OPENING DISCHARGE MEASUREMENT Determination by indirect measurement of peak discharge following a flood through observation of high-water marks and channel and bridge geometry at a bridge constriction.

CONTRACTED WEIR Weir that is shorter that the width of the channel whose flow it measures.

CONTRACTING OFFICER Person designated by the owner as his official representative with specific responsibilities and authority with regard to the project.

CONTRACTING OUT Subcontracting to another for work, material, or supplies.

CONTRACTING REACH Reach of a channel in which flow is accelerating.

CONTRACTION 1. Extent to which the cross-sectional area of a jet, nappe or stream is decreased after passing an orifice, weir or notch; **2.** Reduction in cross-sectional area of a conduit along its longitudinal axis. *See also* **Constriction.**

CONTRACT ITEM Specific unit of work for which a price is provided in a contract. *See also* **Contract.**

CONTRACT LOGGING Operator doing all or part of the logging for a company.

CONTRACTOR One who agrees to supply materials and/or perform certain types of work, to an agreed schedule, and to defined standards, for a specified sum of money. *See also* **Subcontractor,** and **Superintendent.**

CONTRACTOR DEFAULT Failure by a contractor to perform or otherwise comply with the terms of the construction contract.

CONTRACTOR'S AFFIDAVIT Notarized statement by a contractor relating to payment of debts and claims, release of liens, etc. that require specific procedures for the protection of the owner.

CONTRACTOR'S EQUIPMENT FLOATER Insurance against loss or damage to a contractor's tools and equipment that are customarily used away from the contractor's premises.

CONTRACTOR'S LIABILITY INSURANCE Indemnity of a contractor against specified claims that may arise as a result of his, or his subcontractor's work.

CONTRACTOR'S OPTION Provision of a construction contract giving the contractor freedom to select certain materials or use certain procedures of his own option.

CONTRACT OVERRUN (UNDERRUN) Difference between the contract price, including all approved extras and variations to the original specifications, and the final cost of the works.

CONTRACT PAYMENT BOND Security furnished to the contracting agency to guarantee payment of the prescribed debts of the contractor covered by the bond.

CONTRACT PERFORMANCE BOND *See* **Bond.**

CONTRACT PERIOD Number of days between the date of signing of the contract and the date stipulated for completion of the work or, if no completion date is specified, or has been legally changed since commencement of the works, the date that the project is handed over to the owner or his representative.

CONTRACT PRICE Sum of money for which a contractor agrees to complete the work described in the contract documents.

CONTRACT RENT Designated payment for the use of property and facilities as defined in a lease.

CONTRACT SURETY BOND *See* **Bond.**

CONTRACT TIME Number of working days or calendar days allowed for completion of a contract. If a calendar date of

completion is shown in the proposal in lieu of a number of working or calendar days, the contract shall be completed by that date. *Also called* **Period for completion,** and **Time for completion.** *See also* **Contract.**

CONTRACTUAL LIABILITY Responsibilities and liabilities assumed by the signatory to a contract.

CONTRARIES Materials that have a detrimental effect upon reprocessing or recycling.

CONTROL 1. Channel section where bed and bank conditions give measurable indication of flow; **2.** In fighting forest fires, the overall program to control a fire and suppress fire losses (a forest fire is under control when it no longer threatens additional destruction); **3.** Process of comparing actual performance with planned performance, analyzing variances, evaluating possible alternatives, and taking appropriate corrective action as needed; **4.** Exposure of test organisms to dilution water only, or dilution water containing a test solvent or carrier (no toxic agent is intentionally or inadvertently added).

CONTROL CENTER Location to which communication and telemetry lines and/or signals are directed and from which information and/or control signals can be sent.

CONTROL CHARTS Graphic display of the results, over time and against established control limits, of a process. They are used to determine if the process is 'in control' or in need of adjustment.

CONTROL CONSOLE Panel, easily visible from the operator's station, containing dials and gauges that display the principal operating conditions of mechanical equipment and machines.

CONTROL DEVICE 1. Enclosed combustion device, vapor recovery system, or flare; any device, the primary function of which is the recovery or capture of solvents or other organics for use, reuse, or sale; **2.** Any apparatus that reduces the quantity of a pollutant emitted to the air.

CONTROL FACTOR Ratio of the minimum compressive strength to the average compressive strength.

CONTROL FLOAT Float installed in a tank or body of liquid that starts and stops a device according to the level of the liquid.

CONTROL FLUME *See* **Flume.**

CONTROLLED-AIR INCINERATOR *See* **Incinerator.**

CONTROLLED AREA Surface location, to be identified by passive institutional controls, that encompasses no more than 100 km² and extends horizontally no more than 5 km in any direction from the outer boundary of the original location of the radioactive wastes in a disposal system.

CONTROLLED BURNING Use of fire to destroy logging debris, reduce buildup of dead and fallen timber that may pose wildfire hazard, control tree diseases, and clear land. Other functions include clearing a buffer strip in the path of a wildfire. *See also* **Area ignition, Backfire,** and **Prescribed burning.**

CONTROLLED DUMPING *See* **Sanitary landfilling.**

CONTROLLED FILL Suitable inert fill material placed on natural ground in thin layers under controlled compaction after any existing weak and compressible soil has been removed.

CONTROLLED PRODUCT Product which, because of flammability, toxicity or similar hazard-related reasons requires special handling, labelling and a Material Safety Data Sheet.

CONTROLLED STORAGE Portion of the total storage in a reservoir that can be controlled: i.e. from the top of the gates to the bottom of the outlet.

CONTROLLED SURFACE MINE DRAINAGE Surface mine drainage that is pumped or siphoned from the active mining area.

CONTROLLED TIPPING Common method of domestic refuse disposal whereby the refuse is deposited in layers, compacted, and covered at the end of every working day with a layer of suitable material that forms a seal.

CONTROLLED WASTE Waste subject to specific regulations regarding its storage, handling, and/or disposal.

CONTROLLER 1. Device or group of devices that serves to govern, in some predetermined manner, the electric power delivered to the apparatus to which it is delivered; **2.** Device that detects a change in a process variable, and then automatically uses an external source of power to amplify the detected signal and to energize a mechanism that will correct the deviation in the process variable until it returns to a preset value. *See also* **Regulator.**

CONTROL LIMITS Limits relating to personal exposure to hazardous substances.

CONTROL LINE Construction of natural barriers to a forest fire.

CONTROL LOOP Path through a control system between the sensor, which measures a variable, and the controller, which controls or makes adjustments.

CONTROL REACH Section of an open channel where a structure or stretch of rapids make the water level above it a stable index of the discharge.

CONTROL ROD Rod of neutron-absorbing material, commonly cadmium, boron, or hafnium, that is moved in and out of a reactor core to control the number of neutrons available for fission, and thus the power level of a nuclear reactor.

CONTROL ROOM Enclosed room to which telemetry, communications, and control electronics are fed and from which actuating signals emanate.

CONTROL SECTION Cross section of a bottleneck in a waterway which determines the energy head required to produce the flow.

CONTROL STOP Device installed in supply piping to regulate or shut off the flow of water to a flush valve.

CONTROL STRATEGY Combination of measures designated to achieve the aggregate reduction of emissions necessary for attainment and maintenance of national standards.

CONTROL SYSTEM Instrumentation system that senses and controls its own operation on a close, continuous basis.

CONTROL SUBSTANCE Chemical substance or mixture, or any other material other than a test substance, feed, or water, that is administered to a test system in the course of a study for the purpose of establishing a basis for comparison with the test substance for known chemical or biological measurements.

CONTROL VALVE *See* **Valve.**

CONTROL WORKS Structures and mechanisms by which flow is regulated.

CONURBATION Large area occupied by mostly urban development.

CONVECTION Transportation of heat by movement due to the ascension of air or liquid when heated and its descension when cooled.

CONVECTIVE COLUMN Rising warm air above a continuing heat source.

CONVECTION CURRENT Ascending and descending water movement in settling basins and lakes caused by differences in temperature and the force of gravity.

CONVECTIVE EQUILIBRIUM State in which a vertically-displaced parcel of air experiences no buoyancy force.

CONVECTIVE INSTABILITY State in which a vertically-displaced parcel of air experiences a buoyancy force in the direction of displacement.

CONVECTIVE PRECIPITATION Precipitation resulting from the vertical movement of moisture-laden air which, on rising, cools and loses its moisture.

CONVENTIONAL FILTRATION TREATMENT Series of processes including coagulation, flocculation, sedimentation, and filtration resulting in substantial particulate removal.

CONVENTIONAL MINE Open pit or underground excavation for the production of minerals.

CONVENTIONAL TECHNOLOGY 1. Wastewater treatment processes and techniques involving the treatment of wastewater at a centralized treatment plant by means of biological or physical/chemical unit processes followed by direct point source discharge to surface waters; **2.** Wet flue gas desulfurization technology, dry flue gas desulfurization technology, atmospheric fluidized bed combustion technology, and oil hydrodesulfurization technology.

CONVENTIONAL TREATMENT Wastewater treatment involving preliminary, sedimentation, flotation, trickling filter, activated sludge stages, and chlorination.

CONVERGENCE 1. Act or process or meeting at a point; **2.** Compression of a fluid or gas into a smaller volume.

CONVERGING TUBE Tube of gradually reducing diameter along its longitudinal axis from the end at which liquid enters.

CONVERSION 1. Process of changing the use of a property, either from one type of use to another (residential to commercial, for instance) or from one type of occupancy to another (storage to living, for example); **2.** Transition section in a conduit connecting sections, each having different hydraulic elements; **3.** Transformation of wastes into other forms, such as steam, gas, or oil.

CONVERSION FACTOR Number enabling the units of one system to be transformed into the units of another system.

CONVEX BANK Stream bank having its center of curvature away from the channel.

CONVEYANCE LOSS Water lost from a conduit due to leakage, seepage, evaporation, evapotranspiration, or other cause.

CONVEYING SYSTEM Device for transporting materials from one piece of equipment or location to another location within a plant. Conveying systems include but are not limited to: feeders, belt conveyors, bucket elevators and pneumatic systems.

CONVEYOR Device for moving materials, usually a continuous belt, articulated system of buckets, confined screw, or a pipe through which material is moved by air or water. There are several types, including:

> **Apron:** One or more endless chains fitted with interlocking or overlapping plates that transport materials on their upper faces.

> **Belt:** Endless, pulley-driven, rubber-covered fabric belt supported on rollers, that conveys material placed on its upper surface.

> **Decline:** Conveyor that transports and discharges to a lower elevation.

> **Feeder:** Pushing device or short belt that supplies material to a crusher or belt conveyor.

> **Screw:** Revolving auger flights that move bulk materials along a tube or trough.

CONVEYOR SWING ANGLE Measurement in degrees a conveyor will swing left and/or right of a machine centerline.

COOLANT Substance used to draw heat away from a mass.

COOLING COIL Coil surrounded by hot or cold water and through which hot, or cold water is passed to cause an exchange of heat.

COOLING POND Enclosed body of water used to cool an inflow through surface evaporation, convection, and radiation.

COOLING SPRAY Water spray directed into hot gas or air for cooling.

COOLING SYSTEM Any cooling tower or closed cooling water system.

COOLING TOWER Device through which heated water is passed so as to reduce its temperature, in an open or closed system, by convection or conduction.

COOLING TOWER PRECIPITATION Light rain that occurs around water cooling towers that are not equipped with spray eliminators.

COOLING WATER Clean wastewater discharged from a heat transfer system.

COPPER Metal (Cu) used for a wide range of applications, particularly those requiring a high heat transfer ability or electrical conductivity. Copper and its compounds are widely distributed in nature, and copper is found frequently in surface water and some groundwater. Distributed water contains considerably more more copper than the original water supply because of the dissolution of copper from copper piping. The aesthetic objective for copper in drinking water is <1.0 mg/L. This level is below the taste threshold, is protective of health and contributes to minimum nutritional requirements.

COPPERAS Greenish, crystalline, hydrated ferrous sulphate, $FeSO_4 \cdot 7H_2O$.

COPPER–CHROME ARSENATE Wood preservative, applied using vacuum-pressure impregnation, that reacts with cellulose to form compounds lethal to insects and fungi, but safe for humans.

COPPER NAPHTHENATE Wood preservative.

COPPER PIPE STRAPS Straps fabricated of copper, used to secure copper pipe to the structure of a building.

COPPER SULFATE Poisonous blue, crystalline copper salt, $CuSO_4 \cdot 5H_2O$, used to control algal growths. *Also called* **Blue copperas, Bluestone,** and **Cupric sulfate.**

COPPER WATER PIPE *See* **Pipe.**

CORAL Any of various marine polyps (class Anthozoa) having a stony, horny, or leathery external or internal skeleton.

CORAL REEF Reef consisting mainly of coral produced by many colonies of coral polyps over a period of centuries, with new animals building on the skeletons left behind by animals that have died.

CORE 1. Cylindrical sample produced by a core drill; **2.** Cutoff wall of clay within an earth dam or embankment.

CORE CITY Planning concept calling for buildings to be developed very densely in the central zone of a city, to the degree that the total available floor area is approximately equal to the total land areas occupied.

CORE DRILLING 1. Exploratory drilling that involves cutting cylinders of rock or soil and bringing them to the surface for inspection and analysis.

CORE SAMPLE Sample obtained using a core drill. For asphaltic pavement, the core is typically 100 mm (4 in.) in diameter.

CORE WALL Wall of impervious material built inside an earthen dam to reduce percolation.

CORING Act of obtaining cores from soil, rock, or concrete structures for examination and testing.

CORPORATION COCK *See*

CORRASION Mechanical abrasion of rock and soil by the scouring action of rock fragments moved by wind and/or water.

CORRECTED NOISE LEVEL Index of noise emitted from an industrial site, expressed as dB(A), corrected for tonal character, intermittency and duration.

CORRELATION Relationship between two variables, according to a specific range of values.

CORROSION 1. Destruction of metal by chemical, electrochemical, or electrolytic reaction with its environment; **2.** Chemical or galvanic decomposition of the wires in a rope through the action of moisture, acids, alkalines, or other destructive agents; **3.** Chemical attack on a metal or other solid by contaminants in a lubricant. Common corrosive contaminants are:

> **Acids:** Which may form as oxidation products in a deteriorating lubrication oil, or may be introduced into the oil as combustion by-products in piston engines.

> **Water:** Which causes rust.

CORROSION CONTROL 1. Means by which the metallic ions of a conduit are prevented from going into solution; **2.** Chemical treatment causing the sequestration of metallic ions and subsequent formation of protective films on metal surfaces.

CORROSION EMBRITTLEMENT Loss of ductility or workability of a metal due to corrosion.

CORROSION EXPERT Person who, by reason of his knowledge of the physical sciences and the principles of engineering and mathematics, acquired by a professional education and related practical experience, is qualified to engage in the practice of corrosion control.

CORROSION FATIGUE Effect on metal of repeated stress in a corrosive atmosphere characterized by shortened life of the part.

CORROSION TESTING Methods used to determine the susceptibility of materials, especially metals, to corrosion, particularly corrosion from saltwater. There are two principal methods:

> **Fog testing:** *See* **Salt spray testing** (below).

> **Kesternich test:** In this test, the component to be tested is prepared and placed in a Kesternich Test Cabinet in which 2 L (1.69 US gal, 1.40 imp gal) of distilled water has been placed at the bottom. The cabinet is sealed and 2 L (1.69 US gal, 1.40 imp gal) of sulfur dioxide injected and the internal temperature set to 40°C (104°F) for the cycle. Each 24 hour cycle begins with 8 hours of exposure to the acidic bath created in the cabinet, which is then purged and opened, the test specimens rinsed with distilled water and allowed to dry at room temperature for 16 hours. The test specimens are examined for surface corrosion (red rust) at the end of each cycle. The following table compares the relative surface corrosion of various

coatings, platings, and materials after 30 cycles:

Coating/Plating/Material	% Surface Corrosion
Cadmium	100% after 4 cycles
Stainless steel - type 304	5–10% after 30 cycles
Stainless steel - type 316	None after 30 cycles
Stainless steel - type 410	100% after 3 cycles
Zinc w/clear chromates	100% after 3 cycles
Zinc w/yellow dichromate	100% after 3 cycles
Mechanically galvanized	100% after 6 cycles
Zamac 7 alloy	None after 30 cycles

Salt spray test: In this test, the component to be tested is prepared and suspended in a sealed chamber and then subjected to a spray or fog of a neutral 5% salt solution, atomized at a temperature of 35°C (95°F).

CORROSIVE SUBSTANCE Solid, liquid, or gas which when contacting living tissue damages the tissue, or when contacting other materials and certain chemicals, causes fire or accelerated deterioration of the material or chemical.

CORROSIVE WEAR See **Wear.**

COSMIC RAYS Complex system of radiation that reaches Earth from outer space.

COST ACCOUNTING Accounting method concerned with the classification, recording, analysis, reporting, and interpretation of expenditures identifiable with the production and distribution of goods and services.

COST ANALYSIS Review and evaluation of each element of a cost to determine reasonableness, allocability and allowability.

COST APPROACH Determination of the value of a property, including land and buildings, made first by estimating the value of the land as if it were vacant, adding to it the estimated cost to reproduce the buildings as new, then deducting an amount for depreciation of the existing property. See also **Depreciated cost method.**

COST/BENEFIT ANALYSIS See **Analysis.**

COST/BENEFIT RATIO Ratio of the gross costs of a project or activity to the gross benefits, both costs and benefits being calculated over the life of the project at an annual rate of interest.

COST BREAKDOWN Detailed listing or analysis of the constituent costs that contribute to a single cost item.

COST BUDGETING Allocating the cost estimates to individual project components.

COST CENTER Subdivision of an organization with which costs can be identified for the purposes of managerial control.

COST CHECK Periodic estimate review to give an up-to-date analysis of cost in relation to the progress of design or construction.

COST CLASS Any of the subdivisions of the total scope of the work in a project to which costs are assigned.

COST CODE System that allocates codes to defined classes and types of work and materials; an aid to estimating and cost control.

COST CONTROL Accounting procedure designed to produce a range of reports giving information about total expenditure, total income, cash flow, accounts receivable, etc.

COST ESTIMATE Approximation of cost within a given percentage of anticipated actual cost.

COST ESTIMATING Estimating the cost of the resources needed to complete project activities.

COST FLOW Method of allocating costs between inventory and cost of sales. There are several methods, including:

Average: Method where the cost of an item is determined by applying a weighted average of the cost of all similar items at a point of time or over a period.

Base stock: Method where the cost of items sold or consumed during a period is determined by assuming that a predetermined minimum quantity is carried in inventory permanently and at a fixed price.

FIFO: First in, first out, where the cost of items sold or consumed during a period is computed as though they were sold or consumed in order of their acquisition.

LIFO: Last in, first out, where the cost of items sold or consumed during a period is deemed to be at the cost of the most recent acquisitions.

NIFO: Next in, first out, where the cost of items sold or consumed during a period is deemed to be at the cost of the next acquisition.

Specific identification: Where the actual cost of each item is determined separately.

COST FUNCTION Relationship between the degree to which a pollutant emission is reduced and the cost of attaining the reduction.

COSTING Calculation of the cost of work based on the amount done. This actual cost may then be compared to the estimated cost and unit prices of the bid for financial control.

COST OF CAPITAL Investment required to create and maintain productive capital.

COST OF QUALITY Costs incurred to ensure quality, including planning, quality control, quality assurance, and rework.

COST OF WORK Costs incurred in the completion of work described and illustrated in the contract documents.

COST OVERRUN Cost beyond that budgeted for or contracted for.

COST PERFORMANCE INDEX Ratio of budgeted costs to actual costs, a calculation often used to predict the magnitude of a possible cost overrun. *See also* **Earned value.**

COST PER KM/MILE Total cost (including such factors as depreciation, insurance, downtime, repairs and service, replacement parts, etc.) involved in running a vehicle or equipment item for one kilometer or mile, averaged from the costs incurred over as large a number of kilometers or miles as possible.

COST PER TON PER MINUTE Unit used in cost comparisons between transfer and direct-haul operations.

COST PLAN Sets out the total cost limit of a budget, subdivided into meaningful sections, each with its own cost and outline specification stated. It provides the frame of reference required as the first principle of a valid cost control system.

COST-PLUS Form of contract for construction work wherein the construction contractor is reimbursed for the costs he incurs in performing the work, plus a lump sum or proportional fee, hence cost-plus. This type of contract is favored where the scope of the work is indeterminate or highly uncertain and the kinds of labor, material, and equipment needed also are uncertain. Under such an arrangement it is necessary to maintain complete records of all time and materials expended by the contractor on the work.

COST PLUS FIXED FEE Type of contract where the buyer reimburses the seller for the seller's allowable costs (as defined by the contract) plus a fixed amount of profit (fee).

COST PLUS INCENTIVE FEE Type of contract where the buyer reimburses the seller for the seller's allowable costs (as defined by the contract), and the seller earns

135

its profit if it meets defined performance criteria.

COST-RATIO METHOD OF ACCOUNTING See **Method of accounting.**

COST REVIEW Planned, systematic reassessments of the estimated final cost of the scope of work in a cost class. Each design review is accompanied by a reforecast of cost. See also **Forecast to complete.**

COST TYPES Associated with the scope of the work in a cost class are four different cost types, having to do with whether a cost is authorized, contractually committed, expended, or estimated. These types are:

(**a**) authorized appropriation;

(**b**) commitment;

(**c**) expenditure, or actual; and

(**d**) forecast to complete, or estimate to complete, or uncommitted.

Types (a) and (d) contain contingencies. The sum of type (b) and type (d) is the estimated final cost. Before any commitments are made within a cost class, the type (d) cost contains a component called the anticipated award price, that in turn contains an allowance for escalation. Type (c) costs are further subdivided into payments and retentions (holdbacks).

COST VARIANCE Difference between the estimated cost of an activity and the actual cost of that activity.

COULOMB SI unit of electric charge transported in 1 s by a current of 1 A. (1 C=1 A•s). Symbol: C. See also the appendix: **Metric and nonmetric measurement.**

COULOMBMETRIC TITRATION Titration method that measures the quantity of electricity passed during an electron exchange involving the substance being determined.

COUNTERBALANCE VALVE See **Valve.**

COUNTERBALANCE PRESSURE-CONTROL VALVE See **Valve.**

COUNTERDRAIN Drain along the foot of a canal bank or dam to remove leakage.

COUNTERFLOW Liquid flow in the opposite direction to the main flow, as in the return flow from a heat exchanger.

COUNTRY ROCK Body of rock that encloses an igneous intrusion or mineral vein.

COUNTY BUILDING CODES See **Building codes.**

COUPLED NODES Modes of acoustic vibration that influence one another.

COUPLING Pipe fitting containing female threads on both ends.

COVER 1. In reinforced concrete, the least distance between the surface of the reinforcement and the outer surface of the concrete; **2.** Outer component, usually intended to protect the carcass of a product; **3.** Lid or cap that can be removed for access of inspection; **4.** Layer of soil placed over compacted waste at a landfill site at the end of a day's operations; **5.** Device or system placed on or over a waste in a waste management unit so that the entire waste surface area is enclosed and sealed to minimize air emissions. A cover may have openings necessary for operation, inspection, and maintenance of the waste management unit such as access hatches, sampling ports, and gauge wells provided that each opening is closed and sealed when not in use. Example of covers include a fixed roof installed on a tank, a lid installed on a container, and an air-supported enclosure installed over a waste management unit.

COVERAGE 1. Measure of the amount of material required to cover a given surface; **2.** Extent of protection provided by insurance.

COVERED SLUDGE DRYING BED Glass covered enclosure in which sludge is exposed to radiant heating while being protected from rain and snow.

COVER MATERIAL Soil or other suitable material that is used to cover compacted solid wastes in a land disposal site.

CPE MEMBRANE See **Chlorinated poly-**

ethylene membrane.

CPVC (CHLORINATED POLYVINYL CHLORIDE) PIPE *See* **Pipe.**

CRADLE 1. One of a series of piers supporting a pipe, spaced so that the pipe furnishes its own structural bridging; **2.** Continuous concrete bed in the bottom of a trench that supports a pipe and partially encloses it.

CRAMMING Plugging a pipe before making a repair.

CRATON Continental block that has been stable over a relatively long period of the Earth's history and which has undergone only faulting or gentle warping.

CRAWLING Propelling a track-equipped machine or vehicle by causing the movable tracks to rotate.

CREEK 1. Small stream, often a shallow and intermittent tributary to a river; **2.** Small tidal channel through a coastal marsh.

CREEP 1. Movement of aggregate particles under roller pressure during compaction; **2.** Plastic flow of solids under constant stress.

CREOSOTE Oily liquid distilled from wood tar, used as a preservative on wood.

CRESCENT PUMP *See* **Pump.**

CREST 1. Highest part of the water-retaining face of the spillway of a dam, over which water will flow when cresting; **2.** Highest elevation reached by flood waters; **3.** Summit of a wave.

CREST CONTROL Flow control method in which a device on the crest of a spillway dam is used to raise or lower the crest.

CREST GATE Gate built into the crest of the spillway of a dam to maintain or lower the water level.

CRIB 1. Crate-like framing used to support a structure. *See also* **Cellular cofferdam,** and **Cofferdam; 2.** Any of various frameworks of logs or timbers, used in construction work; **3.** Wooden lining on the inside of a shaft; **4.** Openwork of horizontally cross-piled, squared timbers, or beams, used as a retaining wall.

CRIB DAM Barricade formed of bays or cells that are filled with impervious materials.

CRIB HEIGHT Elevation difference between the top of the cap or sill and the bottom of the base log at the center line.

CRIB WALL Retaining wall constructed of rectangular interlocking precast concrete or timber members, forming a cellular structure, laid on top of each other and filled with soil or broken rock.

CRIB WEIR Low diversion weir of log cribs or wire-mesh gabions filled with rock.

CRIBWORK Construction of crib-like structures.

CRITERIA Performance level, standard, or other arbitrary description.

CRITICAL ACTIVITIES OR WORK ITEMS Activities or work items that, if not completed by the indicated time for them, will correspondingly increase the total project duration by the extent of the delay in completion of that activity or work item.

CRITICAL DEPTH Depth of water flowing in an open channel or partially filled conduit that meets a recognized critical velocity.

CRITICAL-DEPTH DISCHARGE MEASUREMENT Indirect measurement of peak discharge following a flood by field survey of high-water marks and channel and control-section geometry.

CRITICAL FLOW Rate at which a fluid flows through an orifice when the stream velocity at the orifice is equal to the velocity of sound in the fluid. Under such conditions, the rate of flow may be increased by an increase in upstream pressure, but it will not be affected by a decrease in downstream pressure.

CRITICAL GRADIENT Channel slope that is exactly equal to the loss of head per unit of measure at a depth that will give uniform flow at critical depth.

CRITICAL GROUP Method of monitoring emissions that relies on the identification of the entities most at risk to a particular discharge.

CRITICAL HEIGHT Height to which a vertical cut made in a cohesive soil will stand without support.

CRITICAL HYDRAULIC GRADIENT Point of saturation where sand becomes quicksand.

CRITICAL MASS Mass of fissile material that will sustain a chain reaction.

CRITICAL MOISTURE Moisture of confined, or unconfined, soil samples attained by increase in the moisture content at a steady rate as a result of deformation under constant load.

CRITICAL ORGAN Most exposed human organ or tissue exclusive of the integumentary system (skin) and the cornea.

CRITICAL PATH Route through the network logic sequence on which all activities must be completed within their expected timings or the schedule for the overall project will change (i.e. activities or work items with zero float).

CRITICAL PATH METHOD Network analysis technique used to predict project duration by analyzing which sequence of activities (which path) has the least amount of scheduling flexibility (the least amount of float). Early dates are calculated by means of a forward pass using a specified start date. Late dates are calculated by means of a backward pass starting from a specified completion date (usually the forward pass's calculated project early finish date). *Also called* **CPM.** *See also* **Arrow,** and **Program evaluation and Review technique (PERT).** *Also called* **Project network analysis.**

CRITICAL POINT State point at which liquid and vapor have identical properties.

CRITICAL PRESSURE *See* **Pressure.**

CRITICAL SILENCER Exhaust silencer that is applied in sensitive noise-control areas.

CRITICAL SLOPE Maximum angle to horizontal that a sloped bank of soil or granular material of given height will stand unsupported.

CRITICAL VELOCITY Rate of flow in a pipe or open channel that will neither deposit or pick up silt.

CROPLAND Land regularly used for the production of agricultural produce.

CROSS Pipe fitting with four female openings at right angles to one another.

CROSSBEARER *See* **Mudsill.**

CROSS BRACE 1. Horizontal member of a shoring system installed perpendicular to the sides of the excavation, the ends of which bear against either uprights or wales; **2.** One of two diagonal scaffold members joined at their center to form an 'X,' used to brace frames and/or uprights.

CROSS BRACING Crossing members usually designed to act only in tension, often used in scaffolding systems, but also between columns to increase support and load-bearing capacity. *See also* **Swaybrace,** and **X-brace.**

CROSS BRIDGING Arrangement of small wooden pieces, 25 x 75 mm, 50 x 50 mm, or 50 x 100 mm (1 x 3 in., 2 x 2 in., or 2 x 4 in.), between joists at intervals through a span, to stiffen the structure. See *also* **Bridging.**

CROSS CONNECTION 1. Any link between a potable water supply and contaminated water. *See also* **Backflow connection**; **2.** Connecting roadway between two nearby and generally parallel roadways.

CROSSCURRENT Current flowing across another current: a conflicting flow.

CROSS-DITCH Shallow channel laid diagonally across the surface of a road so as to lead water off the road and prevent soil erosion. *Also called* a **Water bar.**

CROSSFLOW FILTRATION Water filtration process in which a semipermeable membrane is used to separate waterborne contaminants. The bulk solution flows over and parallel to the filter surface and, under pressure, a portion of the water is forced through the membrane filter.

CROSSING 1. Place at which something, such as a river or highway, may be crossed; **2.** Place where a river passes over a shoal or bar; **3.** Relatively short and shallow length of a river between bends.

CROSS-MEDIA Form of pollution conversion, i.e. a hazardous waste incinerator discharging toxic dust from dust collectors which requires further treatment prior to disposal.

CROSSOVER Connection between two piping runs in the same piping system, or the connection of two different piping systems that contain potable water.

CROWDING FORCE Force that pushes a loader bucket into the bank. It is a function of power available, hydraulic power, and linkage design.

CROWN 1. Upward curve at the middle of a road to allow for water runoff; **2.** Top of a levee or embankment; **3.** Point in a trap where the direction of flow changes from upward to downward; **4.** Highest point of the inside of a drain or sewer.

CROWNED DRAINAGE Where the centerline of a structure, typically a street or highway, is raised above its longitudinal perimeter so as to cast water to the edges. *See also* **Valley drainage.**

CROWN GATE Head gate of a canal lock.

CROWN WEIR Point in the curve of a trap directly below the crown; the point at which the water level will normally remain when the fixture is not discharging through the trap.

CRUDE OIL Liquid petroleum in the form in which it is extracted and prior to it being refined.

CRUDE SEWAGE Untreated domestic waterborne waste.

CRUDE WASTEWATER Mixture of domestic sewage, industrial waste, and stormwater runoff before it receives any treatment.

CRUSHED GRAVEL *See* **Aggregate.**

CRUSHED STONE *See* **Aggregate.**

CRUSHER Machine that reduces rock and other materials to a smaller size. There are several types, including:

> **Gyratory:** Crusher with a central conical member having an eccentric motion in a circular chamber tapering from a wide top opening.

> **Hammermill:** Rock crusher employing hammers or flails on a rapidly rotating axle.

> **Jaw:** Fixed and movable jaw combination, widely spaced at the top and close at the bottom, the movable jaw of which continuously moves toward and away from the fixed jaw.

> **Primary:** Jaw-type crusher that reduces very large rocks to a size that can be processed by a secondary and any subsequent crushers.

> **Roll:** Crusher having two large spring-loaded rolls that counter-rotate, the product size being determined by the space between the rolls.

> **Secondary:** Crusher that receives broken material from a primary crusher and further reduces its size.

CRUSHER-RUN AGGREGATE *See* **Aggregate.**

CRUSHER SCREENINGS Crushed stone or gravel with a gradation from 0 to 6 mm (0 to 0.25 in.).

CRUST Outermost solid shell of the Earth that rests on the mantle and which varies in thickness from an average of about 6 km in ocean areas to 35–70 km in continental regions.

CRYOGENIC SCRAP RECOVERY Technique that employs liquid nitrogen to freeze a mixture of metals and contaminants from scrap automobiles. The very low (-196°C) temperature causes mild steel to fracture like glass when processed in an impactor, allowing it to be easily isolated and removed magnetically.

CRYOGENIC SYSTEM System in which a local temperature is produced that is lower than the surrounding temperature.

CRYOLOGY Science of the physical aspects of snow, ice, hail, sleet, and other forms of water produced at temperatures below freezing.

CRYOTURBATION Disturbance of material by frost action, frost heaving and differential mass movement.

CRYPTOCRYSTALLINE Comprised of crystals too small to be resolved by the naked eye.

CRYSTAL Homogeneous solid that possesses long-range, three-dimensional internal order.

CRYSTALLINE Composed or resembling crystals.

CUBATURE Calculation of the cubic volume of tidal water from the discharges and average currents of a cross section of the waterway from the upstream tidal rises and falls.

CUBIC CENTIMETER Unit of volume equal to 1 mL. Symbol: cm³. Used with the SI system of measurement. Multiply by 0.061 02 to obtain cubic inches, symbol in.³; by 0.035 195 to obtain fluid ounces, symbol: fl oz; by 1.0 to obtain milliliter, symbol: ml; by 0.001 to obtain liters, symbol: L. *See also the appendix:* **Metric and nonmetric measurement.**

CUBIC CONTENT In construction, the cubic measure contained within the walls of a room or combination of rooms, used as a basis for estimating materials, costs, etc.

CUBIC FOOT Nonmetric unit of volume. Symbol: ft³. Multiply by 0.028 316 85 to obtain cubic meters, symbol: m³; by 28.316 85 to obtain liters, symbol: L. *See also the appendix:* **Metric and nonmetric measurement.**

CUBIC FOOT PER HOUR Nonmetric unit of flow rate. Symbol: ft³/h. Multiply by 28.316 85 to obtain liters per hour, symbol: L/h; by 0.007 865 79 to obtain liters per second, symbol: L/s. *See also the appendix:* **Metric and nonmetric measurement.**

CUBIC FOOT PER MINUTE Nonmetric unit of flow rate. Symbol: ft³/min. Multiply by 0.000 471 947 4 to obtain cubic meters per second, symbol: m³/s; by 0.471 947 4 to obtain liters per second, symbol: L/s. *See also the appendix:* **Metric and nonmetric measurement.**

CUBIC FOOT PER SECOND Nonmetric unit of flow. Symbol ft³/sec. Multiply by 0.002 831 684 4 to obtain cubic meters per second, symbol: m³/sec; by 2.831 684 4 to obtain liters per second, symbol: L/s.

CUBIC INCH Nonmetric unit of volume. Symbol: in.³. Multiply by 16.387 064 to obtain cubic centimeters, symbol: cm³; by 16.387 064 to obtain cubic millimeters, symbol: mm³; by 0.016 387 064 to obtain liters, symbol: L. *See also the appendix:* **Metric and nonmetric measurement.**

CUBIC METER A derived unit of volume with a compound name of the SI system of measurement. Symbol: m³. Multiply by 0.2759 to obtain cords (of stacked wood). Symbol: cd; by 1.3080 to obtain cubic yards, symbol: yd³; by 35.3147 to obtain cubic feet, symbol: ft³; by 219.97 to obtain (imperial) gallons, symbol: (imp)gal; by 264.17 to obtain (US) gallons, symbol: (US)gal. *See also the appendix:* **Metric and nonmetric measurement.**

CUBIC METER PER MOLE A derived unit of molar volume with a compound name of the SI system of measurement. Symbol: m³/mol. *See also the appendix:* **Metric and nonmetric measurement.**

CUBIC METER PER SECOND A derived unit of volume flow rate with a compound name of the SI system of measurement. Symbol: m³/s. *See also the appendix:* **Metric and nonmetric measurement.**

CUBIC SCALE Estimate of the cubic-measure volume of wood fiber in a tree, log, or other wood product.

CUBIC YARD Non-SI unit of volumetric

measurement equal to a volume measuring 3 ft by 3 ft by 3 ft, equivalent to 0.765 m³. Multiply by 0.764 555 to obtain cubic meters, symbol: m³. *See also the appendix:* **Metric and nonmetric measurement.**

CUBIC YARD BANK MEASUREMENT Unit of excavation in a cut or natural bed.

CUBIC YARD COMPACTED MEASUREMENT Unit of excavation in a fill or embankment, after compaction.

CUBIC YARD LOOSE MEASUREMENT Unit of excavation in a machine, stockpile, or in uncompacted fill or embankment.

CUBIC YARD STRUCK MEASUREMENT Unit of capacity of the bucket, body, bowl, or dipper of a machine, excluding the teeth, measured by striking off the ends and sides of the container by a straight edge.

CUBING 1. Determining volumes in cubic measurement; **2.** Method of approximating the cost of construction based on a cost per cubic measure of the total volume of the building.

CULLET Broken scrap glass that can be reused.

CULTURE Intentionally developed organic growth produced through the provision of suitable nutrients and environment.

CULVERT Pipe or channel to carry water under a roadway or other obstruction. It is either a short bridge less than 6 m (20 ft) in length, or a structure with enough fill over it so that little of its strength is needed to support the traffic load. *See also* **Drainage ditch.**

CULVERT FLOW DISCHARGE MEASUREMENT Indirect measurement by field survey of high water marks and channel and culvert geometry of peak discharge following a flood

CUMULATIVE BATCHING Measuring more than one ingredient of a batch in the same container by bringing the batcher scale into balance at successive total weights as each ingredient is accumulated in the container.

CUMULATIVE RUNOFF Total volume of runoff of a defined period.

CUMULATIVE TOXICITY Adverse effects of repeated doses occurring as a result of prolonged action on, or increased concentration of an administered substance or its metabolites in susceptible tissues.

CUMULOCIRRUS Cloud that is part cumulus, part cirrus.

CUMULONIMBUS Massive cloud formation having peaks that resemble mountains.

CUMULOSE SOILS Accumulated organic matter.

CUMULOSTRATUS Cumulus cloud with its base spread out horizontally like a stratus cloud.

CUMULUS Cloud formation of rounded heaps having a flat base.

CUNETTE Longitudinal trough or channel, part of a large, flat-bottomed conduit, used to concentrate low flows to develop self-cleansing velocities.

CUNIT Unit of volume consisting of 100 ft³.

CUNETTE CHANNEL Prefabricated manhole channel.

CUPRICHLORAMINE Copper sulfate, ammonia, and chlorine mixture used as an algicide.

CUPRIC SULFATE *See* **Copper sulfate.**

CURB COCK Control valve installed in a house service between the municipal stop and the structure. *Also called* a **Curb stop.**

CURB COLLECTION *See* **Collection.**

CURBSIDE COLLECTION Programs where recyclable materials are collected at the curb, often from special containers color-coded as to type of content, to be brought to various processing facilities.

CURB STOP *See* **Curb cock.**

CURIE Unit of radioactivity, the amount

of any nuclide that undergoes exactly 3.7 x 10^{10} radioactive disintegrations per second.

CURRENT 1. Steady and smooth onward movement of water; **2.** The name given to the apparent transmission or 'flow' of electric force through a conducting body. An electric current is according to its nature called *alternating* or *continuous, intermittent, pulsatory*, or *undulatory*.

CURRENT DENSITY Ratio of the magnitude of current flowing in a conductor to the cross-sectional area perpendicular to the current flow: in desalination, the amperage per unit area flowing through the stack.

CURRENT DIAGRAM Graphic representation of the velocities of flood and ebb currents and times of slack strength over a stretch of the channel of a tidal waterway.

CURRENT DIFFERENCE Difference between the time of slack water or strength of current at a locality and the time of the corresponding phase of the current at a reference station.

CURRENT EFFICIENCY Theoretical power required for a specific desalination task compared to the actual power.

CURRENT FINISH DATE Current estimate of the point in time when an activity will be completed.

CURRENT METER Device for determining the velocity of moving water.

CURRENT POLE Graduated pole used to observe the direction and velocity of a current in flowing water.

CURRENT START DATE Current estimate of the point in time when an activity will begin.

CURRENT TABLES Predictions of the daily times of slack water and of the times and velocities of the flood and ebb maximums.

CURTAIN Vertical barrier to the flow of water.

CURTAIN DRAIN *See* **Intercepting drain.**

CURTAIN WALL Deep cutoff wall below an overflow masonry dam on pervious foundations that prevents underflow.

CURTAIN GROUTING Injection of grout into a subsurface formation to create a water barrier.

CURVATURE FACTOR Numerical quantity expressing the energy loss due to one or more sharp curves in a pipeline.

CURVE 1. Continuous bend; **2.** Graph representing changes in value of physical or statistical quantities.

CUSEC Non-SI unit of volume equal to 1 ft^3/sec of flow in water.

CUT 1. Depth to which material is to be excavated; **2.** Volume of excavation for a given cut; **3.** To lower an existing grade. *See also* **Gross cut,** and **Net cut; 4.** Portion of a land surface or an area from which earth or rock has been or will be excavated; the distance between an original ground surface and an excavated surface.

CUT-AND-COVER Means of developing a sanitary land fill or refuse dump whereby a cut is made, alternate compacted layers of refuse and excavated material from a stockpile are placed in the area of the cut that, when developed to the design profile, is capped with additional fill.

CUT-AND-FILL Construction or development process involving excavation of cuts and using the material for adjacent fills. In a balanced cut-and-fill, the volume of excavated material equals the volume of material required for fill, with allowance for swell or shrink from cut to fill.

CUTBANK Concave wall of a meandering stream that is maintained as a steep, even overhanging, cliff resulting from the action of current flow against its base.

CUTICULAR TRANSPIRATION Direct evaporation of water through the leaf cuticle or outer skin or bark of plants into the atmosphere by transpiration.

CUTOFF 1. Impervious barrier extending into the foundation of a dam; **2.** Natural or artificial channel connecting two points on a stream channel.

CUTOFF DEPTH Depth reached by cofferdam walls or sheet piling below an excavation.

CUTOFF RATIO Ratio of the length of a cutoff to the original length of a river or stream.

CUTOFF TRENCH Trench excavated to impervious material below the bottom of a dam foundation and backfilled with clay or concrete to form a watertight barrier.

CUTOFF WALL Wall or footing constructed to provide resistance to seepage.

CUTTERHEAD Set of revolving blades for fragmenting hard material, fitted to the head of the suction line of a hydraulic suction dredge.

CUTTING SCREEN Device with a sharp leading edge, used to grind, shred, or macerate material removed from wastewater by the screen to which it is attached.

CUTWATER Wedge-shaped upstream face of a pier or a bridge designed to break the flow of water.

CYANIDE Generic name for a salt of hydrocyanic acid (HCN), e.g. potassium cyanide (KCN) used to dissolve gold from crushed rock. Cyanide is an extremely toxic and fast-acting poison. However, because it can be detoxified (through conversion to the relatively non-toxic thiocynate ion) to a certain extent in the human body, cyanide poisoning generally results from acute exposure to high doses, not from chronic ingestion of low doses. The maximum acceptable content for free cyanide in drinking water, based on health effects, is 0.2 mg/L.

CYCLE Complete set of individual operations that a machine performs repetitively.

CYCLE OF FLUCTUATION Time occupied by the rise and following decline of a water table: daily, annual, or secular.

CYCLE TIME 1. Time for a machine to complete one cycle, i.e. load, haul, dump, and return; **2.** *See* **Traffic signal.**

CYCLIC DEPLETION Withdrawal of water from a source in excess of the average rate of supply, or recharge, over a secular cycle.

CYCLIC RECOVERY Rise in elevation of the watertable due to additions of water over a secular period.

CYCLONE 1. Center of atmospheric low pressure around which the air circulates in the same direction as the Earth's rotation; **2.** Cone-shaped air cleaner that removes particulate matter from air by centrifugal separation.

CYCLONIC FLOW Spiralling movement of exhaust gases within a duct or stack.

CYCLONIC PRECIPITATION Precipitation resulting from air movements due to cyclonic storms which transfer moisture from the ocean to land.

CYCLOSTROPHIC FORCE Centrifugal force evident in wind occasioned by the curvature of the path of particles.

CYLINDER GATE Gate installed in a dam to control the outflow from a reservoir.

CYLINDRICAL IMPELLER Round impeller used to pump fluids.

D

DAILY AVERAGE EFFLUENT LIMITA-TION Maximum allowable concentration in a discharge as measured in a representative sample during a sampling day.

DAILY CAPACITY Amount of waste capable of being processed or landfilled each day at a facility.

DAILY COVER Soils placed on top of landfilled waste at the end of each day, or at the completion of a landfill cell, as control against rodents and other disease vectors.

DAILY DISCHARGE Discharge of a pollutant measured during a calendar day or any 24-hour period that reasonably represents the calendar day for purposes of sampling. For pollutants with limitations expressed in units of mass, the daily discharge is calculated as the total mass of the pollutant discharged over the day. For pollutants with limitations expressed in other units of measurement, the daily discharge is calculated as the average measurement of the pollutant over the day.

DAILY FLOOD PEAK Maximum mean daily discharge in a stream during a specific flood event.

DAILY MAXIMUM LIMITATION Value that should not be exceeded by any one effluent measurement.

DAILY ROUTE *See* **Collection method.**

DAILY TYPE OF TIDE Tide with one high and one low in a 24-hour period.

DAM 1. Barrier constructed to hold back water; **2.** Placement of products to hold something back; **3.** Interlocking pairs of cast steel teeth built into the road surface of a bridge expansion joint to allow traffic to pass over the gap.

DAMAGE FUNCTION Relationship between the degree to which a polluting emission is reduced and the value of the damage prevented by the reduction.

DAMP Either moderate absorption or a moderate covering of moisture. Implies less wetness than that connoted by 'wet' and slightly wetter than 'moist.' *See also* **Moist,** and **Wet.**

DAMPING Modulation of energy from an oscillating system or particle using absorption, friction or viscous forces.

DAM SITE Site where topographical and other physical conditions are favorable for the construction of a dam; the place where a dam has been built.

DANGEROUS AIR CONTAMINATION Atmosphere presenting a threat of causing death, injury, acute illness, or disablement due to the presence of flammable and/or explosive, toxic, or otherwise injurious incapacitating substances.

DARCY'S FORMULA Formula used to determine the pressure drop due to flow friction through a conduit.

DART VALVE Drain for a well bailer that opens automatically when rested on the ground.

DATA Records of observations and measurements of physical facts, occurrences and conditions.

DATA GAP Absence of any valid study or studies which would satisfy a specific data requirement.

DATALOGGER Portable electronic device into which signals can be entered, manually or by direct input, and then stored, manipulated, displayed or downloaded, in the field or elsewhere.

DATE OF AGREEMENT *See* **Contract date.**

DATE OF COMMENCEMENT OF WORK Date that work is authorized by the owner's agent to start on site.

DATE OF SUBSTANTIAL COMPLETION Date certified by the owner's agent or consultant when the project may be occupied for the purpose for which it was intended, excepting specific tasks and/or details as noted at that date as requiring to be completed.

DATEOMETER Device attached to equipment that indicates the year/month in which the last maintenance service was performed.

DATUM 1. Position of known and fixed elevation; 2. Position or element relative to which others are determined; 3. Reference or base for other quantities.

DATUM PLANE Reference from which other elevations may be computed.

DAUGHTER PRODUCT Nuclide, which may, or may not be radioactive, resulting from the radioactive disintegration of a 'parent' nuclide.

DAY Non-SI unit of time, equal to 24 hours, 1440 minutes, and 86 400 seconds, permitted for use with the SI system of measurement. Symbol: d. *See also the appendix*: **Metric and nonmetric measurement.** *See also* **Calendar day** and **Work day.**

DAY-DEGREES Sum of the degrees of temperature above a threshold over a specified time period.

DAY LABOR Workers paid on a daily basis for work performed.

DAY–NIGHT SOUND LEVEL Weighted equivalent sound level, in decibels, for any continuous 24-hour period, obtained after addition of 10 decibels to sound levels produced in the hours from 10 pm to 7 am (2200–0700).

DAY TANK Tank in which a known concentration of chemical solution for feed to a chemical feeder, in quantities predicted as sufficient for at least 24 hours of plant operation, is stored.

DDE Metabolite, dichlorodiphenyldichloroethylene, of DDT formed by the action of soil microorganisms on DDT, that is more inert and persistent than DDT and capable of accumulating in organisms in the biosphere.

DDT Organic pesticide, dichlorodiphenyltrichloroethane.

DEACTIVATION Mechanical or chemical removal of dissolved oxygen from water to lessen its corroding power.

DEAD END Branch of a drainage piping system that ends in a closed fitting.

DEAD-END MAIN Water main supplied from only one direction.

DEAD END OF A LINE End of a pipeline that does not lead back to an oil storage tank, so that the oil in the end of the line cannot be recirculated.

DEADMAN 1. Anchor used to keep pipework from separating; 2. Buried log or similar object that acts as an anchor for a guy rope, etc.

DEAD STORAGE Storage area below the lowest outlet levels of a reservoir.

DEAD WATER 1. Standing or still water; 2. Water that fails to circulate to the extent required for proper functioning of equipment upon which it relies.

DEAD WEIGHT RELIEF VALVE *See* **Valve.**

DEAD WELL Shaft driven through an impermeable stratum to allow water to drain to an underlying permeable layer.

DEAERATION Removal of dissolved oxygen from water.

DEATH Lack of opercular movement by a test fish or other higher organism.

DEBRIS Woody material such as bark, twigs, branches that will not pass through a 25.4-mm- (1.0-in.-) diameter round opening and which is present in the discharge from a wet storage facility.

DEBRIS BASIN Basin behind a low dam, or excavated in a stream channel to trap debris, or the bed load carried by mountain torrents.

DEBRIS CONE Fan-shaped deposit of soil, sand, gravel and boulders deposited at the foot of a range of hills by a stream.

DEBRIS DAM Barrier across a stream channel to collect floating debris, and/or to slow the flow causing sand and gravel to accumulate.

DEBRIS JAM Congested debris obstructing the free movement of water in a stream.

DECA Prefix representing 10. Symbol: da. Used in the SI system of measurement. *See also the appendix*: **Metric and nonmetric measurement.**

DECALITER Unit of volume, equal to 10 liters. Symbol: daL. Used in the SI system of measurement. *See also the appendix*: **Metric and nonmetric measurement.**

DECAMETER Unit of length, equal to 10 meters. Symbol: dam. Used in the SI system of measurement. *See also the appendix*: **Metric and nonmetric measurement.**

DECANTATION Separation of liquids from solids, or from a liquid of greater density, by drawing off the upper, lighter, layer from the heavier, settled, material.

DECANTATION TEST Test to determine the actual proportion of loam, silt, or other material in a sample of sand. *Also called* **Silt test.**

DECANTING Pouring off a liquid from a container without disturbing any sediment.

DECAY Reduction of volume, concentration, or other measure of a pollutant due to dilution, erosion, absorption, chemical reaction, transformation, etc.

DECAY HEAT Heat resulting from radioactive decay.

DECENT, SAFE, AND SANITARY DWELLING Dwelling that meets applicable housing and occupancy codes. However, any of the following standards which are not met

by an applicable code shall apply, including: (a) be structurally sound, weathertight, and in good repair; (b) contain a safe electrical wiring system adequate for lighting and other electrical devices; (c) contain a heating system capable of sustaining a healthful temperature (of approximately 70°F), except in those areas where local climatic conditions do not require such a system; (d) be adequate in size with respect to the number of rooms and area of living space needed to accommodate the intended number of persons. There shall be a separate, well-lighted and ventilated bathroom that provides privacy to the user and which contains a sink, bathtub or shower stall, and a toilet, all in good working order and properly connected to appropriate sources of water and to a sewage drainage system. In the case of a housekeeping dwelling, there shall be a kitchen area that contains a fully usable sink, properly connected to potable hot and cold water and to a sewage drainage system, and adequate space and utility service connections for a stove and refrigerator; (e) contains unobstructed egress to safe, open space at ground level. If the dwelling unit is on the second story or above, with access directly from or through a common corridor, the corridor must have at least two means of egress. (f) for a person who is handicapped, be free of any barriers which would preclude reasonable ingress, egress, or use of the dwelling.

DECHLORINATION Removal of excess levels of chlorine from a potable water supply by the addition of sulfur dioxide.

DECI Prefix representing 10^{-1}. Symbol: d. Used in the SI system of measurement. *See also the appendix*: **Metric and nonmetric measurement.**

DECIBEL Non-SI measure of sound level. One-tenth of a bel, the number of decibels denoting the ratio of the two amounts of power and being ten times the logarithm to the base of 10 of this ratio: symbol dB. A unit of measure of noise level in which the faintest sound we can hear, called the threshold of hearing, is 0 dB, and the loudest sound the human ear can tolerate, called the threshold of pain, is 140 dB.

DECIDUOUS Trees, shrubs, etc., that shed their leaves annually.

DECILITER Unit of volume, equal to one tenth of a litre. Symbol: dL. Used in the SI system of measurement. *See also the appendix*: **Metric and nonmetric measurement.**

DECIMETER Unit of length, equal to one tenth of a meter, or 10 centimeters, or 100 millimeters. Symbol: dm. Used in the SI system of measurement. *See also the appendix*: **Metric and nonmetric measurement.**

DECK 1. Horizontal platform or finished surface; **2.** Top of a hollow reinforced-concrete dam.

DECK DRAIN Flat strainer fitted flush with a surface that caps a drain outlet.

DECK PLATE Metal plate with a raised pattern, intended to provide stable footing in exposed and/or hazardous locations.

DECK SCREEN *See* **Screen.**

DECODER Device that converts a digital signal into an analog output.

DECOMMISSIONING Making inoperable; putting permanently out of action.

DECOMPOSERS Organisms, usually bacteria or fungi, that use dead animals and plants as food sources by breaking down the tissue and releasing minerals and nutrients back into the environment.

DECOMPOSITION 1. Bacterial breakdown of organic material; **2.** Chemical alteration of organic material or minerals.

DECONCENTRATOR Apparatus for removing suspended solids.

DECONTAMINATION Process of removing unwanted material or substances; the reduction of contamination to an acceptable level.

DECONTAMINATION AREA Enclosed area adjacent and connected to a regulated area and consisting of an equipment room, shower area, and clean room that is used for the decontamination of workers, materials, and equipment.

DEDUCTIBLE CLAUSE Insurance clause detailing the amount or percentage to be deducted from any loss.

DEDUCTION Amount subtracted from an agreed contract sum by change order.

DECURRENT Especially of a leaf, extending downward from the base as two wings along the stem.

DEEP MANHOLE Sewer inspection chamber connected to the surface by an access shaft.

DEEP PERCOLATION Movement of groundwater to deeply buried permeable rock.

DEEP PERCOLATION LOSS Water that percolates downward through the soil, beyond the reach of plant roots.

DEEP SEAL TRAP Trap with a seal of 100 mm (4 in.) or more.

DEEP SEA That part of an ocean beyond the continental shelf.

DEEP SEA DEPOSIT Deposit on the bed of an ocean, below 182 m (600 ft, 100 fathoms) or more.

DEEP SEEPAGE Fraction of runoff that escapes from a reservoir through the underlying earth or rock stratum.

DEEP SHAFT SEWAGE TREATMENT Technique for sewage treatment in which screened raw sewage is introduced to the bottom of a circular shaft up to 150 m (450 ft) deep and allowed to flow up the shaft with compressed air being introduced at one or more points. The mixing action, coupled with the hydraulic pressure at the point(s) of air injection are said to increase biological reduction of the waste by up to 10 times that of conventional sewage treatment processes.

DEEP-WATER WAVE Wave in water having a depth greater than one-half of its length.

DEEP WELL Well that passes through several impermeable strata, any of which may yield water, but pumps water only from the deepest level drilled.

DEEP-WELL PUMP *See* **Pump.**

DEEP-WELL TURBINE PUMP *See*

Pump.

DEFAULT Failure to meet an obligation; failure to make a sum of money available when contractually required, as with a mortgage payment.

DEFECT 1. Imperfection that, by its size, shape, location, or makeup, reduces the useful service of a part; flaw or blemish; **2.** Work, materials, equipment, or system that is unsatisfactory, defective, deficient or which does not comply with the contract documents; does not meet the requirements of any reference standard, test, or approval; or has been damaged before the agent's recommendation of final payment.

DEFECT IN TITLE Recorded instrument that would prevent a grantor from giving clear title.

DEFECTIVE WORK Work (actually performed on site or incorporated into a supplied product) that does not meet the quality specified in the contract documents.

DEFECTS LIABILITY PERIOD Time after a project is handed over to the owner during which the contractor is responsible for remedying defects.

DEFERRED CHARGE Expenditure not chargable to the period in which it is made, but carried as an asset pending amortization or other disposition.

DEFERRED CREDIT Credit balance or item distributed over a subsequent accounting period(s) or as an expense reduction.

DEFERRED LIABILITY Liability not elsewhere provided for, which is not current, on which payment or other disposition is deferred to a future accounting period(s).

DEFERRED MAINTENANCE Physical deterioration that a prospective purchaser would anticipate having to correct following purchase of a property.

DEFERRIZATION Removal of soluble iron compounds from water.

DEFICIENCY 1. Amount by which a quantity falls short of a given demand; **2.** Amount by which the natural flow of a stream or other source fails to meet an irrigation or other demand; **3.** Amount by which the precipitation over a given period falls short of the normal average for the period; **4.** Item or work missing or incomplete as described in the specifications and contract documents; **5.** Illness caused by the lack of an essential food substance.

DEFINING Within an activated-carbon filter, a process that arranges the carbon particles according to size.

DEFLATION Removal of loose surface material by wind action, leaving bare rock.

DEFOAMANT Substance having a low compatibility with foam and a low surface tension.

DEFOLIATOR 1. Insect that destroys foliage; **2.** Chemical that causes plants to drop their leaves.

DEFOLIANTS Substances or mixtures of substances causing leaves or foliage to drop from plants.

DEFORESTATION Permanent removal of forest and undergrowth.

DEFORMATION 1. Change in the shape or dimensions of a body resulting from stress; strain; **2.** Any change in the original state or size of rock masses, especially as produced by faulting.

DEGASIFICATION 1. Removal of gas from a liquid; **2.** Removal of oxygen from a water source or supply.

DEGENERATIVE DISEASE Illness cause by the deterioration of organs or tissue, rather than by infection.

DEGRADATION 1. Breakdown of matter by biological action; **2.** Process by which various parts of the Earth's surface are worn down and carried away by the action of wind and water.

DEGREASING 1. Removal of grease and oil from wastewater, sludge, or solid waste; **2.** Removal of surface oil and grease from industrial machinery and equipment.

DEGREE ANGLE Unit of plane angle equal

to the circumference of a circle divided by 360. Symbol: °. Multiply by 0.017 453 to obtain radians, symbol: rad. *See also the appendix*: **Metric and nonmetric measurement.**

DEGREE CELSIUS Non-SI unit of measure for temperature where the freezing point of water is 0 and the boiling point 100. Symbol: °C. Permitted for use with the SI system of measurement. Add 273.15 to obtain kelvin, symbol: K; multiply by (°C x 9/5) + 32 to obtain degrees Fahrenheit, symbol: °F. *See also the appendix*: **Metric and nonmetric measurement.**

DEGREE DAY Daily measure of the difference between the average outside temperature and 18°C (64.4°F). The seasonal sum of degree days below 18°C (64.4°F) is used in calculating heating requirements.

DEGREE FAHRENHEIT Nonmetric unit of temperature in which 32 is the freezing point of water and 212 the boiling point. Symbol: °F. Multiply by (°F -32) x 5/9 to obtain degrees Celsius, symbol: °C. *See also the appendix*: **Metric and nonmetric measurement.**

DEGREE-HOUR Measure of strength gain of concrete as a function of the product of temperature multiplied by time for a specific interval. *See also* **Maturity factor.**

DEGREE OF ARC Non-SI unit of angle (180° = π·rad), equal to 60 minutes of arc, and 3600 seconds of arc, permitted for use with the SI system of measurement. Symbol: °. *See also the appendix*: **Metric and nonmetric measurement.**

DEGREE OF COMPACTION Measure of the density of a soil sample, estimated by a standard formula.

DEGREE OF CONSOLIDATION Settlement after time, divided by the final settlement.

DEGREE OF CURVE Number of degrees at the center of a circle, subtended by a chord of 30.48 m (100 ft) at its rim. *See also* **Geometry of circular curves.**

DEGREE OF DENSITY Measure of compaction.

DEGREE OF PURIFICATION Measure of the completeness of removal or destruction of specific impurities or chemicals or organic constituents from water by natural or mechanical means.

DEGREE OF SATURATION Ratio of the weight of water vapor associated with a weight of dry air to the weight of water vapor associated with a similar weight of dry air saturated at the same temperature. *See also* **Saturation.**

DEGREE OF SLOPE *See* **Expression of slope.**

DEGREE OF TREATMENT Measure of the removal or reduction of specific impurities or chemical or organic constituents by mechanical and/or chemical processes.

DEHUMIDIFIER 1. Adsoption or absorption equipment for removing moisture from the atmosphere; **2.** Air cooler or washer for lowering the moisture in air pumped through it.

DEHUMIDIFY Reduction, by any process, of the quantity of water vapor or moisture content in the air of a room.

DEHYDRATION Removal of chemically bound, adsorbed or absorbed water from a material.

DE-INKING Recycling processes that remove ink from printed paper and consisting of pulping the raw product followed by centrifugal cleaning, screening, and washing to remove the ink solids. Yield is up to 75%.

DELAMINATION Mixing of a body of water that has become stratified.

DELAYED DENSITY-DEPENDENT Situation in which the mortality among a host population depends on the population density of the host in successive generations, thus affecting the size of the parasite population.

DELIQUESCE Dissolve and become liquid by absorbing airborne moisture.

DELIVERY Volume of fluid discharged by a pump in a given time, usually expressed in liters per minute (L/min) or gallons per minute (gpm).

DELIVERY BOX Enclosure for the control and measurement of water.

DELTA Body of sediment deposited where a fast-flowing body of water empties into standing, or the slow-current water of a lake, bay, or ocean.

DEMAND Amount of a commodity that can be sold at a profit.

DEMAND FACTOR Ratio of the maximum demand on a system to the load imposed upon the system.

DEMAND HORSEPOWER *See* **Horsepower.**

DEMAND SYSTEM Self-contained breathing apparatus that supplies air to the wearer on demand.

DEMANGANIZATION Removal of compounds of manganese from water.

DEMERSAL Organisms that live in the deepest part of a sea or lake.

DEMINERALIZATION Physical, chemical, or biological reduction of the mineral contents of water or soil.

DEMOGRAPHIC TRANSITION Transition in the pattern of increase in a human population from one characterized by high birth and high death rates to one with low birth and low death rates.

DEMOGRAPHY Science dealing with the statistics of births, deaths, diseases, etc., of a community.

DEMOLITION Wrecking or taking out of any load-supporting structural member of a facility together with any related handling operations or the intentional burning of any facility.

DENATURE 1. Change the nature of something; **2.** Make something unfit for human consumption through the addition of a noxious substance without destroying its usefulness for other purposes.

DENDRITIC CRYSTALS Ice crystals commonly found in snow and characteristically formed when ice particles fall through supersaturated air and which grow by sublimation.

DENDRITIC DRAINAGE Drainage pattern or system in which a main channel connects an expanding array of channels of diminishing capacity.

DENDROCHRONOLOGY Science or technique of dating past events by studying the annual growth rings of timber.

DENITRIFICATION Condition when nitrate ions are reduced to nitrogen gas

DENSITY 1. Ratio of the mass of a substance to its volume, commonly expressed as kg/m^3 (lb/yd^3). *See also* **Material density; 2.** Number of similar things within a given area or volume; **3. Ratio** of the combined weight of soil cover and the underlying solid waste to the combined volume of the solid waste and the soil cover.

DENSITY CURRENT Flow of water through a larger body of water that retains its unmixed identity due to a difference in density.

DENSITY-INDEPENDENT Situation in which the percentage mortality or survival of a species varies independently of population density.

DENSITY FLOW Movement of surface water, or an undertow, resulting from a flow of different density, temperature, salinity, etc.

DENSITY OF SNOW Percentage ratio of the volume that a given quantity of snow would occupy if it were reduced to water divided by the volume of snow.

DENSITY SEPARATION Division of a substance into fractions of varying density by flotation in air or liquid.

DENSITY STRATIFICATION Identifiable layers of different density in a body of water.

DENTATE 1. Edged with toothlike projections, typically in the form of a baffle or weir over which water flows and used to break the force of the water and/or to promote aeration; **2.** Shape of the edge of some leaves.

DENTATED SILL Notched sill at the end of an apron that reduces the force of flowing water and the resulting scour below the apron.

DENUDATION Erosion by natural forces of the organic matter of the soil leaving the underlying strata exposed.

DEOXYGENATION Reduction of the dissolved oxygen level in a liquid.

DEOXYGENATION CONSTANT Rate of biochemical oxidation of organic matter in water under aerobic conditions.

DEPENDABLE CAPACITY Capacity that can be relied on for service during all but exceptional circumstances.

DEPLETION 1. Withdrawal of water from a source at a rate greater than the rate of replenishment; **2.** Exhaustion of natural resources.

DEPLETION CURVE Portion of a hydrograph extending from the point of termination of the recession curve to a subsequent rise or alteration of inflow when additional water becomes available for streamflow.

DEPLETION HYDROGRAPH Surface drainage during prolonged rainless periods, expressed in a hydrograph or discharge from the water bodies maintaining base flow.

DEPOSIT 1. Process of laying down; **2.** Percentage of a bid sum or of a contract sum, or a fixed monetary sum required as a pledge at time of tendering or bidding for a contract; **3.** Percentage of a total price paid in advance of receipt of the goods to secure acquisition of the commodity. *Also called* **Advance.**

DEPOSITION Process or act of causing solid material to settle out from a fluid suspension.

DEPOSITION PROCESSES Removal of particulate matter from an air stream by deposition. There are two principal methods:

Dry deposition – where the particle impacts on to soil, water or vegetation at the Earth's surface; and

Wet deposition – where the pollutant is absorbed into droplets of moisture within, or below clouds, followed by removal by precipitation.

DEPOSITION SEDIMENTATION Laying down of potentially rock-forming material by wind, water, ice, or gravity.

DEPOSITORIAL CLAY Clayey sediment transported by flowing water from its place of origin.

DEPOT Storage or collection center.

DEPRECIATED COST METHOD *See* **Cost approach.**

DEPRECIATION 1. Loss in value or of useful life, for any cause; **2.** In the case of machinery and equipment, loss resulting from wear, obsolescence, inadequacy, or any other cause. *See also* **Accelerated depreciation.**

DEPRECIATION ACCRUAL RATE *See* **Depreciation rate.**

DEPRECIATION FACTOR Percentage of the adjusted tax basis that is depreciated over the depreciation period.

DEPRECIATION RATE Annual percentage by which it is estimated that a loss in value of machinery, equipment, or property occurs. *Also called* **Depreciation accrual rate.**

DEPRESSANT Substance or device that lessens undesirable properties.

DEPRESSED SEWER Section of a sewer that flows below the grade line and therefore runs full and at greater than atmospheric pressure.

DEPRESSION 1. Lowering of amount, force, activity, or quality; **2.** Low place or part; hollow; **3.** Area of low barometric pressure.

DEPRESSION HEAD Extent of the lowering of the water surface in a well, and of the watertable or piezometric surface adjacent to the well, resulting from the withdrawal of water from the well.

DEPRESSION SPRING Water flow from

permeable material where the land surface extends down to or below the watertable.

DEPRESSION STORAGE Water held in natural depressions in the strata over which groundwater flows.

DEPTH FILTER Filter medium that primarily retains contaminants within torque passages.

DEPTH FILTRATION Filtration that primarily retains contaminant within tortuous passages.

DEPTH OF FLOTATION Vertical distance from the surface of the water in which a body is floating to the lowest point of the body.

DEPURATION Elimination of a test chemical from a test organism.

DEPURATION PHASE Portion of a bioconcentration test after the uptake phase during which the organisms are in flowing water to which no test chemical is added.

DEPURATION RATE CONSTANT Mathematically determined value that is used to define the deputation of test material from previously exposed test animals when placed in untreated dilution water, usually reported in units per hour.

DERATE Reduce the normal capacity rating of equipment to allow for some defined circumstance or condition.

DERELICT LAND Land that has been damaged by human activity and which is incapable of beneficial use unless treated.

DERIVED WAVE Tide generated by tidal fluctuations in connected waterways.

DERIVED WORKING LIMIT Rate of discharge of a radioisotope, specific for each isotope, and derived by determining the critical pathway for the radiation release and the critical group of people who are most likely to be affected.

DERMAL CORROSION Production of irreversible tissue damage in the skin following the application of a test substance.

DERMAL IRRITATION Production of reversible inflammatory changes in the skin following the application of a test substance.

DESALINATION Process to convert brackish or salt (sea) water to fresh water.

DESCALING Removal of rust from the inside of cast-iron pipe.

DESERT Dry, barren region, usually sandy and without trees; an area in which one or more of the factors necessary to living organisms is in critically short supply.

DESERTIFICATION Conversion of productive land to a less, or nonproductive state due to human activity such as deforestation, soil exhaustion, salinized or mineralized irrigation, depleted groundwater, overgrazing, etc., or from natural causes such as drought, dune shifting, wind and/or water erosion, etc.

DESERT PAVEMENT Layer of closely spaced stones lying on the surface of silt and sand.

DESERT VARNISH Hard, usually black coating of iron and manganese oxides that forms on the surface of rocks in the desert.

DESICCANT Substance capable of absorbing or adsorbing water or water vapor.

DESICCATION 1. Process in which a region experiences a complete loss of its stored water due to decrease of rainfall, increase in evaporation and/or withdrawal. **2.** Process of shrinkage or consolidation of fine-grained soil, produced by an increase of effective stresses in the grain skeleton accompanying the development of capillary stresses in the pore water.

DESIGN Creation that embodies ideas, aims, and objectives; pictorial depiction from which to work.

DESIGN ALTERNATIVES At each of the design stages (concept, preliminary, and working drawings), identification by the designer of several technical solutions that all satisfy the functional requirements and the standards that constrain his or her choices. The alternative designs may be compared in terms of cost effectiveness, often using an analysis of trade-offs to arrive at an optimum. Ideally, the most economic combination of design alternatives is the one that is finally chosen and implemented.

DESIGN ANALYSIS Tabulation and consideration of the physical data, present and probable future requirements relating to an engineering project.

DESIGNATED FACILITY Hazardous waste treatment, storage, or disposal facility that has received a permit (or interim status).

DESIGNATED PERSON *See* **Authorized person,** and **Competent person.**

DESIGNATED POLLUTANT Air pollutant, emissions of which are subject to a standard of performance for new stationary sources but for which air quality criteria have not been issued.

DESIGNATED REPRESENTATIVE Individual or organization to whom written authorization to act on another's behalf has been given in writing.

DESIGNATED USES Those uses specified in water quality standards for each water body or segment, whether or not they are being attained.

DESIGN–BUILD Construction technique whereby the contractor designs the structures as well as builds them.

DESIGN CAPACITY 1. Maximum volume that something is designed to contain, bear, process, etc.; **2.** Weight of solid waste of a specified gross calorific value that a thermal processing facility is designed to process in 24 hours of continuous operation; usually expressed in tons per day.

DESIGN CHANGE Change in the overall design or a component of the design after the design has been approved by the owner, or following completion of the contract documents.

DESIGN CONTINGENCY Estimated allowance for the cost of changes to the design to make the scope work.

DESIGN DEVELOPMENT STAGE Stage in the design process following conceptual,

sketch, or preliminary designs and prior to final working drawings.

DESIGN DOCUMENTS Documents prepared by the designer (plans, design details, and job specifications).

DESIGN ENGINEER One who conceptualizes the overall design and details its parts.

DESIGN ENGINEERING FIRM Professional organization responsible for the design, plans, and specifications to fulfil the scope of work to be performed to successfully complete the design of a project. The firm may also monitor and observe the construction of the project.

DESIGNER Person responsible for the design.

DESIGN FACTOR Ratio of the nominal strength of a product or component to the total working load.

DESIGN FLOOD Greatest flow that a structure is designed to accommodate or withstand.

DESIGN HOURLY VOLUME *See* **Volume.**

DESIGN LIFE Period of time during which a product or assembly will meet the minimum design standards and requirements.

DESIGN LOAD Factored load. *See also* **Maximum intended load.**

DESIGN RULE Regulation that requires manufacturers to design products so that they will conform to an environmental or other standard.

DESIGN SPECTRUM Consideration of all possible stress ranges, caused by the worst possible load conditions, acting in the most adverse manner.

DESIGN STAGE Stage in the development of drawings that will become part of the contract documents following conceptual and sketch design and prior to working drawings.

DESIGN STORM 1. Storm strength and/or flow for which a hydraulic structure or facil-

153

ity is designed; **2.** Rainfall estimate corresponding to an enveloping depth–duration curve for the selected frequency.

DESIGN STRENGTH 1. Nominal strength of a member multiplied by a strength reduction (Phi) factor. *See also* **Nominal strength,** and **Phi factor**; **2.** Resistance (force, moment, stress, as appropriate) provided by an element or connection; the product of nominal strength and the resistance factor; **3.** Strength of a material, as used in calculations, so that allowable stress is not exceeded under the applicable loading conditions.

DESIGN TEAM Group whose members contribute their individual talents, specialities, and skills toward the resolution of a design problem.

DESIGN VALUE NUMBER Means of rating or classifying materials as to their properties, stability, strength, etc.

DESIGN VOLUME *See* **Volume.**

DESIGN WAVE Type or types of wave of known characteristics against which protection is incorporated in the design of bank protection, harbors, and other marine structures.

DESIGN WEIGHT Maximum weight to which a component may be loaded, without the danger of failure and/or premature wear taking place.

DESIGN WORKING PRESSURE Maximum working pressure for which a system, or part of a system, is designed.

DESILTING BASIN *See* **Sedimentation tank.**

DESILTING WORKS One or more basins installed immediately below the diversion structures of canals to allow for the removal by settlement of silt.

DESLUDGE Removal of accumulated sludge, from a catchpit, catchbasin, etc.

DESMIDIACEAE Large group of unicellular or colonial, freshwater green algae which often form films on mud and aquatic plants.

DESTINATION FACILITY Disposal facility, incineration facility, or facility that both treats and destroys regulated medical waste.

DESTROYED REGULATED MEDICAL WASTE Regulated medical waste that is no longer generally recognizable as such because it has been ruined, torn apart, or mutilated (it does not mean compaction) through: **(a)** processes such as thermal treatment or melting, during which treatment and destruction could occur; or **(b)** processes such as shredding, grinding, tearing, or breaking, during which only destruction would take place.

DESTRUCTION FACILITY Facility that destroys regulated medical waste by ruining or mutilating it or tearing it apart.

DESTRUCTION OR ADVERSE MODIFICATION Direct or indirect alteration of critical habitat which appreciably diminishes the likelihood of the survival and recovery of threatened or endangered species using that habitat.

DESTRUCTIVE DISTILLATION Process similar to pyrolysis in which solid substances are heated within a closed retort in the absence of air with the ensuing volatile gases being condensed to a distillate.

DESTRUCTIVE TESTING Testing of a material, component or sample beyond its capability to withstand stress so as to determine its maximum strength.

DESULFURIZATION Removal from a substance of sulfur or sulfur compounds.

DETACHABLE CONTAINER *See* **Waste container.**

DETACHABLE CONTAINER SYSTEM Partially mechanized self-service refuse removal procedure with specially constructed containers and vehicles. It is mechanized in that special equipment is used to empty the containers and haul refuse to the disposal site. It is self-service when the customer deposits the refuse in the container.

DETAIL DRAWING Large-scale drawing detailing small parts of a larger whole.

DETAILER Draftsman who prepares drawings of details as part of a set of drawings or for shop and site use.

DETAILING Preparation of working drawings.

DETECTION LAG Interval of time between the moment some change occurs or is made in a process and the moment when such a change is detected by an associated sensing or measuring instrument.

DETECTION LIMIT Minimum concentration of an analyte (substance) that can be measured and reported with a 99% confidence that the analyte concentration is greater than zero.

DETECTOR Mechanical or electronic device that senses and signals an event or presence.

DETECTOR CHECK VALVE *See* **Valve.**

DETENTION DAM Small, temporary obstruction to reduce or prevent the flow of surface runoff, and/or to cause the deposition of suspended solids.

DETENTION PERIOD Average time that a unit volume of flow remains in a tank during flow process.

DETENTION RESERVOIR Storage area to which excess flow is directed until such time as a stream can safely carry its normal flow plus the released water.

DETENTION TANK Tank of sufficient capacity to hold the full flow of a stream for sufficient time for physical or chemical reactions to occur.

DETENTION TIME Amount of time that it takes for a flow to pass through a tank or vessel.

DETERGENT Chemical agent that includes petrochemical or other synthetically-derived wetting agents, used to clean and degrease surfaces.

DETRITAL SEDIMENTS Sediments formed from fragments of preexisting minerals and rocks and from the alteration products of rocks, transported to site and then compacted.

DETRITUS Loose material produced by the disintegration of rocks or of the skeletal remains of organisms.

DETRITUS SLIDE Tendency for detritus to move downhill.

DETRITUS TANK Detention chamber where the flow of liquid waste is slowed, allowing sediment and other settleable solids to sink to the bottom where they can be removed by mechanical equipment.

DEUTRIC Alteration of igneous rock by the action of volatiles derived from the magma during the later stages of consolidation.

DEUTERIUM OXIDE An isotopic form of water with composition D_2O, present in natural water as approximately 1 part in 6500 and isolated for use as a moderator in certain nuclear reactors: heavy water.

DEVELOPED LENGTH Length of connected pipe and fittings, measured along the centerline.

DEVELOPED WATER 1. Groundwater artificially brought to the surface; **2.** Artificially induced flow of a stream.

DEVELOPMENT Work done to a dug or drilled well to improve the flow in the aquifer in the immediate vicinity.

DEVELOPMENTAL TOXICITY Property of a chemical that causes *in utero* death, structural or functional abnormalities or growth retardation during the period of development.

DEVELOPMENT-INDUCED DIS-PLACEMENT Population movement due to the industrial development of land formerly used for agriculture and other low-density uses, or the inundation of populated areas due to dam construction, etc.

DEW Water droplets condensed from the air, usually at night, onto cool surfaces.

DEWATER Removal of the water from recently produced tailings by mechanical or evaporative methods such that the water content of the tailings does not exceed 30% by weight.

DEWATERABLE Capable of having water drained from it, with or without chemical aids.

DEWATERING 1. Removal of surface water from an area; **2.** Lowering of the groundwater table to produce a 'dry' area in the vicinity of an excavation that would otherwise extend below water; **3.** Removal of water by filtration, centrifugation, pressing, open-air drying, or other methods.

DEWATERING PUMP See **Pump.**

DEW POINT Temperature at which air becomes saturated with moisture and below which condensation occurs.

DEWPOINT DEPRESSION Difference between the prevailing temperature and the current dew point.

DEWPOINT HYGROMETER Instrument that indicates the dew point, from which the relative humidity can be calculated when the air temperature is known.

DIAGNOSTICS Component-by-component self-check performed on a mechanical, electrical, or electronic device, or series of connected or interdependent components, by a microprocessor programmed to detect irregularities or abnormalities.

DIAGRAM Drawing that illustrates pertinent characteristics, component positions, sizes, interconnections, controls, and actuation of components. There are many types of presentation, including:

> **Cutaway:** Drawing showing principal internal parts of a component, controls and actuating mechanisms, etc., all interconnecting lines and functions of individual components.

> **Graphical:** Drawing or drawings showing each piece of apparatus including all interconnecting lines by means of standard symbols.

> **Line drawing:** Scale drawing in which objects are represented by single lines with no attempt made to show the relative thickness of what is being

shown. In the case of a building, for instance, the lines might indicate the centerlines of walls, or the outside faces of walls. *Also called* **Single-line diagram.**

> **Network:** Graphical display of the sequence, timing and interrelationships of the activities comprising a project.

> **Pictorial:** Drawing showing each component in its actual shape according to the manufacturer's installation drawings.

> **Schematic:** See **Graphical,** above.

> **Schematic wiring:** Drawing that shows a wiring layout using graphic symbols, the electrical connections and functions of a specific circuit arrangement.

> **Single line:** See **Line drawing,** above.

DIAL GAUGE Instrument that shows, by a needle indication on a circular graduated dial, very small displacements of the plunger, indicating pressure, or movement, or temperature.

DIAL INDICATOR 1. Measuring instrument, commonly calibrated in 0.001 in. or 0.01 mm, that shows the reading on a dial; **2.** Device used to determine when the axis of two shafts are in line.

DIALYSATE Stream being depleted of salt by electrodialysis.

DIALYSIS Separation of smaller molecules from larger molecules, or of crystalloid particles from colloidal particles, in a solution by selective diffusion through a semipermeable membrane.

DIAMETER Length of a straight line passing from one side of a circle to the other and passing through the center. *See also* **Inferior diameter,** and **Superior diameter.**

DIAMOND DUST Form of ice crystals.

DIAPHRAGM Flexible partition between two chambers.

DIAPHRAGM PACKING Packing between rigid members in relative motion that is attached to both members and which absorbs the motion through its own deformation.

DIAPHRAGM PUMP *See* **Pump.**

DIAPHRAGM PRESSURE TANK Closed tank divided into compartments by a rubber diaphragm; one space is filled with air or nitrogen, the other with water under pressure.

DIAPHRAGM-TYPE CELLULAR COFFERDAM Structure made of steel sheet piles with each of the inner and outer walls consisting of a series of arc segments that are connected at their intersections with diaphragms that extend through the cofferdam to form a series of cells. The cells are filled with earth, sand, gravel, or rock. *See also* **Circular-type cellular cofferdam.**

DIAPHRAGM-TYPE CONTROLLER Control device that relies upon positive or negative pressure causing movement of a diaphragm for actuation or initiation.

DIATOM Any of various minute, unicellular or colonial algae of the class Bacillariophyceae, having siliceous cell walls consisting of two overlapping, symmetrical parts.

DIATOMACEOUS EARTH Friable earthy material composed of nearly pure hydrous amorphous silica (opal) and consisting essentially of the frustules of the microscopic plants called diatoms. *Also called* **Kieselguhr.**

DIATOMACEOUS-EARTH FILTER Device in which a layer of diatomaceous earth serves as the filtering medium.

DIATOMACEOUS EARTH FILTRATION Process resulting in substantial particulate removal in which (a) a precoat cake of diatomaceous earth filter media is deposited on a support membrane (septum), and (b) while the water is filtered by passing through the cake on the septum, additional filter media known as body feed is continuously added to the feed water to maintain the permeability of the filter cake.

DIATOMITE Fine, powdered diatomaceous earth used as a filtering agent, absorbent, clarifier, and insulator.

DICAMBA Broad-spectrum chlorobenzoic acid herbicide used in large quantities for general weed control in agriculture. It is not strongly adsorbed onto soil particles and is readily leached into groundwater. It has occasionally been detected in public and private water supplies and in some surface waters. The maximum acceptable content for dicamba in drinking water, derived from the acceptable daily intake, is 0.12 mg/L.

DIE-FORMED RING Packing ring, cut to length and mechanically compacted to remove voids prior to being installed in a stuffing box.

DIELDRIN Poisonous insecticide, used for killing wood-destroying insects.

DIELECTRIC Material that does not conduct direct electrical current.

DIELECTRIC UNION Nonconducting plumbing fitting, used to join pipes manufactured of different materials to prevent electrolysis.

DIELECTRIC STRENGTH Ability of insulation to withstand voltage without breaking down: expressed in volts per mil.

DIESEL ENGINE Internal combustion engine that burns crude oil, ignition being brought about by heat compression.

DIESEL FUEL That portion of crude oil that distills out within the temperature range of approximately 200°C to 370°C (392°F to 698°F). There are two grades:

> **Diesel 1:** A kerosene-type fuel, lighter, more volatile, and cleaner burning that Diesel 2, used in engine applications where there are frequent changes in speed and load.

> **Diesel 2:** A fuel used in industrial and heavy mobile service.

DIESEL INDEX Approximation of the cetane number of a distillate fuel: the product of the API gravity and the aniline point (°F) divided by 100.

DIESELING Explosions of mixtures of lubricating oil and air in the compression chambers or other parts of the air system of a compressor. *Also called* **Prefiring,** and **Preignition.**

DIETARY LC$_{50}$ Statistically derived estimate of the concentration of a test substance in the diet that would cause 50% mortality to a test population under specified conditions.

DIFFERENCE IN ELEVATION Vertical dimension between two level surfaces, not withstanding their horizontal displacement.

DIFFERENTIAL DIAPHRAGM Diaphragm between two fluids at substantially different pressures.

DIFFERENTIAL DRY-PIPE VALVE *See* **Valve.**

DIFFERENTIAL GAUGE Gauge that measures the pressure difference between two points in a fluid-filled pipe or receptacle.

DIFFERENTIAL INTAKE Water intake formed by a hollow, low diversion dam having a longitudinal slot for intercepting part of the streamflow, conducting it to a pipeline at one or both ends of the dam.

DIFFERENTIAL PLUNGER PUMP *See* **Pump.**

DIFFERENTIAL PRESSURE *See* **Pressure.**

DIFFERENTIAL PRESSURE INDICATOR *See* **Indicator.**

DIFFERENTIAL PRESSURE VALVE *See* **Valve.**

DIFFERENTIAL SETTLEMENT 1. Where foundations of different parts of the same structure settle at different rates; **2.** In a landfill, the nonuniform subsidence of the landfill surface due to the variety in waste types (biodegradable *vs* inert), decomposition rates, compactive effort, and any voids when the waste was placed.

DIFFERENTIAL SURGE TANK Surge intermediate between a single- and restricted-orifice surge tank, equipped with an internal riser smaller in diameter than the connection to the conduit, with a port at the base of the rise communicating with the tank.

DIFFLUENCE Horizontal flowing apart of air particles.

DIFFRACTED WAVE Wave occurring in an area sheltered by a protecting barrier, different to waves forming in open water outside the barrier.

DIFFRACTION Modification of the behavior of light or other waves resulting from limitation of their lateral extent, as by an obstacle or aperture; **2.** Phenomenon resulting from the propagation of waves into a sheltered area behind a breakwater or similar barrier that produces an interruption of an otherwise regular train of waves.

DIFFRACTION ANALYSIS Use of diffracted electromagnetic radiation to study the structure of solids.

DIFFUSED-AIR AERATION Aeration of a liquid produced by the passage of air through it.

DIFFUSED SURFACE WATER 1. Flood water that has escaped from a stream or channel; **2.** Surface runoff that has not reached or formed a stream or definite channel.

DIFFUSE FIELD Condition where sound pressure is equal at every point in a sound field, and sound waves are likely to be travelling in all directions.

DIFFUSER 1. Inner shell and water passages of a centrifugal pump; **2.** Device through which compressed air is injected in the form of fine bubbles into a liquid waste; **3.** Device, typically a perforated pipe, through which flow is distributed to reduce its velocity.

DIFFUSER PLATE Porous plate through which compressed air is permitted to escape and which, when submerged in a tank, causes air to become entrained in the fluid and to promote mixing.

DIFFUSER TUBE Porous tube used to diffuse air into water and/or wastewater.

DIFFUSING WELL Well into which water is injected to restore an aquifer.

DIFFUSION 1. Spreading or scattering; **2.** Movement of dissolved elements in water or air from areas of higher concentrations to areas of lower concentration.

DIFFUSION VANE Fixed or removable casting in a pump, between the casing and impeller, through which liquid passes, designed to convert velocity head to pressure head.

DIGESTED SLUDGE Sewage sludge that has been subjected to aerobic, or anaerobic digestion to reduce its volatile content.

DIGESTER 1. Vessel in which sewage sludge is processed under anaerobic conditions to produce, following further processing, an inert cake, with methane gas being produced as a by-product; **2.** Heated pressure vessel in which wood chips are reacted with chemicals in the initial stage of pulp production.

DIGESTER COILS System of pipes through which hot water or steam is passed, installed in a sludge digestion tank to heat the contents to within the mesophilic range.

DIGESTER SYSTEM Each continuous digester or each batch digester used for the cooking of wood in white liquor, and associated flash tank(s), chip steamer(s), and condenser(s).

DIGESTION In waste treatment, the biochemical decomposition of organic matter under aerobic or anaerobic conditions.

DIGESTION CHAMBER Lower portion of an Imhoff tank in which sludge is digested.

DIGESTION TANK Sealed tank into which sewage sludge is pumped and in which sludge digestion takes place,

DIGIT 1. Finger or toe; **2.** Numeral from 0 to 9.

DIGITAL Of or pertaining to the general class of devices or circuits whose output varies in discrete steps (i.e. pulses or 'on-off' characteristics).

DIGITAL CIRCUIT *See* **Binary circuit.**

DIGITAL ENCODER Device capable of converting movement into numbers.

DIGITAL INDICATION SYSTEM System that monitors performance by using digital display calibrators.

DIGITAL PROCESSING Electronic data processing technology in which information is expressed in numerical form. It is many times faster than analog processing and requires less space and provides greater reliability.

DIGITAL READOUT Device that uses numbers to indicate the value or measurement of a variable.

DIGITAL-TO-ANALOG CONVERTER Device capable of converting a digital signal to an analog voltage.

DIGITIZER Electronic device that converts an image into Cartesian coordinates.

DIKE, or DYKE 1. Obstacle used to protect land from inundation by water from the sea or a river. *See also* **Seawall; 2.** Raised section built onto the sides of roads to control water runoff and erosion; **3.** Slender rock formation that cuts across the structure of surrounding rock; **4.** Embankment or ridge of either natural or man-made materials, used to prevent the movement of liquids, sludges, solids, or other materials.

DILATANCY Property of silt whereby if it is agitated it exudes water, and if compressed, the water is reabsorbed, leaving a matt surface.

DILUENT Material added to a substance to reduce the concentration of the active ingredient.

DILUTE AND DISPERSE Concept of industrial waste disposal that relies on dilution of the source material, typically into groundwater or a flowing stream or large lake, and its eventual dispersion to migration, current, flow, etc.

DILUTION Reduction of the concentration of a soluble material by the addition of a solvent or water.

DILUTION FACTOR Ratio of a concentrated flow (of raw sewage or treatment process effluent) to the average quantity of diluting water available at the point of disposal.

DILUTION GAUGING Method of flow measurement involving the constant flow of a solution of known concentration for sufficient time at a specific section of a conduit and the determination of the resulting dilution of the solution at a downstream station.

DILUTION WATER Water used to produce a flow-through condition of the test to which a test substance is added and to which a test species is exposed.

DILUVIUM Coarse, detrital material; material resulting from a flood.

DIMENSION Distance in a given direction or along a given line.

DIMENSIONAL ANALYSIS Evaluation of complex design problems by first grouping similar or like elements to reduce the number of variables to be considered. The elements can then be mathematically modelled and their relationship under various situations and conditions calculated.

DIMENSIONAL STABILITY Quality of not changing in size due to any cause: moisture change, heat, cold, pressure, etc.

DIMENSIONAL TOLERANCES Allowable differences in dimensions, squareness or thickness.

DIMENSIONLESS UNIT HYDROGRAPH Unit graph used to compare unit hydrographs of different drainage areas or those resulting from different storm patterns.

DIMENSION LINE On drawings, a line with arrowheads or other limiters at either end, marked with a written dimension, used to show the distance between the two points.

DIMINISHING PIPE Tapered pipe.

DIMINUTION Gradual reduction: of size, intensity, etc.

DIMPLE SPRING Spring occurring in a small depression reaching below the watertable.

DINOSEB Selective, non-systemic herbicide and desiccant that is very toxic to humans and which has been detected in public and private water supplies. The maximum acceptable content for dinoseb in drinking water, derived from the average daily intake, is 0.01 mg/L.

DIOXINS Class of highly poisonous organic compounds that form as a result of incomplete or inefficient combustion of carbon compounds.

DIP 1. Angle at which strata, beds, or veins are inclined from the horizontal; **2.** Lower bend of a trap or the depth of its water seal.

DIPMETER Instrument to record the depth below ground level of the surface of the water in a borehole or piezometric tube.

DIPPER DREDGE Mechanical shovel mounted on a scow, used to remove material from below water level.

DIP PIPE Downward-facing 90° bend with its end below the water surface of an interceptor, to prevent floating gasoline from moving between chambers.

DIP STICK Measuring rod used to gauge the level of liquids within a vessel; of fluids in the reservoirs of machines.

DIQUAT Bipyridyl herbicide generally marketed as a dibromide salt or a dichloride monohydrate. The maximum acceptable content for diquat (measured as the cation) in drinking water, derived from the average daily intake, is 0.07 mg/L.

DIRECT-ACTING PUMP *See* **Pump.**

DIRECT CHARGES Costs that can be identified specifically with a product, service, or activity. *See also* **Charges,** and **Indirect charges.**

DIRECT CIRCULATION Natural condition in which cold air sinks and warm air rises to create flow, and subsequently, wind.

DIRECT-CONNECTED PUMP *See* **Pump.**

DIRECT DISCHARGE Discharge of a pollutant.

DIRECT-DUMP TRANSFER SYSTEM Unloading of solid waste directly from a collection vehicle into an open-top transfer trailer or container.

DIRECT ENERGY One of two components of solar energy arriving at the Earth's surface. *See also* **Diffuse energy.**

DIRECT ENERGY CONVERSION Production of electricity from the conversion of chemical, solar or nuclear energy without involving a mechanical work stage.

DIRECT EXPENSES All expenditures incurred by or attributable to a specific project.

DIRECT-FEED INCINERATOR *See* **Incinerator.**

DIRECT FILTRATION Series of processes, including coagulation and filtration but excluding sedimentation, resulting in substantial particulate removal.

DIRECT FIRE PRESSURE Pressure necessary in a water distribution system to sustain adequate fire streams without the use of pumpers when the fire hose is connected directly to the hydrant.

DIRECT IN-LINE MECHANICAL LINKAGE Metal control rods linked directly to valve sections. This type of linkage provides the operator with positive control as well as parts durability.

DIRECTIONAL CONTROL VALVE *See* **Valve.**

DIRECTION OF IRRIGATION Direction of flow of irrigation water; usually perpendicular to the supply ditch or pipe.

DIRECT IRRIGATION Application of wastewater effluent directly to land by spraying,

DIRECT LABOR Labor directly involved in productive operations, as distinguished from that connected with a support or administrative function.

DIRECT MATERIALS Materials incorporated in a product or project and which can be charged directly to it.

DIRECT OXIDATION Direct combination of substances with oxidants, i.e. by means of

sparging, air diffusion or use of such agents as chlorine, bromine, etc.

DIRECT PHOTOLYSIS Direct absorption of light by a chemical followed by a reaction which transforms the parent chemical into one or more products.

DIRECT RUNOFF Runoff that discharges to stream channels over the ground surface without entering the watertable.

DIRECT SOLAR GAIN Warming of an area by solar energy directly entering the area.

DIRT CAPACITY Weight of a specified artificial contaminant that must be added to the fluid to produce a given differential pressure across a filter under specified conditions. Used as an indication of relative service life. *Also called* **Contaminant capacity,** and **Dust capacity.**

DISC One or more rows of plate-shaped steel wheels that cut into the earth and roll over it to fragment and mix the soil.

DISCHARGE 1. Flow from a pipe, culvert, channel, etc.; **2.** Rated output of a pump; **2.** Includes, but is not limited to, any spilling, leaking, pumping, pouring, emitting, emptying, or dumping.

DISCHARGE ALLOWANCE Amount of pollutant (mg/kg of production unit) that a plant is permitted to discharge.

DISCHARGE AREA Cross-sectional area of a conduit, channel, waterway, etc., at the point where the flow is released.

DISCHARGE CAPACITY Maximum rate of flow that a conduit, channel, waterway, etc., is capable of passing.

DISCHARGE COEFFICIENT Factor used in figuring flow through an orifice. It takes into account the fact that a fluid flowing through an orifice will contract to a cross-sectional area that is even smaller than that of the orifice, and that there is some dissipation of energy due to turbulence.

DISCHARGE CONSENT Permission, usually involving qualifying criteria, given by an authorizing authority for the discharge of a liquid effluent to a receiving watercourse or sewer.

DISCHARGE CONVEYOR Conveyor that transports and deposits material away from the machine.

DISCHARGE CURVE Curve relating the water level of a stream to its rate of discharge.

DISCHARGE HEAD Distance between the intake of a pump and the point at which it discharges freely into the air.

DISCHARGE HOSE Temporary connection between the discharge outlet of a pump and the point of delivery.

DISCHARGE HYDROGRAPH Statistical plot of the discharge or flow of a conduit, channel, stream, etc.

DISCHARGE MEASUREMENT Determination of the quantity of water flowing per unit of time in a conduit, channel, stream, etc.

DISCHARGE OF A POLLUTANT Addition of any pollutant or combination of pollutants to public waters or air from any point source, or addition of any pollutant or combination of pollutants to the waters or air of the contiguous zone or the ocean from any point source other than a vessel or other floating craft which is being used as a means of transportation.

DISCHARGE OF DREDGED MATERIAL Addition from any point source of dredged material into public waters, including the addition of dredged material into the runoff or overflow from contained land or water dredge material disposal area.

DISCHARGE OF FILL MATERIAL Addition of fill material into public waters or lands, including placement of fill that is necessary to the construction of any structure; the building of any structure or impoundment requiring rock, sand, dirt, or other materials for its construction; site-development fills for recreational, industrial, commercial, residential, and other uses, causeways or road fills; dams and dikes; artificial islands; property protection and/or reclamation devices such as riprap, groins, seawalls, breakwa-

ters, and revetments; beach nourishment; levees; fill for structures such as sewage treatment facilities, intake and outfall pipes associated with power plants and subaqueous utility lines; and artificial reefs.

DISCHARGE OF HAZARDOUS WASTES Accidental or intentional spilling, leaking, pumping, pouring, emitting, emptying, or dumping of hazardous waste into or on any land, air or water.

DISCHARGE OF POLLUTANTS Addition of any pollutant to navigable waters or air from any point source or to the waters or air of the contiguous zone or the ocean from any point source.

DISCHARGE PIPE Pipe connecting the discharge outlet of a pump to the point of delivery.

DISCHARGE POINT Point within a disposal site at which dredged or fill material is released.

DISCHARGE RATING CURVE Statistical plot showing the relationship between the discharge of a hydraulic or plumbing fixture or appurtenance and the hydraulic conditions affecting the discharge.

DISCHARGES Includes, but is not limited to, any spilling, leaking, pumping, pouring, emitting, emptying, or dumping.

DISCHARGE STANDARDS Criteria established by a regulating authority setting out the minimum standards that an effluent must meet before its discharge.

DISCHARGE TABLE Table showing the relationship between the gauge height and discharge of a conduit, channel or stream at a gauging station.

DISCHARGE VALVE See **Valve.**

DISCONNECTING SWITCH See **Switch.**

DISCONNECTING TRAP Intercept trap.

DISCONTINUOUS EASEMENT Easement requiring for its exercise an action by one party, as a right-of-way.

DISCONTINUOUS INTERSTICE Small

open space in rock.

DISEASE VECTOR Rodents, flies, and mosquitoes capable of transmitting disease to humans.

DISINFECTANT Any oxidant, including but not limited to chlorine, chlorine dioxide, chloramines, and ozone added to water in any part of the treatment or distribution process, that is intended to kill or inactivate pathogenic microorganisms.

DISINFECTANT CONTACT TIME Time, in minutes, that it takes for water to move from the point of disinfectant application or the previous point of disinfectant residual measurement to a point before or at the point where residual disinfectant concentration is measured.

DISINFECTION Process which inactivates pathogenic organisms in water by chemical oxidants or equivalent agents; sterilization; the destruction of microorganisms that harm humans.

DISINTEGRATION Reduction into small fragments and subsequently into particles. *See also* **Deterioration,** and **Weathering.**

DISK METER Positive displacement-type water meter, used on pipe 50 mm (2 inches) or less in diameter.

DISK SCREEN Flat, bed-like sizing device consisting of rows of disks on driven shafts that rotate the disks to transport material along its length. The up-and-down motion of the material as it travels over the disks separates it, causing smaller portions to fall between the disks to a collector or conveyor below. The spacing of the disks determines the size of the material falling through.

DISMEMBERED STREAM Tributary isolated from its main stream by drowning.

DISPERSANT Material that deflocculates or disperses finely ground materials by satisfying the surface energy requirements of the particles; used as a slurry thinner or grinding aid.

DISPERSE: Distribute or spread material in an orderly manner.

DISPERSION 1. Finely divided particles of a material in suspension in another substance; **2.** Mixing and dilution of one substance or concentration within another, larger volume.

DISPERSION INDEX Measure of the time for short-circuiting of liquid through a continuous-flow tank.

DISPERSION TECHNIQUE Technique that attempts to affect the concentration of a pollutant in the ambient air.

DISPERSIVE CLAY Clay that deflocculates in the presence of water.

DISPLACEMENT 1. Quantity of fluid that can pass through a pump, motor, or cylinder in a single revolution or stroke; **2.** Volume of water displaced by a floating object or vessel, which is the same weight as that of the object or vessel; **3.** Amount of motion associated with waves or vibration, measured in millimeters or inches; **4.** Relative movement of any two sides of a fault measured in any direction.

DISPLACEMENT METER Meter that measures the quantity of a liquid flow by recording the number of times a container of known volume is filled and emptied.

DISPLACEMENT PUMP *See* **Pump.**

DISPLACEMENT TIME Average time that a moving liquid is held, or detained in a tank or channel.

DISPLACEMENT VELOCITY Rate of displacement by inflowing liquids of the contents of a settling tank.

DISPOSABLE CONTAINER *See* **Waste container.**

DISPOSABLE ELEMENT Filter that is discarded and replaced at the end of its service life.

DISPOSABLE FILTER Filter consisting of a filter element encased in a housing that is discarded and replaced in its entirety at the end of the service life of the element.

DISPOSAL 1. Permanent isolation of spent nuclear fuel or radioactive waste from the accessible environment with no intent of recovery, whether or not such isolation permits the recovery of such fuel or waste. For example, disposal of waste in a mined geologic repository occurs when all of the shafts to the repository are backfilled and sealed; **2.** Collection, storage, treatment, utilization, processing, or final disposal of solid waste; **3.** Discharge, deposit, injection, dumping, spilling, leaking, or placing of any hazardous waste into or on any land or water so that it may enter the environment or be emitted into the air or discharged into any waters, including ground water.

DISPOSAL AREA Area where excavated material can be dumped.

DISPOSAL BY DILUTION Effluent disposal by discharge into a body of water.

DISPOSAL FACILITY Facility at which hazardous waste is intentionally placed into or on any land or water, and at which waste will remain after closure.

DISPOSAL FIELD *See* **Leach field.** *Also called* **Tile bed.**

DISPOSAL LEVY Refundable tax imposed on various classes of goods, typically tires, vehicle batteries, etc., at time of purchase, intended to offset their cost of destruction or recycling, that may be claimed only at accredited points of disposal.

DISPOSAL SITE 1. Interim or finally approved and precise geographical area within which ocean or land dumping of wastes is permitted under conditions specified in permits; **2.** Waters where specific disposal activities are permitted and consisting of a bottom surface area and any overlying volume of water. In the case of wetlands on which surface water is not present, the disposal site consists of the wetland surface area.

DISPOSAL SITE DESIGNATION STUDY Collection, analysis and interpretation of all available pertinent data and information on a proposed disposal site prior to use, including but not limited to, that from baseline surveys, special purpose surveys, public data archives, and social and economic studies and records of areas which would be affected by use of the proposed site.

DISPOSAL WELL Well used for the disposal of waste into a subsurface stratum.

DISSIPATER Device that reduces the speed and abrasive power of flowing water.

DISSOCIATION Breakdown of a substance into simpler substances from the effect of heat or the addition of a solvent.

DISSOLVED AIR Air that is dispersed in a fluid to form a mixture.

DISSOLVED OXYGEN Amount of oxygen dissolved in water, expressed as percent saturation, milligrams per litre, or parts per million.

DISSOLVED-OXYGEN SAG CURVE Plot of the uptake of dissolved oxygen associated with biochemical oxidation of organic matter, plus reoxygenation from absorbed atmospheric oxygen and biological photosynthesis, along the course of a stream or channel.

DISSOLVED SOLIDS Anhydrous residues of the dissolved constituents in water.

DISSOLVED WATER Water that is dispersed in a fluid to form a mixture.

DISTANCE Measure of a space or interval.

DISTANCE GAUGE Gauge used to measure a repetitive dimension.

DISTILLATION Process of evaporation and recondensation in which various fractions can be separated according to their boiling points or boiling ranges.

DISTILLED WATER Water formed from the condensation of steam or water vapor.

DISTRESS Soil that is in a condition where a cave-in is imminent or is likely to occur.

DISTRIBUTARIES Channels or conduits that do not return flows to their originating stream but which feed an irrigation system or discharge into another stream or the ocean.

DISTRIBUTARY STREAM Watercourse flowing between two streams, the contents of which may change direction from the stream of higher flow to that of lower flow.

DISTRIBUTED SAMPLE Soil sample that has been thoroughly mixed and is therefore not representative of its original characteristics.

DISTRIBUTION Pipes, ducts and other means by which services are delivered to the various points within a building.

DISTRIBUTION BOX Buried chamber that receives the effluent from a septic tank and distributes it equally to the filter drains.

DISTRIBUTION GRAPH Unit hydrograph in which the ordinates of flow are expressed as percentages of the total volume. *See also* **Unit hydrograph.**

DISTRIBUTION PIPE Pipe carrying water from a storage tank or main to a point of use.

DISTRIBUTION RATIO Ratio of maximum possible 24-hour application of wastewater to a high-rate filter to the actual average daily application.

DISTRIBUTION RESERVOIR Water storage facility that is part of a distribution system.

DISTRIBUTION SYSTEM Complex of conduits and associated appurtenances by which potable water is distributed to consumers throughout a community.

DISTRIBUTION TILE Agricultural tiles, laid in rows to form a drainage bed, that receive the effluent from a septic tank.

DISTRIBUTOR 1. Means of dividing fluid flows from a single path to two or more parallel paths; **2.** Fixed or movable device used to apply liquid over an area at a constant rate.

DISTRICT Specific or defined area of jurisdiction.

DISTRICT HEATING Concept whereby a number of physically separate buildings are heated from a single boiler house or heating plant, that may, or may not be part of one of the structures.

DISTRICT MAP Map showing the boundaries of the districts into which a planning area is divided.

DISTURBED SAMPLE Sample of soil taken without effectively minimizing disturbance of the soil mass.

DISUSE OF LAND Authority granted some public authorities to enforce actions such as removal of debris, basic site development, clearing of unmarketable titles, etc., by the owners of disused land.

DITCH Long narrow excavation, cut or channel in earth, deeper than it is wide, often for conveying drainage or other water but also to permit installation or construction. *See also* **Trench.**

DITCH CHECK Barrier placed in a ditch to reduce the stream velocity of flowing water.

DITCHER Machine used to excavate a ditch. *See also* **Ladder ditcher,** and **Wheel ditcher.**

DITCH LINING Material used to cover the inner face of a ditch to reduce or prevent exfiltration, infiltration, or erosion.

DITCH OXIDATION Activated-sludge wastewater treatment process in which the flow is conducted around an endless ditch by means of paddle wheels or similar devices that promote flow and simultaneously aerate and mix the contents.

DIURNAL 1. Temperature, relative humidity, wind, stability and any other changes between daytime and nighttime; **2.** In tidal hydraulics, having a period or cycle of approximately one tidal day.

DIURNAL FLUCTUATIONS 1. Cyclic rise and fall of the watertable or a stream flow in response to changes in the rate of evaporation; **2.** Any regular cyclic change occurring in 24 h, e.g. tides, leaf movement, photosynthesis, etc.

DIURNAL FORCE Tide-producing force having a period of one lunar day.

DIURNAL INEQUALITY Difference in height between am and pm tides, or the difference in velocity between the two daily flood currents and ebb currents, due principally to lunar declination.

DIURNAL TIDE Tide having only one high and one low water in each 24-hour cycle.

DIURNAL WAVE Regular daily variation in air pressure at one point.

DIURON Substituted urea-based herbicide considered to have high potential for leaching from soil and subsequent groundwater contamination. The maximum acceptable concentration for diuron in drinking water, derived from the average daily intake, is 0.15 mg/L.

DIVE CULVERT Culvert having an outlet lower than its inlet, and in which the middle is at a lower elevation than either. *Also called an* **Inverted siphon.**

DIVER Underwater worker supplied with air, usually under pressure by pipeline from the surface, but also from self-contained equipment that is part of the diver's equipment.

DIVERGENCE Rate of increase of unit volume of a fluid which, in the atmosphere, is approximately proportional to the vertical velocity.

DIVERGING Dividing of a single stream of traffic, water or air into separate streams.

DIVERGING TUBE Conduit, the diameter of which increases, usually at a uniform rate, along its longitudinal axis in the direction of flow.

DIVERSION Channel excavated to make a stream or river bypass, either permanently or during construction.

DIVERSION AREA That portion of an adjacent area beyond the normal groundwater or watershed divide which may, under certain circumstances, contribute water.

DIVERSION CANAL Canal used to divert water from one watercourse to another, or from one point to another.

DIVERSION CHAMBER Chamber on a channel containing a means of diverting all or part of the flow to another channel or channels.

DIVERSION CHANNEL Artificial channel designed to carry excess flows away from the normal course of a stream, usually to prevent flood damage.

DIVERSION CUT Channel around a reservoir connecting the incoming stream to the discharge channel below the dam, used to divert flood flows.

DIVERSION DAM Barrier across flowing water to divert all or part of the water.

DIVERSION DITCH Open waterway used to conduct flows away from their normal course.

DIVERSION DUTY OF WATER Water requirement or duty, including canal or conduit losses, seepage losses, evaporation and transpiration losses, and all waste, measured at the point of diversion. *Also called* **Gross duty of water,** and **Head-gate duty of water.**

DIVERSION GATE Gate used to divert a flow from one channel to another, or to divide a flow between channels.

DIVERSION MANHOLE Manhole containing the means to divert all or part of the incoming flow to another sewer or conduit.

DIVERSION RATE Measure of the amount of waste material being diverted for recycling compared with the total amount that was previously discarded.

DIVERSION TERRACE Terrace built to divert flows from a terrace irrigation system.

DIVERSION VALVE *See* **Valve.**

DIVERSION WORKS Dams, channels, and all appurtenant structures and facilities necessary for the conducting of flows to and through an alternative route.

DIVERSITY FACTOR Factor by which an installed load (water supply, electricity, etc.) is multiplied to produce the probable maximum flow.

DIVERTER VALVE *See* **Valve.**

DIVERTING WEIR Weir positioned so as to cause excess flows in a channel to be diverted.

DIVIDE Separate into parts, equally or unequally, temporarily or permanently, wholly separately or by partitions or dividers.

DIVINING ROD Stick, usually but not necessarily the forked branch of certain species of trees, with which some people appear able to determine the location of the source of underground water. *See also* **Dowsing rod.**

DIVISION BOX Facility for splitting a flow into two or more streams.

DNA Essential compound and the basic structure of all life, deoxyribonucleic acid, found in the nucleus of living cells.

DOCTOR BLADE Fixed blade used to remove excess solids from the outside of a rotating screen.

DOGS Wedges attached to a slide gate and frame that force the gate to seat tightly.

DOMESTIC CONSUMPTION Quantity of water used for all purposes by those connected to a water distribution system, including system losses.

DOMESTIC FILTER Low capacity filter connected to a water supply and used within the home. *Also called* **Household filter.**

DOMESTIC GARBAGE Putrescible and nonputrescible waste originating from a residential unit, and consisting of paper, cans, bottles, food wastes, and may include yard and garden wastes.

DOMESTIC METER Water, electric or gas meter installed on a consumer's residential service.

DOMESTIC WASTE 1. Typical residential-type waste, excluding all commercial, manufacturing and industrial wastes, that requires no pretreatment before discharge into a sanitary sewer system; **2.** *See also* **Domestic garbage.**

DOMESTIC WASTEWATER Waterborne wastes, derived from ordinary living processes in a residential unit, of such character as to permit satisfactory disposal, without special treatment, by conventional publicly-owned treatment works.

DOMINANCE FREQUENCY Proportion of sampling units in which a particular species is most numerous.

DOPPLER SHIFT Apparent change in frequency of sound or electromagnetic waves caused by the relative motion of the source and the observer.

DORMANCY Resting condition in which the growth rate of an organism is halted and its metabolic rate slowed.

DORTMUND TANK Vertical-flow sedimentation tank having a hopper bottom. Wastewater is introduced at low velocity near the bottom of the tank: the solid fraction settles as sludge and is removed at intervals from the hopper; the effluent rises and overflows a weir at the surface.

DOSE 1. Amount of test substance administered, expressed as the weight of test substance (g/mg) per unit weight of test animal (e.g. mg/kg); **2.** In a dermal test, the amount of test substance applied to the skin (applied daily in subchronic tests); **3.** Exposure level, expressed as weight or volume of test substance per volume of air (mg/l), or as parts per million (ppm); **4.** Amount of test substance administered, expressed as weight of test substance (g/mg) per unit weight of test animal (e.g. mg/kg), or as weight of test substance per unit weight of food or drinking water.

DOSE EQUIVALENT Product of the absorbed dose from ionizing radiation and such factors as account for differences in biological effectiveness due to the type of radiation and its distribution in the body as specified by the International Commission on Radiological Units and Measurements (ICRU). The unit of dose equivalent is the rem. (One millirem (mrem)= 0.001 rem.)

DOSE RATE 1. Quantity of a substance per unit of time that is added to a liquid; **2.** Amount of ionizing (or nuclear) radiation to which an individual or a substance is exposed per unit of time.

DOSE RESPONSE Relationship between the dose (or concentration) and the proportion of a population sample showing a defined effect.

DOSE STANDARD Regulatory standard that requires a regulated facility to limit its emissions to the level necessary to ensure that no individual receives an effective dose equivalent greater than the specified level.

DOSIMETER Small, nuclear-radiation detection device that registers the total amount of radiation to which it has been exposed over a specific period of time.

DOSING APPARATUS Device for regulating the flow of wastewater or raw water to filters or other process equipment; **2.** Device for regulating the application of measured quantities of chemicals.

DOSING CHAMBER Storage tank into which an effluent is fed and held until a specific volume has been accumulated, at which point it is automatically siphoned off.

DOSING RATIO Maximum rate of application of wastewater to a given filter area divided by the average rate of application on that area.

DOSING SIPHON Siphon that automatically empties, at a designed rate of flow, the contents accumulated in a tank, commonly to some wastewater treatment device, but also to ensure a self-cleansing velocity of the flow downstream of the siphon.

DOSING TANK *See* **Dosing chamber.**

DOUBLE-ACTING GATE Dump body gate that is hinged at both top and bottom to provide the option of opening the gate for dumping and/or spreading operations.

DOUBLE ACTING PUMP *See* **Pump.**

DOUBLE BLOCK AND BLEED Two safety shutoff valves separated by a vent valve.

DOUBLE CHECK-VALVE Assembly of two single, independently-acting check valves.

DOUBLE COAGULATION Application of a coagulant at two separate points in a treatment process.

DOUBLE CONNECTOR Short pipe connector with a long parallel thread at each end, each fitted with a backnut. It can be removed from a pipe run without dismantling the installation.

DOUBLE FILTRATION Technique of passing a flow through two similar, or dissimilar filters in series.

DOUBLE HUB Cast iron sewer pipe having a bell on both ends.

DOUBLE LOCK Parallel canal-lock chambers separated by a sluice.

DOUBLE OFFSET Piping arrangement that permits one pipe to pass over another and assume its original alignment.

DOUBLE-SEAL MANHOLE COVER Manhole cover with two tongues projecting down into its frame.

DOUBLE-STRENGTH PIPE Specification standard for vitrified clay pipe calling for a minimum strength of 1000 lb/lin ft of pipe/in. of diameter as determined by the three-edge method.

DOUBLE-SUCTION IMPELLER Impeller having suction inlets on each side.

DOUBLE-SUCTION PUMP Centrifugal pump having a suction pipe connected to each side of the casing.

DOUBLE-WALL COFFERDAM Cofferdam enclosed by a wall consisting of two parallel lines of sheeting tied together, and with filling between them, that is usually self-supporting against external pressure.

DOUBLE WASH/RINSE Minimum requirement to cleanse solid surfaces (both impervious and nonimpervious) two times with an appropriate solvent or other material in which PCBs are at least 5% soluble (by weight). A volume of PCB-free fluid sufficient to cover the contaminated surface completely must be used in each wash/rinse. The wash/rinse requirement does not mean the mere spreading of solvent or other fluid over the surface, nor does the requirement mean a once-over wipe with a soaked cloth. Precautions must be taken to contain any runoff resulting from the cleansing and to dispose properly of wastes generated during the cleansing.

DOWNDRAFT Low-pressure zone created on the lee side when wind blows over a promontory, such as a building or a hill.

DOWNGRADIENT Any point that is downslope from a facility in the direction of water flow.

DOWNSPOUT Conductor connecting an eaves trough or gutter to a storm drain or other point of discharge.

DOWNSTREAM In the direction of flow.

DOWNSTREAM FACE Dry side of a dam.

DOWNTIME Operating time lost when a vehicle or item of equipment is not available for any reason, or work is stopped on the project.

DOWSING Imprecise technique for locating subterranean water by which certain people appear able to detect movement in rods or Y-shaped twigs of certain trees that deflect downward when over an underground stream. *See also* **Diving rod.**

DPD METHOD Method of measuring the chlorine residual in water using diethyl phenylene diamine.

DRAFT Process of obtaining water from a static source into a pump that is above the source's level: atmospheric pressure on the water surface forces water into the pump where a partial vacuum has been created.

DRAFT SPECIFICATIONS Specifications that establish the general quality and standards to be maintained but that lack detail as to particular application; outline model against which detailed specifications are measured. *Also called* **Outline specifications,** and **Preliminary specifications.**

DRAFT TUBE Conduit acting as a discharge, or connecting two parts of a process, which by its design contributes to the hydraulic effectiveness of the system components of which it is a part.

DRAFT-TUBE LOSS Energy reduction due to eddies and surface friction.

DRAG Resistance offered by a static surface to a liquid or other substance moving across it.

DRAG COEFFICIENT Ratio of the force per unit area to the stagnation pressure.

DRAG SCARIFICATION Method of site preparation that disturbs the forest floor and prepares logged areas for regeneration. Often carried out by dragging chains or drums behind a skidder or tractor.

DRAIN 1. Pipe and its associated supports and appurtenances, used to conduct fluids from one point to another. *See also* **Combined drain; 2.** To remove fluid from a vessel or system.

DRAINABLE SLUDGE Sludge that is self-draining and does not require pumping, and which can be readily dewatered.

DRAINAGE 1. System for the collection and removal of water, from the ground, or from appliances; **2.** Interception and removal of water from, on, or under an area or roadway.

DRAINAGE AREA 1. Catchment area; **2.** Area served by a continuous system of connected sewers, or by a watercourse.

DRAINAGE BASIN Total area that contributes water to a stream.

DRAINAGE CANAL Canal that drains water from an area having no natural outlet for precipitation or runoff.

DRAINAGE CHANNEL Channel sloped downstream and leading to a side-ditch formed across the path normally taken by surface water, to impede and slow velocity and to divert a portion of the flow.

DRAINAGE DENSITY Expression of natural drainage capacity determined by dividing the total length of drainage channels in an area by the surface area.

DRAINAGE DISTRICT Administrative area, or the organization responsible for designing, financing, constructing and operating a drainage system.

DRAINAGE DITCH Excavation made along the perimeter of an area to which surface water is directed, and that is graded to a slope sufficient to conduct the collected water away from the area. *See also* **Culvert,** and **Drainage swale.**

DRAINAGE DIVIDE Line of highest elevation across which water will not flow.

DRAINAGE EASEMENT *See* **Easement.**

DRAINAGE EQUILIBRIUM Condition where the quantity of water reaching a watertable from all sources is approximately equal to the water drained or removed from the watertable.

DRAINAGE FACILITY Mechanical device or structure (channel, culvert, ditch, pipe, sewer, etc.) designed and placed to intercept, divert, carry, or contain surface water runoff.

DRAINAGE FILL 1. Base course of granular material placed between a floor slab and subgrade to impede capillary rise of moisture. *Also called* **Porous fill; 2.** Lightweight concrete placed on floors or roofs to promote drainage.

DRAINAGE GALLERY Passage in a dam, parallel to its crest, in which water leaking from the upstream face is collected and conducted away from the downstream face.

DRAINAGE HEAD Highest elevation in a drainage area.

DRAINAGE MORPHOMETRY Study of drainage patterns.

DRAINAGE PATTERN Configuration in which surface runoff accumulates and disperses downstream.

DRAINAGE PIPING 1. All or any portion of a drainage piping system; **2.** Pipe that, when fitted together as part of a system, produces a smooth internal surface, especially at joints.

DRAINAGE RIGHT-OF-WAY Land assumed by a public administration for the installation of stormwater channels, sewers, or other facilities for the collection and disposal of runoff.

DRAINAGE RIGHTS Legal right of a landowner to dispose of excess or unwanted water from his land over neighboring land.

DRAINAGE SWALE Excavation that is shallow in proportion to its width, intended to convey surface water away from an area. *Also called* **Drainage ditch.**

DRAINAGE SYSTEM Assembly of pipes, fittings, fixtures, traps, and appurtenances used to convey sewage, clear water waste, or storm water to a public sewer or a private sewage disposal system, but not including subsoil drainage pipes. *Also called* **Building drainage system.**

DRAINAGE TERRACE Terrace, or variable cross-section and grade, constructed on both sides of a ridge and designed to intercept surface runoff.

DRAINAGE TUNNEL Tunnel driven principally to drain underground works.

DRAINAGE WATER Water collected in a drain and discharging to a natural watercourse.

DRAINAGE WELL Borehole downstream of a protected area, used to collect and control seepage so as to reduce uplift pressure.

DRAIN-BACK SYSTEM Type of freeze protection for a solar water-heating system that circulates an antifreeze solution through the collector and into a heat exchanger in the water tank, where heat is transferred from the antifreeze to the water.

DRAIN CHUTE Tapered drain pipe used at the outlet of an inspection chamber to facilitate rodding.

DRAIN COCK Faucet at the lowest point of a water system.

DRAIN-DOWN SYSTEM Freeze protection for a solar water-heating system designed with controls that drain the collector of water when subfreezing temperatures occur.

DRAINED SHEAR TEST Triaxial compression test of cohesive soil that has been consolidated under normal load, conducted in drained conditions slowly enough to allow further consolidation due to shearing during the test.

DRAIN PIPE Pipe, manufactured from a

variety of materials and in a wide range of sizes with accompanying fittings, used underground to convey sewage and stormwater.

DRAIN PLUG Expanding stopper used to seal the ends of a drain being tested.

DRAIN RODS Lengths of flexible fiberglass or other rod that can be joined end-to-end and fitted with various devices for clearing blocked drains.

DRAIN TILE Terracotta, plastic, or concrete pipe, sometimes perforated, laid on a bed of crushed stone to drain water either from waterlogged soil to a filter material, from a drainage system, or from around the footings of a building.

DRAW Natural depression or swale; a small watercourse.

DRAWBAR HORSEPOWER *See* **Horsepower.**

DRAW-DOOR WEIR Weir with vertical gates.

DRAWDOWN 1. Lowering of the level of groundwater, as in dewatering an area in which an excavation is to be dug; **2.** Amount that the water surface in a well is lowered as a result of pumping; **3.** Difference in elevation between the upstream and downstream limits of a continuous water surface with accelerating flow.

DRAWDOWN CURVE Profile of the piezometric surface of the watertable relating drawdown to distance from a well under given pumping conditions.

DRAW-OFF PIPE Pipe leading to a faucet or bib.

DRAW-OFF TAP Water supply tap.

DREDGE Machine for excavating material at the bottom, or at the banks, of a body of water, the excavated material being discharged on the bank or to a scow for transport.

DREDGED BULKHEAD Bulkhead that, after being anchored, has soil excavated or dredged from the front of its base.

DREDGED LEVEL Level to which the ground on the water side of a bulkhead, wharf, or quay has been dredged. *Also called* **Fill bulkhead.**

DREDGED MATERIAL Material that is excavated or dredged from water.

DREDGING Excavating underwater, usually with floating equipment; it may be an elevator ladder, hydraulic suction, grab or dipper bucket, scraper, dragline, clam shell, or backhoe.

DREDGING WELL Opening in a dredge through which the ladder or cutter passes.

DRIFT 1. Superficial deposits caused by wind, ice or water; **2.** Difference between an actual value and the desired, or set-point value.

DRIFT BARRIER Barrier constructed across a stream to catch driftwood and other surface debris.

DRILL Mechanical device for gripping, rotating, and sometimes hammering a cutting tool. There are several types, including:

> **Blasthole:** Air percussion, cable, fusion-type, or rotary drill designed to drill blastholes.
>
> **Churn:** Machine that drills a hole in ground by raising and dropping a string of drilling tools suspended by a reciprocating cable. *Also called* **Cable drill.**
>
> **Core:** Rotary drill equipped with a hollow bit and core lifter, used to obtain samples of the material being drilled through.
>
> **Diamond:** Rotary drill that uses a coring bit studded with industrial diamonds, used chiefly in exploratory drilling.
>
> **Earth:** Auger-type drill with a bucket, used for exploratory drilling in advance of rock-earth excavation.
>
> **Fusion:** Drill that burns out a blasthole by means of fuel, oil, oxygen, and cooling water delivered to a blowpipe within the hole. *Also called* **Jet piercing drill.**
>
> **Hammer:** Pneumatic, compressed air-driven reciprocating drill, commonly

171

used to drill holes in hard rock.

Jet piercing: *See* **Fusion,** above.

Percussion: Drill that hammers and rotates a steel and bit.

Reciprocating: Drill that operates by continuously thrusting a bit against a surface. The bit is held in a chuck that allows it to rotate between blows.

Rock: Pneumatic or electric drill for making holes in rock for blasting or other purposes.

Rotary: 1. Drilling machine that bores holes by the rotation of a kelly bar driving continuous-flight augers or a rotary table, etc.; **2.** A wet rotary drill rig using high-pressure water to open a hole for installation of a mandrel-driven pile.

Spudding: Drill that makes holes by lifting and dropping a chisel bit.

Track: Self-contained rock drill mounted on a self-propelled, tracked chassis.

Turbodrill: Rock drill used at depths below two miles.

Well: Churn drill used to develop a water well, usually truck-mounted.

DRILLABILITY Relative ease with which a hole can be drilled in a certain rock type; factors such as the rock's mineral composition, hardness or compressive strength, and degree of fracturing and weathering all have influence.

DRILL-BLASTER Rock-drill operator who also carries out loading and blasting duties.

DRILL COLLAR Thick-walled drill pipe, used immediately above a rotary bit to provide additional weight.

DRILLED SHAFT *See* **Caisson pile.**

DRILLER'S STROKE Height of fall of a drophammer weight, 63.5 kg (140 lb) or more, used to drive a casing or soil sampling tool. The weight is usually raised by wrapping a rope around a powered and continuously rotating spool or cat-head. The weight is raised by holding the rope taut, thus raising the hammer, then allowing slack in the rope to let the hammer fall to hit either the casing or the sampler rod. The distance from the top of the casing to the height before release is called the fall or driller's stroke. *See also* **N Value.**

DRILL FEED 1. Mechanism that pushes a drill tool into a hole; **2.** Measure of the speed of advance of a drilling tool.

DRILLING AND PRODUCTION FACILITY All drilling and servicing equipment, wells, flow lines, separators, equipment, gathering lines, and auxiliary nontransportation-related equipment used in the production of underground resources.

DRILLING BUCKET *See* **Bucket auger.**

DRILLING CORE Exploratory drilling that includes cutting cylinders and bringing them to the surface for analysis.

DRILLING FLUID Water- or air-based fluid used to decrease friction between the drill string and the hole, cool the bit, and remove cuttings from a drill hole.

DRILLING LINE Cable that supports and manipulates the tools of a churn drill.

DRILLING LOG Detailed daily record of rocks passed through in the drilling of exploratory holes and blastholes.

DRILLING MUD Fluid mixture of clayey soil and water, or commercial mixture of sodium montmorillonite (bentonite) and a clay mineral.

DRILLING OIL Fluid component of mud for rotary drilling, commonly diesel fuel, but also a high-quality mineral oil exhibiting lower viscosity for faster drilling rates, higher flash point for improved safety, and low aromatics content for reduced potential toxicity to workers and the environment.

DRILLING RIG Machine for drilling holes in earth or rock.

DRILLINGS Cuttings produced from the process of drilling.

DRILL JUMBO Movable frame on which drill positioners, jibs, and drills are mounted.

DRILL PIN Round pin protruding from the inside of a lock case to receive a barrel key. *Also called* **Barrel post.**

DRILL PIPE One or several sections of rotary drilling string that connect the kelly with the bit or collars.

DRILL POSITIONER Mechanical control for moving, rotating, and controlling jibs.

DRILL ROD Lengths of steel used to transmit impact energy and rotation from the striker bar to the drill bit; typically 3 m (10 ft) or 3.7 m (12 ft) long and hollow for the flow of flushing air (or water).

DRILL STEEL Hollow steel that connects a percussion drill with the bit.

DRILL STRING All revolving parts below ground of a rotary drill; or the tools hanging from the drilling cable of a churn drill.

D-RING Pipe joint ring with one flat surface, which prevents it from rolling.

DRINKING WATER *See* **Potable water**.

DRINKING WATER STANDARDS Standards for potable water established by competent authorities having jurisdiction.

DRINKING WATER SUPPLY Raw or finished water source that is or may be used by a public water system or as drinking water by one or more individuals.

DRIVING RAIN INDEX Grading of site exposure produced by multiplying annual rainfall (mm) by average wind speed (m/s) and dividing by 1000. 'Severe' exposure is above 7 m²/s, 'moderate' above 3.7 m²/s, and 'sheltered' below 3 m²/s.

DRIZZLE Light rain consisting of relatively uniform droplets, usually less than 0.5 mm (0.02 in.) in diameter, commonly falling under windless conditions and frequently associated with fog or mist.

DROP 1. Structure for vertically dropping water in a conduit to a lower level and simultaneously dissipating energy; **2.** Difference in water surface elevations, upstream and downstream.

DROP CHANNEL Short, steep channel between the inlet and outlet of a drop manhole.

DROP CONNECTION *See* **Drop manhole**.

DROP-DOWN CURVE Longitudinal shape of the water surface as it passes over an obstruction to a lower elevation.

DROP ELBOW Pipe elbow with ears to allow fastening to a surface.

DROP INLET Catch basin with the top set lower than the surrounding pavement.

DROP L L-shaped pipe fitting fabricated with ears that permit it to be attached directly to the building frame.

DROPLET Liquid particle having a very small mass, up to 20μm in diameter, capable of remaining in suspension in a gas.

DROP MANHOLE Vertical branch drain connection at a manhole. *Also called* **Back drop,** and **Drop connection.**

DROP-OFF BOX *See* **Waste container.**

DROPPING HEAD Condition that pertains when the inlet flow to a reservoir or storage facility is less than that of the outflow.

DROP PIPE Suction pipe below a deep-well pump.

DROP TEE Tee-shaped pipe fitting fabricated with ears that permit it to be attached directly to the building frame.

DROUGHT Extended period of dry weather beyond cyclic maximums typical for an area.

DROWNED VALLEY Depression in the ground that extends below the watertable, but which has neither tributary streams or flowing outlet.

DROWNED WEIR Weir whose tailwater level is above the weir crest.

DRUM 1. Cylindrical container, usually with one or more ridges deformed into its perimeter and spaced along its length as reinforcing; **2.** Container having a capacity of less than 230 L (60.7 gal) but more than 30 L (7.92 gal); **3.** Rotating cylindrical member used to transmit compaction forces to soil or other surface materials. *Also called* **Roll.**

DRUM GATE Arc-shaped spillway gate.

DRUM SCREEN Rotating cylindrical screen set in a channel through which raw sewage flows, used to trap solid material, which is then scraped off above the water-line.

DRUM TRAP Trap consisting of a cylinder with its axle vertical.

DRY-BULB TEMPERATURE Air temperature as indicated by any type of thermometer not affected by the water vapor content or relative humidity of the air.

DRY CONNECTION Connection to a water or sewer main completed when the main is dry, i.e. when the main and line are not in service, typically during original construction.

DRY DENSITY Weight of dry material in a soil sample after drying at 105°C (221°F).

DRY-DENSITY/MOISTURE RELATIONSHIP Relationship between the dry density and the moisture content of soil for a specific amount of compaction.

DRY FEEDER Apparatus for feeding chemicals or other fine, granular material in its solid state at a measured rate.

DRY FLUE GAS DESULFURIZATION Sulfur dioxide control system that is located downstream of a steam generating unit and which removes sulfur oxides from the combustion gases by contacting the combustion gases with an alkaline slurry or solution and forming a dry powder material.

DRY ICE Solid carbon dioxide, which sublimes at temperatures above -72°C.

DRY SAMPLING Method of sampling soil by augering a hole in the ground with a sampler or sample spoon attached to the end of an auger. The object is to obtain a complete and undisturbed sample of the natural soil for analysis. *Also called* **Core boring.**

DRY-SEAL PIPE THREAD Pipe threads in which sealing is a function of root and crest interference. *See also* **Pipe thread,** and **Tapered pipe thread.**

DRY SOLIDS Basis for measuring the dry matter in sludge and involving a process that

drives off all water prior to weighing.

DRY SPELL 1. Period of 4 or more days during which the daily maximum temperature remains at least 3°C to 4°C above normal and the relative humidity remains below 50%; **2.** Interval, usually of 14 days duration, during which no measurable rainfall has occurred at a location or within a region.

DRY SUSPENDED SOLIDS Weight of the suspended matter in wastewater after drying for 1 h at 103°C (217.4°F).

DRY VALVE *See* **Valve.**

DRY VENT Any vent that does not carry waste water.

DRY WEATHER FLOW Sewer flow over 24 hours of dry weather.

DRY WEIGHT Actual weight of a substance or organism that normally contains moisture, after all moisture is removed.

DRY WELL 1. Covered pit with open-jointed linings, often filled with rocks, through which drainage from roofs, basement floors, or areaways may seep or leach into the surrounding soil; **2.** Well that was drilled to obtain water, but that did not penetrate an aquifer.

DUAL-FLUSH CISTERN Water closet cistern with a two-button flushing system: one button allows a full flush of 9 L (2 Igal, 2.4 USgal), the other a half flush of 4.5 L (1 Igal, 1.2 USgal).

DUAL POROSITY ELEMENT Element that contains two media of different porosity in parallel.

DUAL POROSITY FILTER Filter that contains two media of different porosity offering parallel flow paths to the fluid.

DUAL SYSTEM Two-pipe plumbing system.

DUAL VENT Vent connecting at the junction of two fixture branches, serving as a back vent for both.

DUCKFOOT BEND 90° pipe bend with a flat seating to carry the weight of vertical pipes, or the thrust due to change of direction.

DUCT Tube, channel or other shape for conveying fluid or air.

DUCTILE Refers to metals that are relatively easily formed into shapes and that can sustain not less than 5% elongation before fracturing. Especially refers to cast ductile iron that has less tendency to crack when subjected to high stress.

DUCTILE IRON PIPE *See* **Pipe.**

DUCTILITY Property of a material by virtue of which it may undergo large permanent deformation without rupture.

DUCTILITY FACTOR Ratio of the total deformation at maximum load to the elastic-limit deformation.

DUFF Partially decomposed organic material of the forest floor beneath the litter of freshly fallen twigs, needles, and leaves.

DUMMY ACTIVITY 1. Activity represented on an arrow diagram of the critical path program to show how progress of one task prevents completion of another task; **2.** Activity of zero duration, used to show a logical relationship in the arrow diagramming method.

DUMP Unmonitored, unprepared land area where unrestricted unloading of refuse is illegally done.

DUMP ANGLE Degree of slope attained by a dump body at the top of its lift, usually measured in degrees from the horizontal plane of the truck axis.

DUMP BODY *See* **Truck.**

DUMPER 1. Rubber-tired vehicle with two large wheels in front and two smaller wheels behind, with a front, or sideways dumping hopper positioned over the front wheels; **2.** Attachment for dumping containers. Various types are available to suit container and operational requirements.

DUMPING Disposition of material provided it does not mean a disposition of any effluent from any outfall structure to any regulated extent.

DUMPSTER Large metal boxlike container used to hold garbage and other refuse at its point of generation, and equipped with means that enable it to be seized, held, lifted, rotated, and emptied into a mobile refuse collection vehicle, itself equipped with the means to effect transfer of the waste.

DUMP TIME Time taken to empty the body, bucket, bowl, etc., of mechanical equipment. *See also* **Hydraulic cycle time.**

DUMP TRAILER *See* **Trailer.**

DUMP TRUCK *See* **Truck.**

DUNE Mound or ridge of loose sand heaped up by the wind.

DUNE STABILIZATION Means used to prevent the windblown migration of sand particles, including fences, and the planting of various grasses to bind the sand among their roots.

DUPLEX Assembly of two filters with valving for selection of either or both filters.

DUPLEX PUMP Reciprocating, continuous-flow pump comprising two cylinders placed side-by-side and connected to the same suction and discharge pipes and arranged so that as one cylinder exerts suction, the other exerts pressure.

DUPLICATION OF PLANT Installation of backup units for essential mechanical services, often with automatic switchover at each start-up.

DURABILITY Ability of a material to withstand conditions of service.

DURABILITY FACTOR Measure of the change in a material property over a period of time as a response to exposure to an influence that can cause deterioration, usually expressed as a percentage of the value of the property before exposure.

DURATION Estimated time to perform an activity.

DURATION COMPRESSION Shortening a project schedule without reducing the project scope.

DUST 1. Particles of solid material, usually 100 microns or less, suspended in air; **2.** Particles formed by other than combustion processes.

DUST BURDEN Weight of dust suspended in a unit volume of a medium.

DUST CAPACITY *See* **Dirt capacity.**

DUST COLLECTOR Device for trapping and storing dust from air and exhaust streams.

DUST DEVIL Rotating convection current made visible because of the particles that it contains, which have been carried off the ground and into the vortex being created by the whirlwind.

DUST-HANDLING EQUIPMENT Equipment used to handle particulate matter collected by a pollution control device.

DUST LOADING Particulate content of air or flue emissions.

DUST PALLIATIVE Asphalt emulsion sprayed onto gravel roads to control dust generation and inhibit erosion.

DUST STORM Wind blowing at speeds between 24 and 48 km/h that has picked up a massive volume of dust, sand, and earth particles from the ground surface. The dust thus raised may be carried to 1500 to 1800 m above the ground.

DUTCH MATTRESS Mattress of timber and reeds, used to protect the river or seabed from scour.

DUTY 1. Load capacity, wear, exposure resistance, etc., of a material or component for a stated service; **2.** Active, as against standby, machine.

DUTY OF CARE Legal concept calling for employers to take all proper precautions to protect employees against foreseeable hazards, and for generators of wastes to take all required measures to prevent uncontrolled and illegal release of the wastes into the environment.

DUTY OF WATER Quantity of water required to satisfy the irrigation requirements of an area of land.

DWELLING Place of permanent or customary and usual residence of a person, according to local custom or law, including a single family house; a single family unit in a two-family, multifamily, or multipurpose property; a unit of a condominium or cooperative housing project; a non-housekeeping unit; a mobile home; or any other residential unit.

DWELL TIME Time required for material to pass through a dryer or drum mixer.

DYKE Embankment that prevents the overflow of water from a natural watercourse.

DYNAMIC DISCHARGE HEAD Total of the static discharge head plus the friction in the discharge line.

DYNAMIC FORCE Force tending to produce motion.

DYNAMIC FORCE APPLIED Vectorial resolution of all the generated forces and the static forces at the interface of the compaction drum and the material being compacted.

DYNAMIC HEAD *See* **Head.**

DYNAMIC LOAD Load that is variable, i.e. not static, such as a moving live load, earthquake, or wind.

DYNAMIC LOADING Loading from units (particularly machinery) that, by virtue of their movement or vibration, impose stresses in excess of those imposed by their dead load.

DYNAMIC PENETRATION TEST Test involving moving parts, as distinguished from a static test.

DYNAMIC PRESSURE Pressure exerted against a surface or body by a moving liquid.

DYNAMIC SEAL Seal that has rotating, oscillating, or reciprocating motion between its components. *See also* **Static seal.**

DYNAMIC STRENGTH Resistance to loads applied suddenly.

DYNAMIC SUCTION HEAD Total of the static suction head and suction line friction.

DYNE Obsolete unit of force, equal to 10 μN, that should not be used with the SI system of measurement. Symbol: dyn. *See also the appendix*: **Metric and nonmetric measurement.**

DYSPHOTIC ZONE Zone of water in a sea or lake that lies between the euphotic zone and aphotic zone, characterized by reduced light levels and usually spanning depths of about 100 to 600 m (328 to 1968.5 ft).

DYSTROPHIC Freshwater bodies that are deficient in calcium, very poor in dissolved plant nutrients (especially nitrates) and that are relatively unproductive.

E

EAR 1. Organ of hearing and balance in the higher vertebrates; **2.** Projection on a plumbing fitting, or on a pipe, by which it may be attached to a wall.

EAR INSERT DEVICE Hearing protective device that is designed to be inserted into the ear canal, and to be held in place principally by virtue of its fit.

EARLIEST EVENT OCCURRENCE TIME Earliest time that all activities that precede an event will be completed.

EARLIEST FINISH Earliest day that the work item can finish if it starts at its earliest start and is completed in its expected time.

EARLIEST START Earliest day that the work item can start provided every preceding work item starts at its earliest start day and is completed in its expected time.

EARLY FINISH DATE In the critical path method, the earliest possible point in time on which the uncompleted portions of an activity (or the project) can finish based on the network logic and any schedule constraints.

EARLY LIFE STAGE TOXICITY TEST Test to determine the minimum concentration of a substance that produces a statistically significant observable effect on hatching, survival, development and/or growth of a fish species continuously exposed during the period of its early development.

EARLY START DATE In the critical path method, the earliest point in time on which the uncompleted portions of an activity (or the project) can start, based on the network logic and any schedule constraints.

EAR MUFF Hearing protective device that consists of two acoustic enclosures which fit over the ears and which are held in place by a spring-like headband.

EARNED VALUE (EV) 1. Method for measuring project performance. It compares the amount of work that was planned with what was actually accomplished to determine if cost and schedule performance is as planned. *Also called* **Earned value analysis.** *See also* **Actual cost of work performed, Cost variance, Cost performance index, Schedule variance,** and **Schedule performance index; 2.** The budgeted cost of work performed for an activity or group of activities.

EARNED VALUE ANALYSIS *See* **Earned value.**

EAR PROTECTORS Ear muffs or ear plugs, used to prevent hearing damage from excessive noise.

EARTH Non-static particulate material lying above bedrock and around rocky outcrops, consisting mostly of the products of rock disintegration such as sand, clay, pebbles, cobbles and boulders, plus organic constituents. *See also* **Material density.**

EARTH BERM Mound of dirt placed to achieve some effect: diversion of water or wind, insulation, impoundment, etc.

EARTH DAM Water barrier made of earth, clay, sand, and gravel and having an impervious core in the form of a vertical curtain. *Also called* **Fill dam.**

EARTH PRESSURE *See* **Pressure.**

EARTH PRESSURE CELL Large, non-ferrous flat cell for the *in situ* measurement of soil pressure.

EARTHQUAKE Vibration or shaking of the ground due to some natural phenomenon within the Earth.

EARTHQUAKE LOADING Load or stress

placed on a wall or floor system due to the earthquake shaking action.

EARTH RESERVOIR Water storage facility formed using excavated material as embankments and with the addition of an artificial waterproof liner or the use of impervious materials.

EARTH RESOURCES TECHNOLOGY SATELLITE Series of Earth-orbiting satellites launched by the National Aeronautics and Space Administration to identify and monitor the natural resources of the world. The satellites complete 14 orbits each day, photographing three strips 185 km wide in North America; 11 strips in the remainder of the world.

EARTH'S SHADOW Darkening of the eastern sky seen just after sunset in suitable conditions of haze.

EARTH TANK Drinking facility for livestock formed by excavating across a natural watercourse and forming an earthen dam at the outlet.

EARTHWORK 1. Moving of surface materials to create a change of landform during site works; **2.** Embankment formed by heaping soil; **3.** Work of excavation.

EASEMENT Strips of land within a larger area over which right of access by a person or public authority other than the owner has been granted for specific purposes, such as the passage of services. There are several distinctions, including:

Drainage: Easement for directing the flow of water.

Planting: Easement for reshaping roadside areas and establishing, maintaining, and controlling plant growth thereon.

Scenic: Easement for the conservation and development of roadside views and natural features.

Sight-line: Easement for maintaining or improving the sight distance.

Slope: Easement for cuts or fills.

EBB Flowing out of the tide.

EBB CHANNEL Channel in an estuary eroded by the river at low tide or low water.

EBB CURRENT Current in a body of water resulting from the movement of tidal flows.

EBBING AND FLOWING SPRING Spring that has cyclic rates of flow.

EBB TIDE Tide occurring at the ebb period of total flow.

EBULLITION Boiling point of a liquid, marked by the rapid formation of bubbles of saturated vapor within the liquid that rise to the surface and escape.

ECCENTRIC FITTING Pipe fitting in which the centerline of the opening is offset.

ECHO Reflected sound that reaches the observer after a time interval long enough for it to be perceived as a separate event.

ECHO LOCATION Method used by some animals and machines to locate and identify objects by emitting sounds and perceiving their echoes.

ECLIPSE Darkening of the sun, moon, etc. when some other heavenly body is in a position that partly or completely cuts off its light as seen from some part of the Earth's surface.

ECLIPTIC Plane of the Earth's orbit about the Sun.

ECOCATATASTOPHY Environmental disaster threatening the quality of life of a community or population.

ECOCLINE Gradient of ecosystems associated with an environmental gradient.

E. coli Abbreviation for the coliform bacterium *Escherichia coli.*

ECOLOGICAL BALANCE Condition of equilibrium among the components of a natural community.

ECOLOGICAL EFFECTS STUDIES Studies performed for the development of information on nonhuman toxicity and potential ecological impact of test substances and their degradation and activation products.

ECOLOGICAL EFFICIENCY Ratio between the amount of energy flow at points along a food chain.

ECOLOGICAL FACTOR Environmental factor that influences living organisms.

ECOLOGICAL INDICATORS Organisms whose presence indicates specific conditions in a sample of soil, water, or air.

ECOLOGICAL NICHE Unique place in the ecosystem occupied by a specific organism.

ECOLOGICAL PYRAMID Diagram depicting the flow of energy through the ecosystem, from generator to final consumer. There are several types, including:

> **Pyramid of numbers** – Shows the number of organisms at each level of the food chain.

> **Pyramid of biomass** – Depicts the total weight of organisms at a point in time at each level of the pyramid of numbers.

> **Pyramid of energy** – Depicts the optimum overall picture of energy flow, i.e. the rate at which food is produced as well as the total amount.

ECOLOGY Study of plants and animals in relation to their physical and biological surroundings.

ECO-MANAGEMENT AND -AUDIT SCHEME Voluntary scheme for individual industrial sites designed to provide recognition for companies that establish a continuously improving program of positive environmental protection action.

ECONOMIC CONSERVATION Management of natural resources, or the environment, so as to sustain a regular yield of a commodity at the highest level feasible.

ECONOMIC DEPRECIATION Loss in value of an asset resulting from external economic conditions affecting the character or degree of utilization.

ECONOMIC DEVELOPMENT Historical process whereby a country changes its economic base from one relying on agriculture and the provision of raw materials, to the industrial processing of materials, to the provision of services, and finally to a reliance on technological development and the provision of high-tech services.

ECONOMIC EFFICIENCY Relationship between the cost of attaining defined economic ends and the value of those ends.

ECONOMIC GROUNDWATER YIELD Maximum rate at which water can be drawn from an aquifer without depleting the supply or altering the character of the remaining water.

ECONOMIC LIFE Period for which a product or structure is designed to remain viable, useful, and/or profitable: usually shorter than physical life.

ECONOMIC OBSOLESCENCE Loss of value due to changing economic circumstances beyond the bounds of a structure or development.

ECONOMIC RATIO Condition when two materials, or materials acting together are each performing at their maximum capability.

ECONOMICS Study of the use of available resources.

ECONOMIZER Device for transferring heat from flue gases to boiler feed water.

ECOSPHERE Sum of the ecological factors that act upon organisms within the biosphere.

ECOSYSTEM Complex ecological community and environment forming a functional whole in nature.

EDAPHIC FACTORS Chemical, physical and biological characteristics of the soil which affect an ecosystem.

EDDY Circular movement occurring in flowing water or air, caused by currents set up by obstructions or changes and irregularities in the banks or bottom of the channel.

EDDY FLOW Uneven or unsteady flow, usually characterized by turbulence.

EDDY LOSS Loss of energy in a flow passing through a conduit, caused by velocity changes, variations in cross-sectional area and/or profile, valves, etc.

EDUCTOR Device for lifting water to a higher level, consisting of concentric tubes that are inserted below the surface, with air under pressure being injected through the central tube.

EFFECTIVE AREA Area of a filter medium through which fluid flows.

EFFECTIVE AREA OF AN ORIFICE Actual area of an orifice times its coefficient of discharge.

EFFECTIVE CORROSION INHIBITOR RESIDUAL Inhibitor residual sufficient to form a passivating film on the interior walls of a pipe.

EFFECTIVE DATE Date indicated as 'the effective date of this agreement' in contract documents. If no date is indicated, it is the date the agreement was signed by the last of the parties to the agreement.

EFFECTIVE DOSE EQUIVALENT Sum of the products of absorbed dose and appropriate factors to account for differences in biological effectiveness due to the quality of radiation and its distribution in the body of reference, man.

EFFECTIVE GROUNDWATER VELOCITY *See* **Actual groundwater velocity.**

EFFECTIVE HEAD Head available for the development of energy.

EFFECTIVE HEIGHT OF EMISSION Height above-ground, and clear of the top of a chimney, at which rising exhaust gases are reckoned to spread horizontally.

EFFECTIVE LEAKAGE AREA Orifice-flow area that will result in the same calculated flow for a given pressure drop as is measured for the seal in question.

EFFECTIVE MODULUS OF ELASTICITY Elastic and plastic effects in an overall stress–strain relationship in a structure.

EFFECTIVE MOMENT OF INERTIA 1. Moment of inertia of the cross section of a member that remains elastic when partial plastification of the cross section takes place, usually under the combination of residual stress and applied stress; **2.** Moment of inertia based on effective widths of elements that buckle locally; **3.** Moment of inertia used in the design of partially composite members.

EFFECTIVE OPENING Cross-sectional area of an opening from which water is discharged from a water supply pipe.

EFFECTIVE POROSITY Ratio of the volume of water or other liquid which a given volume of liquid-saturated rock or soil will yield under any specific hydraulic condition to the total volume of soil or rock.

EFFECTIVE PRECIPITABLE WATER Greatest volume of precipitable water that can be removed from an atmospheric column by convection.

EFFECTIVE PRECIPITATION Precipitation falling during the growing period of a crop and which is available to meet the crop's consumptive moisture requirements.

EFFECTIVE PRESSURE *See* **Pressure.**

EFFECTIVE RAINFALL 1. Rain that produces surface runoff; **2.** In irrigation, that portion of total precipitation which is retained by the soil and is available for crop production.

EFFECTIVE RANGE Portion of the design range in which an instrument has acceptable accuracy.

EFFECTIVE SOIL SIZE Grain size that is larger than 10% by weight of the soil particles as seen on the grading curve of the soil.

EFFECTIVE STACK HEIGHT Actual maximum height above-ground attained by an exhaust plume exiting a chimney or stack.

EFFECTIVE STORAGE Usable contents of a storage vessel.

EFFECTIVE THREAD Includes the complete thread, as well as that portion of the incomplete thread having fully formed roots, but having crests not fully formed.

EFFECTIVE VELOCITY Actual velocity of groundwater percolating through water-bearing material; the volume of groundwater passing through a unit cross-sectional area in a unit of time, divided by the effective porosity.

EFFECTS 1. Direct effects, which are caused by an action and occur at the same time and place; **2.** Indirect effects, which are caused by an action and are later in time or farther removed in distance, but are still reasonably foreseeable. Indirect effects may include growth-inducing effects and other effects related to induced changes in the pattern of land use, population density or growth rate, and related effects on air and water and other natural systems, including ecosystems including ecological (such as the effects on natural resources and on the components, structures, and functioning of affected ecosystems), aesthetic, historic, cultural, economic, social, or health, whether direct, indirect, or cumulative.

EFFERVESCENCE Vigorous release of small gas bubbles from a liquid, especially resulting from chemical action.

EFFICACY Power to produce a desired effect or result, i.e. the capacity of a pesticide product when used according to label directions to control, kill, or induce the desired action in the target pest.

EFFICIENCY Ability to produce the effect wanted without waste of time, energy, etc., i.e. the ability, expressed as a percent, of a filter to remove specified artificial contaminants, at a given concentration, under specified test conditions.

EFFICIENCY FACTOR Percentage of theoretical production obtainable under actual conditions.

EFFLUENT 1. Any solid, liquid, or gas that enters the environment as a by-product of a man-originated process; the substances that flow out of a designated source; **2.** Fluid leaving a component; **3.** Dredged material or fill material, including return flow from confined sites.

EFFLUENT CHARGE Levy imposed on the generator of an effluent for each unit of volume discharged to a public collection system or to a public resource.

EFFLUENT DATA Information necessary to determine the identity, amount, frequency, concentration, temperature, or other characteristics (to the extent related to water quality) of any pollutant

EFFLUENT LIMITATION Limitation established as a condition of a permit issued, or proposed to be issued, by a regulatory agency.

EFFLUENT SEEPAGE Diffuse discharge onto the ground of liquids that have percolated through a mass; may contain dissolved or suspended solids.

EFFLUENT STANDARD Maximum amount of a specified pollutant an effluent is permitted to contain.

EFFLUENT STREAM Stream that receives groundwater. *See also* **Influent stream.**

EFFLUENT TROUGH Channel into which the effluent from a wastewater treatment tank flows, or overflows.

EFFLUENT WEIR Weir forming the outlet to a sedimentation basin or other hydraulic structure.

EFFLUX TUBE Tube inserted into an orifice to provide for outward flow of fluid.

EFFLUX VELOCITY Velocity with which gas leaves a stack, equal to the volume of gas issuing from the stack mouth per second divided by the cross-sectional area of the mouth.

EGG-SHAPED SEWER Sewer pipe with an egg-shaped cross section, installed small end down, used to improve the self-cleaning effect at low flows.

EJECTOR 1. Cleanout device, usually a sliding plate; **2.** Type of pump, similar to an airlift pump, used to lift sewage contained in a pipe by injecting compressed air into it.

EJECTOR WELL POINT Well-point de-watering system whereby a vacuum is created at the well-point tip by means of a high-velocity water flow through a jet nozzle, sucking water from the soil and forcing it up the return pipe.

ELAPSED TIME Time used to complete a defined stage or action, or between stages or actions.

ELASTIC COMPRESSION Movement of soil particles due to the imposition of a load. *See also* **Plastic creep** and **Soil consolidation.**

ELASTICITY Soil property that allows it to return to its approximate original shape when a compressing load is removed. *See also* **Properties of soil.**

ELASTIC LIMIT Maximum stress that can be obtained in a structural material without causing permanent deformation.

ELASTIC MOVEMENT Movement under load that is recoverable when the load is removed.

ELASTIC STRAIN Deformation per unit of length produced by a force acting on a body that disappears when the force is removed.

ELBOW Pipe fitting made to allow a turn in direction of a pipe line.

ELECTRIC FIELD Field in the atmosphere which, in fine weather, has a typical value of about 200 volts/m.

ELECTRIC PUMP *See* **Pump.**

ELECTRIC-RESISTANCE-WELDED PIPE *See* **Pipe.**

ELECTRIC SET Equipment, fixed or portable, designed to generate electricity. There are several distinctions, including:

> **Peaking power:** Electric set that assumes part of the load during peak-load periods.

> **Prime power:** Electric set that is operated by the primary source of power. It may be primary because it is the

sole source or because it provides a special type of power.

> **Standby power:** Emergency electric power system that is on 'standby alert,' ready to assume the load when the normal power source fails.

ELECTRIC SUBMERSIBLE PUMP *See* **Pump.**

ELECTRIC WELL LOG Record of a well developed in rock produced by a travelling electrode. The record is in the form of curves representing the apparent values of the electric potential and resistivity or impedance of the rocks and their contained fluids throughout the uncased portions of the well.

ELECTROCHEMICAL CORROSION Reaction when two metals of different electrical potential are brought into contact in the presence of an electrolyte, the result being that the metal with the higher potential (most noble) will form the cathode, while the metal with the lower potential (the least noble) will form the anode. As current flows from the anode to the cathode, a chemical reaction will take place; the metal forming the anode will corrode and will deposit a layer of material on the metal forming the cathode. As the electric potential between two dissimilar metals increases, the stronger the current flow and corresponding rate of corrosion. *See also* **Chemical corrosion.**

ELECTROCHEMICAL GAUGING Determination of the water flow based on the degree of dilution over time and distance of a salt solution.

ELECTROCHEMICAL SOIL STABILIZATION Stiffening of certain clays produced by the passage of a direct electrical current using an anode of a polyvalent metal such as calcium, aluminum or iron.

ELECTRODE Electric conductor through which a current enters or leaves a medium.

ELECTRODIALYSIS *See* **Membrane processes.**

ELECTRODYNAMIC SEPARATOR Device that incorporates a rotating drum or other moving poles in place of one or more fixed charged plates (poles). Can be used to separate electrically conductive material from nonconductive material.

ELECTROHYDRAULIC SERVO-VALVE Servo-valve that is capable of continuously controlling hydraulic output as a function of electrical input.

ELECTROLYSIS Production of chemical changes by the passage of electrical current through an electrolyte.

ELECTROLYSIS/ELECTROLYTIC PITTING Migration (corrosion, pitting) of metal particles between two dissimilar metals.

ELECTROLYTE Conducting medium in which the flow of electric current takes place by migration of ions. The electrolyte for a lead storage cell is an aqueous solution of sulfuric acid; for alkaline storage cells an aqueous solution of certain hydroxides.

ELECTROLYTIC CHLORINE Chlorine produced by the electrolytic dissociation of hydrochloric acid or one of its salts.

ELECTROMAGNETISM Forces resulting from electrical changes that either attract or repel particles.

ELECTROMETER Instrument used to measure the atmospheric electrical field.

ELECTROMETRIC TITRATION Titration in which the end point is determined from observation of the change of potential of an electrode immersed in the titrated solution.

ELECTRON Negatively charged elementary particle present in the orbital structure of all atoms.

ELECTRON CAPTURE DETECTOR Instrument used to detect the presence of minute quantities of chemical substances in the atmosphere.

ELECTRONIC FILTER Filter that uses an electrically charged plate to attract and retain dust particles and pollen. *Also called* **Electrostatic filter.**

ELECTROOSMOSIS 1. Technique for the removal of certain salts from water by passing it through a series of diaphragms, from cell to cell, containing anode and cathode poles; **2.** Groundwater lowering process in which the flow of water is induced by an electrical current flowing from a positive anode to a negative cathode.

ELECTROPLATING WASTEWATER Process wastewater generated in operations which are subject to regulation.

ELECTROSTATIC FIELD Region in which stationary, electrically-charged particles would be subjected to a force of attraction or repulsion as a result of the presence of another stationary electric charge.

ELECTROSTATIC FILTER *See* **Electronic filter.**

ELECTROSTATIC PRECIPITATOR Device that collects particulates by placing an electrical charge on them and attracting them onto a collecting electrode.

ELECTROSTATIC SEPARATOR *See* **Separator.**

ELEMENT 1. Simple substance consisting of atoms all having the same atomic number, and which cannot be chemically broken down into simpler substances; **2.** Porous device which performs the actual process of filtration. *Also called* **Cartridge**; **3.** *See* **Functional element.**

ELEMENTARY NEUTRALIZATION UNIT Device used for neutralizing corrosive hazardous wastes and which meets the definition of tank, tank system, container, transport vehicle, or vessel.

ELEVATED GRAVITY TANK Water storage tank located at the highest point of the distribution system, to which primary water is pumped, and from which all distribution water flows.

ELEVATED STORAGE SYSTEM System of storing impounded water supplies above the grade level at which the water will be used.

ELEVATED TANK Tank raised above the ground and used to generate a constant head in a water distribution system.

ELEVATION Established point above a known datum (bench mark or other reference).

ELEVATION HEAD Product of the density of a fluid and its height above a point of reference (for water, only the height is stated).

ELEVATION LOSS Loss of pressure caused by raising water through hose or pipe to a higher elevation.

ELL Pipe fitting shaped like a bent elbow or L.

ELLIPTICAL PIPE Pipe having x and y axes of different dimensions.

ELUTRIATION Process wherein materials are separated according to differences in their densities and/or shapes in a countercurrent stream of a fluid (usually water), gas, or air. Typically, the dewatering of sewage sludge.

ELUTRIATOR Classifier used in soil analysis in which large grains sink faster through liquid than small grains of the same material.

ELUVIATION Removal from soil by percolating water of organic material and clay in solution or suspension.

ELUVIUM Deposit of disintegrated material.

EMAGRAM Thermodynamic diagram, having temperature and the logarithm of pressure as its coordinates, used for plotting and analyzing atmospheric soundings of temperature and humidity.

EMBANKMENT 1. Area of fill, the top of which is higher than the surrounding surface; **2.** Structure of soil, soil-aggregate, or broken rock between the embankment foundation and the subgrade.

EMBANKMENT DAM Structure of excavated or imported materials, used to retain water.

EMBANKMENT WALL Retaining wall at the foot of a bank.

EMBAYMENT Deep depression in a shoreline forming a large bay.

EMERGENCY FIELD ORDER *See* **Extra work order.**

EMERGENCY LOCK Air lock designed to hold and permit the quick passage of an entire shift of workers.

EMERGENCY POWER Independent reserve source of electric energy that, upon failure or outage of the normal source, provides electric power. *Also called* **Alternate source of power.**

EMERGENCY SHUTDOWN Manually- or automatically-actuated sequence of events that will cause an engine, motor or system to shut down under certain emergency or detected circumstances.

EMERGENCY SIGNAL 1. Visual or audible signal that warns of a condition beyond that for which a system is designed or that the local environment can tolerate; **2.** *See* **Traffic signal.**

EMERGENCY STOP SWITCH *See* **Switch.**

EMERGENCY WORK ORDER *See* **Extra work order.**

EMISSION 1. Air or gas expelled from an area or mechanical device; **2.** The three major pollutant emissions for which gasoline-powered vehicles are controlled are: unburned hydrocarbons (HC), carbon monoxide (CO), and nitrogen oxides (NOx). Diesel-powered vehicles primarily emit NOx and particulates.

EMISSION CONTROL DEVICE Solvent recovery or solvent destruction device used to control volatile organic compounds (VOC) emissions.

EMISSION DATA Information necessary to determine the identity, amount, frequency, concentration, or other characteristics (to the extent related to air quality) of any emission which has been emitted by the source.

EMISSION FACTOR Amount of a pollutant discharged per unit of time; or the amount of pollutant per unit volume of gas or liquid emitted.

EMISSION INVENTORY Compilation of

185

the rates and amounts of discharge of a pollutant and a diagram or map of its distribution.

EMISSION MEASUREMENT SYSTEM Equipment necessary to transport and measure the level of emissions, including the sample and instrumentation system.

EMISSION SOURCE Point or place where a pollutant originates.

EMISSION STANDARD Rule or measurement established to regulate or control the amount of a given constituent that may be discharged into the outdoor atmosphere.

EMPLOYER Contractor or subcontractor; one who contracts with another, orally or in writing, to pay for the performance of specified acts or duties.

EMPLOYER'S LIABILITY INSURANCE Protection for an employer against claims or common lawsuits by employees for damages arising out of their employment.

EMULSIFIER Agent capable of modifying the surface tension of emulsion droplets to prevent coalescence.

EMULSION Colloidal dispersion of a liquid in another liquid.

EMULSOID Colloid that is readily dispersed in a suitable medium and which may be redispersed after coagulation.

ENCASE Enclose; cover entirely; seal.

ENCLOSED Surrounded by a case, cage, or fence, that will protect the contained equipment and prevent accidental contact of a person with live parts.

ENCLOSED PROCESS Manufacturing or processing operation designed and operated so that there is no intentional release into the environment of any substance present in the operation.

ENCLOSED STORAGE AREA Any area covered by a roof under which metallic minerals are stored prior to further processing or loading.

ENCLOSED TRICKLING FILTER Trickling filter that operates with its own enclosure.

ENCLOSURE 1. Structure that surrounds a volatile organic compound application and drying area, and which captures and contains evaporated VOC and vents it to a control device. **2.** Airtight, impermeable, permanent barrier around asbestos–cement building materials to prevent the release of asbestos fibers into the air.

ENCROACHMENT Seawater that moves inland through the strata to penetrate a freshwater aquifer.

ENCRUSTATION Accumulated dirt, grime and corrosion that is tightly bound to a surface.

ENCUMBRANCES Obligations that reduce the balance of the particular budget account to which they are chargeable. Typically purchase orders and incomplete contracts.

ENDANGERED OR THREATENED SPECIES Any species listed as such in an Endangered Species Act.

END CONTRACTION Contraction of the water area flowing over a measuring weir.

END EVENT Last event in a program of the critical path method.

END MANHOLE Access chamber at the upstream end of a sewer.

ENDOTHERMIC Chemical reaction, process or exchange in which heat is absorbed.

END POINT Stage in titration at which equivalence is attained and revealed by a change that can be seen or measured.

ENDOGENOUS Condition in which living organisms oxidize some of their own cellular mass instead of using new organic matter which they would otherwise absorb or adsorb from their environment.

END-RESULT SPECIFICATION Specification that gives parameters for the completed work but which does not describe or require how the work will be done. *See also* **Compaction specifications.**

END SECTION Flared metal end attachment to a culvert to prevent erosion and improve hydraulic efficiency.

END USE Proposed or potential final use of landfill property after closure.

ENERGY Capacity to do work. There are several definitions, including:

> **Kinetic:** Energy due to motion.

> **Potential:** Energy due to position or condition.

> **Total:** Sum of kinetic energy and potential energy.

ENERGY ANALYSIS Calculation and compilation of the energy consumed at every stage of an operation.

ENERGY AUDIT Accounting of all forms of energy inputs over a period.

ENERGY AVERAGE LEVEL Quantity calculated by taking ten times the common logarithm of the arithmetic average of the antilogs of one-tenth of each of the levels being averaged. The levels may be of any consistent type, e.g. maximum sound levels, sound exposure levels, and day–night sound levels.

ENERGY BUDGET Record of the flow of energy through a system.

ENERGY CONSERVATION Employment of less energy to accomplish the same amount of useful work. *Also called* **Conservation of energy.**

ENERGY DISSIPATION Transformation of mechanical energy into heat energy.

ENERGY EFFICIENCY Rational and economical use of energy as part of an overall conservation policy.

ENERGY FLOW Passage of energy through the trophic levels of a food chain.

ENERGY GRADIENT Slope of the energy line of a body of flowing water relative to a datum plane.

ENERGY-GRADIENT LINE Line representing the gradient that joins the elevations of an energy head.

ENERGY HEAD Elevation of the hydraulic gradient, at any section, plus the head.

ENERGY LINE Line joining the elevations of the energy heads: a line drawn above the hydraulic grade line by a distance equivalent to the velocity head of the flowing water at each section along a channel, conduit or stream.

ENERGY RECOVERY Energy resource recovery where a part, or all, of a waste stream is processed to utilize its heat content to produce hot air, hot water, steam, electricity, synthetic fuel or other useful energy forms.

ENERGY RECOVERY PROCESS Process that recovers the energy content of combustible wastes directly by burning, or indirectly by converting the waste to another fuel form, such as gas or oil.

ENERGY STORAGE DEVICE Rechargeable means of storing tractive energy on board a vehicle, such as storage batteries or a flywheel.

ENGINE 1. Prime mover; **2.** Heat engine driven directly by the expansion of combustion gases, rather than by an externally produced medium, such as steam. Basic versions of the internal combustion engine are:

> **Gasoline engine and gas engine** (spark ignition).

> **Diesel engine** (compression ignition).

> **Gas turbine** (continuous ignition).

Diesel compression-ignition engines are more fuel-efficient than gasoline engines because compression ratios are higher, and because the absence of air throttling improves volumetric efficiency. Gasoline, gas (natural gas, propane), and diesel engines operate on a four-stroke cycle (Otto cycle) or a two-stroke cycle. Most gasoline engines are of a four-stroke type with operation as follows:

> **Intake** – piston moves down the cylinder, drawing in a fuel–air mixture through the intake valve;

Compression – all valves closed, piston moves up, compressing the fuel–air mixture, and spark ignites mixture near top of stroke;

Power – rapid expansion of hot combustion gases drives piston down, all valves remain closed;

Exhaust – exhaust valve opens and piston returns, forcing out spent gases.

The diesel four-stroke cycle differs in that only air is admitted on the intake stroke, fuel is injected near the top of the compression stroke, and the fuel–air mixture is ignited by the heat of compression rather than by an electric spark. The four-stroke engine has certain advantages over a two-stroke, including higher piston speeds, wider variation in speed and load, cooler pistons, no fuel lost through exhaust, and lower fuel consumption.

The two-stroke cycle eliminates the intake and exhaust strokes of the four-stroke cycle. As the piston ascends, it compresses the charge in the cylinder, while simultaneously drawing a new fuel–air charge into the crankcase, which is airtight. (In the diesel two-stroke cycle, only air is drawn in; the fuel is injected at the top of the compression stroke.) After ignition, the piston descends on the power stroke, simultaneously compressing the fresh charge in the crankcase. Toward the end of the power stroke, intake ports in the piston skirt admit a new fuel–air charge that sweeps exhaust products from the cylinder through exhaust ports; this means of flushing out exhaust gases is called 'scavenging'. Because the crankcase is needed to contain the intake charge, it cannot double as an oil reservoir. Therefore lubrication is generally supplied by oil that is pre-mixed with the fuel. An important advantage of the two-stroke-cycle engine is that it offers twice as many power strokes per cycle and, thus, greater output from the same displacement and speed. Because two-stroke engines are light in relation to their output, they are frequently used where small engines are desirable.

Gas turbines differ from conventional internal combustion engines in that a continuous stream of hot gases is directed at the blades of a rotor. A compressor section supplies air to a combustion chamber into which fuel is sprayed, maintaining continuous combustion. The resulting hot gases expand through the turbine unit, turning the rotor and driveshaft.

ENGINEER Registered professional responsible for the design of structural elements within a constructed whole.

ENGINEERING GEOLOGIST Geologist, who may also have engineering training, specializing in the application of geology to engineering problems.

ENGINEERING STUDY Detailed study of the loadings to be accommodated, the structural system to be employed, and the materials suggested for that purpose.

ENGINE RATING Value of engine power output assigned by the manufacturer, to indicate the maximum power level at which the engine should be applied in a given application.

ENGINE SAFETY CONTROL Device that protects against catastrophic damage by shutting the engine down in the event of high coolant temperature, low lube oil pressure, low coolant level, or over-speed.

ENGINE SIDESCREEN Rugged screen that fits on the engine housing of a vehicle used at a sanitary landfill to keep paper and other objects from accumulating and impairing engine performance.

ENGLACIAL Originating in, or carried by the interior of a glacier.

ENGLACIAL LOAD Material embedded in the interior of glacial ice and borne vertically by the ice until finally deposited on the face of the terminal slope.

ENGLACIAL STREAM Stream flowing through the interior of a glacier.

ENLARGEMENT LOSS Head loss in a conduit resulting from eddy losses caused by a sudden velocity change at the point of cross-sectional increase.

Enterococci Group of spherical bacteria normal to the intestines of man or animals.

ENTERIC Of or within the intestine.

ENTERIC BACTERIA Bacteria that inhabit the human gut, including those that may cause disease.

ENTEROBACTER AEROGENES Coliform bacteria, isolated from sewage, feces, soil and dairy products.

ENTRAIN To cause one thing to enter another; to trap bubbles of air in water either mechanically through injection or turbulence, or chemically through a reaction.

ENTRAINED AIR 1. Air induced into a room by the primary air flow, creating a mixed air path; **2.** Microscopic air bubbles, typically between 10 and 1000 μm in diameter and spherical, or nearly so.

ENTRAINMENT Drops of liquid carried over during a given process: as evaporation for distillation; the entrapment and transportation of material by a flowing liquid.

ENTRAINMENT SEPARATOR Apparatus that removes entrapped droplets for a vapor stream in a desalting process.

ENTRANCE HEAD Fluid head required to cause flow into a conduit.

ENTRANCE LOSS Fluid head lost from eddies and friction at the inlet to a conduit.

ENTRANCE WELL Collecting chamber from which surface water is fed to a sewer, or serving as a suction well for a wastewater pump.

ENTROPY 1. In a thermodynamic system, a measure of the energy that is not available for conversion into mechanical work; **2.** Measure of the degree of molecular disorder in a system; **3.** Statistical measure of the predictable accuracy of a system in transmitting information.

ENVIRONMENT 1. Combination of all external conditions which, singularly or in combination, may influence, modify, or change something; **2.** Conditions, circumstances, and influences surrounding and affecting the development of an organism(s); **3.** Includes water, air, and land and the interrelationship which exists among and between water, air, and land and all living things.

ENVIRONMENT(AL) AGENCY Government department or agency responsible for the formulation and administration of public policy on matters concerned with and affecting the natural environment.

ENVIRONMENTAL ASSESSMENT Concise public document that serves to briefly provide sufficient evidence and analysis for determining whether to prepare an environmental impact statement or a finding of no significant impact.

ENVIRONMENTAL AUDIT Assessment of an existing condition affecting the natural environment, or the accounting of effects on the environment of energy use policies, materials use policies, waste output, pollution control measures, etc.

ENVIRONMENTAL DAMAGE ASSESSMENT Methods by which a monetary value can be placed on environmental damage, including:

> **Preventive expenditure** – Amount paid to prevent or reduce unwanted effects.

> **Replacement/restoration** – Amount spent to restore or replace a lost facility.

> **Property valuation** – Difference between the market value of similar properties that reflects the effects of environmental depreciation.

> **Loss of earnings** – Loss through environmentally-caused injury or ill-health.

> **Productivity change** – Value of a reduction in productivity capacity due to environmental damage.

> **Contingent valuation** – Amount people say they would be willing to pay to avoid unwanted effects.

> **Maintenance expenditure** – Cost of maintaining capital structures to repair or offset deterioration due to adverse environmental impact.

ENVIRONMENTAL DESIGN Location and design of a facility that includes consideration of the impact of the facility on the community or region based on aesthetic, ecological, cultural, sociological, economic, historical, conservation, and other factors.

ENVIRONMENTAL EDUCATION AND TRAINING Educational and training activities involving elementary, secondary, and postsecondary students, and environmental education personnel, but not including technical training activities for environmental management professionals or activities primarily directed toward the support of non-educational research and development.

ENVIRONMENTAL ENGINEERING 1. Engineering discipline concerned with measures necessary to contain, treat and dispose of materials capable of injury to the environment; **2.** Engineering design of the mechanical equipment necessary to establish and maintain an enclosed environment suitable for habitation by humans.

ENVIRONMENTAL EVALUATION *See* **Environmental impact assessment.**

ENVIRONMENTAL FORECASTING Technique of predicting the environmental consequences of proposed developments.

ENVIRONMENTAL IMPACT ASSESSMENT Assessment of the impact a development will have on the environment of the site and/or area on which it is to be constructed. *Also called* **Environmental evaluation.**

ENVIRONMENTAL IMPACT STATEMENT Document that identifies and analyses in detail the environmental impacts of a proposed action.

ENVIRONMENTAL INFORMATION DOCUMENT Written analysis prepared by an applicant, grantee or contractor describing the environmental impacts of a proposed action. The document will be of sufficient scope to enable the responsible official to prepare an environmental assessment.

ENVIRONMENTALLY-RELATED MEASUREMENTS Data collection or investigation involving the assessment of chemical, physical, or biological factors in the environment which affect human health or the quality of life, including determina-

tion of (a) pollutant concentrations from sources or in the ambient environment, including studies of pollutant transport and fate; (b) the effects of pollutants on human health and on the environment; (c) the risk/benefit of pollutants in the environment; (d) the quality of environmental data used in economic studies; and (e) the environmental impact of cultural and natural processes.

ENVIRONMENTALLY SENSITIVE AREA In forestry, an area that includes potentially fragile or unstable soils that may deteriorate unacceptably after forest harvesting, and areas of high value to non-timber resources such as fisheries, wildlife, water, and recreation.

ENVIRONMENTALLY TRANSFORMED Chemical substance whose chemical structure changes as a result of the action of environmental processes on it.

ENVIRONMENTAL MANAGEMENT SYSTEMS Procedure by which to evaluate a system for environmental management, consisting of:

a) Definition and documentation of the organizational structure.

b) Drawing up of an inventory of releases, wastes, energy; documentation of raw materials usage .

c) Environmental effects assessments.

d) Establishment of objectives and targets.

e) Drafting of environmental management plans.

f) Management documentation and recording.

g) Environmental audits; audit plans plus reports and follow-ups.

h) Verification and testing.

i) Personnel factors of awareness, training and qualifications.

ENVIRONMENTAL NOISE Intensity, duration, and the character of sounds from all sources.

ENVIRONMENTAL POLICY FORMU-LATION Steps taken by a company wishing to improve its environmental performance.

ENVIRONMENTAL PROTECTION That part of resource management concerned with the release into the environment of substances that might be harmful.

ENVIRONMENT QUALITY STAN-DARDS Maximum limits or concentrations of pollutants that are permitted in specific media.

ENVIRONMENTAL REPORT Compilation and publication by a company of an annual state of health, containing information on waste minimization, materials recycling, energy conservation, emissions reduction, water conservation, etc.

ENVIRONMENTAL REVIEW Process whereby an evaluation is undertaken to determine whether a proposed action may have a significant impact on the environment and therefore require the preparation of an environmental impact statement.

ENVIRONMENTAL STUDIES Ecological effects or chemical fate studies, or both.

ENVIRONMENTAL TEMPERATURE Comfort index calculated from one third of the ambient air temperature plus two thirds the mean radiant temperature.

ENVIRONMENTAL TRANSFORMA-TION PRODUCT Chemical substance resulting from the action of environmental processes on a parent compound that changes the molecular identity of the parent compound.

ENZYME Catalyst produced by living cells.

EPHEMERAL STREAM Stream that flows only following precipitation, or which does not flow continuously.

EPIDEMIOLOGY Study of categories of persons and the patterns of diseases from which they suffer so as to determine the events or circumstances causing the diseases.

EPIGENE Processes that operate in and on the ground.

EPILIMNION Warmer, uppermost layer of water that lies above the thermocline in a lake and which is subject to wind disturbance.

EPIPEDON Organic and leached upper layer of soil.

EPIPELAGIC Seawater zone, or organism inhabitating that zone, at depths of less than 200 m.

EPOCH Unit of geological time, such that two epochs constitute a period, and subdivided into ages.

EQUAL-FALLING PARTICLES Particles of equal specific gravity and equal terminal velocity.

EQUALIZER 1. Piping arranged to maintain a common liquid level or pressure between two or more vessels; **2.** Culvert that allows standing water to rest at a common elevation about a bank.

EQUALIZING BASIN Retention pond or basin, usually between treatment units or stages, used to produce a flow of reasonably uniform volume and composition.

EQUALIZING BED Layer of ballast or concrete on which pipes are laid in the bottom of a trench.

EQUALIZING RESERVOIR Storage tank in a fluid supply system that provides for elasticity of input, operation, and supply.

EQUIPLUVE Line on a rainfall map connecting points having the same value of the Pluviometric coefficient: the ratio of actual mean rainfall for a period to the calculated mean rainfall for the same period at the same location, expressed as a percentage.

EQUIPMENT All machinery and equipment, together with the necessary supplies for upkeep and maintenance and also tools and apparatus necessary for the proper construction and acceptable completion of the work.

EQUIVALENT 1. Alternative designs, materials, or methods that can be demonstrated to provide an equal or greater degree of safety and performance as the item specified; **2.**

Chemical substance or mixture that is able to represent or substitute for another in a test or series of tests, and that the data from one substance can be used to make scientific and regulatory decisions concerning the other substance.

EQUIVALENT DIAMETER Four times the area of an opening divided by its perimeter.

EQUIVALENT DIRECT RADIATION Rate of heat transfer (by both radiation and convection) from a radiator or convector. The equivalent direct radiation is expressed in terms of the surface area of an imaginary standard radiator that would be required to transfer heat at the same rate as does the unit in question. 0.092 m² (1.0 ft²) of EDR gives off 70.32 W (240 Btu/hr) for steam heating units, or 43.9 W (150 Btu/hr) for hot water heating.

EQUIVALENT FLUID PRESSURE Horizontal pressures of soil, or soil and water in combination, that increase linearly with depth, equivalent to that which would be produced by a heavy fluid of a selected unit weight.

EQUIVALENT PER MILLION (epm) Unit of chemical equivalent weight of solute per million unit weights of solution.

EQUIVALENT PIPES Pipes or systems in which the head loss for equal flow rates are the same.

EQUIVALENT TEMPERATURE Index similar to effective temperature, but without consideration for the effects of humidity.

ERA Second largest division (after eon) of geological time, subdivided into periods.

ERADICATION Complete extinction of a species throughout its range.

ERG Obsolete unit of energy, equal to 0.1 μJ, that should not be used with the SI system of measurement. Symbol: erg. *See also the appendix*: **Metric and nonmetric measurement.**

ERGONOMIC DESIGN Design of machines, equipment, and parts to best suit the human form and in a manner to reduce fatigue.

ERGONOMICS Study of the interactions between people and work, especially machines and their component parts.

EROSION 1. Progressive wearing away of land through natural actions of streams, wind, etc.; **2.** Progressive disintegration of a solid by the abrasive or cavitation action of gases, fluids, or solids in motion. *See also* **Cavitation damage.**

EROSION CONTROL Constructions and means used to reduce erosion of land surfaces.

EROSIVE VELOCITY Rate of flow of a stream which, when exceeded, will result in erosion of its banks and/or bed.

ERRATIC Glacially deposited piece of rock, different in composition to the bedrock beneath it, used to reconstruct the movement of ice sheets.

ERROR Difference from a correct value. There are several classifications, including:

> **Absolute:** Errors of position relative to a grid or graticule.

> **Accidental:** Small, unbiased, and unavoidable errors in observations.

> **Compensating:** Combination of two or more errors, usually in calculation or measurement, that combine to cancel each other.

> **Gross:** Large errors, mistakes e.g. misreadings.

> **Relative error:** Errors of position relative to adjacent features or framework point.

> **Systematic:** Small, biased, and sometimes avoidable errors in observations.

ESCALATED BASE PRICE Estimate of the base price of a project at the time of tender.

ESCALATION 1. Anticipated increase in uncommitted costs due to the inflation of prices for resources (labor, materials, equipment); **2.** Component of a cost type; the allowance for escalation is a component within the anticipated award of a cost class.

ESCALATOR CLAUSE Provision in a contract that recognizes anticipated or unexpected cost increases and makes allowance for them.

ESCAPE 1. Organism that was formerly cultivated or in captivity and which has established itself in the wild; **2.** Watercourse for discharging all or part of a stream flow.

ESCAPE CLAUSE Contract provision that allows one or more parties to cancel all or part of a contract under defined conditions, typically if certain events do or do not take place.

ESCAPE ROUTE Pre-planned route along which building occupants can reach safety in an emergency.

ESCAPE STAIR 1. Part of an escape route, usually a stair completely surrounded by protective fire-resisting walls, reached through a lobby with self-closing fire-resisting doors; **2.** Outdoor metal stair connecting all floors with the ground.

ESCARPMENT Steep slope or cliff.

Escherichia coli (E. coli) Species of bacteria in the coliform group: its presence in water or soil is considered indicative of fresh fecal contamination.

ESKER Long, narrow ridge formed by glacial meltwater.

ESSENTIAL MINERALS Mineral constituents of a rock that are necessary to its nomenclature.

ESTIMATE 1. Assessment of the likely quantitative result, usually applied to project costs and duration; **2.** A calculated prediction. *See also* **Cruise.**

ESTIMATE AT COMPLETION Expected total cost of an activity, a group of activities, or of the project when the defined scope of work has been completed.

ESTIMATED COST Prediction of cost based on past experience and/or on information supplied.

ESTIMATED DESIGN LOAD In heating and air conditioning, the sum of the useful heat transfer from, or to the connected piping, plus heat transfer occurring in any auxiliary apparatus connected to the system.

ESTIMATED MAXIMUM LOAD In a heating or air conditioning system, the calculated maximum heat transfer that the system may be called upon to provide.

ESTIMATES Predictions or forecasts of costs that will occur in the future, but that are not yet committed; a cost type. Estimates are described or qualified on the basis of the supporting design information available at the time the estimate is prepared. In ascending order of precision and certainty, the following descriptions are used: (a) order of magnitude, (b) concept, (c) design presentation or appropriation grade, and (d) pretender.

ESTIMATE TO COMPLETE *See* **Cost types.**

ESTIMATING Judging or calculating the amount of material required for a given item of work, including labor content, and extending it by the cost per unit of measurement to give an approximation of the value of the finished product.

ESTIMATOR 1. One who calculates quantities of materials and labor, and their costs, and the costs of construction. **2.** *See* **Cruiser.**

ESTIMATOR'S CONTINGENCY Estimated allowance for price variances due to the inability of the estimator to price any given item.

ESTUARIAL STORAGE Storage of water for domestic water supplies in estuaries.

ESTUARINE Formed in an estuary.

ESTUARY Channel in which the tide meets a river current.

ETHANOL Ethyl alcohol (C_2H_5OH), having a boiling point of 78.4°C and a specific gravity of 0.789.

ETHNOBOTANY Study of the uses to

which plants and plant products are put by peoples of different cultures.

ETHOGRAM Catalogue of the behavior of an animal under conditions as natural as possible and of the contexts in which it occurs.

ETIOLOGY Science of the origin or cause of disease.

EUPHOTIC ZONE Upper zone of a sea or lake into which sufficient light can penetrate for active photosynthesis to occur.

EURYHALINE Organisms able to tolerate a wide range of saline conditions.

EURYTHERMIC Organisms able to tolerate a wide range of temperatures in the environment.

EURYTOPIC Organisms having a wide geographical distribution.

EUSTATIC Worldwide and simultaneous change in sea level, i.e. from the melting of ice sheets and glaciers.

EUTROPHIC Body of water rich in dissolved nutrients and supporting an abundance of plant life.

EUTROPHICATION Natural aging of a body of water that results in the production of an overabundance of organic material and the resulting lowering of the water quality.

EUTROPHIC LAKE Water body rich in dissolved nutrient and producing large quantities of planktonic algae resulting in low water transparency. There is usually a high level of dissolved oxygen in the upper layers; zero in deep layers during warmer weather, and large black- or brown-colored organic deposits. Hydrogen sulfide is often present in the water and deposits.

EVAPORABLE WATER Water present in set cement paste in capillaries or held by surface forces, measured as that removable by drying under specified conditions. *See also* **Non-evaporable water.**

EVAPORATION Change of state from liquid to vapor.

EVAPORATION AREA Surface area of a body of water and of any adjacent wetland that can lose moisture to the atmosphere by evaporation.

EVAPORATION DISCHARGE Gaseous discharge into the atmosphere from the soil and vegetation.

EVAPORATION POND Shallow impoundment in which water may be trapped so as to permit evaporation to deposit minerals, i.e. salt from saline water.

EVAPORATION RATE Quantity of water evaporated from a given surface per unit of time.

EVAPORATION RETARDANT Long-chain organic material such as cetyl alcohol which, when spread on a water film, retards evaporation.

EVAPORATIVE COOLER Air conditioning air by the effect of water evaporation.

EVAPORATIVE EQUILIBRIUM Condition attained when the wetted wick of a wet-bulb thermometer has reached a stable and constant temperature when exposed to moving air in excess of 274.3 m/min (900 ft/min). *Also called* **True wet-bulb temperature.**

EVAPORATIVITY Potential rate of evaporation, as distinguished from the actual rate.

EVAPORATOR Device or apparatus in which a solution is converted to a vapor.

EVAPORIMETER Instrument for measuring evaporation.

EVAPOTRANSPIRATION Portion of precipitation that is returned to the atmosphere through direct evaporation and vegetative transpiration.

EVENT Point in time where certain conditions have been fulfilled, such as the start or completion of one or more activities.

EVENT-ON-NODE Network diagramming technique in which events are represented by boxes (or nodes) connected by arrows to show the sequence in which the events are to occur.

EVENT-ORIENTED SYSTEMS Systems that present information in terms of events (i.e. points in time).

EVERGREEN Plant that bears leaves throughout the year; plants that shed their leaves, but not in any particular season of the year and not all at the same time.

EVOLUTION Process of cumulative change.

EXCAVATION 1. Any digging operation involving the removal of earth; **2.** Space formed by the removal of earth or rock. There are several classifications, including:

> **Classified:** Excavation paid for at a unit price for common excavation plus a unit price for rock excavation.
>
> **Common:** General description of 'soft' excavation, such as earth and residual materials.
>
> **Rock:** Excavation in rocky materials requiring blasting.
>
> **Unclassified excavation:** Excavation paid for at one unit price, whether common or rock excavation.

EXCEEDANCE PROBABILITY Probability of a natural event such as a storm above a stated magnitude occurring during any one year.

EXCESS ACTIVATED SLUDGE Volume of activated sludge beyond that needed for process maintenance and operation which is removed from the system for treatment and disposal.

EXCESS AIR Additional air required to ensure complete combustion of a fuel, stated as a percentage of the theoretical air required.

EXCESS EMISSIONS Emissions of an air pollutant in excess of an emission standard.

EXCESSIVE I/I Quantities of infiltration and inflow that can be economically eliminated from a sewer system by rehabilitation, as determined by cost-effectiveness analysis that compares the cost for correcting the condition with the total cost for transportation and treatment of the infiltration and inflow.

See also **I/I analysis.**

EXCESSIVE PRECIPITATION Rainfall within an area that is greater than limits established as normal from records and other evidence collected over a considerable period.

EXCESS PORE PRESSURE Increment of pore water pressure greater than hydrostatic values, produced by consolidation stresses in compressible materials or by shear strain.

EXCESS PRESSURE PUMP *See* **Pump.**

EXCESS RAINFALL Rain that falls at an intensity exceeding the infiltration capacity and results in direct runoff.

EXCESS SLUDGE Sludge produced as part of the activated sludge process that is not needed to maintain the process and which is withdrawn from circulation.

EXCITATION Exciting or being excited, especially the production of a magnetic field by means of electricity or the raising of an atom or nucleus to a higher level of energy.

EXECUTION Performance of a contract by carrying out the work.

EXFILTRATION Uncontrolled escape of air or water from a structure, system, or vessel.

EXHAUST DUCT Duct through which air is conveyed from a room or space to the outdoors.

EXHAUST EMISSIONS Content of the waste gas that leaves a prime mover through its exhaust system.

EXHAUST FAN Fan that withdraws air under suction.

EXHAUST GAS ANALYZER Instrument that measures the quantities of different gases present in an exhaust stream. The result is a measure of the engine's efficiency.

EXHAUST GAS RECIRCULATION System designed to reduce automotive exhaust emissions of nitrogen oxides (NOx) by routing exhaust gases into the carburettor or intake manifold.

EXHAUST LINE Line returning power or control fluid back to a reservoir or atmosphere.

EXHAUST OPENING Port or void through which air is removed from a space that is being air conditioned.

EXHAUST SYSTEM System of pipes and muffling devices that channels the products of combustion (exhaust gases) from an engine into the atmosphere at a desired location.

EXISTING LAND-USE MAP Map that displays the development and use of an area of land, current at a point in time. It will contain a variety of types of information that may be displayed pictorially, graphically, or texturally, and may include the growth rate of a community, available land, soil types, zoning, community infrastructure (roads, water supply, sewer, electrical distribution, etc.), vegetation, land contours, among others.

EXIT That part of a means of egress from a structure that leads from the floor area it serves, including any doorway leading directly from a floor area, to another floor area, to a public thoroughfare, or to an approved open space. *See also* **Horizontal exit.**

EXIT DOOR Last door on an exit route that opens outward and which must not be locked while the building is occupied, only latched by panic hardware.

EXIT GRADIENT Hydraulic gradient (difference in piezometric levels at two points, divided by the distance between them) near to an exposed surface through which seepage is moving.

EXIT LEVEL Lowest level in an enclosed exit stairway from which an exterior door provides access to a public thoroughfare or to an approved open space with access to a public thoroughfare at approximately the same level either directly or through a vestibule or exit corridor.

EXIT LOSS Loss of head that occurs when a stream of air or water is voided from a restricting structure and assumes a lower velocity.

EXIT STORY Story from which an exterior door provides direct access at approximately the same level to a public thoroughfare or to an approved open space with access to a public thoroughfare.

EXOSPHERE Outermost layer of the Earth's atmosphere, lying beyond the ionosphere, in which air density is such that a molecule moving directly outward has a 50% chance of escaping, rather than colliding with another molecule.

EXOTHERMIC Chemical reaction in which energy in the form of heat is released.

EXPANDING PLUG Adjustable drain plug used for testing pipes.

EXPANDING REACH Channel reach where flow is decelerating due to a change in the channel cross section.

EXPANSION BEND Loop in a pipe that permits the expansion and contraction of the pipe.

EXPANSION CHAMBER Chamber designed to reduce the velocity of the products of combustion and promote the settling of fly ash from a gas stream.

EXPANSION COUPLING Pipe coupling permitting relative movement between the joined pipes.

EXPANSION FIT Fit easily made by placing a cold (subzero) inside component within a warmer outside component and allowing an equalization of temperature.

EXPANSION FLOW Fluid flow in conduits where an increased cross-sectional area results in a decrease in velocity and accompanying increase in pressure.

EXPANSION JOINT *See* **Joint.**

EXPANSION JOINT FILLER Pliable material used to fill gaps left for expansion so as to exclude foreign matter.

EXPANSION PIPE Vent from a hot water system that permits the escape of steam or the relief of excess pressure.

EXPANSION TANK In a hot water or cooling system, a tank designed to allow expansion of the water on heating.

EXPANSION VALVE *See* **Valve.**

EXPANSIVE SOILS Active clay soils that expand on wetting and shrink on drying.

EXPENDING BEACH Beach shaped so as to absorb the energy of waves.

EXPERIMENTAL ANIMALS Individual animals or groups of animals, regardless of species, intended for use and used solely for research purposes and not including animals intended to be used for any food purposes.

EXPERIMENTAL PLOT Area of ground laid out to determine the effects of a certain method of treatment or condition, often divided into subplots.

EXPERIMENTAL START DATE First date the test substance is applied to the test system.

EXPERIMENTAL TECHNOLOGY Technology which has not been proven feasible under the conditions in which it is being tested.

EXPLODED VIEW Drawing depicting an device, apparatus, building, mechanism, etc., with the principal parts separated from each other but remaining in their proper relationship.

EXPLOIT Excavation technique that uses defined materials from a vein or layer, discarding or wasting the remainder.

EXPLORATION Process of research, identification, and classification.

EXPLORATORY DRILLING Drilling performed to obtain data about subsurface materials and conditions.

EXPLORATORY PROGRAM Series of actions or tests carried out to an established plan and designed to reveal information relevant to and typical of an area or situation.

EXPLOSIMETER Instrument capable of detecting a potentially explosive atmosphere.

EXPLOSION-PROOF APPARATUS Electrical apparatus enclosed in a case that is capable of withstanding an explosion of a specified gas or vapor that may occur within it, and of preventing the ignition of a specified gas or vapor surrounding the enclosure by sparks, flashes, or explosion of the gas or vapor within, and that operates at such an external temperature that it will not ignite a surrounding flammable atmosphere.

EXPLOSION VENTING Provision of an opening for the release of pressure and heated (explosive) gases, thus preventing the development of destructive pressure.

EXPLOSION WAVE Wave that travels outward from the center of an explosion to accommodate the extra volume produced.

EXPLOSIVE Any solid, liquid or gaseous mixture or chemical compounds that, by chemical action, suddenly generates large volumes of heated gas, included in the following classifications:

> **Class A Explosives:** Possessing detonating hazard, such as dynamite, nitroglycerin, picric acid, lead azide, fulminate of mercury, black powder, blasting caps, and detonating primers.

> **Class B Explosives:** Possessing flammable hazard, such as propellant explosives, including some smokeless propellants.

> **Class C Explosives:** Including certain types of manufactured articles that contain Class A or Class B explosives, or both, as components, but in restricted quantities.

> *See also* **High explosive** and **Low explosive.**

EXPLOSIVE DUST Mixture of particulate matter and air in such proportions that they can explode, from ignition, compression, spontaneously, etc.

EXPLOSIVE LIMITS Upper and lower limits of petroleum vapor concentration in air outside of which combustion will not occur. *Also called* **Flammability limits.**

EXPONENTIAL GROWTH Increase in a value over a period by a fixed percentage of the original value, such that the increase in each period is equal to the original value plus the interest accumulated in the preceding period.

EXPOSURE 1. Location of a building or structure in relation to the sun, winds, etc.; **2.** Body of bedrock not covered by the regolith. *See also* **Aspect**; **3.** Time that a worker is exposed to a hazardous, or potentially hazardous condition; **4.** Time that a worker is exposed to toxic and hazardous air contaminants.

EXPOSURE: DOSE EFFECT RELATIONSHIP Effect of exposure to a known pollutant over a measured time period.

EXPOSURE HAZARD Probability that a building will be exposed to fire or other threat from surrounding or adjoining property.

EXPOSURE LIMITS Short- and long-term limits to which humans can experience exposure to known pollutants or similar hazards without immediate detrimental effect or long-term hazard.

EXPOSURE PERIOD Five day period during which test birds are offered a diet containing a test substance.

EXPRESSED WARRANTY *See* **Warranty.**

EXPRESSION OF SLOPE Means by which the rise (or fall) of a slope may be defined relative to the horizontal distance. There are several ways of expressing this, including:

> **Degree of slope:** Angle of the face of a slope above or below the horizontal, i.e. a 5° indicates a slope that measures 5° above a horizontal plane.

> **Gradient: 1.** Ascending or descending with a uniform slope; **2.** Degree of slope;

> **Percent grade:** Measure of the rate of ascent of an inclined plane numerically equal to the vertical rise divided by the horizontal length, multiplied by 100.

> **Percent of slope:** Rise of fall of a slope

in m/100 m, i.e. a 3% slope is a rise or fall of 3 m for every 100 m of horizontal run.

> **Slope distance:** Measured distance between the top and bottom of a slope.

> **Slope ratio:** Expression of the total rise of a slope to its total horizontal distance, i.e. a ratio of 0.05:1 gives a vertical rise of 0.05 m for every 1 m of horizontal distance.

EXTENDED AERATION PROCESS Modification of the activated-sludge process which provides for aerobic sludge digestion within the aeration stage.

EXTENSION AGREEMENT Agreement which extends the terms and/or conditions of a contract, signed by the parties to the original document.

EXTENSION OF TIME Additional time granted for the completion of something beyond that stipulated in the contract documents.

EXTENSIONS Extended prices on bid sheets or estimates of cost.

EXTERIOR Not within a shelter or structure; open to the elements.

EXTERNAL COMBUSTION ENGINE Mechanical device that relies on a heat source or fuel combustion outside of the engine, i.e. steam turbines and engines.

EXTERNAL FLOATING ROOF Pontoon-type or double-deck type cover with rim sealing mechanism that rests on the liquid surface in a waste management unit with no fixed roof.

EXTERNAL THREAD Thread on the outside of a part.

EXTERNAL-TOOTH WASHER *See* **Washer.**

EXTERNAL WORKS Works external to the structure or building: site works.

EXTINGUISHER Portable fire fighting appliance designed for use on specific types of fuel and classes of fire.

EXTINGUISHING MEDIA Chemical mixture designed to combat specific types of fire: electrical, oil and grease, organic combustibles, etc.

EXTRA Addition to a work contract after the contract has been awarded.

EXTRACT FAN Wall- or ceiling-mounted fan designed to extract and exhaust air to the outside.

EXTRACTION Process of dissolving and separating out specific constituents of a liquid by treatment of solvents particular to those constituents.

EXTRACTION SITE Place from which dredged or fill material proposed for discharge is to be removed.

EXTRACTOR COLUMN Device used to extract the solute from saturated solutions produced by a generator column.

EXTRAORDINARY STORM Rainfall of an intensity that may be expected to occur only once within a given period: 10 years, 25 years, 100 years, etc.

EXTRA WORK Item of work not provided for in the contract as awarded but found by the engineer (or other senior responsible person) to be essential for the satisfactory completion of the contract within its intended scope.

EXTRA WORK ORDER Unplanned authorization for a construction contractor to perform work beyond the scope of his contract, and hence a commitment document. An extra work order is generally issued in response to an emergency situation encountered during construction. It constitutes an extension to the contract. *Also called* **Emergency work order, Emergency field order** and **EWO.**

EYE Sense organ that detects light.

EYE OF THE STORM Place in a cyclonic storm where the wind velocity approaches zero.

FABRICATE Manufacture, form, assemble, construct, etc.

FABRICATION Preparing, making or assembling materials into elements and components, ready for use in a designed whole, and bringing those elements into use.

FABRICATOR Contractor responsible for furnishing fabricated structural steel.

FACEPIECE Portion of a self-contained breathing apparatus that fits over the face by means of an adjustable harness. It may, or may not have the regulator attached to it.

FACE-TO-FACE DIMENSION Dimension from the inlet face to the outlet face of a plumbing fitting.

FACE WALING Waling across the end of a trench, supported by the ends of the side walings that, together with the end strut also acting as a waling, supports the end face of a trench.

FACTOR One of two or more quantities having a designated product.

FACTORED LOAD Load, multiplied by appropriate load factors, used to proportion members by the strength design method.

FACTOR OF SAFETY 1. Ratio of the forces tending to resist failure to those forces tending to cause failure; **2.** Ratio of load, moment, or shear of a structural member at

the ultimate to that at the service level; **3.** Failure load divided by the design load.

FACTORY FINISHED Complete in all respects, ready for installation as delivered to the site.

FACULTATIVE Capable of response to varying environmental stimulus.

FACULTATIVE BACTERIA Bacteria that can adapt to growth in the presence, as well as the absence, of oxygen.

FACULTATIVE POND Body of water, the upper portion of which is aerobic, the bottom layers of which are anaerobic.

FAHRENHEIT Non-SI temperature measurement scale in which water freezes at 32°F (0°C) and boils at 212°F (100°C).

FAIL-SAFE Process or equipment control system whereby failure of the power supply does not result in process failure or equipment damage.

FAILURE Breakage, displacement, or permanent deformation of a structural member or connection so as to reduce its structural integrity and its supportive capabilities.

FAILURE ANALYSIS Step-by-step process of examination and analysis that allows the determination of the reason for failure of a part, component or system.

FALL 1. Amount of slope given to a item to allow for adequate flow without ponding: to pipe runs, flat roofing, gutters, screed, etc.; **2.** Sudden difference in elevation.

FALLING HEAD PERMEAMETER Apparatus for determining the permeability of soils.

FALLING OBJECT PROTECTION STRUCTURE (FOPS) Attachment to mobile equipment designed to protect operators against falling objects.

FALLING PRESSURE Barometric indication of increasing cloud and the possibility of rain, produced when an upward movement of air, which may lead to cloud formation, produces a convergence of air at lower levels which, in turn, increases the rotation of air near the ground with decreasing pressure at the center.

FALLING TIDE Tide occurring during the ebb period of tidal flow.

FALLOUT 1. Deposition of airborne solid or liquid particles on the ground; **2.** Radioactive debris that resettles to earth following a nuclear explosion.

FALLSTREAK Streak of falling cloud particles, almost always of ice particles that are not evaporating.

FALLSTREAK HOLE Hole in a layer of cloud that is composed of supercooled water droplets.

FALSE BOTTOM Support system above the true floor level: typically, the porous or perforated floor suspended above the true bottom of a water filter to create an underdrain.

FAN Power-driven device that blows or sucks air. There are many types, including:

Attic: Exhaust fan to discharge air from near the top of an attic space while fresh air is drawn in through a louvre at a lower level.

Centrifugal: Fan rotor wheel within a scroll-type housing.

Forced draft: Device that pushes air or gases.

Induced draft: Device that draws or exhausts air or gases.

Overfire air: Device that provides air to the combustion chamber of an incinerator, above the fuel bed.

Propeller: Propeller mounted on an axle or directly-mounted to the shaft of a motor.

Tubeaxial: Propeller or disk-type wheel within a cylinder.

Vaneaxial: Disk-type wheel within a cylinder, usually with air-guiding vanes mounted behind or ahead.

FANNING Behavior of a plume when, in stable air, gases reach their equilibrium level and move horizontally.

FARAD SI unit of electric capacitance of a condenser having a charge of 1 C, across the plates of which the potential difference is 1 V. (1 F=1 C/V). Symbol: F . *See also the appendix:* **Metric and nonmetric measurement.**

FARM DRAIN System of draining water from the surface of fields or grass areas by the use of ditches filled with gravel; perforated pipes may also be used. *Also called* **French drain.**

FARM DUTY WATER Seasonal volume of water delivered for irrigation purposes.

FARM POND Relatively small volume of water, often spring runoff, held behind a dam or contained in a slough.

FASCINE 1. Bundle of long sticks bound together and used for such purposes as filling ditches and making parapets; **2.** Woven willow mattress used along river banks and during river pier construction to minimize scour.

FATAL ACCIDENT RATE Measure of the number of fatalities per 100 million hours spent in one activity.

FATHOM Non-SI unit of length equal to 6 ft (1.8288 m).

FATHOMETER Acoustic measuring device used to measure the depth of water by the time required for a sound wave to travel from the surface to the bottom and its echo to be returned.

FATIGUED Structural failure of a filter medium due to flexing caused by cyclic differential pressure.

FATIGUE FAILURE Cracking or separation of a material when subjected to repeated loadings at a stress substantially less than the static strength of the material.

FATIGUE LIFE Number of cycles of specific loading that a material can sustain before failure occurs.

FATIGUE LIMIT Stress limit below which a material can be expected to withstand any number of stress cycles.

FATIGUE RESISTANCE Ability of a material to withstand repeated or alternating stresses.

FATIGUE STRENGTH Greatest stress that can be sustained for a given number of stress cycles without failure.

FATIGUE TEST Test involving repeated fluctuations or reversals of stress to determine the endurance limit of a material.

FAT Triglyceride ester of fatty acid. *See also* **Grease.**

FAUCET *See* **Valve.**

FAUCET EAR Projection from a faucet permitting nailing of the fixture to a wall.

FAULT Break in a mass of rock with the segment on one side of the break pushed up or down.

FAULT BRECCIA Rock composed of broken fragments lying along a fault plane.

FAULT PLANE Surface movement of a fault.

FAULT SPRING Spring fed by juvenile water and deep groundwater, usually hot and mineralized, which rises through a deep fault or fissure.

FAUNA Animals of a region, country, environment or of a period.

FEASIBILITY PHASE That part of a project life during which the concept design is prepared. *See also* **Construction phase,** *and* **Development objectives.**

FEASIBILITY STUDY/REPORT Investigation of the probability that a project could be completed as proposed, within the terms of reference of the study, that may be limited to finances, or materials, or structural competence, or scheduling, or any combination of these and other considerations.

FECAL *Streptococcus* Group of bacteria, normal to the intestinal tracts of warm-blooded animals other than humans, regarded as indicators of the contamination of water by the feces of these animals.

FECES Waste matter discharged from the intestines.

FEEDBACK Response mechanism between a sensor measuring a process variable and the controller which controls or adjusts the process variable.

FEEDER Channel through which water is fed to a reservoir or canal.

FEEDLOT Concentrated, confined animal or poultry growing operation for meat, milk or egg production, or stabling, in pens or houses where the animals or poultry are fed at the place of confinement.

FEED PIPE 1. Mainline water pipe that carries a supply directly to the point of use, or to secondary lines; **2.** Pipe supplying feed water to a boiler.

FEED PUMP *See* **Pump.**

FEEDSTOCK Crude oil and natural gas liquids; any raw, unprocessed material that is scheduled to become the primary constituent of a refined or finished product.

FEEDWATER Heated water pumped into a water-wall furnace to compensate for steam leaving the furnace.

FEEDWATER HEATERS Heat exchangers used as part of the feedwater system.

FEEDWATER SYSTEM System of pumps, deaerating feedwater heater, piping, and heat exchangers used to optimize the thermal efficiency of a steam cycle.

FEEDWATER TREATMENT Physical or chemical treatment of water in anticipation of its use in heating or processing.

FELDSPAR Most abundant group of rock-forming silicate minerals.

FEMALE Item having an end that encloses a like item.

FEMALE COUPLING Internally threaded pipe.

FEMALE THREAD Internal thread.

FEMTO Prefix representing 10^{-15}. Symbol: f. Used in the SI system of measurement. *See also the appendix*: **Metric and non-metric measurement.**

FEN Area of waterlogged peat that is alkaline or only slightly acid.

FERMENTATION Process of change in organic matter occasioned by the growth of microorganisms in the absence of air.

FERRIC SULFATE Soluble iron salt ($Fe_2(SO_4)_3$), formed by reaction of ferric hydroxide and sulfuric acid or by reaction of iron and hot concentrated sulfuric acid, or in solution by reaction of chlorine and ferrous sulfate.

FERROUS SULFATE Soluble iron salt ($FeSO_4 \cdot 7H_2O$). *Also called* **Copperas, Sugar of iron, Green vitriol,** and **Iron vitriol.**

FERTILIZER An artificial or natural substance added to soil to provide one or more of the nutrients essential to the growth of plants; the principal ones being nitrogen, phosphorus, and potassium.

FESCUE *See* **Grass.**

FETCH 1. Straight-line distance between the rim of a dam and the farthest reservoir shore; **2.** Uninterrupted distance over which wind can travel in raising waves.

FIBER 1. One of the threadlike cells or structures that combine with others to form certain plant or animal tissues; **2.** Slender, threadlike filament of wool, cotton, glass, rayon, nylon, asbestos, etc., used especially for making yarn or cloth.

FIBERGLASS Glass drawn and spun into fine threads or fibers, used for insulation and, bonded with resin, as a strong, light construction material.

FIBROSIS Scarring of lung tissue caused by the inhalation of dust over a prolonged period, usually at the workplace.

FIDELITY 1. Exactness; accuracy; **2.** Ability to reproduce a signal.

FIELD 1. Land yielding some product; **2.** Space throughout which a force operates.

FIELD CAPACITY 1. Water that a soil in place retains, when drainage has become negligible. *Also called* **Field moisture capacity; 2.** Amount of water retained in solid waste after it has been saturated and has drained freely.

FIELD CARRYING CAPACITY Approximate quantity of water that can be permanently retained in the soil in opposition to the downward pull of gravity.

FIELD CHECK Onsite inspection of work, conditions, and/or materials.

FIELD DITCH Smallest lateral of a drainage or irrigation system.

FIELD DUTY WATER Quantity of water delivered to a finite location for irrigation purposes.

FIELD ENGINEER Designated representative at a project site.

FIELD GROUNDWATER VELOCITY Actual velocity of groundwater percolating through water-bearing material. *See also* **Actual groundwater velocity.**

FIELD MOISTURE CAPACITY *See* **Field capacity.**

FIELD ORDER Written order, issued by the owner or owner's agent, that orders a minor change in the work not involving a change in the contract sum or guaranteed maximum price, or the contract time.

FIELD PERMEABILITY COEFFICIENT Rate of flow of water through each measure of thickness of an aquifer over a specific distance. *Also called* **Hydraulic conductivity.**

FIELD REPRESENTATIVE *See* **Owner's representative.**

FIELD SUPERINTENDENT Site engineer.

FIELD SUPERVISION *See* **Site supervision.**

FIELD TEST Experiment conducted under field conditions. Ordinarily less subject to control than a formal experiment; it may also be less precise. *Also called* a **Field trial.**

FIELD TESTING Practical, and generally small-scale testing of innovative or alternative technologies directed to verifying performance and/or refining design parameters.

FIELD TILE Pipe installed as a subsurface drain to collect groundwater; may be perforated along its length or laid in short lengths with a gap at each joint to allow infiltration.

FIELD TREATMENT Onsite application of wood preservative.

FIELD TRIAL *See* **Field test.**

FIELD WASTE Portion of applied irrigation water that is not absorbed or evaporated as it passes over the irrigated land.

FIELDWORK Work done at the jobsite.

FIFTY-YEAR FLOOD Flood that has a 2% chance of occurring in a given year, based on historical data. *See also* **One-hundred year flood.**

FILAMENTOUS Organisms that grow in a threadlike form; typically fungi, *Actinomycetes* and *Thiothrix*, commonly associated with sludge bulking in the activated sludge process.

FILL 1. Any material that is moved or added to the existing terrain to raise its elevation; **2.** Height to which material is to be placed; **3.** Earthwork resulting from the placement of fill; **4.** Coarse grade of earth having low compressibility and good stability, used to fill in around walls or foundations, or to raise the level of ground. *Also called* **Soil fill.** *See also* **Backfill, Compacted fill,** and **Structural fill. 5.** *See* **Fuel tank capacity.**

FILL CAP FILTER Filter that covers the fill opening to a reservoir tank and which filters makeup fluid.

FILL DAM *See* **Earth dam.**

FILL FACE Active area of a landfill, where solid waste is being unloaded and then compacted into the daily cell.

FILLER GATE Openable port in a larger gate by which pressure across the gate may be lessened when it is being opened to admit the passage of water from an already filled section.

FILL MATERIAL Any pollutant which replaces portions of natural water bodies with dry land or which changes the bottom elevation of a water body for any purpose.

FILM BADGE Personal safety device consisting of a small piece of masked photographic film that becomes progressively more 'exposed' on increasing or prolonged exposure to ionizing radiation.

FILM FLOW Movement of water, in any direction, through a system of interconnecting films adhering to the surfaces of solids in, or confining the water.

FILM PRESSURE Inward pull or pressure on water–air surfaces of a film system of suspended water in soils, composed principally of mobile or fringe water.

FILTER Device having a porous medium, whose primary function is the separation and retention of particulate contaminates from a fluid or air stream. *See also* **Bypass filter.**

FILTERABLE SOLIDS Solids retained on a membrane for analysis by weight, count, or observation as a measurement of contamination.

FILTER AID Any of a number of chemicals added to water to help remove fine colloidal suspended solids.

FILTER BAG Device designed to remove particles from a carrier gas or air by passage of the gas through a porous fabric medium.

FILTER BED Fill of rock and pervious soil that receives the effluent from a septic tank.

FILTER BLOCK Hollow, vitrified clay masonry unit, sometimes salt-glazed, designed for trickling filter floors in sewage treatment plants.

FILTER BOTTOM System at the bottom of a filter into which filtered water passes, and through which wash water used to clean the filter by backwashing is fed to the underside of the media.

FILTER BOX Rectangular section of a rapid sand filter complex.

FILTER CAKE Suspended solids deposited and compacted on a porous medium during filtration.

FILTER CLOGGING End result of fine particles filling the voids of a sand filter or biological bed.

FILTER CLOTH Fabric stretched around the drum of a vacuum filter.

FILTER CRIB Crude form of water filter constructed in a stream bed and consisting of a crib in an excavation in the stream bed, filled with sand and gravel and containing the suction of a pump used to draw water for domestic or other uses.

FILTER CROP Dense growth of vegetation across a slope that retards runoff and causes the deposition of soil in suspension.

FILTERED WASTEWATER Wastewater that has passed through a mechanical filtering process, but not through a biological filter.

FILTERED-WATER RESERVOIR *See* **Clear-water reservoir.**

FILTER EFFICIENCY Ability, expressed as percent, of a filter to remove specified artificial contaminants from a specified fluid under specified test conditions.

FILTER ELEMENT Subassembly of a filter that contains the filter medium or media.

FILTER FLOODING Technique of reducing the outflow from a trickling filter to below the rate of inflow so as to raise the level in the filter to a point above that of the media.

FILTER HOUSING Ported enclosure that contains the filter element and directs fluid or air flow.

FILTER GALLERY Type of pipe gallery, but also providing access to the control devices and mechanisms of the individual filters.

FILTER LOADING Weight of biochemical oxygen demand in an applied liquid, per unit area of filter bed or volume per day.

FILTER MEDIUM (MEDIA) Porous material that performs the process of particle separation and retention.

FILTER OPERATING CONSOLE Table mounted on the operating floor of a rapid sand filtration plant from which all equipment controls and adjustments can be made and on which various operating gauges are mounted.

FILTER PACK Clean graded sand and gravel placed in the annulus of a drilled water well to prevent formation material from entering the screen.

FILTER PERFORMANCE Those factors which describe the functions and attributes of a filter element.

FILTER PLANT Complex of installations necessary to remove varying degrees of suspended and/or dissolved solids from raw water or from a wastewater flow.

FILTER PONDING Development of wastewater pools as a result of surface clogging of filters. *Also called* **Filter pooling.**

FILTER POOLING Development of wastewater pools as a result of surface clogging of filters. *Also called* **Filter ponding.**

FILTER PRESS Device for extracting moisture from a slurry to leave a semisolid cake.

FILTER-PRESSED SLUDGE Sludge that has been dewatered by being squeezed in a filter press.

FILTER PRESSURE DIFFERENTIAL Drop in pressure due to flow across a filter or element at any time. The term may be qualified by adding 'initial,' 'final,' or 'mean'.

FILTER RATE Rate that a filter can effectively process a material or, the rate that a material is applied to a process involving filtration.

FILTER RATED FLOW Maximum flow rate of a fluid of specified viscosity for which a filter is designed.

FILTER RATING Measure by which filter performance may be evaluated and rated, including:

> **Absolute:** The diameter of the largest hard spherical particle that will pass through a filter under specified test conditions. This is an indication of the largest opening in the filter element.

> **Filtration (b_x):** The ratio of the number of particles equal to and greater than a given size (x) in the influent fluid to the number of particles equal to and greater than the same size (x) in the effluent fluid.

> **Nominal:** An arbitrary micrometer value, based on weight percent removal, indicated by a filter manufacturer. Due to lack of reproducibility, this rating is depreciated.

FILTER RUN Interval between changing of a filter medium or between backwashing of the filter media.

FILTER STRAINER Device in the underdrain of a rapid sand filter through which filtered water is collected and through which wash water is distributed.

FILTER UNDERDRAINS System for collecting water that has passed through a sand filter or biological bed.

FILTER UNLOADING Flushing technique by which the biological film formed on the filter media of a trickling filter is sloughed off.

FILTER WASH *See* **Backwash.**

FILTRATE Liquid that has passed through a filter.

FILTRATION Process of removing matter, floating or in suspension, from a flow-ing liquid.

FILTRATION PLANT That portion of a raw water or waste water treatment plant in which the flow is subjected to any of various forms of physical filtration.

FILTRATION RATE Rate of application of raw water or waste water to a filter.

FILTRATION RATIO Ratio of the number of particles of a given size entering a filter to the number of particles of the same size leaving a filter.

FILTRATION SPRING Spring water percolating from small openings in permeable material.

FINAL ACCEPTANCE Acceptance by the owner or his agent of a completed construction project from the contractor with an obligation for final payment for the work done. *See also* **Acceptance of work, Interim acceptance,** and **Partial acceptance.**

FINAL CLOSURE Closure of all hazardous waste management units at a facility in accordance with all applicable closure requirements.

FINAL COMPLETION Stage when all work is completed and all contract requirements are fulfilled by the contractor.

FINAL COVER Cover material that serves the same functions as daily cover but, in addition, may be permanently exposed on the surface.

FINAL CURE Time required for a material, or combination of materials subject to chemical reaction, to reach 100% of its rated physical properties.

FINAL DATE *See* **Closing date.**

FINAL DESIGN Project design that is approved by all responsible parties, including the owner, and from which working drawings and other contract documents are prepared.

FINAL EFFLUENT Effluent emanating from the final treatment stage of a wastewater treatment plant and prior to discharge to the receiving water.

FINAL FILTER Last stage of a multistage filter system.

FINAL GRADE Height and shape of a finished landfill after the cap has been installed, or after settlement is complete.

FINAL INSPECTION Final review of a project by the owner's representative that leads to issuance of the final certificate for payment.

FINAL LAYOUT Drawing showing the general arrangement of related components supplied by a subcontractor and approved by the owner's representative.

FINAL PAYMENT Payment by the owner to the contractor, upon issuance of the final certificate for payment by his appointed agent, of the entire unpaid balance of the adjusted contract sum.

FINAL SEDIMENTATION Separation of solids from wastewater in a final settling tank.

FINAL SEDIMENTATION BASIN Tank through which the effluent from a waste effluent oxidizing process is passed to remove settlable solids. *Also called* **Final settling basin.**

FINAL SETTLING BASIN Tank through which the effluent from a waste effluent oxidizing process is passed to remove settleable solids. *Also called* **Final sedimentation basin.**

FINDING OF NO SIGNIFICANT IMPACT Document by a regulatory agency briefly presenting the reasons why an action will not have a significant effect on the human environment and for which an environmental impact statement therefore will not be prepared.

FIN DRAIN Groundwater drain consisting of a geogrid sheet sandwiched between layers of geotextile and inserted in a narrow (as little as 25 mm (1 in.) wide) trench.

FINE GRADING Finish grading that follows rough grading in order to bring a profile to the necessary elevation.

FINE-GRAINED SOIL Soil in which the smaller grain sizes predominate, such as fine sand, silt, and clay.

FINE GRAVEL Rock fragments ranging in size from 2 to 6 mm (0.08 to 0.25 in.)

FINELY GRADED AGGREGATE Sample of aggregate containing a high proportion of small particles.

FINENESS Measure of particle size.

FINENESS MODULUS Factor obtained by adding the total percentages of material in a sample that are coarser than each of a specified series of sieves (cumulative percentages retained) and dividing the sum by 100.

FINE RACK Bar grid having clear spaces of 25 mm (1 in.) or less between the bars.

FINES Finer-grained particles of soil, clay, silt and sand.

FINE SAND Rock fragment ranging in size from 0.06 to 0.2 mm (0.008 to 0.002 in.).

FINE SCREEN Screen generally having clear openings of less that 25 mm (1 in.) in both directions, but may be as little as 1.5 mm (0.625 in.).

FINE SEDIMENT LOAD That part of the suspended load of a stream composed of particles not large enough to be found in shifting portions of the stream bed.

FINISH DATE Point in time associated with an activity's completion, usually qualified by one of the following: actual, planned, estimated, scheduled, early, late, baseline, target, or current.

FINISHED SIZE Overall measurements of any object completely finished and ready for use.

FINISH GRADE Surface elevation of an improved surface following grading.

FINISH GRADING Earthmoving work necessary to bring the site grades to those levels necessary for final drainage patterns and landscaping shapes.

FINISHING Third major stage in plumbing: includes installation of fixtures and other exposed components. *See also* **Rough in**; **4.** Final covering and treatment of surfaces and their intersections.

FIRE Rapid oxidation of combustible materials resulting in light and heat.

FIRE ALARM Audible and/or visible indication of the presence of combustion within a protected area.

FIRE ALARM INDICATOR Panel that shows the location of a fire within a protected area.

FIRE AREA Area of a building enclosed by fire walls and/or exterior walls, within which a fire would be confined for the period of the fire rating of the materials of their construction.

FIRE ALARM Audible and/or visual signal indicating the presence of fire.

FIRE BARRIER Fire resistant walls, doors, and similar construction to prevent spread of a fire in a building. *See also* **Fire stop.**

FIRE BEHAVIOR Manner in which fuel ignites, flame develops, and fire spreads, etc.

FIRE BOOSTER Automatic electric- or diesel-driven pump that is actuated to augment pressure in a fire riser that serves a sprinkler system.

FIRE BOSS Person responsible for all suppression and service activities on a forest fire.

FIREBRAND Source of heat, natural or man-made, capable of igniting a forest fire.

FIREBREAK 1. Area or strip of less flammable fuels that is either natural (standing timber or landslide) or is made in advance (cat trail or road) as a precautionary measure separating areas of greater fire hazard. *Also called* **Fuelbreak**; **2.** Any natural or constructed barrier utilized to segregate, stop, and control the spread of fire or to provide a control line from which to work.

FIRE BREAK Any fire resistant fitting or construction: fire door, closed stairwell, con-

crete floor, division wall, etc.

FIRE BROOM Broom used in suppressing ground cover fire.

FIRE CAMP Temporary camp used to accommodate workers and equipment while suppressing a forest fire.

FIRE CANOPY Horizontal, fire-resistive construction extending beyond the vertical line of an exterior wall.

FIRE CISTERN Tank used to store water to be used in fighting fires.

FIRE CLIMAX Climax community that relies upon fire for its maintenance and the succession of new growth.

FIRE COMPARTMENT Enclosed interior space in a building that is separated from all other parts of the building by enclosing construction providing a fire separation having a required fire resistance rating.

FIRE CONTROL 1. Process of controlling an unwanted forest fire; **2.** Limiting the size of a fire by a) distributing water so as to decrease the rate of heat release, b) pre-wetting adjacent combustibles, and c) controlling the gas temperature near the ceiling.

FIRE CONTROL DAMPER Device that closes an air duct in the event of a fire.

FIRE CONTROL EQUIPMENT Range of tools, machinery, vehicles, etc. used in forest fire control.

FIRE CONTROL IMPROVEMENTS Structures primarily used for forest fire control, such as lookout towers, housing, telephone lines, radio station, roads, etc.

FIRE CONTROL PLANNING Design of organizations, facilities, and procedures to protect forests and similar land areas from fire.

FIRE CUT Angular cut at the end of a joist that is anchored in masonry. In the event of a fire the joist will collapse without forcing the wall to fall outward.

FIRE DAMAGE Loss caused by fire,

present and estimated future.

FIRE DAMPER Closure that consists of a normally-held-open damper installed in an air distribution system or in a wall or floor assembly, and designed to close automatically in the event of a fire in order to maintain the integrity of the fire separation.

FIRE DANGER Measure of the likelihood of a forest fire, based on temperature, relative humidity, wind force and direction, and the dryness of the woods.

FIRE DANGER RATING Management system that integrates a range of fire-related factors into a single index.

FIRE DANGER RATING AREA Geographical area throughout which the fire danger is adequately represented by that measured at a single fire danger station; area relatively homogeneous in climate, fuels, and topography.

FIRE DECK Parting line of the block and head of an engine, surrounding the cylinder bore, where the hottest temperatures and highest pressures of combustion are produced.

FIRE DEMAND Required fire flow and the duration for which it is needed, in volume over time, or the total volume of water needed to deliver the required fire flow for a given number of hours.

FIRE DEMAND RATE Rate of flow needed at a specified pressure for fire suppression at a given location or in a certain area.

FIRE DEPARTMENT CONNECTION Connections at ground level through which a fire department supplies sprinkler systems and/or standpipe systems.

FIRE DETECTOR Device capable of detecting the products of combustion and sending a signal of the fact.

FIRE DEVIL Small-diameter and low-energy cyclone of short duration that develops when heated air or hot gases (as from a fire) rise at a rate faster than the surrounding atmosphere, and that can carry sparks, ash, and burning debris away from the area of an active fire.

FIRE DISTRICT Fire prevention/fighting organization organized and equipped to serve a defined area.

FIRE EDGE Boundary of a fire at a point in time.

FIRE ENDURANCE Ability of a material or assembly to meet defined fire test criteria.

FIRE ENGINE Motorized vehicle equipped to meet specific fire fighting demands.

FIRE ENGINEERING Use of measurement and calculation to provide for fire safety within a structure of building.

FIRE ESCAPE Device used for escape from a building: chute, exterior stairway, ladder, etc.

FIRE EXTINGUISHER Portable device containing chemicals that can be sprayed on a fire to extinguish it.

FIRE FIGHTING SHAFT Dedication of one or more elevator shafts for the exclusive use of fire fighting personnel in the case of an alarm.

FIRE FLOW Quantity of water, additional to that required for normal consumption, required for fire protection in a given area.

FIRE FLOW TEST Procedure to test the rate of flow in L/min (gpm) in a predetermined area.

FIRE GRADING Rating of a material according to its ability to resist fire and flame.

FIREGOUND Area around a fire that is occupied by fire fighters.

FIRE GUARD 1. Man-made barrier (often an area cleared of fuels) constructed at the time of a fire to control it and provide a point from which to carry out fire suppression; **2.** Person trained and assigned to watch for fires and life safety for specific periods or events.

FIRE HAZARD Degree of hazard from fire according to the type of occupancy and contents of a structure, classified as:

Light hazard: Where the quantity and/or combustibility of contents is low and fires with relatively low rates of heat release are expected.

Ordinary hazard, Group 1: Where combustibility is low, quantity of combustibles is moderate, stock piles of combustibles do not exceed 2.4 m (8 ft), and fires with moderate rates of heat release are expected.

Ordinary hazard, Group 2: Where quantity and combustibility of contents is moderate, stock piles do not exceed 3.6 m (12 ft), and fires with moderate rates of heat release are expected.

Ordinary hazard, Group 3: Where quantity and/or combustibility of contents is high, and fires of high rates of heat release are expected.

Extra hazard: Where quantity and combustibility of contents is very high, where flammable liquids, dust, lint, or other materials are present, introducing the probability of rapidly developing fires with high rates of heat release.

FIRE HEADQUARTERS Control center for operations against a particular forest fire.

FIRE HOSE REEL Reel fixed to a wall, often housed in a cabinet, on which is wound a small-diameter fire hose attached to a riser.

FIRE HYDRANT Water supply outlet with a wrench-actuated valve and connection for a fire hose. *Also called* **Plug.**

FIRE INSPECTION Examination of a commercial or industrial building by fire officials prior to issuance of a fire certificate.

FIRE LIABILITY Legal responsibility for loss or damage to the property of others caused by fire.

FIRE LIMITS Boundary line establishing an area in which there exists, or may exist, a fire hazard requiring special fire protection.

FIRE LINE Cleared area extending down to mineral soil that surrounds a forest fire to prevent the fire from reaching fresh fuels.

FIRE LOAD As applied to occupancy, means the combustible contents of a room or floor area expressed in terms of the average weight of combustible materials per square meter (square foot), from which the potential heat liberation may be calculated based on the calorific value of the materials. It includes the furnishings, finished floor, wall and ceiling finishes, and temporary and movable partitions.

FIRE MAIN Water supply pipe dedicated to fire service.

FIREMAN'S PANEL Key diagram near the entrance to a large building that shows a plan of each floor, controls and indicators for fire fighting equipment, a public address microphone, and other aids.

FIRE MODELLING Mathematical simulation used to assess potential fire problems.

FIRE PACK Tools, equipment, and supplies, readied in advance, to be carried by one person to fight a forest fire.

FIRE PARTITION *See* **Wall.**

FIRE PERFORMANCE CHARACTERISTIC Response of a product/material/assembly to a prescribed source of heat or flame under controlled fire conditions.

FIRE POINT 1. Temperature at which a material evolves sufficient vapors that when ignited it will continue to burn; **2.** Lowest temperature at which vapor from asphalt will ignite and remain burning. *Also called* **Flash point.**

FIRE PRESSURE Mains pressure necessary at those times when water is used for fire fighting.

FIRE PREVENTION Planning and activities taken in advance to prevent the outbreak of fire and/or to minimize loss if fire occurs.

FIRE PREVENTION CODE OR ORDI-

NANCE Law enacted by a political jurisdiction to enforce prevention and safety regulations.

FIRE PROGRESS MAP Map that shows the state of development of a large fire and the disposition of forces deployed to fight it.

FIRE PROPAGATION INDEX Rating of the heat output of a material (but not its fire rating) when subjected to fire in a test furnace.

FIREPROOF Something that is not burnable (a redundant term). *See* **Fire resistance.**

FIREPROOFING Material or combination of materials protecting structural members to increase their fire resistance. *See also* **Concrete cover.**

FIRE PROTECTION Procedures, processes, materials and resources developed and used to prevent the outbreak of fire or for its containment and extinguishment.

FIRE PROTECTION DISTRICT Rural or suburban fire protection area that maintains its own fire apparatus or contracts with an adjacent fire department for fire protection.

FIRE PROTECTION ENGINEER Accredited professional who specializes in engineering problems related to fire protection.

FIRE PROTECTION RATING Time in hours or fraction of an hour that a closure, window assembly, or glass block assembly will withstand the passage of flame when exposed to fire under specified conditions of test and performance criteria.

FIRE PUMP Water pump having certain performance characteristics, complete with ancillary equipment such as hoses, nozzles, etc., reserved for and available for use to combat fire.

FIRE PUMPER Motorized vehicle equipped with power-driven pumps, used to boost the pressure of water taken from a hydrant to that needed for fire fighting.

FIRE-RATED SYSTEM Wall, floor, and

roof construction, of specific materials and designs, that has been tested and rated according to fire safety criteria (e.g. flame spread and fire resistance).

FIRE RATING Fire resistance grading.

FIRE RESISTANCE Time that a structural element will perform its normal functions when subjected to standard heating conditions under laboratory test conditions. *See also* **Fireproof.**

FIRE-RESISTANT COATING Paint or coating that will not support combustion and that will provide an effective fire barrier to the surface on which it is applied. *See also* **Fire-retardant coating.**

FIRE-RESISTANT FLUID Fluid, difficult to ignite, that shows little tendency to propagate flame. *Also called* **Nonflammable fluid.** *See also* **Fluid.**

FIRE-RESISTANT GREASE Grease formulated with special flame-retardant additives.

FIRE RESISTANT RATING Time in hours (or major fraction of an hour) that a material will withstand the passage of flame and the transmission of heat at a set distance and temperature.

FIRE RESISTING FINISH Paint formulation based on silicones, polyvinyl chloride, chlorinated waxes, urea formaldehyde resins, casein, borax, or other noncombustible substances that reduces the spread of fire on combustible materials.

FIRE RESISTING FLOOR/WALL Floor or wall construction that has a fire rating appropriate to the occupancy and fire load of the building.

FIRE-RESISTIVE CONSTRUCTION *See* **Construction types.**

FIRE RETARDANT Any substance that, by chemical or physical action, reduces the flammability of forest fuels, usually added to water.

FIRE RETARDANT COATING Paint or coating that will not support combustion, but which will burn, depending on the ma-

211

terial to which it is applied. *See also* **Fire-resistant coating.**

FIRE RETARDANT RATING Rating, based on standard tests of fire-resistive and fire-protective characteristics, of building materials and assemblies.

FIRE-RETARDANT-TREATED Chemical treatment of wood and plywood to retard combustion.

FIRE RISK Chance of a fire starting.

FIRE SAFETY OFFICER Person responsible for identifying the accident and health hazards to forest fire suppression forces and for advising the fire boss on means of keeping the hazards to a minimum.

FIRE SCAR Fresh or healing injury of the cambium of a woody plant, caused by fire.

FIRE SEASON Time(s) of year when fires are most likely to occur, rated according to potential hazard.

FIRE SEPARATION Construction assembly that acts as a barrier against the spread of fire, and that may or may not be required to have a fire-resistance rating.

FIRE SERVICE Fire prevention and fighting organization.

FIRE-SERVICE CONNECTION Pipe extending from a main to supply a sprinkler, standpipe, yard main or other fire-protection device or system.

FIRE-SERVICE DETECTOR CHECK METER Weighted check valve with a disk meter in a bypass used on fire-service connections.

FIRE-SERVICE METER Flow meter having a relatively small loss of head at high rates, used on fire-service connections.

FIRE SETTING Starting of a forest fire, usually with malicious intent.

FIRE SPRINKLER SYSTEM System of water supply pipes, heat detectors, and sprinkler heads, mounted on or near the ceilings of a building, that automatically detects heat or the products of combustion above a pre-

set limit and causes a continuous spray of water to cascade over the area below. There are several types, including:

> **Dense system:** Type of dry-pipe sprinkler system in which all sprinklers are open and where pressure-sensitive devices, rate-of-temperature-rise releases, and/or heat sensitive controls activate the main water supply control valve, causing the entire system to activate flooding of the protected area. *Also called* **Deluge system.**

> **Dry-pipe system:** Automatic sprinkler system that is normally dry, with no water in the sprinkler pipes above the dry-pipe valve; installed where there is a danger of freezing if water were in the pipes.

> **Wet-pipe system:** Automatic sprinkler system that has water in the pipes right up to each sprinkler at all times.

FIRESTORM Violent convection caused by a large continuous area of intense fire, often characterized by destructively violent surface indrafts near and beyond the perimeter, and sometimes by tornado-like whirls.

FIRE STREAM Stream of water from a fire nozzle used to control and combat fires.

FIRE SUPPLY Quantity of water, additional to domestic, industrial, commercial or other public use, needed for fire fighting.

FIRE SUPPRESSION Means taken for the reduction of the rate of heat release of a fire and the prevention of its regrowth.

FIRE SUPPRESSION ORGANIZATION Management structure that enables the line and staff duties of the fire boss to be carried out; supervisory and facilitating personnel assigned to forest fire suppression under the direction of a fire boss.

FIRE SWITCH Switch that allows fire fighters to assume direct control of elevators.

FIRE DISTRIBUTION SYSTEM Independent system of water pipes or mains and their appurtenances installed exclusively to furnish water for fire fighting.

FIRE TEST EXPOSURE SEVERITY
Measure of the degree of fire exposure.

FIRE TOOL CACHE Placing of a supply
of fire tools and equipment at a strategic
location as part of a forest fire prevention or
fire fighting plan.

FIRE TOWER Stairway serving all floors
of a building, plus the exterior at ground
level, accessible only through fire doors,
and enclosed with fire-resistant construction. *Also called* **Lookout.**

FIRE TRAIL Cleared area constructed
around logging slash or other fire hazards in
order to prevent the spread of fire to this
hazardous material.

FIRE TRIANGLE Oxygen, heat, and fuel:
the three factors necessary for combustion
and flame production.

FIRE TRUCK Any motorized vehicle carrying or fitted with fire-fighting equipment.

FIRE VENT Damper for fire venting, usually in the roof and automatically controlled.

FIRE VENTING Inducing hot gas and
smoke to leave a building by fire vents and
smoke outlets.

FIRE WARDEN Person responsible for
fire control within a defined area.

FIRE WATCH Worker who remains at a
logging site for approximately two hours at
the end of the day to watch for possible fires
caused by logging activities.

FIRE WEATHER Weather conducive to
the start and rapid spread of fire.

FIRE WEATHER STATION Meteorological station specially equipped to measure weather elements important to forest
fire prevention and control.

FIRE WHIRLWIND Revolving mass of
air caused by a forest fire; may have sufficient intensity to snap off large trees.

FIRM FIXED PRICE CONTRACT Type
of contract where the buyer pays the seller a
set amount (as defined by the contract) regardless of the seller's costs.

FIRM POWER Electrical or mechanical
power that is always available and dependable to meet demand needs.

FIRM PRICE CONTRACT Contract in
which no variation of bid price is allowed.

FIRST-DRAW SAMPLE One-litre sample
of tap water that has been standing in plumbing pipes at least 6 hours and collected without flushing the tap.

FIRST FOOD USE Use of a pesticide on a
food or in a manner which otherwise would
be expected to result in residues in a food, if
no permanent tolerance, exemption from the
requirement of a tolerance, or food additive
regulation for residues of the pesticide on
any food has been established for the pesticide.

FIRST-ORDER REACTION Reaction in
which the rate of disappearance of a chemical is directly proportional to the concentration of the chemical and is not a function of
the concentration of any other chemical
present in the reaction mixture.

**FIRST-STAGE BIOCHEMICAL OXY-
GEN DEMAND** Oxygen demand associated with biochemical oxidation of carbonaceous (as distinct from nitrogenous) material.

FIRTH Narrow arm of the sea; the opening
of a river into the sea; an estuary.

FISH LADDER Channel containing a series of low weirs or ramps through which
water continuously flows, used to help fish
bypass a dam or waterfall. *Also called* a
Fishway.

FISH SCREEN Barrier that prevents fish
from entering a channel.

FISHWAY Channel containing a series of
low weirs or ramps through which water
continuously flows, used to help fish bypass a dam or waterfall. *Also called* a **Fish
ladder.**

FISSILE 1. Capable of being split along
closely-spaced parallel planes; **2.** Isotopes

of elements that are capable of undergoing nuclear fission upon impact with a slow neutron.

FISSION The splitting that occurs when the nucleus of a heavy atom (uranium, plutonium) under bombardment absorbs a neutron, releasing large quantities of energy.

FISSURE Fracture in a rock with displacement perpendicular to the break.

FISSURED SOIL *See* **Soil types.**

FIT CLASSIFICATIONS Relative tightness of an assembly, classified as:

Class 1: Loose: a large allowance, used where accuracy is not essential.

Class 2: Free: a liberal allowance, used for running fit at 600 rpm or over, and shaft pressure over 4137 kP (600 psi).

Class 3: Medium allowance: used for running fits under 600 rpm, journal pressures under 4137 kP (600 psi), and for sliding fits.

Class 4: Snug, zero allowance: the closest fit that can be assembled by hand. Used when moving parts are not intended to move freely.

Class 5: Wringing, zero, or negative allowance: commonly used where an assembly is not to be interchanged.

Class 6: Tight, slight negative allowance: where light pressure is required for assembly of parts.

Class 7: Medium force, negative allowance: where considerable pressure is required for assembly.

Class 8: Heavy force and shrink, considerable negative allowance: used where material can be highly stressed.

FITMENT An attachment.

FITTING 1. Any functional accessory, including:

Bushing fitting: Short externally threaded connector with a smaller size internal thread.

Closure fitting: Cap or plug.

Compression fitting: Fitting that seals and grips by manual adjustment and deformation.

Connector fitting: Fitting for joining a conductor to a component port or to one or more other conductors.

Flared fitting: Fitting that seals and grips by a preformed flare at the end of a tube.

Flareless fitting: Fitting that seals and grips by means other than a flare.

Reusable hose fitting: Hose fitting that can be removed from a hose and reused.

2. Electrical accessory, such as a lock nut, bushing, or other part of the wiring system, that is intended primarily to perform a mechanical rather than an electrical function; **3.** Part of a piping system, excluding valves, that joins lengths of pipe.

FITTING GAIN Amount of space inside a fitting for which pipe allowance must be made.

FITTING-UP BOLT Bolt used to temporarily hold members together while they are being permanently connected.

FIT UP Assemble prepared or prefabricated units into a whole in a manner that will permit its later disassembly, typically formwork.

FIVE-YEAR LIABILITY Implied or actual warranty effective for five years from date of purchase or acceptance or other defined date.

FIVE-YEAR, SIX-HOUR PRECIPITATION EVENT Maximum 6-hour precipitation event with a probable recurrence interval of once in 5 years.

FIX Position of something determined by reference to two or more fixed entities.

FIXATION Chemical process of changing into a more stable form.

FIXATIVE Chemical substance used to preserve cells without distortion.

FIXED 1. In a stable combined form; **2.** Not subject to change or variation.

FIXED ASSET Tangible, noncurrent asset, such as land, buildings, equipment, etc., held for use rather than for sale.

FIXED-BAR GRILLE Grille, the bars of which cannot be adjusted.

FIXED BUDGET *See* **Budget.**

FIXED CAPITAL Investment in capital assets.

FIXED CAPITAL COST Capital needed to provide all of the depreciable components.

FIXED CARBON Measure of the primary productivity of an ecosystem based on the amount of carbon fixed by photosynthesis per unit area.

FIXED CHARGE Unavoidable expense, e.g. interest, rent, etc.

FIXED COSTS Operation costs that will remain relatively constant for all levels of output.

FIXED DISPLACEMENT HYDRAULIC SYSTEM Hydraulic system that produces a constant flow rate regardless of the actuators engaged at a particular time. Excess hydraulic flow is returned to the reservoir through return lines. Low line pressures provide just enough force to allow the oil to overcome the friction of movement, but are not sufficient to engage hydraulic actuators.

FIXED DISTRIBUTOR Distributor that remains stationary, permitting the discharge of fluid at a constant rate over a specific distance.

FIXED EARTH SUPPORT Pressure distribution assumed in the design of anchored sheet-pile walls where a point of contraflexure is assumed due to the fixed bottom-end support of the wall.

FIXED EXPENSES Property expenses that remain the same regardless of occupancy.

FIXED FEE Provision of services for a stipulated sum.

FIXED FILM DENITRIFICATION Attached growth anaerobic treatment process.

FIXED GROUNDWATER Water within saturated rocks so fine-grained that it is assumed to be permanently attached to the rock particles.

FIXED LADDER Ladder that cannot be readily moved or carried because of being an integral part of a building or structure.

FIXED LIABILITIES Liabilities that are to be paid one year or more after the date of the balance sheet on which they are recorded.

FIXED LIGHT Window that is not made to open.

FIXED PACKER Adjunct of a refuse container system that compacts refuse at the site of generation into a detachable container.

FIXED MOISTURE Moisture held in the soil below the hygroscopic limit.

FIXED NOZZLE Distribution nozzle in a fixed position, used to apply wastewater as a spray.

FIXED-PIN HINGE Hinge in which the pin is fixed permanently in place.

FIXED-PLATFORM TRUCK Industrial truck equipped with a load platform and not capable of self loading.

FIXED PRICE CONTRACT *See* **Stipulated price contract.**

FIXED PRICE INCENTIVE FEE CONTRACT Type of contract where the buyer pays the seller a set amount (as defined by the contract), and the seller can earn an additional amount if it meets defined performance criteria.

FIXED ROOF Cover that is mounted to a tank or chamber in a stationary manner and which does not move with fluctuations in wastewater levels.

FIXED SOLIDS Remaining residue following ignition of volatile suspended or dissolved matter.

FIXED-SPRAY NOZZLE Constant-output spray nozzle used to distribute wastewater over filter media.

FIXED TIME Load, dump, and maneuver times that are relatively constant in a cycle.

FIXING Securing or fastening piece.

FIXTURE 1. Device designed for a particular purpose and fastened securely in place; **2.** Receptacle in which water or other waste may be collected for ultimate discharge into a plumbing system; **3.** Assembly that serves to hold pieces in place while they are worked upon; **4.** Part of fixed assets, usually consisting of machinery or equipment attached to or forming a normal part of a building.

FIXTURE BRANCH Drain from a fixture trap to the junction of the drain with a vent.

FIXTURE DRAIN Drain from a fixture branch to the junction of any other drain pipe.

FIXTURE SUPPLY PIPE Water supply pipe connecting the fixture to the supply system.

FIXTURE UNIT Unit flow from a fixture, usually established as 0.028m³ (1 ft³) or 28 L (7.5 gal) of water per minute.

FLAGELLATA Class of protozoa whose adult members swim by means of flagella.

FLAIL Hammer hinged to an axle so that it can be used to break or crush material.

FLAIL MILL Rotating equipment that uses chains to open or spread solid waste and to shred or crush such material as paper, cardboard, or glass.

FLAME ARRESTER Fine-mesh wire screen inserted in a vent or pipe to resist flame flashback.

FLAME PHOTOMETRY Analytical technique in which substances are strongly heated in a flame, arc, or high-voltage spark to excite their atoms, which emit electromagnetic radiation whose spectrum is indicative of specific elements.

FLAMEPROOF ENCLOSURE Enclosure for electrical apparatus for use in hazardous locations.

FLAMMABLE Subject to easy ignition and rapid flaming combustion.

FLAMMABLE LIQUID Any liquid having a flash point below 60°C (140°F) and having a vapor pressure not exceeding 275.8 kP at 37.7°C (40 psi at 100°F).

FLANGE Rim-like end on a valve, or pipe fitting for bolting another flanged fitting, usually for large diameter or pressurized pipe.

FLANGED JOINT Joint comprising two companion flanges, bolted together and made leakproof by means of a gasket.

FLANGED PIPE Pipe provided with flanges at both ends so that lengths can be joined together with bolts.

FLANGE PACKING *See* **Packing.**

FLANGE UNIT Union secured with nuts and bolts.

FLAP 1. One of two flat plates, joined by a pin to form a hinge; **2.** Any flat surface that is hinged to swing vertically.

FLAP GATE Gate that opens and closes by rotation about a hinge on its top side.

FLAPPER ACTION *See* **Valve.**

FLAP VALVE *See* **Valve.**

FLARE 1. Widening of a tube; **2.** Stack for burning excess quantities of waste combustible gases; the flame produced by burning waste combustible gases..

FLARE ANGLE Angle between the centerline of a structure and a wall.

FLARED FITTING *See* **Fitting.**

FLARED JOINT Reassemblable mechanical joint between two pieces of copper tubing made by flaring the end of one piece so as to receive the face of a grommet passed over the exterior of the other, the grommet then being gripped tightly by fittings previously slid over the ends of both pieces.

FLARELESS FITTING *See* **Fitting.**

FLARING INLET Funnel-shaped inlet to a pipe or conduit that facilitates flow. *Also called* **Bellmouth inlet.**

FLASHBOARD Temporary barrier used to increase the height of a weir or spillway.

FLASHBOARD CHECK GATE Balanced crest gate that opens under a nominated head to relieve pressure on the flashboard.

FLASH DRYER Mechanism for vaporizing water from partly dewatered and finely divided sludge.

FLASH EVAPORATOR Distillation device in which saline water in a superheated state is injected into a vessel under vacuum, in which boiling occurs without a discrete heat-transfer surface.

FLASH FLOOD Flood of short duration accompanied by a relatively high peak rate of flow.

FLASHING RING Collar around a pipe to secure it as it passes through a wall or floor.

FLASH MIXER Device for quickly dispersing chemicals uniformly throughout a liquid.

FLASH POINT Temperature at which a gas, volatile liquid or other substance ignites.

FLASH RANGE Temperature difference from terminal temperature to temperature of brine in the final stage of a flash evaporator.

FLASHY STREAM Stream in which flows collect rapidly due to the steepness of the surrounding slopes of the catchment area.

FLAT Relatively level surface; one without significant elevation, relief or prominences.

FLAT-CRESTED WEIR Weir with a crest that is horizontal in the direction of flow and of significant length relative to the depth of water passing over it.

FLAT RATE Charge for a service without further qualification (such as volume, flow, etc.).

FLAT SLOPE Conduit sloped less than the critical value for a particular discharge.

FLEXIBLE ARMORED REVETMENT Interlocking, precast concrete shapes that can be combined to form a revetment of any length and any height, and that follows the contours of the original bank, and which will adjust to minor changes in the bank shape due to scour.

FLEXIBLE JOINT Pipe joint that permits one member to be deflected without disturbing the other.

FLEXIBLE MEMBRANE Synthetic sheet material, typically butyl-rubber, that can be joined (usually by chemical welding) at the seams to form very large sheets, used to line reservoirs, landfills and similar structures to prevent exfiltration, and in some cases, infiltration.

FLEXIBLE PIPE Pipe that will bend before it breaks, typically plastic, ductile iron or steel.

FLEXIBILITY Degree to which a material will bend without cracking, breaking, or becoming permanently deformed.

FLEXIBILITY FACTOR Degree to which a material or product can be flexed without permanent deformation.

FLEXURAL MODULUS OF STRENGTH Strength of a material in bending, expressed in tensile stress of the outermost fibers of a bent test sample at the instant of failure.

FLEXURAL MOMENT *See* **Positive moment.**

FLEXURAL RIGIDITY Measure of stiffness of a member, indicated by the product of modulus of elasticity and moment of inertia divided by the length of the member.

FLEXURAL STRENGTH Property of a material or a structural member that indicates its ability to resist failure in bending.

FLEXURE *See* **Bending.**

FLIGHT 1. Long, narrow, and relatively thin boards used to collect and move settled sludge or floating scum; **2.** Slow-pitch thread of an auger.

FLIGHT SEWER Sewer laid to a series of steps so as to break up the velocity on a steep grade.

FLOAT 1. Dozer blade resting by its own weight, or held from digging by the upward pressure of a load of dirt against its moldboard; **2.** Object floating on water that opens and closes a valve in a water tank based on water level; **8.** Object that indicates the direction and rate of flow of the water carrying it.

FLOAT ARM Rod connecting the float ball to the inlet valve in a water closet tank.

FLOATATION 1. Ability to float, as with large, low-pressure tires that allow a vehicle or machine to be supported over soft ground; **2.** Separating minerals by floating the lighter components in a fluid.

FLOAT BALL Ball used to control the inlet valve in water closet tanks.

FLOAT GAUGE Device for measuring the elevation of the surface of a liquid in which the elevation of a float is transmitted to a remote-reading gauge.

FLOAT GAUGING Technique, using a series of floats, of measuring the velocity of water flowing in a stream or conduit.

FLOATING BOOM Flexible barrier that floats on the surface, used to separate a body of water for containment or use, to restrict the reach over which waves can develop, to gather floating debris from floating further downstream, etc.

FLOATING COVER Gastight cover to a treatment vessel, kept buoyant by the level of the contained material or by the gas given off by that material.

FLOATING PAN Evaporation pan floating on a body of water.

FLOATING PIPELINE Pipeline supported on or by pontoons, typically used between a floating dredge and land to transport dredged material to be used as hydraulic fill.

FLOATING ROOF Storage vessel cover consisting of a double deck, pontoon single deck, internal floating cover or covered floating roof, which rests upon and is supported by the liquid being contained, and is equipped with a closure seal or seals to close the space between the roof edge and tank wall.

FLOAT RUN Distance over which a float travelling with the flow is timed.

FLOAT SWITCH *See* **Switch.**

FLOAT TUBE 1. Movable intake pipe, the mouth of which is held below the water surface by means of an attached float; **2.** Vertical pipe connected to the wet well of a pumping station and containing a float which actuates a switch for controlling the pumps.

FLOAT VALVE *See* **Valve.**

FLOC Porous mass formed by the agglomeration of suspended particles, in water or in air, often following application of a flocculent.

FLOCCULATION Process to enhance agglomeration or collection of smaller floc particles into larger, more easily settlable particles through gentle stirring by hydraulic or mechanical means.

FLOCCULATION AGENT Substance which, when added to water, promotes coagulation of suspended particulate matter and the formation of a floc that will precipitate.

FLOCCULATION LIMIT Water content of a soil when in the condition of a deflocculated sediment.

FLOCCULATION RATIO Void ratio of a soil when it is in the condition of a deflocculated sediment.

FLOCCULATION TANK Tank in which the formation of floc is propagated by the gentle agitation of liquid suspensions, often with the aid of chemicals.

FLOCCULATOR Mechanical device that enhances the formation of floc in a liquid.

FLOE ICE Ice that has formed on the surface of a body of water and which floats around in large pieces when broken up.

FLOOD Condition where the volume of surface water runoff is greater than the capacity of natural and artificial channels and conduits to adequately and safely contain and conduct the resulting flow, resulting in an accumulation of water over otherwise dry land.

FLOOD BASALT Accumulation of basaltic lava covering many thousands of square kilometers, indicative of a high rate of extrusion of very runny lava from fissures.

FLOOD BASE ELEVATION Elevation of the highest recorded flood for an area, specific to a location and to a point in time.

FLOOD BASIN Part of a river valley outside the natural stream bank and subject to flooding.

FLOOD BENEFITS Assumed value of flood prevention works in terms of safety and non-occurrence of damage.

FLOOD CHANNEL Wetted area that the flood occupies at its peak. *See also* **Channel.**

FLOOD CONTROL Construction (or the prevention of construction) intended to assist in the conveying of excess surface water runoff.

FLOOD-CONTROL STORAGE Water stored during floods for later release.

FLOOD-CONTROL WORKS Structures and other works for the reduction of flood peaks on streams.

FLOOD CURRENT Current in water associated with tidal action.

FLOODED FIXTURE Fixture in which liquid begins to overflow the top or rim.

FLOOD EVENT Series of flows in a stream that culminate in a level above the normal, or calculated maximum carrying capacity of the channel section under observation.

FLOOD FLOW Discharge of a stream during a period of flood.

FLOOD FREQUENCY Frequency with which the maximum flood may be expected to occur at a site in any interval of years.

FLOODGATE Mechanical means by which to regulate or restrain the flow of floodwater through a channel.

FLOOD HAZARD Potential risk due to a sudden and temporary increase of surface water flow.

FLOODING Introduction of water, by gravity, to consolidate backfill surrounding a pipe.

FLOOD IRRIGATION 1. Distribution of flood water over an area of land by means of canals and ditches; **2.** Effluent from wastewater treatment distributed over a suitable area of land as a means of disposal, and of irrigating specific types of vegetation.

FLOOD LEVEL Highest level that flowing water will reach, usually rated as annual, 10-year, 25-year, etc.

FLOOD LEVEL RIM Top edge of a plumbing fixture from which water will overflow.

FLOOD PEAK Maximum rate of flow in a stream during a flood event.

FLOODPLAIN Lowland and relatively flat areas adjoining inland and coastal waters and other flood-prone areas such as offshore islands, including at a minimum, that area subject to a 1% or greater chance of flooding in any given year. The base floodplain is used to designate the 100-year floodplain.

FLOODPLAIN REGULATIONS Building regulations that stipulate the type of construction, or prevent construction, on land within a floodplain.

FLOOD PROBABILITY Based on historic evidence, the probability that a flood of given size and/or duration will occur within a given time period: typically a 1% probability would be a 100-year flood; a 10% probability a 10-year flood.

FLOOD PROFILE Graphic representation of data showing the water surface elevation resulting from a calculated flood condition, relevant to the existing topography.

FLOODPROOFING Modification of individual structures and facilities, their sites, and their contents to protect against structural failure, to keep water out or to reduce effects of water entry.

FLOOD PROTECTION WORKS Engineering structures built to protect property and land from damage by flood.

FLOOD-RELIEF CHANNEL Channel built to carry flood water in excess of the quantity that can be safely carried by a stream.

FLOOD RESERVOIR Ponding area for the temporary impounding of excess surface runoff.

FLOOD ROUTING Use of control channels and temporary storage areas to reduce the force of flood water and contain some of its flow.

FLOOD SOURCE AREA Portion of a drainage basin where conditions of precipitation and cover, topography, or land use favor the development of flood conditions.

FLOOD SPREADING Flooding of relatively pervious lands and the recharge of groundwater basins.

FLOOD STAGE Arbitrarily established gauge height or elevation, above which a rise in water surface elevation would be termed a flood.

FLOOD WAVE Rise in streamflow to a crest in response to runoff.

FLOODWAY Total complex of floodplain areas, streams, and channels, both natural and constructed, that serve to contain and conduct the surface runoff due to excessive inundation.

FLOOR DRAIN Wastewater outlet and trap, usually placed at a low point in a sloping concrete floor.

FLOOR FLANGE Fitting attached to a plumbing system at floor level so that a fixture can be bolted to the drainage piping.

FLOOR HOLE Opening measuring less than 300 mm (12 in.), but more than 25 mm (1 in.) in its least dimension in any floor, roof, or platform through which materials, but not persons, may fall. *See also* **Floor opening.**

FLOOR OPENING Opening measuring 300 mm (12 in.) or more in its least dimension in any floor, roof, or platform, through which persons may fall. *See also* **Floor hole.**

FLOOR OUTLET Gully set flush with the floor surface.

FLOOR PLATE Deck plate, usually removable, set into a structural grid to produce a walkway or complete floor. May be solid, with or without a raised surface pattern, or open pattern.

FLORA Plant of a particular region, country, environment or period.

FLOTATION 1. Condition when a structure can become buoyant due to a rising watertable; **2.** Tendency for an empty but submerged pipeline to float unless weighted down.

FLOURY SOIL Soil that looks like clay when wet but which becomes a powder when dry.

FLOW 1. To move as a liquid does. *See also* **Laminar flow, Metered flow,** and **Turbulent flow**; **2.** Rate of movement of water in a conduit.

FLOWAGE Movement of liquid through a series of facilities, processes or structures.

FLOW BOG Peat bog in which the surface is likely to undulate in response to an increase or decrease of water content.

FLOW CELL Sensor which, when immersed in a fluid, continuously measures some property such as pH, dissolved oxygen, electrical conductivity, etc.

FLOW CHART Program that shows the sequence of performance of different events.

FLOW COEFFICIENT Correction factor used for figuring the volume flow rate of a fluid through an orifice.

FLOW CONTROL (DECELERATION) VALVE See **Valve.**

FLOW CONTROL (FLOW METER-ING) VALVE See **Valve.**

FLOW CONTROL GATE Vertical-acting barrier that can be raised or lowered to regulate a flow passing over, or under it in a control channel.

FLOW CURVE Straight-line graph of the points obtained in a liquid limit test of a soil.

FLOW DEMAND Flow required to satisfy demands on a system.

FLOW DIAGRAM The directions taken by a fluid showing any processes undertaken and noting special equipment.

FLOW DIVIDER See **Valve.**

FLOW DURATION CURVE Duration curve of a streamflow.

FLOW EQUALIZATION Retention of a part of peak flows for release during low-flow periods.

FLOW GRADIENT 1. Slope to which a pipe must be laid to ensure a self-cleansing gradient; **2.** Drainage slope determined by the elevational differences between the inlet and outlet, the distance between those two points, and the required volume and velocity.

FLOW INDEX Slope of a flow curve of a standard liquid-limit soil test, plotted with water content as the ordinate on an arithmetic scale and number of flows as abscissa on a logarithmic scale.

FLOW INDICATOR Device that indicates

whether a flow is present in a stream.

FLOWING WELL Well that naturally discharges water at the surface.

FLOW LINE Line that indicates the direction followed by groundwater toward points of discharge. See also **Equipotential line,** and **Flow mark.**

FLOW-LINE AQUEDUCT Aqueduct placed at an elevation such that the hydraulic gradient is lower than its crown so that the aqueduct never flows full.

FLOW METER Device that indicates either flow rate, total flow, or a combination of both.

FLOWNET 1. Graphical method used to study the hypothetical flow of water through a soil. It is used to indicate the paths of travel followed by moving water and the hydraulic pressures resulting from such water flow; **2.** Pictorial description of the path of water through a dam.

FLOW-NOZZLE METER Differential-medium-type water meter in which flow through the primary element or nozzle produces a pressure differential head, which the secondary element, or float tube, uses as an indication of the rate of flow.

FLOW PRESSURE Pressure in a water supply system measured at or near a faucet or water outlet while it is wide open and water is flowing.

FLOW-PROPORTIONAL COMPOS-ITE SAMPLE Sample composed of grab samples collected continuously or discretely in proportion to the total flow at time of collection, or to the total flow since collection of a previous grab sample. The grab volume or frequency of grab collection may be varied in proportion to flow.

FLOW RATE Volume of fluid per unit of time passing a given cross section of a flow passage in a given direction.

FLOW REGULATOR Installation in a canal or channel to control the flow or level.

FLOW RESISTANCE Measure of the ability of a material to impede the flow of air or water through it.

FLOWSHEET Diagrammatic representation of the series of steps in a process.

FLOW SLIDE Shear failure in which a soil mass moves over a relatively long distance in a fluid-like manner, occurring rapidly on flat slopes in loose, saturated, uniform sands, or in highly sensitive clays.

FLOW STRUCTURE Texture of igneous rocks, especially lavas, in which flow lines are shown by the alignment of prismatic crystals or elongated inclusions and by alternating bands of different minerals or crystal size.

FLOW SWITCH *See* **Switch.**

FLOWTHROUGH CHAMBER Upper compartment of a two-story sedimentation chamber.

FLOWTHROUGH PROCESS TANK Tank that forms an integral part of a treatment process through which there is a steady, variable, recurring, or intermittent flow.

FLOWTHROUGH TEST Toxicity test in which water is renewed continuously in the test chambers, the test chemical being transported with the water used to renew the test medium.

FLOWTHROUGH TIME Time required for a volume of liquid to pass through an identified entity, identified in terms of the characteristics being measures: mean time, modal time, minimum time, etc.

FLUE Passage for conveying hot gases.

FLUE GAS Hot waste gas, usually resulting from combustion.

FLUE-GAS ANALYSIS Statement of the quantities of the various components of a sample of flue gas, usually expressed in percentages by volume.

FLUE-GAS DESULFURIZATION Any of several techniques designed to reduce the sulfur dioxide content of the flue gases emitted from coal-burning furnaces.

FLUE-GAS LOSS Sensible heat carried away by the dry flue gas, and the sensible and latent heat carried away by water vapor in the flue gas.

FLUE-GAS RECIRCULATION Technique for reducing the nitrogen oxide content in exhaust gases by recirculating up to 20% of the exhaust with the combustion air.

FLUE-GAS SCRUBBER Equipment that removes fly ash and other materials from the products of combustion by means of sprays or wet baffles.

FLUE GAS WASHER/SCRUBBER Equipment for removing objectionable constituents from the products of combustion by means of spray, wet baffles, etc.

FLUE LINING Smooth, one-celled hollow tile of fireclay or terracotta, used to protect the masonry of a chimney and to provide a smoother duct through which to exhaust smoke and gases. May be rectangular or round in shape; usually 0.6 m (2 ft) long. *See also* **Chimney lining.**

FLUE PIPE Pipe connecting the flue collar of an appliance to a chimney.

FLUE TERMINAL Cowl or cap on top of a flue to keep out rain and prevent wind gusts disturbing the draft.

FLUE TILE Glazed or unglazed tile about 600 mm (24 in.) long, either round, oblong, or square in section, used to line a chimney flue.

FLUFF Highly combustible shredded rubber, plastics and carpet residues from automobile recycling, sometimes used as a fuel in incineration.

FLUID Any material or substance which flows or moves, whether in a semisolid, liquid, sludge, gas, or any other form or state. *See also* **Fatty-oil fluid, Fire-resistant fluid, Hydraulic fluid,** and **Pneumatic fluid.**

FLUID ANALYSIS Method by which the contamination level of a fluid may be determined. There are a number of methods, including:

Automatic particle count: Technique

that determines the number of particles per milliliter of fluid. It is fast and repeatable, limited to being sensitive to particle concentration and to non-particulate contaminants, e.g. water, air, gels.

Ferrography: Technique that produces a scaled number of the large-to-small particles present in a sample. It provides basic information that will indicate the need for more sophisticated testing upon abnormal results. It is limited by not being capable of detecting nonferrous particles (e.g. brass, copper, silica, etc.).

Gravimetric: Test that determines the milligrams of contaminants per litre of sample. It indicates the total amount of contaminant but cannot distinguish particulate size.

Optical particle count: Technique that determines the number of particulate contaminants per milliliter of sample and that provides an accurate measure of size and quantity distribution.

Patch test and fluid contamination comparator: A field-conducted, visual comparison against a cleanliness code that produces a rapid approximation of system fluid cleanliness levels and that also helps to identify the types of contaminants.

Spectrometry: Technique that identifies and quantifies contaminant material as parts-per-million of the sample being tested. However, it cannot size contaminants, and has limited sensitivity above 5 µm.

FLUID FRICTION Friction due to viscosity of fluid.

FLUIDIC Of or pertaining to devices, systems, assemblies, etc., utilizing fluidic components.

FLUIDICS Engineering science pertaining to the use of fluid dynamic phenomenon to sense, control, process information, and/or actuate.

FLUIDITY Quality of being fluid, or capable of flowing.

FLUIDIZED Mass of solid particles that is made to flow like a liquid by injection of water or gas.

FLUIDIZED BED Solid mass of a filter bed brought to a state of fluid suspension by the contraflow injection under pressure or water and/or air.

FLUIDIZED-BED COMBUSTION Fire-bed configuration where combustion air is blown vertically through a layer of inert particles to which solid fuel or low calorific value waste is added. There are several arrangements, including:

Bubbling – Where the combustion occurs in a conventional bubbling bed.

Circulating – Where the bed medium is entrained and circulated with the combustion gases; a cyclone then separates the inert bed material which is returned to the combustion chamber.

Revolving – Where differential air pressures and bed geometry cause the bed to revolve to ensure effective distribution and efficient combustion of difficult materials.

FLUID LEVEL GAUGE *See* **Gauge.**

FLUID MECHANICS Science of the motion of fluids.

FLUID SAMPLE Small portion of a fluid taken from a system or test apparatus, usually obtained for analyzing fluid properties and/or amount of contamination.

FLUME Artificial channel, often set at an angle, used for measuring the flow of water, or for carrying materials conveyed by water, such as logs. *Also called* **Control flume.**

FLUORESCENCE Light produced by a substance when it is exposed to certain rays (e.g. X-rays or ultraviolet).

FLUORESCING DYES Traceable substances used to measure flow, path of movement, and sedimentation.

FLUORIDATION Addition of fluoride to a water supply to increase the concentration of fluoride ions to a prescribed limit.

FLUORIDE Compound of fluorine with an element or radical that occurs naturally in minerals and soils, concentration of which in natural waters varies widely as it depends on such factors as the source of the water and the geological formations present. The maximum acceptable concentration for fluoride in drinking water is 1.5 mg/L.

FLUORIDES Salts of hydrofluoric acid that are released into the atmosphere as a result of certain industrial processes.

FLUORINE Non-metallic gaseous element, symbol F, forming with chlorine, bromine and iodine the halogen group.

FLUSH Release of a body of water in a controlled manner.

FLUSH BALL Ball-shaped closure that controls the flow of water into the bowl of a water closet.

FLUSH BUSHING Pipe fitting that reduces the diameter of a female-threaded pipe fitting.

FLUSHED Filled up to the surface.

FLUSH HYDRANT Hydrant installed in a pit below ground level (such as near the runway area of an airport, or other locations) where above ground hydrants would be unsuitable.

FLUSHING Release of water in a controlled manner so as to cause the resulting overflow to cleanse the downstream channel.

FLUSHING CHAMBER Tank used to hold water used for flushing.

FLUSHING CISTERN Cistern for a toilet or other sanitary fitting that delivers a measured volume of water.

FLUSHING MANHOLE Manhole equipped with an opening gate so that sewage flows may be accumulated and then released to flush the downstream section of pipe.

FLUSHING MECHANISM Device that empties a cistern of its contents and allows the vessel to refill, either manually controlled, or controlled by a timing device for periodic actuation.

FLUSHOMETER Device that allows a predetermined quantity of flushing water to be accumulated and then released, actuated by water pressure.

FLUSH PIPE Pipe connecting a cistern to a sanitary fitting.

FLUSH TANK Tank that holds a supply of water for flushing one or more plumbing fixtures.

FLUSH VALVE *See* **Valve.**

FLUSH WATER *See* **Wash water.**

FLUVIAL Of or pertaining to rivers, or produced by river action.

FLUVIAL DEPOSIT Sediment deposited by stream action.

FLUVIAL EROSION Erosion caused by stream action.

FLUVIAL INDEX Slope of a flow curve of a standard liquid-limit test of soil, plotted with water content as ordinate on an arithmetic scale, and number of flows as abscissa on a logarithmic scale.

FLUVIAL SOIL Soil whose properties have been the subject of water action; characterized by the roundness of individual particles.

FLUVIOGLACIAL 1. Stream having melted glacial ice as the source of its water and much of its load; **2.** Landform shaped by the action of fluvioglacial streams.

FLUX Rate at which a particular quantity is transferred per unit area.

FLUX STANDARD Regulatory standard that limits the amount of radon that can emanate per square meter of regulated material per second, averaged over a single source.

FLY ASH 1. Suspended particles, charred

paper, dust, soot, and other partially oxidized matter carried in the products of combustion; **2.** Component of coal which results from its combustion, and is the finely divided mineral residue which is typically collected from boiler stack gases by electrostatic precipitators or mechanical collection devices.

FOAM 1. Collection of minute bubbles formed on the surface of a liquid by agitation, fermentation, etc.; **2.** Chemical fire-extinguishing mixture that forms bubbles on application, greatly increasing the mixture volume. There are two types:

> **Air:** *See* **Mechanical foam** (below).

> **Chemical:** Foam formed when an alkaline solution and an acid solution unite to form a gas (carbon dioxide) in the presence of a foaming agent that traps the gas in fire-resistive bubbles.

> **Mechanical:** Type of foam concentrate added to water and agitated or aerated to produce the extinguishing or smothering agent. *Also called* **Air foam.**

FOAM INLET Connection for foam-making equipment by fire fighters.

FOAM-IN-PLACE INSULATION Rigid cellular foam produced by catalyzed chemical reactions that hardens at the site of the work.

FOAM SEPARATION Designed frothing of wastewater or wastewater treatment effluent as means of removing excessive amounts of phosphates.

FOG 1. Cloud of fine drops of water, varying from 2 to 20 micrometers in diameter, that forms just above the Earth's surface; **2.** Jet of fine water spray discharged by spray nozzles, used to extinguish fires.

FOG DRIP Moisture dripping from vegetation having collected from windblown fog.

FOLD Bend in strata or in any planar structure in rocks or minerals.

FOLIATION Layering in rocks caused by parallel orientation of minerals or bands of minerals.

FOLLOWER Sleeve on a pipe die that aligns the die with the pipe.

FOOD CHAIN Series of progressively more sophisticated organisms through which energy is transferred by each consuming the one that precedes it (except for the first, which is herbage).

FOOD-CHAIN CROPS Crops grown for human consumption, and animal feed for animals whose products are consumed by humans.

FOOD/MICROORGANISM RATIO Ratio of the available nutrient in a body of water or wastewater to the growth of bacteria.

FOOD WASTE Organic residues generated by the handling, storage, sale, preparation, cooking, and serving of foods, commonly called garbage.

FOOD WEB Complex of interlocking food chains representing all the separate food chains in a community.

FOOT Nonmetric unit of length equal to 12 inches. Symbol: ft. Multiply by 0.3048 to obtain meters, symbol: m; by 304.8 to obtain millimeters, symbol: mm. *See also the appendix*: **Metric and nonmetric measurement.**

FOOTBATH Sanitary fitting recessed into a floor, used for washing feet.

FOOTBRAKE VALVE *See* **Valve.**

FOOT PER HOUR Nonmetric unit of velocity. Symbol: ft/h. Multiply by 0.084 666 7 to obtain millimeters per second, symbol: mm/s; by 304.8 to obtain millimeters per hour, symbol: mm/h. *See also the appendix*: **Metric and nonmetric measurement.**

FOOT PER MINUTE Nonmetric unit of velocity. Symbol: ft/min. Multiply by 0.005 08 to obtain meters per second, symbol: m/s; by 5.08 to obtain millimeters per second, symbol: mm/s. *See also the appendix*: **Metric and nonmetric measurement.**

FOOT PER SECOND Nonmetric unit of velocity. Symbol: ft/s. Multiply by 0.3048 to obtain meters per second, symbol: m/s. *See also the appendix*: **Metric and nonmetric measurement.**

FOOT-POUND Nonmetric unit of energy, equal to the work done when a mass of one pound, accelerated at a rate of one foot per second per second, has moved one foot. Symbol: ft/lb. Multiply by 1.355 818 to obtain joules, symbol: J. *See also the appendix*: **Metric and nonmetric measurement.**

FOOTPRINT Surface area occupied by an item of equipment or machinery.

FOOT VALVE *See* **Valve.**

FOOTWALL Lower side of an inclined fault or vein, or the ore limit on the lower side of an inclined ore body.

FOPS *Abbreviation for:* falling object protective structure.

FORCE 1. Resultant of distribution of stress of a prescribed area. A reaction that develops in a member as a result of load (formerly called total stress or stress); **2.** That which tends to produce or modify motion, including:

> **Compression:** Force acting on a body, tending to compress it.

> **Shear:** Force acting on a body that tends to slide one portion of the body against its other side.

> **Static:** Force exerted by the weight of a body, structure or equipment.

> **Tensile:** Force acting on a body tending to elongate it.

> **Total applied:** Sum of the various types of force acting on a surface, material, or body.

> **Torsion:** Force acting on a body that tends to twist it.

FORCE CUP Flexible cup attached to a pole-like handle, used for clearing clogged drains and water closets.

FORCE MAIN Pressure pipe connecting a pump discharge at a pumping station with a point of gravity flow.

FORCE PUMP Pump used to deliver fluids to a level considerably higher than the cylinder.

FORCES OF COMPACTION *See* **Compaction.**

FORD Road crossing a stream under water.

FOREBAY Reservoir or pond at the head of a penstock or sluice.

FOREBAY AREA Natural groundwater basin serving as a recharge reservoir to an artesian basin.

FORECAST TO COMPLETE Estimate of the value that remains to be committed within a cost class as of a specified date. Adding this estimate to the committed cost yields the estimated final cost of the scope of work in the class. *See also* **Cost reviews,** and **Cost types.**

FOREIGN WATER Water occurring in a stream, groundwater or other water body that originated in another drainage basin.

FORESHOCK Relatively small Earth tremor that precedes an earthquake by a few days or weeks, originating in the same general area as the subsequent earthquake.

FORESHORE Sloping beach between high and low water marks.

FOREST Concentration of trees and related vegetation in non-urban areas sparsely inhabited by and infrequently used by humans; characterized by natural terrain and drainage patterns.

FOREST ECOLOGY Relationship between forest organisms and their environment.

FOREST ECONOMICS Generally, that branch of forestry concerned with the forest as a productive asset subject to economic principles.

FOREST INFLUENCE Effects on climate, water and soil resulting from forest growth.

FOREST MANAGEMENT Generally, the practical application of scientific, economic, and social principles to the administration and working of a specific forest area for specified objectives.

FOREST MANAGEMENT CYCLE Phases that occur in the management of a forest, including harvesting, site preparation, reforesting, and stand tending.

FOREST MANAGEMENT PLAN General plan for the management of a forest area, usually for a full rotation cycle, including the objectives, prescribed management activities, and standards to be employed to achieve specific goals. Commonly supported with more detailed development plans. *See also* **Managed forest land.**

FOREST PRACTICE 1. Any activity that enhances and/or recovers forest growth or harvest yield, such as site preparation, planting, thinning, fertilization, and harvesting; **2.** Road construction or reconstruction within forest lands for the purpose of facilitating harvest or forest management; **3.** Any management of slash resulting from the harvest or improvement of tree species.

FOREST PROTECTION Prevention and control of any cause of potential forest damage.

FOREST REGENERATION *See* **Reforestation.**

FOREST RENEWAL Renewal of a tree crop by either natural or artificial means.

FOREST RESIDUALS 1. Sum of wasted and unused wood in the forest, including logging residues, rough, rotten and dead trees, and annual mortality; **2.** Unmerchantable material normally left following conventional logging operations other than whole-tree harvesting.

FORESTRY Generally, a profession embracing the science, business, and art of creating, conserving, and managing forest, and forest lands for the continuing use of their resources, materials, and other forest products.

FOREST TYPE 1. Group of forested areas or stands of similar composition (species,

age, height, and stocking) that differentiates it from other such groups; **2.** Classification of forest land in terms of potential cubic-measurement volume growth per ha (acre) at the culmination of mean annual increments in fully stocked natural stands.

FOREST TYPE LABEL Symbol that is used to code information about the forest type on a forest cover map, e.g. site, disturbance, age and height class, species, stocking.

FOREST TYPE LINE Line on a map or aerial photo outlining a forest type.

FORMALDEHYDE Reactive organic compound CH_2O.

FORMATION Rock mass characterized by lithologic homogeneity; mappable on the Earth's surface or traceable in the subsurface.

FORMATION FLUID Fluid present in a formation under natural conditions, as opposed to introduced fluids, such as drilling mud.

FORM LOSS OF HEAD Loss of head attributable to a change in shape of a waterway.

FORMULATION Process of mixing, blending, or dilution of one or more active ingredients with one or more other active or inert ingredients, without an intended chemical reaction, to obtain a product or an end use product.

FOSSIL FUEL Natural gas, petroleum, coal, and any form of solid, liquid, or gaseous fuel derived from such materials for the purpose of creating useful heat.

FOUL AIR Air no longer fit to breath.

FOUL AIR DUCT Suction line of a tunnel ventilating system.

FOULED Anything that hangs up, jams; anywhere the intended movement is restricted.

FOULING Material accumulating in gas passages or on heat absorbing surfaces. *See also* **Slag.**

FOUL WATER Combination of waste and soil water.

FOUNDATION DRAIN Subsoil drain adjacent to the external, and sometimes internal, face of foundations that permits the infiltration of groundwater and conveys it to a sump or tile bed.

FOUNTAIN 1. Spring of water issuing from the ground; **2.** Jet or spray of water pumped into the air, commonly as the centerpiece of an artificial and ornamental pond.

FOUNTAIN AERATOR Device that causes a flow to discharge at the top of an arrangement of basins of gradually decreasing diameter and to cascade from one basin the next.

FOUNTAIN FLOW Characteristic of water flowing from the open end of a vertical pipe, rising a small amount above the rim before spreading out in all directions in a smooth, mushroom shape.

FOUNTAIN HEAD Head in a saturated, confined aquifer.

FOUNTAIN JET Stream of water issuing from, and over, the horizontal rim of a vertical pipe.

FOURNEYRON WHEEL Outward-flow water wheel.

FOUR-WAY VALVE *See* **Valve.**

FRACTURE Break in a mineral that is not along a cleavage plane.

FRACTURE SPRING Spring flowing from a relatively large opening in rocks.

FRACTURE ZONE Zone along which faulting has taken place.

FRAME Any pair of walings on opposite faces of a trench, together with the struts that support them.

FRAZIL ICE Granular or laminar ice that forms on fast-flowing water during long cold spells.

FREE ACCELERATION TEST Procedure for measuring exhaust emissions from vehicles.

FREE AIR Air at normal atmospheric condition for the elevation above sea level. *See also* **Compressed air** and **Standard air.**

FREE AREA Total area across the face of a pipe or duct.

FREE AVAILABLE CHLORINE Chlorine available as dissolved gas, hypochlorous acid, or hypochlorite ion, that is not combined with an amine or other organic compound.

FREE AVAILABLE RESIDUAL CHLORINE Fraction of the total residual chlorine remaining in water and wastewater following a specific contact period and which will react chemically and biologically as hypochlorous acid or hypochlorite ion.

FREEBOARD Vertical distance from a normal water surface to the top of the confining wall.

FREE CHLORINE Chlorine in a liquid or gaseous form and available to combine with water to form hypochlorous and hydrochloric acids, or with wastewater to combine with an amine or other organic compounds to form combined chlorine compounds.

FREE CONVECTION Motion caused by density difference within a fluid. *Also called* **Gravitational convection.**

FREE FIELD Acoustical region in which no significant reflections of sound occur.

FREE FLOW Flow that is not hindered by obstruction or constriction; the maximum velocity possible for the condition.

FREE GROUNDWATER *See* **Groundwater.**

FREE LIQUID Liquid that readily separates from the solid portion of a waste under ambient temperature and pressure.

FREE MOISTURE 1. Moisture having essentially the properties of pure water in bulk; **2.** Moisture not absorbed by aggregate. *See also* **Surface moisture.**

FREE OXYGEN Molecular oxygen available for respiration by organisms; the oxygen molecule, O_2, that is not combined with another element to form a compound.

FREE RADICAL Ion, or molecular fragment that has one or more unpaired electrons, making it highly reactive.

FREE RESIDUAL CHLORINATION Chlorine or chlorine compounds applied to water or wastewater to produce a free chlorine residual directly or through the destruction of ammonia or organic nitrogenous compounds.

FREE SURFACE Surface of a liquid that is in contact with the atmosphere.

FREE SURFACE ENERGY Free energy in a liquid surface produced by the unbalanced inward pull on surface molecules by the underlying molecules. *Also called* **Surface tension.**

FREE WATER 1. Non-chemically combined water; **2.** Water droplets or globules in a system fluid that tend to accumulate at the bottom or top of the system fluid, depending on the fluid's specific gravity.

FREE WAVE Wave in water that continues to exist after the generating force has ceased to act.

FREE WEIR Weir that is now submerged.

FREEZELESS WATER FAUCET Water faucet that discharges on the exterior face of a wall but which has its valve seat on the interior face of the wall.

FREEZE LINE *See* **Frost line.**

FREEZE–THAW CYCLE Cycle from completely frozen to completely thawed and back to completely frozen.

FREEZE-UP 1. Date when outdoor conditions in an area mean that open water can be expected to be frozen, the soil will contain ice crystals, exposed surfaces will be frosted in the early morning; full winter conditions. *See also* **Breakup**; **2.** Condition where a moving part refuses to operate, for whatever reason.

FREEZING INDEX Total number of degree days below freezing for a winter, calculated from the mean night and day air temperatures.

FREEZING LEVEL Elevation of the 0°C isotherm in the atmosphere.

FREEZING NUCLEI Atmospheric particles on to which water freezes.

FREEZING PROCESS Technique of temporarily freezing a water-bearing soil to increase its strength, as well as to eliminate passage of water through the soil.

FREEZING TIME Time for a freezing process to be complete.

FRENCH DRAIN Subsurface trench filled loosely with stone through which water may flow.

FREON Proprietary brand of chlorofluorocarbons.

FREQUENCY Number of vibrations or complete oscillations occurring in one second (designated Hertz or cycles/second). In vibratory compaction, it is a measure of the number of complete cycles or revolutions of the weights around the axis of rotation over a given length of time, usually expressed as vibrations per minute (vpm).

FREQUENCY CURVE Graphical representation of the frequency of occurrence of an event or events.

FREQUENCY DISTRIBUTION Relationship between the magnitude of an observed variable and the frequency of its occurrence.

FREQUENCY RESPONSE Range of frequencies that can be sensed (within certain acceptable limits of error) by a device.

FRESH AIR FILTER Main filter in an air-conditioning system that removes dust and large particles from incoming air.

FRESH AIR INLET Opening for the introduction of atmospheric air.

FRESH AIR RATE Percentage of fresh air that is mixed with recirculating air in a mixing box, replacing some of the return air.

FRESH SLUDGE Sludge in which decomposition is little advanced.

FRESH WASTEWATER Wastewater in which dissolved oxygen can still be measured.

FRESHWATER Uncontaminated source of drinking water; water whose salinity is less than 0.5%.

FRICTIONAL SOIL Silt, sand, or gravel whose shearing strength is mainly decided by the friction between particles.

FRICTION FACTOR Factor used in calculating loss of pressure due to the friction of a fluid flowing through a pipe.

FRICTION HEAD See **Friction loss.**

FRICTION HORSEPOWER See **Horsepower.**

FRICTION LOSS Loss of pressure created by movement of water in a pipe, hose, or fitting. *Also called* **Friction head.**

FRICTION SLOPE Friction head or loss per unit length of conduit.

FRICTION WEIR Weir that is not submerged. *See also* **Free weir.**

FRINGE WATER Water temporarily or permanently held at a level just above the water table.

FRONT Line of separation between air masses of different density and temperature.

FRONTAL ANALYSIS Analysis of weather charts by marking the positions of the fronts between different air masses.

FRONTAL PRECIPITATION Precipitation occurring at a frontal surface.

FRONTAL SLOPE Inclination of the surface of a front to the horizontal.

FRONTAL SURFACE Surface of separation between two adjacent air masses of different characteristics, usually temperature and humidity, normally associated with a belt of clouds and precipitation.

FRONT FOOT One foot (0.3 m) length of land measured along the frontage of a lot. When used as the basis for the sale of commercial property it implies a square measure and includes the ground lying back of the frontage to the rear boundary of the property. *See also* **Front meter.**

FRONT-LOADING REFUSE TRUCK See **Refuse truck.**

FRONT LOT LINE Boundary line of a lot along a street; in the case of a corner lot, either of the boundary lines along a street, the other being considered a side lot line; in the case of a through lot, each of the two shorter boundary lines along streets.

FRONT METER One meter (3.28 ft) of land measured along the frontage of a lot. When used as the basis for the sale of commercial property it implies a square measure and includes the ground lying back of the frontage to the rear boundary of the property. *See also* **Front foot.**

FRONT-MOUNT HOIST Hydraulic cylinder, usually long-stroke telescopic, mounted vertically at the extreme front of a dump body, commonly enclosed in a housing that extends into the front portion of the body. *Also called* **Head lift, Head mount,** and **Vertical hoist.**

FRONT OF LEVEE Side of a levee next to the river. *Also called the* **River side.**

FROST Weather condition during which dew turns to ice.

FROST ACTION Phenomenon that occurs when water in soil is subjected to freezing which, because of the water ice phase change or ice lens growth, results in a total volume increase or the buildup of expansive forces under confined conditions, or both, and the subsequent thawing that leads to loss of soil strength and increased compressibility.

FROST BOIL Softness of soil that has thawed after frost heave.

FROST BOTTOM Portion of a water meter case designed to break easily without damaging the mechanism when water freezes within the device.

FROST BOX Box containing insulation and surrounding a water meter, water pipe or other device containing water that may freeze.

FROST CRACK Radial, longitudinal split in the wood of a tree, generally near the base of the bole, caused by internal stresses due to extremely cold weather.

FROST HEAVE Lifting action of structures, waste material, or surface soil, caused by the expansion of material during the freeze/thaw cycle of water contained in the top soil.

FROST HOLLOW Relatively small, low-lying area that is subject to frequent and severe frosts because of the accumulation of cold air at night.

FROSTLINE Maximum depth to which frost penetrates the ground. This varies in different parts of the country and from year to year. For design purposes it is a figure established for a locality as determined from historical records. It is the depth below which foundations should be placed so as to prevent movement due to frost heave. *Also called* **Freeze line.**

FROST POINT Temperature at which air must be cooled for frost to begin to form on solid surfaces.

FROSTPROOF CLOSET Toilet that has no water in the bowl and a trap and control-valve water supply positioned below the frost line.

FROST WEDGING Pushing up or apart of rock particles due to the action of ice formation.

FROTH Mass of very small bubbles.

FROTH FLOTATION Process that employs a froth of water and oil to effect the separation of finely divided materials.

FROUDE NUMBER In an open channel, a ratio that should be the same for the model analysis as in the full-size project.

FROZEN Condition where a part normally free to rotate cannot move.

FROZEN FOG Fog of cloud composed of ice crystals.

FROZEN SOIL Soil below 0°C (32°F) in which part of the pore water has frozen.

FUEL 1. Substance or combination of substances that can be burned; **2.** In explosive calculations, the chemical compound used to combine with oxygen to form gaseous products and cause a release of heat.

FUEL–AIR RATIO Ratio of fuel supply flow rate to the air supply flow rate when both rates are measured in the same units under the same conditions.

FUEL CELL Device classified as a direct energy converter and consisting of an electrolyte sandwiched by an anode and a cathode.

FUEL CYCLE Sequence of events involved in the location, acquisition, refining or processing, transportation, delivery, use and, where applicable, reprocessing for either reuse or disposal of an energy source.

FUEL EFFICIENCY Relationship between fuel consumption and machine productivity. It is expressed in units of weight carried or material moved per volume of fuel consumed.

FUEL ELEMENT One of the containers of fissile material that is inserted into the core of a nuclear reactor.

FUEL EVAPORATIVE EMISSION Vaporized fuel released into the atmosphere from the fuel system of an engine.

FUEL OIL Broad range of distillate and residual fuels identified by ASTM grades 1 through 6:

> **Grade No. 1:** A light distillate fuel that has the lowest boiling range.

> **Grade No. 2:** Fuel oil, popularly called heating oil, has a higher boiling range and is commonly used in home heating. It is comparable in boiling range to diesel fuel.

> **Grades No. 4, 5, and 6:** Fuel oils that require preheating to facilitate pump-

ing and burning. No. 6 fuel oil is *also called* **Bunker C fuel oil.**

FUGITIVE EMISSIONS 1. Air pollutants emitted to the atmosphere other than from a stack; **2.** Emissions which could not reasonably pass through a stack, chimney, vent or other functionally equivalent opening.

FUGITIVE SOURCE Any source of emissions not controlled by an air pollution control device.

FUGITIVE VOLATILE ORGANIC COMPOUNDS Volatile organic compounds that are emitted from applicators and flashoff areas and which are not emitted in the oven.

FUGITIVE WATER Leakage from an impounding reservoir.

FULL-CELL PROCESS Process for impregnating wood with preservative or chemical in which a vacuum is drawn to remove air from the wood before admitting the preservative.

FULLER'S EARTH Soft, clay-like mixture used to absorb water, coloring matter, grease, and some oils.

FULL-OPEN VALVE *See* **Valve.**

FULL-TIDE COFFERDAM Cofferdam built high enough to exclude water at all tides.

FULL-WAY VALVE *See* **Valve.**

FUMAROLE Volcanic vent that emits only gases, which are at a temperature higher than that of the atmosphere.

FUME Gas, smoke, or vapor, usually offensive and sometimes suffocating.

FUME SCRUBBER Wet air pollution control devices used to remove and clean fumes.

FUME SUPPRESSION SYSTEM Equipment comprising any system used to inhibit the generation of emissions.

FUMIGATION Disinfection with a gas for the destruction of germs, insects, or animal life.

FUNCTIONAL SPACE Room, group of rooms, or homogeneous area (including crawl spaces or the space between a dropped ceiling and the floor or roof deck above), such as classroom(s), a cafeteria, gymnasium, hallway(s), designated by a person accredited to prepare management plans, design abatement projects, or conduct response actions.

FUNCTIONALLY EQUIVALENT COMPONENT Component that performs the same function or measurement and which meets or exceeds the performance specifications of another component.

FUNDAMENTAL FREQUENCY Frequency with which a periodic function reproduces itself; the first harmonic.

FUNDAMENTAL PARTICLES Particles that cannot be demonstrated to contain simpler units.

FUNGI Plantlike organisms that lack chlorophyll and also lack the characteristic plant structures such as leaves, and roots, and including yeasts, moulds, rusts, mildews, mushrooms, etc.

FUNGICIDE Poison, used to kill or retard the growth of fungus.

FUNGUS Group of living plants, like mushrooms, mildew and mold, that feed on decaying plant ingredients and that can attack most organic material under suitable conditions of moisture and temperature.

FUNNEL CLOUD Cloud that appears in the core of a tornado or water spout due to the low pressure.

FURNACE Enclosed structure or vessel in which material is heated to a high temperature, or consumed.

FURRED Pipes and boilers that have become encrusted internally with salts and minerals deposited from the water, usually at elevated temperature.

FURROW IRRIGATION System of small ditches or plowed furrows used to distribute irrigation water.

FUSION 1. Melting together; fusing; **2.**

Combining of two nuclei to create a nucleus of greater mass.

FUSION REACTOR Nuclear reactor whose energy is derived from the fusion of two atoms (deuterium, tritium, lithium, or some combination of these) to form one helium atom, with the simultaneous release of energy.

G

GABION Medium-sized, 200 to 600 mm (8 to 24 in.), rocks confined in a rectangular wire cage, used to restrain the toe of an excavation, as a retaining wall, or for other pressure-resisting purposes.

GAGE *See* **Gauge.**

GAIN Addition of heat, typically from solar gain, that can lessen the load on a heating plant, or increase it on a cooling/ventilating or air-conditioning system.

GAINING STREAM Section of a stream that receives water from groundwater in the saturation zone.

GALL Abnormal growth of plant tissue, produced in response to injury or an invasion of insects, mites, eelworms, fungi, bacteria, or viruses.

GALLERY Underground structure or passageway designed to collect and convey water or wastewater.

GALLON (IMPERIAL) Nonmetric unit of volume, equal to 277.42 cubic inches; 160 fluid ounces; 10.02 pounds of water. Symbol: gal. Multiply by 0.004 546 to obtain cubic meters, symbol m^3; by 4.546 to obtain liters, symbol: L. *See also the appendix*: **Metric and nonmetric measurement.**

GALLON (IMPERIAL) PER DAY Nonmetric unit of flow. Symbol: gpd. Multiply by 0.004 546 to obtain cubic meters per day,

symbol: m^3/d; by 0.000 052 to obtain liters per second, symbol: L/s. *See also the appendix*: **Metric and nonmetric measurement.**

GALLON (IMPERIAL) PER MINUTE Nonmetric unit of flow. Symbol: gpm. Multiply by 0.007 768 to obtain liters per second, symbol: L/s. *See also the appendix*: **Metric and nonmetric measurement.**

GALLON (US) Nonmetric unit of volume, equal to 231 cubic inches; 128 fluid ounces; 8.34 pounds of water. Symbol: gal. Multiply by 0.003 785 to obtain cubic meters, symbol: m^3; by 3.785 to obtain liters, symbol: L. *See also the appendix*: **Metric and nonmetric measurement.**

GALLON (US) PER MINUTE Nonmetric unit of flow. Symbol: gpm. Multiply by 0.063 to obtain liters per second, symbol: L/s. *See also the appendix*: **Metric and nonmetric measurement.**

GALVANIZE Coat iron or steel with zinc by immersion or deposition of molten zinc.

GALVANIZED PIPE *See* **Pipe.**

GALVANIZED STEEL PIPE *See* **Pipe.**

GAMMA RAYS Electromagnetic radiation at the high-energy end of the spectrum, with wavelengths of 10 nanometers or less.

GANG OF WELLS Group of wells from which water is drawn by a single pump. *Also called* **Battery of wells.**

GARBAGE *See* **Solid waste.**

GARBURATOR *See* **Waste disposal unit.**

GAS 1. Any fluid substance that can be expanded without limit; 2. Any gas or mixture of gases except air, including:

Inert – A gas that does not react with materials with which it is in contact.

Liquefied petroleum – A gaseous mixture of light hydrocarbons, including propane, propene, and butenes, and which can be liquefied by increased pressure and/or lowered temperature.

Natural – A naturally-occurring substance frequently, but not necessarily associated with oil reserves.

GAS CAP Accumulation of natural gas above an oil pool.

GAS CHROMATOGRAPHY Analytical technique for separating mixtures of volatile substances.

GAS COCK *See* **Valve.**

GAS-COOLED FAST-BREEDER REACTOR Breeder atomic reactor that uses gas as a coolant.

GAS DOME Steel cover to a sludge digestion tank, floating entirely or in part on the liquid sludge and containing the gas generated by the digestion process.

GASIFICATION Transformation of soluble and suspended organic matter into gas by biological action during waste decomposition.

GAS-LIFT FLOW AREA Area over which the subsurface water is under pressure from the presence of gas. The same water would rise to the surface if it was free to do so.

GAS:OIL RATIO Ratio of oil to gas in a produced volume of crude oil.

GAS-SOLUBILITY FACTOR Quantity of gas absorbed by a unit quantity of water at a given temperature when the water is exposed to a pure atmosphere of the gas under a barometric pressure of 760 mm (29.92 in.).

GAS THERMOMETER Thermometer in which the expanding and contracting medium is gaseous.

GAS TRAP S- or P-shaped section of pipe in which enough water or wastewater is retained to prevent the counterflow of noxious gases.

GAS VENT Opening to the atmosphere from a system or vessel generating or containing lighter-than-air gases.

GATE Hinged metal flap used to control the flow of water.

GATE CHAMBER Underground structure housing a valve or other regulating device and providing access for maintenance.

GATED WYE Hose appliance with one female inlet and two or more male outlets with a gate valve on each outlet.

GATE HOUSE Underground structure housing a valve or other regulating device with an aboveground superstructure providing access for operation and/or maintenance.

GATE LIFT Device for operating a gate in a generally vertical direction.

GATE STEM Rod attached to the top of a gate by means of which it is opened or closed.

GATE-TYPE HYDRANT Hydrant equipped with one main valve consisting of a vertical disk moving vertically across the valve seat and held closed by a wedge nut.

GATE VALVE *See* **Valve.**

GATE YARD Solid waste volumes as calculated at the landfill entrance in the incoming trucks, generally compacted into a space in the landfill equivalent to 50% of that occupied in the truck.

GAUGE (ALSO SPELLED GAGE) 1. Standard measure; **2.** An instrument for measuring, indicating, or comparing a physical characteristic. There are many types, including:

> **Box:** Tide gauge operated by a float in a long vertical box to which the tide is admitted through bottom openings.
>
> **Compound:** Pressure gauge that records the pressure above and below atmospheric pressure.
>
> **Fluid level:** Gauge that indicates a fluid level.
>
> **Manometer:** Differential pressure gauge in which pressure is indicated by the height of a liquid column of known density. Pressure is equal to the difference in vertical height between two connected columns multiplied by the density of the manometer liquid.
>
> **Point:** Sharp point fixed to an attach-

ment which slides on a graduated rod for measuring water level.

Pressure: Gauge that indicates the pressure in the system to which it is connected.

Vacuum: Pressure gauge for pressures less than atmospheric.

Water: Glass U-shaped tube half filled with water, the other end connected by a flexible tube to a system of drains, gas pipes, etc., being tested. It shows whether the pipes are gas tight. *Also called* a **U-gauge.**

GAUGE CORRELATION Stage-discharge relation between gauge height and the discharge of a stream or conduit at the gauging station, as displayed by the rating curve or table for the station.

GAUGE DATUM Elevation of gauge zero above a known datum.

GAUGE GLASS Vertical glass tube that shows a liquid level.

GAUGE HEIGHT Elevation of a water surface as indicated on a gauge that is referenced to a known datum.

GAUGE PRESSURE *See* **Pressure.**

GAUGING Measurement of flow per unit of time in a stream channel, conduit, or orifice at a specific point using current meters, rod floats, weirs, Pitot tubes or other measuring devices.

GAUGING PIG Device used to crudely evaluate a pipeline: a sheet steel or aluminum disc about 95% of the ID of the pipeline that is pulled through the pipe. A deformed or damaged disc indicated an obstruction or fault; the location of the fault can be roughly determined from a pressure chart made of the pig's journey.

GAUGING STATION Point in a channel fitted with a flow gauge.

GAUSSIAN DISTRIBUTION Distribution that shows the maximum number of occurrences at or near to a center or mean point, plus a progressive decrease in occurrences with increasing distance from the center, and symmetrical distribution of occurrences on both sides of the center.

GEAR Two or more toothed wheels meshed together so that the motion of one is transmitted to the other. There are many types, including:

Bevel: Gear made of teeth cut in the surface of a truncated cone, used to transmit power at right angles.

Cluster: Two or more gears of different sizes made in one piece.

Helical: Type of gear where the straight or curved teeth are positioned diagonally across the face of the gear wheel at an angle of less than 90° to the direction of rotation.

Hypoid: Resembles a spiral-bevel gear, except that the pinion is offset so that its axis does not intersect the gear axis.

Herringbone: Gear with V-teeth.

Idler: Gear meshed with two others that does not transmit power to its shaft, used to reverse the direction of rotation in a transmission.

Open: Gear that is exposed to the environment, rather than being housed in a protective gearbox.

Pinion: Small gear that drives a larger gear.

Planetary set: Set of gears consisting of an inner (sun) gear, and outer ring with internal teeth, and two or more small (planetary) gears meshed with both the sun gear and the ring. Used to either speed up or slow down the input *vs* output to gain speed or power.

Rack: Toothed bar.

Ring: Large gear, driven by a worm gear, that is actuated by the drive sheave.

Spiral bevel: Gear with spiral-shaped teeth, used primarily to change the direction of transmitted power.

Sprocket: Metal disk with projecting teeth about its periphery, usually employed in combination with a large chain to transmit rotary power.

Spur: Gear on which the teeth are cut parallel to the axis of the shaft. *Also called* **Straight-tooth gear.**

Straight-tooth: *See* **Spur** (above).

Sun: 1. Central gear in a planetary set. **2.** Planetary gear set consisting of a central gear and an internal-toothed ring gear, and two or more planet gears meshed with both of them.

Worm: Consists of a spirally-grooved screw moving against a toothed wheel.

GEARBOX Casing for gear sets that transmit power from one rotating shaft to another. *Also called* **Gear housing.**

GEAR CASE Enclosure containing a gear train, the shafts on which they are mounted, plus lubricant.

GEARED DRIVE *See* **Driving machine.**

GEAR HOUSING *See* **Gearbox.**

GEAR HYDRAULIC PUMP *See* **Pump, hydraulic.**

GEAR PUMP *See* **Pump.**

GEAR RATIO Number of revolutions a driving gear requires to turn a driven gear through one complete revolution. For a pair of gears, the ratio is found by dividing the number of teeth on the driven gear by the number of teeth on the driving gear.

GEAR REDUCTION UNIT Assembly of meshed gears in which the rpm of the input shaft is higher than that of the output shaft.

GEAR SEGMENT Part of a gear that is segmented to provide a reference for timing, such as combining two teeth together.

GEAR STEP Transmission gear step measured by the proportion of change between successive gears. It is determined by dividing the ratio of the lower gear by the ratio of the next higher gear. The larger the step the greater the rpm drop upon downshifting. Gear steps are usually expressed as percentages.

GEAR TRAIN Assembly of meshed gears in which the related shafts revolve at the same, or different, rpm than that of the input shaft.

GEIGER COUNTER Instrument used to detect radiation by collecting and observing the pulse of ions created in an enclosed volume of gas by the passage of energetic ionizing particles or rays.

GEIGER THRESHOLD Lowest voltage which, when applied to a Geiger counter, will produce pulses of equal magnitude, irrespective of the number of primary ions produced.

GEL Matter in a colloidal state that does not dissolve but remains suspended in a solvent from which it fails to precipitate without the intervention of heat or an electrolyte.

GEL COAT Thin outer layer applied to a moulded shape, usually as a cosmetic.

GELIFLUCTION Mass wasting of thawed material over permafrost.

GELLING Transformation of a liquid into a jellylike consistency.

GEL PORE Void, finely divided from other such voids, left when a cement gel is dried, by heating to 105°C (221°F), and the water gel is driven off.

GEL SPACE RATIO Ratio of the concentration of newly formed hydration products to the original capillary pore space available.

GEL STRENGTH Stress required to break up the gel structure of a bentonite slurry formed by thixotropic buildup, under static conditions.

GEL TIME Time required to change a flowable liquid resin into a non-flowing gel.

GEL WATER Excess water that is bound to a gel but which can be driven off by a strong drying action.

GENE Part of germ plasm that occupies a fixed place on a chromosome and which determines the nature and development of an inherited characteristic: the genes inherited from parents determine traits of plants or animals that develop from a fertilized egg cell.

GENECOLOGY Study of the genetics of plant and animal populations in relation to their environments.

GENE EXCHANGE Sexual reproduction within an ecotype, species or genus, which results in a recombination of parental genes.

GENE FLOW Movement of genes between populations as a result of sexual reproduction between members from each population.

GENE FREQUENCY Frequency with which a certain gene occurs in a population.

GENERAL BENEFIT Advantage accruing from given public works, typically a highway improvement, to a community as a whole, applying to all property similarly situated. *See also* **Benefit,** and **Special benefit.**

GENERAL CIRCULATION Average worldwide system of winds.

GENERAL CONTRACT Contract between an owner and a contractor covering the entire work to be done; one not limited in its scope.

GENERAL CONTRACTOR Owner's designated representative with full responsibility for the completion of the contracted project.

GENERAL DRAWING Drawing showing the plan, elevations, and a cross section of a structure or works, and including overall dimensions, etc.

GENERAL ENVIRONMENT Total terrestrial, atmospheric and aquatic environments.

GENERAL ESTIMATE METHOD Simple estimating method based on a square foot (or equivalent) cost for typical construction types (residential, warehouse, commercial, etc.).

GENERAL FOREMAN Contractor's representative in charge of all labor who coordinates the work of trades foremen, whether directly employed or subcontracted.

GENERATE Act or process of producing.

GENERATED CONTAMINANT Contamination created by the operation of a fluid system or component. *See also* **Artificial contaminant, Built-in contaminant,** and **Contaminant.**

GENERATION CURVE Population density at a given stage of development, plotted on a graph against the generation number over a sequence of generations.

GENERATOR SET Integrated diesel- or gasoline-powered engine and electrical generator. They are rated according to:

> **Prime power:** For continuous electrical service with 10% overload capability for one hour in 12.

> **Standby power:** For continuous electrical service during interruption of normal power.

GENETIC Having to do with genes.

GENETIC DRIFT Change in the genetic composition of a population that occurs by chance and not as a result of natural selection.

GENETIC ENGINEERING Application of genetics involving the interchange of DNA sections between individuals of the same or different species by artificial means.

GENETIC POLLUTION Event in which genes from recombinant DNA organisms escape and become incorporated into wild species.

GENETICS Branch of biology dealing with the principles of heredity and variation in animals and plants of the same or related kinds.

GENOME Genetic material characteristic of a particular species.

GEO- Prefix meaning earth; prefix attached to such words as geology (study of the Earth), geography (depiction of the Earth), etc.

GEOBOTANICAL ANOMALY Indication of enrichment or depletion of particular elements in the soil as evidenced by the presence or absence of certain plant species, or gross physical changes in plants.

GEOBOTANY Study of the global distribution of plants; phytogeography.

GEOCENTRIC 1. As viewed or measured from the Earth's center; **2.** Having or representing the Earth as a center.

GEOCHEMICAL ANOMALY Local enrichment of depletion of an element in soil or rock.

GEOCHEMICAL DISTRIBUTION Dispersion of elements at or near the Earth's surface.

GEOCLINE Gradual and continuous change associated with a geographic gradient.

GEODESIC Shortest possible distance between two points along the Earth's surface.

GEODESY Science or art of measuring the Earth's surface.

GEODETIC Of, having to do with, or involving geodesy.

GEOGNOSY Branch of geology that deals with the structure of the Earth, its rocks and minerals, and the water and air surrounding it.

GEOGRAPHER Person trained in geography, especially one whose work it is.

GEOGRAPHY Science that deals with the Earth's surface and its division into continents and countries, and the climate, animal and plant life, peoples, resources, industries, and products of these divisions.

GEOHYDROLOGIC Having to do with subsurface water and related geologic aspects of surface waters.

GEOLOGICAL ENGINEER Professional engineer who specializes in studies of the Earth's structure.

GEOLOGICAL MAP Map showing the significant physical features of an area identifying the salient materials.

GEOLOGICAL REPORT Report on a specific area or site identifying and locating the materials, and their interrelationship, to prescribed depths and in specified detail, primarily intended to determine load-bearing capabilities, susceptibility to movement, and seismic potential.

GEOLOGICAL SURVEY Investigation of subsurface materials and conditions to meet a requirement and/or specification.

GEOLOGICAL TIME System whereby the history of the Earth from its formation to present is divided into a chronology of definable episodes.

GEOLOGIC EROSION Erosion resulting from geologic processes.

GEOLOGIST Person trained in geology.

GEOLOGY Science of the Earth's crust, its composition, structure, and history of development.

GEOMAGNETIC Of, or having to do with the magnetism of the Earth.

GEOMAGNETIC INDUCTION Induction of the Earth's magnetic field, based on the concept of a diapole at the center of the Earth and measuring the Earth's magnetic induction with reference to three axes mutually at right angles, directed toward geographic north, geographic east, and vertically downward to the Earth's center, the resulting diapole field being overlaid by an irregular, constantly-changing non-diapole field.

GEOMEMBRANE Any synthetic, hydraulically impervious material intended to be used with soil as part of a man-made fluid containment system.

GEOMORPHOLOGY Study of the Earth's surface and its configuration and land forms.

GEOPHONE Microphone that can be lowered below the Earth's surface, or towed behind a ship to record seismic shock waves.

GEOPHYSICS Science that deals with the

relations between the features of the Earth and the forces that produce them; the physics of the Earth.

GEOPHYSICAL EXPLORATION Preliminary subsurface survey using devices located on or above the surface to find the change in wave velocity, electrical resistivity, gravity, magnetism, or other physical variables.

GEOPOLITICS Study of government and its policies as affected by physical geography.

GEOSERE Series of climax plant formations developed through geological time.

GEOSPHERE Solid matter that comprises the Earth.

GEOSTROPHIC WIND Horizontal wind that blows parallel to the isobars, indicating a balance between horizontal pressure-gradient force and the horizontal components of the Coriolis force.

GEOSYNCLINE Elongated basin that has been filled with a great thickness of sediment, usually interlaced with volcanic rocks.

GEOSYNTHETIC Any synthetic material used with soil as an integral part of a manmade system.

GEOTECHNICAL ENGINEER Engineer with specialized training and knowledge of soils and rocks, employed to do soil investigations, design of structure foundations, and provide field observation. *Also called* **Soils engineer.**

GEOTECHNICAL PROCESSES Processes that cause change in soil: freezing/thawing, groundwater levels, etc.

GEOTECHNOLOGY Study of geology, soil, and rock mechanics.

GEOTEXTILE Woven or nonwoven fabric manufactured from synthetic fibers or yarns that is designed to serve as a continuous membrane between soil and aggregate in a variety of earth structures.

GEOTHERMAL ENERGY Energy generated within the Earth's interior.

GEOTHERMAL GRADIENT Change of temperature with depth in the Earth's crust.

GEOTROPIC Responding to gravity.

GERMICIDAL TREATMENT Treatment causing the destruction of microorganisms through use of disinfecting chemicals.

GEYSER Periodically flowing, natural thermal spring that periodically ejects water or steam to the atmosphere.

GIANT Large-diameter nozzle, manually or mechanically controlled, for directing a jet of water under high pressure for hydraulic excavating. *Also called* **Monitor.**

GIGA Prefix representing 10^9. Symbol: G. Used in the SI system of measurement. *See also the appendix*: **Metric and nonmetric measurement.**

GLACIAL DEPOSIT Material deposited by glaciation, usually a wide range of particle sizes.

GLACIAL DRIFT Any rock material, such as alluvia, transported by a glacier and deposited by the ice or by the meltwater from the glacier.

GLACIAL LAKE Lake developed at the bottom of a valley formed by glacation.

GLACIAL STREAM Stream fed by melting glacier ice.

GLACIAL STRIATION Scratches on rocks that have been made smooth from the movement of ice, caused by rock or grit encased in the ice.

GLACIAL TILL Material deposited by glaciation, usually a wide range of particle sizes, not subjected to the sorting action of flowing water. *See also* **Drift.** *Also called* **Moraine.**

GLACIER Stream-like mass of ice, formed by consolidated accumulations of snow at high altitudes and the poles, slowly moving under its own weight to lower elevations.

GLACIER BURST Sudden release of water that has been impounded by a glacier.

GLACIER SNOW Compacted snow that is

in process of being converted to glacier ice.

GLACIOFLUVIAL Stream having melted glacial ice as the source of its water and much of its load. *See also* **Fluvioglacial.**

GLAND 1. Cavity of a stuffing box; **2.** Compressible copper or brass sleeve. *Also called* **End plate**.

GLAND BOLT Bolt used for tightening an unthreaded gland.

GLAND JOINT Pipe joint that allows for movement due to thermal expansion or contraction.

GLAND NUT Nut with one conical face that exerts pressure on the gland packing as it is tightened about the gland spindle.

GLAND RING Part of a stuffing box assembly that exerts a uniform pressure on the packing.

GLAZED FROST Clear, glass-like ice deposited on objects by the impact of super-cooled water droplets in a cloud or fog.

GLAZED WARE Stoneware pipe and fittings, glazed from salt thrown into the kiln during firing.

GLEY Sticky soil layer which develops on ground that is continuously or frequently saturated.

GLOBAL COMMONS Area (land, air, water) outside the jurisdiction of any nation.

GLOBAL SAFETY FACTOR Combined factor of safety resulting from all the partial safety factors.

GLOBE VALVE *See* **Valve.**

GLORY-HOLE SPILLWAY Vertical shaft shaped as an inverted truncated cone and discharging to an outlet tunnel that serves as an overflow to a reservoir.

GLUCOSE Monosaccharide hexose $(C_6H_{12}O_6)$ sugar important to both plants and animals as an energy producer and naturally present in all fruits, and which is the building block of other sugars such as sucrose, lactose, maltose, cellulose and glycogen.

GO-DEVIL 1. Ball of rolled-up burlap or paper or a specially fabricated device put into the pump end of a pipeline and forced through the pipe by water pressure in order to clean the pipeline; **2.** Device used with tremie concrete operations; **3.** Movable hoist-way working platform, used when installing elevator brackets and guide rails.

GOVERNING INSTRUMENTS Legal documents that establish the existence of an organization and define its powers and parameters of operation.

GRABEN Downthrown block between two normal faults.

GRAB SAMPLE Single sample taken without regard for time or flow.

GRADATION 1. Geologic process of aggradation and degradation that results in the Earth's surface adopting a level or uniform slope; **2.** Process of development of a stream bed during which it adopts a gradient at which the water flowing in it is just able to transport the material delivered to it.

GRADE 1. Ground level; **2.** Percentage of rise or fall to the horizontal distance. *See also* **Profile grade**; **3.** Finished design profile of a roadway or other worked ground; **4.** To cut, fill, and smooth a given area to a given profile; **5.** Designation of quality when assigned a letter, number, or description; **6.** Category or rank used to distinguish items that have the same functional use (e.g. hammer) but do not share the same requirements for quality (e.g. different hammers may need to withstand different amounts of force); **7.** Measure of the quality of an ore.

GRADE AND SLOPE CONTROL Automatic system that controls a machine for longitudinal grade and transverse slope to a consistent profile during a planing/milling pass.

GRADE AND SLOPE EQUIPMENT 1. Instruments that help equipment operators maintain the required slope by either noting the plus-or-minus from a predetermined level or by automatically adjusting the equipment so as to compensate grading error; **2.** Attachments that enable a machine to perform fine grading tasks for which it was not primarily designed.

GRADE ASSISTANCE Assistance given a machine or body when descending a grade, caused by the force of the inclined component of its weight.

GRADE BREAK Change in slope from one incline ratio to another.

GRADE CORRECTION Correction applied to a distance measured on a slope to reduce it to a horizontal distance between vertical lines through its end points.

GRADED AGGREGATE Sample of aggregate made up from particles of different, specified sizes.

GRADED BEDDING Sedimentary structure in which the coarsest material is concentrated at the bottom of a bed, and the average grain size decreases toward the top of the bed.

GRADED FILTER Water filter comprising several layers of material, each more coarse that the layer on which it rests.

GRADED RIVER River whose slope and channel have developed to provide the exact velocity needed to transport the sediment load it carries.

GRADED SLOPE Slope that is dynamically stable and which will maintain itself in the most efficient configuration.

GRADED STANDARD SAND Ottawa sand accurately graded between No. 30 (0.59 mm, 0.0232 in.) and No. 100 (0.149 mm, 0.0059 in.) sieves for use in the testing of cements. *See also* **Ottawa sand,** and **Standard sand.**

GRADED STREAM Stream channel that has reached a stable form resulting from flow characteristics.

GRADE LINE Predetermined line indicating a proposed elevation.

GRADE PERCENTAGE Steepness of a grade, measured by dividing the change in elevation by the horizontal distance.

GRADE PIN Steel rod driven into the ground at a surveyor's hub. A string is stretched between the grade pins at the grade indicated on the survey stakes.

GRADE RING Precast concrete ring at the top of a manhole, used to adjust the top of the manhole so that it is set at the proper angle.

GRADE SECTION Portion of the cross section consisting of cut-and-filled sections graded or shaped to the specified elevations.

GRADE STAKE Stake that indicates the amount of cut (-) or fill (+) at a stated point.

GRADE TUNNEL Waterway tunnel with its top elevation above the hydraulic gradient.

GRADIENT 1. *See* **Expressions of slope;** **2.** Rate of change in a variable over time or a distance, as of temperature or moisture.

GRADIENTER Micrometer fitted to the vertical circle of a transit or level that allows it to be moved through a known small angle and with which an angle of inclination is measured in terms of the tangent of the angle, rather than in degrees, minutes and seconds.

GRADIENT OF SLOPE Decimal form of rise and fall in m/m, e.g. 0.05 is the rise or fall of a slope that has 0.05 m of rise of fall for every horizontal meter of run. *See also* **Expression of slope.**

GRADIENT SPEED Theoretical wind speed at 600 m (2000 ft) that can be calculated from isobars.

GRADING 1. Process of changing the lay of the ground, usually to direct the flow of surface water; **2.** Distribution of particles of granular material among various sizes, usually expressed in terms of cumulative percentages larger or smaller than each of a series of sizes (sieve openings) or the percentages between certain ranges of sizes (sieve openings).

GRADING CURVE Graphical representation of the proportions of different particle sizes in a granular material. See *also* **Fuller's curve.**

GRADING LIMITS Allowable maximum and minimum limits, in a sieve analysis test, for a sample of coarse aggregate.

GRADING PLAN Working drawing showing the existing and proposed vertical dimensions of a site layout, by means of contour lines and spot elevations at high and low points.

GRADIOMETER Surveying level with a telescope that can be angled above and below horizontal to assist in setting out a required gradient.

GRADUAL CONTRACTION Systematic and continuous reduction of the cross-sectional area of a stream, channel, conduit or other hydraulic structure.

GRADUATED ACTING Control instrument that gives throttling control, working between fully on or open and fully off or closed.

GRAIN SIZE CLASSIFICATION Classification of soil based on its grain size.

GRAM Unit of mass, equal to 1/1000 kilogram. Symbol: g. Used in the SI system of measurement. Multiply by 0.035 274 to obtain ounces, symbol: oz; by 0.002 679 to obtain pounds, symbol: lb. *See also the appendix*: **Metric and nonmetric measurement.**

GRAM PER CUBIC CENTIMETER Unit of density. Symbol: g/cm³. Used in the SI system of measurement. *See also the appendix*: **Metric and nonmetric measurement.**

GRAM PER LITRE Unit of density. Symbol: g/L. Permitted for use in the SI system of measurement. *See also the appendix*: **Metric and nonmetric measurement.**

GRAM REACTION Bacteriological staining technique in which Gram's stain is used to distinguish between Gram-positive bacteria (e.g. *Streptococcus, Staphylococcus,* etc.) which retain the stain, and Gram-negative bacteria (e.g. *Gonococcus,* etc.) which do not.

GRANULAR ACTIVATED CARBON TREATMENT Process in which nonvolatile and semi-volatile organic compounds are removed from an air stream through adsorption on the surface of carbon particles.

GRANULAR DUST Mineral filler consisting of finely powdered rock dust.

GRANULAR FILL Crushed stone and other non-cohesive material, used as bedding, and sometimes as cover, for pipes and other building services.

GRANULAR SOIL *See* **Soil types.**

GRANULOMAS Condition caused by multiple accumulation of cells distributed in nodules in the lung, caused by inhalation of certain industrial dusts.

GRAPHITE Crystalline form of carbon.

GRAPPLE Clamshell-type bucket, sometimes with heavy tongs, having three or more jaws and generally used to handle rock or large-size demolition rubble.

GRAPPLE DREDGE Floating derrick equipped with a clamshell, orange-peel, or other type of grab bucket, used to remove material at considerable depths below the water level or in confined places.

GRAPPLE LEG Either of the two main legs of a grapple.

GRAPPLE LINE *See* **Holding line.**

GRAPPLER Wedge-shaped spike at the top end of a bracket scaffold that is driven into a brick joint and to which the scaffold is attached.

GRASS Category of plants related to cereals that represents about 10% of the world's flora. A grass leaf typically consists of a sheath and a blade. There are several types having significance in landscaping and construction, including:

Bentgrass: Strain of grasses normally used on golf courses.

Bluegrass: Grass that spreads by underground stems and makes a thick sod, preferring sunny locations and well-drained soils.

Fescue: Shade-tolerant grass, useful for lawns that receive limited maintenance; used on poor and dry soils.

Ryegrass: Fast-growing, rough grass,

either annual or perennial, used to establish a quick grass cover where appearance is not a major factor.

GRASSED WATERWAY Vegetated natural waterway used to convey accumulated runoff from cultivated land in a strip-crop irrigation system.

GRASSLAND Herbaceous vegetation that is dominated by grasses.

GRATE Device used to support solid fuel or solid waste in a furnace during drying, ignition, and combustion. Openings (tuyeres) are provided for passage of combustion air. There are several types, including:

Chain: Stoker that has a massive moving chain as a grate surface; the grate consists of links mounted on rods to form a continuous belt-like surface that is generally pulled by sprockets on the front shaft.

Dead plate: A stationary grate through which no air passes.

Fixed: A stationary grate.

Movable: A grate designed to feed solid fuel or solid wastes into a furnace and discharge combustion residue. *Also called* **Stoker.**

Oscillating: Where the grate surface oscillates to move the fuel and residue from feed end to discharge.

Reciprocating: Stoker grate surface having alternate lateral stationary and moving rows that reciprocate continuously and slowly, forward and backward, for the purpose of stirring the combustible bed of material while conveying it and the resulting residue from the infeeding end to the discharge end of the furnace.

Rocking: An incinerator stoker with moving and stationary grate bars that are trunnion supported. In operation, the moving bars oscillate on the trunnions, imparting a rocking motion to the bars, thus agitating and conveying the solid fuel and resulting residue through the furnace.

Travelling: A travelling grate stoker consisting of a belt-like arrangement of air-admitting grate bars, similar to a chain grate but with grate bars mounted on transverse beams, usually pulled by chains and sprockets through the furnace.

GRATED WASTE Waste outlet that has a grating over its inlet.

GRATE SIFTINGS Materials that fall from a solid waste fuel bed through the grate openings.

GRATING Screen made up of parallel bars, transverse to each other in the same plane.

GRAVEL Granular material, from 2 mm (0.078 in.) to 64 mm (2.5 in.) in diameter, predominantly retained on the No. 4 (4.76 mm, 0.187 in.) sieve, and resulting either from natural disintegration and abrasion of rock or the processing of weakly bound conglomerate. *See also* **Pit run gravel,** and **Soil groups.**

GRAVEL FILL Layer of crushed rock or gravel, placed to a specific thickness at the bottom of an excavation.

GRAVEL FRACTION Fraction of a soil composed of particles between 60 and 2 mm (2.4 and 0.078 in.).

GRAVEL GUARD Screen mounted in front of a rainwater inlet in a curb to prevent gravel from entering. *Also called* **Ballast grating.**

GRAVEL PACK Gravel screen surrounding a perforated well casing.

GRAVEL PUMP Centrifugal pump used to move hydraulically loosened gravel.

GRAVEL-WALL WELL Well developed through a water-bearing formation containing a high proportion of fine-grained material which permits the passage of water at low velocity. Gravel is introduced around the well screen or intake section of the well to increase the specific capacity and to prevent extremely fine material from flowing into the well.

GRAVIMETRIC Of or pertaining to measurement by weight.

GRAVITATIONAL CONVECTION Motion caused only by density difference within a fluid.

GRAVITATIONAL POTENTIAL Work required to move a mass of water occurring in a saturated column of soil from the level water surface to any point a specific distance above the water surface, against the force of gravity.

GRAVITATIONAL WATER Water that enters and flows through the ground by gravity. This water is free-flowing and can be removed from the ground by pumping. *See also* **Absorbed water, Capillary water,** and **Moisture content of soils.**

GRAVITY Force that attracts matter toward the center of the Earth and gives it weight.

GRAVITY CIRCULATION Hot water plumbing circuit that uses temperature differential to move hot water around the piping system.

GRAVITY DAM Dam that relies solely on its own mass to resist lateral pressure.

GRAVITY DRAINAGE Drainage of water by gravity.

GRAVITY FEED Any device that transports materials from one level to another by the force of gravity; a chute.

GRAVITY FILTER Open type, rapid sand filter, the operating level of which is near the hydraulic grade line of the influent and through which water flows by gravity.

GRAVITY FLOW Movement of water in a sloping gutter, channel, drain, etc., which goes downslope under the effect of gravity.

GRAVITY GROUNDWATER Water that is withdrawn from a body of rock or soil in the saturation zone by the force of gravity.

GRAVITY MAIN Pipeline in which water flows downhill from an impounding reservoir to a service reservoir.

GRAVITY PIPE Pipe not intended to take water under pressure.

GRAVITY SEPARATION METHODS Treatment of mineral particles that exploits differences between the specific gravity of each. The separation is usually performed by means of sluices, jigs, classifiers, spirals, hydrocyclones, or shaking tables.

GRAVITY SPRING Spring flow from permeable material or rock openings produced entirely by gravity.

GRAVITY SUPPLY Water distribution system that relies on gravity fall to provide water to all outlets.

GRAVITY SYSTEM Water-distribution system in which no pumping is necessary.

GRAVITY TANK Water storage tank that supplies a flow by gravity pressure.

GRAVITY WATER Water supply that is transported from its source to its place of use by gravity.

GRAY SI unit of absorbed dose of ionizing radiation equal to one joule per kilogram. (1 Gy=1 J/kg). Symbol: Gy. *See also the appendix*: **Metric and nonmetric measurement.**

GRAY WATER Waste water containing any combination of substances excepting human wastes.

GREASE Range of substances including fats, waxes, free fatty acids, calcium and magnesium soaps, mineral oils, etc.

GREASE REMOVAL TANK Tank that promotes the flotation of oil and grease and equipped with a mechanism for its removal.

GREASE SKIMMER Mechanism for removing grease and/or scum from the water surface of wastewater treatment tanks.

GREASE TRAP Drum-type trap installed in a drainage line to separate and collect grease from the flow being conveyed.

GREEN BELT Land, often designated as parkland, near to, and sometimes surrounding a large, developed area, in which further development is either prohibited or severely restricted to a few types in order to preserve its open nature.

GREENHOUSE EFFECT Solar radiation admitted to the Earth's surface that cannot then exit the atmosphere for any reason; within a building, the solar radiation admitted through glass that are transformed to heat waves that cannot pass back through the glass.

GREENHOUSE GASES Gases in the atmosphere that contribute to the greenhouse effect, including carbon dioxide, methane, nitrous oxide and water vapor.

GREEN MANURE Plant crop grown specifically for the purpose of plowing or digging it back into the soil to improve the soil and provide nutrients.

GREENSAND Sand composed mostly of particles of the mineral glauconite, a hydrous potassium iron silicate.

GREEN SLUDGE Sewage sludge in which little bacterial decomposition has taken place.

GREEN VITRIOL Common name for copperas: ferrous sulfate.

GRID Distribution network for public services.

GRIDIRON Water distribution system in which all pipes are connected with all other pipes at street intersections so that fire supply at any location draws from all directions.

GRID ROLLER Soil compaction roller consisting of rolls made up of 38 mm (1.5 in.) diameter steel bars at 130 mm (5 in.) centers.

GRID SYSTEM WATER MAINS Interconnecting system of water mains in a crisscross or rectangular pattern.

GRIP Small channel cut into the uphill side of an excavation to deflect groundwater runoff.

GRIT Heavy suspended mineral matter that will be retained on a #200 mesh having an aperture of 76 μm, present in water or wastewater, and which will precipitate at zero flow and turbulence.

GRIT CATCHER Chamber at the upper end of a depressed sewer, or at other points on combined or storm sewers where wear

from grit is possible, the size and shape of which causes a velocity reduction sufficient for grit to settle out into a sump.

GRIT CHAMBER Chamber in a storm sewer below the invert and discharge levels, used to reduce flow velocity and permit settling of particulate matter. *Also called* **Grit trap.**

GRIT CHANNEL Enlargement in a sewer causing a reduction in rate of flow with the opportunity for the deposition of grit.

GRIT COLLECTOR Mechanism in a grit chamber that removes deposited grit to a collection point.

GRIT COMPARTMENT Section of a grit chamber in which grit is collected and stored prior to removal.

GRIT REMOVAL Mechanical, or more commonly, hydraulic means by which grit in suspension in a stream is isolated and removed.

GRIT TRAP *See* **Grit chamber.**

GRIT WASHER Device for washing organic matter out of grit.

GROIN, or GROYNE Framework or low broad wall run out from a shore to check lateral drifting of the beach and to deter erosion by wave and tidal action. *Also called* **Spur dike.**

GROOVED COUPLING Clamp-type gasketed pipe coupling having a groove machined or cast into its internal circumference; it fits over flanges cast on the exterior circumference of pipe ends.

GROSS ALPHA PARTICLE ACTIVITY Total radioactivity due to alpha particle emissions as inferred from measurements on a dry sample.

GROSS AVAILABLE HEAD Difference in elevation between the point where water is drawn from a stream and the point where it is returned to the same stream, or freely discharged.

GROSS DUTY OF WATER Water requirement or duty, including canal or conduit losses, seepage losses, evaporation and transpiration losses, and all waste, measured at the point of diversion. *Also called* **Diversion duty of water,** and **Head-gate duty of water.**

GROSS HORSEPOWER *See* **Horsepower.**

GROUND AIR Gases in the interstices of the aeration zone that open directly or indirectly to the surface. *Also called* **Soil air,** and **Subsurface air.**

GROUND COVER Vegetation that produces a protective mat on or just above the soil surface.

GROUND DUCT Concrete channel between buildings that contains pipes, cables and building services, etc.

GROUND GARBAGE Garbage that has been shredded prior to disposal.

GROUND ICE Ice formed on the ground by:

a) Fall of drizzle or rain, non-supercooled but subsequently frozen;

b) Layer or part layer of snow, completely or partly melted and frozen again; or

c) Compressed layer of snow, compacted hard by traffic.

GROUND KEY VALVE Valve that can be closed or opened to full flow in a 90° turn of the plug.

GROUND-LEVEL CONCENTRATION Concentration of a pollutant between ground height and approximately 6 ft (2 m) to which a human is normally exposed.

GROUND-LEVEL STORAGE Water distribution storage facility consisting of a shallow tank, the bottom of which is at or below ground level.

GROUND MORAINE Moraine occurring at the bottom of a glacier, consisting of material frozen in the bottom of the glacier and pushed along over its bed. When the glacier melts, this material is deposited in a relatively thin sheet over the area formerly occu-

pied by the ice.

GROUND SILLS Underwater walls built at intervals across the bed of a channel so as to prevent excessive scour or to increase the width of the water surface.

GROUNDWATER Underground water that has collected due to porosity and fissuring of rocks and that is contained by an underlying impervious strata. There are several classifications, including:

Confined: Groundwater that is under pressure substantially greater than atmospheric.

Free: Groundwater in aquifers not bounded or confined by impervious strata.

Juvenile: Geologic water confined in voids in the Earth that are not subject to recharge.

Perched: Unconfined groundwater separated from an underlying body of groundwater by an unsaturated zone.

Unconfined: Water in an aquifer that has a watertable.

GROUNDWATER ARTERY Body of permeable material encased in a matrix of less permeable or impermeable material and saturated with water under pressure, usually artesian.

GROUNDWATER BASIN Pervious formation with sides and bottom of relatively impervious material in which groundwater is held or retained.

GROUNDWATER CASCADE Descent of groundwater on a steep hydraulic gradient to a lower and flatter watertable slope.

GROUNDWATER DAM Underground dam designed to prevent the flow of groundwater from one area to another; to prevent the flow of contaminated groundwater into the downstream aquifer.

GROUNDWATER DECREMENT All groundwater extracted from a subsurface reservoir beneath a given surface area by evaporation, transpiration, spring flow, effluent seepage, pumpage, etc.

GROUNDWATER DEPLETION CURVE
Graphical presentation of recession after passage of the flow created by direct runoff.

GROUNDWATER DISCHARGE Rate at which water flows from a saturated zone to an unsaturated zone or is lost at the surface.

GROUNDWATER DISCHARGE AREA Area where groundwater is discharged by various means.

GROUNDWATER DIVIDE Line representing the underground division between two or more groundwater basins.

GROUNDWATER DRAIN Drain through which groundwater is taken from an area.

GROUNDWATER FLOW Rate of flow of water in an aquifer or soil; that portion of the discharge of a stream that is derived from groundwater.

GROUNDWATER HYDROLOGY Branch of hydrology that deals with groundwater.

GROUNDWATER INCREMENT Water added to a groundwater reservoir from all sources.

GROUNDWATER INFILTRATION Water that enters a treatment facility as a result of the interception of natural springs, aquifers, or runoff which percolates into the ground and seeps into the treatment facility's tailings pond or wastewater holding facility and which cannot be diverted by ditching or grouting the tailings pond or wastewater holding facility.

GROUNDWATER INVENTORY Estimate of the quantities of water stored within a defined subterranean area.

GROUNDWATER IRRIGATION Irrigation by water derived from underground sources.

GROUNDWATER LEVEL Level below which the rock and subsoil are saturated with water.

GROUNDWATER LOWERING Lowering the level of groundwater by pumping to create a dry excavation.

GROUNDWATER MINING Withdrawal of groundwater at a rate exceeding that of recharge.

GROUNDWATER PRESSURE HEAD At a point, the hydrostatic pressure expressed as the height of a column of water that can be supported by the pressure, or the distance that a column of water rises in a tightly cased well where there is no discharge.

GROUNDWATER PROVINCE Area characterized by a general similarity in the occurrence of groundwater.

GROUNDWATER RECESSION Lowering of the groundwater level in an area below an established historic level.

GROUNDWATER RECHARGE Rate at which water is added to the watertable from an unsaturated zone.

GROUNDWATER RESERVOIR Natural or artificial reservoir in which groundwater is stored for future extraction and use.

GROUNDWATER RUNOFF That part of groundwater that is discharged into a stream channel as spring or seepage water.

GROUNDWATER STORAGE Water temporarily stored within permeable rock.

GROUNDWATER-STORAGE CURVE Curve expressing the area under the groundwater-depletion curve showing the volume of groundwater available for runoff at given rates of groundwater flow.

GROUNDWATER TABLE Upper surface of the zone of saturation in permeable rock or soil.

GROUP I STORM WATER DISCHARGE Storm water point source that is subject to effluent limitations guidelines, new source performance standards, or toxic pollutant effluent standards or located at an industrial plant or in plant associated areas.

GROUP VENT Branch vent to a drain that serves two or more traps.

GROWTH Relative measure of increase in mass or size by cell division or enlargement.

GROWTH RATE Experimentally determined constant, typically used to estimate the unit growth rate of bacteria while degrading organic wastes.

GRUBBING Removal of roots and stumps.

GUANO 1. Manure of sea birds; **2.** Artificial fertilizer made from fish.

GUARANTEED MAXIMUM COST Amount established in a contract between an owner and contractor setting out the maximum cost of completing specified work. *Also called* **Upset price.**

GUARANTEED MAXIMUM PRICE Maximum dollar amount for which the owner will be liable to the contractor for performance of all work.

GUARANTY Pledge by which a person commits to the payment of another's debt or the fulfillment of another's obligation in case of default. *See also* **Guarantee.**

GUARANTY BOND *See* **Bond.**

GUARD Any device created or installed to protect a worker or equipment.

GUARD BOARD Board placed on edge at the perimeter of scaffolding to prevent objects from falling off.

GUARDED Covered, shielded, or otherwise protected by means of suitable covers, casings, barriers, rails, screens, mats, or platforms to remove the likelihood of approach to a point of danger or contact by persons of objects.

GUARD GATE Additional pair of gates installed in front of lock gates for emergency use.

GUARD LOCK Lock separating a dock from tidal water.

GUARDRAIL 1. Safety barrier at the edge of an elevated platform or flight of stairs; **2.** Traffic barriers, used to shield hazardous areas from errant vehicles.

GUIDE BANK Protective training embankment to guide a river through its waterway.

GUIDE SPECIFICATION General specification — often referred to as a design standard or design guideline — which is a model standard and is suggested or required for use in the design of all of the construction projects of an agency.

GUIDE VANE Fixed or adjustable device by which the flow of liquid or air can be directed.

GUIDE WALL Temporary, shallow-depth wall constructed in soft or unstable soil in advance of a deeper, permanent wall.

GULLY Outlet into a drain for water from a floor in a wet area.

GUMBO Soil that contains much silt and clay and which becomes very sticky when wet.

GUST Short burst of strong wind.

GUSTINESS Turbulence close to the ground, caused by obstructions such as buildings that prevent the direct flow of air.

GUT Narrow passage or contracted straight between two bodies of water.

GUTTATION Excretion of drops of excess water from glands on the leaves of many plants.

GUTTER Built or formed trough at the edge of pavement to collect and convey runoff.

GUTTER TOOL Tool used to give the desired shape and finish to concrete gutters.

GUYOT Flat-topped sea mount.

GYPSUM Hydrate calcium sulfate $(CaSO_4 \cdot 2H_2O)$.

H

HABIT Characteristic form, mode of growth, etc., of an animal or plant.

HABITAT Place where an animal or plant lives or grows, providing a particular set of environmental conditions.

HABITAT LOSS Significant change in the natural environment that causes the disappearance of flora and fauna, or a change in the natural balance and composition of these elements.

HABITAT MANAGEMENT Management of the forest to create environments that provide habitats (food, shelter) to meet the needs of particular species of wildlife, birds, etc.

HABITUATION Diminishing response to repeated stimulation.

HEMOGLOBIN Iron-containing respiratory pigment found in the blood of vertebrates.

HEMOLYSIS Breaking of blood corpuscles by the action of a poisonous substance.

HAGEN–POISEUILLE LAW Friction factor of Darcey's formula, or the ratio of 64 to the Reynold's numbers when flow is lamina.

HAIL Precipitation in the form of ice pellets or balls formed by the freezing of atmospheric vapor.

HALF-LIFE Time that it takes for a substance to be reduced to one half of its origi-

nal volume, effectiveness, or other measure, including:

> **Radioactivity** – Time it takes for half the atoms in a radioactive substance to decay to atoms of a different mass, which varies from millionths of a second to billions of years.

> **Ingested material – 1.** Time necessary for half the quantity of ingested material to be eliminated from the body naturally; **2.** Time needed for ingested radiobiological material to deliver half of its radiation dose.

> **Environmental** – Time taken for half the quantity of a substance to disappear by any means from the environment.

HALF-ROUND PIPE Pipe, cut in two lengthways.

HALF-TIDE LEVEL Elevation exactly half way between mean high, and mean low water.

HALF-VALUE LAYER Thickness of a material that will reduce the intensity of a beam of radiation to one-half of its value on impacting the material.

HALO Series of colored rings appearing around the sun or moon when it seen through a cloud or ice crystals suspended in the atmosphere.

HALOCLINE Boundary between two bodies of water of differing salinity.

HALOGEN One of the chemical elements chlorine, bromine or iodine.

HALOGENATED FLUOROCARBONS Group name for ethane- or methane-based compounds, in which some or all of the hydrogen in their structures is replaced by chlorine, bromine and/or fluorine.

HALOGENATION Reaction with a member of the halogen group, i.e. bromine, chlorine, fluorine, or iodine.

HALOMORPHIC SOIL Soil containing an excess of salt or an alkali.

HALON Halogenated fluorocarbon containing bromine.

HALOPHYTE Plant that grows in soil containing a high concentration of salt.

HAMADA Desert region where the surface is bedrock.

HAMMERMILL Machine in which swinging hammers are used to crush material against a grid of steel bars.

HAND BORING Site investigation technique of boring a hole in self-supporting soil using a shell and hand auger, down to a maximum depth of 4 m (13 ft).

HANDLING TIGHT Pipe and couplings screwed together by hand so tight, that they require a wrench to loosen.

HAND WHEEL Wheel on top of a valve stem that is turned by hand to open or close it.

HANGER Support specially designed to support pipe lines.

HANGING VALLEY Valley formed by a tributary glacier, which flowed into a main glacier so that the surfaces of the two ice masses were level, but the bed of the tributary was at a higher level than that of the main glacier.

HANG-UP Process of hydrocarbon molecules being adsorbed, condensed, or by any other method removed from a sample flow prior to reaching the instrument detector. It also refers to any subsequent desorption of the molecules into the sample flow when they are assumed to be absent.

HARBOR Sheltered water with access to the open sea, or a channel leading to the open sea.

HARD HAT Protective head gear.

HARDNESS 1. Resistance to deformation of a material; **2.** Total concentration of the calcium and magnesium ions in water expressed as calcium carbonate (mg $CaCO_3$/liter). Depending on the interaction of other factors such as pH and alkalinity, hardness levels between 80 and 100 mg/L are consid-

ered to provide an acceptable balance between corrosion and incrustation.

HARDPAN Dense, hard layer in the subsoil caused by cementation of soil particles with organic matter that obstructs penetration of roots and water.

HARD SUCTION Non-collapsible suction hose, used for drafting water from static sources lower than the pump inlet.

HARD WATER Water containing calcium, and magnesium salts in solution, and iron such as bicarbonates, sulfates, chlorides and nitrates..

HARDWOOD Any hard, compact wood; any tree that has broad leaves or does not have needles.

HARMONIC Of, or having to do with any of the frequencies making up a wave or alternating current, that are integral multiples of the fundamental frequency.

HASS Inside curve of a bent pipe.

HAUL DISTANCE 1. Distance along the most direct and/or practical route between the center of the area to be excavated and the center of the area to be filled; **2.** For refuse collection:

    (a) The distance a collection vehicle travels from its last pickup stop to the solid waste transfer station, processing facility, or sanitary landfill.

    (b) The distance a vehicle travels from a solid waste transfer station or processing facility to a point of final disposal.

    (c) The distance that cover material must be transported from an excavation or stockpile to the working face of a sanitary landfill.

HAUL TIME Time it takes to travel from the load area to the dump area.

HAUNCH Sides and lower third of the circumference of a pipe.

HAUNCHING Concrete support to the sides of a drain or sewer pipe above the bedding.

HAZARD 1. Condition that presents a threat to normal expectations; **2.** Likelihood that use of a pesticide would result in an adverse effect on man or the environment in a given situation.

HAZARD CATEGORY Method of listing actual or potential hazards, including:

(a) Immediate (acute) health hazard, including highly toxic, toxic, irritant, sensitizing, corrosive and other hazardous chemicals that cause an adverse effect to a target organ, and which effect usually occurs rapidly as a result of short-term exposure and is of short duration;

(b) Delayed (chronic) health hazard, including carcinogens and other hazardous chemicals that cause an adverse effect to a target organ, and which effect generally occurs as a result of long-term exposure and is of long duration;

(c) Fire hazard, including flammable, combustible liquid, pyrophoric, and oxidizer;

(d) Sudden release of pressure, including explosive and compressed gas; and

(e) Reactive, including unstable reactive, organic peroxide, and water reactive.

HAZARD INSURANCE Insurance that protects against property damages due to fire, windstorm, and other common hazards.

HAZARDOUS ATMOSPHERE Atmosphere that, by reason of being explosive, flammable, poisonous, corrosive, oxidizing, irritating, oxygen deficient, toxic, or otherwise harmful, may cause death, illness, or injury.

HAZARDOUS MATERIAL In the context of materials transportation, any shipped product that is potentially injurious to people and property.

HAZARDOUS POLLUTANT One to which even slight exposure may cause serious illness or death.

HAZARDOUS SUBSTANCE Substance that, by reason of being explosive, flammable, poisonous, corrosive, oxidizing, irritating, or otherwise harmful, is likely to cause death or injury.

HAZARDOUS WASTE Any waste or combination of wastes which pose a substantial present or potential hazard to human health or living organisms by being nondegradable or persistent in nature or because they can be biologically magnified, or because they can be lethal, or because they may otherwise cause or tend to cause detrimental cumulative effects.

HAZARDOUS WASTE INCINERATION Incineration to values of 99.999% destruction and removal efficiency, under strictly controlled conditions, of wastes deemed to be hazardous.

HAZARDOUS WASTE MANAGEMENT FACILITY All contiguous land, and structures, other appurtenances, and improvements on the land used for treating, storing, or disposing of hazardous waste. A facility may consist of several treatment, storage, or disposal operational units (for example, one or more landfills, surface impoundments, or combination of them).

HAZARDOUS WASTE MANAGEMENT UNIT Contiguous area of land on or in which hazardous waste is placed, or the largest area in which there is significant likelihood of mixing hazardous waste constituents in the same area. Examples of hazardous waste management units include a surface impoundment, waste pile, land treatment area, landfill cell, incinerator, a tank and its associated piping and underlying containment system and a container storage area.

HAZARD REDUCTION Treatment of a hazard that reduces or eliminates threat.

HAZE Small amount of mist, dust, smoke, etc., in the atmosphere, often suspended in water droplets in suspension.

HAZEN NUMBER Unit of measurement for the color of water based on the color produced by 1 mg of platinum per litre in the presence of a cobalt-based compound.

HEAD 1. Pressure resulting from a column of water or an elevated supply of water, within a plumbing system or in the ground, consisting of the sum of static head, velocity head, and friction head. There are several definitions, including:

Dynamic: Maximum elevation that water within a system will reach due to pump pressure, or to induced pressure from a contained source (reservoir or header tank).

Static: Height of a column or body of fluid above a given point.

Static discharge: Static head from the centerline of a pump to the free discharge surface.

Total: Sum of the static and velocity heads of a flowing fluid at the point of measurement. *Also called* **Dynamic head.**

Velocity: Equivalent head through which a liquid would have to fall to attain a given velocity.

2. Back-pressure against a pump from an elevated outlet.

HEAD BAY Widened part of a canal, upstream from the lock gates.

HEAD BOX Chamber in which chemical slurries and liquids can be prepared in anticipation of their distribution into a process stream.

HEADER 1. Water supply pipe to which two or more branch pipes are connected to service fixtures; **2.** Manifold or supply pipe to which a number of branch pipes are connected.

HEAD FLUME Channel formed at the head of a gully or terrace to prevent scouring by running water.

HEAD GATE Upstream gate of a lock, irrigation system, or conduit.

HEAD-GATE DUTY OF WATER Water requirement or duty, including canal or conduit losses, seepage losses, evaporation and transpiration losses, and all waste, measured at the point of diversion. *Also called* **Diversion duty of water,** and **Gross duty of wa-**ter.

HEAD INCREASER Device associated with a draft tube to increase discharge by reducing tailwater pressure.

HEADING Point at which a canal or pipeline diverts water from a stream or other body.

HEAD LOSS Drop in air or water pressure from partly blocked flow.

HEAD PRESSURE Pressure developed by a standing column of water; the pressure a pump must overcome before raising water to a higher level.

HEAD RACE Tunnel or open channel connecting a head pond to a penstock.

HEAD TANK Elevated water storage tank that pressurizes water distribution pipes. *Also called* **Header tank.**

HEADWALL Wall at the end of a culvert or drain that may increase the hydraulic efficiency of the conduit, divert the direction of flow, serve as a retaining wall, or help prevent scour.

HEADWATER 1. Retained water upstream of a structure; **2.** Upper reaches of a stream between its source and a tributary; **3.** Region where groundwater emerges to form a surface stream.

HEADWATER CONTROL 1. Regulation of the depth of water in an impounding structure either to ensure sufficient capacity for potential floodwater flows, or to provide adequate downstream flows; **2.** Component of flood protection involving the control of headwaters, the peripheral upland drainage areas, and the smaller tributaries of a major river system.

HEADWORKS 1. Facilities where wastewater enters a wastewater treatment plant, **2.** Shape at the head of a channel that diverts water into it.

HEALTH AND SAFETY PLAN Plan that specifies the procedures that are sufficient to protect onsite personnel and surrounding communities from the physical, chemical, and/or biological hazards of a site. A health and safety plan outlines:

(a) Site hazards;

(b) Work areas and site control procedures;

(c) Air surveillance procedures;

(d) Levels of protection;

(e) Decontamination and site emergency plans;

(f) Arrangements for weather-related problems; and

(g) Responsibilities for implementing the health and safety plan.

HEALTH AND SAFETY STUDY Study of any effect of a chemical substance or mixture on health or the environment, or on both, including underlying data and epidemiological studies, studies of occupational exposure to a chemical substance or mixture, toxicological, clinical, and ecological or other studies of a chemical substance or mixture.

HEALTH RISK ASSESSMENT Scientific evaluation of the risk to a population's health resulting from certain actions or conditions.

HEALTH HAZARD Chemical for which there is statistically significant evidence based on at least one study conducted in accordance with established scientific principles that acute or chronic health effects may occur in exposed employees.

HEARING Sense by which sound is perceived which, in humans, encompasses frequencies from approximately 20 to 20 000 hertz (Hz).

HEARING LOSS Inability of an individual to detect some or all of the sounds within the normally-audible frequency range, or volume.

HEARING PROTECTIVE DEVICE Device or material, capable of being worn on the head or in the ear canal, that is sold wholly or in part on the basis of its ability to reduce the level of sound entering the ear. This includes devices of which hearing protection may not be the primary function, but

which are nonetheless sold partially as providing hearing protection to the user. This term is used interchangeably with the terms, hearing protector and device.

HEARTWOOD Central part of a tree which no longer contains living cells.

HEAT Temperatures above ambient, as produced by burning or oxidation. There are a number of measures, including:

Humid heat: Ratio increase of enthalpy per pound of dry air with its associated moisture to rise of temperature, under conditions of constant pressure and constant specific humidity.

Latent heat: Heat absorbed or given off by a substance without changing its temperature.

Sensible heat: Heat, the addition or removal of which results in a change in temperature, as opposed to latent heat.

Specific heat: Amount of heat required per unit mass to cause a unit rise of temperature over a small range.

HEAT-ABSORBING GLASS Glass that is substantially opaque to infrared radiation, including shortwave infrared, but having a moderately high degree of transparency to most visible radiation.

HEAT-ACTUATED FIRE DOOR *See* **Fire door.**

HEAT ACTUATING DEVICE Thermostatically-controlled device used to activate fire equipment, alarms, or appliances.

HEAT BALANCE 1. Condition of thermal equilibrium within a space where heat gain is equal to heat loss; **2.** Calculation of the efficiency of a combustion process in which all heat losses (expressed as percentages) are totalled and subtracted from 100; the remaining figure represents the percent efficiency.

HEAT BRIDGE Material capable of conducting heat between two others.

HEAT BUDGET Heat necessary to raise a body of water from the minimum winter temperature to the maximum summer temperature.

HEAT CAPACITY Heat required to raise the temperature of a given mass by one degree.

HEAT CONDUCTOR Material capable of readily conducting heat at a constant rate.

HEAT CONTENT Sum total of latent and sensible heat stored in a substance minus that contained at an arbitrary set of conditions chosen as the base or zero point.

HEAT DETECTOR Device for sensing an abnormally high air temperature or an abnormal rate of heat rise and automatically initiating a signal indicating this condition.

HEAT ENDURANCE Time that a sample can retain its original characteristics and physical properties at a specific temperature.

HEATER Device that produces heat.

HEAT EXCHANGER Device that transfers heat from a warmer to a cooler medium, normally by conduction.

HEAT EXCHANGER TANK Tank used for heating sludge consisting of steam-heated coils, and equipment for diffusion air to pass through the contents, to promote mixing and heat transfer.

HEAT FLUX Rate of flow of heat through a unit area.

HEATING AND VENTILATING ENGINEER One who is professionally qualified to design building services.

HEATING BATTERY Heat exchanger that warms air ducted across hot water pipes.

HEATING CAPACITY Ability of a water heater to raise a given number of liters (gallons) per hour by a specified number of degrees. *Also called* **Recovery capacity.**

HEATING COIL 1. Resistance element that glows red when an electrical current is passed through it; **2.** Coiled pipe through which hot water flows and over which air is blown in a heat exchanger.

HEATING ELEMENT Electrical resistance coil used to heat water or air.

HEATING LOAD Amount of heat required to maintain a desired temperature within the warmed area.

HEATING MEDIUM Any solid or fluid used to convey heat from a source to or through a heat distribution device.

HEATING PLANT Assembly of equipment and its controls designed to produce heat that is subsequently conveyed to the areas of a building, or group of buildings.

HEATING RATE Rate, expressed in degrees per hour, at which the temperature of an object or area being heated increases.

HEATING SEASON Period of the year when it is necessary to operate heating equipment for indoor comfort, sometimes established by local regulation.

HEATING SYSTEM Complete works and installation necessary to generate heat, regulate and deliver it to the areas required to be warmed. There are many types, including:

> **Buried-cable:** Heating system in which electrical resistance cable is buried in a ground-floor concrete slab.

> **Electric:** System where heat is provided from the installation of electric resistance heaters at strategic locations.

> **High-pressure steam:** Steam heating system employing steam at pressures above 1.05 kg/cm² (15 psig).

> **High-pressure water:** Heating system in which water, having a supply temperature above 176°C (350°F), is used as a medium to convey heat from a central boiler, through a piping system, to radiators.

> **Hot water:** Heating system in which water, having supply temperatures less than 121 °C (250°F) is used as a medium to convey heat from a central boiler, through a piping system, to radiators, either by gravity or by a circulating pump.

> **Low-pressure steam:** Steam heating sys-

tem employing steam at pressures between 0 and 1.05 kg.m² (0 and 15 psig).

Medium-temperature water: Heating system in which water having supply temperatures between 121°C and 176°C (250°F and 350°F) is used as a heating medium.

Off-peak, electric: System where electrical resistance cable is buried within a ground-floor concrete slab and operated by a separately metered circuit that operates only during off-peak load hours of the distribution system.

Panel: Heating system consisting of coils or ducts installed in wall, floor, or ceiling panels to provide a large surface of low-intensity heat in which heat is transmitted by both radiation and convection from the panel surfaces to both air and surrounding surfaces.

Perimeter: Heating system where the radiators are placed adjacent to the exterior walls, usually under windows.

Perimeter warm air: Heating system in which warmed air is conducted via ducts cast in a ground slab to registers placed at intervals around the perimeter of the area to be warmed.

Radiant: Heating method usually consisting of coils, pipes, or electric heating elements placed in or mounted on the floor, wall, or ceiling in which only the radiated heat is effective in providing the heating requirements.

Split: System in which heating is accomplished by radiators or convectors supplemented by mechanical air circulation.

Steam: Heating system in which steam is used as a heat transfer medium, at or above atmospheric pressure.

Vacuum: Method of building heating in which a vacuum pump connected to the return main removes moisture and air from the radiators and returns water to the boiler feed tank; a two-pipe heating system that operates below atmospheric pressure.

Vapor: Steam heating system in which the primary supply is at or near atmospheric pressure and the return system conveys the condensate to the boiler by gravity.

Warm air: System in which warmed air is ducted to points of application and vented to the atmosphere.

HEATING VALUE Heat released by combustion of a unit quantity of fuel, measured either in calories or Btu.

HEATING, VENTILATING AND AIR-CONDITIONING Mechanical services that depend upon each other.

HEAT INSULATION Ability of a material to impede heat flow.

HEAT LOAD Sum of the amount of heat that has to be generated and/or transferred to a given volumetric space to meet design requirements and comfort levels.

HEAT LOSS Heat lost through openings around doors, windows, etc. and due to the lack of insulation or the inefficiency of insulation.

HEAT MIRROR Plastic film, fully or partially reflective, placed on or between panes of glass to increase their insulating properties.

HEAT OF COMBUSTION Heat released by combustion of a unit quantity of a fuel.

HEAT OF FORMATION Heat evolved when 1 gram-molecule of a compound is formed from its constituent elements.

HEAT OF FUSION Heat needed to convert a given weight of a substance to its liquid phase.

HEAT OF HYDRATION Heat evolved by chemical reactions with water, such as that evolved during the setting and hardening of Portland cement, or the difference between the heat of solution of dry cement and that of partially hydrated cement. *See also* **Heat of**

solution.

HEAT OF SOLUTION Heat evolved or absorbed when a substance is dissolved in a solvent. *See also* **Heat of hydration.**

HEAT PUMP Heating device that extracts and concentrates usable heat from a medium like air or water. In its reverse mode it can be used for cooling.

HEAT RECOVERY Use of a heat exchanger to recapture heat that would otherwise be lost from ventilation to atmosphere.

HEAT RELEASE RATE Amount of heat liberated in a chamber during complete combustion.

HEAT RESISTANCE Property or ability to resist the deteriorating effects of elevated temperatures.

HEAT TRANSFER Movement and dispersion of heat from its source, by convection, radiation, conduction, or transfer.

HEAT TRANSFER LIQUID Liquid used to transport thermal energy.

HEAT TRANSMISSION Heat flow over time.

HEAT TRANSMISSION COEFFICIENT Coefficient used in the calculation of heat transmission by conduction, convection, or radiation.

HEAT TREATMENT Application of heat in order to change the state or condition of something.

HEAVE 1. Upward movement of soil and/or structures supported on soil caused by swelling of the soil from the freeze/thaw cycle; **2.** Swelling due to increased moisture content, release of overburden pressures, or oxidation of sulfite soils exposed to air; **3.** Displacement of earth adjacent to structures driven into the earth; **4.** Pressure on the underside of basements built into the watertable, or into a zone subject to the influence of tides, that can result in flotation.

HEAVY-DUTY VEHICLE Motor vehicle rated at more than 3056 kg (8500 lb) GVRW or that has a curb weight of more than 2712

kg (6000 lb) or a basic frontal area in excess of 4.18 m² (45 ft²).

HEAVY LIQUIDS Group of dense liquids that are used to separate out accessory minerals.

HEAVY METAL Uranium, plutonium, or thorium placed into a nuclear reactor.

HEAVY METALS 1. Metallic elements of higher atomic weights, including but not limited to arsenic, cadmium, copper, lead, mercury, manganese, zinc, chromium, tin, thallium, and selenium; **2.** Trace emissions that make up the solid particulate emissions and residues from combustion processes of incineration. The emissions are dependent on the composition of the incoming waste stream but can include silver, chromium, lead, tin, zinc, cadmium, mercury, and nickel.

HEAVY MINERALS Detrital accessory minerals having specific gravity of 2.87, or more.

HEAVY SLUDGE Sludge that, while still fluid, has a relatively low moisture content.

HEAVY SPECIALIZED CARRIER Trucking company equipped and authorized to transport equipment and goods that, because of inherent characteristics of size, shape, weight, etc., require special equipment for loading, unloading, or transporting.

HEAVY WATER Water with an isotope of hydrogen having an atomic weight of 2 instead of 1.008; called deuterium oxide or dideutohydrogen oxide.

HECTARE Non-SI unit of area, equal to 10 000 m², permitted for use with the SI system of measurement. Symbol: ha. Multiply by 2.471 to obtain acres, symbol: ac. *See also the appendix*: **Metric and nonmetric measurement.**

HECTO Prefix representing 10². Symbol: h. Used in the SI system of measurement. *See also the appendix*: **Metric and nonmetric measurement.**

HEIGHT OF DAM Difference in elevation between the lowest part of the excavated foundation along the axis of a fixed dam and the highest part of the physical structure, regardless of the level of the impounded water.

HELD WATER Water above the standing watertable. *Also called* **Water of capillarity.**

HELIOTHERMOMETER Instrument that measures the Sun's heat using a thermoelectric junction. *See also* **Pyrheliometer.**

HELIUM Inert gaseous element (He), present in the atmosphere at about 1 part in 200 000, and in some natural gases.

HENRY SI unit of inductance of a closed circuit which gives rise to a magnetic flux of 1 Wb/A. (1 H=1 Wb/A). Symbol: H. *See also the appendix*: **Metric and nonmetric measurement.**

HEPA FILTER High efficiency particulate absolute filter capable of trapping and retaining at least 99.9% of asbestos fibers greater than 0.3 microns in length.

HEPATITIS Disease contracted from blood or the ingestion of excess alcohol or virus-contaminated water and which causes inflammation of the liver.

HERBICIDE Chemical used to kill or retard the growth of plants; a weedkiller.

HERBICIDES Substances or mixtures of substances, except defoliants, desiccants, plant regulators, and slimicides intended for use in preventing or inhibiting the growth of, or killing or destroying plants and plant parts.

HERRINGBONE DRAIN Stone- and gravel-filled trenches laid out in alternating diagonals. *Also called* **Chevron drain.**

HERTZ SI system unit of frequency equal to a periodic occurrence which has a period of one second. (1 Hz=1 s^{-1}). Symbol: Hz. *See also the appendix*: **Metric and nonmetric measurement.**

HETEROTROPHIC ORGANISM Organism that requires sustenance (food) from the environment, i.e. all animals, fungi, yeasts and bacteria.

HICKEY Hand tool with side-opening jaws, used in developing leverage for making bends in bars or pipes in place.

HIGH ALTITUDE Elevation over 1219 meters (4000 feet).

HIGH-CALCIUM LIME Lime containing 95% to 98% calcium oxide.

HIGH-CAPACITY FILTER High-rate filter.

HIGH DUMPER Dump body and hoist combination giving a high angle of elevation, to the extent that retractive power is required for lowering the body rather than the usual gravity return.

HIGH DUTY OF WATER Condition where a given quantity of water will irrigate a relatively large area of land.

HIGH-EFFICIENCY PARTICULATE AIR (HEPA) FILTER Filter capable of trapping and retaining at least 99.97% of all monodispersed particles of 0.3 micrometer in diameter or larger.

HIGH HAZARD INDUSTRIAL OCCUPANCY *See* **Industrial occupancy.**

HIGH-LEVEL INVERSION Atmospheric temperature inversion formed at elevations of 300 m (10 000 ft) or more above the ground by the descent of anticyclonic air.

HIGH-LINE CONDUIT Section of conduit between a head gate and forebay or penstock inlet, used to convey water to be used in developing hydroelectric power.

HIGH-PRESSURE FIRE SYSTEM Separate high-pressure water distribution system reserved for direct hydrant-hose streams.

HIGH-RATE DIGESTION Accelerated anaerobic digestion of sewage sludge promoted by thorough mixing of the digester contents, frequently with the enhancement of thermophilic digestion.

HIGH-RATE FILTER Biological filter operated at a high daily dosing rate.

HIGH-SPEED COMPACTOR Soil and refuse compactor equipped with four or more octagonal wheels, the ridges on the front (steering) wheels corresponding to troughs on the rear (driving) wheels.

HIGH-TEMPERATURE GAS-COOLED REACTOR Nuclear reactor fuelled by enriched uranium and cooled by helium with graphite as a moderator.

HIGH-TEST HYPOCHLORITE Combination of powdered lime and chlorine consisting largely of calcium hypochlorite and having an available chlorine content of greater than 65%.

HIGH-VELOCITY AIR FILTER Air pollution control filtration device for the removal of sticky, oily, or liquid aerosol particulate matter from exhaust gas streams.

HIGH-VOLUME SAMPLING Technique of measuring airborne particulate matter in which large samples of air are drawn through appropriate filters and volumetric or flow metering equipment.

HIGHWALL Face that is being excavated, as distinguished from a spoil pile.

HIGH-WATER LUNITIDAL INTERVAL Time between the moon's meridian passage at a given location and the following high tide at that place.

HIGH-WATER MARK Highest visible mark left by a flood. It is usually the sign of the 10-year (or other arbitrary period) peak flow. *See also* **Low water mark.**

HINDERED SOLIDS SEPARATION Process in which solids settle from a body of water in a thickening, rather than in a clarifying mode.

HISTOGRAM Pictorial method of showing the distribution of various quantities.

HISTOGRAPH Map or chart (of a system, for instance) containing time lines giving data for recorded events (such as flow rates, water levels, etc.).

HISTORIC SITE Building, monument, park, cemetery or other site having public interest and national, regional, or local significance.

HOAR FROST White, crystalline deposit of ice on the surface of solid objects caused

by the direct sublimation of water vapor from the air.

HOLDING VALVE *See* **Valve.**

HOLDING VALVE (INTEGRAL) *See* **Valve.**

HOLDOVER STORAGE Portion of useful storage normally remaining in a reservoir at the end of the drawdown period and available for use in a critically dry year.

HOLLOW DAM Fixed dam consisting of inclined slabs or arched sections supported by transverse buttresses, which transfer the load taken by the slabs or arched sections to the foundation.

HOME RANGE Area habitually occupied by a group of animals or an individual animal.

HOOD Casing on the end of an underwater suction line that causes it to pick up material from the bottom only.

HOODOO Natural pillar of rock, cemented gravel, or clay caused by erosion and often having a fantastic shape.

HOPPER DREDGE Dredge that incorporates a hopper and dump mechanism enabling it to transport and discharge material it has excavated.

HOPPER-TYPE DUMP TRAILER *See* **Trailer.**

HORIZON More-or-less well defined layer or section of the soil profile approximately parallel to the surface and displaying characteristics that have been produced through the soil-building process.

HORIZONTAL BRANCH Drain branch extending laterally from a soil or waste stack or building drain.

HORIZONTAL-FLOW TANK Vessel, with or without baffles, in which the direction of flow is horizontal.

HORIZONTAL PIPE Any fitting or pipe that makes 45° or more with the vertical.

HORIZONTAL PUMP 1. Reciprocating

259

pump in which the piston or plunger moves horizontally; **2.** Centrifugal pump in which the pump shaft is horizontal; **3.** Screw pump where the casing and contained helix is in a more-or-less horizontal plane.

HORIZONTAL WELL Tubular well developed more-or-less horizontally into a water-bearing stratum or under the bed of a lake or stream.

HORSEPOWER 1. Nonmetric unit of power equal to a rate of 33 000 foot-pounds per minute. Symbol: hp. Multiply by 0.745 to obtain kilowatts, symbol: kW. *See also the appendix:* **Metric and nonmetric measurement**; **2.** There are many other ways of measuring horsepower, including:

Boiler: Equivalent of the heat required to change 16 kg (34.5 lb) per hour of water at 100°C (212°F) to steam. It is equal to a boiler heat output of 33 475 Btu per hour.

Brake: Amount of horsepower of an engine, motor, or mechanical device produced as indicated using a brake dynamometer, measured at the end of the crankshaft.

Continuous: 1. Amount of horsepower the engine is capable of developing, at a stated speed and under full load, for more than 24 hours; **2.** Power recommended by a manufacturer for satisfactory operation under specified continuous-duty conditions.

Demand: Amount of horsepower a powertrain needs to propel a specified gross weight at a specified rate of speed; the sum of the power absorbed by the rolling resistance, air resistance, and grade resistance.

Drawbar: Horsepower of an engine, minus friction and slippage loss in the drive mechanisms and in the tracks or tires.

Friction: Power needed to overcome friction within an engine resulting from the pressure of the piston and rings against the cylinder walls, rotation of the crankshaft and camshaft in their bearings, and friction developed by other moving parts.

Gross: Power rating obtained by a dynamometer test of an engine without allowance for power absorbed by engine accessories.

Horsepower: Torque or twisting force produced by an engine over a period of time. A measurement of work: 550 lb/ft/sec = 1 horsepower (76 m-kg/sec), or 746 watts.

Horsepower hour: The equivalent of developing one horsepower for one hour. A heat equivalent of approximately 2454 Btu.

Hydraulic: Horsepower computed from flow rate and pressure differential.

Indicated: Calculated horsepower value using a cylinder pressure gauge and known engine dimensions to provide a reference horsepower value.

Intermittent: Amount of full-load horsepower, at a stated speed, that an engine can maintain on a cycle of 30 minutes under full load, and 60 minutes under no (or reduced) load.

Peak: Horsepower an engine can maintain at maximum load for one minute without speed loss, with a reasonably clean exhaust and with the engine in proper adjustment.

Rated: Horsepower an engine can develop under full load, at a speed recommended by the engine manufacturer, at a measured altitude and temperature.

SAE net: Brake horsepower remaining at the flywheel of an engine to do useful work after the power required by the engine accessories has been provided.

Shaft: Actual horsepower produced by an engine, after deducting the drag of accessories.

Taxable: Arbitrary formula for estimating horsepower that assumes that engines deliver their rated power at a piston speed of 304 m (1000 ft) per min. and that mechanical efficiency will average 75%.

Wheel: Average horsepower delivered at the points of contact between the driving wheels and the ground surface, equal to the flywheel horsepower less the power lost through the power transmission assemblies.

Working: Horsepower delivered at the flywheel with full engine accessories for the working conditions.

HORSEPOWER HOUR See **Horsepower.**

HORSESHOE CONDUIT Conduit having a domed cross-section above the invert.

HOSE BIB See **Bib.**

HOSE COUPLING Joint or connection between a hose and a pipe, or a hose and another hose.

HOSE THREAD Standard screw thread equal to 12 threads per inch on a ¾-in. pipe.

HOST Plant or animal on or in which another lives for nourishment, development, or protection.

HOT SPRING Flow of thermal water having a temperature higher than that of the human body.

HOUR Non-SI unit of time, equal to 3600 seconds or 60 minutes, permitted for use with the SI system of measurement. Symbol: h. See also the appendix: **Metric and nonmetric measurement.**

HOUSE CISTERN Cistern in which rainwater is stored for domestic purposes.

HOUSE CONNECTION Pipe carrying potable water to a residence, or wastewater from it to a common sewer.

HOUSE DRAIN See **Building drain.**

HOUSEHOLD DETERGENT Cleaning product produced for the domestic retail market and usually containing anionic or nonionic surfactants, plus phosphates and such minor ingredients as bleaches, brighteners,

perfumes, etc.

HOUSEHOLD FILTER Low capacity filter connected to a water supply and used within the home. *Also called a* **Domestic filter.**

HOUSEHOLD SOLID WASTE See **Solid waste.**

HOUSEHOLD WASTE Solid waste (including garbage, trash, and sanitary waste in septic tanks) derived from households (including single and multiple residences, hotels and motels, bunkhouses, ranger stations, crew quarters, campgrounds, picnic grounds, and day-use recreation areas).

HOUSE LINE Permanently fixed, private standpipe hoseline.

HOUSE SEWER Pipe conveying sewage from a single residence to a common sewer.

HOUSE SUBDRAIN That portion of a drainage system that cannot drain by gravity flow to a building sewer.

HOUSE TRAP Sanitary seal provided between the building drain and the building sewer. See also **Running trap.**

HOUSING CODE (CANADA) See **Building code, National building code.** *Also called* **Minimum standards bylaw.**

HOUSING CODE (US) See **Building code, housing code (US),** and **Building code, National building code.** *Also called* **Minimum standards bylaw.**

HUB Enlarged end of a hub-and-spigot cast iron pipe.

HUB END Pipe end connections that are leaded and caulked, such as on cast-iron sewer pipe.

HUB HEIGHT Height of a tower-mounted wind turbine measured from ground level to the centerline of the turbine rotor.

HUBLESS PIPE See **Pipe.**

HUMAN ENVIRONMENT Natural and physical environment and the relationship of people with that environment.

HUMAN SUBJECT Living individual about whom an investigator (whether professional or student) conducting research obtains:

(a) Data through intervention or interaction with the individual, or

(b) Identifiable private information.

HUMIC ACID Collective term for a range of organic compounds resulting from the anaerobic decay of vegetable matter.

HUMIDITY Amount or degree of moisture in the air. There are several distinctions, including:

Absolute: Mass of water vapor per unit volume of air.

Percent: Ratio of weight of water vapor in a given weight of dry air to the weight of water vapor that would be present if the same weight of air were saturated, expressed as a percent.

Relative: Ratio of the amount of water vapor present in the air to that which the air would hold at saturation at the same temperature.

Specific: Ratio of the mass of water vapor to the mass of the system: weight of water vapor relative to a given volume of air.

HUMIFICATION Microbial breakdown of organic matter in the soil to form humus.

HUMUS Decomposed and biologically stable residue of plant and animal tissue.

HUMUS SLUDGE Sludge deposited in the final stages of a sewage treatment process which, when dewatered, resembles humus in appearance.

HUNDREDWEIGHT Nonmetric unit of mass, equal to 100 pounds. Symbol: cwt. Multiply by 43.359 to obtain kilograms, symbol: kg. *See also the appendix*: **Metric and nonmetric measurement.**

HUNTING Result of too-sensitive setting of automatic sensors or controls that fluctuate between positive and negative adjustment of the device they are set to regulate.

HURRICANE Wind speed of 120 kph (75 mph) on the Beaufort scale.

HYBRID SOLAR SYSTEM Passive solar system that uses only a few mechanical devices to utilize the collected energy. *See also* **Active solar system.**

HYDRANT Water supply outlet, with its valve located belowground, usually below the frost line, that permits the discharge of water at full bore and full available pressure.

HYDRANT BARREL Upper section of a hydrant covering the operating mechanism and preventing freezing.

HYDRANT BONNET Removable top part of a hydrant.

HYDRANT CAP Casting with a female thread that fits on the male thread of a hydrant outlet nozzle, and having a nut on its outer face that fits a standard hydrant wrench.

HYDRANT DRAIN Drain, operated by the main stem and open when the hydrant is completely closed, for removing water from the barrel of a hydrant following use.

HYDRANT PITOMETER Device for determining the velocity of water discharging freely from one nozzle of a hydrant.

HYDRANT SIZE Net diameter of the valve seat of the main valve of a hydrant.

HYDRANT WRENCH Tool used to open or close a hydrant, and to remove hydrant caps.

HYDRATE Compound produced when any of certain other substances unite with water, represented in formulas as containing molecules of water.

HYDRATED LIME Calcium hydroxide ($Ca(OH)_2$), also called slaked lime.

HYDRATION Chemical process involving the combination or union of water with other substances.

HYDRATION WATER Water that combines with salts when they crystallize.

HYDRAUCONE Draft tube where the

emerging water strikes a shaped plate, causing it to deflect in all directions.

HYDRAULIC Operated by the movement and force of liquids.

HYDRAULIC ACCUMULATOR Device for storing energy by placing a fluid under pressure in a cylinder.

HYDRAULIC ACTUATOR Mechanism that converts hydraulic pressure to mechanical motion.

HYDRAULIC BORE Standing wave that advances upstream in an open conduit from a point where the flow has suddenly been stopped.

HYDRAULIC CONDUCTIVITY Rate of groundwater flow per unit of time through a cross section of unit area in a unit hydraulic gradient at ambient temperature.

HYDRAULIC CONTROL VALVE *See* **Valve.**

HYDRAULIC DREDGE Floating pump used to suck up a mixture of water and sediment, usually for discharge to land, through pipes, but also to another underwater location.

HYDRAULIC ELEMENTS Depth, cross-sectional area, wetted perimeter, velocity, temperature, etc., of water flowing in a channel.

HYDRAULIC ENGINEERING Study and application of the static and dynamic behavior of fluids.

HYDRAULIC EXCAVATION Use of water under high pressure directed against the surface of a material to be moved, the dislodged material being carried away by the resulting flow.

HYDRAULIC FILL Fill built by transporting material by water, either through an open channel or by pumping from a dredge.

HYDRAULIC FILL DAM Dam developed by the deposition of hydraulic fill.

HYDRAULIC FLOW NET Method of calculating pressures and velocities in a curving stream of water where friction, impact, or eddy losses do not exercise a controlling effect.

HYDRAULIC FLUIDS Petroleum-based hydraulic fluids.

HYDRAULIC FRICTION Flow resistance in a pipe or channel caused by surface drag, roughness, or internal obstructions.

HYDRAULIC FRICTION COEFFICIENT Ratio of actual discharge to the theoretical discharge in a frictionless conduit or open channel free of turbulence.

HYDRAULIC GRADIENT Slope of the surface of open or underground water.

HYDRAULIC HEAD Height of the free surface of a body of water above a given point beneath the surface.

HYDRAULIC HORSEPOWER *See* **Horsepower.**

HYDRAULIC JUMP Sudden increase in depth of a stream flowing in a channel of uniform cross-section and even gradient when the stream meets a slower stream downstream.

HYDRAULICKING Earth moving using flowing water.

HYDRAULIC LIFT TANK Tank holding hydraulic fluid for a closed-loop mechanical system that uses compressed air or hydraulic fluid to operate lifts, elevators, and other similar devices.

HYDRAULIC LOAD Total flow over a stated period (hour, day, etc.) that a plant or facility is designed to accommodate or process.

HYDRAULIC LOSS Loss of head due to obstructions, friction, velocity change, and variations in the shape of the conduit.

HYDRAULIC MEAN DEPTH Cross section of the water flowing in a channel, divided by the wetted perimeter of the channel.

HYDRAULIC MODEL Mathematical or physical simulation of a system from which actual conditions and requirements can be predicted and measured.

HYDRAULIC MOLE Microtunneling borer actuated by hydraulic pressure.

HYDRAULIC MONITOR *See* **Monitor.**

HYDRAULIC PERMEABILITY Rate of discharge through a unit cross-sectional area of soil normal to the direction of flow when the hydraulic grade is unity.

HYDRAULIC POWER Transmission of power through fluid under pressure.

HYDRAULIC POWER UNIT Combination of componentry to facilitate fluid storage and conditioning, and delivery of the fluid under conditions of controlled pressure and flow to the discharge port of a pump, including maximum pressure controls and sensing devices where applicable. Circuitry components, although sometimes mounted on the reservoir, are not considered part of the power unit.

HYDRAULIC PROFILE 1. Vertical crosssection of the watertable or piezometric surface of an aquifer or groundwater basin; **2.** Profile along the axis of flow of a stream or conduit showing elevations of the bottom and of the energy line.

HYDRAULIC PUMP *See* **Pump.**

HYDRAULIC RADIUS Cross-sectional area of a stream of water divided by the length of that part of its periphery in contact with its containing conduit.

HYDRAULIC RAM Device for raising water by employing the energy generated by a water hammer created by periodically checking the flow of water in the supply pipe.

HYDRAULIC REACH Section of a river or stream between significant changes in hydraulic character: increase or decrease in flow, bottom, bank height, volume, etc.

HYDRAULIC RELIEF VALVE *See* **Valve.**

HYDRAULIC RESERVOIR Vessel or tank for storing and conditioning liquid in a hydraulic system.

HYDRAULICS Science of transmitting force and/or motion through the medium of a liquid.

HYDRAULIC SIMILITUDE Flow system having characteristics that can be replicated in another system of lesser or greater capacity.

HYDRAULIC SLOPE Slope of the hydraulic gradient.

HYDRAULIC SLUICING Moving materials by water pressure.

HYDRAULIC STRUCTURE Engineering works designed to control water.

HYDRAULIC SYSTEM System designed to transmit power through a liquid medium, permitting multiplication of force in accordance with Pascal's law, which states that 'a pressure exerted on a confined liquid is transmitted undiminished in all directions and acts with equal force on all equal areas.' Hydraulic systems have six basic components:

1: A reservoir to hold the fluid supply.

2: A fluid to transmit the power.

3: A pump to move the fluid.

4: A valve to regulate pressure.

5: A directional valve to control the flow.

6: Working components — such as a cylinder and piston or a shaft rotated by pressurized fluid — to turn hydraulic power into mechanical motion.

HYDRAULIC TEST Test for piping systems in which all conduits are sealed and filled with water to, or slightly above, design pressure.

HYDRAULIC TESTER Device that can measure flow, pressure, and temperature of an active hydraulic circuit, in both forward and reverse flow.

HYDRAULIC TURBINE Prime mover employing the pressure or motion of water for the development of mechanical energy.

HYDRAULIC VALVE Valve operated by means of hydraulic energy.

HYDRAZINE Reagent (N_2H_4) used to scavenge oxygen from boiler water.

HYDRIC 1. Of or containing hydrogen; **2.** Soil that is saturated for sufficient periods of time to produce anaerobic conditions.

HYDRIDE Compound of hydrogen with another element or radical.

HYDRO Of or having to do with water.

HYDROCARBON Class of compounds comprising solely carbon and hydrogen.

HYDRODYNAMICS Engineering science pertaining to the energy of liquid flow and pressure.

HYDROELECTRIC POWER Electrical energy produced using water power as an energy source.

HYDROCHLOROFLUOROCARBONS Chemical compounds, HCFCs, used as a replacement for chlorofluorocarbons, CFCs, in several industrial process, including refrigeration and aerosols.

HYDRODYNAMICS Branch of physics having to do with the forces that water and other liquids exert.

HYDROELECTRIC Of or having to do with the generation of electricity by water power, or by the friction of water or steam.

HYDROELECTRICITY Electrical energy produced from water power.

HYDROFLUORIC ACID Colorless, corrosive, volatile solution of hydrogen fluoride (HF) that reacts with a wide range of materials.

HYDROGEN Colorless, odorless gas (H_2), having an atomic mass of 1.

HYDROGENATION Chemical combination of hydrogen with another substance, commonly through heat and pressure in the presence of a catalyst.

HYDROGEN CHLORIDE Colorless gas (HCl) with a choking odor that is highly soluble in water to form hydrochloric acid.

HYDROGEN CYANIDE Colorless, highly poisonous gas, a compound of hydrogen and cyanide, having an odor of bitter almonds.

HYDROGENESIS Process of natural condensation of the moisture in the air spaces in surface soil or in rock.

HYDROGEN FLUORIDE Colorless fuming gas (HF) having a boiling point of 19.5°C: the aqueous solution is known as hydrofluoric acid.

HYDROGEN ION CONCENTRATION Weight of hydrogen ion per volume of solution, commonly expressed as the pH value.

HYDROGEN PEROXIDE Powerful disinfectant and bleaching agent (H_2O_2) which rapidly gives off oxygen.

HYDROGEN SULFIDE Poisonous gas (H_2S) with the odor of rotten eggs that is produced from the putrefaction of sulfur-containing organic material.

HYDROGEOLOGY Science dealing with the occurrence of surface and groundwater, its utilization, and its functions in modifying the Earth, primarily by erosion and deposition.

HYDROGRAPH Graph showing various conditions of the properties of water with respect to time.

HYDROGRAPHER Person who records measurements of water level, flow, rainfall, runoff, etc.

HYDROGRAPHIC Measurements or other determinations of the characteristics of water.

HYDROGRAPHIC DATUM Baseline reference for depths of water or heights of predicted tides.

HYDROGRAPHIC SURVEY Determination of the configuration of the bottom of a body of water. *See also* **Marine survey.**

HYDROGRAPHY Surveying and charting that which is submerged.

HYDROISOBATHS Contours of similar depth of subsoil watertable below the ground surface.

HYDROISOPLETH MAP Map showing fluctuations of the watertable over time and space.

HYDROKINETICS Engineering science pertaining to the energy of liquids in motion.

HYDROLOGICAL CYCLE General pattern of water movement at, and near the Earth's surface; the means by which water is circulated in the biosphere.

HYDROLOGICAL DATA STATION Location at which data and other hydrological information is obtained and stored.

HYDROLOGIC BUDGET Accounting of the inflow, storage and outflow of a hydrologic unit, typically a drainage basin, reservoir, etc.

HYDROLOGIC INVENTORY Process of obtaining and evaluating data and other information about hydrologic situations and events.

HYDROLOGIC SEQUENCE Series of samples taken as a vertical section from soils derived from the same parent material and which show increasingly poor drainage down a slope, the lowest samples often being waterlogged.

HYDROLOGY Science concerned with the properties, occurrences, distribution, and circulation of water, particularly underground water.

HYDROLYSIS Chemical reaction between a material and the ions of water.

HYDROLYTIC TANK Wastewater tank in which hydrolysis occurs; a sedimentation tank in which biochemical processes are promoted.

HYDROMECHANICS Science of treating of the equilibrium and motion of fluids and of bodies in or surrounding them.

HYDROMETEOROLOGY Meteorology of water in the atmosphere.

HYDROMETRIC PENDULUM Device consisting of a metal ball suspended on a cord which is lowered into flowing water, the current velocity of which can be gauged from the angle from vertical assumed by the cord.

HYDROMETRY Measurement and analysis of the flow of water.

HYDROMORPHIC SOIL Soil containing excess moisture.

HYDROPHYTE Plant typically found in wet areas or in water where oxygen deficiencies occur periodically.

HYDROPHYTE Plant that grows naturally in water or in saturated soils.

HYDROPNEUMATICS Pertaining to the combination of hydraulic and pneumatic fluid power.

HYDROPNEUMATIC TANK SYSTEM Domestic water supply system in which water under pressure is supplied by means of a supply system that incorporates a pressure tank.

HYDROPONICS Growing of plants in water containing the necessary nutrients instead of in soil.

HYDROSEEDING Process of spraying a combination of fertilizer, grass seed, water, and fibrous binder onto prepared ground in the form of a slurry; often used on steep slopes and hard-to-reach places.

HYDROSPHERE Liquid and solid water resting on and invading the outermost zone of the lithosphere.

HYDROSTATIC Pressure or equilibrium of fluids.

HYDROSTATIC EXCESS PRESSURE Pressure in excess of hydrostatic pressure per unit area of soil due to applied loads that exists in the pore water at any time.

HYDROSTATIC HEAD Pressure in a fluid at a given point expressed in terms of the vertical height of the liquid column above that point which would produce the same pressure.

HYDROSTATIC JOINT Bell-and-spigot-type joint.

HYDROSTATIC LEVEL Elevation to which the top of a column of water would rise, if free to do so, from an artesian aquifer basin, or from a conduit under pressure.

HYDROSTATIC PORE PRESSURES Pore water pressures or groundwater pressures exerted under conditions of no flow where the magnitude of pore pressures increases linearly with depth below the ground surface.

HYDROSTATIC PRESS Machine used to magnify force and consisting of linked cylinders of unequal capacity.

HYDROSTATIC PRESSURE *See* **Pressure.**

HYDROSTATIC PRESSURE RATIO Ratio between the pressure on a vertical plane due to active earth pressure and that which would exist at the same point in a liquid of the same density as the soil.

HYDROSTATIC SLUDGE REMOVAL Discharge of sludge from a hopper-bottomed sedimentation tank by means of the hydrostatic pressure of the fluid above the sludge outlet.

HYDROSTATIC STRENGTH Capacity of a vessel, pipe or structure to withstand a specified internal pressure.

HYDROSTATIC TEST Test to determine the capability of a vessel, pipe or structure (or any associated joints) to withstand a prescribed internal hydrostatic pressure.

HYDROSTATIC UPLIFT Upward pressure against the base of an underground structure transmitted by groundwater.

HYDROSULFITE Salt used for reducing chlorine residuals.

HYDROTESTING Testing of piping or tubing by filling with water and pressurizing to test for integrity.

HYDROTHERMAL Geological process involving heat or superheated water.

HYDROXIDE ALKALINITY Alkalinity caused by hydroxyl ions.

HYETAL COEFFICIENT Ratio of actual precipitation in any month to what would have fallen had the rainfall been uniformly distributed throughout the year.

HYETAL REGIONS Division of the world into regions according to rainfall characteristics.

HYETOGRAPH Graphical representation of rainfall characteristics and data.

HYETOGRAPHY Study of the geographic distribution of rain.

HYETOLOGY Science of precipitation.

HYGROGRAPH Recording hygrometer.

HYGROMETER Instrument that measures the degree of moisture suspended in the air.

HYGROMETRIC EXPANSION (CONTRACTION) Expansion and contraction of materials as they absorb or exude moisture.

HYGROMETRY Measurement of the humidity in air.

HYGROSCOPIC Water absorbed from the atmosphere.

HYGROSCOPIC COEFFICIENT Percent moisture that a dry material will absorb in saturated air at a given temperature.

HYGROSCOPICITY Ability to absorb and retain moisture.

HYGROSCOPIC MOISTURE Moisture remaining in an air-dried soil that evaporates if the soil is heated to 105°C (221°F).

HYGROSCOPIC SALTS Salts that absorb moisture.

HYGROSCOPIC WATER 1. Water molecules held on surfaces of particles by forces of adhesion; **2.** Water in the soil that is in equilibrium with atmospheric water vapor.

HYGROSTAT Automatic control responsive to humidity.

HYPERBARIC Pressure above ambient atmosphere.

HYPERCRITICAL FLOW Super-critical flow.

HYPERTROPHIC Condition in which a body of water is grossly enriched with plant nutrients.

HYPOCHLORINATION Application of hypochlorite compounds to water or wastewater for the purpose of disinfection.

HYPOCHLORINATORS Mechanisms used to dispense chlorine solutions made from hypochlorites such as sodium hypochlorite or calcium hypochlorite.

HYPOCHLORITE Chemicals containing available chlorine used for disinfection: calcium, sodium, or lithium hypochlorite.

HYPOCHLORITE OF LIME Combination of slaked lime and chlorine in which calcium oxychloride is the active ingredient.

HYPOLIMNION Colder, non-circulating layer of water in a lake, lying below the thermocline.

I

ICE Solid state of water.

ICE AGE Any of the times in geological history when much of the Earth was covered with glaciers, the most recent of which was the Pleistocene epoch, when most of the northern hemisphere was so covered.

ICE ANVIL Anvil-shaped top of a cumulonimbus cloud, formed of minute ice crystals.

ICE APRON Ramp upstream of a bridge pier, sloping up from below minimum water level, used to lift floating ice, forcing it to break.

ICE CRYSTALS Small, unbranched crystals in the shape of staffs or plates.

ICE EVAPORATION LEVEL Atmospheric level where the temperature is low enough for water to change between the solid and gaseous phase.

ICE FOG Atmospheric suspension of highly reflective ice crystals.

ICE GORGE Stream channel choked with ice that has piled up against an obstruction to form a temporary dam. When the ice is released is often results in a flood of the water impounded behind it.

ICE JAM Ice prevented from moving downstream by an obstruction, or narrowing of the channel, forming a temporary dam.

ICE PACK Ice floe masses that have accu-

mulated and formed a more-or-less solid cake of considerable size and thickness.

ICE PRESSURE Pressure against a natural or artificial obstruction caused by the expansion of ice due to temperature changes.

ICE PUSH Sudden thrust of ice pressure.

ICE SHEET Largest form of glacier: thick layer of ice covering an extensive area, often in excess of several thousand square kilometers.

ICE STORM Storm characterized by the occurrence of glaze.

ICE WEDGE Vertical block of ice in permafrost.

IDEAL CAPACITY *See* **Capacity.**

IDEAL GAS LAW One mole of an ideal gas has a volume of 22.4143 liters at a temperature of 0°C and a pressure of 1 atmosphere.

IDENTIFICATION INDEX Series of numbers and/or letters, used to indicate a sequence of grades, blends, etc.

IDENTIFIED FOR USE Recognized as suitable for the specific purpose, function, use, environment, application, etc., where described as a requirement.

IDLE CAPACITY *See* **Capacity.**

IDLH (Immediately Dangerous to Life or Health) Concentration of a toxic gas representing a maximum level from which a person could escape within 30 minutes without any escape-impairing symptoms or any irreversible health effects.

IGNEOUS Designating rock formed by the solidification of molten matter.

IGNIMBRITE Pyroclastic rock composed of unsorted pumice and other material, formed by the explosive disintegration of flattened pumice.

I/I *See* **Excessive I/I; I/I analysis;** and **Infiltration/inflow.**

I/I ANALYSIS Analysis demonstrating possible excessive or nonexcessive infiltration and/or inflow to a sewer system. *See also* **Excessive I/I.**

ILLUVIATION Gradual accumulation of material by deposition from percolating water.

IMHOFF CONE Cone-shaped graduated glass used to measure the approximate volume of settleable solids in various liquids over time.

IMHOFF TANK Two-storied, anaerobic, sewage treatment tank comprising an upper, continuous sedimentation chamber and a lower, sludge-digestion chamber.

IMMEDIATE BIOCHEMICAL OXYGEN DEMAND 1. Initial quantity of oxygen taken up by wastewater immediately upon being introduced to water containing dissolved oxygen; **2.** The apparent BOD for 15 minutes at 20°C (68°F) as determined in the standard laboratory procedure.

IMMERSE To dip into or cover with a fluid.

IMMERSED TUBE Underwater tunnel made by sinking precast concrete boxes into a channel dredged for the purpose across a waterway from either bank and joining them end-to-end underwater and pumping them dry.

IMMINENT HAZARD Situation that exists when the continued use of a pesticide during the time required for cancellation proceedings would be likely to result in unreasonable adverse effects on the environment or will involve unreasonable hazard to the survival of a species declared endangered.

IMMISSION Reception of a pollutant, or possible pollutant, from a remote source of emission.

IMMOBILIZATION Lack of movement by a test organism except for minor activity of the appendages.

IMPACT Striking of one body against another.

IMPACT LOSS Loss of head in a stream or flowing water resulting from the impact of

particles of water on themselves or on some boundary surface or on some contra or opposing flow.

IMPELLER Component of a centrifugal pump that uses centrifugal force to discharge a fluid into the outlet passages.

IMPELLER EYE Opening through which water flows into the center of an impeller.

IMPELLER PUMP Pump that moves water through the continuous application of power derived from some mechanical agency, typically a centrifugal pump.

IMPERIAL GALLON Nonmetric measure of a fluid gallon equal to approximately 1.2 US gallons. *Also called* **British Imperial gallon.** *See also* **Gallon (Imperial).**

IMPERMEABLE Material that will not permit the passage of water or air, i.e. bituminous pavement.

IMPERMEABLE ROCK Rock having a texture such that water cannot move through it under normal underground pressures.

IMPERMEABILITY FACTOR Ratio of the amount of rain that runs off a surface to that which falls on it.

IMPERVIOUS 1. Incapable of being passed through or penetrated; **2.** Soil in which the spacing of the particles is such as to permit only extremely slow passage of water.

IMPERVIOUS BED Bed or stratum through which water will not move.

IMPERVIOUS CORE *See* **Impervious zone.**

IMPERVIOUSNESS COEFFICIENT Ratio of the effective impervious surface area to the total catchment or tributary area under consideration.

IMPERVIOUS TILE Tile having water absorption of 0.5% or less.

IMPERVIOUS ZONE Clay or silt zone of an earth- or rock-filled dam that provides a water barrier. *Also called* **Impervious core.**

IMPINGEMENT 1. Direct impact of fluid

flow upon or against a surface; **2.** Direct high-velocity impact of a fluid flow upon, or against any internal portion of a filter.

IMPINGEMENT FILTER Coated fabric roll filter that traps dust and particulate matter entering the intake of an air-conditioning system.

IMPLEMENTATION Putting a plan into practice by carrying out planned activities, or ensuring that these activities are carried out.

IMPOUNDING DAM Barrier across a watercourse that serves to regulate the flow and create a reservoir or head pond.

IMPOUNDING RESERVOIR Reservoir to which surplus water is directed, and from which water is drawn through the same system.

IMPOUNDMENT Hollow shape created by an enclosing structure in which water can be stored.

IMPROVED DISCHARGE Volume, composition, and location of a discharge following construction of outfall improvements or of treatment system improvements to treatment levels or discharge characteristics, or implementation of a program to improve operation and maintenance of an existing treatment system or to eliminate or control the introduction of pollutants into the treatment works.

IMPROVED LAND Land that has been partially or fully developed for a higher and better use.

IMPULSE Dynamic pressure of a jet or stream in the direction of its motion when its velocity in that direction is entirely destroyed.

IMPULSE LINE Small-diameter pipe or tube used to convey pressure from a piping system to a diaphragm- or bellows-operated mechanism.

IMPULSE PUMP Pump that raises water by periodic application of a force suddenly applied and suddenly discontinued, typically a hydraulic ram.

IMPULSE-REACTION TURBINE Turbine in which both the impulse and reaction forces are utilized.

IMPULSIVE NOISE Acoustic event characterized by very short rise time and duration.

IMPURITY Chemical substance that is unintentionally present with another chemical substance.

INACTIVE FACILITY Facility which no longer receives solid waste.

INACTIVE WASTE DISPOSAL SITE Disposal site or portion of it where additional waste material has not been deposited within the past year.

INCH Nonmetric unit of length, equal to 1/12 foot. Symbol: in. Multiply by 2.54 to obtain centimeters, symbol: cm; by 0.0254 to obtain meters, symbol: m; by 25.4 to obtain millimeters, symbol: mm. *See also the appendix*: **Metric and nonmetric measurement.**

INCHES OF MERCURY Scale used in measuring negative pressure; used to measure barometric pressure, equal to the pressure exerted by a column of mercury 254 mm (1 in.) high; equivalent to a pressure of 3386.4 newtons/m².

INCH–FOOT CHARGE (FOR WATER SERVICE) Method of charge for fire service equal to the revenue thus obtained minus the hydrant charge, divided by the inch–feet of water mains (diameter of pipe times length).

INCH OF WATER Non-SI unit of pressure equal to the pressure exerted by a column of liquid water 254 mm (1 in.) high at a temperature of 4°C (39.2°F).

INCIDENT RADIATION Energy arriving at the surface of a solar collector or other surface, including both direct and diffuse radiation.

INCINERATION Conversion of matter by burning to ash, carbon dioxide, and water vapor.

INCINERATOR 1. Furnace capable of consuming all of of a range of materials; **2.**

INCINERATOR: Batch fed

Engineered apparatus used to burn waste substances and in which all the factors of combustion (temperature, retention time, turbulence, and combustion air) can be controlled. There are several types, including:

Batch fed: Incinerator that is periodically charged with solid waste; one charge is allowed to burn down or burn out before another is added.

Cell-type: Incinerator, usually batch fed, with grate areas divided into cells, each of which has its own underfire air control and ash dumping grate.

Central: A conveniently located facility that burns solid waste collected from many different sources.

Chute-fed: Incinerator charged through a chute that extends on one or more floors above the furnace.

Continuous-feed: Incinerator into which solid waste is charged almost continuously to maintain a steady rate of burning.

Controlled-air: Incinerator with two or more combustion areas in which the amounts and distribution of air are controlled. Partial combustion takes place in the first zone, and hydrocarbon gases are burned in a subsequent zone(s).

Direct-feed: Incinerator charged through a chute that also functions as a flue to exhaust the products of combustion.

Industrial: Incinerator specifically designed to burn a particular industrial waste.

Multiple-chamber: Incinerator consisting of two or more chambers, arranged as in-line or with retorts, interconnected by gas passage ports or ducts.

Multiple hearth: Incinerator consisting of a series of circular hearths stacked vertically. The combustible material is fed to the uppermost hearth and moved around it by a rotating scrabble arm until it reaches a port or opening where it discharged to the hearth below, where the process is repeated. The upper hearths progressively dry the charge, allowing the lower hearths to more fully oxidize the material before it is discharged as a mixture of ash and clinker.

Onsite: Incinerator that burns waste on the property of the generator of the waste.

Open-pit: Burning device that has an open top and a system of closely spaced nozzles that place a stream of high-velocity air over the burning zone.

Residential: Predesigned, shop-fabricated and assembled unit, shipped as a package for use in individual dwellings.

Single-chamber: Refractory-lined cylindrical furnace charged through a door in the upper part of the chamber. Refuse is batch fed periodically.

INCIPIENT EROSION Early stages of erosion, marked by such developments as gullying.

INCISED RIVER River that has cut its channel through the process of degradation.

INCLUDED GAS Gases in isolated interstices in either the zone of saturation or the zone of aeration.

INCOMPATIBLE FLUIDS Fluids that when mixed in a system, will have a deleterious effect on that system, its components, or its operation.

INCOMPATIBLE WASTE Hazardous waste that is unsuitable for placement in a particular device or facility because it may cause corrosion or decay of containment materials (e.g. container inner liners or tank walls) or comingling with another waste or material under uncontrolled conditions because the comingling might produce heat or pressure, fire or explosion, violent reaction, toxic dusts, mists, fumes, or gases, or flammable fumes or gases.

INCORPORATED INTO THE SOIL Injection of solid waste beneath the surface of the soil or the mixing of solid waste with the surface soil.

INCREASER Short pipe fitting with one end of a larger diameter.

INCRUSTANTS Dense solids that form as a crust on the inside of pipes as a result of the hardness of the water.

INCUBATION Maintenance for growth and reproduction of viable organisms in or on a nutrient substrate at a constant temperature.

INDEX Indicator, usually numerically expressed, of the relation of one thing to another.

INDEX NUMBER Statistical indicator of the changes in magnitude of items in a series of groups of data, in terms of the value of a part of the respective data selected as a base.

INDEX OF LIQUIDITY Figure that is the water content of a sample, minus the water content at the plastic limit of the sample, divided by the index of plasticity. It is the reverse of the consistency index and gives a value of 100% for a clay at the liquid limit, and 0 for a clay at the plastic limit.

INDEX OF PLASTICITY Range of water contents for which a clay is plastic; the difference between the water content of a clay at its liquid and plastic limits.

INDEX ORGANISMS Microscopic organisms, the presence (or absence) of which in water indicates presence or absence and extent of contamination.

INDICATED HORSEPOWER See **Horsepower.**

INDICATED NUMBER In testing of bacterial density by the dilution method, the number obtained by taking the reciprocal of the highest possible dilution (smallest possible sample) in a decimal series.

INDICATING FLOOR STAND Valve operating device that also indicates the extent of opening or closing.

INDICATOR 1. Substance displaying a visible change, commonly of color, in response to or at a desired point in a chemical reaction; **2.** Device or instrument that provides a visual indication or display of a measured parameter. There are many types, including:

Bypass: Indicator that signals that an alternate flow path is being used.

Differential pressure: 1. Indicator that signals the difference in pressure between two points; **2.** Device that indicates continuously during operation the differential pressure across a filter element.

Display: Aid or instrument that provides a visual indication of a measured parameter.

Meter: Type of indicator display or readout. It may be mechanical, hydraulic, electrical, or electronic.

INDICATOR GAUGE Instrument that shows, by means of an index, pointer, dial, display etc., an instantaneous value in response to change.

INDICATOR LIGHT Light that advises of a condition, state, action, need, etc.

INDICATOR PANEL Panel with several lights and/or signals.

INDICATOR POST Post-like attachment to an underground valve that extends above ground and indicates the extent to which the valve is open or closed.

INDICATOR SPECIES Species whose presence is indicative of certain environmental conditions.

INDICATOR VALVE See **Valve.**

INDIGENOUS SPECIES Plant or animal that occurs naturally in, or is native to a region; reciprocal of **Exotic species.**

INDIRECT BANK PROTECTION Works constructed in front of river banks, rather directly on them.

INDIRECT CHARGE Charge that is part of the total charge, but not directly related to it, i.e. overhead.

INDIRECT DISCHARGE Introduction of pollutants into a treatment works from any nondomestic source.

INDIRECT DRAIN Sanitary piping that does not connect directly with the drainage system but discharges liquid wastes into some other fixture or receptacle connected to the drainage system. *Also called* **Indirect waste piping.**

INDIRECT SOLAR Passive solar heating system in which the heat storage unit, a Trombe wall or liquid storage vessel, is physically located between the collector and distributor.

INDIRECT SOLAR GAIN Warming of an area by solar heat gained in an adjacent area: an adjacent sun space, for instance, or via a Trombe wall.

INDIRECT WASTE PIPING Sanitary piping that does not connect directly with the drainage system but discharges liquid wastes into some other fixture or receptacle connected to the drainage system. *Also called* **Indirect drain.**

INDIVIDUAL DRAIN SYSTEM Process drains connected to the first common downstream junction box.

INDIVIDUAL SYSTEM Individual privately owned alternative wastewater treatment works (including dual waterless/gray water systems) serving one or more principal residences, or small commercial establishments. Normally these are onsite systems with localized treatment and disposal of wastewater, but may be systems utilizing small diameter gravity, pressure or vacuum sewers conveying treated or partially treated wastewater. These systems can also include small-diameter gravity sewers carrying raw wastewater to cluster systems.

INDIVIDUAL VENT Branch vent connected to the main vent stack of a plumbing system and extending to a location near a fixture trap to prevent the trap from siphoning. *Also called* **Back vent.**

INDOLE Organic compound (C_8H_7N) containing nitrogen, which has an ammonia odor.

INDUCED-DRAFT WATER-COOLING

TOWER Water-cooling tower having one or more fans located in the saturated air stream leaving the tower.

INDUCED RECHARGE Discharge of water into an aquifer.

INDUCED SIPHONAGE Unsealing of the trap in one sanitary fitting due to air pressure changes in the associated piping caused by flow from another fitting.

INDURATED Process of hardening of sediments through cementation, pressure, or heat.

INDURATED SOIL Soil cemented into a mass hard enough that it will not soften on wetting.

INDURATION Geological process of hardening of sediments or other rock aggregates through cementation, pressure, heat, etc.

INDUSTRIAL CONSUMPTION Water supplied by a public authority for mechanical, trade and manufacturing purposes and accounted for separately.

INDUSTRIAL COOLING TOWER Cooling tower used to remove heat from industrial processes, chemical reactions, or plants producing electrical power.

INDUSTRIAL DISCHARGE Introduction into a publicly-owned wastewater treatment works of a nondomestic pollutant produced by a source subject to a categorical standard or pretreatment requirement, or which contains any substance or pollutant for which a discharge limit or prohibition has been established by any categorical standard or pretreatment requirement.

INDUSTRIAL INCINERATOR *See* **Incinerator.**

INDUSTRIAL OCCUPANCY Occupancy or use of a building for assembling, fabricating, manufacturing, processing, repairing, or storing of goods and materials. There are several subdefinitions, including:

> **High hazard:** Industrial occupancy containing sufficient quantities of highly combustible and flammable or explosive materials that, because of their inherent characteristics, constitute a

special fire hazard.

Medium hazard: Industrial occupancy in which the combustible content is more than 4.5 kg (10 lb) or 105 500 kJ (100 000 Btu) per sq ft of floor area and not classified as high hazard industrial occupancy.

Low hazard: Industrial occupancy in which the combustible content is not more than 4.5 kg (10 lb) or 105 500 kJ (100 000 Btu) per sq ft of floor area.

INDUSTRIAL PARK Area zoned and dedicated for manufacturing and associated activities.

INDUSTRIAL SILENCER Exhaust muffler used to produce the silencing level normally associated within industrial areas.

INDUSTRIAL SOLID WASTE Solid waste generated by manufacturing or industrial processes that is not a hazardous waste.

INDUSTRIAL USER Nongovernmental, nonresidential user of a publicly owned treatment works which discharges more than the equivalent of 11 339 liters (25 000 gallons) per day of sanitary wastes.

INDUSTRIAL WASTE 1. Any liquid, free-flowing waste, including cooling water, resulting from any industrial or manufacturing process or from the development, recovery or processing of natural resources; **2.** *See* **Solid waste.**

INDUSTRIAL WATER Water taken from a source solely for use in an industrial system.

INERT Having inactive chemical properties.

INERT ATMOSPHERE Gas incapable of supporting combustion.

INFECTIOUS AGENT Organism (such as a virus or a bacteria) that is capable of being communicated by invasion and multiplication in body tissues and capable of causing disease or adverse health impacts in humans.

INFECTIOUS WASTE Equipment, instruments, utensils, and fomites of a disposable nature from the rooms of patients who are suspected to have or have been diagnosed as having a communicable disease and must, therefore, be isolated as required by public health agencies; laboratory wastes such as pathological specimens (e.g. all tissues, specimens of blood elements, excreta, and secretions obtained from patients or laboratory animals) and disposable fomites (any substance that may harbor or transmit pathogenic organisms) attendant thereto; and surgical operating room pathologic specimens and disposable fomites and similar disposable materials from outpatient areas and emergency rooms.

INFILTRATION 1. Flow of water from the land surface into the subsurface; **2.** Water entering a sewer system and service connections from the ground through such means as defective pipes, pipe joints, connections, or manhole walls; **3.** Uncontrolled admittance of air through cracks and pores into a building.

INFILTRATION CAPACITY Maximum rate at which a soil can absorb rainfall under given conditions.

INFILTRATION COEFFICIENT Ratio of infiltration to precipitation.

INFILTRATION DITCH Ditch extended into the watertable to collect water and conduct it to a sump, well, canal, reservoir, etc.

INFILTRATION DIVERSION Diversion of stream water using perforated pipes laid under the stream bed.

INFILTRATION GALLERY Gallery developed into a water-bearing strata and having ports or openings at intervals, used to collect and convey groundwater.

INFILTRATION INDEX Rate of infiltration based on recorded rainfall and runoff.

INFILTRATION/INFLOW Total quantity of water entering a sewer system from both infiltration and inflow, without distinguishing the source. *Also called* **I/I.**

INFILTRATION RATE 1. Rate at which water enters the soil or other porous material under specified conditions; **2.** Rate at which

275

water enters an artificial structure from the soil or other porous material.

INFILTRATION WATER Water which permeates through the earth.

INFLATABLE GASKET Gasket whose integrity is dependent on a seal provided by inflation with compressed air.

INFLOW Water other than wastewater that enters a sewer system (including sewer service connections) from sources such as, but not limited to, roof leaders, cellar drains, yard drains, area drains, drains from springs and swampy areas, manhole covers, cross connections between storm sewers and sanitary sewers, catch basins, cooling towers, storm waters, surface runoff, street wash waters, or drainage. Inflow does not include, and is distinguished from, infiltration.

INFLUENCE AREA Surface area surrounding a well or group of wells over which the watertable or piezometric surface is lowered by withdrawal of the water.

INFLUENCE BASIN Basin area surrounding a well or group of wells over which the watertable or piezometric surface is lowered by withdrawal of the water.

INFLUENT Fluid entering a component or system.

INFLUENT SEEPAGE Movement of water from the land surface to the watertable.

INFLUENT STREAM Stream or portion of a stream that contributes water to the groundwater supply.

INFLUENT WATER Subsurface water resulting from migration into the lithosphere of surface water.

INFLUENT WEIR Weir at the influent end of a treatment basin or tank, typically a sedimentation basin.

INFRARED RAYS Wavelengths longer than 7600 Å or 780 nm.

INFRARED SPECTROPHOTOMETRY Measurement of absorption of a light source in the infrared region.

INFRASOUND Low-frequency sound, below 100 Hz.

INGROUND TANK Device meeting the definition of a tank, a portion of which is situated to any degree within the ground, thereby preventing visual inspection of that external surface area of the tank.

INHALABLE DIAMETER Aerodynamic diameter of a particle considered to be inhalable for an organism. It is used to refer to particles which are capable of being inhaled and may be deposited anywhere within the respiratory tract from the trachea to the alveoli. For man, the inhalable diameter is considered as 15 micrometers or less.

INHALATION LC$_{50}$ Concentration of a substance, expressed as milligrams per litre of air or parts per million parts of air, that is lethal to 50% of the test population of animals.

INHERENT ASH Portion of the ash or other material found after combustion that is chemically bound to the molecules of the combustible, as distinguished from extraneous noncombustible materials, that comes from other sources or that may be mechanically entrained with the combustible.

INHIBITION Any decrease in the growth rate of the test organism compared to the control organism.

INHIBITORY SUBSTANCE Material able to kill or restrict the ability of organisms to treat wastes.

INHIBITORY TOXICITY Inhibitory action of a substance on the rate of general metabolism of living organisms.

INITIAL ABSTRACTION RETENTION Amount of rainfall that falls without a significant amount of runoff resulting.

INITIAL CRUSHER Crusher into which nonmetallic minerals can be fed without prior crushing in a plant.

INITIAL DETENTION Volume of water on the ground in depressions or in transit when active runoff beings.

INITIAL LOSS Rainfall that precedes the

commencement of surface runoff due to interception, surface wetting, infiltration, detention, etc.

INJECTION INTERVAL That part of an injection zone in which the well is screened, or in which a waste is otherwise directly emplaced.

INJECTION PUMP *See* **Pump.**

INJECTION WELL Well drilled into an impervious strata, cased as necessary if it passes through permeable strata, into which a (usually waste) liquid is poured and left. *Also called* **Disposal well.**

INJECTION ZONE Geological formation receiving fluids through a well.

INLAND WATERS Those waters of e.g. continental North America in the inland zone, waters of the Great Lakes, and specified ports and harbors on inland rivers.

INLET 1. Port through which matter or material will pass in one direction; **2.** Connection to a closed drain.

INLET CONTROL Control of the upstream inlet to a storage facility so as to influence the relationship between the headwater elevation and discharge.

INLET TIME Time that storm water takes to flow from the most distant point of a drainage system to the point where it enters a storm drain.

INLET WELL Sump or surface opening to which water is conducted and that connects to a sewer.

INNER LINER Continuous layer of material placed inside a tank or container which protects the construction materials of the tank or container from the contained waste or reagents used to treat the waste.

INNINGS Land reclaimed from the sea or a marsh.

INOCULATE Introduction of a seed culture into a system.

INNOVATIVE CONTROL TECHNOL-OGY System of air pollution control that has not been adequately demonstrated in practice, but which would have a substantial likelihood of achieving greater continuous emissions reduction than any control system in current practice, or of achieving at least comparable reductions at lower cost in terms of energy, economics, or non-air quality environmental impacts.

INNOVATIVE TECHNOLOGY 1. Production process, a pollution control technique, or a combination of the two which has not been commercially demonstrated; **2.** Developed wastewater treatment processes and techniques which have not been fully proven under the circumstances of their contemplated use and which represent a significant advancement over the state of the art in terms of significant reduction in life-cycle cost or significant environmental benefits through the reclaiming and reuse of water, otherwise eliminating the discharge of pollutants, utilizing recycling techniques such as land treatment, more efficient use of energy and resources, improved or new methods of waste treatment management for combined municipal and industrial systems, or the confined disposal of pollutants so that they will not migrate to cause water or other environmental pollution.

INORGANIC MATTER Chemical substance of mineral origin.

INORGANIC SOLID DEBRIS Nonfriable inorganic solids contaminated with hazardous wastes that are incapable of passing through a 9.5 mm (0.374 in.) standard sieve; and that require cutting, or crushing and grinding in mechanical sizing equipment prior to stabilization; and, are limited to the following inorganic or metal materials: metal slags (either dross or scoria), glassified slag, glass, concrete (excluding cementitious or pozzolanic stabilized hazardous wastes), masonry and refractory bricks, metal cans, containers, drums, or tanks, metal nuts, bolts, pipes, pumps, valves, appliances, or industrial equipment, scrap metal.

IN-PROCESS CONTROL TECHNOL-OGY Conservation of chemicals and water throughout production operations to reduce the amount of wastewater to be discharged.

INSANITARY Not sanitary.

INSECTICIDES Substances or mixtures intended for preventing or inhibiting the establishment, reproduction, development, or growth of, destroying or repelling any member of the Class Insecta or other allied classes in the Phylum Arthropoda declared to be pests. Insecticides include, but are not limited to: plant protection insecticides intended for use directly or indirectly against insects or allied organisms that attack or infest plants or plant parts, to prevent or mitigate their injury, debilitation, or destruction; animal protection insecticides intended for use directly or indirectly against insects or allied organisms that attack or infest man, other mammals, birds, or certain other animals, to prevent or mitigate their injury, irritation, harassment, or debilitation; premise and indoor insecticides intended for use directly or indirectly against insects or allied organisms to prevent or mitigate their decimation or contamination of man's stored food and animal feeds, injury to raw or manufactured goods, or weakening or destruction of buildings and building materials; and biological insect control agents such as specific pathogenic organisms or entities prepared and utilized by man.

INSERT VALVE Shutoff valve that can be inserted into a pipeline while the line is in service and under pressure.

INSIDE OF LEVEE *See* **Back of levee.**

in situ In its original place; in position.

IN-SITU LEACH METHODS Processes involving the purposeful introduction of suitable leaching solutions into an ore body to dissolve the valuable minerals in place and the purposeful leaching of ore in a static or semistatic condition either by gravity through an open pile, or by flooding a confined ore pile.

IN-SITU SAMPLING SYSTEMS Nonextractive samplers or in-line samplers.

INSOLATION Amount of solar radiation received on a surface.

INSPECTION Examination of work in progress or completed to determine its compliance with contract requirements.

INSPECTION CERTIFICATE Notification that something has been inspected and found to meet the necessary level of installation and/or performance.

INSPECTION CHAMBER Shaft between grade and the invert level of a sewer, giving access for personnel.

INSPECTION COVER Plate that can be removed to allow inspection of whatever lies behind or beyond.

INSPECTION DOOR Panel or trap that can be opened and which will allow a worker to pass through.

INSPECTION FITTING Access eye.

INSPECTION LIST List, made near the completion of work, indicating items to be furnished or work to be done by the contractor or subcontractor to complete the work as specified. *Also called* **Punch list.**

INSPECTOR Authorized representative assigned to make detailed inspections of contract performance.

INSTALL To place or fasten in position ready for use.

INSTALLATION INSPECTOR Person who, by reason of his knowledge of the physical sciences and the principles of engineering, acquired by a professional education and related practical experience, is qualified to supervise the installation of mechanical equipment.

INSTANTANEOUS EFFLUENT LIMITATION Maximum allowable concentration in a wastewater discharge at any time, as measured in a grab sample.

INSTITUTIONAL SOLID WASTE Solid wastes originating from educational, health care, correctional, and other institutional facilities.

INSTITUTIONAL WASTEWATER Wastewater from institutions such as hospitals, sanitoriums, etc.

INSTRUCTIONS TO BIDDERS Requirements contained in the bidding documents describing procedures for submitting bids.

INSTRUCTIONS TO CONTRACTOR *See* **Jobsite instructions.**

INSTRUMENT PANEL Board on which measuring and recording instruments, switches, and controls are mounted.

INSTRUMENT SHELTER Naturally or artificially ventilated structure, used to shield temperature-measuring instruments from direct sunshine and precipitation.

INSULATED FLANGE Coupling used with metal pipes to interrupt electrical conductivity.

INSULATED STREAM Stream or section of a stream that is separated from the saturation zone through which it passes and from or to which it neither receives nor contributes water.

INTAKE Portion of a pipe, pump, or structure through which water or air enters from the source of supply.

INTAKE AREA Surface area on which water that eventually reaches an aquifer or groundwater basin is initially absorbed.

INTAKE HEADING Point where a channel enters a control works.

INTAKE HOSE Noncollapsible hose used to duct water into a pump.

INTAKE MANIFOLD Pipes and fittings that connect a source to several destinations; in an internal combustion engine, connecting the air cleaner outlet air pipe to each cylinder inlet.

INTAKE PIPE 1. Pipe at the head of a pipeline or system through which water enters; **2.** Pipeline conveying water by gravity from a source to an intake well.

INTAKE SCREEN Screen used to prevent foreign objects from entering the intake of a pump.

INTAKE WORKS 1. Structures at the location where water is taken from a source into a conduit for transportation to other locations; **2.** Structures used to adjust elevation and flow rates at the point where a sewer system enters a wastewater treatment facil-ity.

INTANGIBLE ASSET Element of non-physical value applied to permanent property.

INTEGRAL HOLDING VALVE *See* **Valve.**

INTEGRAL VISTA View perceived from within the mandatory area of a specific landmark, or panorama located outside such a boundary.

INTEGRATED CURBSIDE COLLECTION Residential area service combining collection of source-separated, recyclable materials and refuse simultaneously.

INTEGRATED FLOW CURVE Curve representing the summation of all preceding quantities.

INTEGRATED POLLUTION CONTROL Concept whereby all major emissions are considered simultaneously, rather than in isolation, with limitations and controls measures designed in concordance.

INTEGRATED SOLID WASTE MANAGEMENT Solid waste management strategy that ranks the preferred alternatives, typically in the following order: source reduction, recycling, resource recovery, and landfill disposal.

INTEGRATED WASTE MANAGEMENT Strategy for the management of all types of waste involving a range of proven environmentally sound techniques and processes.

INTEGRATOR Instrument that indicates the total quantity of flow through a measuring device.

INTENSIFIER Device that converts low-pressure fluid to a higher-pressure fluid power.

INTENSITY Amount or degree of strength of heat, light, sound, etc., per unit of area, volume, etc.

INTERBEDDED Laid down in sequence: between one layer and another, or in a repetitive sequence of layers.

INTERCEPTING CHANNEL Channel excavated so as to intercept surface runoff or flow.

INTERCEPTING DITCH Ditch that prevents the flow of surface water over the slopes of a cut or against the foot of an embankment.

INTERCEPTING DRAIN Drain constructed between a source of ground, or surface water and the area to be protected. *Also called* **Curtain drain.**

INTERCEPTING SEWER Sewer that receives the dry-weather flow from a number of transverse sewers or outlets, with or without a determined quantity of storm water.

INTERCEPTION 1. Process by which precipitation is caught and held by vegetation and lost by evaporation; **2.** Process of diverting wastewater from a main sewer into a conduit leading directly to a treatment facility.

INTERCEPTOR Receptacle that is installed to prevent oil, grease, sand, or other materials from passing into a drainage system.

INTERCEPTOR TRENCH Trench filled with stone or gravel that intercepts excess water runoff before it reaches a building.

INTERCEPTOR SEWER Sewer designed for one or more of the following purposes:

(a) To intercept wastewater from a final point in a collector sewer and convey such wastes directly to a treatment facility or another interceptor;

(b) To replace an existing wastewater treatment facility and transport the wastes to an adjoining collector sewer or interceptor sewer for conveyance to a treatment plan;

(c) To transport wastewater from one or more municipal collector sewers to another municipality or to a regional plant for treatment; or

(d) To intercept an existing major discharge of raw or inadequately treated wastewater for transport directly to another interceptor or to a treatment plant.

INTERCONNECTION *See* **Backflow connection.**

INTERCONNECTOR Pipe that joins two or more water supply systems.

INTERFACE Boundary layer between two fluids of dissimilar characteristics.

INTERFERENCE 1. Overlap of influence between two wells pumping from the same aquifer; **2.** Discharge that, alone or in conjunction with a discharge or discharges from other sources, inhibits or disrupts a waste treatment processes or operations, or its sludge processes, use or disposal.

INTERFERENCE ANGLE Difference in angle between a valve seat and its mating valve.

INTERFLOW That portion of rainfall that infiltrates into the soil and moves laterally through the upper soil horizons until intercepted by a stream channel or until it returns to the surface at some point down slope from its point of infiltration.

INTERFLUVE Area of land between two adjacent streams.

INTERGLACIAL Period of ice retreat between Ice Ages.

INTERIM MAXIMUM ACCEPTABLE CONCENTRATION (IMAC) Substances for which there are insufficient toxicological data to derive a maximum acceptable concentration with reasonable certainty, and which are recommended, taking into account the available health-related data, but employing a larger safety factor. Such concentrations may also be established for those substances for which estimated lifetime risks of cancer are greater than essentially negligible.

INTERIOR PLAIN Plain distant from an ocean and separated from it by mountains.

INTERLOCK Mechanism in which the action of one device is dependant upon the action or setting of a coupled device.

INTERLOCKING Binding of particles one with another.

INTERMEDIATE BELT Portion of the zone of aeration lying between the soil-water belt and capillary fringe.

INTERMEDIATE COVER Soil material placed on a completed landfill lift to act as a layer between the completed lift and a planned lift on top of it.

INTERMEDIATE GROUNDWATER Water in the zone of aeration between fringe and soil water.

INTERMEDIATE HANDLER Facility that either treats regulated medical waste or destroys it, but which does not do both.

INTERMEDIATE-LEVEL WASTES Radioactive wastes produced by the nuclear industry and consisting of substances used to clean gases and liquids prior to their discharge, sludges from cooling ponds where spent fuel is stored, and material contaminated with plutonium.

INTERMEDIATE TREATMENT Removal from wastewater of a high percentage of suspended solids and a substantial percentage of colloidal matter, but relatively little dissolved matter.

INTERMEDIATE VADOSE WATER Subsurface water lying between the soil–water belt and capillary fringe.

INTERMEDIATE WATER Water lying between the upper surface of the capillary fringe and the soil–water belt.

INTERMITTENT CONTROL SYSTEM Dispersion technique which varies the rate at which pollutants are emitted to the atmosphere according to meteorological conditions and/or ambient concentrations of the pollutant, in order to prevent ground-level concentrations in excess of applicable ambient air quality standards.

INTERMITTENT FILTER Natural or artificial bed of fine-grained material to which wastewater is intermittently applied in flooding doses and through which it passes.

INTERMITTENT HORSEPOWER *See* **Horsepower.**

INTERMITTENT INTERRUPTED

STREAM Stream with intermittent stretches separated by ephemeral stretches.

INTERMITTENT PERIODIC SPRING Period spring that discharges only intermittently.

INTERMITTENT SAMPLING Sampling successively for limited periods and/or at intervals throughout an operation or for a predetermined period.

INTERMITTENT SPRING Spring that discharges only during certain predictable periods.

INTERMITTENT STREAM Stream that flows only in direct response to precipitation.

INTERMITTENT VAPOR PROCESSING SYSTEM Vapor processing system that employs an intermediate vapor holder to accumulate total organic compound vapors, and treats the accumulated vapors only during automatically controlled cycles.

INTERNAL COMPACTION TRANSFER SYSTEM Reciprocating action of a hydraulically powered bulkhead contained within an enclosed trailer that packs and compresses solid waste.

INTERNAL CONTROL Plan of organization and other coordinate methods and measures adopted to safeguard assets, produce accurate and reliable accounting data, promote operational efficiency, and encourage adherence to prescribed managerial policies.

INTERNAL DIAMETER Size of a pipe bore; the usual description of a metal pipe, which includes the wall thickness; the usual description of a plastic pipe, which does not include the wall thickness.

INTERNAL FLOATING ROOF Cover that rests or floats on the liquid surface inside a waste management unit that has a fixed roof.

INTERNAL INSPECTION Part of a sewer system evaluation survey that involves inspecting, by physical, photographic, and/or TV methods, sewer lines that have previously been cleaned.

INTERNAL PIPEWORK Building services that supply hot and cold water, gas, sprinklers, plus sanitary pipework.

INTERNAL THREAD Thread on the inside of a pipe or part.

INTERNAL TREATMENT Water treatment chemicals fed into a boiler, rather than into the water before it enters the boiler.

INTERNAL WATER Water within the lithosphere at depths where the pressure of overlying rocks prevents the existence of interstices.

INTERNATIONAL NUCLEAR EVENT SCALE (INES) System that classifies nuclear events at seven levels, the lowest three of which apply to incidents of increasing seriousness and the upper four to accidents of increasing seriousness.

INTERNATIONAL SYSTEM OF UNITS International metric system (Système International d'Unités) or SI. *See* appendix **Metric and nonmetric measurement** for complete details.

INTRINSICALLY SAFE EQUIPMENT Equipment and associated wiring in which any spark or thermal effect, produced either normally or in specified fault conditions, is incapable, under certain prescribed conditions, of causing ignition of a mixture of flammable or combustible material in air in its most easily ignitable concentration.

INTRINSIC ENERGY VALUE Energy either directly consumed in the manufacture of a product or the energy needed to produce feedstock.

INTERRUPTED WATERTABLE Watertable with a pronounced dip along a groundwater dam.

INTERSTICE Pore or open space in rock or granular material.

INTERSTITIAL ICE Ice below the surface of the lithosphere, formed from frozen saturated or unsaturated rock or soil.

INTERSTITIAL WATER Water within the interstices of rock.

INTERSTREAM GROUNDWATER RIDGE Ridge in the watertable between two effluent streams.

INTRAPERMAFROST WATER Water in unfrozen layers, lenses or veins within the permafrost.

INTRINSIC RATE OF INCREASE Rate of population increase, measured by deducting the instantaneous death rate from the instantaneous birth rate.

INUNDATE Cover with water, as in the case of flooding.

IN VACUUM SERVICE Equipment operating at an internal pressure that is at least 5 kPa below ambient pressure.

INVENTORY Detailed list showing descriptions, quantities and values of property.

INVERSE SLUDGE INDEX Reciprocal of the sludge volume index multiplied by 100. *Also called* **Sludge density index.**

INVERSION Horizontal layer of air through which temperature increases with increasing height.

INVERT Inside bottom level of a pipe, trench, or tunnel.

INVERTED CAPACITY Maximum rate at which a well will dispose of water admitted at or near its upper end by discharge through openings at lower levels.

INVERTED DRAINAGE WELL Well installed to drain swampy land or dispose of stormwater or wastewater at or near the surface.

INVERTED SIPHON Pipe, usually a gravity-flow sewer, that dips below an obstacle, typically a road, causing the flow to continue at almost the same level. *See also* **Dive culvert.**

INVERTED WELL Well in which the movement of water is down, rather than up.

INVERT LEVEL Level of the lowest part of the invert.

IN VHAP SERVICE Equipment that either

contains or contacts a fluid (liquid or gas) that is at least 10% by weight a volatile hazardous air pollutant (VHAP).

INVITATIONAL BIDDING See **Selective bidding.**

INVITATION FOR BID Advertisement for proposals for all work or materials on which bids are required. Such advertisements will indicate, with reasonable accuracy, the quantity and location of the work to be done or the character and quantity of the material to be furnished and the time and place of the opening of proposals.

INVITATION FOR PROPOSALS See **Call for proposals.**

INVITED BIDDERS Bidders selected by the owner and his agent as the only ones from whom bids will be received. *Also called* **Closed list of bidders,** and **Selected bidders.**

IN VOC SERVICE Equipment that contains or contacts a process fluid that is at least 10% volatile organic compound (VOC) by weight.

INVOICE Document issued as a request for payment; by a contractor in accordance with the terms of his contract for works performed by him.

INVOLUNTARY LIEN Lien imposed against property without consent of the owner.

INWARD-FLOW TURBINE Reaction turbine in which water or steam enters the runners from outside and flows toward the axis of the runner.

IODINE-131 Radioactive, poisonous and very mobile in the environment where it is found mainly as a result of nuclear explosions and releases from reactors and fuel reprocessing facilities, this radionuclide has a half-life of 8 days and a maximum acceptable concentration in drinking water of 9 Bq/L.

IODOMETRIC CHLORINE TEST Determination of residual chlorine in water or wastewater by the addition of potassium iodide and titration of the liberated iodine with a standard solution of sodium thiosulfate, with starch solution as a colorimetric indicator.

ION Electrically charged atom or group of atoms.

ION EXCHANGE Chemical exchange of a dissolved substance in water for another, typically, zeolites used in base exchange water softening that take in calcium and magnesium in exchange for the sodium they give out.

ION EXCHANGE RESINS Synthetic zeolites used in base exchange water conditioning.

ION EXCHANGE TREATMENT Use of ion-exchange materials to remove undesirable ions from a liquid in substitution of acceptable ions.

IONIC CONCENTRATION Concentration of any ion is solution.

IONIZATION Process of adding electrons to, or removing them from, atoms or molecules, thereby creating ions.

IONIZING RADIATION Radiation that is capable of energizing atoms sufficiently to remove electrons from them.

IONOSPHERE Region of the upper atmosphere beginning at a height of 90 km (56 miles) above the Earth's surface which includes the high-ionized Appleton and Kennelly–Heavyside layers which enable intercontinental radio transmission.

ION-SELECTIVE ELECTRODE Electrode having a high degree of selectivity for one ion over other ions which may be present in a sample.

IRON Fourth most abundant element in the Earth's crust; a malleable tenacious metallic element, symbol Fe. The aesthetic objective for iron in drinking water is <0.3 mg/L.

IRON BACTERIA See **Bacteria.**

IRON PIPE THREAD Standard system of threads for connecting various types of rigid piping.

IRON VITRIOL Common name for copperas: ferrous sulfate.

IRRADIATION Exposure to electromagnetic waves.

IRREGULAR WEIR Weir having a nonstandard crest.

IRRIGABLE AREA Portion of the arable area of an irrigation project suitable for irrigated farm use.

IRRIGABLE LAND Land physically suited for sustained irrigation and for which adequate and suitable water can be provided at reasonable cost.

IRRIGATED AREA Gross area to which water is artificially applied for the production of crops.

IRRIGATING HEAD Flow of water used for a particular area of land or specific irrigation ditch, or which is rotated among a group of irrigators.

IRRIGATING STREAM Flow used for irrigation.

IRRIGATION Artificial distribution of water to promote plant growth.

IRRIGATION CANAL Channel built and used primarily to transport water to be used to irrigate crops.

IRRIGATION DISTRICT Administrative area established for the purpose of financing, constructing and operating an irrigation system.

IRRIGATION DITCH Ditch that conveys water used for irrigation of crops.

IRRIGATION EFFICIENCY Fraction of irrigation water delivered to the point of distribution and available in the soil for consumptive use by crops.

IRRIGATION REQUIREMENT Quantity of water, including wastes but excluding rain, needed to grow a crop.

IRRIGATION ROTATION Technique of delivering the entire flow of irrigation water in a lateral ditch or pipeline that serves several users as a large flow to each user in rotation according to a negotiated formula.

IRRIGATION STRUCTURE Physical works required for the efficient conveyance, control, measurement, or application of irrigation water.

IRRIGATION WATER Water artificially delivered to and applied to lands to meet the moisture needs of growing plants.

IRRIGATION-WATER REQUIREMENTS Quantity of water required to be delivered and applied by artificial means for crop production.

ISLAND Area of land surrounded by water.

ISOBAR Line of equal barometric pressure.

ISOBATH Line of a map connecting all points having the same vertical distance above a plane of interest (which, unlike the reference for a contour, need not be in a horizontal plane).

ISOCHION Line of equal snow depths or equal water content of snow. *Also called* **Isonival.**

ISOCHLOR Line on a map connecting all points of equal concentrations of chlorides — in groundwater, for instance, tidal estuarial, soil, etc.

ISOCHRONE Plot of the drainage of a river or sewer system on which the times of various incidents are placed showing the history of water as is originates at one place and is discharged from the system at another.

ISOELECTRIC POINT Point of electrical neutrality.

ISOHEL Line drawn on a map to connect points of equal average hours of sunlight.

ISOHYET Line on a map connecting all points of equal precipitation.

ISOHYETRAL MAP Map containing isohyets from which the distribution of precipitation over an area for a given period can be read.

ISOLATED INTERSTICE Small, open

space in otherwise homogeneous rock.

ISOLATED SOLAR GAIN Passive solar heating system in which heat is collected at one location for use at another.

ISOLATING VALVE *See* **Valve.**

ISOLATING SWITCH *See* **Switch.**

ISOMERS Compounds identical with respect to number and kind of constituent atoms, but of different arrangement.

ISONIVAL Line of equal snow depths or equal water content of snow. *Also called* **Isochion.**

ISOPERCENTIL Having equal percentages.

ISOPHYTOCHRONE Line drawn on a map that connects points where the growing season for plants is the same length.

ISOSEISMIC LINE Imaginary line that joins points of equal magnitude of earthquake shock.

ISOTACH Line of equal wind speed.

ISOTHERMAL Having the same temperature.

ISOTHERMAL COMPRESSION Compression of air at constant temperature.

ISOTOPE Atoms of the same element that have the same number of protons in their nuclei but differing numbers of neutrons, giving them the same atomic number but different mass numbers.

ISOTROPIC Having the same physical properties in all directions.

ISOTROPIC SOIL Soil mass having essentially the same properties in all directions at any given point.

ISOTROPY Behavior of a medium having the same properties in all directions.

ITEM Nomenclature for a particular kind of work to be performed or material to be supplied under the contract documents.

J

JACKET Insulation that covers exposed heating and cooling pipes.

JACKETED PUMP Pump having jackets around its cylinders, heads, and stuffing boxes, through which steam or other heat may be forced to permit the handling of materials that, when cold, exhibit high viscosity.

JACKETING Surrounding a prime flow with a confined bath or flow of fluid for temperature control or heat absorption.

JACKING PIPE Pipe installed using pipe jacking techniques to form part of a conduit.

JAR TEST Laboratory procedure that simulates coagulation/flocculation with differing chemical doses.

JET 1. Nozzle containing a small-diameter exit orifice to which a pipe or tube is fitted, having the effect of controlling the air or fluid being expelled; **2.** Stream coming from a nozzle.

JET ACTION Valve design type in which flow effect is controlled by the relative position of a nozzle and a receiver.

JET HEIGHT Vertical distance to which an unconfined jet of water rises above the orifice or opening from which it issues.

JET PUMP *See* **Pump.**

JET STREAM Current of air travelling at very high speed (often more than 350 km/hr (430 mph)) from west to east at altitudes between 13 and 20 km (8 and 12.4 miles).

JETTING 1. Use of a water jet to aid in the placing or driving of a well through hydraulic displacement of parts of the soil. *Also called* **Wash boring**; **2.** Process of injecting great amounts of water through a hose into soil and granular material to speed the process of compaction.

JETTY Structure or long body of fill extending into a body of water from the shore, that serves to aid in access to deeper water so as to load and unload vessels, and to change the direction or velocity of the water flow.

JOB ACCOUNT Account specific to a repetitive type of work, or to a specific part of a project, giving all details of expenditure for that purpose.

JOB LAYOUT Factors taken into account prior to pricing or commencing a soil compaction project. There are two approaches:

Project method: Small jobs are best suited to this method. Fill material is moved into the area and spread in thick lifts. Compaction then proceeds over the entire area until density is reached. Then, another lift is spread and compacted, with the process repeated until correct grade is reached.

Progressive method: Here, with continuous operation of the haul and compaction equipment, material is spread progressively in front of the compactor(s) until required density and finish grade is reached.

JOB SITE Location of the project; place where the work is to be done. *Also called* **Construction site, Project site,** and **Site.**

JOBSITE INSTRUCTIONS Written clarification or interpretation of the terms, conditions, specifications, or drawings contained in a contract. Such an interpretation is rendered by the owner or his agent in response to a request for an urgent reply from the site of the work. *Also called* **Instructions to contractor.**

JOB SPECIFICATION Specification specific to a project that may have to be read along with a standard specification.

JOB SUPERINTENDENT Person, on site, responsible for the works.

JOGGING Frequent stop/start cycling of an electric motor.

JOINT 1. Fracture in rock, often across bedding planes, along which little or no movement has taken place; **2.** Where two pipes are connected either by bolting, welding, or by screwed connection.

JOINTING COMPOUND Mastic smeared over the threads of a pipe joint to make it leakproof.

JOINT RING Rubber or plastic ring used to make a push-fit joint in a drain pipe.

JOINT RUNNER Tool composed of asbestos rope and a clamp, used in leading (leding) joints in horizontal runs of bell-and-spigot cast iron pipe.

JOINT SEALANT Compressible material used to exclude water and solid foreign materials from joints.

JOINT-SEALING COMPOUND Impervious material used to fill joints.

JOINT TAPE Polytetrafluorethylene tape used on the thread of pipe fittings to ease the fitting together and help create a tight joint.

JOULE SI unit of energy, work done when the point of application of a force of 1 N is displaced through a distance of 1 m in the direction of the force. (1 J=1 N.m). Symbol: J. Multiply by 0.000 9 to obtain British thermal units, symbol: Btu; by 0.737 56 to obtain foot-pounds, symbol ft/lb; by 0.3725 x 10^6 to obtain horsepower-hour, symbol: hp/h; by 0.2778 x 10^6 to obtain kilowatt hour, symbol: kW.h; and by 1.0 to obtain newton meters, symbol N.m. *See also the appendix*: **Metric and nonmetric measurement.**

JOULE PER KELVIN A derived unit of heat capacity with a compound name of the SI system of measurement. Symbol: J/K. *See also the appendix*: **Metric and nonmetric measurement.**

JOULE PER KILOGRAM A derived unit of specific latent heat of the SI system of measurement. Symbol: J/kg. *See also the appendix*: **Metric and nonmetric measurement.**

JOULE PER KILOGRAM KELVIN A derived unit of specific heat capacity of the SI system of measurement. Symbol: J/(kg.K). *See also the appendix*: **Metric and nonmetric measurement.**

JOULE PER MOLE A derived unit of molar internal energy of the SI system of measurement. Symbol: J/mol. *See also the appendix*: **Metric and nonmetric measurement.**

JUNCTION Purpose-made section of drain pipe made to join a branch to a main run. It has a short tail off the main run at a 45° angle, or can be a wye.

JUNCTION CHAMBER Access manhole in which one or more branch sewers are joined to a main sewer.

JUNCTION MANHOLE Access manhole formed over the junction of two or more sewers.

JUNK FILL Fill that has been dumped over a long period and that contains refuse of all kinds.

JUVENILE WATER *See* **Groundwater.**

K

KAME Long mound of glacial detritus: an eskar.

KANAT Horizontal tunnel collection system developed within an aquifer at the bottom of a water well.

KAOLIN Type of clay having a high aluminum content.

KAPLAN TURBINE Water turbine, the propeller blade pitch of which is automatically varied according to the demand load.

KARST Geologic setting where cavities are developed in massive limestone beds by solution in flowing water. *Also called* **Karstic limestone.**

KATABATIC WIND Wind caused by cold air flowing downhill.

KEEL Plumbers colored marking crayon.

KELVIN SI unit of thermodynamic temperature equal to the fraction 1/273.16 of the thermodynamic temperature of the triple point of water. Symbol: K. The Kelvin scale has its origin, or zero point, at absolute zero (the point at which all atomic vibration ceases). Its fixed point of 273.16 K or 0°C is the temperature at which water exists in vapor, liquid, and solid state. *See also the appendix:* **Metric and nonmetric measurement.**

KESSENER BRUSH Large cylinder with paddle-shaped protrusions throughout its surface which, when mounted across the width of an aeration ditch and forced to rotate, promotes flow and agitates the contents of the ditch to entrain oxygen.

KEY EVENT Planning and scheduling terms that correspond to a stage of work progress.

KEY EVENT SCHEDULE *See* **Master schedule.**

KEY INTERLOCK Mechanism that permits insertion, operation, or removal of a key to a piece of equipment only if certain conditions have been met, or prescribed sequences of operations has been completed.

KEY-OPERATED SWITCH *See* **Switch.**

KEY SWITCH *See* **Switch.**

KEY VALVE *See* **Valve.**

KILO Prefix representing 10^3. Symbol: k. Used in the SI system of measurement. *See also the appendix:* **Metric and nonmetric measurement.**

KILOGRAM SI unit of mass equal to the mass of the international prototype of the kilogram, a cylinder of platinum–iridium alloy, kept by the International Bureau of Weights and Measures in Paris, France. Symbol: kg. Multiply by 35.274 to obtain ounces, symbol oz; by 2.205 to obtain pounds, symbol lb; by 0.000 98 to obtain long tons (2240 lb); and by 0.001 to obtain short tons (2000 lb). *See also the appendix:* **Metric and nonmetric measurement.**

KILOGRAM-FORCE Obsolete unit of force, equal to 9.8 Newtons, that should not be used with the SI system of measurement. Symbol: kg·f. *See also the appendix:* **Metric and nonmetric measurement.**

KILOGRAM METER PER SECOND A derived unit of momentum of the SI system of measurement. Symbol: k·m/s. *See also the appendix:* **Metric and nonmetric measurement.**

KILOGRAM METER SQUARED A derived unit of the moment of inertia of the SI system of measurement. Symbol: kg·m². *See also the appendix:* **Metric and nonmetric measurement.**

KILOGRAM METER SQUARED PER SECOND A derived unit of angular momentum of the SI system of measurement. Symbol: kg·m²/s. *See also the appendix:* **Metric and nonmetric measurement.**

KILOGRAM PER CUBIC METER A derived unit of density (mass density) of the SI system of measurement. Symbol: kg/m³. *See also the appendix:* **Metric and nonmetric measurement.**

KILOGRAM PER LITRE A derived unit of density of the SI system of measurement. Symbol: kg/L. *See also the appendix:* **Metric and nonmetric measurement.**

KILOGRAM PER MOLE A derived unit of molar mass of the SI system of measurement. Symbol: kg/mol. *See also the appendix:* **Metric and nonmetric measurement.**

KILOGRAM PER SECOND A derived unit of mass flow rate of the SI system of measurement. Symbol: kg/s. *See also the appendix:* **Metric and nonmetric measurement.**

KILOHERTZ Unit of frequency, equal to 1000 hertz, of the SI system of measurement. Symbol: kHz. *See also the appendix:* **Metric and nonmetric measurement.**

KILOKELVIN Unit of temperature, equal to 1000° kelvin, of the SI system of measurement. Symbol: °kK. *See also the appendix:* **Metric and nonmetric measurement.**

KILOJOULE Unit of energy, equal to 1000 joules in the SI system of measurement. Symbol: kJ. *See also the appendix:* **Metric and nonmetric measurement.**

KILOLITER Unit of volume, equal to 1000 liters. Permitted for use with the SI system of measurement. Symbol: kL. *See also the appendix:* **Metric and nonmetric measurement.**

KILOMETER Unit of length, equal to 1000 meters in the SI system of measurement. Symbol: km. *See also the appendix:* **Metric and nonmetric measurement.**

KILOMETER PER HOUR Unit of speed, equal to 1000 meters covered in one hour. Permitted for use with the SI system of measurement. Symbol: km/h. *See also the appendix:* **Metric and nonmetric measurement.**

KILONEWTON Unit of force, equal to 1000 newtons in the SI system of measurement. Symbol: kN. *See also the appendix:* **Metric and nonmetric measurement.**

KILOPASCAL Unit of pressure, equal to 1000 pascals in the SI system of measurement. Symbol: kPa. *See also the appendix:* **Metric and nonmetric measurement.**

KILOVOLT-AMPERE Electrical unit of power, 1000 volt amperes (apparent power), equivalent to about 0.89 kW.

KILOVOLT AMPS REACTIVE 1000 volt amps reactive (reactive power).

KILOWATT Unit of power, equal to 1000 watts in the SI system of measurement. Symbol: kW. *See also the appendix:* **Metric and nonmetric measurement.**

KILOWATT HOUR Non-SI unit of energy, equal to 3.6 megajoules, permitted for use with the SI system of measurement for a limited time. Symbol: kW·h. Multiply by 3412 to obtain British thermal units, symbol: Btu; by 1.3405 to obtain horsepower hours, symbol: hp/h. *See also the appendix:* **Metric and nonmetric measurement.**

kin. *Abbreviation for* kinetic.

KINEMATIC VELOCITY Ratio of absolute velocity to density.

KINEMATIC VELOCITY COEFFICIENT Ratio of the coefficient of absolute viscosity of a fluid to its unit weight.

KINETIC ENERGY Energy possessed by a body resulting from its motion.

KINETIC FLOW FACTOR Degree of turbulence or of tranquillity that prevails in flowing water.

KINETIC FRICTION COEFFICIENT Factor used as an index of the force necessary to keep a body sliding at a uniform rate on the surface of another body.

KINETIC HEAD Theoretical vertical height

through which a liquid body must be raised by virtue of its kinetic energy, equal to the square of the velocity divided by two times the acceleration due to gravity.

KIP Nonmetric unit of mass, equal to 1000 lb. Symbol: kip. Multiply by 4.448 to obtain kilonewtons, symbol: kN. *See also the appendix:* **Metric and nonmetric measurement.**

KIP PER SQUARE INCH Nonmetric unit of force. Symbol: kip/in.^{2}. Multiply by 6.895 to obtain megapascals, symbol: MPa. *See also the appendix:* **Metric and nonmetric measurement.**

KJELDAHL NITROGEN Nitrogen in the form of organic proteins or their decomposition product, ammonia, as measured by the Kjeldahl method.

KNIFE-BLADE SWITCH *See* **Switch.**

KNIFE SWITCH *See* **Switch.**

KOLK Rising turbulent current that lifts denser water, or coarse material, to the surface where it spreads and moves laterally: a vortex action.

KNUCKLE BEND Pipe bend that turns as sharply as possible.

KRAUS PROCESS Modification of the activated sludge process in which aerobically conditioned supernatant liquid from anaerobic sludge digesters is added to activated sludge aeration tanks to improve the settling characteristics of the sludge and to add oxygen in the form of nitrates.

L

boom. *Also called* **Ladder trencher.** *See also* **Ditcher** and **Wheel ditcher.**

LADDER DREDGE Endless chain fitted with buckets and mounted on a scow.

LAG To wrap pipes either to reduce heat losses or prevent freezing.

LAGGING Planks placed horizontally between soldier piles in a shored or braced excavation.

LAGOON 1. Shallow lake near a river or the sea, due to the infiltration or overflow of water from the larger body; **2.** Pond in which aerobic or anaerobic stabilization of raw or partially treated wastewater occurs.

LAGOONING Retention of raw or partially-treated wastewater prior to, or as an intermediate step, in a treatment process.

LAHAR Mud flow of water-saturated volcanic ash.

LAID LENGTH Length of a completed pipeline.

LAKE Large sheet of water, entirely surrounded by land.

LAMBERT Non-SI unit of brightness equal to 10 lumens/mm², 1000 millilamberts, 10^4/ μ candelas/m², 929 footcandles; the average brightness of a surface emitting or reflecting one lumen per square centimeter.

LAMINAR FLOW Water flow in which the stream lines remain distinct and in which the flow direction at every point remains unchanged with time. *See also* **Flow.**

LAMINAR (STREAMLINE) FLOW Flow situation in which fluid moves in parallel lamina or layers. *See also* **Reynold's number.**

LAMINAR VELOCITY Speed in a specific horizontal stream of water, distinct from those above or below it, in which streamline or turbulent flow may occur.

LAMINATION Layer or sheet of material of similar substance throughout, and within which similar physical phenomena occur.

LABELLED Equipment or materials to which has been attached a label, symbol, or other identifying mark of a qualified testing laboratory, indicating compliance with appropriate standards of performance in a specified manner.

LABOR Cost of services of employees performing physical or mechanical tasks, usually paid as wages rather than salaries.

LABOR AND MATERIAL PAYMENT BOND *See* **Bond.**

LABORATORY SAMPLE Representative portion of a gross sample received by a laboratory for further analysis.

LABOR CONSTANT Amount of labor necessary to perform a defined unit of work.

LACUSTRINE Sedimentary deposits laid down underwater in the bed of a lake.

LACUSTRINE PLAIN Plain originally formed as the bed of a lake from which the water has disappeared.

LADDER Digging-boom assembly of a hydraulic dredge or chain bucket-type excavator.

LADDER DITCHER Trench digger that operates by means of buckets mounted on a pair of chains travelling on the outside of a

LAMP HOLE Vertical shaft, 225 mm (9 in.) in diameter, within an inspection chamber down which an inspection lamp can be lowered to the invert of the sewer, the light from which, if seen from the next manhole up- or downstream, will indicate a sewer free of major obstruction or damage.

LAND Solid portion of the Earth's surface.

LAND ABUSE *See* **Land-use control.**

LAND ACCRETION *See* **Land reclamation.**

LAND AREA *See* **Land-use classes (forest management).**

LAND ASSEMBLY Acquisition of, or option to acquire, contiguous parcels of land with the intent to create an area for development, or reserve against development.

LAND BANKING Acquisition of land, usually by a public authority, with the intent of reselling the whole or part of the area for designated types of development at some future date.

LAND BATTURE Portion of a river bank which may be inundated at comparatively infrequent intervals.

LAND BOUNDARY Demarcation line between adjoining parcels of land.

LAND-CLEARING RAKE Heavily constructed equipment attachment, used to cut and collect brush and grub small roots from a site, or a logged-over area.

LAND DEVELOPMENT Work done on land either to bring it to a stage of higher or better use or in anticipation of a future construction project. *See also* **Development.**

LAND DISPOSAL 1. Disposal of raw or treated wastewater onto land; **2.** Placement of solid or liquid wastes in or on the land, including landfill, surface impoundment, waste pile, injection well, land treatment facility, salt dome formation, salt bed formation, underground mine or cave, or placement in a concrete vault or bunker intended for disposal purposes.

LAND DRAIN Drain for drawing off surface and subsurface water from land.

LAND DRAINAGE Process of removing water from on top of and within the ground.

LANDFILL 1. Area of land or an excavation in which wastes are placed for permanent disposal; **2.** Disposal facility where regulated medical waste is placed in or on the land and which is not a land treatment facility, a surface impoundment, or an injection well; **3.** Disposal facility where hazardous waste is placed in or on land and which is not a pile, a land treatment facility, a surface impoundment, an underground injection well, a salt dome formation, a salt bed formation, an underground mine, or a cave.

LANDFILL CELL Engineered volume in which waste is placed, compacted, and covered daily to form a portion of the landfill.

LANDFILL COMPACTOR *See* **Loader.**

LANDFILL GAS Natural by-product of the decomposition of organic material in the waste stream following disposal. Anaerobic decomposition produces the gas, which consists of approximately equal proportions of methane and carbon dioxide.

LANDFILL MACHINE Any machine that is used on a sanitary landfill; generally considered to be dozers, tractors, loaders, compactors, and/or scrapers.

LAND FILTRATION Distribution of wastewater through numerous ditches which divide agricultural land into beds which are kept moist from horizontal seepage from the ditches.

LANDFORM Physical feature of the surface of the land, such as a hill or a valley, whether resulting from natural processes or man-made.

LAND IMPROVEMENT Work done and expenditures incurred in the process of putting land into usable condition, e.g. clearing, grading, landscaping, paving, installing utilities and services, etc.

LAND RECLAMATION Gaining usable land by removing surface water, preventing ingress of water (diking), lowering groundwater, or otherwise making stable land that was previously incapable of supporting loads. *Also called* **Land accretion.**

LAND RIGHTS Rights and/or obligations associated with the ownership, occupation or use of land.

LANDSCAPING General use of plant and man-made materials on a site (as opposed to landscape architectural design).

LAND SIDE OF LEVEE *See* **Back of levee.**

LANDSLIDE Sudden downward movement of a mass of surface soil and/or rock, usually at a shear plane, often triggered by excess moisture content or seismic vibration, and frequently contributed to by changes in the ground profile in the immediate vicinity.

LANDSLIDE SPRING Spring flowing at the base of a landslide.

LAND SUBSIDENCE General lowering in elevation of a significant area of land surface due to the gradual removal of the underlying supporting material, from natural causes such as the removal of soluble material by water, or such artificial causes as withdrawal of water from artesian storage, or underground mining.

LAND TREATMENT Disposal of the effluent, and also of the digested sludge, from sewage treatment plants by distribution over arable land.

LAND TREATMENT FACILITY Facility at which hazardous waste is applied onto or incorporated into the soil surface.

LAND USE Division of available land into defined categories under such headings as use, class, ground cover, mineralization, etc.

LAND-USE ANALYSIS Study and classification of an existing land-use pattern, and the potential of the area.

LAND-USE CONTROL Public control over the manner in which land is used and/or developed. This control is exercised in several distinct ways:

> **Land abuse:** Abortive subdivision, cutover lands, etc.

> **Land-use prescription:** Zoning, building restrictions, etc.

> **Nonuse or disuse:** Taxing to enforce development and/or maintenance, clearing, etc.

> **Public protection:** Building standards, density regulations, etc.

> **Reuse control:** Urban redevelopment, slum clearing, etc.

LAND-USE DESIGNATION Type and extent of development established for an area of land defined on an Official Plan.

LAND-USE INTENSITY Expression of the intensity to which an area of land is developed in relation to an established zoning plan.

LAND-USE MAP Overall map of a community showing the character and density of land use.

LAND-USE PLAN Official determination for the future uses of land contained within an Official Plan, showing the public and private improvements to be made on the land and assumptions and reasons for arriving at the indications.

LAND-USE PLANNING Development of plans intended to produce the highest and best use of land, together with proposals as to how such use can be achieved.

LAND-USE PRESCRIPTION *See* **Land-use control.**

LAND-USE REGULATION Zoning, official maps, and subdivision regulations to guide or control land development.

LAND-USE SURVEY Record of the ways in which an area of land is used at a point in time, usually classified as commercial, industrial, public, residential, etc.

LANGELIER'S INDEX Hydrogen ion concentration that a water must contain to be in equilibrium with its calcium carbonate content.

LAPSE RATE Rate of change of air temperature with increasing elevation.

LAP WELD PIPE *See* **Pipe.**

LARGE ROUTE *See* **Collection method.**

LARGE WATER SYSTEM Water system that serves more than 50000 persons.

LATE FINISH In the critical path method, the latest possible point in time that an activity may be completed without delaying a specified milestone (usually the project finish date).

LATENT ENERGY Energy necessary for a change of state at constant temperature: melting of ice, for instance, or the vaporization of water.

LATERAL To the side.

LATERAL CANAL Canal that parallels a river.

LATERAL EARTH PRESSURE *See* **Pressure.**

LATERAL EROSION Erosion of the side walls of the banks of a stream.

LATERAL-FLOW SPILLWAY Spillway in which the initial and final flows are perpendicular to each other.

LATERAL MORAINE Moraine occurring along the sides of a valley glacier.

LATERAL SEWER Sewer that discharges into a branch or other sewer and has no other sewer tributary to it.

LATE START DATE In the critical path method, the latest possible point in time that an activity may begin without delaying a specified milestone (usually the project finish date).

LATEST EVENT OCCURRENCE TIME Deadline by which an event must be completed so as not to delay the project.

LATEST FINISH DATE Latest day a work item can finish without affecting the project duration, assuming that all subsequent work items start as soon as they are able and are completed in their expected time.

LATEST START DATE Latest day that a work item can start without affecting the final project duration, assuming that it is completed in its expected time and that all subsequent work items start as soon as they are able and are completed in their expected times.

LATRINE Multiseat toilet discharging to a single trough.

LAUNDER Clarify water by passing it through a sedimentation basin, the discharge from which passes over weir plates into discharge channels.

LAVA Molten rock flowing from a volcano or fissure in the Earth, or the rock formed by the cooling of this molten material.

LAVATORY 1. Basin for washing hands and face; **2.** Place providing sanitary facilities.

LAY BARGE Vessel from which a submerged pipeline is assembled and progressively lowered to the bottom from the stinger.

LAYOUT OF CONSTRUCTION PLANT Disposition on site of the physical properties, fixed equipment, and materials necessary to complete a project.

LC50 Lethal concentration, 50% mortality; a measure of inhalation toxicity. It is the concentration in air of a volatile chemical compound at which half the test population of an animal species dies when exposed to the compound. It is expressed as parts per million by volume of the toxicant per million parts of air for a given exposure period.

LD50 Lethal dose, 50% mortality; a general measure of toxicity. It is the dose of a chemical compound that, when administered to laboratory animals, causes death in one-half of the test population. It is expressed in milligrams of toxicant per kilogram of animal weight. The route of administration may be oral, epidermal, or intraperitoneal.

LEACH 1. To cause liquid to filter through a material; **2.** Extraction of a soluble material through immersion in water or by flowing water.

LEACHATE Liquid that has percolated through a material mass and contains dissolved or suspended mineral and/or microbial constituents.

LEACHATE RECIRCULATION Recycling or reintroduction of leachate into or on a disposal facility.

LEACH BED System of subsoil piping that permits absorption of fluids into the earth. *Also called* **Disposal field,** and **Tile bed.**

LEACHING Continuous removal of soluble matter by the dissolving action of water.

LEACHING BASIN Sand-filled pit with gravel sides into which drainage water is fed.

LEACHING FIELD System of subsoil piping that permits absorption of fluids into the earth. *Also called* **Disposal field** and **Tile bed.**

LEACHING REQUIREMENT Fraction of irrigation water entering the soil that must pass through the root zone in order to prevent salinity from exceeding a specific value.

LEAD (led) Soft, malleable, ductile, bluish-white, dense metallic element, symbol Pb. Lead is present in tap water as a result of dissolution from natural sources or from household plumbing systems containing lead in pipes, solder, or service connections to homes. The World Health Organization has established a provisional tolerable weekly intake for lead for children of 25 µg/kg bw, equivalent to an average daily intake of approximately 3.5 µg/kg bw. The maximum acceptable concentration for lead in drinking water, derived from the average daily intake, is 0.010 mg/L.

LEAD (led) ARSENATE White crystalline compound, $Pb_3(AsO_4)_2$, used in insecticides and herbicides.

LEAD (led) CAULKED JOINT Pressure-tight joint between cast-iron pipes made by filling the space between the socket and spigot with molten lead poured in behind a joint runner, pipe-jointing clip, or caulking.

LEADER Conductor that takes water from a roof and discharges it to a storm drain or other means of disposal.

LEAD (led) JOINT Molten lead poured into the joint between sections of cast-iron pipe.

LEAD (led) LINE Weighted rope used to measure the depth of water in hydrographic survey.

LEAD (led) POISONING Illness resulting from the toxic accumulation of lead in the body.

LEAD (leed) TIME Time taken in preparation for something to be done: delivery of goods to site, ordering and making special equipment, putting up temporary works, curing of concrete, etc.

LEAD (led) WOOL Lead in the form of thin strands, used to make lead caulking for joints where it does not have to be melted but can be rammed to form a watertight joint.

LEAKAGE Uncontrolled loss from, or movement between water or gas being stored or conveyed, due to hydrostatic pressure.

LEAKAGE CURRENT Unwanted flow of electricity through liquid passages, rather than through membranes and cells.

LEAK-DETECTION SYSTEM System capable of detecting the failure of either a primary or secondary containment structure or the presence of a release of hazardous waste or accumulated liquid in the secondary containment structure.

LEAK DETECTOR Device capable of determining the presence of water leaking from a pipe or vessel, pinpointing its source and estimating the rate of flow.

LEAKER Crack or hole in a tube that allows fluid to escape.

LEAN COMBUSTION Technique for the reduction of exhaust emissions from internal combustion engines which makes use of gasified petroleum as the engine fuel as compared to the conventional method of fine droplets, enabling the fuel to be burned in air-to-gasoline ratios approximating 20:1, as opposed to the conventional 15:1. This reduces the emission of NO_x and CO.

LEAPING WEIR Opening or gap in the invert of a combined sewer through which the dry-weather flow will fall to a sanitary sewer and over which a portion or the majority of the storm flow will 'leap'.

LEASEHOLD Right to use land for a specific purposes, for which consideration has been paid for a specified time period.

LEAT Channel excavated along a contour to conduct a flow to be used to generate power.

LEDGE DRAIN Type of drainage system installed in the upstream face of a concrete dam.

LEE Side away from current or flow.

LEFT BANK Bank bounding flowing water that is on the left when looking downstream.

LEGAL WATER LEVEL Elevation of a body of water that defines a riparian boundary.

LEGIONNAIRE'S DISEASE Severe pneumonia, caused by the bacterium *Legionella,* which occurs in small numbers in soil, and in water (including potable water). It is spread through the air inside sprayed droplets of water between 20°C and 50°C (68°F and 122°F) that is oxygen deficient, has iron present, and contains algae and protozoa. Such conditions can occasionally occur in large, air-conditioned buildings with incorrectly installed, or incorrectly operated cooling towers, but are also associated with evaporative condensers, whirlpool spa baths, and hot water taps and showers having a secondary circulation system.

LENGTH OF DAM Distance between end abutments, measured along the centerline axis of the crest.

LENGTH OF RUN 1. Distance that water must flow in furrows, or over the surface, between head ditches; **2.** Period of time for which one user is allowed to use the irrigating head in a rotational irrigation system.

LETHAL CONCENTRATION Concentration of a substance in air or in water that can cause death, i.e., 50% of a sample within a fixed time, typically abbreviated to LC_{50}.

LEVEE 1. Rock or earth embankment to prevent inundation or erosion; **2.** Landing place on the bank of a river.

LEVEE GRADE Slope of the centerline of the crown of a levee.

LEVEL CONTROL Device that senses and maintains the water level in a storage vessel.

LEVEL RECORDER Pressure- or float-operated device that continuously senses and records the level of a fluid in a channel or vessel.

LICHENS Compound plant formed by the symbiotic association of two organisms: a fungus and an alga.

LIDAR Method for the detection of cloud patterns and atmospheric pollutants using the particle scatter radiation in a tuned laser beam.

LIEN Claim on the property of another against payment of a debt. *See also* **Judgement lien,** and **Mechanic's lien.**

LIEN BOND *See* **Bond.**

LIEN RELEASE Legal document that assures that materials and services furnished to a project have been paid for.

LIEN WAIVER Undertaking by a person or entity who has, or may have, a right of mechanic's lien against the property of another, to relinquish such right. *Also called* **Release of lien.**

LIFE-CYCLE ANALYSIS Method for evaluating the whole life cycle of a product.

LIFE-CYCLE COST Total of the costs incurred during the design, or actual life of a product, typically including capital cost, maintenance costs, labor (operating) costs, running (fuel, supplies) costs, etc.

LIFE EXPECTANCY Length of time a device, material, construction or other fabrication may be expected to remain operating under the circumstances for which it was designed.

LIFELINE Rope, suitable for supporting one person, to which a lanyard or safety belt (or harness) is attached.

LIFE TEST Laboratory procedure used to determine the resistance of a product to a specific set of destructive forces or conditions. *See also* **Accelerated life test.**

LIFT 1. Vertical distance between a static water source and the suction chamber of a pump; **2.** Height of a column or body of fluid below a given point, expressed in linear units; **3.** In sanitary landfill, a compacted layer of solid wastes and the top layer of cover material; **4.** Vertical distance a vessel rises or falls when passing through a lock.

LIFT-AND-CARRY *See* **Waste container.**

LIFT BRIDGE Bridge, part of whose deck can be raised vertically within a fixed framework so as to provide increased clearance.

LIFT DEPTH Vertical thickness of a compacted volume of solid wastes and the cover material immediately above it in a sanitary landfill.

LIFT LOCK Canal lock serving to lift or lower a vessel from one reach of water to another.

LIFT PUMP *See* **Pump.**

LIFT STATION Wastewater pumping station positioned on a gravity sewer to raise the flow to a higher elevation.

LIGHT-DUTY DUMP BODY *See* **Truck, dump body.**

LIGHT-GAUGE COPPER TUBE Thinwall copper pipe, used for domestic plumbing (water distribution), connected with fittings and solder joints.

LIGHT RAIN Rain falling with an intensity between a trace and 2.5 mm/h (0.10 in./h).

LIGHT-WATER REACTOR Nuclear reactor that uses light water (as distinct from heavy water) as a coolant and moderator.

LIGNIN Organic phenolic polymer resistant to biological attack, that holds the cell walls and cellulose fiber of plants together.

LIME Limestone burned in a kiln until the carbon dioxide has been driven off.

LIME AND SODA-ASH PROCESS Process for softening water by the addition of lime and soda ash to form the insoluble compounds of calcium carbonate and magnesium hydroxide.

LIME BLOOM Unwanted chalky-white powder appearing on the surface of hardened concrete, usually due to the formation of insoluble calcium carbonate formed from rainwater flowing over the concrete surface immediately after striking the formwork.

LIME POWDER Powder obtained by air slaking lime.

LIME ROCK Naturally occurring calcium carbonate containing varying quantities of silica, that hardens on exposure to the atmosphere.

LIME-SODA SOFTENING Process in which calcium and magnesium ions are precipitated from water by reaction with lime and soda ash.

LIME STABILIZATION Use of burned lime products, quicklime, or hydrated lime, as an additive to plastic clayey soils and granular materials to improve water resistance and cohesive properties of the particles.

LIMESTONE Sedimentary rock consisting primarily of calcium carbonate.

LIMIT CONTROL Device that activates, or deactivates another mechanism within preset limits.

LIMITED WATER-SOLUBLE SUBSTANCES Chemicals that are soluble in water at less than 1000 mg/L.

LIMITING FACTOR Environmental factor that restricts the distribution or activity of an organism or population.

LIMITING PERMISSIBLE CONCENTRATION Liquid phase of a material which, after allowance for initial mixing, does not exceed applicable marine water quality criteria or, when there are no applicable marine water quality criteria, that concentration of waste or dredged material in the receiving water which, after allowance for initial mixing will not exceed a toxicity threshold defined as 0.01 of a concentration shown to be acutely toxic to appropriate sensitive marine organisms in a bioassay.

LIMIT OF WORKS Point or place where work by one contractor, or work in one trade, begins or ends.

LIMITS Maximum permitted dimensions of a part.

LIMITS OF OSCILLATION Distance over which a river has ranged within historic time.

LIMIT SWITCH *See* **Switch.**

LIMNETIC Living in the open water of a lake or pond.

LIMNIC Sediments deposited in freshwater lakes and the environment of their deposition.

LIMNOLOGY Study of the biological productivity and characteristics of freshwater bodies.

LINE Pipe or hole for conducting fluid.

LINEAL Distance measured in one direction.

LINEAL SHRINKAGE Decrease in one dimension experienced by a soil mass when the moisture content is reduced from an amount equal to the field moisture equivalent to the shrinkage limit.

LINEAR ALKYLATE SULFONATE (LAS) Biodegradable surfactant, the major surfactant component of household synthetic detergents.

LINEAR DESIGN AND CONSTRUCTION Process in which each stage of design and construction follows in logical sequence and is not commenced until the preceding stage is completed.

LINEAR ENERGY TRANSFER Linear rate of energy dissipated by the passage of particulate or electromagnetic radiation through an absorbing material.

LINEAR FORCE 1. Force exerted in a straight line; **2.** For a static roller, the vertical force produced directly below the width of the drum or wheels that creates the shear stresses for compaction.

LINEAR GROWTH Increase in value of an asset based on interest calculated on the original value only, so that the sum added at the end of each period remains constant.

LINEARITY Ability of an instrument to measure the actual values of a variable through its effective range.

LINED CANAL Canal, the sides and bottom of which have been lined or covered with an impervious material to prevent leakage and erosion, and to improve carrying capacity and flow characteristics.

LINE LOSS Loss of power occurring in lines carrying electrical current due to resistance offered by the line, and which varies with the length of line, size of conductors, voltage, draw, etc.

LINE MANHOLE Manhole in a sewer located for reasons other than permitting connection of a branch sewer, such as change of direction, service access, change of grade, etc.

LINE PIPE Welded or seamless pipe, with plain, bevelled, grooved, expanded, flanged or threaded ends, principally used to convey gas, oil or water.

LINER 1. Something placed inside another object and covering it's inside surface; **2.** Insulating fabric insert worn underneath safety caps or hats; **3.** Continuous layer of natural (e.g. earthen) or man-made material beneath, or on the sides of a surface impoundment, landfill, or landfill cell, that restricts the downward or lateral escape of hazardous waste, hazard waste constituents, or leachates.

LINE SPINNING Method of rotating a metal pipe by wrapping a chain or line around it; commonly used to make and break a threaded connection.

LINE SQUALL Belt of severe thunderstorms accompanying a cold front.

LINING Material placed on the bottom of a reservoir, channel or canal to reduce scour, friction or leakage.

LINK 1. Key event in an arrow diagram; a point of articulation, where lines representing several activities come together, indicating that they all have to be completed before the work program can proceed; **2.** One hundredth of an engineer's measuring chain, or 12 in. (7.92 in. of a Gunter's chain); **6.** See **Logical relationship.**

LINKED SWITCH See **Switch.**

LIPOIDAL MATTER Substance found in wastewater and consisting mainly of fatty acids and their salts, mineral oils, vegetable oils, and animal fats.

LIP UNION In plumbing, a union with a ringlike inner projection that restrains the gasket in proper alignment.

LIQUEFACTION Change of state to liquid: can be from a previously gaseous state, or from a solid state. In the latter it is a loss of strength occurring in saturated, fine-grained, cohesionless soil when exposed to shock or vibrations. Under such circumstances, the soil particles momentarily lose contact due to pore pressure increase. The material then behaves as a fluid without shear strength. *See also* **Quicksand.**

LIQUID Fluid; flowing or capable of flowing; substance whose molecules are incompressible and inelastic and which move freely among themselves.

LIQUID CHLORINE Elemental chlorine placed in a liquid state through compression and refrigeration of dry, purified chlorine gas.

LIQUID COLLECTOR Solar collector that uses water or other liquid as the heat transfer medium.

LIQUID LIMIT Minimum moisture content that will cause soil to flow if jogged. *See also* **Soil limits.**

LIQUID–METAL FAST-BREEDER RE-ACTOR Breeder-type nuclear reactor that uses plutonium dioxide or uranium dioxide as a fuel and molten sodium and potassium as a coolant.

LIQUID-MOUNTED SEAL Foam or liquid-filled primary seal mounted in contact with the liquid between a tank wall and a floating roof continuously around the circumference of the tank.

LIQUID SLUDGE Sludge containing sufficient water to permit flow by gravity or pumping.

LIQUID SPECIFIC GRAVITY Ratio of the weight of a given volume of liquid to an equal volume of water. *See also* **Specific gravity.**

LIQUID TRAP Sumps, well cellars, and other traps for the purpose of collecting oil, water, and other liquids.

LIQUID WASTE Fluid discharge that must be contained (in a pipe or other vessel) during transport and treated before discharge or release to the environment.

LIQUID WASTE HAULER Person or company equipped and qualified (and licenced, where necessary) to collect, transport and discharge wastewater or liquid wastes.

LIQUID WEIGHT Water content, expressed as a percentage of the dry weight of soil, at which the soil passes from the plastic to the liquid state under standard test conditions.

LIQUOR Liquid phase in which other phases are present.

LISTED Equipment or materials included in a list published by a qualified testing laboratory whose listing states either that the equipment material meets appropriate standards, or has been tested and found suitable for use in a specified manner.

LITER Non-SI unit of volume, equal to a cubic decimeter, permitted for use with the SI system of measurement. Symbol: L. Multiply by 0.035 to obtain cubic feet, symbol: ft^3; by 61.024 to obtain cubic inches, symbol: in.^{3}; by 0.001 to obtain cubic meters, symbol: m^3; by 0.22 to obtain (Imp) gal, symbol: (I)gal; by 0.264 to obtain (US) gal, symbol: (US)gal; by 35.195 to obtain fluid ounces, symbol: fl oz; and by 0.88 to obtain quarts, symbol: qt. *See also the appendix*: **Metric and nonmetric measurement.**

LITHIUM HYPOCHLORITE Dry powder comprising a combination of lithium and chlorine mixed so that, when it is dissolved in water, active chlorine is released.

LITHIFICATION Process in which sediments are consolidated to form sedimentary rock.

LITHOLOGY Study of rocks. *See also* **Petrography** and **Petrology.**

LITHOSPHERE Rocky portion of the Earth's crust composed mainly of solid materials such as rock, clay, earth, gravel, etc.

LITMUS Organic chemical indicator of acidity or alkalinity; red in color for pH values below 4.5 and blue above 8.3.

LITTER Vegetable matter on the ground surface.

LITTORAL Pertaining to the shore; the zone between high- and low-water marks.

LITTORAL CURRENT Current that moves along the shore in a direction parallel to the shore line.

LITTORAL DEPOSIT Material laid down in the littoral zone and in the shallow water adjacent to the low-water mark.

LITTORAL DRIFT Moving of beach material along a coast by wave action.

LITTORAL ZONE Shore line between high- and low-water marks.

LIVE BOTTOM BIN Storage bin for shredded or granular material whereby controlled discharge is through a mechanical or vibrating device across the bin bottom.

LIVE BOTTOM PIT Storage pit, usually rectangular, receiving truck-unloaded material, and using a push platen or bulkhead, reciprocating rams or mechanical conveyor across the pit floor for controlled discharge (retrieval) of the material.

LIVE-BOTTOM TRAILER Transfer trailer whereby controlled discharge is by a mechanical or vibrating device across the bottom of the trailer.

LIVE MAIN Gas or water pipe under pressure.

LIVE TAPPING Connection of a water pipe to a live main using a saddle.

LOAD ALLOCATION Portion of a receiving water's loading capacity that is attributed either to one of its existing or future non-point sources of pollution or to natural background sources.

LOAD CURVE Curve describing a variation of load over time.

LOAD-DIVIDING PRESSURE CONTROL VALVE *See* **Valve.**

LOADED FILTER Filter at the foot of an earth dam. It stabilizes the toe of the dam by its weight and permeability since water cannot exist in it under pressure.

LOAD FACTOR Ratio of the average load carried by an operation to the maximum load carried, during a given period of time.

LOAD GATE *See* **Elevating gate.**

LOAD HEIGHT Height a bucket, attached to an excavator, loader or other equipment, can be raised to dump a load into a truck.

LOADING 1. Introduction of waste into a waste management unit but not necessarily to complete capacity (also referred to as filling); **2.** Ratio of fish biomass (grams, wet weight) to the volume (liters) of test solution in a test chamber or passing through it in a 24-hour period; **3.** Measure of the application of a factor to a device, or the demand upon a system at a point in time.

LOADING CAPACITY Greatest amount of loading that a water can receive without violating water quality standards.

LOADING PIPE Filling a pipe with a suitable material, or a device, to prevent distortion or collapse of the pipe wall during bending.

LOAM Earth having a relatively even mixture of clay, silt, and sand plus a considerable proportion of organic matter. *Also called* **Topsoil.** *See also* **Material density** and **Soil types.**

LOCK Chamber in a canal or river with gates on each end through which vessels can pass up- or downstream and be raised or lowered to a new elevation.

LOCKAGE Water passed from the upper to the lower reach of a canal by operating a lock.

LOCK BAY Enclosed water space within a lock chamber.

LOCK CHAMBER Section of the main waterway of a lock, equipped with gates at each end, into which a vessel passing through the lock is moved.

LOCK CUT Short canal beside a river that enables vessels to bypass a weir and move through a lock to a higher or lower elevation.

LOCK GATE Gate at either end of a lock permitting passage of vessels into or out of the chamber, or within the length of the chamber, dividing the lock into two or more compartments.

LOCKOUT RELAY Electrically reset or hand-reset auxiliary relay, used to hold associated devices inoperative until it is reset.

LOCK PADDLE Sluice used to fill or empty a lock chamber.

LOCK SEAM Spiral or longitudinal seam in corrugated or other pipe or sheet metal, formed by overlapping or folding the adjacent edges.

LOCK SILL Part of the floor of a lock chamber against which the gates rest when closed.

LOESS Fine, porous mineral filler deposited by the wind.

LOG BOOM Floating structure of logs, used to protect the upstream face of a dam or other structure from damage by wave action or floating debris.

LOG CHUTE Channel through or beside a dam for passage of logs and driftwood.

LOGICAL RELATIONSHIP Dependency between two project activities, or between project activity and a milestone. *See also*

Precedence relationship. There are four possible types of logical relationship:

> **Finish-to-start:** Where the 'from' activity must finish before the 'to' activity can start.

> **Finish-to-finish:** Where the 'from' activity must finish before the 'to' activity can finish.

> **Start-to-start:** Where the 'from' activity must start before the 'to' activity can start.

> **Start-to-finish:** Where the 'from' activity must start before the 'to' activity can finish.

LOGIC BOARD Assembly of decision-making circuits on a printed-circuit board.

LOGIC DEVICE One of a general category of components that perform logic functions; for example AND, NAND, OR, and NOR. They can permit or inhibit signal transmission with certain combinations of control signals.

LOGIC DIAGRAM *See* **Project network diagram.**

LOGIC PROBE Instrument used to determine the logic level at a tested point.

LOGIC STATE Signal levels in logic devices characterized by two stable states, the logical 1 (one) state and the logical 0 (zero) state. The designation of the two states is chosen arbitrarily. Commonly, the logical 1 state represents an 'on' signal, and the 0 state represents an 'off' signal.

LOGIC SYMBOL Symbol used to represent a logic element.

LONGITUDINAL DRAINAGE Technique in which the conduits are placed approximately parallel to the steepest slope of the land to be drained.

LONG NIPPLE Pipe nipple having an unthreaded length at its center.

LONG-NOSE PLIERS *See* **Pliers.**

LONG PIPE Pipeline 500 times its diameter in which the loss of head due to entrance and to velocity head is negligible.

LONG-QUARTER BEND 90° pipe fitting with one end longer than the other.

LONG-RADIUS ELBOW Pipe fitting having a larger-than-normal radius for the 90° through which it turns.

LONG SCREW In plumbing, a threaded connector, 150 mm (6 in.) or longer, with one end containing a longer thread than the other.

LONGSHORE CURRENTS Currents in water that are generated by waves, winds and tides and which move parallel to the coast.

LONGSHORE DRIFT Movement of bottom material in longshore currents along a coast.

LONG-SWEEP FITTING Pipe fitting that has a longer than normal radius curve.

LONG TON Nonmetric weight equal to 2240 lb (1016 kg). *See also* **Metric ton** and **Short ton.**

LONG TUBE Tube inserted in an orifice, having a length greater than three times its diameter.

LONG-TUBE EVAPORATOR Evaporator in which multiple stages are interconnected by continuous tubes.

LOOP A network path that passes the same node twice (and which cannot be analyzed using traditional network analysis techniques such as CPM and PERT; but which are allowed in GERT).

LOOPED WATER MAIN Water main arranged in a complete circuit so water will be supplied to a given point from more than one direction.

LOOPING Behavior of a chimney plume when large-scale thermal eddies briefly direct puffs of concentrated gases to the ground before carrying them aloft again.

LOOP VENT Vent that loops back and connects with a waste stack vent.

LOOSE APRON Stones or blocks laid loosely on the berm of a river embankment to prevent erosion.

LOOSE-BOUNDARY HYDRAULICS Study of the movement of material due to the flow of fluid along unstable boundaries such as the seashore.

LOOSE CUBIC METER/CUBIC YARD One cubic meter or cubic yard of material after it has been removed from natural conditions. *See also* **Bank cubic meter/cubic yard.**

LOOSE FIT Fit between mating parts that allows considerable freedom.

LOOSE GROUND Material having a relative compaction less than 90%.

LOOSE-MEASURE VOLUME Volume of earth once it is moved from its original position and deposited in another location, either for transport or for storage.

LOOSE-ROCK DAM Dam built up of rocks, loosely dumped in place and without mortar.

LOOSE YARD Cubic yard loose measurement, a unit of volume of excavation.

LOSING STREAM Stream which in whole or in part contributes water to the saturation zone.

LOSS OF HEAD Hydraulic friction. *Also called* **Lost head.**

LOSS-OF-HEAD GAUGE Gauge on a rapid sand filter that indicates the loss of head resulting from the filtering process.

LOST ENERGY Heat energy of water flowing in a waterway, the result of friction, lost through absorption into the stream.

LOST RIVER Stream that, through secular increase in aridity, has lost its trunk.

LOUDNESS Intensity of a sound at the point of detection, i.e. the human ear.

LOW ALTITUDE Elevation equal to or less than 1219 meters (4000 feet).

LOW ALTITUDE CONDITIONS Test altitude less than 549 meters (1800 feet).

LOW DOSE Should correspond to 1/10 of the high dose.

LOW DUTY OF WATER Given quantity of water that will serve to irrigate a small area of land.

LOWERING VALVE *See* **Valve.**

LOWEST-OBSERVED-ADVERSE-EF-FECT LEVEL Lowest dose in a toxicity study that results in an observed adverse effect (usually one dosage level above the no-observed-adverse-effect level).

LOWEST-OBSERVED-EFFECT LEVEL Lowest dose in a toxicity study that results in an observed, but not adverse, effect (usually one dosage level above the no-observed-effect level.

LOW GROUND PRESSURE Version of some tracked equipment that combines an increase in track length with wider-than-normal track shoes. A machine so equipped distributes its weight over a greater area with a resulting drop in pressure exerted on the ground.

LOW HAZARD INDUSTRIAL OCCUPANCY *See* **Industrial occupancy.**

LOW-LEVEL GROIN Groin built up slowly by natural causes from the bed of a stream by progressive and uniform reclamation over a large area.

LOW-LEVEL WASTE Radioactive wastes associated with clothing and equipment from hospitals and laboratories where radioactive substances have been used, soil and rubble from demolished buildings that demonstrates slight radioactive contamination, etc.

LOW-LYING AREA TECHNIQUE *See* **Sanitary landfilling, wet or low-lying area technique.**

LOW-PRESSURE STORAGE TANK Storage tank designed to operate at pressures between 3.5 kPa (0.51 psi) and 100 kPa (14.5 psi) gauge.

LOW-RATE TRICKLING FILTER Trickling filter designed to receive a relatively small load of BOD per unit volume of filtering material and to have a low dosage rate per unit of surface area.

LOW-VELOCITY ZONE Area of the world where seismic shock waves travel at significantly reduced speeds: the top of the zone varies between 7 and 150 km (4.3 and 93 miles) in depth; the bottom between 200 and 300 km (124 and 186 miles).

LOW VOLUME WASTE SOURCE Taken collectively as if from one source, wastewater from all sources except those for which specific limitations are otherwise established. Low volume wastes sources include, but are not limited to: wastewaters from wet scrubber air pollution control systems, ion-exchange water treatment systems, water treatment evaporator blowdown, laboratory and sampling streams, boiler blowdown, floor drains, cooling tower basin cleaning wastes, and recirculating house service water systems. Sanitary and air conditioning wastes are not included.

LOW-WATER MARK Average surface elevation of the expected low water. *See also* **High-water mark.**

LOW-WATER LINE Line on a tidal shore which is ordinarily reached at low water.

LOW-WATER LUNITIDAL INTERVAL Interval between the moon's meridian passage at a given place and the following low water at that place.

LOW-WATER REGULATION Stream flows adjusted to a desirable, necessary, or imposed condition.

LOW-WATER SLACK Condition of zero current velocity at the end of an ebb tide.

LUGGED PIPE BRACKET Fixing for attaching pipe to a wall consisting of a lug and two-piece cleat.

LUGGER BODY Used for commercial/industrial waste collection systems, dropping off and retrieving trough-shaped containers of 4.6 to 15.3 m³ (6 to 20 yd³) capacity.

LUMEN SI unit of luminous flux emitted in a cone of solid angle of 1 sr by a spherical point source of uniform luminous intensity of 1 cd. (1 lm=1 cd·sr). Symbol: lm. *See also the appendix*: **Metric and nonmetric measurement.**

LUMEN SECOND A derived unit of the quantity of light of the SI system of measurement. Symbol: lm·s. *See also the appendix*: **Metric and nonmetric measurement.**

LUMP SUM BID Single sum bid without breakdown.

LUMP SUM CONTRACT Contract in which the owner will pay the contractor a specified sum of money for a project completed to the terms of the contract document. *Also called* **Lump sum agreement.** *See also* **Stipulated price agreement.**

LUNAR DAY Time of rotation of the Earth with respect to the moon, or the interval between two successive upper transits of the moon over the meridian of a place: 24.84 solar hours long, or 1.035 times the mean solar day.

LUX SI unit of illumination (or illuminance) of 1 lm uniformly over an area of 1 m². (1 lx = 1 lm/m²). Symbol: lx. *See also the appendix*: **Metric and nonmetric measurement.**

LUX-SECOND Derived unit of light exposure of the SI system of measurement. Symbol: lx·s. *See also the appendix*: **Metric and nonmetric measurement.**

LYSIMETER Structure containing a mass of soil, designed to permit measurement of water flowing through the soil.

M

MACHINE 1. Apparatus for applying mechanical power; **2.** *See* **Driving machine.**

MACHINE AVAILABILITY *See* **Machine time.**

MACHINE BASE Foundation for mechanical equipment, which may be isolated from any surrounding floor slab and which commonly projects above the slab as a plinth.

MACHINE BEAM Horizontal steel beam that supports machinery.

MACHINED END–END Pipe that has been milled on each end and left rough in the center. *Also called* **MEE.**

MACHINED OVER ALL Pipe that has been milled end-to-end to allow easier joining of the pipe if the length must be cut to fit. *Also called* **MOA.**

MACHINE DOWNTIME *See* **Machine time.**

MACHINE DRAWING Drawing of a mechanical part, including dimensions and notes.

MACHINE EFFICIENCY Ratio of the rate of horsepower energy output to rate of horsepower energy input over a given period.

MACHINE LANGUAGE Series of binary numbers for interpretation by a computer.

MACHINE OPERATING EFFICIENCY Ratio of actual work time to available work time, expressed as minutes worked per hour.

MACHINE RATE Cost per unit for owning and operating a machine or other piece of equipment. The rate is composed of fixed costs such as depreciation, interest, taxes, and license fees, and variable costs including fuel, lubricants, and repairs and replacement of components such as tires.

MACHINE RATING Amount of load, or power, an electrical machine can deliver without overheating.

MACHINE RESISTING MOMENT Moment of the dead weight of a crane or derrick, less boom weight, about the tipping fulcrum; the moment that resists overturning. *Also called* **Machine moment** and **Stabilizing moment.**

MACHINE ROOM Room where the driving machine for an elevator is located.

MACHINE TIME Time allocated to a machine or piece of equipment under a range of cost or profit headings, including:

> **Active repair time:** Time during which actual repair work is carried out on the machine itself or a dismantled part of the machine.
>
> **Actual productive time:** Time spent using a machine to carry out an actual task.
>
> **Delay time:** Time when, for any reason, a machine, ready, available, and capable of performing its assigned task, is unable to do so.
>
> **Disturbance time:** Examples are: time spent for securing a load, towing, detail planning, talking to supervisor, waiting for a load, and waiting for better weather.
>
> **Idle time:** Scheduled nonoperating time during which a machine is not working, moving, under repair, or being serviced.
>
> **In-shift moving time:** That part of non-mechanical delay time during which

a machine is moving or being transported. Includes the time taken to move or transport the machine between operating sites or between base and site, assuming the machine is not under repair. It does not include time spent moving between adjacent working positions on any one site.

In-shift repair time: Part of mechanical delay time when a machine is actually undergoing repair plus the time during which a machine is waiting to be repaired or for repair parts, mechanics, or facilities.

In-shift service time: Part of mechanical delay time when a machine is waiting for service parts, mechanics, or repair facilities.

Machine availability time: Percent of the scheduled operating time during which a machine is not under repair or service.

Machine downtime: Time during which a machine cannot be operated in production or auxiliary work because of breakdown, maintenance requirements, or power failure.

Machine utilization time: Percentage of the scheduled operating time that is productive time. It is computed by productive time divided by the scheduled operating time and multiplied by 100.

Mechanical delay time: Part of scheduled operating time spent in repair or service during which a machine cannot work. It does not include replacement of oil filters and spark plugs as scheduled in a preventive maintenance program. Servicing is fuelling, lubricating, and doing the work specified in a scheduled preventive maintenance program. When a machine is being serviced while under repair, the time involved is classified as repair time, not service time. Repair and service time occur in both scheduled operating and nonoperating times.

Operating time: Productive time.

Operational lost time: Time during which production is halted due to such factors as operating conditions, non-availability of auxiliary equipment, or using the machine or equipment in a nonproductive manner to assist other machines.

Other productive time: Time when a machine is carrying out tasks other than those for which it is intended.

Out-of-shift repair time: Part of nonoperating time when a machine is actually undergoing service time. Does not include waiting time.

Out-of-shift service time: Part of nonoperating time during which a machine is actually undergoing repair. Waiting time is not included here as an in-shift repair time element.

Personnel time: Part of nonmechanical delay time in which a machine lacks an operator or any other member of the machine crew.

Productive machine hour: Time during scheduled operating hours when a machine performs its designated function (time exclusive of such things as machine transport, operational or mechanical delays, and servicing or repair).

Productive time: Part of scheduled operating time in which a machine is performing a function for which it was intended. Also, time spent in carrying out the task; the sum of actual productive and other productive time.

Repair time: Sum of active repair time, waiting repair time, and time spent servicing the machine while undergoing repair.

Scheduled machine hour: Allocation of productive or nonproductive machine time to an accountable classification.

Scheduled nonoperating time: Time when no production is scheduled for a machine or item of equipment.

Scheduled operating time: Time when

a machine is scheduled to do productive work. Time during which a machine is on standby as a replacement is not considered as scheduled operating time. When a machine is replaced, the scheduled operating time of the replaced machine is considered as ending when the replacement arrives on the job. Scheduled operating time of the replacement commences when it starts to move to the location of the machine it is replacing. Extension of the regular shift operation into overtime is considered as scheduled operating time.

Service time: Time for normal service and maintenance of machines and equipment.

Waiting repair time: Time during which the machine is waiting for a mechanic, spare parts, or repair equipment. Includes time for transporting the machine to and from the workshop.

MACHINE UTILIZATION *See* **Machine time.**

MADE GROUND Usable land created by dumping and/or filling.

MAGNESIUM Light, silvery, moderately-hard metallic element, symbol Mg, an essential element in human nutrition and one of the major contributors to water hardness. There is no evidence of adverse health effects specifically attributable to magnesium in drinking water.

MAGMATIC WATER Water existing in magma or molten rock.

MAGNETIC FRACTION That portion of municipal ferrous scrap remaining after the nonmagnetic contaminants have been removed and the magnetic fraction washed with water and dried at ambient temperature.

MAGNETIC SEPARATOR *See* **Separator.**

MAGNETIC SWITCH *See* **Switch.**

MAIN CANAL Principal irrigation conduit that receives its water from the supply and delivers it to the laterals.

MAIN CHANNEL Middle, deepest, or most navigable channel.

MAIN CONTRACTOR *See* **Prime contractor.**

MAIN DRAIN *See* **Trunk sewer.**

MAINLINE METER Water meter installed on the main line of a water distribution system.

MAIN SEWER 1. Large-diameter sewer to which building drains (and, in a combined system, storm drains) are connected; **2.** Sewer that receives sewage from two or more branch sewers as tributaries.

MAINTENANCE Process of sustaining the level of physical quality of an existing building, machine, site, system, etc., and usually involving a program of inspection, cleaning, and repair activities.

MAINTENANCE CONTRACT Contract, not necessarily with the supplier of equipment or builder of a structure, for a range of maintenance procedures over a specified period.

MAINTENANCE PERIOD Time, following completion of a project, during which the contractor is still responsible for his work.

MAINTENANCE SCHEDULE Detailed list of maintenance procedures required to be completed according to an established timetable.

MAINTENANCE STANDARD Formally established criterion for a specific operation that encompasses elements usually found in quality, quantity, and performance standards.

MAIN VENT Principal soil or waste stack in a plumbing system that connects the system to the open sky.

MAIN WATER LINE *See* **Water main.**

MAJOR AND MINOR CONTRACT ITEMS Major contract items are listed as such in a bid schedule or in the special provisions; all other original contract items are considered as minor items; or in cases where the major contract items are not listed as such, the original contract item of greatest cost, computed from the original contract price and estimated quantity, and such other original contract items next in sequence of lower quantities of not less than x% (60% suggested) of the original contract cost considered as a major item or items; or any item having an original contract amount is considered as a major item or items.

MAJOR DISASTER Hurricane, tornado, storm, flood, high water, wind-driven water, tidal wave, earthquake, drought, fire, or other catastrophe that is or threatens to become of sufficient severity and magnitude to warrant disaster assistance.

MAJOR MODIFICATION Any physical change in or change to the method of operation of a major stationary source that would result in a significant net emissions increase of any pollutant.

MAKEUP WATER Water that is added to a boiler, tank, or some other container to replace water that has been lost, thus maintaining the proper water level.

MALATHION An organophosphorus insecticide and acaricide that does not leach to groundwater. The maximum acceptable concentration for malathion in drinking water is 0.19 mg/L.

MALE COUPLING Threaded hose nipple that fits in the thread of a female swivel coupling of the same pitch and appropriate diameter.

MALE THREAD External pipe thread.

MALFUNCTION 1. To function other than in the designed manner; **2.** Sudden failure of a control device or a hazardous waste management unit so that organic emissions are increased; **3.** Unanticipated and unavoidable failure of air pollution control equipment or process equipment.

MALLEABLE CAST IRON Cast iron made by annealing white cast iron while the metal undergoes decarburization, graphitization, or both, thus eliminating all or most of the cementite. *See also* **Iron** and **Wrought iron.**

MALLEABLE IRON PIPE *See* **Pipe.**

MANDATORY RECYCLING Programs requiring that residents and/or businesses keep secondary materials from their solid wastes.

MANDATORY STANDARD Standard with which it is obligatory to comply, established by an authority having the necessary legal authority to enforce it.

MANEUVER TIME Part of cycle time, and for mobile equipment, includes basic travel and four changes of direction at full throttle.

MANGANESE Gray-white or silvery, brittle metallic element, symbol Mn, that occurs in over 100 common salts and mineral complexes that are widely distributed in rocks, in soils and the floors of lakes and oceans. It is generally present in natural surface waters at concentrations below 0.05 mg/L; and more prevalent in groundwater supplies owing to the reducing conditions that exist underground. High manganese concentrations are also found in some lakes and reservoirs as a result of acidic conditions. The aesthetic objective for manganese in drinking water is <0.05 mg/L.

MANGANESE BACTERIA Bacteria capable of utilizing dissolved manganese and depositing it as hydrated manganic hydroxide.

MANHOLE 1. Chamber giving access to an underground service; **2.** Covered access hole to a tank or boiler.

MANHOLE COVER Removable cast-iron plate mounted in a metal frame that is set flush with grade or the finished road surface.

MANHOLE HEAD Two-part cast-iron fixture: a frame that rests on the shaft of the manhole, and a removable cover. The frame may be equipped for height adjustment.

MAN HOUR Unit of work performed by one worker in one hour.

MANIFOLD 1. Chamber or tube having several inlets and one outlet, or one inlet and several outlets. *See also* **Vented manifold; 2.** Filter assembly containing multiple ports and integral related components which services more than one fluid circuit.

MAN LOCK Chamber through which workers pass from one air pressure environment to another.

MANING ROUGHNESS COEFFICIENT Roughness coefficient for determining the discharge coefficient in the Chezy formula.

MANOMETER See **Gauge.**

MANTLE Soil, sand, and other loose materials covering bedrock.

MANUAL SEPARATION Separation of materials from waste by hand sorting.

MANUFACTURERS' IDENTIFICATION Code symbol used on some products to indicate the manufacturer.

MANUFACTURING WASTES Solid and liquid wastes produced by a manufacturing process.

MARIGRAPH Gauge that records tide levels.

MARINA Basin offering dockage and other services for small boats.

MARINE BAYS AND ESTUARIES Semi-enclosed coastal waters that have a free connection to the territorial sea.

MARINE CLIMATE Climate whose meteorology is primarily influenced by the ocean.

MARINE DEPOSIT Deposit laid down in ocean water.

MARINE ENVIRONMENT Territorial seas, the contiguous zone and the oceans.

MARINE SANITATION DEVICE Equipment for installation onboard a vessel and designed to receive, retain, treat, or discharge sewage.

MARINE SOIL Soil formed from materials deposited by or in the waters of oceans and seas and exposed by elevation of the land or reduction in the water level.

MARITIME PLANTS Saltwater plants that grow on tidal flats and the foreshore and which help reduce scour.

MARSH Soft, wet, spongy land, usually vegetated by reeds, grasses and small shrubs.

MASKING AGENT Substance used to overcome or disguise an unpleasant odor.

MASONRY CHECK Barrier built of rock and stones laid in mortar, having a flush surface and conforming to the upstream cross-section of the ditch in which it is built.

MASONRY DAM Dam built of stone set in mortar or concrete.

MASS BALANCE Quantitative accounting of the distributions of chemicals in plant components, support medium, and test solutions. It also means a quantitative determination of uptake as the difference between the quantity of gas entering an exposure chamber, the quantity leaving the chamber, and the quantity adsorbed to the chamber walls.

MASS CURVE Form of mass diagram.

MASS DIAGRAM Diagram, curve, or graph plotted with rectangular coordinates and representing an integration of all preceding quantities — each ordinate equalling the sum of preceding terms in the series; the abscissa representing elapsed time or other variable.

MASS MOVEMENT Unit movement of a portion of the land surface due to creep, landslide, earth flow, subsidence, etc.

MASS RUNOFF Total volume of runoff from a specific area over a specific period.

MASS SPECTROMETRY Classification of ions by measuring their masses according to their ratios of charge to mass.

MASTER SCHEDULE Summary-level schedule which identifies the major activities and key milestones. See also **Milestone schedule.**

MASTER SPECIFICATIONS Specifications that serve as the principal reference and guide for more specific requirements particular to a location or circumstance, the intent of which may be amplified but may not be altered or diminished.

MASTER SWITCH See **Switch.**

MATERIAL

MATERIAL Substance used to form products or construction works. There are several classes, including:

Direct: Material used in a manufacturing process that will form an integral part of the final product.

Indirect: Material in a manufacturing process that is necessary to the production of the final product but which does not form an integral part of it.

Raw: Goods acquired for the purpose of being consumed or changed in form in the manufacturing process.

MATERIAL DENSITY Mass of a unit volume of a substance (weight is synonymous with mass in the case of unit volume). Some typical material densities are:

Caliche: 1250 kg/m^3 (2100 lb/yd^3).

Clay (natural bed): 1600 kg/m^3 (2800 lb/yd^3).

Clay (dry): 1480 kg/m^3 (2500 lb/yd^3).

Clay (wet): 1660 kg/m^3 (2800 lb/yd^3).

Clay (with gravel, dry): 1420 kg/m^3 (2400 lb/yd^3).

Clay (with gravel, wet): 1540 kg/m^3 (1600 lb/yd^3).

Earth (dry, packed): 1510 kg/m^3 (2550 lb/yd^3).

Earth (wet, excavated): 1600 kg/m^3 (2700 lb/yd^3).

Granite (broken or large crushed): 1660 kg/m^3 (2800 lb/yd^3).

Gravel (dry): 1510 kg/m^3 (2550 lb/yd^3).

Gravel (pit run to gravelled sand): 1930 kg/m^3 (3250 lb/yd^3).

Gravel (dry to 13-50 mm (0.5-2 in.): 1690 kg/m^3 (2850 lb/yd^3).

Gravel (wet to 13-50 mm (0.5-2 in.): 2020 kg/m^3 (3400 lb/yd^3).

Limestone (broken or crushed): 1540 kg/m^3 (2600 lb/yd^3).

Loam: 1250 kg/m^3 (2100 lb/yd^3).

Sand (dry): 1420 kg/m^3 (2400 lb/yd^3).

Sand (wet): 1840 kg/m^3 (3100 lb/yd^3).

Sand (with gravel, dry): 1720 kg/m^3 (2900 lb/yd^3).

Sand (with gravel, wet): 2020 kg/m^3 (3400 lb/yd^3).

Sandstone (broken): 1510 kg/m^3 (2550 lb/yd^3).

Shale: 1250 kg/m^3 (2100 lb/yd^3).

Slag (broken): 1750 kg/m^3 (2950 lb/yd^3).

Stone (crushed): 1600 kg/m^3 (2700 lb/yd^3).

Topsoil: 950 kg/m^3 (1600 lb/yd^3).

MATERIAL HANDLER Range of mechanical equipment designed to handle different types of materials (as distinct to processing material, or simply moving it by pushing (as with a dozer) or digging it (as with an excavator or backhoe)). Includes forklifts, telescopic boom equipment, etc.

MATERIAL SAFETY DATA SHEET Information data sheet listing the components of a product, their hazard level, the hazard level of the product when used, and how to extinguish fires should the material catch fire.

MATERIAL SPECIFICATION Stipulation of the character of certain materials to meet necessary performance criteria.

MATERIALS RECOVERY Concept of resource recovery where emphasis is on separating and processing waste materials for beneficial use or reuse.

MATTRESS Blanket made up of natural or artificial materials, or both, placed on the bank of a stream and weighted down, used to prevent scour.

MATURE RIVER Stream, the slope of which is so established that the tractive forces are just sufficient to carry debris delivered by its tributaries having steeper slopes than the main stream.

MATURE SHORELINE Shoreline that is relatively stable and not subject to wave erosion.

MATURE VALLEY Stream valley so developed and established that cutting of the bottom has practically ceased.

MAXIMUM ACCEPTABLE CONCENTRATION (MAC) Concentrations established for certain substances known or suspected to cause adverse effects on health. *See also* **Interim maximum acceptable concentration.**

MAXIMUM ACCEPTABLE PRESSURE Highest water pressure in a distribution system that will not result in premature or accelerated damage to any of the components of the system.

MAXIMUM ACCEPTABLE TOXICANT CONCENTRATION Maximum concentration at which a chemical can be present and not be toxic to the test organism.

MAXIMUM ALLOWABLE CONCENTRATION Upper limit for the concentration of noxious or toxic emissions in the workplace.

MAXIMUM ALLOWABLE COST Maximum cost for a project undertaken on a unit cost or cost-plus basis.

MAXIMUM AVAILABLE WATER 1. Quantity of water that can be withdrawn from an aquifer without lowering the groundwater table; **2.** Quantity of water that can be readily extracted by a plant for the purposes of growth: the difference between field capacity and wilting coefficient.

MAXIMUM COMPUTED FLOOD Largest momentary flood discharge from a watershed computed from factors such as probable maximum rainfall and snow cover, and geomorphic conditions such as stream gradients or land slope.

MAXIMUM   CONTAMINATION

LEVEL Maximum permissible level of a contaminant in water which is delivered to any user of a public water system.

MAXIMUM DISCHARGE Maximum rate of flow that a stream, conduit, channel or other hydraulic structure or appliance is capable of passing.

MAXIMUM EXCURSION Maximum pressure deviation from an operating pressure after an abrupt disturbance.

MAXIMUM EXPOSURE LIMIT Maximum level of exposure to a harmful substance allowed to a human.

MAXIMUM FLOOD FLOW Maximum rate of discharge of flood water from a drainage basin resulting from a high-intensity rainfall, melting snow, failure of a hydraulic structure, etc.

MAXIMUM INDIVIDUAL RISK Maximum additional cancer risk of a person due to exposure to an emitted pollutant for a 70-year lifetime.

MAXIMUM INLET PRESSURE Maximum rated pressure applied to the inlet port of a device.

MAXIMUM MINING YIELD Total volume of groundwater in storage in a particular source that can be extracted and used.

MAXIMUM PROBABLE PRECIPITATION Maximum precipitation, based on historic data, that can be expected to occur in a drainage basin.

MAXIMUM PROBABLE RAINFALL Maximum precipitation, based on historic data, that can be expected to occur over a given period in a drainage basin.

MAXIMUM RATED RPM Engine speed measured in revolutions per minute (rpm) at which peak net brake power is developed.

MAXIMUM SOUND LEVEL Greatest A-weighted sound level in decibels measured during a designated time interval or during an event.

MAXIMUM STREAM FLOW Maximum rate of discharge from a stream at a point over a specified period.

MAXIMUM SUSTAINED YIELD Maximum rate at which groundwater can be withdrawn perennially from a source.

MAXIMUM 30-DAY AVERAGE Maximum average of daily values for 30 consecutive days.

MAXIMUM WATER-HOLDING CAPACITY Approximate volume of water that can be permanently retained in the soil in opposition to gravity.

MAXIMUM WORKING PRESSURE *See* **Pressure.**

MEAN 1. Intermediate between extremes, equidistant to each; **2.** Average of a number of measurements of the same variable.

MEAN ANNUAL PRECIPITATION Average over a period of years of the annual amounts of precipitation.

MEAN ANNUAL RUNOFF Average over the years of the annual amounts of runoff discharged to, or by, a stream.

MEAN CELL RESIDENCE TIME Average time that a microorganism will spend in the activated sludge process.

MEAN DEPTH Cross-sectional area of a stream divided by its surface width.

MEANDER Tortuous or intricate course or bend.

MEANDER BELT Portion of a valley floor across which a stream occasionally shifts its bed as meanders move their position.

MEANDERING CHANNEL Bed of a slow-flowing stream in easily erodible material, on flat ground, with many curves and abandoned channels. *See also* **Channel.**

MEANDERING LINE Survey line at the high watermark on navigable lakes and streams; the line at which continuous vegetation ends and sandy or muddy shore begins.

MEAN FILTRATION RATING Measurement of the average size of the pores of a filter medium.

MEAN FLOW Average discharge at a given point or station on the line of flow for a specific period.

MEAN GRADIENT Average slope or inclination.

MEAN HIGH TIDE Average daily maximum water surface elevation. *See also* **Mean high water, Mean low tide** and **Mean low water.**

MEAN HIGH WATER Average high elevation to which the surface of a body of water rises. *See also* **Mean high tide Mean low tide,** and **Mean low water.**

MEAN LOW TIDE Average daily minimum water surface elevation. *See also* **Mean high tide, Mean high water** and **Mean low water.**

MEAN LOW WATER Average low elevation to which the surface of a body of water falls. *See also* **Mean high tide, Mean high water** and **Mean low tide.**

MEAN RANGE OF TIDE 1. Average of the differences in height of high and low water at a particular place; **2.** Difference in elevations of mean high and mean low water.

MEAN SEALEVEL Average 19-year height of the surface of the sea for all stages of the tide.

MEAN TIDE LEVEL Plane midway between mean low water and mean high water.

MEAN VELOCITY Average velocity of a stream flowing in a channel or conduit at a given cross-section or in a given reach, equal to the discharge divided by the cross-sectional area of the section, or the average cross-sectional area of the reach.

MEAN VELOCITY POSITION In flow with a free surface, the point on a vertical section at which the actual velocity is equal to the mean velocity.

MEASURED VARIABLE Characteristic or component part that is sensed and quantified by a primary element or sensor.

MEASUREMENT CONTRACT Contract in which payment is based on the measurement of completed work.

MEASUREMENT FOR PAYMENT Determination of work done, materials supplied, or services provided to a point in time for authorization of payment under a contract.

MEASUREMENT STANDARD Prescribed procedure for taking a measurement, including the degree of accuracy required to obtain reliable, reproducible results.

MEASURING Determination against an established standard.

MEASURING WEIR Weir that measures the flow of water passing over it.

MECHANICAL AERATION 1. Mechanical agitation of the surface of a body of water, or of the whole contents of a storage basin by mechanical means; **2.** Use of a mechanical device to bring fresh surfaces of liquid into contact with the atmosphere.

MECHANICAL AERATOR Mechanical device used to introduce atmospheric oxygen into a liquid.

MECHANICAL AGITATION Use of a mechanical device to introduce atmospheric oxygen into a liquid.

MECHANICAL ANALYSIS Determination of the distribution of grains of a granular material in accordance with size.

MECHANICAL AND THERMAL INTEGRITY Ability of a converter to continue to operate at its previously determined efficiency and light-off time and be free from exhaust leaks when subject to thermal and mechanical stresses representative of the intended application.

MECHANICAL COLLECTOR Device that separates entrained dust from gas through the application of inertial and gravitational forces.

MECHANICAL CONTROL Use of artificial structures so as to reduce erosion.

MECHANICAL DELAY TIME *See* **Machine time.**

MECHANICAL EFFICIENCY 1. Ratio of an engine's useful horsepower available at the flywheel or power takeoff, to the horsepower developed in the engine cylinders, expressed in percent; **2.** Ratio of energy or work of the output of a machine to the energy of input.

MECHANICAL ENGINEER Person qualified to design mechanical services.

MECHANICAL ENGINEERING Design and construction of engines, machines, and mechanical equipment of all kinds.

MECHANICALLY-CLEANED SCREEN Screen equipped with an apparatus that mechanically removes retained solids.

MECHANICAL OPERATOR Device which, when activated, causes some other mechanism to actuate.

MECHANICAL PIPE JOINT Flexible joint involving lugs and bolts and a flexible gasket.

MECHANICAL RAKE Machine-operated device used to clean debris from racks located at the intake of conduits supplying water to a power-generating or treatment system.

MECHANICAL SEPARATION Separation of waste into various components using mechanical means, such as cyclones, trommels, and screens.

MECHANICAL SERVICE PIPE Welded steel pipe, available in standard, extra-strong, and double-extra-strong weights in sizes up to 300 mm (12 in.) internal diameter.

MECHANICAL SERVICES Building services such as heating, ventilating, air conditioning, plumbing and electrical distribution, etc.

MECHANICAL SPECIFICATIONS Detailed descriptions applicable to the supply, installation, operation, and maintenance of mechanical systems.

MECHANICAL UNIT Plumbing, heating, air conditioning, or electrical system that may be assembled on- or off-site and then installed.

MECHANIC'S LIEN Charge placed against a project for satisfaction of unpaid debts on work performed or materials supplied. *See also* **Lien**. *Also called* **Material lien**.

MEDIA Any of a range of materials selected and placed to achieve a calculated result, such as trapping and retaining suspended solids from a flow passed through it, or encouraging the accumulation of organisms in a biological contactor.

MEDIA MIGRATION Material passed into an effluent stream composed of the materials making up a filter medium.

MEDICAL EMERGENCY Unforeseen condition which a health professional would judge to require urgent and unscheduled medical attention. Such a condition is one which results in sudden and/or serious symptom(s) constituting a threat to a person's physical or psychological well-being and which requires immediate medical attention to prevent possible deterioration, disability, or death.

MEDICAL WASTE Any solid waste generated in the diagnosis, treatment (e.g. provision of medical services), or immunization of human beings or animals, in research, or in the production or testing of biologicals.

MEDIUM-HAZARD INDUSTRIAL OCCUPANCY *See* **Industrial occupancy**.

MEDIAN In a statistical array, the value having as many cases larger in value as cases smaller in value.

MEDIAN STREAM FLOW Rate of stream discharge for which there are equal numbers of greater and lesser flow occurrences during a given period.

MEDIAN TOLERANCE LIMIT In toxicological studies, the concentration of pollutants at which 50% of the test animals can survive for a specified period of exposure.

MEDICINAL SPRING Water for which some therapeutic value is claimed.

MEDIUM SAND Sediment particles having a diameter between 0.25 and 0.5 mm (0.01 and 0.02 in.).

MEDIUM-SIZE PLANT Plant that processes between 3720 kg/day (8200 lbs/day) and 10 430 kg/day (23 000 lbs/day) of raw materials.

MEDIUM-SIZE WATER SYSTEM Water system that serves between 3300 and 50 000 persons.

MEGA Prefix representing 10^6. Symbol: M. Used in the SI system of measurement. *See also the appendix*: **Metric and nonmetric measurement**.

MEGAGRAM Unit of mass, equal to one million grams in the SI system of measurement. Symbol: Mg. *See also the appendix*: **Metric and nonmetric measurement**.

MEGAHERTZ Unit of frequency, equal to one million hertz in the SI system of measurement. Symbol: MHz. *See also the appendix*: **Metric and nonmetric measurement**.

MEGAJOULE Unit of energy, equal to one million joules in the SI system of measurement. Symbol: MJ. *See also the appendix*: **Metric and nonmetric measurement**.

MEGANEWTON Unit of force, equal to one million newtons in the SI system of measurement. Symbol: MN. *See also the appendix*: **Metric and nonmetric measurement**.

MEGALOPOLIS City of great or overpowering size, especially thought of as a center of power, wealth and influence of a country.

MEGASCOPIC Visible to the unaided eye.

MEGAWATT Unit of power, equal to one million watts in the SI system of measurement. Symbol: MW. *See also the appendix*: **Metric and nonmetric measurement**.

MEGOHM Unit of resistance equal to one million ohms. Symbol: MΩ.

MEGOHMMETER Instrument for measuring resistance, giving a reading in ohms.

MELTDOWN Theoretical condition in a nuclear reactor where the coolant escapes and the fuel becomes sufficiently hot that it melts and falls onto the reactor base resulting in uncontrollable heat and radioactive release.

MELTING POINT Constant temperature, at a given pressure (commonly atmospheric pressure), at which the solid and liquid phases of a substance are in equilibrium.

MELTWATER Water formed from melting snow, rime, or ice.

MEMBRANE 1. Layer of usually impervious material sandwiched between two other materials to prevent the transmission of moisture or vapor; **2.** Thin sheet of ion-exchange material used in electrodialysis to remove impurities from brackish water; in reverse osmosis the membrane is semipermeable.

MEMBRANE BARRIER Thin layer of material impermeable to the flow of gas or water.

MEMBRANE FILTER Filter having a filtration surface fabricated from plastic with a defined pore diameter.

MEMBRANE POISONING Absorption of ions of low mobility into a membrane, resulting in its inability to transport desired salt ions.

MEMBRANE PROCESSES Any of several processes that use a filtering skin, including:

> **Electrodialysis:** Process relying on an ion-selective membrane impervious to water that permits the passage of ions in the form of electric current to desalt water down to approximately 200 ppm.

> **Microfiltration:** Process for the desalination of water that employs very dense filters under pressure.

> **Reverse osmosis:** Process incorporating a permeator (a cartridge filter of cellulose ester, nylon, polyphenylene oxide, or other material) to reverse the osmosis of water through the membrane to reduce the salt content, or to trap other dissolved solids, bacteria, viruses and pyrogens.

> **Ultrafiltration:** Low-pressure, low-energy process capable of straining particles down to 0.002 microns.

MEMBRANE SELECTIVITY Ability of a membrane to permit passage only of cations or anions.

MEMBRANE WATERPROOFING Coating the external faces of the underground portions of basement walls with a waterproof substance.

MEMORANDUM OF UNDERSTANDING Communication that sets out the basic framework of an agreement, that is, the basis for a more detailed text and possibly a contract, but that is not binding at that stage.

MENISCUS Curved edge of a liquid at its plane of contact with a perpendicular surface.

MENSURATION Branch of mathematics dealing with length, area, or volume.

MERCAPTANS Sulfur-containing compounds that emit a strong, and usually offensive odor.

MERCURY Heavy metal (Hg) that exists as a liquid at normal temperatures and having an atomic mass of 200.59. Mercury is a toxic element and serves no physiological function in humans. Elevated mercury levels have been found in all freshwater fish taken from areas with suspected mercury contaminating resulting from industrial and agricultural activities. The maximum acceptable concentration of mercury is drinking water is 0.001 mg/L.

MERCURY GAUGE Gauge in which the pressure of a fluid is measured by the height of a column of mercury which the fluid pressure will sustain.

MERCURY SWITCH *See* **Switch.**

MEROMIXIS Permanent stratification of water in a lake.

MESA Isolated, flat-topped hill or upland with steep sides.

MESH Opening or space in a screen — the value of a mesh is generally the number of openings per linear measure.

MESH SCREEN Screen composed of woven fabric of any of various materials.

MESOPHILIC BACTERIA Group of bacteria that thrive in a moderate temperature range between 20°C and 45°C (68°F and 113°F).

MESOCLIMATE Local climate, occurring over an area of several square kilometers and 100 to 200 m (328 to 656 ft) above sea level.

MESOHALINE Brackish waters in which salinity is between 5 to 15 parts per thousand.

MESOPHILIC DIGESTION Sewage sludge digestion by biological action at temperatures between 27° and 32°C (80°F and 90°F).

MESOPELAGIC Group of organisms that live below the epipelagic zone (200–1000 m) in water where little light can penetrate.

MESOPHILIC MICROORGANISMS Group of microorganisms whose optimum temperature for growth lies between 20°C and 45°C (68°F and 113°F).

MESOPHILIC RANGE Temperature range between 27° and 32°C (80°F and 90°F) at which mesophilic bacteria are most effective in the digestion of sewage sludge.

MESOPHILOUS Plants that thrive in neutral soil.

MESOPHREATOPHYTE Long-rooted plants that are neither alkali-resistant nor resistant to drought.

MESOPHYTE Plant that grows under normal conditions of atmospheric moisture supply, as distinct from one that grows under very dry or extremely wet conditions.

MESOSAPROBIC Body of water in which organic matter is rapidly decomposing, significantly reducing the level of dissolved oxygen.

MESOSPHERE Layer of the Earth's atmosphere lying above the stratosphere, extending from about 50 km (31 miles) to about 85 km (52 miles) above the Earth's surface, and characterized by a decrease in temperature with increasing altitude, to about -90°C at the point where the thermosphere begins.

MESOTROPHIC Body of freshwater containing a moderate amount of plant nutrient.

METABOLISM Continuous chemical change going on in living matter, either constructive, by which nutritive material is built up into complex living matter, or destructive, by which it is broken down into simpler substances.

METABOLITE Chemical entity produced by one or more enzymatic or nonenzymatic reactions as a result of exposure of an organism to a chemical substance.

METACENTER Point of intersection of a vertical line drawn through the center of buoyancy of a floating body and the axis of equilibrium of such a body when some external force causes the axis of equilibrium to depart from a vertical position. When the metacenter lies above the center of gravity of the body, the latter is said to be in stable equilibrium and will return to its original position when the external force is removed.

METACENTRIC HEIGHT Vertical distance between the center of gravity of a floating body and its metacenter.

METAL DETECTOR Device for detecting concealed metal objects by radiating a high-frequency electromagnetic field and detecting the change produced in the field by the presence of ferrous objects.

METAMORPHIC WATER Water driven from rocks by the process of metamorphism.

METAMORPHISM Change in the structure of rocks caused by heat.

METASOMATISM Geologic process of practically simultaneous solution and deposition through interstices by which a new mineral of partly or wholly different composition may grow in the body of an old mineral or mineral aggregate.

METEORIC WATER Water in, or derived from the atmosphere.

METEOROGRAPH Instrument capable of recording measurements of two or more meteorological elements on a single printout.

METEOROLOGICAL STORM Meteorological expression for precipitation, synoptic conditions, pressure condensation, pressure, and specific humidity.

METEOROLOGICAL WATER Water in or derived from the atmosphere.

METEOROLOGIC ELEMENTS Six quantities that specify the state of the weather at any given time and place: air temperature, barometric pressure, wind velocity (direction and speed), humidity, cloudiness (type and amount), and precipitation. In addition, any other meteorological phenomena which distinguish the particular condition of the atmosphere, such as sunshine, visibility, radiation, halos, thunder, mirages, lightning, etc. may be noted.

METEOROLOGY Science of the atmosphere and its phenomena.

METER 1. SI unit of length equal to 1 650 763.73 wavelengths in vacuum of the radiation corresponding to the transition between the $2p_{10}$ and $5d_5$ of the krypton-86 atom. Symbol: m. Derived units of area are the square meter, symbol m^2, and the hectare (10 000 m^2), symbol ha; and of volume and capacity the cubic meter, symbol m^3, and the litre, symbol L. Multiply by 39.37 to obtain inches, symbol in.; by 3.28 to obtain feet, symbol ft; and by 1.0936 to obtain yards, symbol yd. *See also the appendix*: **Metric and nonmetric measurement; 2.** Device used for measuring the amount of a material consumed, passed or conveyed; **3. See Indicator.**

METER BOX Receptacle for the enclosure and protection of a water or other meter, installed outside of the building whose consumption is being recorded, with facilities for direct or remote reading.

METER BYPASS Bypass pipe with a stopvalve that allows an unmeasured full flow of water for firefighting.

METER PER SECOND SI unit of speed or velocity. Symbol: m/s. Multiply by 196.85 to obtain foot per second, symbol: fps; by 3.28 to obtain foot per minute, symbol: fpm; by 2.237 to obtain mile per hour, symbol: mph. *See also the appendix*: **Metric and** nonmetric measurement.

METER PER SECOND SQUARED SI unit of acceleration. Symbol: m/s^2. Multiply by 3.28 to obtain foot per second squared, symbol: ft/s^2. *See also the appendix*: **Metric and nonmetric measurement.**

METERED FLOW Flow at a controlled rate. *See also* **Flow.**

METERED SYSTEM Water distribution system in which meters are located at all strategic points on main supply lines, pumping stations, storage outlets, and each consumer service.

METERING PIN Valve plunger that regulates the rate of flow of a liquid or gas.

METER RATE Charge for water or other commodity based on the quantity used as measured by a meter at the point of connection to the distribution system.

METER STOP Valve on a water service between the street main and the water meter.

METHANE Odorless, colorless, flammable and explosive gas (CH_4), typically resulting from the anaerobic decomposition of organic waste matter; the major component of natural gas.

METHANE FERMENTATION Fermentation of organic matter into methane gas.

METHANOGENS Organisms producing methane, requiring completely anaerobic conditions for growth.

METHANOL Chemical, methyl alcohol or wood alcohol (CH_3OH) produced by the destructive distillation of coal or wood, and which can be synthesized from carbon monoxide and hydrogen or from methane.

METHOXYCHLOR Organochlorine insecticide having a maximum acceptable concentration in drinking water of 0.9 mg/L.

METHYL-ORANGE ALKALINITY Measure of the total alkalinity of an aqueous suspension or solution as measured by the quantity of sulfuric acid required to bring the water to a pH of 4.3 as indicated by the change in color of methyl orange.

METHYL TERTIARY BUTYL ETHER
Nontoxic, unleaded fuel additive.

METOLACHLOR Chloroacetanilide herbicide that has been detected in private and public water supplies, for which the maximum acceptable concentration in drinking water is 0.05 mg/L.

METRIBUZIN Triazine herbicide that has been detected in private and public water supplies and for which the maximum acceptable concentration in drinking water is 0.08 mg/L.

METRIC MEASUREMENT *See appendix* **Metric and nonmetric measurement.**

METRIC TON Non-SI unit of weight equivalent to 1000 kg, 2205 lb, or 1.102 short tons. *See also* **Long ton,** and **Short ton.**

METROPOLIS Most important city of a country or region.

MICRO Prefix representing 10^{-6}. Symbol: μ. Used in the SI system of measurement. *See also the appendix*: **Metric and nonmetric measurement.**

MICROBE Microscopic organism: commonly, bacteria, some of which are pathogenic. While these organisms are essential scavengers of organic material, breaking down dead plant and animal remains, some are associated with food poisoning and disease in man, e.g. *Salmonella.*

MICROBIAL ACTIVITY Chemical changes resulting from the metabolism of living organisms.

MICROBIAL FILM Gelatinous film of zoogleal growths covering a filtration medium, or spanning the interstices of a biological bed. *Also called* **Biological slime.**

MICROBIAL METALLURGY Use of bacteria to separate valuable metals from some ores.

MICROBIOLOGICAL GUIDELINES Guidelines to control the presence of pathogenic or disease-causing microorganisms that occure in polluted surface water, including protozoa, bacteria and enteric viruses; protozoa are not commonly found in groundwater.

MICROBIOLOGY Study of very small units of living matter and their processes.

MICROBIOTA Microscopic plants and animals.

MICROCHEMICAL Very small-scale chemical reactions.

MICROCLIMATE Localized climate of a given area, that may differ from surrounding general climatic conditions, being influenced in natural surroundings by topography, drainage, vegetation, and orientation to the sun, and within urban environments by the mass and disposition of large buildings, airways created by roads and streets, nonporous surfaces such as paved parking lots, etc.

MICROCLIMATOLOGY Study of the effects on climate of the relationship of structures, artificial land forms, and other man-made obstructions to the free flow of air.

MICROFARAD One millionth of a farad.

MICROFILTRATION *See* **Membrane processes.**

MICROHENRY One millionth of a henry.

MICROMETER 1. Precision measuring device that measures thicknesses and diameters; **2.** Device for interpolating accurately between two graduations of a scale, typically on the horizontal or vertical circles of a transit, by the rotation of a graduated drum actuating a moving mark.

MICROMETEOROLOGY Study of variations in meteorologic conditions over very small areas.

MICROMETER Unit of length, equal to one millionth of a meter in the SI system of measurement. Symbol: μm. *See also the appendix*: **Metric and nonmetric measurement.**

MICRON Obsolete metric unit of length, equal to 1 μm, that should not be used with the SI system of measurement. Symbol: μ. *See also the appendix*: **Metric and nonmetric measurement.**

MICRONUTRIENT Element or compound required in very small amounts by living organisms.

MICROORGANISM Living creature, too small to be seen with the naked eye. The smallest class are viruses, followed by bacteria, with microalgae being the largest.

MICRORELIEF Small-scale topography in which the vertical difference is measured in centimeters.

MICROSAND Fine aggregate passing the No. 100 (0.149 mm, 0.0059 in.) sieve, and essentially free of clay and shale.

MICROSCOPIC Object having dimensions between 0.5 and 100µm and visible only by magnification with an optical microscope.

MICROSCREEN Straining media with openings in the range of 20 to 60 µm.

MICROSITE Small area that exhibits localized characteristics different from the surrounding area.

MICROSTRAINER Rotating drum supporting a finely-woven, stainless-steel mesh through which a flow is passed to remove particulate matter..

MICROSWITCH Electrical switch that reacts to a very small physical movement of the actuator.

MICROTUNNEL Machine-made shaft or tunnel too small for a human to work in.

MICROTUNNELLING Technique for conduit installation using a steerable, remote-controlled tunnel boring machine, typically by pipe jacking, the excavated material being removed either by mechanical auger or as a slurry.

MIGRATION Contaminant released downstream.

MIL Nonmetric unit of length, equal to 0.001 in. (0.0254 mm). Symbol: mil. Multiply by 25.4 to obtain micrometers, symbol: µm. *See also the appendix*: **Metric and nonmetric measurement.**

MILDEW Any of various fungi that attack plants or grow on food, paper, cloth, leather, etc., especially in damp conditions.

MILD SLOPE Conduit laid at less than the critical slope for a particular discharge; a slope less than the critical slope.

MILE Nonmetric unit of length, equal to 1760 yards. Symbol: mi. Multiply by 1.609 to obtain kilometers, symbol: km; by 1609.344 to obtain meters, symbol: m. *See also the appendix*: **Metric and nonmetric measurement.**

MILE (NAUTICAL) Nonmetric unit of length, equal to 1852 meters. Symbol: n.mi. Multiply by 1.852 to obtain kilometers, symbol: km. *See also the appendix*: **Metric and nonmetric measurement.**

MILE PER HOUR Nonmetric unit of velocity. Symbol: mph. Multiply by 1.609 to obtain kilometers per hour, symbol: km/h; by 0.447 to obtain meters per second, symbol: m/s. *See also the appendix*: **Metric and nonmetric measurement.**

MILESTONE Key or important intermediate goal in the Network system of project analysis and control.

MILESTONE SCHEDULE Summary-level schedule which identifies major milestones. *See also* **Master schedule.**

MILE YARD Measure of payment for excavation representing the movement of 1 yd^3 of material through a horizontal distance of 1 mile.

MILLED REFUSE Solid waste that has been mechanically reduced in size.

MILLI Prefix representing 10^{-3}. Symbol: m. Used in the SI system of measurement. *See also the appendix*: **Metric and nonmetric measurement.**

MILLIBAR Non-SI unit of pressure, equal to 1000 dynes per cm^2. Symbol: mbar. The millibar is used as a measure of atmospheric pressure; a standard atmosphere being equal to 1013.25 mbar. *See also the appendix*: **Metric and nonmetric measurement.**

MILLIFARAD Unit equal to 1/1000 of a farad. Symbol: mF. *See also the appendix*:

Metric and nonmetric measurement.

secutively.

MILLIGRAM Unit of mass, equal to 1/1000 of a gram in the SI system of measurement. Symbol: mg. *See also the appendix*: **Metric and nonmetric measurement.**

MILLIGRAMS PER LITER Unit of concentration commonly applied to water and wastewater: equal to 0.001 g or the constituent in 1.0 L of the whole; also equal to the more historic and still used parts per million (ppm).

MILLIHENRY Unit equal to 1/1000 of a henry. Symbol: mH. *See also the appendix:* **Metric and nonmetric measurement.**

MILLILITER Unit of volume, equal to 1/1000 of a litre, permitted for use with the SI system of measurement. Symbol: mL. *See also the appendix*: **Metric and nonmetric measurement.**

MILLIMETER Unit of length, equal to 1/1000 of a meter in the SI system of measurement. Symbol: mm. *See also the appendix*: **Metric and nonmetric measurement.**

MILLIMETER OF MERCURY Obsolete metric unit of pressure, equal to 133.32 pascals, that should not be used with the SI system of measurement. Symbol: torr. *See also the appendix*: **Metric and nonmetric measurement.**

MILLIMICRON Unit equal to 1/1000 of a micron, 1/1 000 000 of a millimeter, or 10 angstroms. Symbol: mμ. *See also the appendix:* **Metric and nonmetric measurement.**

MILLION GALLONS PER DAY Nonmetric unit of volume. Symbol: mgd. Multiply by 4546.09 to obtain cubic meters per day, symbol: m^3/d; by 0.052 6 to obtain cubic meters per second, symbol: m^3/s. *See also the appendix*: **Metric and nonmetric measurement.**

MILLIPASCAL Unit of pressure, equal to 1/1000 of a pascal in the SI system of measurement. Symbol: mPa. *See also the appendix:* **Metric and nonmetric measurement.**

MILLISECOND DELAY Electric detonation caps that have a built-in delay element, usually 25/1000th of a second apart, con-

MILLRACE *See* **Tailrace.**

MILL TAIL *See* **Tailrace.**

MIMIC DIAGRAM Electronically controlled panel displaying diagrams of buildings, usually floor plans divided into zones, with indicator lights and lines that represent equipment, circuits, status and other key data.

MINE Area of land, surface or underground, actively used for or resulting from the extraction of a mineral from natural deposits.

MINERAL Any inorganic, naturally occurring, solid chemical element or compound having a crystalline structure.

MINERAL DEPOSIT Impurity in water that precipitates on metal and other surfaces, such as the interior of pipes and storage vessels.

MINE DEWATERING Pumping or draining of any water that is impounded or that collects in a mine.

MINE DRAINAGE Water drained, pumped or siphoned from a mine.

MINERAL FIBER INSULATION Insulation (rock wool or fiberglass) which is composed principally of fibers manufactured from rock, slag or glass, with or without binders.

MINERAL SPRING Spring water containing large quantities of mineral salts.

MINE WATER Water that enters or accumulates in underground mining operations.

MINIMUM ACCEPTABLE PRESSURE Lowest pressure permitting safe, efficient, and satisfactory operation of the most remote hydraulic fixture or component in a water distribution system.

MINIMUM ANNUAL FLOOD Smallest flood recorded.

MINIMUM BURST PRESSURE Lowest pressure at which rupture occurs under prescribed conditions.

MINIMUM FACTOR In forestry, the factor which limits distribution and tree growth by the intensity of its occurrence.

MINIMUM FLOW Stream flow during the driest period of the year.

MINIMUM GRADE Grade at which a gravity sewer will maintain minimum self-cleaning velocity.

MINING WASTE Residues that result from the extraction of raw materials from the earth, or residues left after ore benefication.

MINOR CHANGE Change of a minor nature to the work under contract, that does not involve an adjustment in the contract sum or time of completion, but that must be confirmed by a field order or other written order signed by an appropriate authority.

MINOR CONTRACT ITEMS *See* **Major and minor contract items.**

MINUTE Non-SI unit of time, equal to 60 seconds, permitted for use with the SI system of measurement. Symbol: min. *See also the appendix*: **Metric and nonmetric measurement.**

MINUTE OF ARC Non-SI unit of angle, equal to 60 seconds of arc, permitted for use with the SI system of measurement. Symbol: '. *See also the appendix*: **Metric and nonmetric measurement.**

MIST Very fine drops of moisture of such size, less than 2 micrometers, that gravity separation is hindered.

MISUSE OF LAND Use of land that might injuriously affect the interests of a community.

MITIGATION Steps taken to lessen risk by lowering the probability of a risk event's occurrence or reducing its effect should it occur.

MIXED-FLOW PUMP Centrifugal pump that develops head partly through centrifugal force and partly by the lift imparted on the flow by the vanes on the impeller.

MIXED LIQUOR Sewage effluent containing floc from an aeration tank.

MIXED LIQUOR SUSPENDED SOLIDS Concentration of suspended solids (mg/L) in the mixed liquor of an aeration tank.

MIXED LIQUOR VOLATILE SUSPENDED SOLIDS Concentration (mg/L) of organic or volatile suspended solids in the mixed liquor of an aeration tank: a measure or indication of the volume of microorganisms present.

MIXED MUNICIPAL REFUSE *See* **Solid waste.**

MIXER Machine used for blending materials.

MIXING BASIN Vessel in which agitation is applied to a liquid content.

MIXING CHAMBER Chamber, usually placed between the primary and secondary combustion chambers of an incinerator, in which the products of combustion are thoroughly mixed by turbulence created by increased velocities of the gases.

MIXING CHANNEL Channel in which two or more flows are blended, or in which chemicals are introduced to and completely mixed with a flow.

MIXING CYCLE Time taken for a complete cycle in a batch mixer, i.e., the time elapsing between successive repetitions of the same operation (e.g. successive discharges of the mixer).

MIXING FAUCET *See* **Valve.**

MIXING LAYER Lower part of the atmosphere within which air movement is affected by the proximity and shape of the Earth's surface and where airborne pollutants are dispersed and mixed.

MIXING RATIO **1.** Amount of one substance relative to the amount of another; **2.** Amount of the mass of water vapor to the mass of dry air with which it is associated.

MIXING SPEED Rotation rate of a mixer drum or the paddles in an open-top, pan, or trough mixer.

MIXING TANK Tank having a configuration, and equipped with a mechanism designed to provide a thorough mixing of two or more liquids of different characteristics, or of chemicals introduced into liquids.

MIXING VALVE *See* **Valve.**

MIXING ZONE For a discharge into the ocean, the zone extending from the sea's surface to seabed and extending laterally to a distance of 100 meters (328 feet) in all directions from a discharge point(s) or to the boundary of the zone of initial dilution as calculated by a plume model.

MIX-IN-PLACE Soil stabilization technique in which soil is mixed by a travelling mixer without being removed from the site.

MIXTURE 1. Combination of two or more elements and/or compounds in solid, liquid, or gaseous form except where such substances have undergone a chemical reaction so as to become inseparable by physical means; **2.** Heterogeneous association of substances where the various individual substances retain their identities and can usually be separated by mechanical means. Includes solutions or compounds but does not include alloys or amalgams.

MOBILE BELT Region of the Earth's crust that is undergoing uplift and subsidence, earthquakes and volcanic activity, often characterized by folding and faulting involved in mountain building.

MOBILE COMPACTOR *See* **Compactor.**

MOBILE CONVEYOR Self-propelled, wheel-mounted bulk material handling/delivery system.

MOBILE PACKER *See* **Refuse truck.**

MOBILE WATER Water free to move in any direction under the pull of gravity or the effect of some other force.

MODEL Simplified description of a system used to prove theories and/or develop complete designs.

MODERATE RAIN Rainfall having an intensity between 2.8 and 7.6 mm/h (0.11 and 0.30 in./h).

MODERATOR Substance that when added to two or more others in a mixture will slow a reaction or process, or change it.

MODIFICATION 1. Written amendment to an agreement, signed by both (all) parties; **2.** Change order signed by all appropriate parties; **3.** Field order for a minor change in the work, issued by the owner or owner's agent.

MODIFIED AERATION Variation of the activated sludge process calling for a reduced period of aeration accompanied by a corresponding reduction in the quantity of suspended solids in the mixed liquor.

MODIFIED MERCALLI SCALE Scale used to measure the intensity of earthquake shock, ranging from instrumental (I) to catastrophic (XII), based on damage to human structures caused by the event.

MODIFIED VELOCITY Flow velocity corrected for drift and angularity.

MODIFIER Additive that changes some or several aspects of a substance without causing it to loose its essential characteristics.

MODULATE Regulate or adjust.

MODULATED CONTROLS Automatic controls that can gradually change an mechanical, hydraulic or electrical system continuously, increasing or decreasing supply, speed, flow, etc., to match demand.

MOISTURE Condensed or diffused liquid.

MOISTURE CONTENT Quantity of water present in a substance.

MOISTURE CONTENT OF SOIL Water is present in all soils in their natural state, and appears in one of three forms:

> **Gravitational:** Water that is free to move downward due to the force of gravity and which can drain naturally from a soil.

> **Capillary:** Water held in a soil by small pores or voids; considered free water, but which can be removed only by lowering the watertable or by evaporation.

Hygroscopic: Water present in soil after gravitational and capillary water are removed, held by individual soil grains in the form of a very thin film having physical and chemical affinity for the soil grains.

MOISTURE DENSITY Mass of water per unit volume of space occupied by soil, air, and water.

MOISTURE EJECTION VALVE *See* **Valve.**

MOISTURE FILM COHESION In soil stabilization, the resistance of particles to separation due to surface tension of the moisture film surrounding them.

MOISTURE GRADIENT Rate of change of moisture content of soil with depth.

MOISTURE HOLDING CAPACITY Quantity of water held by compacted solid wastes beyond which the application of additional water will cause it to drain rapidly to underlying material.

MOISTURE MOVEMENT Movement of moisture through a porous medium.

MOISTURE PENETRATION Depth to which moisture penetrates soil before the rate of downward movement becomes negligible.

MOISTURE PERCENTAGE Water content of a semiliquid material, such as sludge, expressed as the ratio of the loss in weight after drying for a defined period at 103°C (217°F), to the original weight of the sample.

MOISTURE REGAIN Reabsorption following a loss of moisture.

MOISTURE TENSION 1. Measure of the energy with which water is held in the soil; **2.** Equivalent negative or gauge pressure to which water must be subjected in order to be in hydraulic equilibrium, through a porous membrane wall or membrane, with the water in the soil.

MOLD *See* **Mould.**

MOLE 1. Tunnel-boring machine; **2.** Egg-shaped device pulled behind the tooth of a subsoil plow to open passages for drainage or for the insertion of piped services or cable; **3.** *See* **Breakwater**; **4.** SI unit of substance of a system that contains as many elementary entities as there are atoms in 0.012 kg of carbon-12. Symbol: mol. *See also the appendix*: **Metric and nonmetric measurement.**

MOLECULAR OXYGEN Oxygen molecule, O_2, that is not combined with another element to form a compound.

MOLECULAR WEIGHT Sum of the atomic weights of the elements in a compound.

MOLECULE Smallest division of a compound that still retains or exhibits all the properties of the substance.

MOLE DRAIN Drainage channel cut in stiff clay by drawing a mole plough through it.

MOLE FRACTION Ratio of the number of moles of a component of a mixture to the total number of moles in the mixture.

MOLE PER CUBIC METER A derived unit of concentration in the SI system of measurement. Symbol: mol/m^3. *See also the appendix*: **Metric and nonmetric measurement.**

MOLE PER KILOGRAM A derived unit of molality in the SI system of measurement. Symbol: mol/kg. *See also the appendix*: **Metric and nonmetric measurement.**

MOLE PLOUGH Vertical knife blade carrying a bullet shape at its lower end, drawn by a tractor so that the blade slices through the ground and the plough creates an underground channel through the soil, to drain the soil, or to bury pipe or cable that is automatically fed to the plough from reels carried on the tractor.

MOLING Installing services with the aid of a mole plough.

MOMENTARY-CONTACT SWITCH *See* **Switch.**

MONITOR Nozzle mounted in a swivel that generates extremely high pressures in water jetted through it. *See also* **Giant,** and

Hydraulic monitor.

MONITORING 1. Capture, organization, and reporting of measures of performance in comparison to standards for project scope, time, and cost; **2.** Process in the project management system.

MONITORING DEVICE Equipment used to measure and record (if applicable) process parameters.

MONITORING SYSTEM System used to sample and condition (if applicable), to analyze, and to provide a record.

MONITORING WELL Well developed specifically to permit monitoring of water levels and/or water quality.

MONOCHLORAMINE By-product of drinking water chlorination, or can be added to maintain residual disinfection activity in potable water distribution systems.

MONOCHLOROBENZENE Solvent for adhesives which, being very volatile, is ingested by humans from the atmosphere in higher quantities than from food or drinking water. The maximum acceptable concentration for monochlorobenzene in drinking water, derived from the average daily intake, is 0.08 mg/L.

MONOMER Chemical substance that has the capacity to form links between two or more other molecules.

MONOMOLECULAR LAYER Layer that is one molecule thick, and which may form a boundary between two fluids, preventing the natural transfer of molecules between them.

MONTHLY AVERAGE Arithmetic average of individual data points from effluent sampling and analysis during any calendar month.

MOOR Open area of acid peat, usually hilly or high altitude and having low plant growth.

MORAINE Debris of rocks brought down by glaciers.

MORPHOLOGY Study of the form of organisms or of the Earth's physical features.

MORTALITY RATE Incidence of death in a population.

MOSS Any of a class of multicellular, chlorophyll-containing, flowerless plants in the Bryophyta; usually growing as low, dense, carpet-like masses on tree trunks, rocks, moist ground, etc.

MOST PROBABLE NUMBER That number of organisms per unit volume that, in accordance with statistical theory, would be more likely than any other number to yield the observed test result with the greatest frequency: in water quality analysis the results are computed from the number of positive findings of coliform-group organisms resulting from multiple-portion decimal-dilution platings.

MOTHER LIQUOR Solution substantially freed from undissolved matter by filtration, centrifuging, or decantation.

MOTILE Capable of self-propelled movement.

MOTOR Rotating machine that transforms chemical, electrical or explosive energy into mechanical energy.

MOTOR CIRCUIT SWITCH *See* **Switch.**

MOTOR CONTROL CENTER Assembly of one or more enclosed sections having a common power bus, and principally containing motor control units.

MOTOR GENERATOR Electrical generator propelled by an electric motor.

MOTOR-GENERATOR SET Machine that converts mechanical energy into electrical energy by electromagnetic induction, typically an AC drive motor coupled directly to a DC generator.

MOTORIZED VALVE *See* **Valve.**

MOULD 1. Wooly or furry growth of multicelled filamentous fungi that appears on food and other animal or vegetable substances in warm, moist conditions; **2.** Soft, rich, crumbly soil, earth mixed with decaying leaves, manure, etc. **3.** Hollow shape in which anything is formed or cast.

MOUND BREAKWATER Rocks heaped into a furrow to serve as a breakwater.

MOUTH 1. Entry point to a shaft, tunnel, etc.; **2.** Exit point of discharge of a stream into another stream or body of water.

MOVABLE BED 1. Stream bed composed of materials readily transportable under some conditions of flow; **2.** Bed composed of materials readily movable under certain hydraulic conditions established in a tank or chamber.

MOVABLE BULKHEAD REFUSE TRUCK *See* **Refuse truck.**

MOVABLE DAM Barrier that may be opened in whole or in part, usually through the movement of vertically hinged gates that provide an adjustable weir.

MOVABLE DISTRIBUTOR Mechanism consisting of rotating or reciprocating pipes or troughs from which liquid is discharged.

MOVABLE WEIR Temporary weir that can be removed from a stream channel as conditions necessitate.

MOVE LATERALLY IN SOIL Transfer through soil generally in a horizontal plane by physical, chemical, or biological means.

MOVING VANE Surface offering resistance to flow and which will be moved by the dynamic pressure exerted by an impinging jet or stream.

MUCK 1. Soft mud containing vegetable matter; **2.** General term for all kinds of excavated material.

MUD Wet, loose mixture of particles less than 60μm in diameter.

MUDBALLS Balls of accumulated sediment found in debris-laden flow and channel deposits.

MUD BLANKET Flocculent layer that forms on the surface of a sand filter.

MUD BOIL Sudden upsurge of watery subsoil through an opening in a less yielding surface soil, caused by soil pressure change.

MUD FLAT Muddy area submerged by the rise of the tide.

MUD FLOW Flow of water so heavily charged with earth and debris that the mass is thick and viscous.

MUD LINE River bed at the interface of water and soil.

MUD PUMP *See* **Pump.**

MUD VALVE Plug valve in the bottom of a settling tank through which accumulated sediment is drained.

MUD WAVE Front face of a mud slide.

MUFFLER Expansion chamber used to muffle the noise of engine combustion, oil, or gas flow noise.

MULCH Straw, leaves, loose earth, etc., spread on the ground around trees or plants, used to protect roots from cold or heat, to prevent evaporation of moisture from the soil, to control weeds, or to enrich the soil.

MULTILEVEL INTAKE Hydraulic structure by which water may be drawn into or through a dam from various depths below the surface of the head pond.

MULTIMEDIA FILTER Water filtration multibed filter containing several media.

MULTIPASS FILTER PERFORMANCE TEST Test designed to obtain consistent and repeatable information on a filter's ability to control specific size particles.

MULTIPHASE FLOW Underground movement of different flows (sometimes of different substances) through the same strata.

MULTIPLE-ARCH DAM Dam consisting of a number of similar arches whose axes slope at about 45° to the horizontal and which are carried on parallel buttress walls.

MULTIPLE-CHAMBER INCINERATOR *See* **Incinerator.**

MULTIPLE CORRELATION Relationship between several sets of data or quantities, one representing values of a dependent variable, the others values of independent variables.

MULTIPLE-DOME DAM Buttress dam in which widely-spaced buttresses support a dome arched horizontally and longitudinally, on which the water rests.

MULTIPLE DOSING TANKS Two or more identical dosing tanks through which a flow can be passed in rotation.

MULTIPLE-EFFECT EVAPORATOR Battery of single-effect evaporators working in series.

MULTIPLE OUTLETS Array of branching pipes leading to widely separated outlets through which wastewater or effluent can be discharged under water over an extended area.

MULTIPLE-STAGE PUMP *See* **Pump.**

MULTIPLE-STAGE SLUDGE DIGESTION Digesting of sewage sludge progressively in two or more treatment tanks in series.

MULTIPLE-TRAY CLARIFIER Clarifier consisting of several trays positioned vertically above each other so that the effluent of one becomes the influent of the next below it.

MULTIPLE-USE MANAGEMENT Management of land resources with the objective of achieving optimum yields of products and services from a given area without impairing the productive capacity of the site.

MULTIPLE-USE RESERVOIR Reservoir used to store water intended for more than one purpose: raw water supply and fire protection, for instance.

MUNICIPAL COLLECTION *See* **Collection.**

MUNICIPAL ENGINEERING Design and construction of publicly owned services and facilities.

MUNICIPAL SERVICES Services provided to the residents and visitors to a municipality and including water supply, sanitary and storm sewerage, roads and streets, curbs, sidewalks, parks, garbage collection,

etc., and may include electrical distribution, hospitals, local airports, harbor facilities, etc.

MUNICIPAL SOLID WASTE LAND-FILL Discrete area of land or an excavation that receives household waste, and that is not a land application unit, surface impoundment, injection well, or waste pile.

MUNICIPAL SOLID WASTES Garbage, refuse, sludges, wastes, and other discarded materials resulting from residential and non-industrial operations and activities, such as household activities, office functions, and commercial housekeeping wastes.

MUNICIPAL STOP Valve installed in a building water service line at the water main.

MUNICIPAL-TYPE SOLID WASTE Household, commercial/retail, and/or institutional waste. **Household waste** includes material discarded by single and multiple residential dwellings, hotels, motels, and other similar permanent or temporary housing establishments or facilities. **Commercial/retail waste** includes material discarded by stores, offices, restaurants, warehouses, nonmanufacturing activities at industrial facilities, and other similar establishments or facilities. **Institutional waste** includes material discarded by schools, hospitals, non-manufacturing activities at prisons and government facilities and other similar establishments or facilities. Household, commercial/retail, and institutional waste do not include sewage, wood pallets, construction and demolition wastes, industrial process or manufacturing wastes, or motor vehicles (including motor vehicle parts or vehicle fluff). Municipal-type solid waste does include motor vehicle maintenance materials, limited to vehicle batteries, used motor oil, and tires. It does not include wastes that are solely segregated medical wastes. However, any mixture of segregated medical wastes and other wastes which contains more than 30% waste medical waste discards, is considered to be municipal-type solid waste.

MUNICIPAL WASTE COMBUSTOR Device that combusts, solid, liquid, or gasified municipal solid waste including, but not limited to, field-erected incinerators (with or without heat recovery), modular incinerators (starved air or excess air), boilers (i.e. steam generating units), furnaces (whether suspension-fired, grate-fired, mass-fired, or fluidized bed-fired) and gasification/combustion units. This does not include combustion units, engines, or other devices that combust landfill gases collected by landfill gas collection systems.

MUSHROOM VALVE Valve consisting of a flat disk that can be raised or lowered without rotation about the valve opening.

MUSKEG Swamp or bog in an undrained or poorly drained area of alluvium or glacial till, commonly a rocky basin.

MUTAGENIC Property of a substance or mixture to induce changes in the genetic complement of either somatic or germinal tissue in subsequent generations.

MUTATION Random alterations of reproductive cells of organisms that result in changes in other cells of the organisms.

N

NAMEPLATE RATING 1. Durable plate that lists critical installation and operating conditions of the equipment to which it is attached; **2.** Full load continuous rating of an electric generator and its prime mover, or other electrical or mechanical equipment, used under specific conditions as designated by the manufacturer.

NANO Prefix representing 10⁻⁹. Symbol: n. Used in the SI system of measurement. *See also the appendix*: **Metric and nonmetric measurement.**

NANOPLANKTON Smallest of the phytoplankton.

NAPPE Unbroken film of water flowing over the crest of a weir or dam.

NARROW-BASE TERRACE Terrace having a base width of from 1.2 to 2.4 m (4 to 8 ft).

NARROWS 1. Contracted part of a stream or body of water; **2.** Strait connecting two bodies of water.

NATIONAL STANDARD HOSE COUPLING Fire hose coupling having universally adopted thread dimensions.

NATIVE ELEMENT Element that occurs in the uncombined state as a mineral.

NATRIC Soil enriched with sodium.

NATURAL CONDITIONS Includes naturally occurring phenomena that reduce visibility as measured in terms of visual range, contrast, or coloration.

NATURAL FLOW Unregulated stream flow.

NATURAL FREQUENCY Frequency at which a system oscillates freely.

NATURAL GAS Any gas found in the earth, as opposed to manufactured gas. Naturally occurring mixture of gaseous saturated hydrocarbons, consisting of 80% to 95% methane (CH_4), lesser amounts of propane, ethane and butane, and small quantities of non-hydrocarbon gases (e.g. nitrogen, helium), having a calorific value of about 8.9 kilocalories/m³ (1000 Btu/ft³) of gas.

NATURAL GAS LIQUIDS Hydrocarbons, such as ethane, propane, butane, and pentane, that are extracted from field gas.

NATURAL GRADE (GROUND) Original ground profile and elevation before any excavation is done.

NATURAL HARBOR Sheltered inlet in a coastline.

NATURAL LEVEE Low, alluvial ridge confining the channel of a stream.

NATURALLY DEVELOPED WELL Water well in which the screen is in direct contact with the aquifer materials, no filter pack having been installed.

NATURAL MOISTURE CONTENT *In situ* moisture at the time of measurement.

NATURAL OUTLET Outlet into a watercourse, ditch, or other body of surface or groundwater.

NATURAL PURIFICATION Natural processes occurring in a stream or other active body of water that results in the reduction of bacteria, satisfaction of the BOD, stabilization of organic constituents, replacement of depleted dissolved oxygen, deposition of sediments, and return of the stream biota to natural levels.

NATURAL RESOURCES Land, fish, wildlife, biota, air, water, ground water, drinking water supplies, and other such resources belonging to, managed by, held in trust by appertaining to, or otherwise controlled by a public authority.

NATURAL SLOPE Steepest angle to horizontal that can be adopted by an unsupported granular semisolid or semifluid material.

NATURAL SUBIRRIGATION Water delivered to plant roots by the capillary fringe.

NATURAL TURNOVER Annual throughput of material processed by a natural system in equilibrium.

NATURAL VENTILATION Ventilation obtained without the use of mechanical devices, using flues, vents, windows, etc.

NATURAL WATERCOURSE Naturally created and/or developed surface or underground watercourse.

NAVIGABLE DEPTH Governing maximum safe depth in a navigable channel, commonly less than the actual depth.

NAVIGABLE WATER Waters that have been surveyed, mapped and marked for the passage of defined classes of shipping.

NAVIGATION CANAL Canal built and used primarily for navigation or transport by ships.

NEAP TIDES Semimonthly (lunar month) tides having a decreased range because of the moon being in quadrature.

NEAR-CRITICAL ACTIVITY Activity that has low total float.

NEAR THE FIRST SERVICE CONNECTION At one of the 20% of all service connections in an entire system that are nearest the water supply treatment facility, as measured by water transport time within the distribution system.

NECK 1. Narrow channel connecting two bodies of water; **2.** Narrow stretch of land connecting two larger areas.

NECROSIS 1. Death of body cells or of a portion of tissue of an organ, resulting from the irreversible damage caused by an outside agent; **2.** Any of various bacterial diseases of plants characterized by spots of decayed tissue.

NEEDLE Plank-shaped object set on end either vertically or inclined, used in guides to regulate the flow of water through a control structure.

NEEDLE DAM Movable dam consisting of upright plank-like pieces, the lower ends of which bear against a sill, the upper ends against a lintel.

NEEDLE VALVE *See* **Valve.**

NEEDLE WEIR Fixed frame weir supporting heavy vertical timbers in contact. The timbers can be withdrawn vertically to lower the water level.

NEGATIVE CONFINING BED Bed that retards or prevents the downward movement of groundwater where the overlying water has sufficient head to produce a downward pressure.

NEGATIVE HEAD 1. Partial vacuum; **2.** Condition of negative pressure produced in a rapid sand filter near the end of a filter run due to clogging of the filtering media.

NEGATIVE PRESSURE Air pressure below ambient atmospheric.

NEGATIVE PRESSURE VALVE Air-relief valve that releases accumulated air before it accumulates in quantities required to actuate an automatic air valve, by increasing the water through partial closure of a downstream gate.

NEGATIVE WELL Shaft or well driven through an impermeable stratum to permit water to drain through to a permeable one.

NEGLIGENCE Omission of any reasonable precaution, care, or action.

NEGLIGIBLE RESIDUE Amount of a pesticide chemical remaining in or on a raw agricultural commodity or group of raw agricultural commodities that would result in a daily intake regarded as toxicologically insignificant on the basis of scientific judgment of adequate safety data.

NEGOTIATED CONTRACT Construction contract in which the contract sum is discussed, rather than arrived at by competitive bid.

NEMATICIDES Substances or mixtures intended for preventing or inhibiting the multiplication or establishment of, preventing or mitigating the adverse effects of, repelling or destroying any members of the Class Nematoda of the Phylum Nemathelminthes declared to be pests.

NEPHELOMETER Instrument used to compare the turbidity of solutions by passing a beam of light through a transparent tube and measuring the ratio of the intensity of the scattered light to that of the incident light.

NEPHOLOMETRIC Means of measuring turbidity in a sample using light.

NERITIC Those parts of the sea lying over the continental shelf, usually less than 200 m (656 ft) deep.

NESSLERIZATION Method for determining the presence of ammonia from its reaction with a mercury complex in alkaline solution.

NESSLER TUBES Matched cylinders with stain-free, clear-glass bottoms for comparing color density or opacity.

NET AVAILABLE HEAD Difference in pressure elevation between the water in a power conduit before it enters a water wheel and the first free water surface in the conduit below the wheel.

NET CONSUMPTIVE USE OF WATER Consumptive use required for the production of irrigated crops, minus the estimated contribution due to rainfall.

NET DUTY OF WATER Irrigation water requirement, measured at the point of application to crops, excluding canal, conduit, or transpiration loss, seepage, evaporation, or wastage.

NET HEAD Hydraulic head available for production of energy in a hydroelectric plant after deduction of all system losses except those chargeable to the turbine.

NET HORSEPOWER *See* **Horsepower, SAE net.**

NET PEAK FLOW Maximum stream flow, minus corresponding base flow.

NET POSITIVE SUCTION HEAD Greatest distance below the suction inlet of a pump from which it can raise water.

NET POWER Power output of a 'fully equipped' engine.

NET RAINFALL Rainfall, minus infiltration and other surface runoff.

NETWORK Tool used for planning and scheduling a project. The Network is a schematic display of the sequential relationships among the activities that a project comprises. Two popular drawing conventions or notations for scheduling are arrow networks and precedence networks. *See also* **Project network diagram.**

NETWORK PATH Continuous series of connected activities in a project network diagram.

NETWORK PLANNING Identification of the sequence, timing, and interrelationships of the activities comprising a project, including a graphical display of this information.

NETWORK PLANNING TECHNIQUE Planning tool to provide information with which to plan, monitor, and control the time to accomplish specified goals.

NEUTRAL ATMOSPHERE Atmosphere in which the adiabatic lapse rate matches the environmental lapse rate.

NEUTRAL DEPTH Depth of water in an open conduit corresponding to uniform velocity for a given flow.

NEUTRALIZATION Acid or alkyl reaction with an opposite reagent that results in the hydrogen and hydroxyl ion in solution being approximately equal.

NEUTRAL PRESSURE Hydrostatic pressure in the pore water of the soil.

NEUTRAL SHORELINE Shoreline resulting from the building forward of alluvial deposits into a body of water.

NEUTRAL SOIL Soil in the range from slightly acid to slightly alkaline, usually in the range of pH 6.6 to 7.3.

NEUTRAL ZONE Zone between the underside of surface scum and the top of deposited sediments in a sedimentation or settling tank.

NEUTRON Elementary particle that has no electrical charge; a component of atomic nuclei except hydrogen.

NEVE Granular mass, somewhere between snow and ice, that forms in a snow field as the result of compaction due to the weight of overlying material under conditions of alternate thawing, when water percolates down from surface layers, and subsequent refreezing.

NEW SOURCE Source discharging a toxic pollutant, the construction of which is commenced after proposal of an effluent standard or prohibition applicable to such source.

NEWTON SI unit of force which, when applied to a mass of 1 kg gives it an acceleration of 1 m/s². (1N=1 kg·m/s²). Symbol: N. *See also the appendix*: **Metric and nonmetric measurement.**

NEWTONIAN BEHAVIOR Property of a simple liquid in which the shear rate or flow rate is directly proportional to the shear stress or the pressure. This constant proportion is the viscosity of the liquid. *See also* **Non-Newtonian behavior.**

NEWTON METER A derived unit of the moment of force in the SI system of measurement. Symbol: N·m. *See also the appendix*: **Metric and nonmetric measurement.**

NEWTON PER METER A derived unit of surface tension in the SI system of measurement. Symbol: N/m. *See also the appendix*: **Metric and nonmetric measurement.**

NEW WATER 1. Groundwater artificially brought to the surface which would otherwise have continued to flow to an underground reservoir or to have appeared in some other source; **2.** Water from any discrete source such as a river, creek, lake or well

which is deliberately allowed or brought into a plant site.

NICKEL Silvery-white magnetic metal (Ni) that resists corrosion and having an atomic mass of 58.91.

NIGHT SOIL Human excrement.

NIMBY Acronym for 'Not In My Backyard'.

NIPHER SHIELD Cone-shaped wind shield for precipitation gauges.

NIPPLE 1. Short length of pipe threaded on both ends to allow for joining pipe elements. *See also* **Close nipple**; **2.** Small valve at the high points of a hot-water distribution system through which air can be released; **3.** Cylindrical, pipe-like attachment, one end of which is securely inserted and retained in the end of a hose, serving the same purpose as a hose coupling.

NITRATE Salts of nitric acid (NO_3) formed naturally in the soil by microorganisms from protein breakdown and nitrites and available as a plant nutrient.

NITRATE/NITRILE Naturally occurring ions that are ubiquitous in the environment and for which the maximum acceptable concentration in drinking water is 45 mg/L, expressed as nitrate (equal to 10 mg/L as nitrate-nitrogen.

NITRIFICATION 1. Conversion of nitrogenous matter into nitrates by bacteria; **2.** Treatment of a material with nitric acid.

NITRILOTRIACETIC ACID An important chelating agent that has been detected in private and public water supplies and for which the maximum acceptable concentration in drinking water is 0.4 mg/L.

NITROBACTERIA *See* **Bacteria.**

NITROGEN Colorless, odorless, tasteless gaseous chemical element (N) that forms about four-fifths of the Earth's atmosphere by volume.

NITROGEN CYCLE Depiction of the stages through which all matter passes, from living animals through to dead organic matter, including various stages of decomposition, plant life, and the return to living animals.

NITROGEN FIXATION Utilization of free nitrogen and fixation by bacteria associated with legumes.

NITROGEN-FIXING PLANTS Vegetation capable of assimilating and fixing the free nitrogen in the atmosphere using bacteria living symbiotically in their root nodules.

NITROGEN OXIDES (NOx) Nitric oxide (NO), with minor amounts of nitrogen dioxide (NO_2). NOx is formed whenever fuel is burned at high temperatures in air, from nitrogen in the air as well as in the fuel.

NITROGENOUS Chemical compound, usually organic, containing nitrogen in combined forms.

Nitrosomonas Genus of bacteria that oxidize ammonia to nitrite.

NITROUS FUMES Poisonous, reddish fumes produced when nitroglycerine explosives burn instead of detonating.

NODE In CPM networks, events indicating a point in time when all activities leading up to the event have been completed and when all activities leaving the node may start (i.e. junction points); one of the defining points of a network — a junction point joined to some or all of the other dependency lines.

NO DETECTABLE EMISSIONS Less than 500 ppm above background levels, as measured by a detection instrument.

NO-HUB PIPE Pipe having smooth ends (without bell or spigot).

NOISE Sound that is not musical or pleasant; loud, harsh sound that is socially or medically undesirable.

NOISE ABATEMENT ZONE Area designated by a local authority as one in which noise levels may not exceed specified levels throughout designated periods of the day.

NOISE AND NUMBER INDEX Index of air traffic noise based on the number of aircraft passing a point and the sound level

measured at that moment.

NOISE CERTIFICATION Noise levels that newly-designed aircraft must not exceed on takeoff or landing, based on engine power and unladen weight.

NOISE CONTROL Means by which noise, once generated, is abated or reduced, at its point of origin, in the intervening space between the generator and receiver, or by the listener.

NOISE CONTROL BOUNDARY Perimeter of a noise abatement zone on which registered levels are fixed.

NOISE FOOTPRINT Area below a flightpath during takeoff or landing where maximum noise levels will be registered.

NOISE INDICES Measures of the disturbing qualities of noise-loudness.

NOISE-INDUCED HEARING LOSS Damage to the ear not caused by ageing or disease.

NOISE RATING Use of graph curves to relate sound levels in octave bands to their acceptability for particular applications and/or environments.

NOISE REDUCTION COEFFICIENT Average absorption coefficient of a surface or material at 250 Hz, 500 Hz, 1 kHz, and 2 kHz.

NOISE REDUCTION RATING Single number noise reduction factor in decibels, determined by an empirically derived technique which takes into account performance variation of protectors in noise reducing effectiveness due to differing noise spectra, fit variability and the mean attenuation of test stimuli at the one-third octave band test frequencies.

NOISE ZONING Classification of areas according to the noise levels generated by their use.

NOMINAL Dimensional value assigned for the purpose of convenient designation; existing in name only.

NOMINAL CONCENTRATION Amount

of an ingredient that is expected to be present in a typical sample of a pesticide product at the time the product is produced, expressed as a percentage by weight.

NOMINAL DIMENSION Dimension equal to the actual dimension.

NOMINAL FILTRATION RATING Arbitrary value indicated by a filter manufacturer. Due to lack of reproducibility, this rating is deprecated.

NOMINALLY HORIZONTAL In plumbing, at an angle less than 45° with the horizontal.

NOMINAL STRENGTH Published strength calculated by a standard procedure.

NOMINAL VALUE Reference value selected to establish equipment ratings.

NOMINATED SUBCONTRACTOR Subcontractor chosen by the owner, not by the general contractor.

NOMINATED SUPPLIER Supplier selected by the owner, not by the general contractor.

NOMOGRAM (NOMOGRAPH) Diagram or chart consisting of two or more lines (straight or curved) marked with scales, against which measurements and calculations can be made. The straight-line version is *also called* **Alignment chart.**

NONABSTRACTIVE USE Use of a resource that does not permanently remove it from its natural environment or alter its state, i.e. the use of a natural flow of water to generate electricity.

NONARTESIAN WELL Non-flowing well in which the water does not rise above the zone of saturation.

NONBIODEGRADABLE Substances not susceptible to biological degradation.

NONCARBONATE HARDNESS Harness in excess of carbonate hardness.

NONCLOGGING IMPELLER Pump impeller having large passages for the passing of solids.

NONCOMBUSTIBLE WASTE *See* **Solid waste.**

NONCONDENSABLE Gaseous matter not liquefied under ambient conditions.

NONCONFORMING USE Use for which land or a building is lawfully occupied at the time a new regulation becomes effective but that does not comply with the new regulation.

NONCONFORMING WORK Work that does not meet the standards or requirements of regulations or specifications applicable to the site or application.

NONCONTRIBUTING AREA Portion of a drainage having physical characteristics or topography that prevent surface runoff flowing into a river system.

NONCORRODIBLE Material that resists corrosion.

NONCRITICAL ACTIVITIES OR WORK ITEMS Activities or work items that have positive float (i.e. within defined limits, can take longer to complete than is planned without affecting the overall project duration).

NONCONCUSSIVE ACTION Action of a slow-closing automatic valve or tap to prevent water hammer.

NONDEPLETABLE ENERGY SOURCE Renewable energy source, such as wind, solar energy, water power, etc.

NONDEPOSITING VELOCITY Flow velocity that will maintain sediment in suspension.

NONDETECTABLE EMISSIONS Less than 500 parts per million by volume (ppmv) above background levels.

NONERODING VELOCITY Flow velocity that will maintain silt in movement that will not scour a stream or canal bed or banks.

NON-FLOWING WELL Well that does not discharge water at the surface except through use of a pump.

NONFRONTAL PRECIPITATION Precipitation that may occur in any kind of barometric depression.

NONGOVERNMENTAL ORGANIZATION Voluntary body that is accorded official status by the United Nations Organization.

NONIONIC SURFACTANT Surfactant that when in solution, the entire molecule remains associated.

NONMANIPULATIVE JOINT Compression joint in pipework requiring no work other than cutting the ends of the pipe square.

NONPARAMETRIC Statistical test that does not involve a specific form of distribution.

NONPERFORMANCE Situation where a contractor has not done work described in the contract documents, and is therefore in default of the contract obligations. This may entitle the owner to cancel the contract.

NONPOSITIVE DISPLACEMENT PUMP *See* **Pump.**

NONPRESSURE DRAINAGE Drainage system constructed to handle design capacities by gravity flow.

NONPRESSURE PIPE Pipe designed for use, or in a system conveying liquid only by gravity; has no pressure rating.

NONREGENERATIVE CARBON Series, over time, of non-regenerative carbon beds applied to a single source or group of sources.

NONRENEWABLE RESOURCE Resource such as a mineral or fossil fuel that are generated across geological time.

NONRETURN VALVE *See* **Valve.**

NONRISING-STEM VALVE Valve where the stem does not rise when opening the valve.

NONSETTLEABLE MATTER Matter that remains in suspension for longer than one hour.

NONSETTLEABLE SOLIDS Wastewater matter that will stay in suspension for one hour or longer.

NONSILTING VELOCITY Flow velocity that will maintain silt in movement.

NONSIPHON TRAP Plumbing trap whose seal is not easily broken by virtue of physical design and/or quantity of water (usually not less than 0.95 L (0.24 gal)) held in the trap.

NONSPLITTABLE ACTIVITY Activity on a program that must be fully completed before the next activity can begin.

NONTIDAL CURRENT Current induced in water wholly independently of tidal action: caused by wind, temperature, runoff, density differences, etc.

NONUNIFORM FLOW Condition of flow in which velocity varies either in magnitude or direction, or both, from place to place along a stream or conduit due to such factors as variation in cross-section, discharge, slope, roughness, etc.

NONUSE OF LAND Declaration by a public authority that land is not being used, or is disused, empowering it to levy a tax to enforce development, to clear unmarketable titles, to restrain owners of occupied dwellings from discontinuing their use, etc.

NONUSE, OR DISUSE *See* **Land-use control.**

NO-OBSERVED-ADVERSE-EFFECT LEVEL Highest dose in a toxicity study that does not result in any observed adverse effect (one that significantly alters the health of the target animal for a sustained period of time or reduces survival).

NO-OBSERVED-EFFECT LEVEL Highest dose in a toxicity study that results in no observed effects.

NORMAL According to rule, standard, established — the usual state; the average mean value of observed quantities.

NORMAL AMBIENT VALUE Concentration of a chemical species reasonably anticipated to be present in a water column, sediments, or biota in the absence of disposal activities at the disposal site in question.

NORMAL CONSOLIDATION Condition that exists if a soil deposit has never been subjected to an effective stress greater than the existing overburden pressure, and if the deposit is completely consolidated under the existing overburden pressure.

NORMAL DEPLETION CURVE Graphical representation of the mean values obtained from a number of groundwater-depletion curves.

NORMAL DEPTH 1. Depth of water measured perpendicular to the bed; **2.** Depth of water in an open conduit that corresponds to uniform velocity for a given flow.

NORMAL DISTRIBUTION Common symmetrical distribution found in statistics that can be employed as a working approximation for predictive purposes.

NORMAL EROSION Land erosion due to natural activities.

NORMAL FLOW Mean or average flow that prevails for the majority of time.

NORMAL HAUL Haul, the cost of which has been included in the cost of excavation.

NORMAL MOISTURE CAPACITY Approximate quantity of water that the soil will permanently retain in opposition to the drag of gravity.

NORMAL POSITION Default position of a switch; its position prior to being operated.

NORMAL PRESSURE 1. Pressure exerted by any substance in equilibrium with the surface against which it is pressing; **2.** Mean atmospheric pressure at any location; **3.** Pressure exerted most of the time in a fluid system.

NORMAL SOLUTION Solution containing one gram equivalent weight of reactant (compound) per liter of solution.

NORMAL TEMPERATURE AND PRESSURE Temperature of 273.15°K (0°C) and a pressure of 101.325 KPa, used as a reference standard for gas volumetric measurements.

NORMAL THREAD ENGAGEMENT Amount of overlap necessary to insure a tight connection between threaded pipes and fittings.

NORMAL WEAR AND TEAR Physical depreciation resulting from age and normal use of a property.

NORMAL YEAR Year during which the precipitation or streamflow approximates the average measured over a long period.

NOTCH Small measuring weir with its upper edge above water level and a V-shaped, graduated notch cut to allow flow.

NOTCHED WEIR Wide weir, the upper surface of which is broken by a regular series of v-shaped notches of uniform shape and known hydraulic performance.

NOTCH PLATE 1. Horizontal metal plate cut to form V- or U-shaped notches and acting as a weir over which a contained body of water is permitted to discharge at a calculated rate; **2.** Small weir used in laboratory models of hydraulic structures.

NOTICE OF AWARD 1. Formal notification of the awarding of a contract to a successful bidder; **2.** Public notification of the company to which a contract has been awarded, usually as the result of competitive bidding for a publicly funded project.

NOTICE OF CHANGE *See* **Change order.**

NOTICE TO BIDDERS Formal notification to prospective bidders giving the bidding requirements and procedures.

NOTICE TO PROCEED Written notice to a contractor to begin contract work; when applicable, includes the date of beginning of contract time.

NOZZLE Spout, projecting mouthpiece, end of a pipe of hose.

NOZZLE AERATOR Shaped nozzle through which water is sprayed into the air.

NOZZLE JET Stream of water issuing from a nozzle.

NPDES PERMIT National Pollutant Dis-

NUCLEAR DENSITY METER TEST

charge Elimination System permit: the regulatory agency document designed to control all discharges of pollutants from point sources into waterways.

NUCLEAR DENSITY METER TEST *See* **Soil compaction test.**

NUCLEAR ENERGY Energy derived from fission or fusion: fission uses the energy released in a controlled nuclear reaction whereby an isotope of uranium-235 is split by the capture of neutrons; fusion relies on the energy release when a heavier element is formed by the fusion of lighter ones.

NUCLEAR FISSION Splitting of the nucleus of a heavy atom, with the consequent emission of neutrons and energy.

NUCLEAR FUSION Nuclear reaction in which the nuclei of light atoms are fused to form heavier atoms with an accompanying release of energy.

NUCLEAR POWER Electricity generated by equipment powered by a controlled nuclear reaction.

NUCLEAR RADIATION Particulate and electromagnetic radiation emitted from atomic nuclei in various nuclear processes.

NUCLEAR REACTOR Facility for generating heat by the controlled nuclear fission of atoms of uranium-235, or by the nuclear fusion of light atoms to form heavier atoms.

NUCLEAR REACTOR WASTES Waste products produced by the controlled release of energy from fissile fuel and, to a lesser degree, by neutron bombardment of otherwise neutral materials in a reactor, including reactor metals, coolant fluids, air, carbon dioxide and other gases, all of which fall under one of three classifications:

High level – Radioactivity greater than 1 curie/4.5 L.

Intermediate level – Radioactivity in the range 1 x 10^{-6} curie/4.5L.

Low level – Radioactivity less than 1 x 10^{-6} curie/4.5L.

NUCLEIN Constituent of the nuclei of cells.

NUCLEUS Central part about which aggregation, accretion or growth goes on; the charged center of an atom consisting of protons and neutrons, about 1/10 000th its diameter but containing nearly all the mass.

NUCLIDE Isotope of an atom (if it is radioactive, it is a radionuclide).

NUISANCE Someone or something that annoys, troubles, offends, or is disagreeable.

NUISANCE THRESHOLD Lowest concentration of an air pollutant that can be considered objectionable.

NULL Zero set on an automatic sensing mechanism.

NUMBER Measure of discrete quantity.

NUTRIENT Raw material necessary for life and which is consumed during the metabolic process of nutrition.

NUTRIENT CYCLE Transformation or change of a nutrient from one form to another until the nutrient has returned to its original form.

NUTRIENTS Substances necessary to support the growth of organisms, e.g. compounds containing carbon, hydrogen, oxygen, sulfur, nitrogen, and phosphorus.

NYMPH Larval stage of aquatic insects that can be a useful biological indicator of the quality of the water in which they are hatched.

O

OBLIGATE AEROBES Bacteria that depend upon atmospheric or dissolved molecular oxygen.

OBLIGATION In a building contract, whatever the owner and contractor agree on and as shown and described in the contract documents.

OBSERVATION WELL Perforated pipe installed in the ground for monitoring groundwater levels.

OBSERVED EFFECT CONCENTRATION Lowest tested concentration in an acceptable early life stage test:

(a) Which caused the occurrence of any specified adverse effect (statistically different from the control at the 95% level); and

(b) Above which all tested concentrations caused such an occurrence.

OCCASIONAL STORM Rainfall intensity that may be anticipated to occur once in every 10 to 25 years.

OCCLUDED FRONT Boundary of a warm front when it has been overtaken and lifted by an advancing cold front.

OCCUPATIONAL EXPOSURE LIMIT Time-weighted average concentration of a material in air for an 8-hour workday, 40-hour workweek to which nearly all workers may be exposed repeatedly without adverse effect.

OCCUPATIONAL EXPOSURE STANDARD Concentration of an airborne substance, averaged over a reference period, for which there is no evidence that it is likely to be injurious to employees if they are exposed on a regular basis at the workplace.

OCEAN Waters of the open seas lying seaward of the baseline from which the territorial sea is measured

OCEAN CURRENT Nontidal current forming part of the overall oceanic circulation.

OCEANIC Waters of oceans deeper than 200 m (656 ft).

OCEANIC CLIMATE Climate primarily influenced by the ocean, i.e. relatively high humidity and fairly uniform temperatures.

OCEANIC TRENCH Narrow, elongated depression at the margin between an ocean basin and continental shelf.

OCEANOGRAPHY Science of the oceans, their form, physical features and phenomena.

OCTANE NUMBER Percent of isooctane by volume in a mixture of isooctane and normal heptane that has the same antiknock character in a standard variable-compression test engine as the fuel under test. A mixture having 75% octane and 25% heptane is said to have an octane rating of 75.

OCTAVE Interval between a tone and another tone having twice or half as many vibrations.

ODOR Smell or scent detectable by the human olfactory senses. Although odor in drinking water can often be attributable to a specific chemical, it is usually impractical and often impossible to isolate and identify the odor-producing chemical. Odor is rarely indicative of the presence of harmful chemical substances, but may indicate an undesirably high level of biological activity in a drinking water source, contamination of the water supply or treatment and distribution inadequacies.

ODORANT Distinctive, sometimes unpleasant, odor added to odorless, usually gaseous, substances to give warning of its presence.

ODOR CONTROL *See* **Air contaminant control.**

ODOR PANEL Group of people assembled to detect and evaluate odors.

ODOR THRESHOLD Lowest concentration of a vapor that can be detected by smell.

OFFENSIVE INDUSTRY Manufacturer whose operations are noxious or offensive by reason of the process it employs, the materials is uses, its products or the wastes it discharges.

OFF-GAS Gaseous products from chemical decomposition of a material.

OFF-PEAK POWER Electrical energy that can be produced at off-peak hours, outside the load curve, when combined primary and secondary load has fallen below plant capacity.

OFFSET 1. Difference between an actual and desired value or set-point; **2.** In plumbing, a combination of elbows or bends that brings one section of a pipe out of line but parallel with the other section.

OFFSET ELBOW S-shaped pipe fitting, used to offset a straight pipe.

OFFSET FITTING Pipe fitting having two changes of direction, one offsetting the other.

OFFSHORE BAR Beach more-or-less parallel with the shoreline but some distance from the low water mark.

OFFSHORE TERRACE Sand built up in deep water by the combined action of waves and currents, and running more-or-less parallel to the shoreline.

OHM SI unit of electric resistance between two points of a conductor when a constant difference of potential of 1 V applied between these two points produces in this conductor a current of 1 A, the conductor not being the source of any electromotive force. Symbol: Ω ($1\Omega = 1$ V/A). *See also the appendix*: **Metric and nonmetric measurement.**

OHMMETER Meter used to measure electrical resistance.

OHMS-PER-VOLT Rating showing the sensitivity and accuracy of an electric meter.

OIL FIELD Area underlain by deposits of oil or natural gas which are related to a single geological feature that is either structural or stratigraphic.

OIL SHALES Large oil-bearing shale deposits in the Green River formation of Colorado, Wyoming and Utah, and in other areas of the world.

OIL SLICK Floating layer of oil on a water body, usually resulting from a spill or intentional discharge.

OIL-WATER SEPARATOR Waste management unit, generally a tank or surface impoundment, used to separate oil from water and consisting of a separation unit, forebay, and other separator basins, skimmers, weirs, grit chambers, sludge hoppers, and bar screens that are located directly after the individual drain system and prior to additional treatment units such as an air flotation unit, clarifier, or biological treatment unit.

OILY WASTEWATER Wastewater that contains oil, emulsified oil, or other hydrocarbons.

OLD RIVER River in which all channels are subject to aggradation.

OLD VALLEY Stream valley that is fully developed, being very wide with a relatively flat bottom and low sloping sides.

OLIGOTROPHIC LAKE Body of low nutrient water characterized by low quantity of planktonic algae, high water transparency with dissolved low organic deposits, and an absence of hydrogen sulfide in the water and deposits.

OMBROMETER Type of rain gauge.

ONCE-THROUGH OPERATION Treatment process in which none of the flow is passed back for any reason.

ONE-HUNDRED-YEAR FLOOD Flood that has a 1% chance of occurring in any given year, based on historical records for a given location. *See also* **Fifty-year flood.**

ONE HUNDRED-YEAR FLOODPLAIN Land area that is subject to a 1% or greater chance of flooding in any given year from any source.

ONE-PIPE SYSTEM In drainage, two vertical pipes with waste and soil water flowing down the same pipe, and all branches connected to the same antisiphon pipe.

ON-LINE REPLACEMENT The breaking out of an existing service and the installation of a new conduit in the same place.

ON/OFF VALVE *See* **Valve.**

ONSHORE All land areas landward of the territorial seas.

ONSITE DISPOSAL Includes all means of disposal, but more commonly the volume reduction of refuse on the premises before removal for ultimate disposal.

ONSITE INCINERATOR *See* **Incinerator.**

OPACITY 1. Ratio of transmitted to incident light; **2.** Degree to which emissions reduce the transmission of light and obscure the view of an object in the background.

OPEN Not closed, covered, or clogged.

OPEN BIDDING/TENDERING System of public advertising of construction and material supply requirements. *Also called* **Open tendering.**

OPEN-BOTTOMED WELL Well in which the bottom of the casing is left open to allow an inflow of water.

OPEN BURNING Combustion of any material without:

(a) Control of combustion air to maintain adequate temperature for efficient combustion;

(b) Containment of the combustion reaction in an enclosed device to pro-

vide sufficient residence time and mixing for complete combustion; and

(c) control of the emission of the combustion products.

OPEN-CENTER VALVE *See* **Valve.**

OPEN CENTRIFUGAL PUMP Centrifugal pump having an impeller built with independent vanes.

OPEN CHANNEL Conduit in which the water surface is not in contact with the crown of the pipe.

OPEN-CHANNEL DRAINAGE Drainage by means of open channels or ditches.

OPEN-CHANNEL FLOW Flow in a channel where the upper surface of the fluid is exposed to the atmosphere.

OPEN-CIRCUIT APPARATUS Breathing apparatus in which the wearer's exhalations are vented to the atmosphere.

OPEN CUT Excavation where the working area is kept open, as against cut-and-cover and underground work.

OPEN-CUT MINE Any form of recovery of ore from the earth except by a dredge.

OPEN-CUT TRENCHING Traditional method for conduit installation involving the excavation of a trench, laying the conduit, refilling, and reinstating the surface.

OPEN DUMP Land disposal site at which solid wastes are disposed of in a manner that does not protect the environment, are susceptible to open burning, and are exposed to the elements, vectors, and scavengers.

OPEN IMPELLER Impeller without attached side walls.

OPEN-PIT INCINERATOR *See* **Incinerator.**

OPEN SPOOL VALVE *See* **Valve.**

OPEN TENDERING *See* **Open bidding/tendering.**

OPEN-TOP CULVERT Culvert under a road with openings in the top to allow for drainage of the road surface.

OPEN WELL Well large enough in diameter to allow access for a worker.

OPERABLE TREATMENT WORKS Treatment works that will treat waste water, transport waste water to or from treatment, or transport and dispose of waste water in a manner that will significantly improve an objectionable water quality situation or health hazard.

OPERATING BAND Range of pressures above and below the operating pressure within which it is desired to keep a supply output.

OPERATING COSTS Costs that result when a machine is being used, i.e. fuel, repair, wear items, operator wages, etc. *See also* **Variable costs.**

OPERATING DEVICE Push button, lever, or other manual device used to actuate a control.

OPERATING ENGINEER Person certified to operate static machinery.

OPERATING EXPENSE Expense necessary for the maintenance and operation of a device or facility.

OPERATING FLOOR Level in a control building from which controls can be operated and where indicating devices are commonly located.

OPERATING INSTRUCTIONS Details of how something should be operated under specific conditions, a condition of warranty protection.

OPERATING PRESSURE *See* **Pressure.**

OPERATING TIME *See* **Machine time.**

OPERATIONAL LOST TIME *See* **Machine time.**

OPERATIONAL WASTE Water lost from an irrigation system through spillways and other devices.

OPERATION SELECTOR SWITCH *See* **Switch.**

OPERATOR Mechanism that actuates another device.

OPERATOR PROTECTIVE STRUCTURE Equipment for mobile equipment that protects the operator from falling objects and if the machine rolls over. *See also* **Rollover protective structure** and **Falling object protective structure.**

OPERATOR'S MANUAL Book that accompanies equipment, a machine, or mechanical tools that lays out operating conditions, methods, sequences and limits, gives maintenance schedules and procedures, describes recommended safety measures, and lists guarantee or warranty coverage.

OPTICAL DENSITY Method of expressing degrees of contamination of a fluid by removing contaminants through filtration and measuring change in optical transmission of the filter disk or fluid, or both.

OPTIMAL CORROSION CONTROL TREATMENT Corrosion control treatment that minimizes the lead and copper concentrations at users' taps while insuring that the treatment does not cause the water system to violate any national primary drinking water regulations.

OPTIMUM AIR SUPPLY Quantity of air that will give the greatest thermal efficiency under actual conditions. With perfect mixing of fuel and air, the optimum air supply is equal to the chemically correct amount of air.

OPTIMUM MOISTURE CONTENT Moisture content in percent by weight of dry rock and earth that results in the least voids and greatest density when the material is compacted.

OPTIMUM POINT OF COAGULATION Hydrogen-ion concentration in a coagulation process at which the best floc occurs in the shortest time.

OPTIMUM POPULATION Number of individuals that can be accommodated within an area to the maximum advantage of each, and without continuous depletion of the local resources.

OPTION Available equipment or feature not standard equipment on a model.

ORDINARY STORM Rainfall intensity that may be anticipated to occur once each 5 to 10 years.

ORE Naturally-occurring mineral containing a valuable substance such as metal.

ORE BODY Mass of ore that is capable of commercial exploitation.

ORE DRESSING Conversion of raw material obtained by mining into a marketable commodity.

ORGAN Part of a plant or animal that forms a structural unit.

ORGANIC DEPOSIT Vegetable and mineral substances laid down under water.

ORGANIC FARMING Crop production without the use of industrial fertilizers or pesticides.

ORGANIC MATTER Chemical substances consisting of hydrocarbons and their derivatives; organic fraction of the sediment or soil; it includes plant and animal residues at various stages of decomposition, cells and tissues of soil organisms, and substances synthesized by the microbial population.

ORGANIC MATTER DEGRADATION Biological conversion of organic matter to inorganic forms.

ORGANIC NITROGEN Nitrogen that is combined in organic molecules such as amines, amino acids, and proteins.

ORGANIC SILT Mineral particles ranging in size from 0.05 to 0.074 mm (0.002 to 0.003 in.) containing appreciable quantities of organic materials.

ORGANIC SOIL Soil containing an appreciable content of decayed vegetable matter. *See also* **Soil groups.**

ORGANIC WASTE Waste material originating from animals or plants.

ORGANOCHLORINES Class of organic chemicals characterized by persistence, mobility and intense biological activity.

ORGANOMERCURY FUNGICIDES Poisonous compounds that contain compounds of mercury.

ORGANOPHOSPHORUS COMPOUNDS Group of pesticides that includes azodrin, malathion, parathion, diazonon, trithion and phosdrin.

ORIFICE 1. Literally, an opening, but more commonly used to designate a constriction in a passage, circular in shape unless otherwise specified. **2.** Small hole in the tip of a nozzle.

ORIFICE FEED TANK Vessel that receives a chemical solution from a large chemical-storage tank, and which in turn is used to apply the solution to a flow.

ORIFICE FLOW Flow through the orifice of a pipe, the inlet of which is completely submerged; the outlet either discharging freely, partially or completely submerged.

ORIFICE METER Pierced plate placed across a pipeline to create a pressure differential, the amount of which is related to the flow through the pipe.

ORIFICE PLATE Plate having an orifice of smaller area than the cross section of the pipe to which it is attached.

ORIFICE RESTRICTOR Restrictor, the length of which is relatively small with respect to its cross-sectional area. The orifice may be fixed or variable. Variable types are noncompensated, pressure-compensated, or pressure- and temperature-compensated.

OROGENY Extended geological period of mountain building, involving the formation of folds and faults, metamorphism and igneous intrusion.

OROGRAPHIC PRECIPITATION Precipitation caused by rising land in the path of a moisture-laden wind, causing the air flow to rise, with subsequent dynamic cooling and expansion.

ORTHOPHOSPHATE Acid salt containing phosphorus as PO_4.

ORTHOTOLIDINE Colorimetric indicator of chlorine residual.

ORTHOTOLIDINE-ARSENITE TEST Technique using orthotolidine, sodium arsenite, and colorimetric standards to differentiate between free available chlorine, combined chlorine, and color due to interfering substances.

ORTHOTOLIDINE REAGENT Solution prepared from orthotolidine crystals, used to determine colorimetrically the amount of available chlorine in a solution.

OS&Y Valve See **Valve.**

OSCILLATION Periodic movement across a neutral point.

OSMOSIS Diffusion of a liquid through a skin (permeable in one direction only) into a more concentrated solution.

OSMOTIC PRESSURE Pressure in osmosis exerted by a solvent when its entry through a fabric into a more concentrated solution is prevented.

OSMOTIC WATER TRANSPORT Movement of water between separated compartments through a membrane.

OTHER PRODUCTIVE TIME See **Machine time.**

OTTAWA SAND Silica sand produced by processing a material obtained by hydraulic mining of massive quartzite situated in open-pit deposits near Ottawa, IL, composed almost entirely of naturally rounded grains of nearly pure quartz, used in mortars for testing of hydraulic cement. See also **Standard sand** and **Graded standard sand.**

OUNCE Nonmetric unit of mass, equal to 1/16 pound. Symbol: oz. Multiply by 28.349 to obtain grams, symbol: g. See also the appendix: **Metric and nonmetric measurement.**

OUTAGE Difference between the rated capacity of a storage tank and what it will contain, allowing for expansion of the contents.

OUTBURST BANK Middle part of the slope of a sea embankment, between the footing and the highest point that a wave will reach.

OUTCROP Exposure of a rock body or a rock surface.

OUTFALL Discharge end of a drain or sewer.

OUTFALL SEWER Sewer that receives flow from a collector system and conducts it to a point of final discharge.

OUTGO Pipe tail from a sanitary fitting that connects to the drain.

OUTHOUSE Detached privy.

OUTLET Portion of a pipe, pump, or structure through which water or air exits.

OUTLET CHANNEL Structure used to carry water away from a storage facility.

OUTLET CONTROL Means by which to regulate the relationship between a headwater elevation and discharge.

OUTLET HOSE NOZZLE Small hose connection on a hydrant, commonly 62 mm (2.5 in.) in diameter.

OUTLET PIPE Pipe that conveys the effluent from a sewage treatment plant to its point of final disposal.

OUTLINE SPECIFICATIONS See **Draft specifications.**

OUT-OF-SHIFT REPAIR TIME See **Machine time.**

OUT-OF-SHIFT SERVICE TIME See **Machine time.**

OUTSIDE AIR Air drawn from the atmosphere and not previously circulated through an air handling or air conditioning system.

OUTSIDE STANDPIPE Standpipe riser on the outside of a building and equipped with a fire department Siamese connection.

OUTWARD-FLOW TURBINE Reaction turbine in which water or steam enters the runners and flows away from their axis.

OUTWASH DEPOSIT Silt carried by running water from a glacier and deposited beyond the marginal moraine.

OUTWASH PLAIN Stratified deposit of material initially eroded by glaciers and then transported and deposited by meltwaters beyond the glacial margin.

OVALITY Shape that a buried pipe tends to assume due to the compressive force of the cover.

OVERALL REDUCTION Percentage reduction in a measurable constituent between an inflow and outflow.

OVERBAND MAGNET Electromagnet positioned laterally above a conveyor carrying solid waste or incinerated residue so as to recover ferrous metals.

OVERBANK FLOW Percentage of a streamflow in excess of the normal carrying capacity of its channel and which overflows to the adjoining flood plain..

OVERBURDEN Surface material, usually top soil, that must be removed and disposed of prior to excavation of a site.

OVERBURDEN PRESSURE Vertical pressure at a point in a soil mass due to the bulk of material above it.

OVERCHUTE Flume passing transversely over a canal.

OVERCONSOLIDATION Condition that exists if a soil deposit has been subjected to an effective stress greater than the existing overburden pressure.

OVERDEVELOPMENT Condition where the volume of water pumped from a groundwater supply continually exceeds the rate of recharge.

OVERFALL That part of a dam over which water can be discharged; the overpouring water. *Also called* **Overtopping.**

OVERFALL DAM Dam having a crest over which water may flow.

OVERFALL WEIR Impounding structure over which a flow may pass.

OVERFLOW 1. Flow or spread over or across; **2.** To flow over an edge or brim; **3.** Amount that overflows.

OVERFLOW CHANNEL Hydraulic structure through which water overflowing a storage facility is directed to absorb excess energy and to prevent scour of a natural channel.

OVERFLOWED LAND Land that is periodically submerged by tidal or seasonal flows.

OVERFLOW RATE One of the design criteria of settling tanks; the volume of overflow per day per surface area of the storage vessel.

OVERFLOW SIPHON Pipe arrangement that allows excess volume to be discharged other than by overflowing the upper perimeter.

OVERFLOW STAND Standpipe in which water rises and overflows at the hydraulic gradient.

OVERFLOW STREAM Effluent channel of a lake or pond.

OVERFLOW TOWER Outlet from a pressurized water conduit, positioned at elevational summits to limit pressure on the pipe.

OVERFLOW TUBE Vertical tube in the tank of a water closet that prevents overflow.

OVERHEAD Indirect expenses that cannot be charged to individual costs or bid items except by proration.

OVERIRRIGATION Application to agricultural land of more water than is needed by a crop.

OVERLAND FLOW Surface runoff.

OVERLAND RUNOFF Uncontrolled flow of water over the surface before it reaches an established stream channel.

OVERPOUR HEAD GATE Head gate to

a canal through, or over which excess water from a river discharges.

OVERPUMPING 1. Extraction from a groundwater source of more water than the source naturally receives; **2.** Technique for increasing the capacity of a well by pumping from it for brief periods at volumes higher than the normal recharge rate.

OVERRIDE PRESSURE *See* **Pressure.**

OVERRIDE SWITCH *See* **Switch.**

OVERSIZE WASTE *See* **Solid waste.**

OVERSPEED SWITCH *See* **Switch.**

OVERSTORY Uppermost layer in a forest, typically the trees that form a canopy.

OVERTOPPING Flow of water over the top of a dam or embankment. *Also called* **Overfall.**

OVERTURN Mixing of a thermally stratified body of water due to temperature changes.

OVER-YEAR STORAGE Gradual accumulation of water stored in a reservoir during times of excess supply, used to offset lower availability during dry years.

OWNER–CONTRACTOR AGREE-MENT Contract to complete a described project between an owner and contractor.

OWNER DEFAULT Failure by an owner to pay a contractor under the terms of a construction contract.

OWNER OF RECORD Person or persons stated in the public records as the owner(s) of a properly described property.

OWNING COSTS Costs that result from owning a machine, i.e. depreciation, interest cost, taxes, storage, insurance, replacement cost escalation. These costs accrue whether a machine works or not.

OXBOW Abandoned part of a meandering stream.

OXBOW LAKE Lake formed in the closed-off meander of a stream.

OXIDATION Reaction of oxygen on an organic substance, usually evidenced by a change in the appearance or feel of the surface or by a change in physical properties.

OXIDATION DITCH Open channel, circular or oval in plan, with one or more paddle wheels mounted on a bridge and partially submerged in the contents that promote flow, agitate the wastewater and promote aeration and thus oxidation of the organic fraction.

OXIDATION LAGOON Type of oxidation pond.

OXIDATION POND Basin in which pretreated sewage is stored and where biological oxidation of organic material is effected by natural or artificially accelerated transfer of oxygen to the contents from the air.

OXIDATION PROCESS Any process by which putrescible organic matter in wastewater is oxidized through exposure to dissolved or atmospheric oxygen.

OXIDATION RATE Rate at which the organic fraction present in wastewater is stabilized.

OXIDATION–REDUCTION POTENTIAL Potential necessary to transfer electrons from an oxidant to a reductant, used as a qualitative measure of the oxidation process in wastewater treatment.

OXIDATION TREATMENT Conversion of organic matter to a stable mineral form through the agency of living microorganisms in the presence of oxygen.

OXIDE FUEL Nuclear fuel manufactured from the oxides of the fissile material.

OXIDIZABLE SALT Salt, such as ferrous sulfate or carbonate or the corresponding salts of iron and manganese, in solution in groundwater and which may be oxidized to other forms and deposited from solution on exposure to dissolved or atmospheric oxygen.

OXIDIZED ORGANICS Organic material that has been broken down in a biological process.

OXIDIZED SEWAGE Waterborne wastes that have been aerated to the point where organic matter has become stabilized.

OXIDIZED SLUDGE Liquid and solid product resulting from the wet air oxidation of wastewater sludge.

OXIDIZED WASTEWATER Wastewater in which the organic fraction has been stabilized through exposure to dissolved or atmospheric oxygen.

OXIDIZER Supplier of oxygen.

OXIDIZING Combining oxygen with any other substance.

OXIDIZING AGENT Chemical that either gives up oxygen in a chemical reaction or supplies an equivalent element, such as chlorine, to combine with a reducing agent.

OXIDIZING SUBSTANCE Product or substance that:

(a) Causes or contributes to the combustion of another material by yielding oxygen or other oxidizing substances, whether or not the product or substance is itself combustible; or

(b) Is an organic compound that contains the bivalent O-O structure.

OXYGEN Colorless, odorless gas (chemical symbol O, formula O_2, molecular weight 32) essential for all forms of life and for combustors and which forms 21% by volume of the atmosphere.

OXYGENATION CAPACITY Measure of the ability of an aerator to supply oxygen to a liquid.

OXYGEN BALANCE 1. Relation between the biochemical oxygen demand of a liquid and the dissolved oxygen available in the receiving stream; **2.** Level of dissolved oxygen at any point in a stream.

OXYGEN CONSUMED Measure of the oxygen uptake capability of organic and inorganic matter present in water.

OXYGEN DEFICIENCY State of insufficient oxygen to support life or flame; 16% oxygen is needed for flame production and human life.

OXYGEN DEFICIT Difference between the dissolved oxygen level at saturation and the actual dissolved oxygen concentration in water.

OXYGEN DEMAND Quantity of oxygen required by the biochemical oxidation of organic matter, for a specific period, at a specific temperature, under specific conditions.

OXYGEN DEPLETION Loss of dissolved oxygen from water due to biochemical or chemical action.

OXYGEN SAG Decline and subsequent rise in percentage saturation of dissolved oxygen in a river downstream of an effluent discharge containing biodegradable material.

OXYGEN SATURATION Maximum quantity of dissolved oxygen that a liquid of given chemical characteristics, in equilibrium with the atmosphere, can contain at a given temperature and pressure.

OXYGEN UTILIZATION 1. Oxygen consumed in an aerobic biological process; **2.** Oxygen used to support combustion.

OZONATION Application of ozone to water for the purpose of disinfection.

OZONATOR Device for injecting ozone into water.

OZONE Poisonous blue gas (O_3) formed from oxygen on exposure to UV light in the stratosphere or as a pollutant from breakdown of nitrogen oxides, used to disinfect potable water.

OZONE CRACKING Surface cracks, checks, or crazing caused by exposure to an atmosphere containing ozone.

OZONE DEPLETION POTENTIAL Measure of the potential for depletion of the Earth's stratopheric ozone layer.

OZONE LAYER Layer of ozone surrounding the Earth between 15 and 30 km (9.3

and 18.6 miles) above the surface, formed by ultraviolet radiation which splits molecular oxygen (O_2) into two atoms of oxygen, which are highly reactive and forms ozone (O_3), which acts as a barrier to radiation and protects the biosphere.

OZONE RESISTANCE Ability to withstand the deteriorating effects of ozone (generally cracking).

OZONOMETER Instrument for measuring the amount of ozone present in the atmosphere.

P

the stem; **2.** Stuffing of water-resilient material to prevent leaking at a valve stem; **3.** Product intended for use as a mechanical seal in circumstances involving rotary, reciprocating, and helical motions, and which are intended to restrict fluid or gas leakage between moving and stationary surfaces including seals in pumps, valves, compressors, mixers, swing joints, and hydraulic cylinders.

PACKING GLAND Seal round the stem of a faucet or valve, filled with packing and made watertight by screwing down the gland nut.

PACKING NUT Nut on the stem of a valve that compresses the packing.

PAD FOOT Tapered-shank pad with a circular cross section that projects from the roller of a soil compactor. *See also* **Sheepsfoot roller** and **Tamping foot.**

PADDLE Wooden panel used to close a water passage in a lock and to close a sluice or culvert.

PADDLE AERATOR Rotating wheel equipped with projecting paddles that dip into the water contained in an aeration ditch or lagoon, promoting circulation and aeration.

PADDLE HOLE Hole used to allow water into or out of a lock.

PADDLE WHEEL Water-wheel-like device that imparts motion to the stream into which its paddles dip.

PANAMETRIC ESTIMATING Estimating technique that uses a statistical relationship between historical data and other variables (e.g. square footage in construction, lines of code in software development) to calculate an estimate.

PAN COEFFICIENT Ratio of evaporation from a large body of water to that measured in an evaporation pan.

PANEL An access trap.

PANEL MOUNTING Panel on which a number of components may be mounted.

PACKAGE BUILDER Contractor who provides a range of services, in addition to construction, toward the completion of a project. This may include all or any of such items as financing, site selection, site provision planning and design, supply of structures and materials, project management, etc. *Also called* **Turnkey contractor.**

PACKAGE PLANT Prefabricated, and usually partially preassembled, plant designed to process up to a given volume per hour/day. Could be a raw water plant, pumping station, lift station, sewage plant, etc.

PACKED TOWER Pollution control system device that forces dirty air through a tower packed with crushed rock or wood chips while liquid is sprayed over the packing material. The pollutants in the air stream either dissolve or chemically react with the liquid.

PACKER Device which, on being lowered in the tubes lining a well, swells automatically, or can be expanded to produce a watertight joint against the borehole or casing.

PACKER TRUCK Type of solid waste collection vehicle, usually used for residential collection, that compacts refuse into a high density mass for maximum collection efficiency; can incorporate a rear loading or top loading device.

PACKING 1. Material used in the stuffing box of a valve to keep a leakproof seal around

347

PARABOLIC WEIR Weir with a notch shaped as a parabola with a vertical axis.

PARALIC Sediments formed on the landward side of a coast in an area subject to marine incursion.

PARALLEL OPERATION 1. Two or more generators, or other power sources, of the same phase, voltage, and frequency characteristics supplying the same load; **2.** Operation of a multistage pump when each of its impellers receives water from a common source and contributes volume directly to the discharge; **3.** Raw- or wastewater treatment plant in which all processes are duplicated and the flow is divided equally to pass through each.

PARALLEL THREAD Screw thread of uniform diameter, commonly used on machine bolts.

PARAMETER Quantity that is constant in a particular calculation or case but which varies in other cases.

PARAQUAT Bipyridyl contact herbicide most commonly marketed as paraquat dichloride and for which a maximum acceptable concentration in drinking water is 0.01 mg/L.

PARASITE Organism that lives attached to or within another living organism, gaining food and often shelter, but contributing nothing to the host, which usually suffers as a result.

PARASITIC BACTERIA See **Bacteria.**

PARATHION Organophosphorus insecticide and acaricide for which the maximum acceptable concentration in drinking water, derived from the average daily intake, is 0.05 mg/L.

PARENT MATERIAL Weathered rock from which soil is formed.

PARSHALL FLUME Type of Venturi flume, used in hydraulics for measuring flows in open channels.

PARTIAL ACCEPTANCE Acceptance from the contractor of a portion of a construction project that has been completed according to requirements, while waiting for final completion of the entire project. *See also* **Acceptance of work, Final acceptance,** and **Interim acceptance.**

PARTIAL DIVERSION Taking, or removing water from one location in a natural drainage area and discharging it somewhere else in the same drainage area.

PARTIAL PRESSURE *See* **Pressure.**

PARTIAL VACUUM Space condition at less than one atmosphere.

PARTICLE Minute mass of matter that while still having inertia and attraction is treated as a point without length, breadth, or thickness; one of the fundamental units of matter, as the electron, neutron, photon, or proton.

PARTICLE COUNT BLANK Allowance for the determinable background contamination.

PARTICLE-SIZE ANALYSIS Mechanical determination of the proportion by weight of different particle sizes in soil or sand.

PARTICLE SIZE DISTRIBUTION Tabular or graphical listing of the number of particles according to particle size ranges.

PARTICLE VELOCITY Velocity at which a water particle or molecule travels through a formation of porous material.

PARTICULATE CONTAMINATION Foreign matter in a fluid taken in the form of solid particles.

PARTICULATE MATTER 1. Airborne finely divided solid or liquid material with an aerodynamic diameter smaller than 100 micrometers; **2.** Finely divided solid or liquid material, other than uncombined water.

PARTICULATE SAMPLING Methods by which to determine the volume of particles present in a gas stream, most commonly using a device that shines a light beam across the duct or chimney and measuring the reduction in light intensity.

PARTICULATES Atmospheric particles made up of a wide range of natural materials (e.g. pollen, dust, resins), combined with man-made pollutants (e.g. smoke particles, metallic ash).

PARTLY SUBMERGED ORIFICE Orifice with its top above the water discharge surface and its bottom below the surface.

PARTS PER BILLION (PPB) Parts of a substance per billion parts of the whole, equal to 0.000001%

PARTS PER MILLION (PPM) Parts of a substance per million parts of the whole, equal to 0.0001%

PASCAL SI unit of pressure or stress resulting from a force of 1 N acting uniformly over an area of 1 m². (1 Pa = 1 N·m²). Symbol: Pa.

PASCAL SECOND A derived unit of dynamic viscosity in the SI system of measurement. Symbol: Pa·s. *See also the appendix*: **Metric and nonmetric measurement.**

PASCAL'S LAW Principle that states that pressure applied to a confined fluid at rest is transmitted with equal intensity throughout the fluid.

PASS 1. Single progression of a compactor over the work area (number of passes equals the number of progressions); **2.** One-way working trip or passage of a machine. A round trip in the same path is two passes; **3.** Opening through a sand bar or between two bodies of water.

PASSIVE SOLAR ENERGY Collection of solar energy through fixed, south-facing panels, thermal storage devices and a transfer medium; one in which no pumps are used to accomplish the transfer of thermal energy.

PASS THROUGH Discharge from a wastewater treatment plant of a volume of effluent that alone, or in combination with other discharges, is a violation of established quality levels.

PATCH TEST Any method of evaluating fluid contamination wherein the sample is passed through a standardized laboratory filter, and the change in color, reflectivity, etc. of the laboratory filter is compared with previously established standards.

PATH Set of sequentially connected activities in a project network diagram.

PATH CONVERGENCE In mathematical analysis, the tendency of parallel paths of approximately equal duration to delay the completion of milestones where they meet.

PATHOGENIC BACTERIA *See* **Bacteria.**

PATHOGENS Pathogenic or disease-producing organisms.

PATHOLOGICAL WASTE *See* **Solid waste.**

PAYBACK Ratio of the annual income or savings derived from a project to its capital cost.

PAYLOAD 1. Total weight of the commodity being carried on a truck at a given time; **2.** Excavated load for which payment is made.

PAYLOAD AND BODY ALLOWANCE Payload capacity of a truck with allowance for the weight of the truck body.

PAYMENT REQUEST Request made by a contractor or supplier for payment for work done or materials supplied.

PAYMENT SCHEDULE Sequence and times of partial payments made as part of a total contract sum.

PAYNE'S PROCESS Method of fireproofing lumber by treating it first with an injection of sulfate of iron, then with a solution of sulfate of lime or soda.

PEA GRAVEL Screened gravel, most of the particles of which pass a 10-mm (0.38-in.) sieve and are retained on a No. 4 (4.76-mm, 0.197-in.) sieve.

PEAK Maximum quantity or demand that occurs over a relatively short period.

PEAK DEMAND Maximum momentary load placed on a supply facility or system.

PEAK FLOOD LEVEL Water surface elevation of a design flood.

PEAK HORSEPOWER *See* **Horsepower.**

PEAK LOAD Maximum average load placed on a supply facility or system.

PEAK-LOAD PLANT Electric power generating plant used primarily to meet peak system loads.

PEAK SOUND PRESSURE LEVEL Value, in decibels, of the maximum sound pressure.

PEAK TORQUE Speed at which an engine develops maximum torque.

PEAT Lightweight mixture of decomposed plant tissue in which parts of the plant are easily recognized.

PEATLAND *See* **Muskeg.**

PEBBLE Sediment particles between 2 mm (0.06 in.) and 65 mm (2.5 in.) in diameter.

PEDESTAL BASIN Bathroom washbasin mounted on a matching, freestanding pedestal, through which the hot and cold water supply pipes and waste outlet and trap are concealed.

PEDESTAL URINAL Urinal which is not connected to a wall but is mounted on a single pedestal.

PEDIMENT Eroded wash slope at the foot of a mountain.

PEDOGENESIS Origin and development of soils.

PEDON Smallest vertical column of soil that contains a sample of all the soil horizons present at that point.

PELAGIC Organisms of the plankton and nekton that inhabit the open water of a sea or lake.

PELLETED FUEL Combustible material, often waste derived, compressed into solid fuel pellets.

PELLICULAR FRONT Even front, developed only in pervious material, on which pellicular water, depleted by evaporation, transpiration, or chemical action, is regenerated by influent seepage.

PELLICULAR WATER Film of water surrounding each grain of water-bearing material after gravity water has been drained off.

PELLICULAR ZONE Maximum depth below the surface down to which water will evaporate.

PENALTY FOR DELAY Contractual clause that imposes a penalty, usually financial, if a contract is not completed on the date stipulated in the contract documents.

PENDANT SWITCH *See* **Switch.**

PENSTOCK Pipeline or pressure shaft in a dam, leading from the headrace or reservoir to the turbines.

PENTACHLOROPHENOL Toxic, oil-soluble chemical, used as a wood preservative and insect repellent.

PENTOSANS Polysaccharides, $(C_5H_8O_4)_x$, present with cellulose in plant tissue.

PERCEIVED NOISE DECIBELS Frequency-weighted noise unit (PNdB) used for aircraft noise measurement.

PERCENTAGE REDUCTION Ratio of material removed to the material originally present.

PERCENT COMPLETE Estimate, expressed as a percent, of the amount of work which has been completed on an activity or group of activities.

PERCENT FINES 1. Amount, expressed as a percentage, of material in an aggregate finer than a given sieve, usually the No. 200 (0.074 mm, 0.0029 in.); **2.** Amount of fine aggregate in a concrete mixture expressed as a percent by absolute volume of the total amount of aggregate.

PERCENT GRADE *See* **Expressions of slope.**

PERCENTILE Any value in a series of points on a scale arrived at by dividing a group into a hundred equal parts in order of magnitude.

PERCENT LOAD Fraction of the maximum available torque at a specified engine speed.

PERCENT REMOVAL Percentage expression of the removal efficiency across a treatment plant for a given pollutant parameter, as determined from the 30-day average values of the raw wastewater influent pollutant concentrations to the facility and the 30-day average values of the effluent pollutant concentrations for a given time period.

PERCENT SATURATION Amount of a substance that is dissolved in a solution compared with the amount that could be dissolved in the solution, expressed as a percent.

PERCENT OF SLOPE *See* **Expressions of slope.**

PERCHED GROUNDWATER Groundwater separated from the main body of groundwater by an aquiclude.

PERCHED SPRING Spring rising from a perched zone of saturation.

PERCHED WATER *See* **Groundwater.**

PERCOLATING FILTER Type of aerobic sewage treatment bed consisting of a bed of inert material through which water is allowed to trickle in order to purify it.

PERCOLATION Movement of gas or liquid through the void spaces of soil without a definite channel.

PERCOLATION PATH Course followed by water percolating through a permeable material.

PERCOLATION RATE Speed at which water moves under hydrostatic pressure through a permeable material.

PERCOLATION TEST Standard test to measure the rate at which the soil of an area will absorb and disperse moisture, used to determine the suitability of the area to receive the effluent from a domestic waste disposal system. *Also called* **Perk test.**

PERENNIAL INTERRUPTED STREAM All-year-flowing stream having stretches in which there is no surface flow.

PERENNIALLY-FROZEN SOIL Soil that has been continuously frozen for more than three years.

PERENNIAL SPRING All-year-flowing spring.

PERENNIAL STREAM All-year-flowing stream.

PERFORATED CASING Well casing in which holes have been cut or punched so as to allow the inflow of water.

PERFORMANCE Legal term used in contract documents; the contractor's principal obligation.

PERFORMANCE CURVE Graphic representation of an operating characteristic of equipment.

PERFORMANCE EVALUATION SAMPLE Reference sample provided to a laboratory for the purpose of demonstrating that it can successfully analyze the sample within limits of a performance specification. The true value of the concentration of the reference material is unknown to the laboratory at the time of the analysis.

PERFORMANCE FACTOR Ratio of the useful output capacity of a system to the input required to obtain it. Input and output units need not be consistent. *See also* **Tractive effort.**

PERFORMANCE SPECIFICATION Specification that states the desired operation or function of a product but does not specify the materials from which the product must be constructed.

PERFORMANCE TEST *See* **Service test.**

PER HECTARE FACTOR Number used to convert sample plot information to per-hectare information (as in converting a plot volume to be a per-hectare volume).

PERICLINE Anticline that plunges in both directions.

PERIGLACIAL Area surrounding the limit of glaciation and subject to intense frost action.

PERIOD Interval required for completion of an event; a particular duration of time.

PERIOD FOR COMPLETION *See* **Contract time.**

PERIODIC Occurring, appearing, or done again and again at regular intervals.

PERIODIC APPLICATION OF COVER MATERIAL Application and compaction of soil or other suitable material over disposed solid waste at the end of each operating day or at such frequencies and in such a manner as to reduce the risk of fire and to impede vectors access to the waste.

PERIODIC SPRING Spring that has periods of different, but predictable rates of flow that occur at more-or-less regular intervals.

PERIODIC TABLE Table in which the elements, arranged in the order of their atomic weights, are shown in the following related groups:

> **Group 1** Li, Na, K, Rb, Cs, Fr: alkali metals.
>
> **Group II** Be, Mg, Ca, Sr, Ba, Ra: alkaline earth metals.
>
> **Group VI** O, S, Se, Te, Po: chalcogenides.
>
> **GroupVII** F, Cl, Br, I, At: halogens.
>
> **Group O** He, Ne, Ar, Kr, Xe, Rn: noble gases.

PERIPHERAL FLOW Flow parallel to the circumference or periphery of a circular tank.

PERIPHERAL PUMP Pump having an impeller that develops head by recirculating the liquid through a series of rotating vanes.

PERIPHERAL WEIR Continuous weir around the inside edge of the circumference of a circular tank.

PERISTALTIC HYDRAULIC PUMP *See* **Pump, hydraulic.**

PERISTALTIC PUMP *See* **Pump.**

PERK TEST *See* **Percolation test.**

PERM Measure of water vapor movement through a material: one grain per square foot per hour per inch of mercury difference.

PERMANGANATE VALUE Value used as a rapid effluent treatment test, consisting of the amount of oxygen absorbed from a standard potassium permanganate solution during 4 hours at 27°C.

PERMANENT HARDNESS Hardness of water attributable to metallic ions other than calcium or magnesium.

PERMANENTLY ABSORBED WATER Water that percolates beyond the root zone to become part of groundwater within the saturation zone.

PERMANENT SNOW LINE Line of lowest elevation of a permanent snow field, above which snow remains on the ground throughout the entire year.

PERMANENT STREAM All-year-flowing stream.

PERMEABLE Soil or material that permits the passage of water or gas.

PERMEABILITY Rate of diffusion of a gas or vapor through a porous material. *See also* **Properties of soil.**

PERMEABILITY COEFFICIENT Rate of flow through a cross section of permeable material under a hydraulic gradient.

PERMEAMETER Instrument for measuring the coefficient of permeability of a soil sample: a constant-head unit is used for permeable materials like gravel; a falling-head unit for impermeable materials like clay.

PERMEANCE Ratio of the rate of water vapor transmission through a material or assembly to the vapor pressure differential between the surfaces.

PERMEATE To spread or flow throughout; pass through the openings or interstices.

PERMEATION GROUTING Low-pressure grouting of soil in a manner that creates minimum disturbance.

PERMEATING TUBE Used for calibrating instruments and in analytical methods for measuring concentrations of pollutants in air, a sealed polymer tube containing a liquefied sample of a gas to be measured which, at a fixed temperature, diffuses through the walls of the tube at a constant rate.

PERMISSIBLE VELOCITY Greatest speed for which a structure is designed to carry a flow without damage.

PERMIT Written authority to do something, or to have something.

PERMIT Document issued by a regulatory authority permitting the bearer to take a defined action.

PERMIT TO WORK Authorization to enter and work within areas controlled or restricted due to security or hazard.

PEROXYACETYL NITRATES Any of a number of complex compounds present in photochemical smog that cause irritation to eyes and which are toxic to plants.

PERSISTENCE Capacity of a substance to remain chemically stable.

PERSONAL PROPERTY Property that is not real property and that is movable or not attached to the land. Personal property can be a single or multicomponent or multi-material product, it may be tangible (having physical existence), such as equipment and supplies, or intangible (having no physical existence), such as patents, inventions, and copyrights.

PERSONAL PROTECTIVE EQUIPMENT Chemical protective clothing or device placed on the body to prevent contact with, and exposure to, an identified chemical substance or substances in the work area.

PERSONAL SAMPLER Device that can be worn and which samples air in the immediate vicinity to record exposure to pollutants, or which progressively records exposure to radiation.

PERT. *See also* **Arrow** and **Critical-path method**.

PERT CHART Specific type of project net-

work diagram. *See also* **PERT**.

PERT SCHEDULE Project Evaluation and Review Technique (PERT). The charting of activities and events anticipated in a work process.

PERVIOUS Permeable by virtue of being mechanically discontinuous.

PERVIOUS BED Layer or stratum containing voids through which water will flow.

PERVIOUSNESS Property of permitting the passage of water once saturated.

PERVIOUS SOIL Soil that allows relatively free movement of water.

PEST Insect, rodent, nematode, fungus, weed, or any other form of terrestrial or aquatic plant or animal life or virus, bacterial organism or other microorganism (except viruses, bacteria, or other microorganisms on or in living man or other living animals) which are injurious to health or the environment.

PEST CONTROL Methods used to limit the presence and procreation of pests that affect public health or attack resources of use to man.

PESTICIDE Substance or mixture intended for preventing, destroying, repelling, or mitigating any pest, or intended for use as a plant regulator, defoliant, or desiccant.

PESTILENCE Contagious disease that is epidemic and deadly.

PEST PROBLEM Pest infestation and its consequences, or any condition for which the use of plant regulators, defoliants, or desiccants would be appropriate.

PETCOCK *See* **Valve**.

PETROCHEMICAL Obtained from petroleum or relating to such chemicals.

PETROLEUM Range of distillate and residual liquid fuels derived from oil

PETROLOGY Study of the origin, structure and mineralogy and chemical composition of rocks.

pH Logarithm of the reciprocal of the weight of hydrogen ions in grams per litre of solution: a simplified system of measuring acidity or alkalinity irrespective of the acid or alkali involved; in which neutrality is 7.0; e.g. mineral acid solution is 1.0 to 2.8, acetic acid solution or citric acid solution is 3.0 to 4.0, ammonia is 9.0, and lime water is 12.0. Corrosion effects may become significant below pH 6.5, and the frequency of incrustation problems my be increased above pH 8.5. There is also a progressive decrease in the efficiency of chlorine disinfection processes with increasing pH levels. An acceptable range for drinking water pH is from 6.5 to 8.5.

PHASE Portion of a physical system separated by an interface from the rest of the system.

PHASED CONSTRUCTION *See* **Accelerated design and construction.**

PHASED DESIGN AND CONSTRUCTION *See* **Accelerated design and construction.**

PHASED DISPOSAL Method of tailings management and disposal that uses lined impoundments which are filled and then immediately dried and covered.

PHENOL Carbolic acid; any of various weakly acid organic compounds derived from benzine, and containing a hydroxyl group, used in the manufacture of epoxy resins, phenol formaldehyde resins, plasticizers, plastics, and wood preservatives.

PHENOL FORMALDEHYDE Water-resistant thermosetting resin system.

PHENOLPHTHALEIN ALKALINITY Measure of the hydroxides plus one half of the normal carbonates in aqueous suspension.

PHORATE Organophosphorus insecticide and acaricide for which the maximum acceptable concentration in drinking water, derived from the average daily intake, is 0.002 mg/L.

PHOSPHATE Salt of phosphoric acid.

PHOSPHORUS Element (P) that plays an essential role in the growth and development of both plants and animals.

PHOTIC ZONE Surface waters of a sea or lake into which sunlight can penetrate and in which photosynthesis occurs.

PHOTOCHEMICAL Of, having to do with, or resulting from the chemical action of light or other radiant energy.

PHOTOCHEMICAL SMOG Presence in the atmosphere of substances that condense to form a haze of minute droplets containing chemical compounds that can cause irritation of the eyes and lungs as well as cause severe damage to some leafy plants.

PHOTOCONDUCTIVE CELL Photoelectric cell whose resistance is proportional to the intensity of light striking it.

PHOTODEGRADABLE Process where ultraviolet radiation degrades the chemical bond or link in the polymer or chemical structure of a plastic.

PHOTODIODE Diode that can switch and regulate electrical current in proportion to the intensity of the light striking the PN junction.

PHOTOELASTICITY Examination by transparent model analysis of the distribution of stresses in an object through the use of polarized light that reveals isochromatic and isoclinic lines. The lines indicate the direction of axis of principal stresses at any point and the magnitude of the difference of principal stresses.

PHOTOELECTRIC CELL Switch that is held in one position by a fixed beam of light and actuated by an interruption to that beam.

PHOTOELECTRIC EFFECT Flow of electrical energy in some conductors caused by exposure to light.

PHOTOELECTRIC LOAD INDICATOR Device comprising a cylinder of optical glass that is placed in a hole in a steel cell body, the load being applied through end plates. When the glass is strained under load, photoelastic interference fringe patterns are visible when the glass is illuminated with polarized light.

PHOTOGRAMMETRY Surveying technique employing a photo-theodolite or other photographic equipment to take a successive series of photographs from a fixed station along a fixed camera axis to record movement.

PHOTO-IDENTIFICATION Precise determination of the position of the image of a ground point or feature on an air photograph or map.

PHOTO INDEX Small-scale map that shows the layout of a series of aerial photographs.

PHOTO-INTERPRETATION Determination of the nature of a feature seen on an air photograph.

PHOTOMETER Instrument that measures photometric quantities such as luminance, luminous intensity, luminous flux, and illumination, etc. *See also* **Flame photometer.**

PHOTOMETRY Measurement of quantities associated with light.

PHOTOMICROGRAPH Photograph of an object taken through a microscope.

PHOTO MAP Map in which the detail has been reproduced directly from air or satellite photography.

PHOTON Quantum or unit particle of light, having a momentum equal to its energy and moving with the velocity of light.

PHOTOPERIODISM Response of plants and animals to the relative duration of light and dark throughout an annual cycle.

PHOTOSYNTHESIS Process by which chlorophyll-containing cells in green plants convert incident light to chemical energy and synthesize organic compounds from inorganic compounds, especially carbohydrates from carbon dioxide and water, with the simultaneous release of oxygen.

PHOTOSYNTHETIC BACTERIA *See* **Bacteria.**

PHOTOSYNTHETIC EFFICIENCY Percent of total energy arriving at the Earth's surface that is fixed by plants, estimated at 6%.

PHOTOTHEODOLITE Calibrated camera combined with a theodolite for taking precise terrestrial photographs from a ground station.

PHOTOTROPIC Capable of obtaining energy from sunlight.

PHOTOVOLTAIC ARRAY Group of interconnected photovoltaic modules acting as a composite unit.

PHOTOVOLTAIC CELL Device that converts light into electricity through the excitation of electrons.

PHOTOVOLTAIC EFFECT Production of an electromotive force by radiant energy falling on the junction of two dissimilar materials.

PHREATIC Relating to the subsurface layer in which groundwater is present.

PHREATICOLOUS Organisms that live in subterranean fresh water.

PHREATIC CYCLE Time between regular fluctuations in the surface level of groundwater.

PHREATIC DECLINE Downward shift of the upper level of the watertable.

PHREATIC DIVIDE Boundary in a body of groundwater separating flows to individual streams.

PHREATIC FLUCTUATION Upward, or downward movement of the upper surface of groundwater.

PHREATIC WATER Groundwater.

PHREATIC ZONE Saturated ground below the watertable.

PHREATOPHYTE Plant that obtains its moisture from the saturation zone.

PHYSICAL DETERIORATION Depreciation of a structure due to the passage of time, or the action of the elements, or the wear and tear to which it has been subjected in use, or any combination.

PHYSICAL INVENTORY Inventory

shown to be present by observation and counting and/or sampling. *See also* **Book inventory.**

PHYSICAL LIFE Estimated life expectancy of a structure.

PHYSICAL MAINTENANCE Preservation and upkeep of a property, equipment, machinery, structures, a highway, etc., including all of its elements, in as nearly as practicable its original, as-constructed condition or its subsequently improved condition.

PHYSICAL PROPERTIES Properties or qualities, other than mechanical properties, that have to do with the physics of a material.

PHYSICAL RESTRAINT Physical activity or work item that must be completed before the next activities or items in a sequence can begin (i.e. concrete must be cured before forms can be removed).

PHYSICAL STABILITY Ability of a material, product, assembly, etc., to maintain its physical dimensions and properties when exposed to conditions normally encountered in its service environment.

PHYSICAL SURVEY 1. On-ground activity that determines or confirms the physical state and/or condition of an area, facility, or structure; **2.** Activity of a sewer system evaluation survey designed to determine specific flow characteristics, groundwater levels, and the physical condition of the system.

PHYSICAL TESTING Examination of a material to determine its physical properties.

PHYSICAL WASTE TREATMENT PROCESS Any of a number of mechanical devices used to reduce or remove the solid fraction from a waste stream: racks, screens, clarifiers, comminutors, etc.

PHYSIOGRAPHIC EQUILIBRIUM Balance at a locality between surface landforms and vegetation, and precipitation and erosion.

PHYSICOCHEMICAL EFFLUENT TREATMENT Nonbiological effluent treatment: precipitation, setting, etc.

PHYTOMETER Instrument for measuring vegetal transpiration.

PHYTOPLANKTON Free-floating microscopic plants, mainly one-celled algae, that are a constituent of plankton.

PICKING TABLE OR BELT Table or belt on which solid waste is manually sorted and certain items are removed.

PICO Prefix representing 10^{-12}. Symbol: p. Used in the SI system of measurement. *See also the appendix*: **Metric and nonmetric measurement.**

PICLORAM Chloropicolinic acid-derived herbicide for which the maximum acceptable concentration in drinking water is 0.19 mg/L.

PIEDMONT ALLUVIAL DEPOSIT Alluvial fans built up by streams emerging from mountains.

PIEDMONT ALLUVIAL PLAIN Series of alluvial fans merging one into the other. *Also called* **Compound alluvian fan.**

PIESTIC INTERVAL Difference in altitude between two isopiestic lines or lines of equal pressure.

PIEXOELECTRIC EFFECT Interaction between electrical and mechanical stress-strain variables in a medium, whereby energy is converted from one form to another.

PIEXOMETRIC LEVEL Level to which water in a confined aquifer will rise under its own pressure in a borehole.

PIEZOMETER Device for measuring the pressure head of pore water at a specific point within a soil mass.

PIEZOMETER NEST Two or more piezometers installed in a common borehole or in close proximity to measure the vertical hydraulic gradient

PIEZOMETER TUBE Standpipe used to measure moderate pressures.

PIEXOMETRIC HEAD Elevation plus pressure head.

PIEZOMETRIC MAP Map depicting isopestic lines showing the shape of the piezometric surface of an aquifer.

PIEZOMETRIC SURFACE Hypothetical surface above or within the ground at which the water would settle in a piezometer tube whose lower end passes below the water table: the level to which water from an artesian well could rise.

PILLOW Plastic, pillow-shaped tube containing a chemical or reagent used in test procedures.

PILOT CIRCUIT Control circuit in electrical work.

PILOT LIGHT Low-output electric lamp indicating a closed and energized circuit.

PILOT-OPERATED CHECK VALVE *See* **Valve.**

PILOT-OPERATED VALVE *See* **Valve.**

PILOT PLANT Process plant that replicates in small scale all of the techniques and equipment proposed for a full-size plant, used to evaluate efficiencies and costs.

PILOT PRESSURE *See* **Pressure.**

PILOT TUNNEL Exploratory, small-diameter tunnel or shaft, often driven to full length, that confirms the geology of the route and enables the full-size tunnel to be driven from both ends.

PILOT VALVE *See* **Valve.**

PINGO Dome-shaped body in permafrost, caused by lenses of ice accumulating beneath the surface.

PIONEER PLANTS Plants capable of invading bare sites, e.g. a newly exposed soil surface, and persisting there, i.e. colonizing until supplanted by invader or other succession species.

PIPE Cylindrical tube. Pipe is fabricated from a wide range of materials, including:

ABS: Rigid plastic pipe made of acrylonitrile-butadiene-styrene (ABS)

Alloy: Steel pipe fabricated with one or more elements in addition to carbon that give the metal greater strength and a higher resistance to corrosion than carbon steel pipe.

Aluminized steel: Steel pipe made of helically-wound corrugated sheet hot-dipped in aluminum.

Armored: Large-diameter plastic pipe reinforced by steel wire embedded in its wall.

Asbestos–cement: Pipe fabricated from concrete containing an amount of asbestos fibers.

Banded steel: Steel pipe, around the shell of which, bands have been shrunk to increase its strength for use under extremely high heads.

Black iron: In fact, steel pipe that has not been galvanized.

Brass: Pipe composed of 84% to 86% copper and up to 0.06% lead, plus zinc.

Butt-weld steel pipe: Pipe intended for low-pressure steam, water, gas, or air service.

Cast iron: Cast pipe fabricated from a composite of alloys, primarily of iron, carbon, and silicon.

Cement lined: Cast- or wrought-iron pipe, or steel pipe, the interior of which is lined with a smooth, dense layer of cement mortar.

Clay: Pipe made of clay and burned in a kiln.

Concrete: Pipe formed by casting, extruding, or spinning a concrete mix, made in diameters ranging from approximately 75 mm to 2 m (3 in. to 6 ft) or more.

Copper pipe: Solid copper pipe produced in three types:

Type K: Heavy-wall tube identified by a 6.4 mm (0.25 in.) green stripe along the entire length of the pipe, used primarily for underground services, plumbing, heating and cooling systems, and for steam, oil, air, oxygen, and hydraulic lines.

Type L: Medium-wall tube, identified by a blue stripe, used for underground services, interior plumbing, heating and cooling systems, snow-melting systems, and steam, oil, and oxygen lines.

Type M: Light-wall tube, identified by a red stripe, used for interior water distribution, heating and cooling systems, steam lines, interior waste lines, and drainage lines.

CPVC: Semirigid plastic pipe formed of chlorinated polyvinyl chloride; a type of plastic pipe that will carry hot water and some chemicals.

Ductile iron: Cast iron pipe that can sustain not less than 5% elongation before fracturing. Especially refers to cast ductile iron that has less tendency to crack when subjected to high stress.

Electric-fusion-welded: Pipe where coalescence is produced in the longitudinal seam of the preformed tube by manual or automatic electric-arc welding.

Electric-resistance-welded: Individual or continuous lengths of pipe produced from skelp with the longitudinal butt joint coalesced by pressure and an electric-resistance circuit of which the pipe is one element.

Galvanized steel: Steel pipe produced in three grades of wall thickness (standard-weight, extra-strong, and double-extra-strong), galvanized to resist corrosion.

Hubless: Plain-end pipe without built-in facility for mating with adjacent lengths.

Lap weld: Pipe formed by welding along a scarfed longitudinal overlapping seam.

Malleable iron: Cast-iron pipe, heat treated to reduce its brittleness.

Polybutylene (PB): Pipe formed of a light-colored, liquid, straight-chained aliphatic hydrocarbon polymer.

Polyethylene (PE): Thermoplastic high-molecular-weight organic compound used in formulating pipe.

Polypropylene (PP): Pipe formed by the polymerization of high-purity propylene gas in the presence of an organometallic catalyst at relatively low pressures and temperatures. It has exceptional flex life, good surface hardness, scratch and abrasion resistance, and high chemical resistance. It will not stress crack and will take high temperature fluids.

Polyvinyl chloride (PVC): Flexible pipe formed of synthetic resin prepared by the polymerization of vinyl chloride.

Seamless: Pipe or tube formed by piercing a billet of steel and then rolling.

Seamless carbon-steel: Drawn pipe having a high carbon and manganese content giving it greater tensile strength but less ductability.

Solid drawn: Metal tube made by extrusion; seamless pipe.

Spiral: Pipe made by coiling a plate into a helix of uniform diameter and riveting or welding the overlapping edges.

Spiral weld: Strip steel or plate turned into an open pipe with the longitudinal, spiral joint welded or crimped into a tight seam.

Steel: Pipe manufactured in any of a large number of steel alloys, either extruded (seamless) or welded. Its wall thickness ranges from Schedule 10 (lightest) to Schedule 160 (heaviest).

358

Spun: Centrifugally-cast concrete pipe.

Stainless steel: Corrosion-resisting alloy-steel pipe high in nickel and chromium.

PIPE BRACE Brace extending between opposite hangers on a spring-type suspension. *Also called* **Angle brace.**

PIPE BRACKET Support for a pipe, fixed to a wall or ceiling.

PIPE BRANCH *See* **Branch fitting.**

PIPE BUFFER Worker who buffs the ends of pipe to be joined prior to welding.

PIPE BURSTING Technique for conduit installation in which an existing, defective pipeline is first used as a pilot and is then broken up and removed, after which an enlarged excavation is made for a larger pipeline.

PIPE CLAMP *See* **Clamp.**

PIPE CLIP Pipe support having a removable section. *See also* **Saddle clip,** and **Two-piece cleat.**

PIPE CLOSER Noncombustible sleeve that closes the gap between a pipe and the opening made through a wall to allow its passage.

PIPE COUPLING Short threaded collar, used to connect the threaded ends of two pipes.

PIPE COVERING Wrapping around a pipe which acts as a thermal insulation and/or vapor barrier.

PIPE CROSS Pipe fitting having four openings in the same plane at right angles to each other.

PIPE CUTTER Tool for cutting pipes that leaves a smooth end.

PIPE DIAMETER Nominal or designated inside diameter of a pipe bore.

PIPE DRILL Drill bit that cuts round holes in masonry.

PIPE DUCT Enclosed interior space reserved for pipework.

PIPE ENAMEL Asphaltic coating given to buried pipes.

PIPE FINDER Instrument that uses magnetic attraction or electronic detection to locate buried metal pipes.

PIPE FITTER Worker skilled in the installation of piping systems for water, steam, gas, oil, chemicals, etc.

PIPE FITTING Fitting used to connect lengths of pipe, to support them and permit them to change direction.

PIPE FLOW Factors that affect the flow of a fluid in a pipe, including:

a) Velocity of the fluid (mean velocity).

b) Viscosity of the fluid.

c) Density of the fluid.

d) Diameter of the pipe.

e) Friction where the fluid is in contact with the pipe.

PIPE FREEZING Technique of freezing pipe so that work can be done on a section without the necessity of draining the system. There are several methods, including the use of liquid nitrogen at -196°C (-320°F), or liquid carbon dioxide, which changes into 'dry ice' when released.

PIPE GALLERY Space reserved for the installation of pipe runs, usually with access for maintenance.

PIPE HANGER Device to support from above a pipe or group of pipes.

PIPE HOOK Hook-shaped support with a spiked extension that can be driven into a mortar joint or a timber, used to support a horizontal, or near horizontal run of pipe.

PIPE INSPECTION CHAMBER Manhole in which an open channel connects two lengths of drain, used for inspection and pipe rodding, commonly placed at 30 m (100 ft) intervals on a drain line where no other type of manhole is required.

PIPE JACKING Technique of installing a line of pipes through the ground in a previously excavated bore by means of hydraulic jacks from the drive shaft. After pushing a pipe length into the ground a new pipe is positioned and the process repeated. *Also called* **Pipe pushing.**

PIPE JOINT *See* **Pile joint.**

PIPE KEY *See* **Barrel key.**

PIPE KOOK Spiked fastener, driven into a stud or masonry joint, with a curved end.

PIPE LAYER 1. Tracked equipment fitted on one side with a hinged boom and on the other with a counterweight and winch. The equipment can lift, support, and transport sections of pipe, or an assembled pipeline (working in series with other units), and lower the pipe into a trench offset to the side; **2.** Worker skilled in installing all types of pipe in trenches, laying to correct levels, making joints and connections.

PIPE LEAD *See* **Lead.**

PIPELINE Lengths of pipe joined together.

PIPE LINER Tube of flexible synthetic sheet that is drawn into an existing pipeline and expanded to form a new internal lining.

PIPELINE TRANSPORT Use of a pipeline to convey a range of materials, separated one from the other by a volume of material immiscible to each.

PIPE LINING Application of a protective or flow enhancing lining to the interior of pipe prior to its installation.

PIPE PLUG Solid pipe fitting with a male thread, used to close the end of a ferrule or other fitting having a female thread.

PIPE PUSHING *See* **Pipe jacking.**

PIPE RAMMING Technique for conduit installation involving a casing driven through the ground by a percussive hammer.

PIPE RUN Path or route taken by a pipeline.

PIPE SCHEDULE Detailed description of the lengths and sizes of pipe and fittings necessary for a system.

PIPE SLEEVE Cylindrical insert, placed in a concrete form at the location where a pipe will pass through the completed wall.

PIPE STOPPER Screw plug for sealing a drain.

PIPE STRAP Metal strap used to support or position a pipe.

PIPE TAIL Short length of pipe, usually from a fitting, for connection to a branch pipe.

PIPE THREAD Screw threads for joining pipe. *See also* **Dry-seal pipe thread** and **Tapered pipe thread.**

PIPE VISE Vise with specially shaped jaws, used to grip pipe without damaging it so that it can be worked on.

PIPEWORK Plumbing systems inside buildings, above grade, for water supply, heating/ventilating/air-conditioning, and drainage.

PIPEWRAP Plastic or wax-coated tape wound around pipe to be buried to protect against corrosion.

PIPE WRENCH *See* **Wrench.**

PIPING 1. Arrangement of pipes of the same or different diameters as part of a scheme or layout; **2.** Movement of soil particles as a result of unbalanced seepage forces produced by percolating water, leading to the development of boils or erosion channels. *See also* **Blowing.**

PIPING SYSTEM Complex of pipes, fittings, and appurtenances, not necessarily of the same size or material, through which a fluid flows.

PISTON PUMP *See* **Pump.**

PITCH 1. Black, sticky substance obtained from the distillation of tar, petroleum, etc.; **2.** Determination of the key in which a sound falls based upon its frequency, sound pressure and waveform.

PITCHBLENDE Mineral consisting largely

of uranium oxide, occurring in black, pitch-like masses: a source of radium, uranium, and actinium.

PITCHER FITTING Plumbing fitting, tee, elbow, etc., with a gently curved turn instead of the sharp angles of conventional tees and elbows

PITLESS ADAPTER Threaded or welded fitting to a well casing that provides underground connection between the well and the service pipe with ready access to the drop pipe and well shaft.

PITOMETER Device used to measure the velocity of flowing fluids.

PITOMETRIC SURVEY Survey using a pitometer to determine the velocity of water flow at various points in a distribution system.

PITOT TUBE Tube having an opening that is inserted into a stream of water or air and to which a gauge is attached indicating the discharge pressure of the stream.

PIT TECHNIQUE See **Sanitary landfilling.**

PLACER (plasser) Deposit of heavy minerals concentrated mechanically, as by stream flow, wave action, or wind.

PLACER (plasser) DIGGINGS Areas where placer mining has overturned or removed the earth and left a rough, eroded, and scarred surface.

PLACER (plasser) MINING Process of washing loose sand or gravel for gold or other minerals.

PLAIN Extensive, level, treeless land region.

PLAIN SEDIMENTATION Natural, unaided sedimentation of suspended matter from a liquid.

PLAIN SETTLING TANK Tank or basin of sufficient volume to retain a flow for the time necessary for settleable solids to precipitate without mechanical or chemical aid.

PLANE OF SATURATION Ground within the watertable.

PLANKTON Plant and animal organisms, generally microscopic, that float or drift in great numbers in fresh or salt water.

PLANNED START DATE See **Scheduled start date.**

PLANOSOL Soil with a compact, clayey layer than develops into hardpan.

PLANS Contract drawings that show the location, character, and dimensions of the prescribed work, including layouts, profiles, cross sections, and details. See also **Standard plans** and **Working drawings.**

PLANT Generally, power-operated machinery and equipment used in engineering and construction.

PLANT AND EQUIPMENT Mechanical devices necessary to provide the services required to operate a building.

PLASTICITY 1. Complex property of a material involving a combination of qualities of mobility and magnitude of yield value; **2.** Property of soil that allows it to deform beyond the point of recovery without cracking or appreciable volume change.

PLASTICITY INDEX Range in water content through which a soil remains plastic.

PLASTICIZER Substance incorporated in a material to improve its processability and/or flexibility.

PLATE COUNT Number of colonies of bacteria grown on selected solid media at a given temperature and incubation period.

PLATE PRESS Filter press comprising a series of parallel plates lines with filter cloth through which filtrate squeezed from a sludge is drained.

PLATE SCREEN Screen comprising one or more perforated plates through which a liquid flows.

PLATE TECTONICS Study of the structure of the Earth's crust on the basis of the theory that the crust is made up of huge segments, called plates, that float on the mantle below, and whose individual movement is responsible for continental drift, mountain building, etc.

PLATFORM Stable, flat area on which a thin sequence of sediments may accumulate.

PLAYA Lake that exists temporarily following rainfall, or its dried-up bed.

PLOW Attachment, mounted to the rear of such equipment as a trencher, that is pushed vertically into the ground to a predetermined depth under hydraulic pressure, and that is used to draw pipe, cable, and other small-diameter flexible lines into the ground as the prime mover travels forward. A common variation is to provide the plow with a vibratory force, the amplitude of which may often be varied to best suit the ground being worked.

PLUG 1. Cap used for shutting of a tapped opening.

PLUG COCK *See* **Valve.**

PLUG FLOW Hydraulic phenomenon in which a 'slug' of water or wastewater moves through a tank without dispersing or mixing with the tank's contents.

PLUGGED Filter element that has collected a sufficient quantity of insoluble contaminants such that it can no longer pass rated flow without excessive differential pressure.

PLUGGING RECORD Systematic listing of permanent or temporary abandonment of water, oil, gas, test, exploration and waste injection wells, and which may contain a well log, description of amounts and types of plugging material used, the method employed for plugging, a description of formations that are sealed and a graphic log of the well showing formation location, formation thickness, and location of plugging structures.

PLUG PRESSURE Water pressure available at a fire hydrant: generally refers to static pressure, but could refer to residual pressure when in use.

PLUG VALVE Valve having as its movable control element a cylindrical or conical plug.

PLUMBER Worker skilled in all aspects of plumbing.

PLUMBER'S DOPE Soft, nonhardening compound used to seal pipe threads.

PLUMBER'S FURNACE Heating source, used to heat soldering irons, melt lead or solder, etc.

PLUMBER'S RULE Measure with a standard scale on one side and a scale for measuring the length of 45° offsets on the other.

PLUMBER'S SOLDER Solder with a relatively low melting point containing approximately two parts of lead to one part of tin; used in making wiped joints. *See also* **Solder.**

PLUMBER'S SOIL Mixture of glue and lampblack, used in lead work to prevent lead from bonding to lead pipe and fittings in unwanted areas.

PLUMBING 1. Pipes, fixtures, and other apparatus and appurtenances for the supply of water and the removal of waterborne wastes; **2.** Transferring a point established at one level to a point vertically below or above it.

PLUMBING FIXTURE Device that receives and discharges water and liquid wastes.

PLUMBING INSPECTOR Person authorized to inspect plumbing works for compliance with a regulatory code.

PLUMBING SYSTEM Drainage system, a venting system, and a water system.

PLUMBING TILE (PIPE) Glazed tile, with bell joints, for below-grade drains.

PLUME Stream of gases issuing from a chimney or stack prior to its dispersion into the surrounding atmosphere.

PLUME RISE Manner in which a stream of gases issuing from a stack behaves, and which involves factors and conditions including the temperature and velocity of the rising stream, and a range of meteorological conditions of the atmosphere into which the plume is venting.

PLUNGE POINT Final breaking point of a wave when its crest overruns the wave body.

PLUNGE POOL Pool below a vertical fall of water, filling a basin scoured out of the bed.

PLUNGER Hand-operated suction cup, used for clearing plugged drains.

PLUNGER PUMP Reciprocating pump having a plunger that does not come in contact with the cylinder wall, but enters and withdraws from it through packing glands.

PLUTONIUM Extremely toxic, radioactive metallic element (Pu) found naturally in trace amounts in uranium ores and produced in great amounts from uranium in nuclear reactors. The most important isotope of plutonium is plutonium-239, used as a fuel in nuclear fission, which has a half-life of 24,360 years.

PLUVIAL Of or pertaining to rain.

PLUVIAL INDEX Quantity used as an index of anticipated total rainfall for a period at a location.

PLUVIOGRAPH Theoretical hydrograph that would result from a storm if the runoff were 100% of the rainfall and if the proportions fixed by the distribution graph were applicable to the rainfall.

PLUVIOMETER Rain gauge.

PLUVIOMETRIC COEFFICIENT Ratio of actual precipitation for a period to what would have fallen had the rainfall been uniformly distributed throughout the year.

PM10 Airborne particles having a diameter less than 10μm and which can be inhaled beyond the larynx: those with a diameter less than 2.5μm are called respirable particles and can penetrate to the inner lungs.

PNEUMATIC EJECTOR Means of raising the level of a liquid by admitting it through a check valve into the bottom of a chamber and then ejecting it through another check valve into a discharge pipe by admitting compressed air to the chamber above the liquid.

PNEUMATIC PUMPING Pumping using an airlift pump.

PNEUMATICS Engineering science pertaining to gaseous pressure and flow.

PNEUMATIC VALVE *See* **Valve.**

PNEUMATIC WATER SUPPLY SYSTEM Water supply system, usually confined to one dwelling, in which a cistern connected to the main supply is pressurized by air that forces water through the internal distribution system.

PNEUMOCONIOSIS Incapacitating disease of humans caused by the retention of inhaled dusts in the lungs.

PODZOL Relatively acid, surface soil from the A-horizon of temperate climate zones, from which much soluble material has been leached into underlying soils.

POINT Terminal to a bored well having a drive point surmounted by a fine-mesh screen.

POINT BAR Sand or gravel deposit built up by rivers on the inner curve of a meander.

POINT GAUGE Device used to measure the surface elevation of water and consisting of a needle sliding along a graduated staff. The point is lowered until the tip causes a streak in flowing water and a meniscus jump in still water.

POINT OF APPLICATION Location with a process or series of processes where a specific treatment chemical is introduced.

POINT OF DISINFECTANT APPLICATION Point where a disinfectant is applied and water downstream of that point is not subject to recontamination by surface water runoff.

POINT-OF-ENTRY TREATMENT DEVICE Treatment device applied to the drinking water entering a house or building for the purpose of reducing contaminants.

POINT OF WASTE GENERATION Location where samples of a waste stream are collected for the purpose of determining the waste flow rate.

POINT RAINFALL Rainfall recorded at only one out of a series of stations in a region.

POINT SOURCE Discernible, confined, and discrete conveyance, including any pipe, ditch, channel, tunnel, conduit, well, discrete fissure, container, rolling stock, concentrated animal feeding operation, landfill leachate collection system, vessel or other floating craft from which pollutants are or may be discharged. This term does not include return flows from irrigated agriculture or agricultural storm water runoff.

POISE Obsolete unit of dynamic viscosity, not permitted for use with the SI system of measurement. Symbol: P. *See also the appendix*: **Metric and nonmetric measurement.**

POLARIZATION Partial or complete polar separation of positive and negative electric charge in a nuclear, atomic, molecular or chemical system.

POLAROGRAPHY Automatic electroanalysis of a solution by measuring current flow with voltage changes.

POLLUTANT OR CONTAMINANT Element, substance, compound, or mixture, including disease-causing agents, which after release into the environment and upon exposure, ingestion, inhalation, or assimilation into any organism, either directly from the environment or indirectly by ingestion through food chains, will or may reasonably be anticipated to cause death, disease, behavioral abnormalities, cancer, genetic mutation, physiological malfunctions (including malfunctions in reproduction) or physical deformations, in such organisms or their offspring.

POLLUTION Presence in a body of water, soil, or air of substances of such character and in such quantities that the natural quality of the body (water, soil, or air) is degraded so that it impairs the body's usefulness or renders it offensive to the senses of sight, taste, or smell.

POLLUTION CONTROL Collective term for the laws and regulations defining pollution and mandating its control, and the equipment and devices used for compliance.

POLLUTION CONVERSION Transformation of one type of pollutant into another (burning of a solid waste with the generation of airborne pollutants), or the partial elimination of a pollutant that leaves a different type of contaminant (sorting of municipal solid waste may remove ferrous metals, most types of paper, plastics, and glass, but leave putrescible matter not suitable for landfilling or incineration).

POLLUTION HAZARD APPRAISAL Evaluation of a pollutant to determine its potential as a hazard, particularly to those who must process it or be exposed to its effects, including the following characteristics:

a) Hazardous substance.

b) Techniques for prevention and minimization.

c) Techniques for abatement.

d) Scale of process.

e) Location.

f) Frequency of operation.

g) Offensive substance in the process.

POLLUTION INDEX Formula used in chimney height calculation to determine a limiting pollutant.

POLLUTION INDICATORS Range of organisms and plants that react in a predictable manner to one or more defined pollutants.

POLLUTION PERMIT Permission by a regulator authority to generate and/or discharge a specified volume or specified rate of a defined pollutant(s).

POLLUTION PREVENTION Source reduction and other practices that reduce or eliminate the creation of pollutants through increased efficiency in the use of raw materials, energy, water or other resources, or the protection of natural resources by conservation.

POLLUTIONAL INDEX Criterion by which the degree that something is polluted may be measured.

POLLUTIONAL LOAD Volume of material carried by a stream or in a body of water that exerts some detrimental influence.

POLYBUTYLENE PIPE *See* **Pipe.**

POLYCHLORINATED BIPHENYLS Chlorinated hydrocarbons (PCBs) formerly used as plasticizers and in transformer-cooling oils and whose use has been banned.

POLYELECTROLYTE High-molecular-weight substance formed by natural or a synthetic process, used as flocculent aids in raw and wastewater clarification.

POLYETHYLENE Thermoplastic polymer of ethylene having good flexibility.

POLYETHYLENE PIPE *See* **Pipe.**

POLYMER Long-chain molecule formed of many monomers, used as a flocculent aid in raw and wastewater clarification.

POLYMERIZATION Joining together of monomer molecules by additional polymerization (where the polymer is a simple multiple of the monomer molecule) or by condensation polymerization (where the resulting polymer does not have the same empirical formula as the basic monomer constituent).

POLYPROPYLENE Lightweight thermoplastic, similar to, but harder than, polyethylene, commonly used for moulded articles, insulation, etc.

POLYPROPYLENE PIPE *See* **Pipe.**

POLYSACCHARIDE Any of a large group of natural carbohydrates, including starch, cellulose, and glycogen, whose molecules consist of more than two molecules of simple sugars linked together.

POLYSAPROBIC Body of water in which organic matter is decomposing rapidly and in which there is little or no dissolved oxygen.

POLYSTYRENE Synthetic organic polymer formed from styrene: a rigid, colorless, thermoplastic resin that is resistant to acids, alkalis, and many solvents.

POLYSYNTHESIS Synthesis of several elements.

POLYUNSATURATED Having to do with or designating a class of vegetable and animal fats whose molecules consist of long carbon chains with many double bonds.

POLYURETHANE Any of a group of synthetic organic polymers that may be rubbery, resinous, or fibrous.

POLYVINYL Of or having to do with a group of thermoplastic resins formed from the polymerization of vinyl.

POLYVINYL CHLORIDE Colorless, synthetic thermoplastic material produced by the polymerization of vinyl chloride and which can be softened for shaping by raising its temperature and then can be hardened again by cooling without any chemical change taking place.

POLYVINYL-CHLORIDE PIPE *See* **Pipe.**

POND 1. Small lake. *See also* **Settling pond**; **2.** Stretch of water between two canal locks operating in series; **3.** Area where discharge water from a dredge is held long enough for particulate matter to settle out.

PONDAGE 1. Water that is being retained for later release; **2.** Storage capacity available for the temporary retention of water from a flow.

PONDAGE FACTOR Percent of power produced or water available during heavy demand hours at a hydroelectric generating plant to the quantity available continuously during a 24-hour period.

PONDING 1. Condition occurring in trickling filters when hollow spaces become plugged to the extent that adequate throughflow is prevented; **2.** Gathering of water at low or irregular areas of an ostensibly flat surface.

POND WATER SURFACE AREA Area within an impoundment for rainfall, and the actual water surface area for evaporation.

PONTOON Flat-bottomed boat or other structure used to support a floating bridge.

POOL 1. Small pond, often artificial and used as an ornamental feature in a garden; **2.** Deep area of a river or stream; **3.** Swimming pool.

POOL PRICE Base price paid by regional electrical generating and/or distribution companies for marginal supplies of electricity.

POP-OFF VALVE See **Valve.**

POPPET VALVE See **Valve.**

POPULATION 1. People of a defined area or zone; **2.** Part of the inhabitants distinguished in any way from the rest; **3.** Total number of organisms of a specific kind in a given area; **4.** Total number of individuals or things from which samples are taken.

POPULATION DENSITY Numerical size of a population relative to the area it occupies.

POPULATION DYNAMICS Study of changes in population density with time.

POPULATION EQUIVALENT Expression of the strength of organic material in wastewater which averages 77 g (0.17 lb) of oxygen per capita per day.

POP-UP WASTE Lever-operated captive plug for wash basin.

PORCELAIN ENAMEL Lustrous finish used on sanitary ware, similar to vitreous enamel but fired at a higher temperature.

PORE Tiny passage in material, usually microscopic, through which water may be absorbed or discharged.

PORE PRESSURE See **Pressure.**

PORE SIZE DISTRIBUTION Ratio of the number of holes of a given size to the total number of holes per unit area, expressed as a percent and as a function of hole size.

PORE SPACE Naturally-occurring space or void in a body of rock or sediment.

PORE TREATMENT Use of chemicals such as silicone resin brushed on the exterior of masonry and other porous materials to reduce the infiltration of moisture.

PORE WATER Water present in the interstices of granular permeable rock.

PORE-WATER-PRESSURE CELL Instrument for measuring pore-water pressures due to load changes.

POROSITY 1. Ratio, usually expressed as a percentage, of the volume of voids in a material to the total volume of the material including the voids; **2.** Ability of a material to absorb a fluid; **3.** Ability of an aggregate to absorb a liquid, generally asphalt.

POROUS Admitting the passage of gas or liquid through interstices in porous material.

POROUS FILL See **Drainage fill.**

POROUS PIPE Pipe which lets in water, through the material from which it is made, or via holes in its diameter, used for subsoil drainage.

POROUS TUBE Physical condition of a hose tube due to the presence of pores.

PORT Opening in a storage or process vessel.

PORTABLE CONTAINER Reusable container that has a capacity of 30 L (7.92 gal) or less, but excluding a container that is integral with, or permanently attached to, any appliance, equipment, or vehicle.

PORTABLE GENERATING SET Any wheel-, skid-, truck- or railroad car-mounted, but not self-propelled, equipment designed to supply electric current. This consists of an electric generator and a prime mover mounted on a common frame with all equipment necessary to constitute a complete, self-contained unit. In larger sources, the fuel source may be separate from the unit.

PORTABLE TANK Closed container having a liquid capacity more than 230 L (60 gal), and not intended for fixed installation.

POSITION HEAD Elevation head of a fluid.

POSITIVE CONFINING BED Layer that prevents or retards the upward movement of groundwater.

POSITIVE DISPLACEMENT PUMP See

Pump.

POSITIVE-DISPLACEMENT HYDRAU-LIC PUMP *See* **Pump.**

POSITIVE HEAD Energy per unit weight of a fluid, due to its elevation above some datum.

POSITIVE PRESSURE Pressure greater than atmospheric for the specific elevation.

POSITIVE-PRESSURE FABRIC FIL-TER Fabric filter with the fans on the up-stream side of the filter bags.

POSITIVE ROTARY PUMP Displacement pump in which the rotating elements fit closely against the case.

POSTCHLORINATION Application of chlorine as a sterilant to an effluent after all biological waste treatment processes have been completed.

POSTCONSUMER RECOVERED PA-PER Paper, paperboard and fibrous wastes from retail stores, office buildings, homes and so forth, after they have passed through their end-usage as a consumer item including: used corrugated boxes; old newspapers; old magazines; mixed waste paper; tabulating cards and used cordage.

POSTCONSUMER RECYCLING Reuse of materials generated from residential and commercial waste, excluding recycling of material from industrial processes that has not reached the consumer, such as glass broken during manufacturing processes.

POSTCONSUMER WASTE Material or product that has served its intended use and has been discarded for disposal after passing through the hands of a final user.

POST INDICATOR VALVE *See* **Valve.**

POTABLE WATER Water that is safe for human consumption. *Also called* **Drinking water.**

POTAMOLOGY Branch of hydrology dealing with streams and rivers.

POTASSIUM PERMANGANATE Dark purple crystalline compound, $KMnO_4$, used as an oxidizing agent, disinfectant, and deodorizer.

POTENTIAL 1. Work required to move a unit mass against a unit force from one point to another; **2.** Work required to bring a unit electric charge, magnetic pole, or mass from an infinitely distant position to a designated point in a static electric, magnetic, or gravitational field, respectively.

POTENTIAL HEAD Height of a particle of water above a datum.

POTENTIAL SWITCH *See* **Switch.**

POTENTIOMETRIC SURFACE Level to which water will rise in tightly cased wells.

POUND Nonmetric unit of mass, equal to 16 ounces. Symbol: lb. Multiply by 453.592 to obtain grams, symbol: g; by 0.453 592 to obtain kilograms, symbol: kg. *See also the appendix*: **Metric and nonmetric measurement.**

POUND PER CUBIC FOOT Nonmetric unit of density. Symbol: lb/ft^3. Multiply by 16.018 to obtain grams per litre, symbol: g/L, or kilograms per cubic meter, symbol: kg/m^3. *See also the appendix*: **Metric and nonmetric measurement.**

POUND PER CUBIC INCH Nonmetric unit of density. Symbol: $lb/in.^3$. Multiply by 27.68 to obtain grams per cubic centimeter, symbol: g/cm^3, or kilograms per litre, symbol: kg/L. *See also the appendix*: **Metric and nonmetric measurement.**

POUND (FORCE) PER SQUARE INCH Nonmetric unit of pressure. Symbol: psi. Multiply by 6.895 to obtain kilopascals, symbol: kPa. *See also the appendix*: **Metric and nonmetric measurement.**

POUND (MASS) PER SQUARE INCH Nonmetric unit of force. Symbol: psi. Multiply by 703.07 to obtain kilograms per square meter, symbol: kg/m^2. *See also the appendix*: **Metric and nonmetric measurement.**

POUR POINT Lowest temperature at which a liquid will flow.

POWER Strength or force exerted or capable of being exerted.

POWER CAPACITY OF A STREAM Hydroelectric generating potential of a stream.

POWER FACTOR Ratio of the power passing through an electric circuit to the product of the volts times the amperes in the circuit.

POWER HEAD Actuating mechanism of a deep-well pump.

POWER UNIT 1. Combination of pump, pump drive, reservoir, controls and conditioning components that may be required for its application.

PRACTICAL QUANTITATION LIMIT Concentration of a substance that can be measured in a laboratory within reasonable limits of precision and accuracy.

PRAIRIE Large area of level or rolling land with grass but very few or no trees.

PREAERATION Injection of air into a wastewater stream in advance of formal treatment processes, to remove gases, add oxygen, promote flotation of grease, and aid coagulation.

PREALARMS Warning prior to actual actuation of automatic engine safety measures to indicate impending shutdown.

PRECAUTIONARY PRINCIPLE Reduction of risk to the environment by taking avoiding action prior to a problem arising.

PRECEDENCE DIAGRAMMING METHOD (PDM) Network diagramming technique in which activities are represented by boxes (or nodes). Activities are linked by precedence relationships to show the sequence in which they are to be performed. *Also called* **Activity-on-node.**

PRECEDENCE OF DOCUMENTS Order of authority of contract documents if there should be an inconsistency. A typical sequence would be: contract agreement; conditions of contract; specifications; large-scale drawings; small-scale drawings; specifications; and finishings schedules.

PRECEDENCE PROGRAMMING Activity-oriented system to more effectively display the logic and interrelationships of work

items than is possible using arrow diagramming.

PRECEDENCE RELATIONSHIP Logical relationship in the precedence diagramming method. In current usage, however, precedence relationship, logical relationship, and dependency are widely used interchangeably regardless of the diagramming method in use.

PRECHLORINATION Chlorination of a raw water supply prior to any treatment necessary for it to be used as a potable water, so as to oxidize any microbiological pollutants that might contaminant treatment filters, etc.

PRECIPITATABLE WATER Total water vapor contained in an atmospheric column of unit cross-section.

PRECIPITATE Physical separation from a fluid as a result of a chemical or physical change.

PRECIPITATION Total volume, measured over a stated period, of moisture in the form of rain, snow, hail, and sleet.

PRECIPITATION GAUGE Device for catching and measuring the amount of precipitation.

PRECIPITATION NUMBER The number of milliliters of precipitate formed when 10 ml (0.35 oz) of lubricating oil are mixed with 90 ml (3.15 oz) of a precipitation naptha and centrifuged under prescribed conditions.

PRECISE SETTLEMENT GAUGE Instrument that monitors long-term ground settlement.

PRECISION Relates to the quality of procedures used, fineness of measurement, and the repeatability of a result.

PRECISION WORK Work that is required to be exact in measurement, finish, etc.

PRECOAT Application of a free-draining, noncohesive material to a filtering material to reduce the frequency of media washing and facilitate cake discharge.

PRESCRIBED PROCESS Regulation that requires that a defined waste be treated in a defined manner, possibly with described equipment, to produce a specific result.

PRESCRIBED SUBSTANCE Releases to air, water and/or land that are subject to regulation and control.

PREDECESSOR ACTIVITY 1. In the arrow diagramming method, the activity which enters a node; **2.** In the precedence diagramming method, the 'from' activity.

PREDECESSOR WORK ITEM Work item that directly precedes one or more work items in the logic sequence of a precedence network.

PREDRAINAGE Dewatering of a site by lowering the water table by pumping in the vicinity of a proposed excavation.

PREEXCAVATION 1. Advance excavation of a general site; **2.** Removal by augering of soil that may heave; **3.** Removal of soil by driving and cleaning out an open-end pipe.

PREFILL VALVE *See* **Valve.**

PREFILTER Fine-mesh screen or other device placed ahead of a filter to trap large particles.

PREFORMED STRUCTURAL GASKET Flexible, preformed shape that holds a material in its intended position and that forms a watertight seal.

PRELIMINARY FILTER Filter, usually rapid sand, ahead of the final filtration stage in a water treatment plant.

PRELIMINARY TREATMENT Any process or stage applied to a wastewater stream that has the effect of reducing the treatment load on the plant to which the flow is directed.

PRELOAD To temporarily heap fill to a calculated depth on a site to be developed so as to compress the natural surface materials. *See also* **Compaction.**

PREQUALIFICATION Assessment of a potential bidder or supplier prior to an invitation to bid. Prequalification is based on past record, present capability and availabil-

ity, etc.

PRESERVATIVE Any substance that, for a reasonable length of time, will prevent the action of wood-destroying fungi, borers, and similar destructive agents.

PRESERVATIVE TREATMENT Treatment of wood that prevents decay from moisture or bacteria.

PRESETTLING Sedimentation stage immediately ahead of a further treatment process.

PRESSURE Equals force per unit area, usually expressed in kilograms per square meter (kg/m^2), or in pounds per square inch (psi). There are many ways of determining pressure, including:

Absolute: Ordinary outside air pressure (barometric pressure); the sum of atmospheric pressure and gauge pressure.

Active earth: *See* **Earth pressure** (below).

Atmospheric: Pressure of air at sea level, usually 101.325 Kp (14.7 psia) (1 atmosphere), or 0 psig.

At rest earth: *See* **Earth pressure** (below).

Back: (a). Measured loss in efficiency in an engine exhaust system due to friction or pumping losses; **(b).** Earth pressure exerted on the vertical walls of an abutment by the fill, or the lateral loading of a piling or face log in an abutment; **(c).** In a plumbing system, compressing of trapped air, that resists the flow of waste through the drain, waste, and vent piping.

Burst: Pressure at which rupture occurs.

Charge: Pressure at which replenishing fluid is forced back into a fluid power system.

Critical: Vapor pressure corresponding to the critical state of the substance at which the liquid and vapor have identical properties.

Differential: Difference in pressure between two points in a system.

Earth: Thrust from retained soil. There are several classifications, including:

Active: The minimum earth pressure, which is the force from soil tending to overturn a free retaining wall.

At rest: The thrust from earth on to a fixed retaining wall.

Lateral: Lateral forces interacting between an earth-retaining structure and the retained earth mass.

Passive: Earth pressure that results from the resistance of an earth outface to deformation by other forces. *Also called* **Lateral pressure.**

Effective: Pressure in a soil between the points of contact of the soil grains.

Gauge: Pressure measured by a gauge and indicating the pressure exceeding atmospheric.

Hydrostatic: 1. Pressure at any point in a liquid at rest; **2.** State of stress in which all the principal stresses are equal (and there is no shear stress), as in a liquid at rest; the product of the unit weight of the liquid and the difference in elevation between the given point and the free water elevation; **3.** Pressure exerted on the underside of a concrete slab when the watertable is above its elevation.

Lateral earth: *See* **Earth pressure** (above).

Maximum working: Pressure as stated by the machine manufacturer as the maximum at which a hydraulic circuit shall be operated. This pressure may be limited by a relief valve or other means.

Operating: Maximum pressure permitted to be developed in a hydraulic system, as limited by the setting of a main relief valve to the desired pressure.

Override: Difference between the cracking pressure of a valve and the pressure reached when the valve is passing at full flow.

Partial: Portion of total gas pressure of a mixture attributable to one component.

Passive earth: *See* **Earth pressure** (above).

Pilot: 1. Pressure in a pilot circuit; **2.** Hydraulic pressure used to actuate or control hydraulic components.

Pore: Water pressure developed in the voids of a soil mass.

Precharge: Pressure of a compressed gas in an accumulator prior to admission of a liquid.

Proof: Specified pressure that exceeds the manufacturer's recommended working pressure applied to a hose to indicate its reliability at normal working pressure. (Proof pressure is usually twice the working pressure.) *Also called* **Test pressure.**

psi: Pounds per square inch (kilopascals); measure of the pressure within a system or appliance, or produced by a device.

psia: Pounds per square inch, absolute (kilopascals); absolute pressure is equal to atmospheric pressure plus the gauge pressure.

psig: Pounds per square inch, gauge (kilopascals); the pressure above atmospheric pressure — 0 at sea level.

Rated: Qualified operating pressure recommended for a component or system by the manufacturer.

Rated fatigue: Pressure that a pressure-containing component is represented to sustain 10 million times without failure.

Rated static: Pressure that a component

can withstand without failure.

Saturation: For a pure substance, that pressure at which vapor and liquid, or vapor and solid, can coexist in stable equilibrium.

Service: *See* **Working pressure,** below.

Static: 1. Water pressure head available at a specific location; **2.** Force resulting from the imposition of a static load; *See also* **Compaction.**

Suction: Operating pressure measured in the suction line at a compressor inlet.

System: Pressure that overcomes the total resistance in a system. It includes all losses as well as useful work.

Test: *See* **Proof pressure,** above.

Total: Pressure on a horizontal plane of soil from the mass of material above it, plus any superimposed load.

Vapor: 1. Component of atmospheric pressure, caused by the presence of vapor, expressed in inches, centimeters, or millimeters of height of a column of mercury, or in pascals; **2.** Pressure, at a given fluid temperature, in which the liquid and gaseous phases of the fluid are in equilibrium.

Velocity: In moving fluid, the pressure capable of causing an equivalent velocity, if applied to the same moving fluid through an orifice such that all pressure energy expended is converted into kinetic energy.

Working: Maximum pressure to which a component will be subjected, including the momentary surges in pressure that can occur during service. *Also called* **Service pressure.**

PRESSURE-ACTUATED SEAL Sealing device in which sealing action is aided by fluid pressure.

PRESSURE BALANCE Pressure in a system or container equal to that which exists outside.

PRESSURE CONTROL VALVE *See* **Valve.**

PRESSURE CUTOUT Sensor that actuates a switch or a control at a preset pressure.

PRESSURE-DIFFERENTIAL SWITCH *See* **Switch.**

PRESSURE DROP Reduction in pressure between two points in a line or passage due to the energy lost in maintaining flow.

PRESSURE EQUALIZING Permitting high- and low-side pressures to equalize during idle periods through the use of an unloading valve or vapor lock liquid control; or nearly equalizing inlet and discharge pressures on a compressor to reduce starting torque load.

PRESSURE FILTER Rapid sand filter to which incoming water is introduced under pressure.

PRESSURE GAUGE *See* **Gauge.**

PRESSURE HEAD Amount of force or pressure created by a depth of one meter (or one foot, or other dimension) of water.

PRESSURE INTENSITY Force exerted by a material or body per unit area of contact surface.

PRESSURE-LIMITING DEVICE Pressure-responsive mechanism designed to arrest operation of the pressure-imposing element at a predetermined level.

PRESSURE LINE Line conducting pressurized fluid to a working device or devices.

PRESSURE LOSS Reduction in pressure due to expenditure of pressure energy required to move water or air through a line, including loss from back pressure, elevation, friction, etc.

PRESSURE METER Instrument for *in-situ* testing of the mechanical properties of soil or rock by hydraulically expanding a probe in a bore hole and measuring the volume changes produced by successive increments of pressure.

PRESSURE POTENTIAL Work necessary to move a unit mass of water occurring in a saturated soil column from the surface, against any static pressure in the column.

PRESSURE-PRESERVATIVE TREATED Wood treated with preservative or fire retardants by pressure-injecting treating solutions into wood cells, e.g. creosote, pentachlorophenol, or ammoniacal copper arsenate or chlorinated copper arsenate.

PRESSURE-REDUCING (WATER) VALVE *See* **Valve.**

PRESSURE-REDUCING VALVE *See* **Valve.**

PRESSURE REGULATING VALVE Valve used to regulate the pressure in a water or gas line either upstream, or downstream of its position.

PRESSURE REGULATOR Device used to maintain a constant pressure within a system or device. *See also* **Valve.**

PRESSURE RELEASE Emission of materials resulting from a system pressure being greater than the set pressure of the pressure relief device.

PRESSURE-RELIEF CONE Depression in the piezometric surface of a confined artesian aquifer created when a well is used to extract water at a given rate.

PRESSURE-RELIEF DAMPER Damper installed in a system which relieves pressure in excess of a preset limit.

PRESSURE-RELIEF VALVE *See* **Valve.**

PRESSURE RIDGE Ridge created on floating ice by buckling under lateral pressure of wind and tide.

PRESSURE SENSING VALVE *See* **Valve.**

PRESSURE STRAINER Device on a pressure line with two or more compartments holding removable screens or strainers.

PRESSURE SWITCH *See* **Switch.**

PRESSURE TANK Storage or process vessel capable of holding its contents under a maximum design pressure.

PRESSURE TRANSDUCER Transducer that converts pressure to a proportional electrical signal.

PRESSURE-UNLOADING VALVE *See* **Valve.**

PRESSURE VESSEL Container designed to operate at pressures greater than 100 kPa (14.5 psi) gauge.

PRESSURE ZONE Area within a water distribution system within which mains pressure is maintained within a specified range.

PRETREATMENT Reduction of the amount of pollutants, the elimination of pollutants, or the alteration of the nature of pollutant properties in wastewater prior to or in lieu of discharging or otherwise introducing such pollutants into a treatment works. The reduction or alteration may be obtained by physical, chemical or biological processes, process changes or by other means.

PREVAILING WIND Direction from which the wind blows most often during a specific season of the year.

PREVENTIVE MAINTENANCE Maintenance measures taken in advance to avoid breakdowns.

PRIMARY AIR Air supplied to a fuel in its early stages of combustion, which may be present with the fuel at time of combustion, or entrained with the fuel.

PRIMARY CONTROL SYSTEM Air pollution control system designed to remove gaseous and particulate matter.

PRIMARY DRAIN Single sloping connection from the base of a soil or waste stack to its junction with the main building drain or with another branch.

PRIMARY ELEMENT Hydraulic structure used to measure flow.

PRIMARY EMISSION CONTROL SYSTEM Hoods, enclosures, ducts, and control devices used to capture, convey, and remove particulate matter from exhaust gases which are captured directly at the source of generation.

PRIMARY FILTER First stage of a multi-stage filter system.

PRIMARY MANUFACTURING RESIDUE Waste products that results from the improvement of a raw material; sawdust, chips, slabs, and the like from lumber conversion, for instance.

PRIMARY MINERALS Minerals formed directly from cooling magma and persisting as the original mineral even in sedimentary rocks.

PRIMARY POLLUTANT Pollutant emitted or introduced directly into the environment.

PRIMARY RECYCLING Return of secondary material to the same industry from which it came and processing of that secondary material so that it will yield the same or similar product as that to which it was a secondary material.

PRIMARY SETTLING TANK First treatment tank in a sewage treatment plant where settleable solids are removed.

PRIMARY SLUDGE Sludge obtained from a primary settling tank.

PRIMARY SPILLWAY *See* **Spillway.**

PRIMARY (SEWAGE) TREATMENT *See* **Sewage treatment.**

PRIME 1. Provision of means to start a process; **2.** To supply sufficient fluid (water, gasoline, etc.) to start a pump; **3.** Manual injection of a volatile fuel into an engine to help start the mechanical process.

PRIME CONTRACT Contract to the party having principal responsibility for his own actions and work, plus that of contractors and suppliers to whom he subcontracts.

PRIME CONTRACTOR Contractor having a contract directly with the owner; the contractor responsible for the project until its completion. *Also called* **Main contractor.**

PRIME MOVER 1. Machine used to pull or push other machines; **2.** Engine or motor, used to drive a machine or machines.

PRIMING Initiating a process through the addition of a small volume of material that will stimulate reaction: priming a pump to wet all surfaces, exclude air and induce flow, for instance.

PRIMING PUMP *See* **Pump.**

PRIMITIVE WATER Water that has been captive within the interior of the lithosphere since the Earth's formation.

PRIORITY CIRCUIT Electrical circuit providing power for essential services, either from an uninterruptible power supply or standby generator. Normal loads are unloaded by automatic load shedding.

PRIORITY WATER QUALITY AREAS Specific stream segments or bodies of water where municipal discharges have resulted in the impairment of a designated use or significant public health risks, and where the reduction of pollution from such discharges will substantially restore surface or groundwater uses.

PRIVATE SEWAGE DISPOSAL SYSTEM Privately owned plant for the treatment and disposal of sewage (such as a septic tank with an absorption field).

PRIVATE SEWER Privately owned, maintained and operated sewer serving one or more properties and discharging to either a private sewage disposal plant or to a public sewer.

PRIVATE UTILITY Enterprise owned by individuals or a corporation and operated for the purpose of rendering utility service.

PRIVATE WATER Surface water that is not navigable or that is not on public land or publicly owned.

PRIVATE WATER SUPPLY Water supply developed by the owner for private enjoyment and not for public benefit.

PRIVY Portable or fixed structure sited over a pit or vault, equipped with seating and used as a latrine.

PROBABILITY CURVE Expression of the cumulative frequency of an event based on an extended record of its occurrence.

PROBABLE MAXIMUM FLOOD Flood that may be expected from the most severe combination of meteorologic and hydrologic conditions possible for a region.

PROBABLE MAXIMUM PRECIPITA-TION Maximum amount and duration of precipitation that can be expected to occur on a drainage basin.

PROBABLE RESERVES Assumed reserves of a substance based on exploration and/or initial development, i.e. a mineral, or oil, or natural gas.

PROBE Device that can be used to investigate an area for one or more specific conditions or states (temperature, humidity, resistance to penetration, etc.).

PROBING Sensing a state or condition using a probe sensitive to the required data. *Also called* **Piercing.** *See also* **Sounding.**

PROBLEM WASTE *See* **Solid waste.**

PROCESS Series of actions, changes, or functions that bring about an end or result.

PROCESS VARIABLE Physical or chemical quantity, the adjustment and regulation of which affects an operation or sequence of operations.

PROCESS WASTE Waste material from an industrial process.

PROCESS WASTE WATER Water which, during manufacturing or processing, comes into direct contact with or results from the production or use of any raw material, intermediate product, finished product, by-product, or waste product.

PROCESS WATER Water in contact with, or incorporated into, an end product.

PRODUCTION WELL Well used for the production of water from an aquifer or water-bearing formation.

PRODUCTIVE MACHINE TIME *See* **Machine time.**

PRODUCTIVE TIME *See* **Machine time.**

PRODUCT PIPE Permanent pipe used in a conduit installation.

PRODUCTS OF COMBUSTION Gases, vapors and solids that result from the combustion of a material.

PRODUCT STANDARD Standard that states how a product must be manufactured.

PROFILE 1. Vertical section through an object; **2.** Graphical representation of elevation plotted against distance; **3.** Longitudinal section through a pipeline, channel, conduit, stream, etc.; **4.** Graph or table representing numerically the extent to which something changes according to various tested characteristics.

PROGRAM 1. Criteria and/or requirements that must be met; **2.** Plan of operations by which to meet an objective; **3.** Instructions in logical sequence that directs a computer to perform a given task; **4.** Group of related projects managed in a coordinated way, usually including an element of ongoing activity.

PROGRAM EVALUATION AND RE-VIEW TECHNIQUE Event-oriented system generally used in the research and development field where, at the planning stage, activities and their durations between events are difficult to define, but where completion of these activities by a specific date is essential to the success of the project. Typically, these projects involve massive programs, many large organizations, and extensive operations in many different locations.

PROGRAMMABLE LOGIC CONTROL-LER Solid-state device used to control equipment variables.

PROGRESS CHART Diagrammatic presentation of a project represented by a number of stages and elements with an indication of the stage of completion of each represented on a time line.

PROGRESSIVE SYSTEM Well-pointing system, used in trench work, that dewaters the ground prior to excavation. As each trench section is completed, the well points are extracted and relocated in a new section.

PROGRESSIVE TIDAL WAVE Wave having a crest that advances from one body of water to another at times of high and low water.

PROGRESSIVE WAVE Wave in which the water particles move in circles or ellipses.

PROGRESS PAYMENT Payment in response to an invoice submitted by a construction contractor, asking for payment for works completed, or in progress, or materials required for the project that are in inventory and on site, to a specific date. The invoice is usually supported by a progress payment certificate. *Also called* **Partial payment.**

PROGRESS REPORT Regular report that describes the progress made to that point in time of a construction project, and that describes any variation from an established construction program. *Also called* **Construction progress report, Construction status report, Project status report,** and **Status report.**

PROGRESS SCHEDULE Diagram, graph, or other presentation showing proposed and actual times of starting and completing described events.

PROJECT 1. Entire sequence of activities associated with satisfying a client request through new construction or renovation; **2.** Two or more activities or tasks which, when performed, lead to a common goal, within the concept that a project has a single starting point and a single ending point.

PROJECT ACCOUNTANT Title of an organizational position designating the person responsible for financial accounting and reporting of a project.

PROJECT ANALYSIS Evaluation of a project, at any stage from conceptualization to final completion, according to any criteria established for the purpose.

PROJECT BRIEF Capsulated description of a project defining its purpose, scope, anticipated completion date, cost, etc.

PROJECT BUDGET Sum established by the owner as available for the entire project.

PROJECT CHARTER Document issued by senior management that provides the project manager with the authority to apply organizational resources to project activities.

PROJECT CLOSEOUT Full completion of a project.

PROJECT COMMUNICATIONS MANAGEMENT Subset of project management that includes the processes required to ensure proper collection and dissemination of project information, and consisting of communications planning, information distribution, performance reporting, and administrative closure.

PROJECT COMPLETION Stage at which a project is completed as described and illustrated in the contract documents. *Also called* **Completion of project.**

PROJECT CONTROL A management function that consists of taking actions (or directing or motivating others to take actions) that will yield an outcome in accordance with a predetermined performance standard.

PROJECT COST Any of the cost types (appropriation, commitment, expenditure, or estimate to complete) associated with the total scope of work in a project; that is all cost classes. The project cost comprises:

(a) The costs of the feasibility phase;

(b) Construction costs; and

(c) Management and overhead costs during the construction phase.

PROJECT COST MANAGEMENT Subset of project management that includes the processes required to ensure that the project is completed within the approved budget, and consisting of resource planning, cost estimating, cost budgeting, and cost control.

PROJECT DEFINITION Mission statement and overall scope of a project.

PROJECT DRAWING One of the drawings that, along with the project specification, completely describe the construction of the work required or referred to in the contract documents.

PROJECT ENGINEER Professional engineer responsible for onsite works.

PROJECT EVALUATION AND REVIEW TECHNIQUE A probabilistic, event-oriented control technique, events are identified within the nodes themselves. *Also called* **PERT.**

PROJECT HANDOVER Acceptance of a building or project by the operator or owner for occupancy or use, even though construction may not be fully complete.

PROJECT HUMAN RESOURCES MANAGEMENT Subset of project management that includes the processes required to make the most effective use of the people involved with the project, and consisting of organizational planning, staff acquisition, and team development.

PROJECT INSPECTOR Person responsible to the prime or general contractor for inspecting materials supplied to a job and work completed on the job to ensure that they meet the appropriate specifications and local regulations.

PROJECT INTEGRATION MANAGEMENT Subset of project management that includes the processes required to ensure that the various elements of the project are properly coordinated, and consisting of project plan development, project plan execution, and overall change control.

PROJECT LIFE CYCLE Collection of generally sequential project phases whose name and number are determined by the control needs of the organization or organizations involved in the project.

PROJECT MANAGEMENT Application of knowledge, skills, tools, and techniques to project activities in order to meet or exceed stakeholder needs and expectations from a project.

PROJECT MANAGEMENT BODY OF KNOWLEDGE Sum of knowledge within the profession of project management, and which includes proven, traditional practices which are widely applied as well as innovative and advanced ones which have seen more limited use.

PROJECT MANAGEMENT PROFESSIONAL Individual certified as such by the Project Management Institute.

PROJECT MANAGEMENT SOFTWARE Class of computer applications specifically designed to aid with planning and controlling project costs and schedules.

PROJECT MANAGEMENT TEAM Members of a project team who are directly involved in project management activities.

PROJECT MANAGER Person given specific duties and assignments on a project.

PROJECT MANAGEMENT Planning, scheduling, measuring physical progress and controlling the entire project: engineering, purchasing, and construction as one coordinated entity.

PROJECT MANUAL Volume assembled by the architect for the work and including, among other things, bidding instructions and requirements, sample forms, the agreement, any special or supplementary conditions, the specifications, contract, and other documents.

PROJECT NETWORK DIAGRAM Schematic display of the logical relationships of project activities, always drawn left to right to reflect project chronology.

PROJECT PHASE Collection of logically related project activities, usually culminating in the completion of a major deliverable.

PROJECT PLAN Formal, approved document used to guide both project execution and project control, and whose primary uses are to document planning assumptions and decisions, facilitate communications among stakeholders, and document approved scope, cost, and schedule baselines.

PROJECT PLAN DEVELOPMENT Taking the results of other planning processes and putting them into a consistent, coherent document.

PROJECT PLAN EXECUTION Carrying out the project plan by performing the activities included therein.

PROJECT PLANNING Development and maintenance of a project plan.

PROJECT PROCUREMENT MANAGE-MENT Subset of project management that includes the processes required to acquire goods and services from outside the performing organization, and which consists of procurement planning, solicitation planning, solicitation, source selection, contract administration, and contract closeout.

PROJECT QUALITY MANAGEMENT Subset of project management that includes the processes required to ensure that the project will satisfy the needs for which it was undertaken, and which consists of quality planning, quality assurance, and quality control.

PROJECT RISK MANAGEMENT Subset of project management that includes the processes concerned with identifying, analyzing, and responding to project risk, and which consists of risk identification, risk quantification, risk response development, and risk response control.

PROJECT SCHEDULE Timetable specifying the dates of key project events including public notices of proposed procurement actions, subagreement awards, issuance of notice to proceed with building, key milestones in the building schedule, completion of building, initiation of operation and certification of the project.

PROJECT SCOPE MANAGEMENT Subset of project management that includes the processes requires to ensure that the project includes all of the work required, and only the work required, to complete the project successfully, and which consists of initiation, scope planning, scope definition, scope verification, and scope change control.

PROJECT SPECIFICATIONS Written documents that describe the requirements for a project in accordance with the various criteria established by the owner.

PROJECT STATUS REPORT *See* **Progress report.**

PROJECT TEAM MEMBERS People who report either directly or indirectly to the project manager.

PROJECT TIME MANAGEMENT Subset of project management that includes the processes required to ensure timely completion of the project, and which consists of activity definition, activity sequencing, activity duration estimating, schedule development, and schedule control.

PROLIFERATION Growth or production by multiplication of parts rapidly and repeatedly.

PROOF PRESSURE *See* **Pressure.**

PROPAGATE Increase or multiply; spread.

PROPELLANT Gas or volatile liquid used in a pressurized product for the purpose of expelling the contents of the container.

PROPELLER PUMP Centrifugal pump that develops head by a propelling or lifting action of the vanes of its impeller.

PROPERLY SHREDDED GARBAGE Garbage that has been shredded to such a degree that all particles will be carried freely under flow conditions normally prevailing in public sewers, with no particle greater that 6 mm (0.25 in.) in any dimension.

PROPORTIONAL CONTROL Mode of control in which there is a continuous linear relationship between the value of the controlled variable and the position of the final control element.

PROPORTIONAL SAMPLING Sampling at a rate that produces a constant ratio of sampling rate to flow rate.

PROPORTIONAL WEIR Weir in which the discharge is directly proportional to the head.

PROPORTIONER Device, possibly adjustable, used to blend two or more substances or compounds into a flow or stream.

PROSPECTING Detailed exploration and evaluation of the Earth's surface and crust. *See also* **Seismic exploration.**

PROTECTED WASTE PIPE Waste pipe from a plumbing fixture not directly connected to a drain, soil pipe, vent pipe, or waste pipe.

PROTECTION Measures against damage

and/or injury to buildings or their occupants.

PROTECTIVE EQUIPMENT 1. Personal safety equipment, including hard hats, safety goggles and shoes, ear plugs, etc.; **2.** Electrical switches and circuits that protect a machine and its operator from faults or overload.

PROTEIN Any of a group of complex nitrogenous organic compounds of high molecular weight that contain amino acids as their basic structural units and that occur in all living matter and are essential for the growth and repair of animal tissue.

PROTEOLYSIS Breaking down of proteins into simpler, soluble substances, as in digestion.

PROTEOLYTIC ENZYMES Enzymes that have the power to decompose or hydrolyze proteins.

PROTON Nuclear particle carrying one unit of positive electric charge, equal to and opposite to that of an electron, found in the nucleus of every kind of atom. An element is identified and classified according to the number of protons in the nucleus of each of its atoms; the number of protons gives the element its atomic number.

PROTOPLASM Complex, typically colorless and semifluid substance considered the physical basis of life and having the power of spontaneous motion and reproduction.

PROTOTYPE Original type, form, or instances that serves as a model on which later stages are based or judged.

PROTOZOAN Single-celled, usually microscopic organisms of the phylum or subkingdom Protozoa.

PROVEN RESERVES Reserves of a substance based on exploration and/or initial development, i.e. a mineral, or oil, or natural gas.

PROXIMATE ANALYSIS Method of measuring the percentage of free moisture, volatile matter, ash, and fixed carbon in a solid fuel.

Psychoda Small gray fly found around trick-

ling filters and whose larvae live in the zoogleal film on the filter stones.

PSYCHROMETER Instrument consisting of two thermometers, one with a dry bulb and the other with a wet bulb, used to determine relative humidity.

PSYCHROMETRIC CHART Graphic representation of the thermodynamic properties of moist air.

PSYCHROMETRY Branch of physics relating to the measurement or determination of atmospheric conditions, particularly regarding moisture mixed with air.

PSYCHROPHILIC BACTERIA Group of bacteria that grow and thrive in temperatures below 20°C (68°F).

P-TRAP Sanitary fitting that provides a water seal where the inlet is vertical and the outlet horizontal (or up to 5° below horizontal). *See also* **S-trap.**

PUBLIC AUTHORITY Any applicable governmental entity having jurisdiction over the work.

PUBLIC CONTRACT Contract for works or the supply of materials by a public authority in which the terms of the contract and scope of the work are open for public inspection and the awarding of a contract is public knowledge.

PUBLIC HEALTH Matters concerning the mental and physical well being of people, both individually, and as a community.

PUBLIC HEARING Open discussion (in this context) about a proposed project, some aspects of which will either affect members of the community or may be at variance with existing regulations or codes.

PUBLICLY-OWNED TREATMENT WORKS Treatment works and connecting sewer collection systems that are owned and/or operated, in whole or in part, by a public authority or its designated agency.

PUBLICLY-OWNED TREATMENT WORKS Any device or system used in the treatment (including recycling and reclamation) of municipal sewage or industrial wastes of a liquid nature which is owned by a municipality or other public authority. This includes sewers, pipes, or other conveyances only if they convey wastewater to the plant providing treatment.

PUBLICLY-OWNED TREATMENT WORKS RESIDUALS Liquid effluent and/or solids, including sludge, scum, screenings and grit, that are the by-product of wastewater treatment operations and which must be discharged to the environment for ultimate disposal and/or reuse.

PUBLIC SEWER Publicly owned and maintained sewerage system to which private sewers may be connected under specific conditions.

PUBLIC USE OF WATER Acquisition, treatment, and distribution of raw and/or potable water for the public use by an agency, corporation, or other organization.

PUBLIC UTILITY Private business organization that is subject to governmental regulation because it provides an essential service or commodity, such as water, electricity, transportation, or communication, to the consuming public.

PUBLIC WATER Water available for public enjoyment.

PUBLIC WATER MAIN Water supply pipe for public use, administered by a public authority.

PUBLIC WATER SUPPLY Water supply from which water is available for the public enjoyment, the quality of which is subject to regulation.

PUBLIC WATER SYSTEM System for the provision to the public of piped water for human consumption having at least 15 service connections or regularly serving at least 25 individuals daily at least 60 days out of the year. The term includes any collection, treatment, storage and distribution facilities that are used primarily in connection with the system.

PUBLIC WORKS Works owned by the members of a community for their collective benefit and enjoyment, constructed and maintained from the tax revenues.

PUDDLE 1. Small pool of (usually dirty) water; **2.** To mix a soluble material with a fluid until they are of a uniform consistency.

PUDDLE CLAY Clay having certain properties that make is suitable for use as a watertight lining or backing.

PUDDLED CORE Core to an earth dam formed of puddled or compacted clay.

PUFF PIPE Antisiphon pipe.

PUG MILL Mechanical device with rotating paddles or blades, used to mix and blend materials.

PULL-ON CONTAINER Detachable container system in which a large container (around 15 to 23 m^3 (20 to 30 yd^3)) is pulled onto a service vehicle or tilt frame or hoist truck by mechanical or hydraulic means and carried to a disposal site for emptying.

PULMONARY IRRITANTS Airborne pollutants that affect the mucous lining of the respiratory tract.

PUMP Mechanism receiving power from an engine or motor to create a flow. There are many types, including:

> **Air:** Pump that exhausts, or compresses air, or forces it through something.

> **Air chamber:** Displacement pump equipped with an air chamber in which air is alternately compressed and expanded by the water displaced by the pump, resulting in a more even rate of discharge.

> **Air displacement:** Displacement pump in which compressed air, rather than pistons or plungers, is used as the displacement agency in the cylinders.

> **Airlift:** Pump consisting of two pipes hanging in a tank or well (one pipe may be inside the other). Compressed air is passed out of the bottom of the smaller pipe with the resulting increase in pressure and uplift of the released air forcing the fluid in the well or tank to rise up the larger pipe.

> **Automatic fire:** Pump that provides the

required water pressure in a fire stand-pipe or sprinkler system should the system pressure fall below a preset value.

Axial flow: Centrifugal pump that develops most of its head by the propelling or lifting action of the vanes. *Also called* **Propeller pump.**

Booster: Pump that operates in the discharge line of another pump, either to increase pressure or to restore pressure lost between the prime pump and booster pump for any cause.

Borehole: Electrically-driven centrifugal pump driven by mechanical linkage from the surface, or equipped with a submersible electrical motor and located at the foot of the shaft to be pumped, used to dewater shafts.

Bucket: Pump consisting of a series of buckets attached to a belt or chain that are continuously filled with water, raised to an elevated position, emptied, and returned for refilling.

Centrifugal: Pump that uses centrifugal force developed by the rapid rotation of an impeller to increase flow or pressure of a fluid. *See also* **Closed centrifugal pump, Self-priming centrifugal pump,** and **Trash-handling centrifugal pump,** all below.

Centrifugal screw: Axial-flow, combined axial- and radial-flow centrifugal pump having a screw-type impeller.

Chain: Pump consisting of an endless chain, fitted at intervals with disks or cups, which moves within a pipe or sleeve, used to raise sludge or viscous materials.

Circulating: Centrifugal pump used in central heating systems to ensure continuous distribution of heated water and flow of return water to the boiler.

Close coupled: Pump directly connected to its power unit without reduction gearing or shafting.

Closed centrifugal: Centrifugal pump having its impeller built with the vanes enclosed within circular disks.

Concrete: Apparatus that forces concrete to the placing position through a pipeline or hose.

Continuous flow: Displacement pump within which the direction of flow is not changed or reversed.

Crescent: Half-moon-shaped pump used in hydraulic and lubrication systems.

Deep well: Centrifugal pump used to raise water more than 7.62 m (25 ft).

Deep-well turbine: Centrifugal-type pump equipped with stages, each of which comprises a set of vanes, and which increase in number according to the operating head.

Dewatering: Pump used to remove ponded water and to counter inflow to a site or excavation. A rugged, heavy-duty unit capable of tolerating a high silt content in the water being pumped. May be self-priming centrifugal, trash-handling centrifugal, electric submersible, or diaphragm type.

Diaphragm: Pump that develops pressure by the reciprocating motion of a diaphragm in a chamber having inlet and outlet check valves.

Direct-acting: Reciprocating pump having two opposing pistons connected to the same crankshaft: one piston is powered by steam, compressed air, etc., the other pumps water at each cycle.

Displacement: Ram-operated or piston pump in which compressed air displaces the water to be pumped.

Double-acting: Reciprocating pump in which the piston does work in both directions.

Electric: Any pump driven by a direct-coupled electric motor.

Electric submersible: Vertical, close-coupled, motor-driven pump designed to be totally immersed in the fluid it is designed to pump.

Excess pressure: Pump required to maintain a higher pressure than municipal pressure in a wet-pipe automatic sprinkler system, located above the alarm valve to prevent false alarms.

Feed: Pump that provides feed water to a boiler.

Fuel transfer: Integrally mounted and driven pump on an engine that supplies fuel to the operating system.

Gear: Hydraulic pump that creates hydraulic flow by moving oil from the inlet to the outlet through the spaces between the teeth of the pump's meshing gears. Gear pumps are considered to be more reliable than other forms of hydraulic pump due to their ability to handle higher levels of contamination.

Hydraulic: Device that converts mechanical force and motion into hydraulic fluid power by means of producing flow. There are several types, including:

Axial-piston: Pump comprising several pistons in a rotating cylinder block. Each piston makes one stroke per revolution of the block.

Gear: Pump housing containing one or more sets of meshed gears that, as they rotate, force the oil between them in the direction of rotation.

Peristaltic: Pump that employs two diametrically-opposed rollers inside a circular chamber subjected to subatmospheric pressure.

Positive-displacement: Pump that for each cycle or revolution, positively displaces (usually by mechanical means) a specific amount (volume) of fluid.

Vane: Pump housing an offset, rotating cylinder having a number of sliding radial vanes mounted around its perimeter. The vanes slide in and out to form a seal with the inner face of the housing, creating a chamber that changes in volume as the drum rotates.

Injection: Device that meters fuel and delivers it under pressure to the nozzle and holder assembly of an engine.

Jet: Water pump that develops very high discharge pressure.

Lift: Suction pump that discharges slightly above or below the elevation of the suction inlet.

Multiple-stage: Pump with more than one impeller.

Mud: Circulating pump supplying fluid to a rotary drill.

Nonpositive displacement: Pump that displaces a variable amount each revolution.

Peristaltic: Pump that transmits force through the regular pulsing of a diaphragm, inducing a wavelike reaction.

Piston: Available in radial and axial designs for both fixed- or variable-displacement systems. All piston pumps operate on the principle that a piston reciprocating in a bore will draw in fluid as it is retracted and expel fluid on the forward stroke. By varying the stroke length of the piston, the pump displacement varies, thus varying pump output.

Positive displacement: Any of several piston or rotary-gear pump designs that move a given quantity of fluid through the pump chamber with each stroke or cycle.

Priming: Small-capacity pump used to prime another pump.

Propeller: *See* **Axial flow pump** (above).

Ram: Single action reciprocating pump employing a ram in place of a piston.

Reciprocating: Pump utilizing a piston within a cylinder.

Rotary: Pump employing geared wheels to drive fluids between the gear teeth.

Rotary gear: Positive-displacement pump employing closely fitting rotors or gears to force water through the pump chamber.

Sand: Tube, fitted with a check valve at its lower end, lowered into a borehole to extract mud and cuttings.

Screed: Machine for mixing the materials for a screed with water and pumping the liquid mix through a hose to where it is placed.

Self-priming centrifugal: High-volume, high-head pump with the capability of priming itself provided the suction is within a defined height/distance of the water surface. May be powered either by an engine or electric motor; often designed as a portable unit. When engine driven the pump speed can be adjusted to vary pump flow rate.

Sewage: Any type of pump specifically designed to handle raw sewage and its constituent components.

Shell: Type of pump used to evacuate sand from boreholes.

Single-acting: Reciprocating pump, or compressed air or steam engine, in which only one side of the piston works; every second stroke is a power stroke.

Single suction: Pump having a spiral-shaped case in which water enters the impeller from one side only.

Sludge: Pump designed to handle highly viscous materials.

Spout-delivery: Pump designed to deliver water to the level at which it is working.

Submersible: Pump that can operate while partially or wholly submerged.

Sump: Pump, usually electrically operated, to remove water that collects in a sump.

Supply: Pump for transferring fuel from a tank and delivering it into the injection pump of an engine.

Trash-handling centrifugal: Centrifugal pump capable of handling small sticks, stones, sand, and other solid particles, up to 25% by volume, carried in the water being pumped.

Vacuum: Device that uses mechanical force and motion to evacuate gas from a connected chamber to create subatmospheric pressure.

Vane: Pumps that generate volume by employing small vanes that force a fluid through the opening between the oval pump housing and circular vane housing, normally used only with closed, pressurized hydraulic systems due to their lower tolerance to contamination.

Vertical: Small-diameter, long, multistage, electrically-powered pump designed primarily to pump water from a deep drilled well.

Wellpoint: Centrifugal pump that can move a variable mixture of air and water, used to dewater a subterranean area or to temporarily lower the watertable over a limited area.

PUMPABILITY Ability of a lubricating grease to flow under pressure through the lines, nozzles, and fittings of grease dispensing systems.

PUMPAGE Volume pumped in a given period.

PUMP CAN Five-gallon water can with an attached, hand-operated pump, used to extinguish small fires.

PUMP CHARACTERISTIC CURVE Graphic representation showing the interrelation of such factors as horsepower, efficiency, head, capacity, speed, etc. of a pump.

PUMP DISCONNECT Mechanism used to interrupt a pump drive, typically for roading a mobile crane or easier cold-weather starting.

PUMPED CONCRETE Concrete that is transported through hose or pipe by means of a pump.

PUMPED STORAGE RESERVOIR Reservoir filled entirely with water pumped from outside its natural drainage area.

PUMP EFFICIENCY 1. Ratio of energy converted into useful work to the energy applied to a pump shaft; **2.** Energy difference in the water at the discharge and suction nozzles of a pump, divided by the power input to its shaft.

PUMPER OUTLET NOZZLE Hydrant outlet, usually 112 mm (4.5 in.) in diameter, to which direct connection to a fire pumper can be made.

PUMPING HEAD Sum of the static head and friction head on a pump discharging a specific volume of water.

PUMP HOUSE Structure housing pumps and their appurtenances.

PUMPING LEVEL Level to which water rises in a well when pumping is in progress.

PUMPING LINE Discharge line from a pump.

PUMPING MAIN Pipe on the outlet side of a pump.

PUMPING STATION Structure housing relatively high-capacity pumps and their associated appurtenances.

PUMP PIT Sump or underground chamber in which a pump is installed.

PUMP PRIMER Vacuum pump attached to the suction of a pump and used to prime it.

PUMP SETTING Distance a well pump is positioned below the pump head.

PUMP SLIPPAGE Difference between the amount of water taken into the suction of a pump and that which is discharged.

PUMP STAGES Series of impellers in a centrifugal pump, i.e. a single-stage pump has one impeller; a two-stage pump has two impellers, etc.

PUMP STRAINER Device at the end of a pump suction, used to strain out suspended matter from entering the pump.

PUMP STROKE Distance travelled by the piston or plunger of a reciprocating pump in one half of its operating cycle.

PUMP SUBMERGENCE Distance between the bottom of a pump suction and the water level in the pump pit or afterbay.

PUMP-VALVE CAGE Number of water valves mounted in a single casting on the end of a pump suction.

PUMP VALVES Openings through which water enters and leaves the cylinders of a displacement pump.

PUMP WELL Well whose only point of discharge is through the suction of a pump.

PURGE 1. Remove or clean out one substance prior to the introduction of another; or one color or type of material before using a second color or material; **2.** Removal of air from a hydraulic or fuel system. *Also called* **Bleed.**

PURE OXYGEN ACTIVATED SLUDGE PROCESS Activate sludge sewage treatment process that employs pure oxygen instead of ambient air.

PURIFICATION Act or process of removing objectionable matter and cleansing water by natural or artificial means.

PUTREFACTION Partial decomposition of organic matter by microorganisms, producing foul-smelling matter.

PUTRESCIBILITY Tendency of organic matter to decompose in the absence of oxygen.

PYCNOMETER Instrument for measuring the density of a liquid.

PYRETHROIDS Active insecticidal constituents of pyrethrum flowers (*Chrysanthemum cinerariaefolium* and *Chrysanthemum coccineium*).

PYRHELIOGRAPH Instrument for measuring and recording the intensity of the Sun's radiation entering the Earth's atmosphere.

PYROCLASTIC Material blown out by an explosive volcanic eruption and subsequently deposited.

PYROLYSIS Volatilzation of organic matter by heating in a closed retort in the absence of air and which produces combustible gases, a low-calorific value combustible char, a mixture of oils, and liquid effluent.

PYROLYTIC GAS AND OIL Gas or liquid products that possess usable heating value that is recovered from the heating of organic material (such as that found in solid waste), usually in an essentially oxygen-free atmosphere.

PYROMETER Apparatus used to measure high temperature,

Q

Q_{10} Amount by which the growth or activity of an organism or an enzymic reaction increases per 10°C rise in temperature.

QUAD One quadrillion (1×10^{15}) Btu.

QUADRAT Sampling area used to study the composition of an area of vegetation.

QUAGMIRE Soft, muddy ground.

QUALIFICATION OF BIDDER Means of determining if a prospective bidder is capable of, and competent to complete a contract within the terms specified.

QUALIFICATION TEST Examination of samples from a typical production run to determine adherence to a given specification; performed for approval as a supplier.

QUALIFIED BIDDER Bidder for a contract who has met the requirements for prequalification at the time of invitation to bid.

QUALIFIED GROUNDWATER SCIENTIST Scientist or engineer who has received a baccalaureate or postgraduate degree in the natural sciences or engineering, and has sufficient training and experience in groundwater hydrology and related fields as may be demonstrated by registration, professional certifications, or completion of accredited university courses that enable that individual to make sound professional judgements regarding groundwater monitoring and con-

taminant fate and transport.

QUALIFIED PERSON One who, by possession of a recognized degree, certificate, or professional standing, or who, by extensive knowledge, training, and experience, has successfully demonstrated the ability to solve or resolve problems relating to the subject matter, the work, or the project.

QUALIFIED TESTING LABORATORY Properly equipped and staffed testing laboratory that has capabilities for, and that provides, the following services:

(a) Experimental testing for safety of specified items of equipment and materials to determine compliance with appropriate test standards of performance in a specified manner;

(b) Inspecting the run of such items of equipment and materials at factories for product evaluation to assure compliance with test standards;

(c) Service-value determinations through field inspections to monitor the proper use of labels on products, and with authority for recall of the label in the event a hazardous product is installed;

(d) Employing a controlled procedure for identifying the listed and/or labelled equipment or material tested; and

(e) Rendering creditable reports or findings without bias of the test methods employed.

QUALITY ASSURANCE 1. Actions taken by an owner or his representative to provide assurance that what is being done and what is being provided are in accordance with the applicable standards of good practice for the work; **2.** Process of evaluating overall project performance on a regular basis to provide confidence that the project will satisfy the relevant quality standards; **3.** Organizational unit that is assigned responsibility for quality assurance.

QUALITY CHARACTERISTICS Property that is actually measured to determine conformation of a unit or product with a given requirement.

QUALITY CONFORMANCE INSPEC-TION OR TEST Examination of samples from a production run to determine adherence to a given specification, for acceptance of that production run.

QUALITY CONTROL 1. Actions taken by a producer or contractor to provide control over what is being done and what is being provided so that the applicable standards of good practice for the work are followed; **2.** Process of monitoring specific project results to determine if they comply with relevant quality standards and identifying ways to eliminate causes of unsatisfactory performance; **3.** Organizational unit that is assigned responsibility for quality control.

QUALITY FACTOR Weighting between 1 and 20 of the biological effects that ionizing radiation has on human tissue.

QUALITY STANDARD 1. Formally established criterion for a specific activity that (a) describes a deficiency, condition, or schedule that established the need for work, (b) outlines the work involved, (c) tells how to achieve good workmanship, and (d) lists expected end results; **2.** Formally established criterion for a specific activity that (a) outlines the work involved, and (b) lists the number of work units that are usually required to meet the quality standards for various categories.

QUALITY STANDARDS Stipulations of measurable physical properties or characteristics that materials, equipment, or construction items must have as a minimum.

QUANTITATIVE ANALYSIS Process of determining the amount or proportion of each chemical component of a substance.

QUANTITY TAKEOFF Estimation and measurement of the quantities of separate materials and tasks, including labor, required for a project.

QUANTUM Smallest amount of energy capable of existing independently.

QUANTUM THEORY Theory that whenever radiant energy is transferred the transfer occurs in pulsations or stages rather than continuously, and that the amount of energy transferred during each stage is of a definite quantity.

QUARANTINE Isolation from others for a time, especially to prevent the spread of an infectious disease.

QUARK Any of the fundamental particles from which the composite particles called hadrons (including protons and neutrons) are formed, and which, together with leptons, are believed to be the basic constituents of all matter.

QUARTER BEND Pipe fitting passing through an arc of 90°.

QUARTERING Method of obtaining a representative sample by dividing a circular pile of a larger sample into four equal parts and discarding opposite quarters successively until the desired size of sample is obtained.

QUARTERNARY (SEWAGE) TREATMENT *See* **Sewage treatment.**

QUARTZ Crystalline form of silica $(Si)_2$, usually white, but can be colorless.

QUICKLIME Burnt lime; calcium oxide.

QUICK-OPERATING VALVE Valve having a lever attached to a revolving plug permitting it to be suddenly opened or closed.

QUICKSAND Semifluid mass of silt and fine sand, thoroughly saturated with water. *See also* **Liquefaction.**

QUOTATION Offer to supply something or do something for a given price.

R

RABBLE ARM Radial rotating arm used to transport material within a multiple hearth furnace.

RABBLING Process of moving or plowing the material inside the combustion chamber of a furnace.

RACE 1. Inner or outer ring of a ball or roller bearing; **2.** Channel to or from a head-race, tailrace, or turbine.

RACHION Line on a lake shore where wave action causes the most disturbance.

RACK 1. Parallel, evenly-spaced metal bars within a frame; **2.** Trash rack in a waterway.

RACK RAKE Manual or mechanical device used to clear accumulated debris from the face of a trash rack.

RAD Measure of absorbed radiation dose equal to 0.01 joules/kg.

RADIAL DRAINAGE River systems that form a radial pattern.

RADIAL EXPANSION Tendency of communities experiencing rapid growth to expand radially from their urban center.

RADIAL FLOW Flow to, or from the center of a tank from its periphery.

RADIAL-FLOW TANK Circular tank configured so that the flow is to, or from the center to its periphery.

RADIAL GATE Dam gate with a curved water face and horizontal pivot axis.

RADIAL IMPELLER Impeller set perpendicular to the impeller shaft; material being pumped flows at a rightangle to the impeller.

RADIAL WELL System in which a number of well strainers, each connected to a central shaft or sump, are driven horizontally into a water-bearing stratum.

RADIAN SI unit of plane angle equal to the measure of a plane angle with its vertex at the center of a circle and subtended by an arc equal in length to the radius. Symbol: rad. *See also the appendix*: **Metric and nonmetric measurement.**

RADIAN PER SECOND Unit of angular velocity in the SI system of measurement. Symbol: rad/s. *See also the appendix*: **Metric and nonmetric measurement.**

RADIAN PER SECOND SQUARED Unit of angular acceleration in the SI system of measurement. Symbol: rad/s². *See also the appendix*: **Metric and nonmetric measurement.**

RADIATION Any or all of the following: Alpha, beta, gamma, or X-rays; neutrons; and high-energy electrons, protons, or other atomic particles; but not sound or radio waves, nor visible, infrared, or ultraviolet light.

RADIATION DOSE EQUIVALENT Measure of the effect of ionizing radiation on the substance or tissue that absorbs it.

RADIATION SICKNESS Effects, possibly lethal, caused by a large dose of ionizing radiation to the whole body.

RADIATION WINDOW Band in the radiation spectrum extending from 8.5 to 11.0 micrometers in wavelength in which little absorption by water vapor occurs.

RADICAL Atom or group of atoms with at least one unpaired electron.

RADIDE Portion of a plant embryo that develops into the primary root.

RADIOACTIVE DECAY Progressive decrease in the number of radioactive atoms in a substance by spontaneous nuclear disintegration or transformation.

RADIOACTIVE MATERIAL Material that spontaneously emits radiation.

RADIOACTIVE-SOLUTION GAUGING Flow measurement technique in which a constant flow of a radioactive solution of known concentration is introduced to a stream and the degree of dilution measured at a downstream point.

RADIOACTIVE SPRING Spring in which the water has an abnormally high and easily measurable radioactivity.

RADIOACTIVE WASTE Waste that consists of, or is contaminated with, radionuclides.

RADIOACTIVITY Spontaneous emission of radiation, either directly or from unstable atomic nuclei, or as a consequence of a nuclear reaction.

RADIOACTIVITY SHIELDING Construction or material that absorbs or deflects radiation.

RADIOBIOLOGY Study of the effects of radiation on living organisms.

RADIOCHEMISTRY Chemistry of radioactive materials.

RADIO CONTROL Use of radio signals to activate devices.

RADIOISOTOPES Naturally or artificially produced radioactive isotope of an element.

RADIOMETRIC AGE Time, measured in years before the present, that it has taken for a particular ratio of 'daughter' to 'parent' atoms to be formed by the radioactive decay of the parent atom.

RADIONUCLIDE Type of atom that spontaneously undergoes radioactive decay.

RADIUS OF INFLUENCE Distance from the center of a wellbore to the point where there is no lowering of the watertable when the well is being pumped.

RADON Poisonous, radioactive, heavier-than-air gas released from some granites, and from naturally-occurring or discarded radioactive materials, that can seep into basement and subgrade areas. It has a half-life of 3.28 days, producing radionuclides that decay within hours. Radon is the major source of naturally-occurring radiation exposure for humans — via the ingestion of radon dissolved in water and the inhalation of airborne randon. Because of radon's volatility, water drawn from surface supplies does not generally contain appreciable levels of the radionuclide, usually on the order of 0.01 Bq/L. Groundwater generally contains higher levels, on the order of 10 to 75 Bq/L. The radon level in outdoor air for continental locations in temperate lattitudes is about 9 Bq/m³. Indoor radon levels (in tap water, building materials and natural gas) in addition to that entering from the substructure averages 50 Bq/m³.

RAIN Water condensed from atmospheric vapor, falling to earth in drops that are generally larger than 0.5 mm (0.02 in.) and which descend at velocities greater than 3 m/sec (10 fps).

RAINFALL Precipitation in the form of water.

RAINFALL AREA Geographic surface over which rain has fallen.

RAINFALL DISTRIBUTION COEFFICIENT Product of the maximum rainfall at any point within a storm area divided by the mean for the drainage area or basin.

RAINFALL INDEX Average rainfall intensity above which the volume of rainfall equals the volume of observed runoff.

RAINFALL INTENSITY Volume of rain falling over a given period, converted to its equivalent in mm/h (in./h) at the same rate.

RAINFALL INTENSITY CURVE Curve that represents the rates of rainfall to their duration for a given period.

RAINFALL SIMULATION Determination of the impact of rainfall and/or runoff on a sewer system.

RAIN-FED STREAM Stream whose source

is primarily runoff from rainfall.

RAIN GAUGE Instrument that collects falling rain and indicates the total fallen over a given period.

RAINOUT Removal of particulate matter from the atmosphere by formation of water droplets on the particles which act as condensation nuclei, followed by rain.

RAINWASH Decomposed rock and similar matter washed down into the soil by rain.

RAINWATER Water resulting from precipitation.

RAISE Small tunnel excavated upward from a drive or level.

RAISED BEACH Former beach formed when local sea or lake levels were higher than at present.

RAKINGS Matter removed from bar screens.

RAMP Length of drainpipe laid more steeply than the remainder.

RAMP TECHNIQUE *See* **Sanitary landfilling.**

RAM PUMP *See* **Pump.**

RANDOM ERROR Error that distorts particular statistical results, but which balances overall.

RANDOM FILL Fill placed in an embankment that may consist of a variety of materials.

RANDOM INCIDENT FIELD Sound field in which the angle of arrival of sound at a given point in space is random in time.

RANDOM NOISE Fluctuating sound whose amplitude distribution is gaussian.

RANDOM SAMPLE Small part of a group that is used to represent the whole, so chosen that each portion of the group has an equal probability of being selected.

RANGE 1. Extensive area upon which livestock wander and graze; **2.** Geographical region in which a specific type of plant or animal normally lives or grows; **3.** Totality of points in a set established by mapping; **4.** Measure of dispersion equal to the difference of interval between the smallest and the largest of a set of quantities; **5.** An extended group or series, especially of mountains; **6.** Single series or row of townships, each six miles square, extending parallel to, and numbered east and west from, a survey base meridian line.

RANGE MANAGEMENT Planned use and management of grazing land so as to optimize maximum stock production with conservation of the range resource.

RANGE OF TIDE Difference in height between a high water and a preceding or following low water.

RANKINE Non-SI temperature scale in which freezing point is 491.6°, boiling point is 671.6°, and absolute zero is 0°.

RAPID Extremely fast-moving part of a river, caused by a rapid descent of the river bed, where the surface of the stream is turbulent and obstructed.

RAPID FILTER Sand or pressure filter.

RAPID FLOW Supercritical flow.

RAPID SAND FILTER Filter used to remove particulates from water and consisting of layers of increasingly smaller media built up on a supporting grid. The flow is passed through the media into an underdrain until such time as the head resulting from plugging of the of the media reaches a predetermined level, at which time the flow is reversed with the backwash water being drained to a sewer.

RASPING SYSTEM Procedure in which refuse is ground through a screen partly covered with steel pins, that have the effect of a rasp.

RATE BASE Equity on which a public utility is entitled to a fair return and on which rates my be based.

RATED FATIGUE PRESSURE *See* **Pressure.**

RATED HORSEPOWER *See* **Horse-**

power.

RATED POWER 1. Power specified by a manufacturer for a given application at a given (rated) speed; **2.** Stated or nameplate net electric output that is obtainable from a generator set when it is functioning at rated conditions.

RATED PRESSURE *See* **Pressure.**

RATED SPEED Revolutions per minute at which an engine or apparatus is designed to operate.

RATED STATIC PRESSURE *See* **Pressure.**

RATE MAKING Establishing charges for a public service or commodity.

RATE OF ANNUAL DEPLETION Average annual rate at which water is withdrawn from a groundwater reservoir.

RATE OF FLOW Amount of a liquid that passes a given point over a period of time.

RATE-OF-FLOW CONTROLLER Device used to control the rate of flow of a fluid.

RATE-OF-FLOW INDICATOR Device that indicates and/or records the rate of flow of a fluid for a point in time.

RATE-OF-FLOW RECORDER Device that registers the rate of flow of a fluid relative to time.

RATING 1. Nominal maximum capacity of a material, assembly or item of equipment; **2.** Relationship between the water level and discharge of a stream, well, or aquifer; **3.** Greatest volume of water that can be removed from a stream, well or aquifer without damage to it.

RATING CURVE Graph of the water level of a stream against its flow rate.

RATING FLUME Open conduit designed to maintain a consistent regimen in which a flow may be measured, or flow-measuring devices rated.

RATING TABLE Table showing the relationship between the gauge height at a gauging station and the discharge of a stream or conduit.

RATIO Relationship between two similar things in respect to the number of times the first contains the second, either integrally or fractionally.

RATIONAL METHOD Method of estimating the runoff in a drainage basin at a specific point and time using the rational runoff formula.

RATIONAL RUNOFF FORMULA Formula used in drainage computations to determine the amount of runoff within a watershed.

RATIO REGULATOR Proportional control device that regulates the downstream pressure in a pipeline in which it is located, used to maintain proportional pressures in fuel and air lines in a pressure control system.

RAVINE Deep, more-or-less-linear depression or hollow worn by running water.

RAW DATA Laboratory worksheets, records, memoranda, notes, or copies that are the result of original observations and activities of a study and are necessary for the reconstruction and evaluation of a report.

RAW LAND Unimproved land, i.e. without landscaping, drainage, utilities, structures, etc.

RAW SEWAGE SLUDGE Solids concentrated by various methods in wastewater treatment plants, usually containing 90% to 95% water.

RAW WASTEWATER Influent to a wastewater treatment facility, before any treatment has commenced.

RAW WATER Water as obtained from its source, particularly a surface source such as a lake, river, etc.; untreated water.

REACH 1. Stretch of water visible between bends in a river or channel; **2.** Section of a stream between two gauges; **3.** Length of a canal in which the hydraulic elements remain uniform.

REACTION 1. Reverse or opposing action; **2.** Chemical change or transformation in which a substance decomposes, combines with another substance, or interchanges constituents with other substances; **3.** Dynamic backward pressure of a jet or stream, in the line of its motion.

REACTION TANK Vessel in which the reaction of chemicals added to a liquid takes place.

REACTION VELOCITY Speed at which a chemical reaction occurs.

REACTIVE Substance that is normally unstable and which readily undergoes violent change; or reacts violently with water; or generates toxic gases, vapors, or fumes in quantity sufficient to present a danger to human health or the environment when mixed with water.

READILY ACCESSIBLE Capable of being reached quickly for operation, renewal, or inspection without requiring those to whom ready access is required to climb over or remove obstacles. *See also* **Accessible.**

READILY-AVAILABLE WATER Quantity of water that can be extracted by a plant and used for its normal growth: the difference between field capacity and wilting coefficient.

READILY WATER-SOLUBLE SUBSTANCES Chemicals that are soluble in water at a concentration equal to or greater than 1000 mg/L.

READOUT Visual display of data and/or information derived, collected, relayed, or transmitted by a mechanical, electric, or electronic device.

READY BIODEGRADABILITY Substances that, in certain biodegradation test procedures, produce positive results that are unequivocal and which lead to the reasonable assumption that the substance will undergo rapid and ultimate biodegradation in aerobic aquatic environments.

READY CONDITION Any equipment, device, system, etc. that is in a condition whereby it can be activated or used without further preparation or adjustment.

REAERATION Absorption of oxygen into oxygen-deficient water.

REAERATION CONSTANT Expression of the rate at which oxygen is dissolved into oxygen-deficient water based on a specific unit of time (commonly a day), temperature, depth, and flow velocity.

REAGENT Pure chemical substance from which new products are made, or which is used in chemical tests to measure, detect, or examine other substances.

REAL DENSITY Mass per unit volume of an oven-dried soil sample, pore space excluded.

REAL-EAR PROTECTION AT THRESHOLD Mean value in decibels of the occluded threshold of audibility (hearing protector in place) minus the open threshold of audibility (ears open and uncovered) for all listeners on all trials under otherwise identical test conditions.

REAL PROPERTY Land, and generally whatever is built upon it, including all water within its boundaries, minerals and other deposits of value beneath its surface, and the air above it.

REAL-TIME MEASUREMENT Measurement made simultaneously with the event that is measured.

REAR-END SYSTEM Chemical, thermal, or biological system and its supplementary facilities used for the conversion of preprocessed wastes into useful or nonhazardous products.

REAR LOADER, DETACHABLE CONTAINER Detachable container system in which roll-out containers, typically 0.8 to 2.3 m^3 (1 to 3 yd^3) capacity, are hoisted at the rear of the collection vehicle and mechanically emptied; the container being left with the customer.

REAR-LOADING REFUSE TRUCK *See* **Refuse truck.**

REAR PACKER *See* **Refuse truck.**

REASONABLY AVAILABLE CONTROL TECHNOLOGY Devices, systems process modifications, or other apparatus or techniques that are reasonably available taking into account the necessity of imposing such controls in order to attain and maintain an established ambient emission standard, the social, environmental and economic impact of such controls, and alternative means of providing for attainment and maintenance of such standards.

REAUMUR Temperature scale in which freezing point is 0°, boiling point is 80°, and absolute zero is -218.5°.

RECALCINATION Lime-recovery process in which the calcium carbonate fraction of a complex such as sewage sludge is converted to lime by heating at 980°C (1800°F).

RECARBONATION Introduction of carbon dioxide as a final stage in the lime–soda ash water-softening process so as to convert carbonates and stabilize the solution against their precipitation.

RECEIVER Device that indicates the result of a measurement.

RECEIVING BASIN Vessel into which an effluent is discharged.

RECEIVING WATER Stream, estuary, lake, sea, etc., that receives effluent.

RECEPTION PIT Pit dug at the opposite end from the launch pit of a microtunnel or to which a pipe is jacked.

RECEPTION SHAFT Excavation into which trenchless technology equipment is driven and recovered following conduit installation. *Also called* **Exit shaft.**

RECESSION VELOCITY Mean velocity of water flowing in a channel or conduit immediately downstream of a structure.

RECHARGE Process, natural or artificial, by which water is added to the saturated zone of an aquifer.

RECHARGE AREA Area in which water reaches the zone of saturation (groundwater) by surface infiltration; in addition, a major recharge area is an area where a major part of the recharge to an aquifer occurs.

RECHARGE BASIN Excavation into which streams or storm drains discharge, and which empties into an aquifer.

RECHARGE RATE Rate at which water is added beneath the surface of the ground to replenish or recharge groundwater.

RECHARGE WELL Well used to conduct surface water or an artificial discharge into an aquifer.

RECIPROCATING COMPRESSOR Equipment that increases the pressure of a gas by positive displacement, employing linear movement of the driveshaft.

RECIPROCATING PUMP *See* **Pump.**

RECIRCULATED COOLING WATER Water that is passed through the main condensers for the purpose of removing waste heat, passed through a cooling device for the purpose of removing such heat from the water and then passed again, except for blowdown, through the main condenser.

RECIRCULATION Directing all or part of a flow to once again move through a process.

RECIRCULATION OPERATION Distilling desalination plant in which all or part of the feedwater is returned to each stage more than once.

RECLAIMED Material that is processed to recover a usable product, or which is regenerated.

RECLAMATION Restoration to a better or more useful state, such as land reclamation by sanitary landfilling, or the obtaining of useful materials from solid waste. *See also* **Road reclamation.**

RECLAMATION AREA Typically, the surface area of a coal mine that has been returned to required contours, and on which revegetation (specifically, seeding or planting) work has commenced.

RECOMBINANT DNA TECHNOLOGY Techniques in molecular biology that enable sections of DNA to be transferred between individuals of the same or different species.

RECORD DRAWING Drawing that shows something as constructed or fabricated and which records any changes made since first delivery or completion.

RECORDER Device that produces a visible record of an operation or process.

RECORDING GAUGE Gauge that continuously plots the value it is reading.

RECORDING RAIN GAUGE Device that automatically records the cumulative amount of rainfall over time.

RECORD KEEPING Log of all required actions, setting, events, or other data.

RECOVERABLE RESOURCE Material that still has useful physical or chemical properties after serving its original purpose and that can be reused or recycled for the same or other purposes.

RECOVERED MATERIALS Waste material and by-products that have been recovered or diverted from solid waste, but excluding materials and by-products generated from, and commonly reused within, an original manufacturing process.

RECOVERED SOLVENT Solvent captured from liquid and gaseous process streams that is concentrated in a control device and which may be purified for reuse.

RECOVERY Process of retrieving materials or energy resources from waste.

RECOVERY DEVICE Individual unit of equipment, such as an absorber, carbon adsorber, or condenser, capable of and used for the purpose of recovering chemicals for use, reuse, or sale.

RECTANGULAR WEIR Measuring weir with a rectangular notch.

RECTILINEAR DISTRIBUTOR Movable distributor spanning a rectangular vessel (typically a trickling filter bed), mounted on side rails along which it moves back and forth.

RECTILINEAR TIDAL CURRENT Characteristic of tidal estuaries of rivers and streams in which the direction of flow is reversed between ebb and flow and flow and ebb tides.

RECURRENCE PERIOD Interval between occurrences of repeating events.

RECYCLABLE PAPER Paper separated at its point of discard or from the solid waste stream for utilization as a raw material in the manufacture of a new product.

RECYCLED MATERIAL Material that can be utilized in place of a raw virgin material in manufacturing a product, consisting of materials derived from postconsumer waste, industrial scrap, material derived from agricultural wastes, or other items, all of which can be used in the manufacture of new products.

RECYCLE Process of passing all or part of a flow back through all or part of a process or processes.

RECYCLING Separating a given waste material (e.g. glass) from a waste stream and processing it so that it may be used again as a raw material for products which may or may not be similar to the original.

RECYCLING AGENT Organic material with chemical and physical characteristics selected to restore aged asphalt to desired specifications. *Also called* **Rejuvenating agent,** and **Softening agent.**

RECYCLING BODY Multicompartmented dumping body, typically installed on a low-profile chassis, used primarily for residential pickup of recyclable refuse.

RECYCLING TRAIN Succession of mobile processing units used to recycle old pavement on site.

REDUCE 1. Lessen in amount, concentration, size, etc.; **2.** Deoxidize: the action of a substance to decrease the positive valence of an ion; **3.** Change the form of a mathematical expression without changing the value.

REDUCER 1. Bacteria or fungi that break down dead organic matter into simpler compounds; **2.** Pipe fitting with a smaller opening at one end.

REDUCING AGENT Substance that will

readily donate electrons: typically a base metal such as iron, or the sulfide ion.

REDUCING TEE T-shaped pipe fitting having two or more different sizes of opening.

REDUCING VALVE Spring- or lever-loaded valve that permits a constant pressure to be maintained on one side.

REDUCTION Addition of hydrogen, removal of oxygen, or the addition of electrons to an element or compound.

REDUCTION CONTROL SYSTEM Emission control system that reduces emissions from sulfur recovery plants by converting them to hydrogen sulfide.

REED BEDS Sealed soil or gravel pits containing reeds (commonly *Phragmites australis*) for the treatment of effluent.

REEF Narrow ridge of rock, sand, or coral at or near the surface of the water.

REEF ECOSYSTEM Community of organisms living on and close to a coral reef.

REEL CARRIER Attachment, commonly to a trencher, used to support and discharge the reel carrying the pipe or cable being buried.

REFERENCE Physical or chemical quantity whose value is known and unchanging.

REFERENCE SUBSTANCE Chemical substance or mixture, or analytical standard, or material other than a test substance, feed, or water, that is administered to or used in analyzing a test system in the course of a study for the purposes of establishing a basis for comparison with the test substance for known chemical or biological measurements.

REFILL TUBE Tube from the ball cock to the overflow tube in a water closet tank.

REFLECTED WAVE Wave that is returned when a wave rides up a steep beach or hits an obstruction.

REFLUX Flow back.

REFLUX VALVE Check valve.

REFRACTION Deflection of a propagating wave, as of light or sound, at the boundary between two mediums with different refractive indices, or in passage through a medium of nonuniform density.

REFRACTORY MATERIALS Materials difficult to remove entirely from wastewater, such as nutrients, color, taste- and odor-producing substances, and some toxic materials.

REFRIGERANT Fluid medium (gas or liquid) for removing heat rapidly in refrigerating equipment.

REFUSE *See* **Solid waste.**

REFUSE BURNER *See* **Burner.**

REFUSE CHUTE 1. Vertical or near-vertical pipe for the disposal of building refuse generated at above-grade levels into dumpsters on the ground; **2.** Vertical pipe connecting the several floors of a high-rise to a basement storage area from where refuse sent through the chute is held prior to disposal or incineration.

REFUSE COMPACTOR Mechanical device equipped with a ram that reduces the volume of waste material by forcing it into a removable container or package.

REFUSE-DERIVED FUEL Shredded refuse fuel, used principally as a supplement in industrial or utility boilers that have ash handling capabilities. Using a separation system, much of metal, glass, and other inorganics are first removed, the remaining organic fraction is processed to relatively uniform size particles that are classified as:

RDF-1: Waste used as a fuel in as-discarded form with only bulky wastes removed.

RDF-2: Waste processed to coarse particle size with or without ferrous metal separation.

RDF-3: Combustible waste fraction processed to particle sizes, 95% passing 50-mm (2-in.) screening.

RDF-4: Combustible waste fraction processed into powder form, 95% passing 10-mm mesh screening.

RDF-5: Combustible waste fraction densified (compressed) into the form of pellets, slugs, cubettes, or briquettes.

RDF-6: Combustible waste fraction processed into liquid fuel.

RDF-7: Combustible waste fraction processed into gaseous fuel.

REFUSE HANDLING What is done to prepare refuse for disposal or for processing, which is conversion of wastes into something useful.

REFUSE TRUCK Vehicle specially designed for the collection and transport of refuse. Includes the following types:

Batch loader: Type of enclosed compactor truck equipped with a loading hopper at the rear end and a large mechanized panel that sweeps the solid wastes into the body of the unit.

Front loading: Used for commercial and industrial pickup; capacities range from 15 to 38 m^3 (20 to 50 yd^3). Usually installed on a tilt-cab tandem-axle chassis. Front fork arms lift steel containers over the cab to empty the waste into a hopper atop the body; a hydraulic ram and push-plate system compacts the waste rearward inside the body.

Mobile packer: An enclosed vehicle provided with special mechanical devices for loading, compressing and distributing refuse within the body.

Movable bulkhead: Type of side-loading, enclosed compactor truck equipped with a movable bulkhead that pushes the solid wastes from the front loading area to the rear of the vehicle.

Rear packer: Primarily used for residential refuse collection by two- or three-man crews. Capacities range from 7.65 to 24.5 m^3 (10 to 32 yd^3) of ram-compacted waste; hopper swings up for unloading and waste is pushed out by a movable front wall; may include a container-attachment and winch system for collecting waste from apartment complexes and small businesses in containers of up to 7.65 m^3 (10 yd^3).

Side loading: Designed for residential and some commercial applications using a one-man crew. Sizes from 5.4 to 30.1 m^3 (7 to 40 yd^3).

Stationary packer: An adjunct of a refuse collection system that compacts refuse into a pull-on, detachable container at the site of generation.

REGAIN OF MOISTURE Amount of moisture absorbed by a material as a percent of its original weight.

REGELATION Refreezing of moist ice under pressure at a temperature above freezing.

REGENERATED WATER Water taken from a source for irrigation purposes that, not having been consumed, passes directly back to the same or another body of water.

REGENERATE Form, construct or create anew; to restore or refresh.

REGENERATION 1. Process of restoring an ion-exchange material to the state employed for adsorption; **2.** Periodic restoration of some aspect or material associated with a process to the state necessary for completion of that process.

REGENERATION EFFICIENCY Regeneration level of an ion exchange unit divided by its breakthrough capacity.

REGENERATION LEVEL Quantity of regenerant used in an ion exchange process.

REGENERATIVE CARBON ADSORBER Carbon adsorber applied to a single source or group of sources, in which the carbon beds are regenerated without being moved from their location.

REGENERATIVE-CYCLE GAS TURBINE Stationary gas turbine that recovers thermal energy from exhaust gases and uses it to preheat air prior to entering the combustor.

REGIMEN Condition in a stream or canal where the flowing water neither picks up or deposits material.

REGISTER 1. Device that counts, records, or indicates variations in magnitude; **2.** Book or document of original entry.

REGISTERED LAND Land recorded in a public register.

REGOLITH Loose, unconsolidated and broken rock covering bedrock.

REGOSOL Weakly developed soil.

REGRESSION 1. Tendency of one variable that is correlated with another to revert to the general type and not to equal the amount of deviation of the second variable; **2.** Destruction of existing vegetation and the subsequent colonization of the area by a lesser order of growth.

REGULATED FLOW Streamflow that is subject to control through the use of reservoirs, diversions, or other hydraulic works and/or equipment.

REGULATION 1. Government order having the force of law; **2.** Actions or works that result in an artificial influence on a natural streamflow or its discharge.

REGULATOR Device installed to control, modify or maintain a condition established elsewhere.

REHABILITATION All aspects of maintaining or upgrading the performance of existing pipeline systems.

REJUVENATED STREAM Stream that, having approached its base level, had its gradient and velocity increased from geological uplift resulting in a new cycle of erosion.

REJUVENATED WATER Water returned to the terrestrial supply as a result of geological compaction or metamorphism.

RELATIVE COMPACTION Dry density of a soil sample divided by the maximum dry density of the same sample as determined by a standard compaction test.

RELATIVE DENSITY State of compaction of a granular soil mass relative to the loosest and most dense conditions possible.

RELATIVE ERROR *See* **Error.**

RELATIVE EVAPORATION Ratio of the actual to the potential rate of evaporation of a land or water surface in contact with the atmosphere.

RELATIVE HUMIDITY *See* **Humidity.**

RELATIVE PHYSICAL INTENSITY SCALE Instrument showing the relationship between decibels and physical intensity.

RELATIVE SETTLEMENT Differential settlement.

RELATIVE STABILITY Percentage ratio of dissolved oxygen, nitrate and nitrate oxygen to the total oxygen required to satisfy the biochemical oxygen demand.

RELATIVE TRANSPIRATION Rate of transpiration per unit area from a plant's surface, divided by the rate of evaporation from an equivalent area of open water surface under similar climatic conditions.

RELATIVE UTILIZATION FACTOR Ratio of the utilization efficiency of a fuel to the utilization efficiency of the base (comparative) fuel.

RELATIVE VELOCITY Velocity of a body relative to some other body that may itself be in motion.

RELAY 1. Valve or switch that reflects, amplifies, or restores initial strength to a hydraulic, pneumatic, or electrical impulse; **2.** Electrical magnetic switch employing an armature to open or close contacts.

RELAY EMERGENCY VALVE *See* **Valve.**

RELAY VALVE *See* **Valve.**

RELEASE Spilling, leaking, emitting, discharging, escaping, leaching, or disposing of a pollutant into groundwater, surface water, surface or subsurface soils, or to the atmosphere.

RELICTION Gradual, and relatively permanent, subsidence of water leaving dry land.

RELIEF Difference in elevation between high and low parts of a land surface.

RELIEF SEWER Standby sewer built to carry flows in excess of the capacity of an existing sewer.

RELIEF VALVE *See* **Valve.**

RELIEF VENT Branch from a vent stack, connected to a horizontal branch between the first fixture branch and the soil or waste stack, whose primary function is to provide for circulation of air between the vent stack and the soil or waste stack.

RELIEF WELL Borehole drilled at the toe of an earthen dam to relieve pressures created by the dam's weight.

RELIQUEFACTION Return of a gas to the liquid state.

REMOTE SENSING Technique of obtaining information at a distance from that which is being examined.

RENEWABLE ENERGY SOURCE Source of energy, such as wind power and solar heat, that is inexhaustible or which is derived from organic matter that reproduces itself continually, such as wood or moss.

RENOVATION 1. To make new or like new; **2.** Methods by which a new pipeline is constructed replacing the original fabric on the same line.

REOXYGENATION Oxygen replenishment of a stream from dilution water entering it, biological activity by certain oxygen-producing aquatic plants, and atmospheric aeration.

REPAIR Element of maintenance, as distinct to replacement or retirement.

REPEATABILITY Closeness of agreement between successive results obtained with the same test method, test material, and under the same conditions.

REPLACEMENT New or alternate equipment installed in place of existing equipment

of the same size, capacity, rating, etc.

REPLACEMENT COST Actual or estimated cost of duplicating something with property of equal utility and desirability.

REPLACEMENT UNIT Landfill, surface impoundment, or waste pile unit from which all or substantially all of the waste is removed and which is subsequently reused to treat, store, or dispose of hazardous waste.

REPLICATE Two or more duplicate tests, samples, organisms, concentrations, or exposure chambers.

REPRESENTATIVE IMPORTANT SPECIES Species that are representative, in terms of their biological needs, of a balanced, indigenous community of shellfish, fish and wildlife in a body of water or other defined space.

REPRESENTATIVE SAMPLE Sample of a universe or whole (e.g. waste pile, lagoon, groundwater) which can be expected to exhibit the average properties of the universe or whole.

REPROCESSING Industrial process whereby a material is reused to form a product, or where a reclaimed or otherwise scrap material is combined with others to form a new product.

REPRODUCIBILITY Closeness of agreement between individual results obtained with the same method on identical test material but under different conditions.

REQUEST FOR PAYMENT Application for interim or final payment under the terms of a contract.

REQUEST FOR PROPOSAL Type of bid document used to solicit proposals from prospective sellers of products or services. *Also called* **Request for quotation.**

REQUEST FOR QUOTATION *See* **Request for proposal.**

REQUIRED FIRE FLOW Rate of flow at a specified residual pressure for firefighting purposes at a location or within an area.

RESEALING TRAP Trap on a plumbing

fixture drain pipe in which the rate of flow at the end of a discharge from the fixture seals the trap but does not cause self-siphonage.

RESEARCH Systematic investigation, including research development, testing and evaluation, designed to develop or contribute to general knowledge.

RESERVES Identified or obtained quantities of a substance or product, the initial portion of which has been processed or utilized.

RESERVOIR Place where anything (usually a fluid) is collected and stored, especially a natural or artificial impoundment.

RESERVOIR CAPACITY Total storage space in a reservoir between designated elevations.

RESERVOIR CLEARING Cutting and grubbing of vegetation from the site of a proposed water impoundment area.

RESERVOIR DEPOSITION Accumulated sediment deposited in a reservoir.

RESERVOIR LINING Waterproof membrane or lining to prevent or reduce seepage through the bottom and sides of a reservoir.

RESERVOIR ROUTE *See* **Collection method.**

RESET Return something to its normal operating state after it has been switched, manually or automatically, to some other state, such as standby.

RESIDENCE Home, house, apartment building, or other place of dwelling which is occupied during any portion of the relevant year.

RESIDENCE TIME Time that a substance or condition remains unchanged from a given state.

RESIDENTIAL/COMMERCIAL AREAS Areas where people live or reside, or where people work in other than manufacturing or farming industries. Residential areas include housing and the property on which housing is located, as well as playgrounds, roadways, sidewalks, parks, and other similar areas within a residential community. Commercial areas are typically accessible to both members of the general public and employees and include public assembly properties, institutional properties, stores, office buildings, and transportation centers.

RESIDENTIAL INCINERATOR *See* **Incinerator.**

RESIDENTIAL SILENCER Exhaust muffler used to produce the silencing level usually associated with or required for residential areas.

RESIDENTIAL SOLID WASTE Garbage, rubbish, trash, and other solid waste resulting from the normal activities of households.

RESIDENTIAL WASTE *See* **Solid waste.**

RESIDUAL CHLORINE Amount of chlorine remaining in potable water for purposes of disinfection following dechlorination of the supply as it leaves a treatment plant.

RESIDUAL CLAY Clay that has formed in place from the decomposition of rock.

RESIDUAL DIRT FACTOR Dirt capacity remaining in a service loaded filter element after use, but before cleaning, measured under the same conditions as the dirt capacity of a new filter element.

RESIDUAL DISINFECTANT CONCENTRATION Concentration of disinfectant measured in mg/L in a representative sample of water.

RESIDUAL DRAWDOWN Distance that the watertable present in a well is below the initial static water level following a period of pumping.

RESIDUAL FUEL OIL Residue from the distillation of petroleum from oil which it is not economical to process further.

RESIDUAL OXYGEN Dissolved oxygen content of a stream following the commencement of deoxygenation.

RESIDUAL PRESSURE Pressure remaining in a water distribution system when a specified flow at a designated pressure is being withdrawn, usually for firefighting purposes.

RESIDUAL WASTE Those materials (solid or liquid) that still require disposal after the completion of a resource-recovery activity, e.g. slag and liquid effluents, plus the discards from front-end separation systems.

RESIDUE **1.** Active ingredient(s), metabolite(s) or degradation product(s) that can be detected in crops, soil, water, or other components of the environment, including man, following the use of a pesticide; **2.** Solids that remain after completion of thermal processing, including bottom ash, fly ash, and grate siftings.

RESILIENT CONNECTOR Flexible connector used to join pipes, or connect pipes to fittings, that are subject to vibration.

RESILIENT HANGER Pipe hanger incorporating a spring to dampen vibration.

RESISTANCE In fluid flow, the opposition to flow that makes it inevitable that there will be a pressure drop when a fluid is flowing.

RESISTANCE LOSSES Head loss of a flow resulting from resistance of a pipe or channel.

RESOLUTION Minimum incremental measurement possible in a given measuring device.

RESONANCE **1.** Reinforcement and prolongation of sound due to reflection or vibration of other objects; **2.** Condition of a circuit adjusted to allow the greatest flow of current at a certain frequency.

RESONANT Resounding; continuing to sound.

RESONANT FREQUENCY Actual frequency at which the combination of a vibratory roller drum and the material exhibits the greatest amplitude; that is when the generated frequency coincides with the natural frequency of the material being compacted.

RESOURCE Any supply that will meet a need.

RESOURCE LEVELLING Any form of

network analysis in which scheduling decisions (start and finish dates) are driven by resource management concerns (e.g. limited resource availability or difficult-to-manage changes in resource levels).

RESOURCE-LIMITED SCHEDULE Project schedule whose start and finish dates reflect expected resource availability.

RESOURCE PLANNING Determining what resources (people, equipment, materials) are needed, and in what quantities, to perform project activities.

RESOURCE RECOVERY General term used to describe the extraction of materials or energy from wastes.

RESOURCE RECOVERY FACILITY Physical plant that processes residential, commercial, and/or institutional solid wastes biologically, chemically, or physically, and which recovers useful products, such as shredded fuel, combustible oil or gas, steam, metal, glass, etc. for recycling.

RESOURCES Personnel, equipment, and material available for application to a given situation; anything that is of use to man.

RESPIRABLE DUST Airborne dust in sizes capable of passing the upper respiratory system to reach the lower lung passages.

RESPIRATION Process by which an animal, plant, or living cell secures oxygen from the air or water, distributes it, combines it with substances in the tissues, and gives off carbon dioxide.

RESPIRATOR Filtering mask for individual protection against smoke and fumes.

RESPIRATORY QUOTIENT Ratio of the volume of carbon dioxide given off to the volume of oxygen used up by an organism during respiration.

RESPONSE Act of responding to a stimulus or alarm.

RESPONSE TIME Time required for effective transition.

REST MASS Mass of a body when it is at rest relative to the observer.

RESTORATION CLAIM Preauthorized or emergency claim for restoring, rehabilitating, replacing or acquiring the equivalent of any natural resources injured by the release of a hazardous substance.

RESTORE Reestablish a setting or environment in which the natural functions of a floodplain can again operate.

RESTRICTED-ORIFICE SURGE TANK Storage vessel with a restricted opening, used to create an increased head differential when water is flowing to or from the tank.

RESTRICTED WASTES Those categories of hazardous wastes that are prohibited from one or more methods of land disposal either by regulation or statute.

RESTRICTION Reduced cross-sectional area in a fluid- or vapor-carrying line that produces a pressure drop.

RESTRICTOR Device that reduces the cross-sectional flow area. *See also* **Orifice restrictor.**

RESURGENT WATER Magmatic water of external origin.

RESUSCITATOR Approved mechanical device for assisting the respiration of an unconscious person.

RETAINAGE Amount withheld from progress payments in accordance with the terms of the construction contract, paid following completion and acceptance of the project.

RETAINED DIRT CAPACITY In fluid filter evaluation, the amount of dirt that is captured by the filter in a test system before the terminal differential pressure is reached.

RETAINED WATER Water suspended in an aeration zone.

RETARDING BASIN Storage basin used to reduce flood flows of a stream.

RETARD SWITCH *See* **Switch.**

RETENTION Difference between total precipitation and total runoff for a given area over a given period.

RETENTION CAPACITY Maximum retention capability under stated conditions.

RETENTION CHAMBER Structure within a flowthrough test chamber which confines the test organisms, facilitating their observation and eliminating their loss.

RETENTION PERIOD Theoretical time required to displace the contents of a tank or storage vessel at a given rate of discharge.

RETENTION POND Basin in which sudden influxes of surface runoff are held temporarily before being released gradually into a drainage system.

RETENTION TIME Time that a liquid waste takes to pass through a process being performed upon it.

RETREAT VELOCITY Mean velocity of water flowing in a stream or channel immediately downstream of a hydraulic structure.

RETROGRESSION Lowering of the bed level of a river.

RETURN BEND Pipe fitting providing a 180° change in direction.

RETURNED SLUDGE Settled activated sludge returned to mix with incoming raw or primary settled wastewater.

RETURN FLOW Water returned to a stream channel after diversion for irrigation or other purposes.

RETURN OFFSET Double offset that allows a pipe to return to its original line having passed over an obstacle. *Also called* **Jumpover.**

RETURN SEEPAGE Water that percolates from canals and irrigated areas to the underlying groundwater, returning eventually to natural channels.

RETURN SYSTEM Pipes or ducts that return air or water from a warmed or cooled area to the processing equipment.

RETURN TIME Time it takes to travel from the dump area to the load area.

REUSE Return of a commodity or product

into the economic stream for use in exactly the same form and kind of application as before, without any change in its identity.

REVERBERATION Sound that echoes back.

REVERBERATION TIME Time required for the mean square sound pressure level, originally in a steady state, to fall 60 dB after the source is stopped.

REVERBERATORY FURNACE Furnace in which the roof is heated by flames and the resulting heat radiated down to the material to be consumed.

REVERSE FLOW Flow opposite to the normal direction.

REVERSE OSMOSIS *See* **Membrane processes.**

REVERSE ROTARY DRILLING Similar to rotary drilling but where waste material is carried away by water or mud forced up the inside of the drill pipe.

REVERSIBLE MOTOR Motor whose direction of rotation can be changed by varying the electrical connections or by mechanical means.

REVERSING CURRENT Characteristic of tidal waterways where the flow is alternately upstream (flood) and downstream (ebb).

REVERSION Hydrolytic phenomenon of the higher polymeric or condensed phosphates wherein in aqueous media they revert to the uncondensed form, orthophosphate.

REVETMENT Layer of solid material (rock, aggregate, concrete block, etc.) placed on the bottom or banks of a river or waterway to minimize erosion.

REVETMENT WALL Wall built along the toe of an embankment to protect the slope against erosion.

REVOLUTIONS PER MINUTE Non-SI unit of rotational speed. It is equal to the number of complete cycles, typically of an engine, pulley, shaft, etc., completed in one minute; the customary term used to describe engine speed. Symbol: rpm.

REVOLUTIONS PER SECOND Unit of rotational speed of the SI system of measurement. Symbol: r/s. *See also the appendix:* **Metric and nonmetric measurement.**

REVOLVING DISTRIBUTOR Movable distributor having a rotary motion.

REVOLVING SCREEN Circular trash rack that is turned mechanically or by the force of the water passing through it.

REVOLVING VANES Vanes attached to an axis around which they move as a result of pressure upon their larger face.

REYNOLD'S CRITICAL VELOCITY Velocity in a conduit or channel at which flow changes from laminar to turbulent, and at which friction ceases to be proportional to the first power of the velocity and becomes proportional to the higher power, almost the square.

REYNOLD'S NUMBER Numerical ratio of the dynamic forces of mass flow to the shear stress due to velocity. Flow usually changes from laminar to turbulent between Reynold's numbers 2000 and 4000. *See also* **Turbulent flow.**

RHEOLOGY Science dealing with the flow of materials. *See also* **Flow.**

RICHTER SCALE Logarithmic scale that measures the energy released by movement of the ground from the center of a seismic shock.

RIDER SEWER Shallow sewer to which house connections are made, and which connects at intervals to a parallel main sewer at a greater depth.

RIDGE Long, narrow upper section or crest of something.

RIFFLE Rocky shoal or sandbar lying just below the surface of a waterway.

RIFT VALLEY Linear valley bounded by normal faults.

RIGGING SWITCH *See* **Switch.** *Also called* **System override switch.**

RIGHT BANK Bank bounding flowing water that is on the right when looking downstream.

RIGHT-OF-ACCESS Lawful right to enter private property for a described reason.

RIGHT-OF-WAY 1. Route that is lawful to use; **2.** Strip of land acquired or used for utility installation and service; **3.** Land or water rights necessary for construction; **4.** General term denoting land, property, or interest therein, usually in a strip, acquired for or devoted to transportation purposes.

RIGHT-OF-WAY APPRAISAL Expert opinion of the market value of property including damages, if any, as of a specified date, resulting from an analysis of facts.

RIGHT-OF-WAY ESTIMATE Approximation of the market value of property including damages, if any, in advance of an appraisal.

RIGHT-OF-WAY LINE Line marking the limit between land secured for public use and adjacent private property.

RIGHT-OF-WAY MAP Plan of a highway improvement showing its relation to adjacent property, the parcels or portions thereof needed for highway purposes, and other pertinent information.

RIGID PIPE Pipe fabricated of an inflexible material, such as concrete, glazed or unglazed clay, grey cast iron, etc.

RILL Small stream; a rivulet.

RIM Unobstructed edge, usually in a single plane.

RIME Frost of granular ice crystals.

RINGELMANN CHART Set of five charts that emulate smoke densities in percentages of black, used to assess the opacity of smoke issuing from stacks, etc.

RING NOZZLE Nozzle containing a ring at its smaller end, which significantly reduces its diameter.

RING SYSTEM Well pointing system around the periphery of the area in which construction is to take place that creates a

dry zone.

RIP Stretch of broken water in a river, estuary, or tidal channel.

RIPARIAN LAND Land abutting the banks of a natural body of water.

RIPARIAN OWNER One who owns land bounding upon a body of water with right of access to the water and its enjoyment.

RIPARIAN RIGHTS Rights of a land owner to water on or bordering his property, including the right to prevent upstream diversion or misuse.

RIPARIAN ZONE Timber left standing on lake and river banks to give shade and protection.

RIP CURRENT Current of water disturbed by an opposing current, especially in tidal waters, or by passage over an irregular bottom.

RIPPLE 1. Small wave on the surface of otherwise still water, usually forming in concentric circles about a point of disturbance; **2.** Series of small more-or-less parallel undulations with a flat upstream face and steep downstream face in silt and sand caused by slow flow slightly above the least flow of water required to move the sand.

RIPRAP Rough stones of various sizes, from about 150 kg (330 lb) up to 10 t (9.07 tons), placed irregularly and compactly to prevent scour by water.

RIP TIDE Surface current that briefly flows out from a shore during certain phases of an ebb tide.

RISER 1. Vertical water supply pipe extending from a horizontal water supply pipe to a fixture; **2.** Vertical water pipe used to carry water for fire protection systems above grade.

RISE RATE Ratio of pressure rise to time.

RISER PIPE Vertical pipe rising from one story to another.

RISING MAIN Electrical, gas, or water supply service that passes through two or more floors.

RISING SLUDGE Condition where settled and compacted sludge present at the bottom of a secondary clarifier in the activated sludge process, rises to the surface as a result of denitrification.

RISING STEM VALVE Valve whose stem rises when the valve is opened.

RISK Chance of harm or loss.

RISK ESTIMATION Actuarial technique used to assess the relative costs and benefits of a particular circumstance or technique in which the actual recorded incidence of death or injury to humans is extended over an extended period.

RIVER Large natural stream of water emptying into an ocean, lake, or other body of water, and usually fed along its course by converging tributaries.

RIVER BASIN Land area drained by a river and its tributaries.

RIVER BED Area covered or once covered by water, between the banks of a river.

RIVER CAPTURE Evolutionary process in which a river system erodes the slopes that provide its drainage to the point where the it envelopes an adjacent drainage basin.

RIVER FORECASTING Hydrologic and meteorologic forecasting of the stages of a river.

RIVER GAUGE Any of a number of devices used to measure the height of a river's surface water above a datum: the river stage.

RIVER GRAVEL Gravel having a rounded shape and found at the site of a current or former river bed.

RIVER HEAD Source of a river.

RIVERINE Pertaining to or resembling a river.

RIVERSIDE Bank of a river.

RIVER STAGE Height of the surface of a river at a given point above a datum.

RIVER TERRACE Bench bordering a stream, formed by previous meanderings at a period when the stream carried a higher flow or when the land was at a lower elevation.

RIVER TRAINING Engineering works and hydraulic structures employed to direct or lead the flow to and/or through a defined channel, or away from some point.

RIVER VALLEY Valley created primarily by the erosive activities of flowing water.

RIVER WASH Alluvial deposits carried downstream by moving water and subject to erosion and deposition during periods of flood.

RIVULET Small brook or stream.

ROCK Any naturally formed mineral mass or aggregate that constitutes a significant part of the Earth's crust; any relatively-hard, naturally-formed mass of mineral or petrified matter; stone.

ROCK CRUSHING AND GRAVEL WASHING Facilities that process crushed and broken stone, gravel, and riprap.

ROCKFILL DAM Dam constructed with a central core of loose rock, faced with rolled earth, concrete, or other impervious surfacing.

ROCK FLOWAGE AND FRACTURE ZONE That part of the lithosphere lying between the zone of rock fracture and the zone of rock flowage.

ROCK FLOWAGE ZONE That part of the lithosphere in which all rocks are under stresses exceeding their elastic limits and therefore subject to permanent deformation: interstices are insignificant or absent.

ROCK FLOUR Finely comminuted rock debris, especially that produced by the grinding effect of glaciers.

ROCK FRACTURE ZONE Upper section of the lithosphere in which rocks are under stresses less than those required to close their interstices.

ROCK WOOL INSULATION Insulation composed principally from fibers manufactured from slag or natural rock, with or without binders.

RODDING Use of a flexible section of metal rod capped with one of several devices to clear a blocked drain.

RODDING EYE Access in a pipe fitting for cleaning purposes. *See also* **Access eye,** and **Cleanout.**

RODENTICIDES Substances or mixtures of substances intended for preventing, destroying, repelling, or mitigating animals belonging to the Order Rodentia of the Class Manunalia, and closely related species.

ROD FLOAT Wooden rod, weighted at one end to float vertically with most of its length submerged so as to average out the velocity of the stream throughout its submerged depth.

ROENTGEN Unit of exposure to radiation based on the capacity to cause ionization and equal to 2.58×10^4 coulomb/kg in air.

ROLLED-EARTH DAM Dam constructed of succeeding layers of selected material of proper moisture content placed in thin lifts and compacted by rolling.

ROLLER 1. Heavy compacting machine, self-propelled or towed, that may be equipped with rubber-tired wheels, rubber-tired wobble wheels, steel rolls, or special types of rolls such as sheepsfoot or grid, etc. Vibratory or oscillating equipment often is added.

ROLLER GATE Hollow cylindrical crest gate used in dam spillways. It operates by being driven up and down geared racks in the side piers or walls. *See also* **Sector gate.**

ROLL-OFF BODY Truck body used for heavy-duty, high-volume commercial and industrial pickup. There are two basic types:

   Hook-arm: Uses hydraulic boom arms with hook ends to grab and pull containers or other body types onto the chassis.

   Tilt frame: Employs subframe assemblies that rise and roll rearward for lowering containers by cable and winch.

ROLL-OFF CONTAINER Steel box with wheels used to collect waste at a site, such as a construction site, that can be rolled onto a truck using a winch and then taken to a disposal facility for discharge. The empty container can then be returned or taken to another site.

ROLL-ON/ROLL-OFF *See* **Waste containers.**

ROLL-OVER PROTECTION STRUCTURE Structure on mobile equipment that protect the operator in the event that the machine rolls over. *Also called* **ROPS.**

ROLLWAY Spillway of a dam carrying the overflow.

ROOF DRAIN Vertical pipe installed to receive water drained from the surface of a roof and to discharge it to a drainage system.

ROOT Section of a dam that merges into the ground.

ROOT CROPS Plants whose edible parts grow below the surface of the soil.

ROTAMETER Device used to measure the flow rate of gases and liquids.

ROTARY DISTRIBUTOR Movable distributor consisting of a number of near-horizontal arms that revolve about, and receive a flow from, a central post, and which distribute the liquid over a circular bed through orifices spaced along their length.

ROTARY DRILL Rock drill in which the bit it rotated in the hole, the rock being cut or abraded by knives or teeth, or by hard material set in the bottom of the bit.

ROTARY DRYER Long steel rotating cylinder, mounted at a slight angle to horizontal, through which the material to be dried is passed against a countercurrent of hot air.

ROTARY GEAR PUMP *See* **Pump.**

ROTARY PLUG VALVE *See* **Valve.**

ROTARY PUMP *See* **Pump.**

ROTARY SCREEN *See* **Screen.**

ROTARY SWITCH *See* **Switch.**

ROTARY TIDAL CURRENT Offshore ocean tidal current that does not slack and reverse, but which changes direction continually throughout 360°.

ROTARY VALVE Spherical valve containing a circular gate mounted on trunnions that permit it to rotate through 90°.

ROTATING-BALL FAUCET Single-handle faucet that regulates and mixes the flows of hot and cold water through a single spout.

ROTATING BIOLOGICAL CONTACTOR Paddle wheel that propels, lifts and aerates the wastewater contained in an aeration ditch.

ROTIFER Any of various minute, multicellular aquatic organisms of the phylum Rotifera, having at the anterior end a wheel-like ring of cilia.

ROTOR Rotating part of a machine.

ROUGHING FILTER Any of several types of filter used to remove relatively coarse material from wastewater as a form of pretreatment.

ROUGHING TANK Settling tank in which wastewater is held to remove coarse suspended matter and grease.

ROUGHNESS COEFFICIENT Effect on water flow of the degree of roughness of the inner surface of a conduit.

ROUND-CRESTED WEIR Weir with a crest that is upwardly convex in the direction of normal flow.

ROUND-HEADED BUTTRESS DAM Concrete dam consisting of parallel buttresses thickened at their inner faces until they touch, the spillway being a curved slab overlapping the exterior or downstream face of the buttresses.

RUBBISH General term for solid waste, excluding food wastes and ashes, taken from residences, commercial establishments, and institutions.

RUBBISH CHUTE Pipe, duct, or trough through which waste materials are conveyed by gravity from above to a storage area preparatory to burning, compaction, or removal.

RUBBLE Fragments of rock or masonry crumbled by natural or man-made forces.

RUBBLE DAM Dam constructed of rocks laid without mortar.

RUBBLE DRAIN Trench filled with stones selected to allow water to flow between them.

RUBBLE-MOUND BREAKWATER Breakwater constructed on heaped stones weighing up to 10 tons each.

RUN 1. Migration of fish ascending a river to spawn; **2.** Discharge, or flow; **3.** That part of a pipe or fitting that continues in the same straight line as the direction of flow.

RUNNING GROUND Water-bearing, or very dry sand that will flow down even a very slight slope. *See also* **Running sand.**

RUNNING SAND Sand below the natural groundwater level that flows into excavations. *See also* **Running ground.**

RUNNING TRAP Sanitary fitting that provides a water seal where both the inlet and outlet are horizontal. *See also* **House trap.**

RUNOFF Excess surface water that flows over a site instead of percolating through the soil after precipitation.

RUNOFF COEFFICIENT Ratio of the volume of runoff from a drainage basin to the volume of rainfall.

RUNOFF CYCLE Portion of the hydrologic cycle between periods of precipitation over a land area and subsequent discharge through stream channels.

S

SACRIFICIAL ANODE Anode, or easily corroded material, connected electrically to a cathode, used for cathodic protection.

SACRIFICIAL PROTECTION Use of a sacrificial metal, in the form of an anode, or a coating such as a zinc-rich paint, to protect steel or a steel alloy. In the presence of an electrolyte, a galvanic cell is established and the anode or coating corrodes instead of the steel.

SADDLE 1. Pipe fitting; **2.** Bends in a pipe allowing it to pass over other pipes in the same plane or similar obstacles; **3.** Natural depression in a ridge.

SADDLE FITTING Plumbing fitting used to install a branch from an existing run of pipe.

SADDLE REEF Ore body occupying the space between relatively competent beds in the hinge zone of a fold.

SAE NET HORSEPOWER *See* **Horsepower.**

SAFE BEARING CAPACITY Load per unit area, inclusive of a factor for safety, that the soil can carry. *Also called* **Admissible load.**

SAFETY BELT 1. Single belt or harness-type system with means for securing it about the waist or body and for attaching to a lanyard or lifeline; **2.** Flexible means to per-mit an operator or helper engaged in 'order picking' or building maintenance freedom to perform those tasks while offering restraint from a free drop to the ground in the event of a fall.

SAFETY CAN Approved, closed fuel con-tainer, maximum 18.5-L (5-gal) capacity, having a flash-arresting screen, spring-closed lid, and spout cover, designed to safely re-lieve internal pressure when subjected to fire exposure.

SAFETY DEVICE Any device for the pro-tection of operators, equipment, and machines in case of accident.

SAFETY FENCING Temporary wood, metal, or plastic fencing intended to prevent unauthorized entrance or proximity to a haz-ardous or restricted area for a limited period.

SAFETY FLOORING Flooring materials that, through their surface design or type of materials used, or both, incorporate slip-re-sistance or sanitation features, or both. Other features that may be built into such flooring include underfoot comfort, impact and cut resistance, etc.

SAFETY GATE Gate installed in a high-level navigation canal to prevent loss of wa-ter in an emergency.

SAFETY GEAR 1. Mechanical devices de-signed to arrest movement, or right an out-of-normal condition that might otherwise pose a hazard to life, the structure or the equipment itself; **2.** Range of equipment, gear and materials which, when used in an appro-priate manner under the conditions for which they were designed, can help prevent per-sonal injury.

SAFETY GLASSES Spectacles with plain or prescription lenses made of shatter-resis-tant material.

SAFETY GOGGLES Safety device, used to protect the eyes from dust, chips, and other small particles by enclosing the eyes (and spectacles being worn by the user) in a transparent, ventilated shield made of shat-terproof material.

SAFETY HARNESS Harness worn by a user to restrain a fall.

SAFETY-OPERATED SWITCH *See* **Switch.**

SAFETY SHUTOFF VALVE *See* **Valve.**

SAFETY SWITCH *See* **Switch.**

SAFETY VALVE *See* **Valve.**

SAFETY WORK SURFACE Surface intended to reduce the possibility of foot slippage.

SAFE VELOCITY Flow velocity that will maintain solids in movement without scouring the conduit.

SAFE WORKING PRESSURE Maximum working pressure of a pressure vessel (boiler, flask, cylinder, etc.) permitted under a regulation, code, etc., and stamped on the unit.

SAFE YIELD Sustained pumping rate from an aquifer or water-bearing formation in which the yield does not exceed the aquifer storage and recharge rate over a defined period.

SALINE Containing salt.

SALINE CONTAMINATION Depreciation of usable water through the intrusion of salt water.

SALINE ESTUARINE WATERS Semienclosed coastal waters that have a free connection to the territorial sea, undergo net seaward exchange with ocean waters, and have salinities comparable to those of the ocean. Generally, these waters are near the mouth of estuaries and have cross-sectional annual mean salinities greater than 25 parts per thousand.

SALINE SPRING Spring water that contains a considerable quantity of sodium chloride or other minerals.

SALINE WATER Water containing from 10,000 to 33,000 mg/L of dissolved salts.

SALINITY Relative concentration of dissolved mineral substances in water.

SALINOMETER Hydrometer calibrated for salt solutions.

SALT 1. White crystalline compound, sodium chloride, found in the earth and in salt water; **2.** Crystalline chemical compound derived from an acid by replacing the hydrogen wholly or partly by a metal or an electrochemical radical.

SALTMARSH Marsh periodically flooded by salt water.

SALT STABILIZATION *See* **Soil stabilization.**

SALVAGE Quantity of materials, sometimes of mixed composition, no longer useful in its present condition or at its present location, but capable of being recycled, reused, or used in other applications. *See also* **Salvage scrap.**

SALVAGE AND RECLAMATION Refuse disposal process in which discarded material is separated mechanically or by hand into various categories.

SALVAGE SCRAP Materials, products, or equipment beyond repair that have to be sold or disposed of as scrap. *See also* **Salvage.**

SALVAGE VALUE Probable sale price of an item, if offered for sale, on the condition that it will be removed from the property at the buyer's expense, allowing a reasonable period of time to find a person buying with knowledge of the uses and purposes for which it is adaptable and capable of being used, including separate use of serviceable components and scrap when there is no reasonable prospect of sale except on that basis.

SALVAGING Controlled removal of waste materials for utilization.

SAME LOCATION Same or contiguous property that is under common ownership or control, including properties that are separated only by a street, road, highway, or other public right-of-way. Common ownership or control includes properties that are owned, leased, or operated by the same entity, parent entity, subsidiary, subdivision, or any combination, including any municipality or other governmental unit, or any quasi-governmental authority (e.g. a public utility district or regional waste disposal authority).

SAMPLE Either a group of units, or a portion of material taken respectively from a larger collection of units or a larger quantity of material that serves to provide information that can be used as a basis for action on the larger collection or quantity or on the production process.

SAMPLER Device used to obtain a single (grab), composite, continuous, or periodic sample of flowing or still water.

SAMPLING SPOON Split shell, 50 mm (2 in.) OD, 35 mm (1.375 in.) ID, 600 mm (24 in.) long, for taking earth samples.

SAMPLING TRAIN Series of devices into which a known volume of exhaust or stack gas is drawn for collection and analysis.

SAND *See* **Soil types**, and **Soil groups.**

SANDBAG Hessian or plastic sack averaging 300 x 700 mm (12 x 28 in.), holding between 7 and 13.5 kg (15 and 30 lb) of sand; fabric bags are closed with a draw string, plastic bags by stapling. Both types are used for a wide range of tasks, principally to form a low dike or water barrier to exclude floodwater.

SANDBAR Offshore shoal of sand built up by the action of waves or currents.

SAND BOIL Sudden upwelling of water through sand caused by unbalanced hydrostatic pressure.

SAND CATCHER 1. Hydrographic instrument through which water flows and deposits sand, used to determine the volume of sand in suspension; **2.** Chamber within a storm or combined sewer designed to reduce flow velocity and permit the settling out of sand and grit.

SAND:COARSE AGGREGATE RATIO Ratio of fine to coarse aggregate in a batch, by mass or volume.

SAND DRAIN Vertical sand columns installed to speed drainage and rapid consolidation of marshy land. *See also* **Pile, sand, and Sand or earth wick.**

SAND DUNE Mound or ridge of sand, blown by prevailing winds, and exhibiting a long windward slope and steeper leeward slope.

SAND EQUIVALENT Measure of the relative proportions of detrimental fine dust or claylike material, or both, in soils or fine aggregate.

SAND FILTER Large-scale equipment for the filtration of potable water consisting of layers of increasingly fine aggregate ranging from coarse stone at the bottom to fine quartz grains at the top. Raw water is fed to the bottom of the tank, flows through the graded media, and decants over weirs at the top. When the filter bed becomes sufficiently clogged with trapped material the flow is reversed with the wash water fed to a drain.

SAND GATE Sluice gate through which sand and sediment collected in a sand trap can be ejected.

SAND-GRAIN METER Hydrographic instrument used to measure the quantity of sand in flowing water.

SAND ISLAND Temporary structure formed in a body of water by creating a perimeter of sheet piling into which bottom silts are pumped.

SAND LINE In well drilling, the wire rope that operates the bailer that removes water and drill cuttings.

SAND OR EARTH WICK Ribbonlike woven strip that is inserted into small-diameter holes drilled vertically into water-bearing surface strata. Used to raise the water to the surface by capillary action for subsequent removal.

SAND PUMP *See* **Pump.**

SAND PUMP DREDGE Suction dredge.

SAND TRAP Trap in a water line that permits sand and other heavy particles to settle out from the flow.

SAND WASHER Machine used to wash sand used in a slow sand filter.

SANDY CLAY *See* **Soil types.**

SANDY LOAM *See* **Soil types.**

SANITARY APPLIANCE Fixed appliance, normally supplied with water and connected to a drain.

SANITARY DRINKING FOUNTAIN Drinking fountain that delivers potable water in a manner that permits an individual to drink without coming into contact with the equipment.

SANITARY ENGINEERING Professional responsibility for the design and maintenance of facilities for the provision of potable water and the collection and disposal of liquid and solid wastes.

SANITARY LANDFILL Disposal facility employing an engineered method of disposing of solid wastes on land in a manner which minimizes environmental hazards by spreading the solid wastes in thin layers, compacting them to the smallest practical volume, and applying cover material at the end of each working day.

SANITARY LANDFILL COMPACTOR *See* **Compactor.**

SANITARY LANDFILLING *Also called* **Controlled dumping.** An engineered method of disposing of solid waste on land in a manner that protects the environment, by spreading the waste in thin layers, compacting it to the smallest practical volume, and covering it with soil by the end of each working day. There are several methods, including:

> **Area technique:** Where refuse is deposited on the ground level or upon an earlier lift of solid wastes.

> **Canyon technique:** An area method in a depression where cover material is obtained within the depression.

> **Quarry or pit technique:** An area method in a depression where the cover material generally is obtained from within the depression.

> **Ramp technique:** An area method where cover is obtained by excavating in front of the working face.

> **Trench technique:** A method where a trench is excavated specifically for

placement of solid wastes and the excavated soil used as cover material.

> **Wet or low-lying area technique:** A method of operating in swampy ground where precautions are made to avoid water pollution before proceeding with area landfill.

SANITARY PLUMBING Internal pipework for taking away discharges from sanitary fittings, which may be separated into soil water and wastewater, and which connects to the drainage system.

SANITARY SEWAGE Waterborne wastes containing biological matter.

SANITARY SEWER Sewer intended to carry only sanitary and industrial wastewater from residences, commercial buildings, industrial plants, and institutions and to which storm, surface and groundwater is not intentionally admitted.

SANITARY SURVEY 1. Investigation of any condition that may affect public health; **2.** Study of conditions and potential hazards associated with the collection, treatment, and disposal of solid, liquid, and airborne wastes.

SANITARY WASTEWATER Public water supply of a community after it has been used and discharged to a system of sewers.

SAPROBE Organism that derives its nourishment from nonliving or decaying matter.

SAPROBIC CLASSIFICATION Classification of river organisms according to their tolerance to organic pollutants.

SAPROGENIC Resulting from decay or putrefaction.

SAPROLITE Clay, silt, or other rock remnants unmoved from the site of disintegration.

SAPROPEL Aquatic sludge rich in organic matter.

SAPROPHAGOUS Feeding on decaying matter.

SAPROPHYTIC BACTERIA *See* **Bacte-**

ria.

SAPROZOIC Pertaining to nutrition by absorption of dissolved organic and inorganic materials, as in protozoans and some fungi.

SAPWOOD Sap-carrying tissue between the bark and the heartwood of most trees.

SATELLITE COMMUNITY Relatively self-contained community, dependent upon an adjacent town or city for many services and facilities.

SATELLITE VEHICLE Small solid waste collection vehicle that transfers its load into a larger vehicle operating in conjunction with it.

SATURATE Fill to capacity or beyond; to cause a solution or compound to be saturated.

SATURATED AIR Condition where air contains all the water vapor it is capable of holding at a given temperature and pressure.

SATURATED FLOW Flow of water through a porous material under saturated conditions.

SATURATED LIQUID Liquid at a given temperature containing as much of a solute as it can retain in solution in the presence of an excess of that solute.

SATURATED ROCK Rock having all of the interstices and void spaces filled with water.

SATURATED SOIL *See* **Soil types.**

SATURATED SOLUTION Solution in which the dissolved solute is in equilibrium with an excess of undissolved solute; or a solution in equilibrium such that at a fixed temperature and pressure, the concentration of the solution is at its maximum value and will not change even in the presence of an excess of solute.

SATURATED ZONE Subsurface zone in which all voids are completely filled with water.

SATURATION 1. Degree of difference of a color from a gray of the same lightness; the degree of a color's chromatic purity; **2.** Relative humidity of 100%.

SATURATION CAPACITY Condition of an ion-exchange column when fully saturated.

SATURATION DEFICIT Difference between the quantity of dissolved oxygen present in a stream or body of water and the quantity of oxygen required to create a condition of saturation.

SATURATION LINE Line through a cross-section of an earth structure used to contain or retain water that marks the uppermost limit of flowthrough.

SATURATION PRESSURE *See* **Pressure.**

SATURATION TEMPERATURE Air temperature at which, for any given water vapor content, the air is saturated; any further temperature reduction results in condensation.

SATURATION ZONE That portion of the lithosphere in which the functional interstices and voids in permeable rock are filled with water under hydrostatic pressure.

SCABLAND Areas where the earth cover over bedrock is nonexistent or so shallow as to cause serious runoff and erosion problems.

SCALE 1. Marked plate against which an indicator or recorder reads; **2.** Solid material precipitated out of water and forming a solid sheath on interior surfaces.

SCARIFICATION 1. Method of seedbed preparation that consists of exposing patches of mineral soil through mechanical action; **2.** Shallow loosening of the soil surface.

SCARIFIER Mobile equipment accessory attachment used to loosen a compacted surface to a shallow depth.

SCARIFY Make scratches or cuts in; break up and loosen a surface.

SCARP Steep slope; a cliff, abrupt declivity.

SCATTERED DEVELOPMENT Unorganized development of land for a variety of purposes.

SCAVENGING Uncontrolled and unauthorized removal of materials at any point in a solid waste management system.

SCHEDULE 1. Sequence of events arranged over time; **2.** List of similar things with special details, e.g. doors or windows and their locations.

SCHEDULED MAINTENANCE Adjustment, repair, removal, disassembly, cleaning, or replacement of components or systems which is performed on a periodic basis to prevent part failure or malfunction.

SCHEDULE OF DEFECTS List of items remaining to be completed before a certificate of practical completion or final certificate can be issued. *Also called* **Inspection list,** and **Punch list.**

SCENIC EASEMENT *See* **Easement.**

SCENIC OVERLOOK Roadside area provided for motorists to stop their vehicles beyond the shoulder, primarily for viewing the scenery in safety.

SCHEDULED FINISH DATE Point in time work is scheduled to finish on an activity.

SCHEDULED MACHINE HOUR *See* **Machine time.**

SCHEDULED NONOPERATING TIME *See* **Machine time.**

SCHEDULED OPERATING TIME *See* **Machine time.**

SCHEDULED START DATE Point in time work is scheduled to start on an activity.

SCHEDULE OF WORK *See* **Construction schedule.**

SCHEDULER Person who organizes the timing of tasks and the flow of materials.

SCHEDULE VARIANCE Difference between the scheduled completion of an activity and the actual completion of that activity.

SCHEMATIC Diagram showing the general principles of construction or operation, usually without accuracy of scale or mechanical representation.

SCHEMATIC DESIGN PHASE Interpretation in graphic form, supported by various types of documentation, of a client's physical requirements for a project.

SCHEMATIC DIAGRAM *See* **Diagram.**

SCHEMATIC WIRING DIAGRAM *See* **Diagram.**

SCHEME Preliminary design and description for the resolution or solution of a proposal.

SCHOKLITSCH FORMULA Mathematical means of determining the volume of water lost to seepage through a dam.

SCINTILLATION COUNTER Radiation detector in which radiations cause individual flashes of light in a scintillator, the intensity of the flash being relative to the energy of the radiation.

SCOPE Sum of the products and services to be provided as a project.

SCOPE CHANGE Any change to the project scope, and which almost always requires an adjustment to the project cost and schedule.

SCOPE CHANGE CONTROL Controlling changes to project scope.

SCOPE DEFINITION Decomposing the major deliverables into smaller, more manageable components to provide better control.

SCOPE OF WORK Work content of a project or any component of a project, such as a work package of a cost class. Scope is fully described by naming all activities performed, the end product(s) that result, and the resources consumed.

SCOPE PLANNING Developing a written scope statement that includes the project justification, major deliverables, and project objectives.

SCOPE VERIFICATION Ensuring that all

identified project deliverables have been completed satisfactorily.

SCORIA Dark-colored, pyroplastic material ejected by a volcano.

SCOUR 1. Erosion caused by fast-flowing water containing abrasive particles or solids; **2.** Removal of sand, earth, or silt from the bottom or banks of a river.

SCOURING SLUICE Opening in the foot of a dam, regulated by a gate, through which accumulated silt, etc., may be ejected.

SCOURING VELOCITY Minimum flow velocity necessary to dislodge stranded material from within a waterway, channel, or pipeline.

SCOUR PROTECTION Mechanical means (sheet piling, riprap, revetments, etc.) to protect submerged organic silts.

SCOW END Raised portion of the floor in the rear of a dump body, particularly on rock bodies, that replaces the conventional tailgate in preventing load spillage.

SCRAP Material or articles discarded as useless and fit only to be broken down, melted, etc. and reprocessed.

SCRAPER 1. Device that can be pushed or pulled through a pipeline so as to dislodge accumulated organic and/or mineral deposits; **2.** Device installed in the bottom of a sedimentation tank and used to move settled sludge to a discharge port; **3.** Blade used to remove accumulated matter from filter or screen surfaces; **4.** Digging, hauling, and grading machine equipped with a cutting edge, carrying bowl, movable front wall, and a dumping mechanism, used to dig, transport, and spread soil. *Also called* **Pan,** and **Tractor scraper.** There are a number of designs, including:

> **Auger:** Self-loading system incorporating a hydraulically-powered auger located in the center of the scraper bowl. As material flows over the scraper's cutting edge, it is lifted by the auger.

> **Bottom-dump:** Carrying scraper that ejects its load over the cutting edge.

Drag: 1. Digging and hauling equipment consisting of a bottomless bucket working on the cable of a mast and anchor, corresponding to headworks and tailworks; **2.** Towed bottomless scraper used for levelling and maintaining haul roads.

Rear-dump: Two-wheel scraper that discharges its load at the rear.

Self-powered: Tractor and scraper built and operating as a single unit.

Two-axle: Scraper mounted on or built into a full trailer.

SCRAP LOADER Machine equipped with a magnet or grapple attachment for handling metal in scrap yards.

SCREEFING Removing weeds and small plants, together with most of their roots, from the area immediately surrounding a planting hole.

SCREEN 1. Device with rows of uniform-size openings, used to trap and retain floating or suspended matter in flowing water or wastewater; **2.** Barrier, usually vertical, fabricated of a wide range of materials and in a variety of forms, from solid embankments to open-weave mesh, each appropriate to its taks of providing a visual or acoustic baffle; **3.** Production equipment for separating granular material according to size using woven-wire cloth or other similar device with regularly spaced apertures of uniform size. There are many types, including:

> **Deck:** Two or more screens, commonly vibrating, placed one above the other, used to classify the same run of material.

> **Rotary:** An inclined, meshed cylinder that rotates on its axis and screens material placed in its upper end.

> **Scalping:** Vibrating grizzly.

> **Shaking:** Screen that is moved back-and-forth, or in rotary motion, causing material to move along it or through it.

> **Vibrating:** An inclined screen that is

vibrated to move material along it and/or through it.

SCREEN ANALYSIS *See* **Sieve analysis.**

SCREEN CHAMBER Place where a fixed or mobile screen is installed.

SCREENED MATERIAL Material that has passed through or over a screen.

SCREENED WELL Water well in which water enters the bore through screens (as distinct to, or in addition to, perforations in the casing).

SCREENING Process of removing floating or suspended solids from water or wastewater.

SCREENINGS Undersized or oversized rejects from a screening process.

SCREENINGS GRINDER Mechanism for shredding or macerating material removed from wastewater screens to reduce its size.

SCREW AUGER Auger with a threaded, sharp point at its lower end.

SCREWED PIPE Pipe having a male thread on each end, joined by means of a threaded sleeve, coupling, or fitting.

SCREW-FEED PUMP Pump equipped with a helix rotating within a casing to either move or compress material fed to its inlet.

SCREW IMPELLER Helical impeller of a screw pump.

SCRUB Low, stunted trees or shrubs.

SCRUBBER Equipment for removing fly ash and other objectionable materials from the products of combustion by means of sprays, wet baffles, wetted packing, etc. Also reduces elevated temperatures.

SCRUBBING Washing of impurities from any process gas stream.

SCUM Filmy layer of extraneous or impure matter that forms on or rises to the surface of a liquid.

SCUM BAFFLE Board that dips below the surface at the discharge end of a tank to prevent the passage of floating matter.

SCUM BOARD Baffle that dips below the surface at the discharge end of a tank to prevent scum from floating out.

SCUM BREAKER Mechanical device used to break up accumulated scum in a sludge-digestion tank.

SCUM CHAMBER Area within a sludge-digestion tank where scum rising as part of the digestion process is collected.

SCUM COLLECTOR Mechanical device that continuously skims the surface of a wastewater treatment tank to collect and remove accumulated scum.

SCUM TROUGH Channel into which scum is deposited for collection and transport to a separate processing stage.

SCUPPER 1. Opening in the base of a wall or parapet that allows surface water to drain away; **2.** Catch basin at the low point of a bridge slab to collect and drain water.

SEA Any large body of salt water, smaller than an ocean, partly or wholly enclosed by land.

SEAFLOOR SPREADING Result of new material from the earth's mantle being added to the Earth's crust at the site of oceanic ridges.

SEAL 1. To permanently close off; **2.** Material used to cover or close the joint between materials to prevent passage of dust, moisture, wind, etc.; **3.** Depth of water held in a trap to prevent the passage of air or gas; **4.** Airtight or watertight joint; **5.** Gland, stuffing box, or similar mechanical device between a static and moving part.

SEALABLE EQUIPMENT Equipment enclosed in a case or cabinet that is provided with a means of sealing or locking so that electrically live parts cannot be made accessible without opening the enclosure.

SEALANT An elastomeric material with adhesive qualities that joins components of similar or dissimilar nature to provide an effective barrier against the passage of the elements.

SEALANT RESERVOIR Cavity, indentation, channel, or formed joint into which a sealant is placed.

SEAL CASE Rigid member to which a seal lip is attached.

SEALED BEARING Bearing with seals on both sides to retain lubricant within the device and prevent the ingress of foreign matter.

SEALED BID Contract for which interested parties submit written bids at the time and place specified.

SEALED HYDRAULIC SYSTEM Leakproof hydraulic system with static oil pressure.

SEALED SYSTEM Pipe circuit for low-pressure hot-water heating that contains a constant volume of water, kept separate from air, and consequently exhibiting a low rate of corrosion.

SEA LEVEL Level of the ocean's surface; especially, the mean level halfway between high and low tide, used as a standard in reckoning land elevations or sea depths.

SEALING WATER 1. Water either positioned, or under sufficient pressure to prevent the passage of sewer gas, or the backflow of water; **2.** Water used to prevent contaminated water from reaching a mechanical device.

SEAMLESS CARBON-STEEL PIPE *See* **Pipe.**

SEAMLESS PIPE *See* **Pipe.**

SEA MOUNT Isolated, usually conical, submarine mountain rising at least 1000 m (3280 ft) above the sea floor.

SEASON One of the four equal natural divisions of the year: spring, summer, autumn (fall), and winter, indicated by the passage of the Sun through an equinox or solstice and derived from the apparent north–south movement of the Sun caused by the fixed direction of the Earth's axis in solar orbit.

SEASONAL DEPLETION Withdrawal of ground- or surface water at a rate beyond that of seasonal supply, but not in excess of the average supply over a secular cycle; a condition preceding seasonal recovery.

SEASONAL EFFICIENCY Ratio between solar energy collected and used to that striking a solar collector, measured over a heating season.

SEASONAL RECOVERY Cyclical recharge of groundwater and the accompanying rise in the elevation of the watertable during or after a wet season and following seasonal depletion.

SEASONAL STORAGE Water diverted to, or retained in a reservoir at times when the available supply exceeds the average demand.

SEAWALL Work constructed along a shore line consisting of loose mounds or heaps of rubble, or masonry walls supplemented with treated timber, steel, or reinforced concrete sheet piling driven into the beach and strengthened by wales and guide- and brace-piles. Intended as a barrier to prevent the encroachment of the sea upon land by direct wave action. *See also* **Dike,** and **Pile bulkhead.**

SEA WATER Raw water as found under natural conditions in the sea and generally containing from 33,000 to 36,000 mg/L of total dissolved solids.

SECCHI DISC Flat, white disc of specific diameter and thickness, lowered into the water by a rope until it is just visible, which gives indication of the transparency of the water.

SECOND 1. SI unit of time equal to the duration of 9 192 631 770 periods of the radiation corresponding to the transition between two hyperfine levels of the ground state of the caesium-133 atom. Symbol: s. Derived units are the hertz (symbol: Hz), a unit of frequency representing one period (cycle) per second; speed or velocity, measured in meters per second; and acceleration, measured in meters per second per second. *See also the appendix*: **Metric and nonmetric measurement**; **2.** One of the divisions of a minute of an angle or a minute of time into 60 equal parts; **3.** Second-quality material.

SECONDARY AIR Air introduced during the combustion process to promote further burning and complete combustion of volatile material.

SECONDARY BRANCH Branch off the primary branch of a drain.

SECONDARY CIRCULATION Small-diameter, insulated pipe used to constantly circulate hot water in a large system serving many appliances or appliances that are widely separated and from which short dead legs connect to individual faucets.

SECONDARY CONSOLIDATION Reduction in volume of a soil mass by the temporary application of a sustained load to the mass resulting in an adjustment of the internal structure of the soil by the displacement of water and a closer alignment of the solid particles.

SECONDARY EMISSION CONTROL SYSTEM Combination of equipment used for the capture and collection of secondary emissions (e.g. an open hood system for the capture and collection of primary and secondary emissions, with local hooding ducted to a secondary emission collection device such as a baghouse).

SECONDARY EMISSIONS Emissions that occur as a result of the construction or operation of an existing stationary facility but which do not come from the existing stationary facility.

SECONDARY ENRICHMENT Natural enrichment of an ore by material of later origin, deposited from ascending aqueous solutions.

SECONDARY FILTER Second stage of a multistage filter system.

SECONDARY INTERSTICE Interstice in rock that developed after the rock was formed.

SECONDARY LIQUID FUEL Organic chemicals no longer suitable for their original use or recovery, blended to form a combustible fuel that may be mixed with conventional fuels in certain industrial processes.

SECONDARY MATERIALS 1. Materials that have fulfilled their primary function and are available for another use; **2.** Useful materials produced as a by-product of a manufacturing or fabricating process.

SECONDARY POLLUTANT Pollutant formed as a result of chemical combinations and changes resulting from exposure to an external influence, i.e. photochemical changes resulting from exposure to sunlight.

SECONDARY PROCESS Where components separated from solid waste may be further processed to allow reuse in their original form or use in an entirely different form.

SECONDARY RECYCLING Use of a secondary material in an industrial application other than that in which the material originated.

SECONDARY SEDIMENTATION Sedimentation that follows secondary, or biological treatment of sewage.

SECONDARY SETTLING TANK Vessel through which the effluent from some prior sedimentation process flows for the purpose of removing settleable solids.

SECONDARY SOILS Previously existing organic soils and rock debris that has been transported to its present location through geologic and hydrologic processes.

SECONDARY (SEWAGE) TREATMENT *See* **Sewage treatment.**

SECONDARY USE Use of a material in an application other than that in which it originated; however, the material is not changed significantly by processing and retains its identity.

SECOND-FOOT In hydraulics, a unit of flow; one cusec (ft^3/sec).

SECOND OF ARC Non-SI unit of angle permitted for use with the SI system of measurement. Symbol: ". *See also the appendix:* **Metric and nonmetric measurement.**

SECOND-STAGE BIOCHEMICAL OXYGEN DEMAND Oxygen required for the biochemical oxidation of nitrogenous material, following satisfaction of the first-stage demand necessary for the oxidation of carbonaceous material.

SECRETION Substance that is secreted by some part of an animal or plant.

SECTION 1. View obtained at an intersecting plane surface; **2.** Shape produced in a continuous process to a definite cross section that is small in relation to its length; **3.** Area equal to 259 ha (640 a or 1 mile^2); **4.** Portion of an area being worked; **5.** Subsubdivision of a specification dealing with the work of a single trade.

SECTIONALIZING VALVE One of several valves installed in the main lines of a water distribution system so as to allow the isolation of specific portions for inspection and/or repair.

SECTIONAL TANK Storage vessel made up of modular panels.

SECTOR GATE Roller-type crest gate in which the roller consists of sectors of a circle instead of a cylinder.

SECULAR Long time – up to, and even beyond, a century.

SECULAR CYCLE Period of a continuous group of years during which precipitation is considerably above, or below average.

SECURITY FENCE Fence of sufficient height, and of materials designed to prevent unauthorized entry.

SEDENTARY SOIL Soil remaining in the place at which it was formed.

SEDIMENT Regolith that has been transported and deposited by water, air, or ice; predecessor of sedimentary rocks.

SEDIMENTARY GROIN Groin developed slowly from the river or sea bed.

SEDIMENTARY ROCK Rock that results from the decomposition, disintegration, and deposition of preexisting rocks, minerals, plants, and animals.

SEDIMENTATION Process for removal of solids, before filtration, by gravity or separation; act or process of depositing sediment.

SEDIMENTATION TANK Treatment tank sized so that the flow passing through it is slowed to the point where suspended sediment will gravitate to the bottom, where it can be removed by scraping or pumping. *Also called* **Desilting basin.**

SEDIMENT TRAP Device, with a means for evacuating, educting, or cleaning, placed in a conduit to slow the flow and allow sediment to precipitate and collect.

SEEDING MATERIAL Digested sewage sludge that is used to 'prime' or 'seed' sludge digestion tanks after emptying, etc.

SEEPAGE Water percolating through a soil deposit or soil structure.

SEEPAGE BOIL Uprising of water on the landward side of a dike, dam or levee, the result of infiltration.

SEEPAGE COLLAR Projecting collar around a pipe, tunnel, or conduit under an embankment dam that lengthens the seepage path along the exterior of the conduit.

SEEPAGE FORCE *See* **Capillary pressure.**

SEEPAGE LOSS Loss of water through the banks of a canal.

SEEPAGE PIT Below-grade soakaway: pit filled with large granular material. It will fill with water which will then slowly soak into the surrounding soil.

SEEPAGE SPRING Outflow of water resulting from percolation from numerous small openings in permeable material.

SEEPAGE TRENCH One or more interconnected small trenches excavated at a slight gradient and filled with crushed stone, and used to distribute effluent from a treatment facility.

SEGMENTAL SLUICE GATE Radial hydraulic gate.

SEGREGATED STORMWATER SEWER SYSTEM Drainage and collection system designed and operated for the sole purpose of collecting rainfall runoff at a facility, and which is segregated from all other individual drain systems.

SEICH Wave that oscillates in lakes, bays, or gulfs from a few minutes to a few hours as a result of seismic or atmospheric disturbances.

SELECTIVE BIDDING System of restricted bidding for a contract whereby those being invited to bid are selected according to some criteria (prequalification, previous experience, etc.). *Also called* **Closed bidding, Invitational bidding,** and **Limited tendering.**

SELECTIVE DIGGING Separating two or more types of soil while excavating them.

SELECTIVE MEMBRANE Sheet material that is preferentially selective to passage of either cations or anions in solution.

SELECTIVITY 1. Property of a circuit, instrument, etc., by virtue of which it responds to electric oscillations of a particular frequency; **2.** Characteristic of plants and insects to adapt to their environment by means of genetic selection; **3.** Characteristic of radiation, toxic chemicals and heavy metals to affect certain functions and organs of the body to a much greater extent than the body as a whole.

SELECTOR Reactor or basin consisting of a number of compartments, the environment and resulting microbial population of which can be controlled, used to select specific organisms to produce predictable results or conditions.

SELECTOR RECYCLE Recycling of return sludge or oxidized nitrogen to provide a desired environment.

SELENIUM Nonmetallic, toxic chemical element (Se) resembling sulfur in chemical properties and having an atomic number of 34 and an atomic weight of 78.96. An essential micro-nutrient, the maximum acceptable concentration of 0.01 mg/L has been established for selenium in drinking water.

SELF-CLEANSING GRADIENT Gradient at which the periodic flow in a pipe is sufficient to prevent the deposition of solids along its invert.

SELF-CLEANSING VELOCITY Minimum flow in a drain pipe necessary to initiate a scrubbing action.

SELF-CONTAINED BREATHING APPARATUS Device enabling an individual to have air or oxygen independent of the atmosphere in which he is working.

SELF-LOADING Capability of a powered industrial truck to pick up, carry, and deposit its load without the aid of external handling means.

SELF-MULCHING SOIL Surface layer sufficiently well aggregated that it does not crust or seal under the impact of rain; a soil that breaks into dust on cultivation, forming a mulch.

SELF-PRIMING CENTRIFUGAL PUMP *See* **Pump.**

SELF-PURIFICATION Natural process of flowing water that results in the deposition of sediments, reduction of bacteria, satisfaction of the biochemical oxygen demand, stabilization of organic constituents, replacement of depleted dissolved oxygen, and return of the stream biota to normal.

SELF-SIPHONING Condition where a partial vacuum in an unventilated or inadequately ventilated plumbing system causes the seal to be drawn from a trap(s), allowing the passage of air or gases through the fitting. *Also called* **Siphonage.**

SEMIARID Land that is neither humid nor arid, with a tendency toward dryness.

SEMIANNUAL Happening or issued twice a year.

SEMIAQUATIC Adapted for living or growing in or near water.

SEMIAUTOMATIC BATCHER *See* **Batcher.**

SEMIBOLSON Topographic basin having a centripetal or inward-flowing drainage basin and drained by an intermittent stream.

SEMIDIURNAL Occurring or coming once every 12 hours.

SEMIHYDRAULIC-FILL DAM Earth dam, some of which material has been transported by water.

SEMIPERCHED Groundwater, or a watertable, having a greater pressure head than an underlying body of water not separated from it by an unsaturated rock.

SEMIPERMANENT SNOW LINE Elevation of the lower limit of a snow field, above which snow accumulates and remains on the ground.

SEMIPERMEABLE MEMBRANE Material that is permeable only in one direction.

SENSING UNIT Device that senses the condition, state, or position of something.

SENSITIVE SWITCH *See* **Switch.**

SENSITIVENESS Responsiveness of an instrument to the signal or condition being recorded or monitored.

SENSITIVITY Responsiveness to change.

SENSITIVITY LEVEL Least concentration that can be detected and quantified by an instrument or test method.

SENSOR Device designed to respond to a physical stimulus (as temperature, illumination, motion, etc.) and transmit a resulting signal for interpretation, measurement, or for operating a control.

SEPARATE COLLECTION Collecting recyclable materials that have been separated at the point of generation and keeping those materials segregated from other collected solid waste in separate compartments of a single collection vehicle or through the use of separate collection vehicles.

SEPARATE SEWER Sewer intended to receive only wastewater.

SEPARATE SYSTEM Drainage system in which rainwater and sewage are carried in separate pipes.

SEPARATE SLUDGE DIGESTION Process of digesting sludge in tanks that are separate from those in which it had been allowed to accumulate and settle.

SEPARATION Tendency for solids to separate from water by gravitational settlement.

SEPARATION LINE Interface of clarified liquid and the underlying sludge blanket in a solids-contact unit.

SEPARATOR 1. Component placed between membranes in a desalting stack to form flow passages and promote fluid turbulence; **2.** Device whose primary function is to isolate contaminants by physical properties other than size. There are several types, including:

Absorbent: Separator that retains certain soluble and insoluble contaminants by molecular adhesion.

Ballistic: Device that drops mixed materials having different physical characteristics onto a high-speed rotary impeller; they are hurled off at different velocities and land in separate bins.

Centrifugal: Separator that removes nonmiscible fluid and solid contaminants that have a different specific gravity than the fluid being purified by accelerating the fluid in a circular path and using the radial acceleration component to isolate these contaminants.

Coalescing: Separator that divides a mixture or emulsion of two nonmiscible fluids, using the interfacial tension between the two liquids and the difference in wetting of the liquids on a particular porous medium.

Electrostatic: Separator that removes contaminants from dielectric fluids by applying an electrical charge to the contaminant that is then attracted to a collection device of different electrical charge.

Inertial: 1. Device that relies on ballistic or gravity separation of materials having different physical characteristics; **2.** Device that removes particles from a gaseous stream by imparting a centrifugal motion with fixed mechanical parts.

Magnetic: Separator that uses a magnetic field to attract and hold ferromagnetic particles.

Vacuum: Separator that utilizes subatmospheric pressure to remove certain gases and liquids from another liquid because of their difference in vapor pressure.

SEPARATOR ENTRAINMENT Removal of entrapped droplets from the vapor stream of a desalinization device.

SEPARATOR TANK Device used for separation of two immiscible liquids.

SEPTAGE Aerobic wastewater originating from a domestic source (residential, commercial, or industrial facility) that is not hazardous and which is compatible with biological wastewater treatment processes.

SEPTIC BED Area adjacent to a septic tank in which an underground network of perforated pipes distributes the effluent from the tank into the surrounding soil.

SEPTICITY Condition produced by the growth of anaerobic organisms.

SEPTIC SEWAGE Sewage from which all dissolved oxygen has been absorbed by the organic matter.

SEPTIC SLUDGE Undigested or partially digested sludge.

SEPTIC TANK Tank, embedded in the earth, into which sewage is allowed to drain, divided into compartments and of an internal configuration that promotes anaerobic decomposition of organic wastes suspended in water, and discharging to a tile bed or drainage field; sludge from settled solids is retained for sufficient time to secure satisfactory decomposition of organic solids by bacterial action.

SEPTIC TREATMENT SYSTEM Small-scale sanitary treatment system consisting of a septic tank and tile bed in which sewage flows are subjected to bacterial treatment in an anaerobic environment. The resulting sludge settles to the bottom of the tank, the liquid effluent decants over a weir to the tile bed, and a foamy by-product of the bacterial action produces a blanket within the tank that prevents oxygen from approaching the contents.

SEPTIC WASTEWATER Anaerobic wastewater undergoing putrefaction.

SEQUENCE Order in which operations take place or are scheduled.

SEQUENCE ARROW Indicator of an arrow diagram.

SEQUENCE VALVE *See* **Valve.**

SEQUENT DEPTHS Elevations before and after a hydraulic jump.

SEQUENTIAL STARTING Starting a number of electric motors drawing power from the same source in an order that smooths the starting load on the circuit.

SEQUESTERING AGENT Chemical that causes the coordination complex of certain phosphates with metallic ions in solution so that they may no longer be precipitated.

SEQUESTRATION Inhibition or prevention of normal ion behavior by combination with added materials, especially the prevention of metallic ion precipitation from solution by formation of a coordination complex with a phosphate.

SERIAL BONDS Bonds having an established date of maturity.

SERIAL DISTRIBUTION Group of absorption trenches so arranged that the total effective absorption of one is utilized before the liquid flow into the next.

SERIES Group or number of similar things arranged in a row.

SERIES OPERATION Operation of a multistage centrifugal pump in which the first impeller provides its water volume and pressure to the second impeller, thus building pressure until the final impeller delivers the same volume of water at increased pressure to the discharge.

SERIOUS ACUTE EFFECTS Human injury or human disease processes that have a short latency period for development, resulting from short-term exposure to a chemical substance, and which are likely to result in death or severe or prolonged incapacitation, disfigurement, or severe or prolonged loss of the ability to use a normal bodily or intellectual function with a consequent impairment of normal activities.

SERIOUS CHRONIC EFFECTS Human injury or human disease processes that have a long latency period for development, result from long-term exposure to a chemical substance, or are a combination of these factors, and which are likely to result in death or severe or prolonged incapacitation, disfigurement, or severe or prolonged loss of the ability to use a normal bodily or intellectual function with a consequent impairment of normal activities.

SERULE Miniature succession composed of minute or microscopic organisms engaged in the breaking down of organic matter to its simpler constituents.

SERVICE Water or sewer pipes, electrical or gas lines that connect public services to a building.

SERVICE AGE Time since a physical unit was commissioned.

SERVICE AREA Geographic area over which service will be provided under specific conditions. This includes permanent and fixed service, such as water supply or electrical energy, as well as mobile service, such as that offered on site to owners of machinery and equipment.

SERVICE BOX *See* **Curb box.**

SERVICE CHARGE Basic charge for providing service, to which a consumption charge may be added.

SERVICE CLAMP Pipe saddle.

SERVICE CONNECTION Water connection between a street main and the building meter, or the building line in an unmetered connection.

SERVICED LOT Surveyed lot provided with domestic water, sanitary sewers, storm drainage and proximity to electrical distribution lines, plus a curb crossing where roads and curbs have been developed. *Also called* **Fully-serviced lot.**

SERVICE ELL *See* **Street elbow.** *Also called* **Service L.**

SERVICE L *See* **Street Elbow.**

SERVICE LIFE 1. Period for which a product is guaranteed, warranted, or can be reasonably expected to perform in the manner for which it was designed; **2.** In fluid filtration, the length of time that a filter will survive in an actual system before the terminal differential pressure is reached.

SERVICE METER Recording meter installed on a service branch to register the amount consumed.

SERVICE PIPE Water or gas supply pipe connecting the public supply main to the building.

SERVICE PRESSURE *See* **Pressure.**

SERVICE RESERVOIR Reservoir sited within a service area and capable of storing water during hours of low demand in preparation for periods when demand exceeds the available supply.

SERVICE ROOM Room or space provided in a building to accommodate building service equipment such as air-conditioning or heating appliances, electrical services, pumps, compressors, and incinerators.

SERVICES Systems comprising equipment, pipes, cables, ducts, fittings, appliances, and appurtenances that supply drainage, water, steam, electricity, gas, and communication services to a building or structure.

SERVICE SPACE Space provided in a building to facilitate or conceal the installation of building service facilities such as chutes, ducts, pipes, shafts, or wires.

SERVICE SWITCH *See* **Switch.**

SERVICE SYSTEMS Heating, ventilating, air conditioning, water supply, electrical supply, and gas supply to a building.

SERVICE T *See* **Street T.**

SERVICE TEST Test in which a product is used under actual service conditions. *Also called* **Performance test.**

SERVICE TIME *See* **Machine time.**

SERVICING VALVE *See* **Valve.**

SERVOCIRCUIT Closed-loop circuit that is controlled by some type of feedback, i.e. the output of the system is sensed or measured and is compared with the input. The actual output and the input control the circuit. The system output may be position, velocity, force, pressure, level, flow rate, temperature, etc. *See also* **Circuit.**

SERVOCONTROL Control actuated by a feedback system that compares the output with a reference signal and makes corrections to reduce the difference.

SERVOPISTON Piston that moves another part on command from a servovalve.

SERVOVALVE *See* **Valve.**

SERVOVALVE OVERLAP Lap condition that results in a decreased slope of the normal flow curve in the null region.

SERVOVALVE UNDERLAP Lap condition that results in an increased slope of the normal flow curve in the null region.

SERVOVALVE NULL Condition where the servovalve supplies zero control flow at zero load pressure drop.

SET Direction of the flow of water.

SETBACK LINE Line outside the right of way, established by public authority or private restriction, on the highway side of which the erection of buildings or other permanent improvements is controlled.

SET POINT Position at which a control or controller is set.

SETTLEABILITY TEST Determination of the ability of solids in suspension to settle under measurable conditions.

SETTLEABLE SOLIDS Solid fraction of a wastewater that will not remain in suspension following a specified period under prescribed conditions.

SETTLED WASTEWATER Wastewater from which most of the settleable solids have

been removed by sedimentation.

SETTLING Gravity process that promotes the subsidence and deposition of suspended matter carried by water.

SETTLING CHAMBER Any chamber designed to reduce the velocity of a material (fluid, gas, products of combustion, etc.) to promote the settlement of particulate matter.

SETTLING POND Natural or artificial depression where water is held long enough to allow the precipitation of particulates.

SETTLING TANK Container that gravimetrically separates oils, grease, and dirt, together with the piping and ductwork used in its installation.

SETTLING VELOCITY Terminal rate of fall of a particle through a fluid, or a given dust from a dust-laden gas, as induced by gravity or other external force.

SEVEN-DAY AVERAGE Arithmetic mean of pollutant parameter values for samples collected in a period of seven consecutive days.

SEVERE EXPOSURE Conditions of high wind, heavy rain, heavy frost or snowfall.

SEWAGE Liquid waste that contains animal, mineral, or vegetable matter in suspension or solution, conveyed in a sewer.

SEWAGE CHARGE Service charge levied for providing wastewater collection and/or treatment service.

SEWAGE COLLECTION SYSTEM Common lateral sewers, within a publicly owned treatment system, that are primarily installed to receive wastewaters directly from facilities, which convey wastewater from individual structures or from private property, and which include service connection 'Y' fittings designed for connection with those facilities.

SEWAGE DISPOSAL Process of collecting, conveying, and treating waterborne organic wastes originating from domestic, commercial, and industrial sources, but not necessarily including industrial wastes, or stormwater runoff.

SEWAGE DISTRIBUTOR Mechanism that applies sewage, or sewage effluent to the top of a filter.

SEWAGE EJECTOR Device for raising sewage to a higher elevation by entraining it in a high-velocity jet of water or air.

SEWAGE FROM VESSELS Human body wastes and the wastes from toilets and other receptacles intended to receive or retain body wastes that are discharged from marine vessels, except that with respect to commercial vessels on the Great Lakes. This term includes graywater (galley, bath, and shower water).

SEWAGE GAS By-product of bacterial action on organic material, predominantly methane.

SEWAGE PUMP *See* **Pump.**

SEWAGE RATE Charge made to users of a wastewater collection and treatment system based on the volume and composition of the wastewater they discharge to the system.

SEWAGE SLUDGE Solid, semisolid, or liquid residue removed during the treatment of municipal wastewater or domestic sewage. Sewage sludge includes, but is not limited to, solids removed during primary, secondary, or advanced wastewater treatment, scum, septage, portable toilet pumpings, type III marine sanitation device pumpings, and sewage sludge products. Sewage sludge does not include grit or screenings, or ash generated during the incineration of sewage sludge.

SEWAGE TREATMENT Processes that reduce and/or eliminate the organic and pathogenic content of raw sewage to produce a hazard-free effluent that will not impose a burden upon the receiving water. The treatment is divided into several stages, not all of which may be necessary, including:

Primary: Straining and settlement to remove the largest of the floating and suspended solids.

Secondary: Introduction by various means of oxygen in sufficient quantities to cause oxidation of the organic fraction and reduction of the biochemical oxygen demand, followed by settling and removal of the resulting sludge.

Tertiary: Chemical dosing to causing any remaining suspended particles to floc and settle, followed by filtration.

Quarternary: Chemical treatment to promote the precipitation, oxidation, or conversion of various undesirable chemical fractions.

Sterilization: Injection of a sterilant such as chlorine, ozone, bromine, etc., in sufficient quantities to ensure the elimination of any remaining pathogens.

SEWAGE TREATMENT PLANT Processing facility that receives untreated waterborne wastes from a community, and that discharges treated effluent into a receiving body of water and treated sludge cake to a disposal area.

SEWAGE TREATMENT RESIDUE Coarse screenings, grit, and dewatered or air-dried sludge solids from sewage treatment plants, or pumpings of cesspool or septic tank sludges, which require disposal as putrescible wastes.

SEWAGE TREATMENT WORKS Municipal or domestic waste treatment facilities of any type; a combination of facilities for collecting, pumping, treating and disposing of sewage that are publicly owned or regulated to the extent that feasible compliance schedules are determined by the availability of public funding.

SEWER Pipe or other construction, and its associated manholes, vents, and other appurtenances, forming part of a sewerage system.

SEWERAGE System for the collection and conveyance of surface water and wastewater from an area.

SEWER APPURTENANCES Structures, equipment, fittings, devices, etc., excepting pipe or conduit, associated with a sewer system.

SEWER BRICK *See* **Brick.**

SEWER CONNECTION Connection of

drainage pipes to a sewer.

SEWER DISTRICT Land area, or organization, created for the purpose of collecting, treating, and disposing of wastewater.

SEWER GAS Mixture of gases, odors, and vapors found in a sewer and emanating from sewage, some of which, and in certain concentrations, may be explosive and/or poisonous.

SEWER LINE Lateral, trunk line, branch line, ditch, channel, or other conduit used to convey wastewater to downstream components of a wastewater treatment system.

SEWER MANHOLE Shaft or chamber giving access for the maintenance or inspection of a sewer.

SEWER OUTFALL Point of discharge of final effluent from a sewer.

SEWER OUTLET Hydraulic structure through which flows the final effluent of a sewage treatment plant.

SEWER PIPE Any of several types of pipe used to convey sewage.

SEWER ROD One of a series of interconnecting rods used to propel an object through a sewer to dislodge an obstruction.

SEWER SYSTEM All of the interconnected pipes, structures, appurtenances, equipment and plant installed to collect, transport, treat, and dispose of wastewater.

SEWER SYSTEM EVALUATION SURVEY Systematic examination of tributary sewer systems or their subsections to determine the location, flow rate, and cost for correction for each definable element of a demonstrated total infiltration and inflow problem.

SEWER TAP Wye, saddle or other device placed on a public sewer to receive a building connection.

SEWER TILE *See* **Tile.**

SEWER TRAP *See* **Trap.**

SEWER UTILITY Company or entity formed to install and administer wastewater collection and disposal.

SHAFT HORSEPOWER *See* **Horsepower.**

SHAFT SPILLWAY Vertical shaft, flared at the top and bottom, through which overflow water from a reservoir is discharged.

SHALE *See* **Soil types.**

SHALLOW MANHOLE Inspection chamber not containing branch connections and of the same cross-section for all of its depth.

SHALLOW-WATER DEPOSIT Ocean sediments at depths up to 100 fathoms (600 ft, 200 m).

SHALLOW-WATER WAVE Progressive gravity wave in water having a depth of up to one half of the wave's length.

SHALLOW WELL Well less than 6 m (20 ft) deep.

SHARP-CRESTED WEIR Measuring weir topped with a thin metal plate.

SHARP-EDGED ORIFICE Orifice through which water will touch only the line of the edge.

SHEAR Machine or attachment used to cut sheet metal, often to reduce large waste items to a size suitable for handling, transportation and/or smelting.

SHEAR GATE Pivoted slide, held closed in one direction only by the pressure of water and the seating lugs.

SHEAR SHREDDER Size reduction machine that cuts material between two large blades or between a blade and a stationary edge.

SHEAR SLIDE Landslide where a mass of earth separates and moves down a slope.

SHEAR STRESS Force per unit area in fluid flow tending to produce shear, generated at the surface as a result of turbulence and viscosity.

SHEEP'S FOOT Projection from the drum of a compactor that leaves an imprint resem-

423

bling that of a sheep's foot. *Also called* **Foot.**

SHEEP'S FOOT ROLLER Compacting roller with feet expanded at its outer face, used in compacting soil. The sheeps foot pad is cylindrical, usually 203 mm (8 in.) long, and ranges in size from 0.45 to 1.16 m^2 (7 to 18 in.^{2}). *See also* **Tamping foot,** and **Pad foot.**

SHEET DRAINAGE Where water flowing on the surface stays in a thin layer, uniformly covering the entire area.

SHEETED PIT Excavation where all the sides have been sheeted.

SHEET EROSION Lowering of land elevation by the almost uniform removal of soil particles by flowing water.

SHEETERS Light steel vertical polling boards for protecting trench sides, driven prior to excavation.

SHEET FLOW Flow in the form of a relatively thin sheet of generally uniform thickness.

SHEETING 1. Members of a shoring system that retain the earth in position and which in turn are supported by other members of a shoring system; **2.** Tongue-and-groove board, 22 mm (0.875 in.) thick, used in shoring and bracing.

SHEETING DRIVER Air hammer attachment that encloses plank ends, or shaped steel sections, enabling them to be driven without splintering.

SHEETING JACK Push-type turnbuckle, used to set trench bracing.

SHELL ICE Ice under which the water has receded. *Also called* **Cat ice.**

SHELTERBELT Windbreak, usually a stand of trees.

SHIELD Structure that is able to withstand the forces imposed on it by a cave-in and thereby protect workers within a structure. *Also called* **Trench box.**

SHINGLE BEACH Self-draining foreshore covered with stones of more-or-less equal size.

SHOCK LOAD Introduction to a process of a body of material having characteristics in excess of those which the process was designed to handle.

SHOCK WAVE Travelling wave in which the thermodynamic properties of the air change suddenly to a different value.

SHOE TILE Box towed behind a ditcher from which tile can be laid in the ditch bottom.

SHORE 1. Land bordering a body of water; **2.** Lumber used as a prop, support or brace.

SHORE CURRENT Current in water adjacent to a shoreline, caused by waves striking at an oblique angle.

SHORE DRIFT Sand, pebbles and other material carried along the shoreline by shore currents.

SHORE EROSION Wind or water action that results in the removal of sand and other material from the land adjacent to the shoreline.

SHORE HEAD Wood or metal horizontal member placed on and fastened to vertical shoring members. *See also* **Raker.**

SHORELINE Zone along the edge of a body of water where the land surface comes in contact with wave action above and below the surface of the water.

SHORE PROTECTION Various types of construction aimed at protecting the shore at its waterline.

SHORING 1. Materials used to support a vertical, or near-vertical face cut in earth as part of an excavation. *See also* **Bracing,** and **Horizontal shoring**; **2.** Props or posts of timber or other material in compression used for the temporary support of formwork, or unsafe structures; **3.** Support for structures, especially underpinning; **4.** Process of erecting shores.

SHORING LAYOUT Drawing prepared prior to erection showing the arrangement of equipment for shoring.

SHORT CIRCUIT Hydraulic condition where the time of travel of water through a tank is less than the flowing-through time.

SHORT NIPPLE Pipe nipple having a total length only slightly greater than that of the two threaded ends.

SHORT PIPE Pipeline shorter than 500 times its diameter, for which special hydraulic factors apply.

SHORT-TERM TEST INDICATIVE OF THE POTENTIAL TO CAUSE A DE-VELOPMENTALLY TOXIC EFFECT Either any *in vivo* preliminary development toxicity screen conducted in a mammalian species or any *in vitro* developmental toxicity screen, including any test system other than the intact pregnant mammal, that has been extensively evaluated and judged reliable for its ability to predict the potential to cause developmentally toxic effects in intact systems across a broad range of chemicals or within a class of chemicals that includes the substance of concern.

SHORT-TERM TEST INDICATIVE OF CARCINOGENIC POTENTIAL Either any limited bioassay that measures tumor or preneoplastic induction, or any test indicative of interaction of a chemical substance with DNA (i.e. positive response in assays for gene mutation, chromosomal aberrations, DNA damage and repair, or cellular transformation).

SHORT TON Measure equal to 907 kg (2000 lb). *Also called* **Net ton.**

SHOULDER DITCH Ditch cut at the top of a slope or embankment to convey surface runoff and reduce erosion.

SHOULDER NIPPLE Pipe nipple with a space of approximately 19 mm (0.75 in.) between threads at the middle.

SHOWER Brief fall of rain, hail, or sleet.

SHOWER HEAD Water outlet mounted on the wall of a shower enclosure, approximately 2 m (6 ft 6 in.) above the base. It may create a spray, or be adjustable to produce a range of effects from fine spray to a jet or a pulsed output.

SHOWER RECEPTOR Floor and side walls of a shower up to and including the curb of the shower.

SHOWER ROOM Room with more than one shower head, used for communal bathing.

SHOWER TRAY Waterproof tray set in or on the floor of a cubicle, on the wall of which is mounted a shower head. The tray contains a waste outlet that is connected to the drainage system.

SHREDDER Size reduction machine that tears or grinds materials to a smaller and more uniform particle size.

SHREDDING Mechanical process that reduces large pieces to smaller pieces.

SHRUB Woody plant smaller than a tree, usually with many spreading stems starting from near the ground.

SHUTDOWN 1. Work stoppage for any reason; **2.** To shut off a machine.

SHUTOFF VALVE *See* **Valve.**

SHUTTER Swinging flashboard that opens under the pressure of water.

SHUTTER WEIR Movable weir consisting of panels hinged at the top and inclined downstream toward the top when the weir is closed.

SHUTTLE 1. Repeated movement back and forth over a predetermined route; **2.** Back and forth motion of a machine that continues to face in one direction.

SHUTTLE HAULING Use of preloaded trailers to reduce truck turnaround time.

SHUTTLE VALVE *See* **Valve.**

SIAL Upper layer of the Earth's continental crust, composed of comparatively light rocks, such as grabite, that are rich in silica and aluminum.

SIAMESE CONNECTION Hose fitting for combining the flow from two or more lines of hose into a single stream; one male coupling to two female couplings.

SICK BUILDING SYNDROME Building in which the occupants suffer from a higher-than-normal number or range of illnesses. Said to be caused by such factors as insufficient air changes, exudation from materials used, type or intensity of lighting levels, etc.

SIDECASTING Piling spoil alongside the excavation from which it is taken.

SIDE CHANNEL *See* **Spillway**.

SIDE DITCH Open drain alongside a road, designed to receive water from it and the adjacent ground.

SIDE-ENTRANCE MANHOLE Deep manhole having a horizontal access shaft to an inspection chamber.

SIDEFILL Aggregate laid each side of a buried pipe and which becomes part of the pipe surround.

SIDEHILL CUT Excavation in a hill involving one cut slope and, usually, one fill slope.

SIDE-LOADER DETACHABLE CONTAINER System similar to a rear loader except that it is loaded at the side of the collection vehicle.

SIDE-LOADING PACKER *See* **Refuse truck.**

SIDE OUTLET Fitting having an outlet or opening in the side.

SIDE SEAL Longitudinal seam of a filter medium in a filter element.

SIDESTREAM Waste flows discharged from storage or treatment facilities that may, or may not require supplemental or special treatment.

SIDETRACKING Drilling a branch hole in directional drilling.

SIDE VENT Vent connected to a drain at an angle of 45° or less.

SIDE WATER DEPTH Depth of water measured at a vertical exterior wall.

SIEMEN SI unit of electrical conductance between two points of a conductor when a constant current of 1 A in the conductor produces a potential difference of 1 V, and when the conductor itself is not the source of any electromotive force. (1 S=1 A/V=1 Ω^{-1}). *See also the appendix*: **Metric and nonmetric measurement.**

SIEVE Metallic plate or sheet, a woven wire cloth, or other similar device with regularly spaced apertures of uniform size, mounted in a suitable frame or holder for use in separating granular material according to size. The following are the opening sizes of US standard sieves:

COARSE SERIES

	mm	in.
4 in.	101.6	4.0
3-1/2 in.	88.9	3.5
3 in.	76.2	3.0
2-1/2 in.	63.5	2.5
2 in.	50.8	2.0
1-3/4 in.	44.4	1.75
1-1/2 in.	38.1	1.5
1-1/4 in.	31.7	1.25
1 in.	25.4	1.0
7/8 in.	22.2	0.875
3/4 in.	19.1	0.75
5/8 in.	15.9	0.675
1/2 in.	12.7	0.5
3/8 in.	9.52	0.375
1/4 in.	6.35	0.25

FINE SERIES -

	mm	in.
No. 4	4.76	0.187
No. 5	4.00	0.157
No. 6	3.36	0.132
No. 8	2.238	0.0937
No. 10	2.00	0.0787
No. 16	1.19	0.0469
No. 20	0.84	0.0331
No. 30	0.59	0.0232
No. 40	0.42	0.0165
No. 80	0.177	0.0070
No. 100	0.149	0.0059
No. 200	0.074	0.0029
No. 500	0.029	0.0011

SIEVE ANALYSIS Particle size distribution, usually expressed as the weight percentage retained upon each of a series of standard sieves of decreasing size, and the percentage passed by the sieve of the finest size. *Also called* **Screen analysis.**

SIEVE CORRECTION Correction of a sieve analysis to adjust for deviation of sieve performance from that of standard calibrated sieves.

SIEVE FRACTION That portion of a sample that passes through a standard sieve of specified size and is retained by some finer sieve of specified size.

SIEVE NUMBER Number used to designate the size of a sieve, usually the approximate number of openings per linear inch, applied to sieves with openings smaller than 6.3 mm (0.25 in.).

SIEVERT SI unit of radiation dose equivalent, equal to 1 joule of energy per kilogram of absorbing tissue. Symbol: Sv.

SIEVE SHAKER Mechanically vibrated table on to which a bank of standard sieves is clamped.

SIEVE SIZE Nominal size of openings between the cross wires of a testing sieve.

SIEVE TEST Test used to determine quantitatively the percent of asphalt present in a mix in the form of relatively large globules.

SIGHT GLASS Glass tube inserted into piping to show the liquid level in pipes, tanks, bearings, etc.

SIGNIFICANT ADVERSE ENVIRONMENTAL EFFECTS Injury to the environment by a chemical substance that reduces or adversely affects the productivity, utility, value, or function of biological, commercial, or agricultural resources, or which may adversely affect a threatened or endangered species.

SIGNIFICANT ADVERSE REACTIONS Reactions that may indicate a substantial impairment of normal activities, or long-lasting or irreversible damage to health or the environment.

SIGNIFICANT BIOLOGICAL TREATMENT Use of an aerobic or anaerobic biological treatment process in a treatment works to consistently achieve a 30-day average of at least 65% removal of BOD.

SIGNIFICANT ENVIRONMENTAL EFFECTS Either:

(a) Any irreversible damage to biological, commercial, or agricultural resources of importance to society;

(b) Any known or reasonably anticipated loss of members of an endangered or threatened species.

SIGNIFICANT HAZARD TO PUBLIC HEALTH Any level of contaminant that causes or may cause an aquifer to exceed any maximum contaminant level established in a Primary Drinking Water Standard.

SIGNIFICANT SOURCE OF GROUNDWATER Aquifer that:

(a) Is saturated with water having less than 10 000 mg/L of total dissolved solids;

(b) Is within 2500 ft of the land surface;

(c) Has a transmissibility greater than 200 gpd/ft, provided that any formation or part of a formation included within the source of groundwater has a hydraulic conductivity greater than 2 gpd/ft; and

(d) Is capable of continuously yielding at least 10,000 gpd to a pumped or flowing well for a period of at least a year.

SIGNIFICANT INDUSTRIAL USER 1. User of a public wastewater collection and treatment system that is subject to pretreatment regulations; **2.** User of a public wastewater collection and treatment system that discharges 94 635 L (25,000 gal) per day or more of process wastewater, excluding sanitary, noncontact cooling, and boiler blowdown wastewater; **3.** User of a public wastewater collection and treatment system that contributes a process wastewater stream which makes up 5% or more of the average dry weather hydraulic or organic capacity of the treatment plant; **4.** User of a public wastewater collection and treatment system that is designated as such on the basis that it has a reasonable potential for adversely affecting the treatment plant operation or for violating any pretreatment requirement or standard.

SIGNIFICANTLY MORE STRINGENT LIMITATION BOD5 and suspended solids limitations necessary to meet the percent removal requirements of at least 5 mg/L more stringent than the otherwise applicable concentration-based limitations (e.g. less than 25 mg/L in the case of the secondary treatment limits for BOD and suspended solids).

SIGNIFICANT NONCOMPLIANCE Industrial user of a wastewater collection and treatment system who:

(a) Chronically violates wastewater discharge limits by exceeding, by any magnitude, the daily maximum limit, or the average limit for the same pollutant parameter in 66% or more of all measurements taken over a six-month period.

(b) Following a review of technical criteria, equals or exceeds the product of the daily maximum limit or the average limit, times the applicable TRC (TRC = 1.4 for BOD, TSS, fats, oil, and grease, and 1.2 for all other pollutants except pH) in 33% or more of all of the measurements for each pollutant parameter taken during a six-month period.

(c) Violates any pretreatment effluent daily maximum or long-term average limit to the extent that it alone, or in combination with other discharges, causes interference with or pass-though of the treatment plant.

(d) Discharges a pollutant that has caused immanent endangerment to human health, or to the environment, or has caused the treatment plant to exercise emergency authority to halt or prevent the discharge.

(e) Has failed to meet, within 90 days after a scheduled date, a compliance schedule milestone contained in a permit or enforcement order for starting construction, completing construction, or attaining final compliance.

(f) Has failed to provide, within 30 days after the due date, required reports such as baseline monitoring reports, 90-day compliance reports, periodic self-monitoring reports, and reports on compliance with relevant schedules.

(g) Has failed to accurately report noncompliance.

(h) Is in violation of regulations or requirements to a degree that will adversely affect the operation or implementation of a local pretreatment program.

SILAGE Green fodder for farm animals which has molasses added to promote fermentation and preservation in the silo in which it is fermented and stored.

SILENCER Device for reducing gas flow noise; noise is decreased by tuned resonant control of gas expansion.

SILENCING PIPE Vertical tube in a cistern from the ball valve to the bottom, with its outlet always below water level. It has a small antisiphon hole near the top, which prevents completely silent filling.

SILICOSIS Lung disease, principally caused from inhaling rock dust.

SILL 1. Horizontal overflow line of a dam, spillway, weir, etc.; **2.** Submerged structure across a river to control upstream water levels.

SILL COCK See **Bib.**

SILT Soil particles 5 μm and less in size.

SILTATION Deposition of silt from water.

SILT BASIN Small chamber in a storm sewer, used to retard flow velocity and promote the deposition of sand and grit for subsequent removal.

SILT BOX Metal box at the bottom of a road gully for collecting sand and other debris, that can be removed and emptied.

SILTING Filling with soil or mud conveyed by water.

SILT LOAM *See* **Soil types.**

SILT TEST *See* **Decantation test.**

SILT TRAP Settling basin that prevents waterborne soil or mud from entering a drainage system or waterway.

SILVA Forest trees of a particular region or time.

SILVICULTURAL POINT SOURCE Discernible, confined and discrete conveyance related to rock crushing, gravel washing, log sorting, or log storage facilities that are operated in connection with silvicultural activities and from which pollutants are discharged into local waters. The term does not include nonpoint source silvicultural activities such as nursery operations, site preparation, reforestation and subsequent cultural treatment, thinning, prescribed burning, pest and fire control, harvesting operations, surface drainage, or road construction and maintenance from which there is natural runoff.

SIMA Lower part of the Earth's crust.

SIMAZINE A triazine soil sterilant and pre-emergence herbicide that has been detected in private and public water supplies and for which the maximum acceptable concentration of 0.01 mg/L has been established for drinking water.

SIMPLE SURGE TANK Surge tank without a restricted orifice.

SINGLE-ACTING PUMP *See* **Pump.**

SINGLE-ARCH DAM Curved masonry dam that spans between sidewalls in a single arc.

SINGLE-BAG COMPACTOR Semiautomatic refuse compactor in which the refuse is crushed against a front-opening door into a specified volume.

SINGLE-CELL PROTEIN Protein manufactured by microbes from organic substances.

SINGLE CENTRIFUGAL PUMP Centrifugal pump in which the suction inlet admits water to one side of the impeller.

SINGLE-CHAMBER INCINERATOR *See* **Incinerator.**

SINGLE-EFFECT EVAPORATOR Device in which liquid is subjected to only one evaporative stage.

SINGLE-HUB PIPE Pipe having a bell at one end and a spigot at the other.

SINGLE-PASS SOIL STABILIZER Machine equipped with several wheels or paddles that rotate rapidly in contact with the soil, pulverizing it to a measured depth, and mixing it with any materials spread ahead of the machine, such as binders, stabilizers, conditioners, etc.

SINGLE-PIECE CONVEYOR One-piece conveyor that transports and deposits material away from the machine.

SINGLE-STACK DRAINAGE SYSTEM Sanitary pipework with a large-diameter vertical stack which conducts wastewater and provides ventilation.

SINGLE-STAGE DIGESTION Complete sludge digestion carried out within a single vessel.

SINGLE-STAGE PUMP Centrifugal pump having one set of impellers.

SINK 1. Bowl-like depression in the land surface; **2.** Surface hollow which, in limestone regions, may connect with a cavern or subterranean passage; **3.** Water basin fixed to the wall and equipped with a water supply and drain; **4.** A cesspool.

SINK BIB Tap designed for a kitchen sink.

SINK DRAIN Conduit that carries liquid wastes from kitchen and slop sinks.

SINKHOLE Natural depression in a land surface communicating with a subterranean passage, formed by the collapse of a cavern roof.

SINKING PUMP Pump designed to operate in a well or shaft under construction to keep it dry.

SIPHON 1. Tube or pipe through which a fluid flows over a high point by gravity; **2.**

429

SIPHONAGE

See **Siphon spillway,** and **Spillway.**

SIPHONAGE *See* **Self-siphoning.**

SIPHON BREAK Small groove to arrest capillary action between adjacent surfaces.

SIPHON SPILLWAY Spillway constructed as a siphon. Water levels must rise to the crest of the siphon, which then primes itself and flows until the water level falls below its inlet, that is, below the crest. *See also* **Spillway.**

SIPHON TRAP *See* **S-trap.**

SIREN Warning device that makes a loud, high-pitched, or an alternating high- and low-pitched wailing sound.

SITE Location where construction takes place; the area to be occupied by the project and all adjacent and related areas to be used by the contractor during performance of the work including easements, rights-of-way, buildings not scheduled for demolition, and storage, staging, production, and disposal areas. *Also called* **Job site.**

SITE ACCOMMODATION Space for offices, storage, etc., provided on site during the construction phase.

SITE ASSEMBLY Putting together on site components fabricated elsewhere.

SITE BOUNDARY Edge of a building site, commonly the fence line, that delimits on-site from off-site.

SITE CLASS *See* **Land classification (forest management).**

SITE CLEANUP Removal of construction debris, refuse, and other unwanted materials from a site following completion of the works and prior to handover.

SITE CLEARING Stripping a site of unwanted vegetation, rubble, debris, and structures ready for further development. *Also called* **Clearing the site.**

SITE CONDITIONS Overall description of the site for a project: ground cover, relative elevations, surrounding environment, presence of surface water and the normal watertable, soil borings, known history, as well as available access and egress, etc.

SITE CONSTRAINTS Conditions and limitation that are part of the contract documents that limit or otherwise condition the use of a site during the construction phase.

SITE DESIGN Planning of an area of land in anticipation of development.

SITE DEVELOPMENT Stage of construction that sees the land on which a project is to be built prepared in anticipation of the construction phase.

SITE DEVELOPMENT PLAN Detailed plan illustrating the proposed arrangement of a site, including site layout, grading, hard materials, and planting. *Also called* **Site plan,** and **Plot plan.**

SITE DIARY Record kept by the project engineer or his deputy of what happens each day on site.

SITE DRAINAGE Removal of surface water from a site by natural runoff or through a storm sewer system.

SITE EXPLORATION *See* **Site investigation.**

SITE FENCE Temporary fence around the perimeter of a construction site, not necessarily on its boundary.

SITE INVESTIGATION Detailed survey and physical determination of the characteristics of a site, on and below grade, according to terms of reference or a specification particular to the site. *Also called* **Site exploration.**

SITE MANAGER Person responsible for onsite organization and administration; an agent of the general contractor.

SITE MEETING Meeting held on site between representatives of the contractor, the owner, and other interested parties, to review progress, etc.

SITE OFFICE Location of person responsible for the site and works. *Also called* **Shack,** and **Site shack.**

SITE ORGANIZATION Disposition on site of entities, trades, personnel, materials, etc., plus establishment of a line of authority and responsibility.

SITE PLAN *See* **Site development plan.**

SITE PREPARATION Disturbance of an area's topsoil and ground vegetation to create conditions suitable for construction, development, forest regeneration, etc.

SITE REHABILITATION Conversion of the existing unsatisfactory cover on highly productive forest sites to a cover of commercially valuable species.

SITE SECURITY Measures and personnel necessary to ensure the safety of equipment and structures within the site boundary, and the responsibility of the general contractor.

SITE SERVICES Temporary connections to public services, such as electricity, water, drainage, etc., necessary for completion of the works proposed for the site.

SITE SHACK *See* **Site office.**

SITE SIGN Sign visible from outside the site fence stating, as a minimum, the name and location of a firm responsible for the project. In some cases, it may be required that details of the owner, architect (and other professionals), and general contractor, together with information about the applicable building, development, or planning permits also be displayed. The sign may also list the names of some or all of the entities contributing to the development, including principal suppliers, etc.

SITE SPECIFIC Data such as surface or subsurface soil and water testing taken on the property of interest.

SITE SUPERVISION Responsibility, on site, for day-to-day management of operations and supervision of construction. *Also called* **Field supervision.**

SITEWORK Work done on site, as against prefabrication or supply of complete or partially completed units or the supply of raw materials.

SIZE 1. Rated capacity in tonnes (tons) per hour of a crusher, grinding mill, bucket elevator, bagging operation, or enclosed truck or railcar loading station; **2.** Total surface area of the top screen of a screening operation; **3.** Width of a conveyor belt; and the rated capacity in tons of a storage bin.

SKATOLE Organic compound (C_9H_9N) containing nitrogen and having a fecal odor.

SKIM Remove floating matter.

SKIMMING DETRITUS TANK Tank designed to promote the settling of solids for subsequent removal, while, at the same time, allowing buoyant materials to float, where they can be skimmed by mechanical rakes.

SKIMMINGS Floating materials removed from the surface of wastewater in settling tanks.

SKIMMING TANK Tank that promotes buoyant material to float and remain on the surface of wastewater while permitting the liquid to discharge continuously under baffle or scum boards.

SKIMMING WEIR Weir with an adjustable crest that will allow restriction of the depth of overflow water.

SKIN OF WATER Thin layer on the free surface of water.

SLACK WATER 1. Period at high or low tide when there is no visible flow of water; **2.** Area in a sea or river unaffected by currents; still water.

SLAG More-or-less completely fused and vitrified matter separated during the reduction of a metal from its ore.

SLAKE Combine (lime) chemically with water or moist air.

SLAKED LIME Common name for calcium oxide (CaO) to which water has been added converting it to slaked lime, $Ca(OH)_2$.

SLANT Sewer pipe connecting a house to a common drain.

SLAVE UNIT Machine that is controlled by or through a similar machine.

SLAVE VALVE *See* **Valve.**

SLEET Precipitation consisting of generally transparent frozen or partially frozen raindrops.

SLEEVE 1. Metal adapter used to splice pipe by driving the two pipes into, or onto the sleeve. An inside sleeve reduces the ID at the splice while maintaining the pipe OD; an outside sleeve increases the OD while maintaining the pipe ID.

SLEEVE PIECE Short, thin-walled brass or copper tube used in plumbing to solder pipes of different metals.

SLICK 1. Smooth, glossy and slippery, as if covered with oil or ice; **2.** Floating patch of oil or scum.

SLICKENS Thin layer of extremely fine silt.

SLICKENSIDE Polished and striated rock surface caused by one rock mass sliding over another in a fault plane.

SLIDE 1. Small landslide; **2.** Fresh tile wall that has buckled or sagged.

SLIDE COUPLING Slip joint.

SLIDING GATE Valve or gate that slides directly on its bearings or seats.

SLIDING-PANEL WEIR Frame weir containing panels that slide between grooved uprights.

SLIME Soft viscous deposit or coating.

SLIMICIDES Substances or mixtures, except antimicrobial agents, fungicides, and herbicides, intended for use in preventing or inhibiting the growth of, or destroying biological slimes composed of combinations of algae, bacteria or fungi.

SLIP 1. Smooth crack at which rock strata have moved on each other; **2.** Difference between optimal and actual output of a mechanical device; **3.** Movement between two parts where non should exist.

SLIP COUPLING Pipe coupling without a stop, permitting it to slip over a pipe.

SLIP JOINT Inserted joint in which the end of one pipe is slipped into the flared or swaged end of another.

SLIPLINING Insertion of new pipe, usually of polyethylene (PE), within an existing defective pipe. Developments provide for the new PE line to be temporarily reduced in diameter before insertion, and subsequently enlarged to provide a tight fit into the original pipeline.

SLIP NUT Nut used on P traps and similar connections where a gasket is compressed around the joint to form a watertight seal.

SLIP-ON FLANGE Flange that slips on to the end of a plain pipe and is soldered or welded in place.

SLIP PLANE Potentially hazardous sloping surface along a fault plane.

SLOPE Incline upward or downward; lie on a slant.

SLOPE ANGLE Natural angle of repose of fill material, measured from a horizontal plane.

SLOPE AREA DISCHARGE MEASUREMENT Determination of the peak discharge, based on visual observation, of the result of a flood.

SLOPE DISCHARGE CURVE Graphical representation of the discharge at a gauging station allowing for the slope of the water surface plus the gauge height.

SLOPE GAUGE Instrument used to determine the slope of a water surface.

SLOPE OF FRONT Gradient of an approaching weather front.

SLOP OIL Floating oil and solids that accumulate on the surface of an oil–water separator.

SLOPS *See* **solid waste.**

SLOP SINK Large, rectangular deep sink for service or janitorial use.

SLOT DRAIN Gutter concealed under a narrow gap between paving slabs.

SLOUGH (slew) 1. Secondary river channel through which flow is usually sluggish; **2.** Shallow depression created to store water; **3.** Small muddy tidal waterway that is often dewatered at low tide; **4.** Natural or artificial depression in which water is stored, commonly for watering livestock.

SLOUGH (sluff) 1. Sliding of overlying material such as overburden upon rock. **2.** Release of contaminant from the upstream side of a filter element to the downstream side of the filter enclosure or into the effluent stream.

SLOUGHINGS (sluffings) Slimes that have been washed off the media of trickling filters.

SLOW BEND Bend about a wide radius.

SLOW SAND FILTER Column or layer of fine sand through which raw water is passed by gravity to an underdrain, the dirty or clogged filter media being periodically removed and replaced.

SLOW SAND FILTRATION Process involving passage of raw water through a bed of sand at low velocity resulting in substantial particulate removal by physical and biological mechanisms.

SLUDGE 1. Aggregate of oil, or oil and other matter of any kind in any form, other than dredged oil, having a combined specific gravity equivalent to or greater than water; **2.** Any solid, semisolid, or liquid waste generated from a municipal, commercial, or industrial wastewater treatment plant, water supply treatment plant, or air pollution control facility or any other such waste having similar characteristics and effect; **3.** Waste from a wet grinding process; **4.** Particulate contaminant or a mixture of particulate and liquid contaminant separated from a fluid in an unconsolidated state. *See also* **Solid waste**.

SLUDGE AGE Time that a particle of suspended solids has been subjected to aeration in the activated sludge process.

SLUDGE BLANKET Accumulated mat of sludge hydrodynamically suspended within or on an enclosed body of water or wastewater.

SLUDGE BOIL Upwelling of sludge and water caused by release of gases developed in the decomposing sludge.

SLUDGE BULKING Condition in an activated sludge process in which the sludge expands beyond normal limits and resists concentration.

SLUDGE CAKE Dewatered, compressed product of a sludge digester.

SLUDGE CIRCULATION Mechanical or hydraulic overturning of the contents of a sludge digestion tank.

SLUDGE COLLECTOR Device that moves across the bottom of a settling tank to move accumulated sludge to a drawoff sump.

SLUDGE COMPARTMENT Separate section of a settling tank in which accumulated sludge is digested.

SLUDGE CONCENTRATION One or more processes used in sequence to dewater sludge.

SLUDGE CONDITIONING Mechanical and/or chemical treatment of digested sludge to enhance drainability and facilitate dewatering.

SLUDGE DENSITY INDEX Reciprocal of the sludge volume index multiplied by 100.

SLUDGE DEWATERING Means taken to lower the fluid content of digested sludge: may be hydraulic, mechanical, or chemical, or any combination.

SLUDGE DIGESTION Treatment of sludges collected from the primary, secondary and tertiary stages of sewage treatment by anaerobic digestion in closed tanks with the resulting production of sludge cake, liquor, and methane gas.

SLUDGE DIGESTION TANK Tank to which waste sludge is pumped and in which sludge digestion is promoted.

SLUDGE DRYER Enclosed thermal treatment device that is used to dehydrate sludge and that has a maximum total thermal input, excluding the heating value of the sludge itself, of 2500 Btu/lb of sludge treated on a

433

wet-weight basis, heating to temperatures above 65° C (150° F) directly with combustion gases.

SLUDGE DRYING BED Shallow tank with a porous bed into which digested sludge is pumped and allowed to dry by drainage and evaporation.

SLUDGE EXCESS Sludge produced in the activated sludge treatment process that is surplus to the process and which is drawn off for treatment and disposal.

SLUDGE FILTER Device in which wet sludge is conditioned by the addition of a chemical coagulant and partly dewatered by means of vacuum or pressure, or both.

SLUDGE FOAMING Formation of a froth caused by the generation of excess quantities of gas in a sludge blanket.

SLUDGE GAS Product of the process of decomposition of the organic fraction of sludge removed from wastewater and allowed to decompose under anaerobic conditions.

SLUDGE GAS HOLDER Storage vessel used to hold gas collected from sludge digestion tanks.

SLUDGE LAGOON Basin in which processed sludge is held prior to disposal.

SLUDGE MOISTURE CONTENT Net weight of water present in a sludge.

SLUDGE PRESS Device that partially dewaters sludge using mechanical or hydraulic pressure.

SLUDGE PUMP *See* **Pump.**

SLUDGE REAERATION Injection of oxygen into partially treated sludge as a means of reconditioning.

SLUDGE SEEDING Use of biologically active sewage sludge as an inoculant to help start, or to accelerate the biological treatment of wastewater.

SLUDGE SHREDDER Device used to break down lumps in air-dried sludge.

SLUDGE STRIPPER Device for removing air-dried sludge from drying beds.

SLUDGE THICKENER Tank with, or without, mechanical equipment, designed to concentrate wastewater sludge.

SLUDGE THICKENING Process of increasing the solids content of sludge by mechanical, hydraulic, or chemical means, or any combination.

SLUDGE TREATMENT Processing of wastewater sludge to render it innocuous.

SLUDGE UTILIZATION Use of treated wastewater sludge for a productive purpose as an alternative to disposal to land fill.

SLUDGE VOLUME INDEX Calculation that indicates the tendency of activated sludge solids to thicken or become concentrated during the sedimentation/thickening process.

SLUDGE VOLUME RATIO Volume of a sludge blanket divided by the daily volume of sludge pumped from the thickener.

SLUG Relatively small quantity of fluid in a pipeline, isolated from the main fluid in the pipe.

SLUG DISCHARGE Discharge of a nonroutine, episodic nature, including but not limited to an accidental spill or a noncustomary batch discharge.

SLUGGISH STREAM Stream in which the flow velocity is not sufficient to prevent the deposition of silt and sediment.

SLUGS Intermittent volumes of a discharge released into a stream.

SLUICE Steep, narrow waterway.

SLUICE GATE Device used to control the rate of flow through a sluice.

SLUICEWAY Means by which water can be drained from a reservoir, head pond, channel, etc., under controlled conditions.

SLUICING Moving granular materials and soil by flowing water, often under pressure.

SLUMP Minor landslide in which a relatively small mass of material, usually topsoil, moves as a body to a lower elevation.

SLURRY Thin mixture of a liquid, usually water, and any of several finely divided substances.

SLURRY TRENCH Narrow trench with vertical, unshored walls, in which caving or sloughing of the earth walls is prevented by the hydrostatic pressure of the slurry or mud with which the trench is filled.

SLURRY TUNNEL-BORING MACHINE Type of microtunneling machine in which the soil is turned to slurry and is used to counterbalance water pressure to stabilize the face, before being pumped to the surface.

SLURRY WALL Wall constructed from the ground surface to an impermeable layer; consisting of slurry, usually soil, bentonite, and/or cement. *Also called* **Diaphragm wall.**

SLUSH 1. Broken or crushed ice mixed with water; **2.** Mixture having a high water content.

SMOG Collective name for a fog comprising air pollutants lying relatively close to the ground through a temperature inversion; made up of sulfur dioxide, unburned hydrocarbons and nitrogen oxides, characterized by a dull yellowish haze, and which can cause eye and respiratory irritation.

SMOKE Suspended particles and aerosols of the products of combustion, mostly incomplete combustion, less than 0.1 μm in size.

SMOKE ALARM Electrical device that sounds an alarm when sensing the products of combustion; a combined smoke detector and audible alarm device.

SMOKE CONTROL AREA Area in which the generation and emission of smoke is either regulated or prohibited.

SMOKE DETECTOR Device for sensing the presence of visible or invisible particles produced by combustion, and automatically initiating a signal indicating this condition.

SMOKE ROCKET Smoke-producing incendiary used in tightness testing of drains.

SMOKE SHADE Degree of darkness of a smoke plume measured using a comparative device such as a Ringelmann chart.

SMOKE STACK Industrial chimney designed to remove airborne emissions from an industrial process.

SMOKE TEST Test of a pipe system to determine its tightness by filling it with smoke and sealing all inlets and outlets.

SMOTHER Deny oxygen to a fuel.

SNAKE Spring-steel wire or tape used to travel through pipes and conduit when pulling wires or to clear an obstruction. *See also* **Fish.**

SNOW Solid precipitation in the form of white or translucent crystals of various shapes originating in the upper atmosphere as frozen particles of water vapor.

SNOWBANK Snow banked up by the wind.

SNOW COVER Accumulated snow and ice that obscures the ground.

SNOWFALL Amount of snow that falls during a given period or in a specified area.

SNOW FENCE Temporary fencing used to prevent snow from drifting.

SNOWFIELD Area where snow accumulations remain on the ground throughout the year.

SNOW FLAKE Single flake or crystal of snow.

SNOW GAUGE Device used to measure the depth of snow, or the amount of snow that has fallen over a specified period.

SNOW LINE Lower altitudinal boundary of a snow-covered area, especially one that is perennially covered.

SNOW PACK Accumulation of snow on the ground at a point in time.

SNOW PLOW Device or vehicle used to move or remove snow.

SNOW RIPENING Stage of melting snow

when the crystals tend to become granular; the spaces between the crystals becoming filled with water, and the water content of the snow tends to become uniform at all depths.

SNOW SAMPLE Core removed from an accumulation of snow.

SNOW SAMPLER Device used to obtain a sample of snow in as undisturbed condition as possible.

SNOW SCALE Measuring rod from which the depth of accumulated snow can be read from a reasonable distance.

SNOW SHED Roofed structure designed to permit unobstructed passage beneath an accumulated mass of snow, or to allow passage of sliding snow across something: a road or railway track, for instance.

SNOW SHOE Racket-shaped frame containing interlaced thongs which, when attached to the foot, distributes a person's weight and facilitates walking on the surface of snow.

SNOW STORAGE Volume of snow accumulated in a drainage basin which will eventually become available water.

SNOW STORM Storm marked by heavy snowfall and high winds.

SNOW SURVEY Evaluation of fallen snow to determine its depth, water content, density, etc.

SNOW TRAP Opening cleared in a dense forest to trap snow and retard melting.

SNOWY Abounding in snow; covered with or subject to snow.

SOAP Product that induces surface activity in water.

SOCKET Enlarged end, shaped to receive an unenlarged end, as in a pipe joint.

SOD Matting of grass and soil, that is cut just below the roots and then used on a new site to provide quick grass cover.

SODA ASH Crude anhydrous sodium carbonate.

SODIUM Soft, light, extremely malleable silver-white metallic element, symbol Na, that is the most abundant cation in extracellular fluid. The minimum daily requirement for sodium is approximately 50 mg for the average adult. The aesthetic objective for sodium in drinking water is <200 mg/L.

SODIUM ADSORPTION RATIO Expression of the relative activity of sodium ions in the exchange reactions with soil.

SODIUM ALUMINATE Coagulating chemical and softening agent used in water treatment.

SODIUM ARSENITE Salt used to test for residual chlorine.

SODIUM CARBONATE Soda ash.

SOFT DETERGENT Synthetic detergent that responds to biological degradation.

SOFT GROUND Ground unable to bear even a moderate imposed load due to such factors as a high water content, lack of cohesion between soil particles, etc.

SOFTWARE Instructions that tell a computer how to perform a given task or tasks.

SOFT WATER Water, free of magnesium or calcium salts, in which soap readily dissolves, forming a lather without being precipitated.

SOFTWOOD Tree that has needles or does not have broad leaves.

SOIL *See* **Soil types.**

SOIL ABSORPTION CAPACITY Ability of soil to absorb an effluent flow.

SOIL ADHESION Sticking of soil to foreign materials such as soil implements, tracks, or wheels.

SOIL ANALYSIS Investigation of earth and the preparation of a report giving conclusions and recommendations, a fundamental stage prior to the design of foundations for most commercial and industrial structures. The analysis covers, at least, soil density, moisture content, load-bearing capacity, shear strength, plasticity index, organic content, and grain size. *See also* **Soil types.**

SOIL BORING Small-diameter hole drilled into the soil for the purpose of obtaining earth samples and exploring the subsurface conditions. *Also called* **Test boring.**

SOIL BRANCH Sewer branch leading to a soil pipe.

SOIL CLASSIFICATION SYSTEMS (*See also* **Soil groups, Soil limits,** and **Soil types.**) Method of categorizing soil and rock deposits in a hierarchy in decreasing order of stability. The categories are determined based on an analysis of the properties and performance characteristics of the deposits and the environmental conditions of exposure. Such a classification includes:

> **Stable rock:** Natural solid mineral matter that can be excavated with vertical sides and remains intact when exposed.

> **Submerged soil:** Soil that is under water or is free seeping.

> **Type A:** Cohesive soils with an unconfined compressive strength of 144 kPa (1.5 tons/ ft^2) or greater.

> **Type B:** Cohesive soil with an unconfined compressive strength greater than 48 kPa (0.5 tons ft^2) but less than 144 kPa (1.5 tons ft^2); granular cohesionless soils including angular gravel, silt, silt loam, sandy loam, and in some cases silty clay loam and sandy clay loam; previously disturbed soils (except those classed as Type C); soil that meets the unconfined compressive strength or cementation requirements for Type A, but is fissured or subject to vibration; dry rock that is not stable; or material that is part of a layered system where the layers dip into an excavation on a slope less steep than 4:1.

> **Type C:** Cohesive soil with an unconfined compressive strength of 48 kPa (0.5 tons ft^2), or less; granular soils, gravel, sand and loamy sand; submerged soil or soil from which water

is freely seeping; submerged rock that is not stable; or material in a sloped, layered system where the layers dip into an excavation on a slope of 4:1 or steeper.

The American Association of State Highway and Transportation Officials has developed the following soil classification system based on field performance of soils for highway construction. It is also known as the **AASHTO Classification system** and as the **Bureau of Public Roads Soil Classification:**

A-1-a Mostly gravel with or without well-graded fines.

A-1-b Mostly sand with or without well-graded fines.

A-2-4 Granular material with silty fines.

A-2-6 Granular material with clayey fines.

A-3 Poorly graded sand with almost no fines or gravel.

A-4 Mostly silty fines.

A-5 Uncommon soil type with silty fines, usually elastic and hard to compact.

A-6 Have either silty or clayey fines with low liquid limit.

A-7-5 The more plastic clays and silts.

The **Unified soil classification system** is used by the U.S. Army Corps of Engineers and the U.S. Bureau of Reclamation. It uses texture as the descriptive terms, plus a system of modifiers, as follows:

SYMBOLS:

G - Gravel, smaller than 76 mm (3 in.) and larger than 6 mm (0.25 in.)

S - Sand, smaller than 6 mm (0.25 in.) and larger enough to see.

M - Silt, fine-grained soils with

individual grains.

C - Clay, with grains too small to see with the naked eye.

MODIFIERS (sand and gravel)

W - Well-graded having large, medium and small grains.

P - Poorly-graded and having a uniformly sized grain.

C - Clayey.

M - Silty.

MODIFIERS (silt and clay)

L - Low plasticity.

H - High plasticity.

SOIL COHESION The mutual attraction exerted on soil particles by molecular forces and moisture films.

SOIL COMPACTION Compression of soil as a result of heavy equipment traffic. *See also* **AASHTO test, Nuclear density meter test, Proctor test, Sand-cone test,** and **Water balloon test method.**

SOIL COMPACTION TEST Field or laboratory test used to determine the degree to which soil has been compacted following densification. There are several types, including:

AASHTO (T-99): Procedure by which to determine the density of compacted soil in which a 2.5 kg (5.5 lb) hammer is dropped freely 25 times from a height of 305 mm (12 in.) to compact 0.00935 m^3 (0.33 ft^3) of the soil, in three layers, in a 102-mm- (4-in.-) diameter confined mold. The test imparts a total 16 812 Nm (12,400 ft/lbs) of compactive effort.

AASHTO (T-180): Procedure by which to determine the density of compacted soil in which a 4.5 kg (10 lb) hammer is dropped freely 25 times from a height of 457 mm (18 in.) to compact 0.00935 m^3 (0.33 ft^3) of the soil, in five layers, in a 102-mm- (4-in.-) di-

ameter confined mold. The test imparts a total 76 196 Nm (56,200 ft/lbs) of compactive effort.

Nuclear density meter: Means of measuring the density of compacted soil or asphalt pavement at the job site that employs the principal of either backscatter or direct transmission gauging in which gamma rays emitted by a radioisotope source contained in the gauge body penetrate the pavement and are scattered and/or absorbed. A counter in the device establishes the number of rays that return, the number of which are proportional to the moisture content and the density of the pavement.

Modified Proctor: Moisture-density test of more rigid specifications than Proctor. The basic difference is the use of a heavier weight that is dropped from a greater distance in laboratory determinations.

Proctor: Method developed for determining the optimum water content and the corresponding maximum dry density of compacted soils. Similar in most respects to the AASHTO test.

Sand cone: Multistep test procedure for determining the compaction level of soil in which a test site away from operating equipment is selected and levelled and on which the unit's base plate is laid. Material is excavated through the hole in the plate to a depth of approximately 150 mm (6 in.), dried in an oven and weighed to determine moisture content. The volume of the hole is measured by filling it with dry, free-flowing sand from a special sand-cone cylinder. The density (wet unit weight) of the compacted sample is found by dividing the weight of the material by the volume of the hole. Dry unit weight can be found by dividing the wet unit weight by 1 plus the moisture content (expressed as a decimal). For example: if the moisture content is 9%, the wet unit weight would be divided by 1.09 to find dry density.

Water balloon: Method by which to test

the density of soil in which the first three steps – excavating a sample, weighing it and drying it – are the same as performed in the sand-cone method and produce a measure of the moisture content. Next, balloon is suspended from the base plate into the hole and filled with water from a Washington Densometer to give an accurate measure of the volume. The density (wet unit weight) is found by dividing the weight of the sample by the volume of the hole; dry unit weight by dividing the wet unit weight by one plus the moisture content.

SOIL CONSERVATION Management of soil so as to minimize erosion while maintaining or enhancing its ability to sustain crops.

SOIL CONSOLIDATION 1. Techniques and materials that assist in the removal of moisture from a soil mass, or that prevent the entry of moisture to the mass; **2.** Moving together of soil particles as water and air are squeezed out due to an imposed load. *See also* **Elastic compression,** and **Plastic creep.**

SOIL COVER Covering of lightweight plastic film, roll roofing, or similar material used over the soil to minimize moisture permeation.

SOIL CREEP Tendency of a cohesive mass of soil to move a relatively short distance downslope under the force of gravity.

SOIL DISCHARGE Evaporation of groundwater lifted by capillarity to the surface from the zone of saturation.

SOIL DRAINAGE Extraction from soil of water that is surplus to its requirements, either naturally by gravitation and evaporation, or artificially through ditching, pumping, etc.

SOIL EROSION Detachment and movement of soil from the land surface by wind or water.

SOIL FAILURE Alteration or destruction of the soil structure by mechanical forces such as in shearing, compression, or tearing.

SOIL GROUPS Five fundamental groups into which soils are divided (*See also* **Soil classification system, Soil limits,** and **Soil types**):

Gravel: Where individual grains vary in size from 2.0 to 76.2 mm (0.08 to 3.0 in.) in diameter and have a rounded appearance.

Sand: Small rock or mineral fragments smaller than 2.0 mm (0.08 in.) in diameter and semi-sharp.

Silt: Fine grains that appear soft and floury when dry; when moist, silt pressed between the thumb and forefinger will have a broken appearance.

Clay: Very fine textured soil that forms hard lumps of clods when dried; when moist, clay is very sticky and can be rolled into a ribbon between the thumb and forefinger.

Organic: This matter consists of either partially decomposed vegetation (peats) or finely divided vegetable matter (organic silts and clays).

SOIL HORIZON Any layer of soil that may be distinguished from adjacent layers because it differs in physical, chemical, or biological characteristics, usually designated as A, B, and C horizons.

SOIL INFILTRATION RATE Maximum rate at which a soil, under specific conditions, can absorb water.

SOIL INVESTIGATION Study of the earth in the area of a foundation consisting of sampling, classification, preparation of logs of borings, and a report setting forth conclusions and recommendations; a basic practice preparatory to the design of foundations as required under most building codes.

SOIL LIMITS (*See also* **Soil classification system, Soil groups,** and **Soil types**). Means for differentiation between highly plastic, slightly plastic and nonplastic soils, as measured by the following tests:

Liquid limit (LL): Moisture content at which a soil passes from plastic to a liquid state.

Plastic limit (PL): Condition when a soil changes from a semisolid to a plastic state.

Plasticity index (PI): Numerical difference between a soil's plastic limit and liquid limit.

Shrinkage limit (SL): Point where a soil, dried below the plastic limit, shrinks and becomes brittle to the point where all the particles are in contact.

SOIL MAP Map that shows the distribution of soil, by types, volumes, etc., in relation to other features of the land.

SOIL MECHANICS Investigation of the composition of soils, their classification, consolidation, strength, etc.

SOIL MOISTURE Pellicular water in the soil zone.

SOIL MOISTURE DEFICIT Difference between the moisture in soil at a point in time and the field moisture capacity.

SOIL ORGANIC MATTER Organic fraction of the soil; it includes plant and animal residues at various stages of decomposition, cells and tissues of soil organisms, and substances synthesized by the microbial population.

SOIL-OR-WASTE PIPE Pipe in a sanitary drainage system.

SOIL-OR-WASTE STACK Vertical soil-or-waste pipe that passes through one or more story and that includes any offset that is part of the stack.

SOIL PARTICLE SIZE Classification into which soils can be grouped by particle size, i.e.:

	International (mm)	US (mm)
Coarse sand	0.2–2.0	0.2–2.0
Fine sand	0.002–0.2	0.05–0.02
Silt	0.002–0.02	0.002–0.05
Clay	<0.002	<0.002

SOIL pH Value obtained by sampling the soil to the depth of cultivation or solid waste placement, whichever is greater, and analyzing by the electrometric method (the negative logarithm to the base 10) of the hydrogen ion activity of a soil as determined by means of a suitable sensing electrode coupled with a suitable reference electrode at a 1:1 soil:water ratio.

SOIL PHASE Local variation of the soil classification in use, based on some unusual condition.

SOIL PIPE Vertical drainage stack into which branch lines drain plumbing fixtures.

SOIL PLASTICITY Property that allows soil to be deformed or molded in a moist condition without cracking or falling apart.

SOIL POROSITY Percentage of soil or rock not occupied by solid particles.

SOIL PRESSURE *See* **Contact pressure.**

SOIL PROFILE Vertical section showing the succession of soils at a location.

SOIL REPORT Report based on samples, tests, and analysis done to determine the various material composition and structural capabilities of soils in a given area.

SOIL SAMPLE Representative specimen.

SOIL SAMPLER Equipment used to extract soil samples from borings or test pits made in subsurface investigations.

SOIL SATURATION Stage where all the pore spaces of a soil are filled with water.

SOILS ENGINEER *See* **Geotechnical engineer.**

SOIL SERIES Basic unit of soil classification, consisting of soils that are alike in all major profile characteristics save texture of the surface layer, and that have similar horizons.

SOIL STABILIZATION Chemical or mechanical treatment designed to either increase or maintain the stability of a mass of soil or to otherwise improve its engineering properties. There are several techniques, including:

Cement-modified soil: Soil mixed with a relatively small amount of Portland cement; less than would be necessary to produce a hardened compound. *Also called* **Cement stabilization.**

Flyash stabilization: Technique of mixing controlled quantities of fly ash with soil during the process of preparation, compaction of base courses and subbases for road construction, embankments and other earth works, to limit or control the rate of moisture loss.

Salt stabilization: Use of sodium chloride to lower the rate of evaporation of water during the compaction of soil.

Soil–cement: Mixture of soil and measured amounts of Portland cement and water, compacted to a high density.

SOIL STACK General term for the vertical main of a system of soil, waste, or vent piping.

SOIL TEST Sampling of an area to determine the characteristics of its soils and to map their location; usually accomplished by borings and subsequent laboratory analysis. *See also* **Soil types.**

SOIL TEST BORING Subsurface sample-taking program under controlled conditions using equipment that obtains samples of predictable size and shape to predetermined depths, maintaining the sample size and shape for transport to a laboratory for analysis.

SOIL TEXTURE Classification of soils based on the relative proportions of the various soil separates present. The soil textural classes are: clay, sandy clay, silty clay, clay loam, silty clay loam, sandy clay loam, loam, silt loam, silt, sandy loam, loamy sand, and sand.

SOIL TYPES (*See also* **Soil classification system, Soil groups,** and **Soil limits**). Classification of soils into identifiable groups or types. A typical grouping is:

Bank sand: Sand having sharp edges.

See also **Lake sand** and **Sand** (below).

Cemented: Soil in which the particles are held together by a chemical agent, such as calcium carbonate, such that a hand-size sample cannot be crushed into powder or individual soil particles by finger pressure.

Clay: Natural cohesive soils having plastic properties and composed of very fine particles that are firmly coherent, compact, and hard when dry, but usually stiff, viscid and ductile when moist. *See also* **Material density.**

Clay loam: Mixture of sand, clay, and silt, having a large percentage of clay.

Cohesive: Clay (fine-grained soil) or soil with a high clay content, that has cohesive strength. Cohesive soil does not crumble, can be excavated with vertical sideslopes, and is plastic when moist. Cohesive soil is hard to break up when dry, and exhibits significant cohesion when submerged. It includes clayey silt, sandy clay, silty clay, clay, and organic clay.

Compact coarse sand: Soil consisting of coarse particles, 0.063 mm (0.0024 in.) or less in diameter, that have been compacted by the weight of overburden or weather. The grains are generally spherical or angular in shape, depending on the extent of weathering and/or decomposition.

Compact fine sand: Sand predominantly retained on the No. 200 (0.074 mm, 0.0029 in.) sieve that, when confined, whether wet or dry, will bear heavy loads. Fine sand has a lower bearing value than coarse sand, since the fine particles can be rearranged and tend to squeeze together, However, water will make fine sand flow more readily than coarse sand.

Dry: Soil that does not exhibit visible signs of moisture content.

Fissured: Soil material that has a tendency to break along definite planes of fissure with little resistance, or material that exhibits open cracks, such as tension cracks, in an exposed surface.

Granular: Gravel, sand, or silt (coarse-grained soil) with little or no clay content. Granular soil has no cohesive strength. Some moist granular soils exhibit apparent cohesion. Granular soil cannot be molded when moist and crumbles easily when dry.

Hardpan: 1. Dense, heterogeneous mass of clay, sand and gravel of glacial drift origin; **2.** Hard layer of consolidated or cemented earth underlying surface soil.

Lake sand: Sand having rounded, waterworn edges. *See also* **Bank sand,** (above) and **Sand** (below).

Layered system: Two or more distinctly different soil or rock types arranged in layers. Micaceous seams or weakened planes in rock or shale are considered layered.

Loam: *See* **Topsoil,** below.

Moist soil: Condition in which a soil looks and feels damp. Moist cohesive soil can easily be shaped into a ball and rolled into small-diameter threads before crumbling. Moist granular soil that contains some cohesive material will exhibit signs of cohesion between particles.

Plastic: A property of a soil that allows it to be deformed or molded without cracking or appreciable volume change. A test is to roll it into 3-mm (0.125-in.) diameter strings without it crumbling.

Sand: 1. Small grain of mineral, largely quartz, that is the result of disintegration of rock; **2.** Granular material passing the 9.6-mm (0.375-in.) sieve and almost entirely passing the No. 4 (4.76-mm, 0.187-in.) sieve and predominantly retained on the No. 200 (0.074-mm, 0.0029-in.) sieve. *Also called* **Fine aggregate.** *See also* **Material density.**

Sandy clay: Soil type characterized by sand containing enough clay to act as a binder, thus exhibiting the properties of a sandy soil but lacking the movement of loose sand, and without the slipping or shearing qualities of clays.

Sandy loam: Soil type containing enough clay and silt to render it cohesive.

Saturated: Soil in which the voids are filled with water. Saturation does not require flow. Saturation, or near saturation is necessary for the proper use of instruments such as a pocket penetrometer or sheer vane.

Shale: Laminated and fissile sedimentary rock, the constituent particles of which are principally in clay and silt sizes. *See also* **Material density.**

Sharp sand: Coarse sand consisting of particles of angular shape.

Silt: Granular material resulting from the disintegration of rock, with grains passing a No. 200 (0.074-mm, 0.0029-in.) sieve. *See also* **Mineral filler.**

Silt loam: Sandy soil having a moderate amount of clay and sand, in which over 50% of the sand is composed of extremely fine particles.

Soil: Generic term for unconsolidated natural surface material above bedrock.

Topsoil: 1. Uppermost layer of soil; **2.** Adequately drained soil containing humus and capable of supporting good plant growth. *Also called* **Loam.** *See also* **Material density.**

SOIL-WATER BELT *See* **Belt of soil water.**

SOLAR Of, pertaining to, or proceeding from the Sun.

SOLAR ACCESS 1. Exposure to, and collection of sunlight; **2.** Right to maintain exposure to sunlight already enjoyed.

SOLAR BATTERY System consisting of a large number of connected solar cells.

SOLAR CELL Device that converts sunlight into electrical energy, or that captures and transmits the thermal component of sunlight.

SOLAR COLLECTOR Device that transforms solar radiation to usable heat. There are various types, including:

Concentrating: Curved collector, that may either increase the available surface area of the collecting surface, or reflect the collected sunlight and focus it on a solar cell.

Flat-plate: Fixed flat-shape collector, oriented to receive maximum exposure to sunlight.

Tracking: Moving collector that orients itself toward the sun.

SOLAR CONSTANT Average solar radiation reaching the Earth's atmosphere per minute.

SOLAR-CONTROL GLAZING Window glass that reduces the transmission of the Sun's heat and glare through tinting or partial reflection. The effect produced also reduces the transmission of daylight.

SOLAR COOLING SYSTEM System that converts solar energy into other forms of energy, which is then used for cooling.

SOLAR DAY Interval between two successive meridian passages of the Sun.

SOLAR DEGRADATION Deterioration in a material's properties caused by exposure to solar energy.

SOLAR ENERGY Heat that is derived from sunlight.

SOLAR ENERGY SYSTEM Building subsystem used to convert solar energy into thermal energy for heating and/or cooling.

SOLAR FLARE Temporary outburst of solar gases from a small area of the Sun's surface; a source of intense radiation.

SOLAR FRACTION Percentage of a building's heat energy requirement provided through a solar system.

SOLAR FURNACE Heat exchanger by which the solar energy focused through a concentrating collector is transferred to another medium: rocks, water, etc.

SOLAR GAIN Percentage of a building's heating or cooling load contributed by solar radiation striking the structure or entering it.

SOLAR HEATING Use of the Sun's heat to warm the interior of buildings and/or to heat water.

SOLAR HOUSE House having large quantities of heat-absorbing material behind large glass areas, designed to supplement or replace conventional heating methods.

SOLARIMETER Instrument used to measure the flux of solar radiation through a surface.

SOLAR INSOLATION Total available solar radiation, composed of direct, diffuse, and reflected radiation.

SOLAR IRRADIANCE IN WATER Related to the sunlight intensity in water, it is proportional to the average light flux (in the units of 10^{-3} Moles cm^2 day^{-1}) that is available to cause photoreaction in a wavelength interval over a 24-hour day at a specific latitude and season date.

SOLARIUM Room, gallery, or glassed-in porch exposed to the Sun.

SOLARIZE Affect by exposing to the Sun's rays.

SOLAR LIGHT Floodlight powered by batteries that are recharged during daylight hours by solar energy via a photovoltaic array.

SOLAR LOAD Cooling load attributable to heat from the Sun.

SOLAR MONTH One-twelfth of a tropical year.

SOLAR NOON Moment of the day that divides daylight hours exactly in half; half of the time between sunrise and sunset.

SOLAR ORIENTATION Positioning and design of a building to take advantage of maximum exposure to the winter Sun.

SOLAR POWER Useful energy extracted from solar radiation.

SOLAR PROTECTION Substances added to materials, or materials themselves that wholly or partially block ultraviolet rays from the Sun.

SOLAR PUMP Mechanical device driven by energy obtained through a solar collector.

SOLAR RADIATION Sun's energy as received on the Earth's surface.

SOLAR REFLECTING SURFACE Exterior finish to a roof or wall intended to reduce the effects of solar heating.

SOLAR RIGHTS Right to continue to enjoy the ability to receive direct sunlight.

SOLAR STILL Desalination plant driven by solar heat.

SOLAR STORAGE Medium used to hold heat obtained through a solar collector.

SOLAR SYSTEM Assembly of equipment designed to collect solar radiation and convert it to another form of usable energy.

SOLAR YEAR Time interval between two successive passages of the Sun through the vernal equinox; the calendar year of 365.2422 mean solar days: a tropical year.

SOLENOID Magnetically-operated mechanical device.

SOLID DRAWN TUBE *See* **Pipe.**

SOLIDS CONCENTRATION Density of the microorganism-carrying solids in an aeration tank.

SOLIDS-CONTACT CLARIFIER Unit in which liquid passes up through a solids blanket and discharges at or near the surface.

SOLIDS CONTENT Percentage by weight or by volume of nonvolatile components in a solution.

SOLID SLEEVE Fitting used to join two pipes of the same nominal diameter in a straight line.

SOLIDS RETENTION TIME Average retention time of suspended solids in a biological waste treatment process.

SOLID WASTE 1. Garbage, refuse, sludges, and other discarded solid materials, including solid waste materials resulting from industrial, commercial, and agricultural operations, and from community activities, but not including solids or dissolved materials in domestic sewage or other significant pollutants in water resources, such as silt, dissolved or suspended solids in industrial wastewater effluents, dissolved materials in irrigation return flow, or other common water pollutants; **2.** Sludge from a wastewater treatment plant, water supply treatment plant, or air pollution control facility and other discarded material, including solid, liquid, semisolid, or contained gaseous material resulting from industrial, commercial, mining, and agricultural operations, and from community activities; **3.** Refuse, more than 50% of which is municipal-type waste consisting of a mixture of paper, wood, yard wastes, food wastes, plastics, leather, rubber, and other combustibles, and noncombustible materials such as glass and rock. It can be grouped under the following categories:

Ash: Residue from the burning of combustible materials; may include extraneous noncombustibles, unburned carbon, as well as mineral matter inherent in the combustible material.

Bulky waste: Large discarded materials: appliances, furniture, scraped automobile parts, diseased trees, large branches, stumps, etc.

Combustible waste: The organic content of solid waste, including paper, cardboard, cartons, wood, boxes, excelsior, plastic, textiles, bedding, leather, rubber, paints, yard trimmings, leaves, and household waste, all of which will burn.

Commercial waste: From businesses, office buildings, apartment houses, stores, markets, theatres, and hospitals and institutional facilities.

Domestic waste: Putrescible and nonputrescible waste originating from a residential unit, and consisting of paper, cans, bottles, food wastes, and may include yard and garden wastes.

Food waste: Animal and vegetable discards from handling, storage, sale, preparation, cooking, and serving of foods.

Garbage: *See* **Food waste,** above.

Hazardous waste: Any waste material, or combination thereof, which poses a substantial present or potential hazard to human health or living organisms because such wastes are nondegradable or are persistent in nature, or because they can be biologically magnified, or because they can be lethal, or because they may otherwise cause or tend to cause detrimental cumulative effects.

Household solid waste: *See* **Domestic waste,** above.

Industrial waste: Discarded waste materials from industrial processes and/or manufacturing operations.

Infectious waste: Waste materials from a medical facility, hospital, or laboratory which may contain pathogens or other disease infected wastes which could be transmitted to another human during the collection, transportation, or disposal cycle.

Mixed municipal refuse: *See* **Municipal solid waste,** below.

Municipal solid waste: Domestic refuse and some commercial waste.

Noncombustible waste: Inorganic content of solid waste, including glass, metal, tin cans, foils, dirt, gravel, brick, ceramics, crockery, and ashes.

Oversize waste: *See* **Bulky waste,** above.

Pathological waste: *See* **Infectious waste,** above.

Problem waste: A general term used to describe bulky wastes, dead animals, abandoned vehicles, construction and demolition waste, hazardous and infectious waste, and any other waste that requires special considerations in the collection, handling/transportation, or disposal cycle.

Refuse: Term sometimes used in place of **Solid waste.** Used to define general community waste, which includes kitchen/food wastes and rubbish.

Residential waste: Discarded materials originating from residences.

Rubbish: Nonputrescible materials collected from residences, commercial establishments, and institutions.

Rubble: Rough stones of irregular shape and size, broken from large masses either naturally or artificially, as by weathering action or by demolition of buildings, pavements, roads, etc.

Slops: *See* **Swill,** below.

Sludge: A semiliquid sediment resulting from the accumulation of settleable organic/inorganic solids deposited from wastewater or other fluids in tanks or basins.

Special wastes: Wastes that require special care and consideration in the storage, collection, handling/transportation, and disposal cycle by reason of their pathological, explosive, or toxic/hazardous nature.

Street refuse: Material collected by manual and mechanical sweeping of streets and sidewalks; litter from public receptacles and dirt removed from catch basins.

Swill: Semiliquid waste material consisting of food waste and free liquids.

Trash: Larger (nonputrescible) residential solid wastes unsuitable for routine pickup by refuse collection.

Unconventional waste: Hazardous

445

wastes by reason of their pathological, explosive, radioactive, or toxic nature.

White goods: Discarded appliances such as stoves and washing machines.

Yard waste: Plant clippings, prunings, grass clippings and leaves, and other discarded material from yards and gardens.

SOLID WASTE BOUNDARY Outermost perimeter of the solid waste (projected in the horizontal plane) as it would exist at completion of the disposal activity.

SOLID WASTE-DERIVED FUEL Fuel derived from solid waste that can be used as a primary or supplementary fuel in conjunction with, or in place of, fossil fuels.

SOLID WASTE DISPOSAL Disposal of all solid wastes through landfilling, incineration, composting, chemical treatment, and any other method that prepares solid wastes for final disposition.

SOLID WASTE GENERATING UNIT Device that combusts any fuel or by-product/waste to produce steam or to heat water or any other heat transfer medium. This term includes any municipal-type solid waste incinerator with a heat recovery steam generating unit or any steam generating unit that combusts fuel and is part of a cogeneration system or a combined cycle system.

SOLID WASTE MANAGEMENT A planned program for effectively controlling the generation, storage, collection, transportation, processing, reuse, conversion or disposal of solid wastes in a safe, sanitary, aesthetically acceptable, environmentally sound, and economic manner.

SOLID WASTE STORAGE CONTAINER Receptacle used for the temporary storage of solid waste while awaiting collection.

SOLID-WASTE-DERIVED FUEL 1. Solid, liquid, or gaseous fuel derived from solid fuel for the purpose of creating useful heat which includes, but is not limited to, solvent refined coal, liquefied coal, and gasified coal; **2.** Fuel that is produced from solid waste and which can be used as a primary or supplementary fuel in conjunction with or in place of fossil fuels. The solid-waste-derived fuel can be in the form of raw (unprocessed) solid waste, shredded (or pulped) solid waste, or classified solid waste, gas or oil derived from pyrolyzed solid waste, or gas derived from the biodegradation of a solid waste.

SOLIFLUCTION Gradual movement of saturated soil and other material from high ground to a lower level.

SOLUBLE BOD Biochemical oxygen demand of water that has been filtered in the standard suspended solids test.

SOLUM Soil mantle, comprising the organic layer, the leached layer and the layer of deposition, but not including the parent material.

SOLUTE Substance dissolved in another substance, usually the component of a solution present in a lesser amount.

SOLUTION Spontaneously forming homogeneous mixture of two or more substances, retaining its constitution in subdivision to molecular volumes, displaying no settling, and having various possible proportions of the constituents which may be solids, liquids, gases, or intercombinations; the state of being dissolved.

SOLUTION CHANNEL Opening in solid rock created by the dissolving, usually by water, of material that formerly occupied the opening.

SOLUTION FEEDER Device for dispensing an additive in liquid or solid state at a predetermined rate.

SOLUTION MINING Technique of mineral extraction in which a chemical solution is used to leach the metal *in situ,* the leachate then being processed to recover the metal.

SOLVENT Material capable of dissolving another substance.

SOLVENT EXTRACTION Operation or method of separation in which a solid or solution is contacted with a liquid solvent (the two being mutually insoluble) to preferentially dissolve and transfer one or more components into the solvent.

SOLVENT RECOVERY SYSTEM Air pollution control system by which VOC solvent vapors in air or other gases are captured and directed through a condenser(s) or a vessel(s) containing beds of activated carbon or other adsorbents. For the condensation method, the solvent is recovered directly from the condenser. For the adsorption method, the vapors are adsorbed, then desorbed by steam or other media, and finally condensed and recovered.

SOMATIC MUTATION Mutation arising in a nonreproductive cell.

SONDE Instrument used to obtain direct measurements of the conditions of the atmosphere at various altitudes consisting of a set of transducers and either a recorder or transmitter, lifted aloft by a balloon or a rocket.

SONIC Of or relating to audible sound.

SONIC BOOM Loud transient explosive sound caused by the shock wave preceding an aircraft travelling at supersonic speed.

SOOT Finely-divided carbon particles resulting from the incomplete combustion of carbon fuels.

SORPTION Process of adsorption or absorption of a substance on or in another substance.

SORTED Divided into similar kinds, classes, classifications, etc.

SORTING Process of dividing into kinds, classes, classifications, etc.

SOUND 1. Long, relatively wide body of water connecting larger bodies of water; **2.** To determine the depth of water; **3.** Vibratory disturbance in the pressure and density of a fluid, or in the elastic strain in a solid, with frequency in the approximate range of 20 and 20,000 cycles per second, and capable of being detected by the organs of hearing.

SOUND ABSORPTION COEFFICIENT Fraction of sound energy absorbed by a material. *Also called* **Acoustic absorptivity.**

SOUND ATTENUATING INSULATION Unfaced mineral or glass-fiber material used to impede the transmission of sound waves.

SOUND ATTENUATION Reduction of objectionable noise to acceptable limits. *See also* **Attenuate.**

SOUND BARRIER Any structure or material that is relatively opaque to sound and which serves to absorb, dissipate, or otherwise diminish and obstruct the continuation of a sound wave.

SOUND BOARD Reflective surface designed and placed to direct or deflect sound toward a specific point.

SOUND EXPOSURE LEVEL Level in decibels calculated as ten times the common logarithm of time integral of squared A-weighted sound pressure over a given time period or event divided by the square of the standard reference sound pressure of 20 µPa and a reference duration of one second.

SOUND FOCUS Relatively small area in a room where the sound level is significantly higher than elsewhere.

SOUNDING 1. Determination of the measurement between the surface of a body of water and the bottom immediately below; **2.** Method of examining soil to 6 to 10 m (20 to 30 ft) depth by driving or hydraulically pushing a cone, steel rod, or small-diameter pipe (gas pipe) into the ground with a hammer or maul. With experience, the movement of the rod or pipe under each hammer blow can give an indication of the approximate types of soil materials that are being penetrated. *Also called* **Probing.**

SOUNDING LINE Weighted rope, wire, or cable used for measuring the depth of water.

SOUNDING WELL Vertical conduit in the mass of coarse aggregate for preplaced aggregate concrete, provided with continuous or closely spaced openings to permit the entrance of grout.

SOUND INSULATION Means taken to reduce, or eliminate the transmission of sound from one area to another.

SOUND ISOLATION Various means to

447

prevent transmission of noise and vibration.

SOUND LEVEL 20 times the logarithm to base 10 of the ratio of pressure of a sound to the reference pressure. The reference pressure is 20 µPa (20 µN/m²); the value of the sound level pressure, in psi or decibels.

SOUND-LEVEL METER Instrument designed to indicate, and possibly record the amplitude of sound, sometimes displaying the magnitude of the various frequencies.

SOUND LOCK Acoustically treated vestibule intended to prevent or diminish the transmission of sounds from one area to another; similar to an air lock.

SOUNDNESS 1. Freedom from flaws; **2.** Solid, free from cracks, flaws, fissures, or variations from an accepted standard.

SOUND POWER LEVEL Total energy per second emitted by a sound source, expressed in decibels.

SOUND PRESSURE Minute fluctuations in atmospheric pressure that accompany the passage of a sound wave and give rise to the sensation of hearing.

SOUND PRESSURE LEVEL Effective value of pressure fluctuations above and below atmospheric pressure caused by the passage of a sound wave, expressed in decibels.

SOUNDPROOFING Construction techniques and materials used exclusively to minimize or eliminate the transmission or reflection of sound waves.

SOUND REDUCTION FACTOR Value, usually in decibels, of the measure of sound transmission intensity reduction of a material. *Also called* **Acoustical reduction factor.**

SOUND TRANSMISSION CLASS Numerical rating of the ability of an assembly of materials to resist the transmission of airborne sound.

SOUND WAVE Pressure disturbance in air proceeding at approximately 341 m/sec (1120 ft/sec).

SOURCE Place from which anything comes

or is obtained.

SOURCE SEPARATION 1. Sorting at point of generation of specific discarded materials into specific containers for separate collection; **2.** Practices that reduce the amount of a pollutant or contaminant entering a waste stream or otherwise released to the environment.

SOUR GAS Natural gas containing high levels of hydrogen sulfide.

SPAN Scale or range of values an instrument is designed to measure and/or record.

SPARGER Air diffuser that produces large bubbles.

SPECIAL AQUATIC SITES Geographic areas, large or small, possessing special ecological characteristics of productivity, habitat, wildlife protection, or other important and easily disrupted ecological values. These areas are generally recognized as significantly influencing or positively contributing to the general overall environmental health or vitality of the entire ecosystem of a region.

SPECIAL ASSESSMENT Direct tax levy assessed to meet a specific condition beyond the range of the regular scale of charges.

SPECIALLY-DESIGNATED LANDFILL Landfill at which complete long-term protection is provided for the quality of surface and subsurface waters from pesticides, pesticide containers, and pesticide-related wastes deposited therein, and against hazard to public health and the environment. Such sites are located and engineered to avoid direct hydraulic continuity with surface and subsurface waters, and any leachate or subsurface flow into the disposal area is contained within the site unless treatment is provided. Monitoring wells are established and a sampling and analysis program conducted.

SPECIAL WASTES Nonhazardous solid wastes requiring handling other than that normally used for municipal solid waste.

SPECIES Fundamental category of taxonomic classification, ranking after a genus, and consisting of organisms capable of interbreeding. In chemistry, a kind of atom, molecule, or radical which has a distinct chemical structure and composition.

SPECIFIC ABSORPTION Capacity of water-bearing material to absorb water after gravity water has been removed.

SPECIFICATION Compilation of provisions and requirements for the performance of prescribed work, material, or product.

SPECIFICATION OF WORKS Written document describing all aspects of the construction to be carried out, the materials to be used, and the manner of their finishing. *Also called* **Architectural specification,** and **Book of specifications.**

SPECIFIC CAPACITY OF A WELL Amount of water that can be pumped from a well per unit measure of drawdown.

SPECIFIC DISCHARGE Rate of discharge of groundwater per unit cross-sectional area measured at right angles to the direction of flow.

SPECIFIC ENERGY Energy contained in a stream of water, expressed in terms of head, referred to the bed of the stream.

SPECIFIC GRAVITY Comparison of the density of one material with a reference material under specific test conditions. The reference material for specific gravity is the maximum density of water, that is, water at 4°C (39.2°F). The device used for this, typically a comparison of fuel, or coolant, or battery electrolyte with water, is a hydrometer. *Also called* **Apparent specific gravity.** *See also* **Baumé scale, Bulk density, Bulk specific gravity, Density (dry),** and **Liquid specific gravity.**

SPECIFIC HEAT Ratio of the amount of heat necessary to raise the temperature of a given weight of a given substance to one unit of temperature to the amount of heat required to raise the temperature of a similar mass of a reference material, usually water, by the same amount.

SPECIFIC HUMIDITY Ratio of the mass of water vapor to the total mass of the mixture of air and water vapor.

SPECIFIC LEVEL Level of the water surface at a particular site for a given discharge.

SPECIFIC RETENTION Ratio of the volume, or weight of water that a soil will retain against the force of gravity, having once been saturated, to its dry volume or weight.

SPECIFIC STORAGE Volume of water released from, or taken into storage, per unit volume of the porous medium.

SPECIFIC VOLUME Volume of unit mass.

SPECIMEN Material derived from a test system for examination or analysis.

SPIGOT Plain end of a bell and spigot pipe joint.

SPECIFIC YEAR FLOOD Flow of a stream that is equalled or exceeded, on average, once in a designated period.

SPECIFIC YIELD Volume of water that a unit volume of saturated permeable rock or soil will yield when drained by gravity.

SPECTRAL UNCERTAINTY Possible variation in exposure to the noise spectra in the workplace.

SPECTROGRAPH Spectroscope equipped to photograph spectra.

SPECTROHELIOGRAM Photograph of the Sun taken in a narrow wavelength band centered on a selected wavelength.

SPECTROHELIOGRAPH Instrument used to make a spectroheliogram.

SPECTROHELIOSCOPE Instrument used to observe solar radiation.

SPECTROMETER Spectroscope equipped with scales for measuring the positions of spectral lines.

SPECTROPHOTOMETER Instrument used to determine the distribution of energy in a spectrum of luminous radiation.

SPECTROSCOPY Study of spectra, especially the experimental observation of optical spectra.

SPECTRUM Distribution of a characteristic of a physical system or phenomenon.

SPECTRUM OF TURBULENCE Variation of turbulence intensity with frequency.

SPEED OF SOUND In air, the speed of propagation of sound waves is approximately 332 m/sec at 0°C

SPENT NUCLEAR FUEL Fuel that has been withdrawn from a nuclear reactor following irradiation, the constituent elements of which have not been separated by reprocessing.

SPHAEROTILUS **BULKING** Type of sludge bulking that occurs when a genus of filamentous bacteria, *Sphaerotilus*, is present in large numbers.

SPIGOT Plain end of a bell and spigot pipe joint.

SPILL Intentional or unintentional spills, leaks, and other uncontrolled discharges where the release results in any quantity of a contaminant running off or about to run off an external surface.

SPILL AREA Area of soil on which visible traces of a spill can be observed plus a buffer zone of 0.30 m (1 ft) beyond the visible traces.

SPILL BOUNDARIES Actual area of contamination as determined by post-cleanup verification sampling or by precleanup sampling to determine actual spill boundaries.

SPILL EVENT Discharge of oil or other contaminant into or upon soil or water.

SPILLWAY Overflow channel or chute. When part of a dam, it may be of several types:

> **Fuse plug:** Auxiliary or emergency spillway comprising a low embankment or saddle that is overtopped only during exceptionally high water.

> **Ogee:** Overflow whose channel, in longitudinal section, is in the shape of an S or ogee curve.

> **Primary:** Principal spillway over which water flows during flood flows.

> **Shaft:** Vertical or inclined shaft passing through, under, or around a dam and into which floodwater or surplus water is spilled.

> **Side channel:** Spillway whose crest is approximately parallel to the downstream channel.

> **Siphon:** Spillway having one or more siphons at crest level.

SPIRAL PIPE *See* **Pipe.**

SPIRAL AIRFLOW DIFFUSION Use of air, diffused into the aeration tank of the activated sludge process in such a manner as to induce a spiral or helical movement to the tank contents.

SPIRAL-RIVETED PIPE Pipe formed of steel sheets curved to form a cylinder with the edges overlapping and riveted together to form a helical seam.

SPIRAL-WELD PIPE *See* **Pipe.**

SPIT Narrow point of land extending into a body of water.

SPLASH PAD Construction that protects bare soil from erosion by splashing or falling water.

SPLIT PIPE Pipe cut along its length to form a channel.

SPLIT SPOON SAMPLER Type of drill core used in soil exploration.

SPLITTER BOX Chamber that divides a flow into two or more streams.

SPOIL Refuse material removed from an excavation.

SPOON Pipe-shaped tool split in half for part of its length, used to obtain soil samples when driven into the ground.

SPOON BLOW Blow of a 63.5-kg (140-lb) hammer falling 760 mm (30 in.) onto a 50-mm (2-in.) diameter OD by 35-mm (1.375-in.) ID split spoon sampler. *See also* **N Value.**

SPORE Microorganism, as a bacterium, in a dormant or resting state.

SPOUT Water outlet of a faucet.

SPOUT-DELIVERY PUMP *See* **Pump.**

SPOUTING VELOCITY Theoretical velocity of water issuing from an orifice under a stated head, when the effect of friction is eliminated.

SPRAY Liquid droplets larger than 10 μm, created mechanically by disintegration.

SPRAY AERATOR Device comprising a pressure nozzle through which water is propelled into the air in a fine spray.

SPRAY DRYER Type of dryer in which the liquid containing the solids to be dried is sprayed or atomized into a heated chamber.

SPRAY-IN-PLACE INSULATION Insulation material that is sprayed onto a surface or into cavities.

SPRAY IRRIGATION *See* **Direct irrigation.**

SPRAY NOZZLE Nozzle that produces a spray of water, as against a jet.

SPRAY POND Basin over which water is sprayed from nozzles, generally to reduce temperature and/or to oxygenate.

SPRAY TOWER Structure that permits the removal of particulate matter from an exhaust steam by spraying the rising plume with an alkaline solution.

SPRING Point where water spontaneously issues from rock or soil

SPRING TIDE Tide occurring when its range is at maximum, immediately following the new and full moon.

SPRING WATER Water derived from a natural spring.

SPRINKLE Light rain of scattered drops.

SPRINKLER Device used to scatter or spray wastewater onto a filter bed, or to distribute water over a designated area.

SPRINKLER CONNECTION Siamese connection used to increase the water supply and pressure to a sprinkler system.

SPRINKLERED Building or part of a building equipped with a system of automatic fire extinguishing sprinkler heads.

SPRINKLER IRRIGATION Use of sprinklers set at specific intervals on pipes spaced apart at specific distances to distribute irrigation water over a large area.

SPRINKLER NOZZLE Irrigation nozzle that ejects a spray through a 360° arc in segments, actuated by water pressure on an internal mechanism.

SPRINKLER SYSTEM 1. Fire extinguishing system consisting of pipes mounted in or under the ceilings of rooms and equipped with nozzles. The pipes are connected to a water main (but not necessarily charged with water) and the nozzles will eject a spray of water when the system is activated by heat or some other criteria; **2.** System, usually underground, by which lawns and beds can be irrigated.

SPRINKLER TONGS Tool used to stop the flow of water from a sprinkler head.

SPRINKLER WEDGE Device used to temporarily shut off the flow of water from a sprinkler head.

SPUD Movable vertical pipe, H-section or pile, placed through a frame on a floating pile driver or dredge and driven into the bottom silt of a waterway to hold the vessel in position; **2.** Steel tube, pointed at the bottom and fitted with lifting tackle at the top; spuds are the means by which a dredge is held in position while operating.

SPUDDING Act of opening a hole through dense material by dropping or driving a spud.

SPUD KEEPER Framework on the back of a dredge that holds spuds or legs dropped down to anchor the vessel while dredging.

SPUN PIPE *See* **Pipe.**

SPUR DIKE Riprap or rock fill shaped as a quarter of an ellipse, built in front of or behind abutments or piers to prevent scouring. Its length is more than 1.5 times its width. *See also* **Groin.**

SPUR VALLEY Short branch valley.

SQUALL Brief, sudden and violent wind storm accompanied by rain or snow.

SQUALL LINE Zone of squalls and other violent changes in weather that marks the replacement of a warm air current by cold air.

SQUARE Non-SI unit of measurement, such as 100 ft^2 (10 x 10 ft) (9.29 m^2 3.048 x 3.048 m) usually applied to roofing materials (that may also be used as sidewall covering).

SQUARE CENTIMETER SI unit of area. Symbol: cm^2. *See also the appendix*: **Metric and nonmetric measurement.**

SQUARE FOOT Nonmetric unit of area. Symbol: ft^2. Multiply by 0.092 9 to obtain square meters, symbol: m^2. *See also the appendix*: **Metric and nonmetric measurement.**

SQUARE INCH Nonmetric unit of area. Symbol: in.^{2}. Multiply by 6.4516 to obtain square centimeters, symbol: cm^2; by 645.16 to obtain square millimeters, symbol: mm^2. *See also the appendix*: **Metric and nonmetric measurement.**

SQUARE KILOMETER SI unit of area. Symbol: km^2. *See also the appendix*: **Metric and nonmetric measurement.**

SQUARE METER SI unit of area. Symbol: m^2. Multiply by 0.0001 to obtain hectares, symbol: ha; by 10.7639 to obtain square feet, symbol: ft^2; by 1.196 to obtain square yards, symbol: yd^2. *See also the appendix*: **Metric and nonmetric measurement.**

SQUARE METER PER SECOND A derived unit of kinematic viscosity with a compound name of the SI system of measurement. Symbol: m^2/s. *See also the appendix*: **Metric and nonmetric measurement.**

SQUARE MILE Nonmetric unit of area. Symbol: mi^2. Multiply by 258.999 to obtain hectares, symbol: ha; by 3.589 999 to obtain square kilometers, symbol: km^2. *See also the appendix*: **Metric and nonmetric measurement.**

SQUARE YARD Nonmetric unit of area.

Symbol: yd^2. Multiply by 0.836 to obtain square meters, symbol: m^2. *See also the appendix*: **Metric and nonmetric measurement.**

STABILITY Resistance of material in a cut or fill to movement downslope due to inherent characteristics or to the weight of a superimposed load.

STABILIZATION Process by which wastes are rendered relatively inert, uniform, biologically inactive, nuisance-free, or harmless.

STABILIZATION LAGOON Shallow pond for the storage of wastewater before its discharge.

STABILIZATION POND Oxidation pond in which biological oxidation of organic matter is encouraged by natural or artificial accelerated transfer of oxygen to the water from the air.

STABILIZE To make soil firm and prevent it from moving.

STABILIZED Condition of equilibrium.

STABILIZED CHANNEL Earth channel which, over a period of time, has not experienced appreciable erosion of deposition of silt.

STABILIZED WASTE Waste which, if discharged or released, will have no deleterious effect on the stream into which it is emptied.

STABLE EFFLUENT Wastewater treatment effluent that has no adverse effect or demand upon the receiving water.

STABLE ROCK *See* **Soil classification system.**

STACK 1. Any structure that contains a flue or flues for the discharge of gases; **2.** Vertical run of drain-waste-and-vent piping.

STACK EFFECT Rise of air through a tall space caused by temperature and/or pressure difference.

STACK EMISSIONS Particulate matter captured and released to the atmosphere through a stack, chimney, or flue.

STACK GAS Conglomerate of gaseous, solid and liquid particles generated by a source and contained within a ventilating stack.

STACK PIPE Waste pipe to which waste branches are connected.

STACK SAMPLING Collection of representative samples of gaseous and particulate matter that flows through a duct or stack.

STACK VENT Extension of the waste pipe above the highest horizontal connecting drain, through the roof to atmosphere.

STAFF GAUGE Graduated scale by which the height above a datum of a fluid can be measured.

STAGE 1. Stages in the life of a project, separated by milestone events that are sign-offs or approvals by authorities. The seven stages in the life-cycle of a construction project are:

(a) Investigations.

(b) Preliminary analysis.

(c) Conceptual design.

(d) Project brief.

(f) Control and drawings.

(g¹) Construction.

(g²) Construction closeout.

(h) Operations.

Of these, stages **(c)** and **(d)** are the development process or feasibility phase, and stages **(e)**, **(g¹)**, and **(g²)** are the construction process or construction phase; **2.** Suspended structure that supports the working load of a single, two-point or multiple-point scaffold; **3.** Hydraulic amplifier used in a servovalve: may be single stage, two stage, three stage, etc.; **4.** Space designed primarily for theatrical performances with provision for quick-change scenery and overhead lighting, including environmental control for a wide range of lighting and sound effects and which is traditionally, but not necessarily, separated from the audience by a proscenium

wall and curtain opening; **5.** Water level measured from any chosen reference.

STAGE AERATION Division of the activated sludge process into stages allowing for intermediate settling of solids and the return of sludge to each stage.

STAGE DIGESTION Processing of waste sludge in two or more stages in series; primary and secondary digestion.

STAGE-DISCHARGE RELATION Relation between gauge height and discharge of a stream or conduit at a gauging station.

STAGE TREATMENT Any process in which a number of treatment stages are arranged in series.

STAGED TRICKLING FILTRATION System where the effluent from one trickling filter becomes the influent of another unit of the same basic design and capacity, with or without intermediate sedimentation.

STAGNATION POINT Phenomenon on the upstream side of the point where a fluid flow divides, to pass by an object, or a point on the surface of the object, at which the velocity of flow is zero.

STAINLESS STEEL PIPE *See* **Pipe.**

STALE WASTEWATER Wastewater containing little or no oxygen and in which organic matter is close to putrefaction.

STAND 1. Volume of standing timber per unit area; **2.** Aggregation of plants of more-or-less uniform species composition, age and condition.

STANDARD Document, or an object for physical comparison, for defining product characteristics, products, or processes, prepared by a consensus of a properly constituted group of those substantially affected by, and having the qualifications to prepare the standard for use.

STANDARD ABSORPTION TRENCH Trench, 0.3 to 1.2 m (12 to 36 in.) wide, containing 300 mm (12 in.) of clean, coarse aggregate and a distribution pipe covered with a minimum 300 mm (12 in.) of earth.

STANDARD AIR Air at a temperature of 20°C (68°F), a pressure of 1.03 kg/cm² (14.70 psi) absolute, and a relative humidity of 36%. (In gas industries, the temperature of standard air is usually given as 21.1°C (70°F).) *See also* **Compressed air,** and **Free air.**

STANDARD ATMOSPHERE Non-SI unit of pressure, equal to 101.325 kPa, permitted for use with the SI system of measurement for a limited time. Symbol: atm. *See also the appendix:* **Metric and nonmetric measurement.**

STANDARD BIOCHEMICAL OXYGEN DEMAND Biochemical oxygen demand as determined under standard laboratory procedures.

STANDARD CONDITION 1. Atmospheric pressure at sea level of 760 mm (29.92 in.) of mercury; **2.** Reference for compressible fluids of 15.6°C (60°F) at 1.0 atmosphere (total pressure) at dry-gas conditions.

STANDARD CUBIC FEET OF AIR PER MINUTE Volume of air delivered under standard conditions of temperature, pressure, and humidity, i.e. 0°C, 14.7 psia, and 50% relative humidity.

STANDARD DEVIATION Constant allowance due to mechanical misadjustment or other permanent condition. *See also* **Coefficient of variation,** and **Deviation.**

STANDARD DIMENSION Manufacturer's designated dimension.

STANDARD DRYING DAY Day that produces the same net drying as experienced during a 24-hour period under laboratory conditions where the dry-bulb temperature is maintained at 32°C (90°F) and the relative humidity at 20%.

STANDARD-DUTY DUMP BODY *See* **Truck.**

STANDARD ERROR Standard deviation of many samples of the mean that shows the mount of inconsistency between the sample and the mean of the whole.

STANDARD FALL DIAMETER Diameter of a sphere having a specific gravity of 2.65 and the same standard fall velocity as the particle in question.

STANDARD FALL VELOCITY Average rate of fall that a particle would attain if falling alone in quiescent distilled water of infinite extend and at a temperature of 24°C (75.2°F).

STANDARD FORM OF CONTRACT Generic contract documents with conditions typical of most forms of construction that require the addition of many details specific to the job on hand.

STANDARD INSIDE DIAMETER DIMENSION RATIO Ratio of the average specified inside diameter to the minimum specified wall thickness of a pipe.

STANDARDIZE Reduce to or compare to a standard.

STANDARD METHODS Procedures established by a recognized authority.

STANDARD ORIFICE Orifice with a sharp edge, in which water passing touches only the line of the edge.

STANDARD OUTSIDE DIAMETER DIMENSION RATIO Ratio of the average specified outside diameter to the minimum specified wall thickness of a pipe.

STANDARD PENETRATION RESISTANCE Number of blows of a 63.5-kg (140-lb) hammer falling 750 mm (30 in.), required to advance a 50-mm (2-in.) OD, split-barrel sampler 300 mm (12 in.) through a soil mass.

STANDARD PENETRATION TEST Number of blows required to drive a 50-mm (2-in.) O.D., 35 mm (1-3/8 in.) I.D., 600-mm (24-in.) long, split, soil-sampling spoon 300 mm (1 ft) with a 63.5-kg (140-lb) weight falling freely 750 mm (30 in.). The count is recorded for each of three 150-mm (6-in.) increments. The sum of the second and third increments is taken as the N value in blows/ mm (blows/ft).

STANDARD PLAN Drawing approved for repetitive use, showing details to be used, where appropriate. *See also* **Plan,** and **Working drawings.**

STANDARD PRESSURE Working pressure of 861.8 kPa (125 psi); the minimum pressure for which plumbing fixtures are designed.

STANDARD PROVISION Contract and construction clause that is common to many types of project and that is used unless the situation requires some special consideration or treatment.

STANDARD-RATE FILTER Trickling filter in which both hydraulic and organic loadings are relatively low, commonly designed to operate without recycling or recirculating the flow.

STANDARD RATING Rating based on tests performed under standard rating conditions.

STANDARD SAMPLE Aliquot of finished drinking water that is examined for the presence of coliform bacteria.

STANDARD SAND Ottawa sand accurately graded to pass a No. 20 (0.84-mm, 0.0331-in.) sieve and be retained on a No. 30 (0.59-mm, 0.0232-in.) sieve. *See also* **Graded standard sand** and **Ottawa sand.**

STANDARD SHORT TUBE Tube having a diameter about one third its length.

STANDARDS OF PROFESSIONAL PRACTICE Statements of ethical principles promulgated by professional societies to guide their members in the conduct of professional service.

STANDARD SOLUTION Solution in which the exact concentration of a chemical or compound is known.

STANDARD SPECIFICATIONS Book of specifications approved for general application and repetitive use.

STANDARD TEMPERATURE AND PRESSURE Temperature of 0°C (32°F) and a barometric pressure of 760 mm (29.9 in.) of mercury.

STANDARD THREAD National Standard Hose threads.

STANDARD WIPE TEST For spills of high concentration on solid surfaces, a cleanup to numerical surface standards and sampling by a standard wipe test to verify that the numerical standards have been met. This definition constitutes the minimum requirements for an appropriate wipe testing protocol. A standard-size template (100 x 100 mm) is used to delineate the area of cleanup; the wiping medium is a gauze pad or glass wool of known size which has been saturated with hexane. It is important that the wipe be performed very quickly after the hexane is exposed to air.

STANDBY Workers or equipment ready to undertake some activity without further preparation or delay; for machinery, it is often a condition between start/stop and run.

STANDBY EQUIPMENT Equipment that comes into use following a breakdown in supply of normal services.

STANDBY POWER SUPPLY Power supply that is selected to furnish electric energy when the normal power supply is not available.

STANDING CROP Amount of living matter present in a population of one or more species within a given area: the biomass.

STANDING WATER LEVEL Level at which groundwater stands in a hole or pit left open for a prolonged period.

STANDING WAVE Wave on the surface of a body of water formed when a stream enters it at high velocity, or because of a hydraulic jump, or because a wave travelling in one direction has met another wave of equal force travelling in the opposite direction.

STANDPIPE 1. Vertical water storage pipe used to establish uniform pressure in a water distribution system; **2.** Pipe or tank connected to a closed conduit and extending to or above the hydraulic grade line; **3.** Fire protection system consisting of a piping arrangement either wet or dry to take water to upper floors or remote areas of buildings where fire department outlets and private hoselines are provided.

STAND-UP TIME Time an unsupported excavation can be maintained in a tunnel or drill pier.

STAPHYLOCOCCUS Any of gram-positive, spherical parasitic bacteria of the genus *Staphylococcus,* occurring in grapelike clusters and which can be pathogenic for man. *Also called* **'staph.'**

START DATE Point in time associated with an activity's start, usually qualified by one of the following: actual, planned, estimated, scheduled, early, late, target, baseline, or current.

STARTING MATERIAL Substance used to synthesize or purify a technical grade of active ingredient (or the practical equivalent of the technical grade ingredient if the technical grade cannot be isolated) by chemical reaction.

START TIME Commencement of an operation or action in the critical path method of project management.

START-UP Initiation of a mechanical device or system.

START-UP COST Aggregation of the costs, excluding acquisition costs, incurred to bring a new facility into production.

START-UP TIME Period of time needed to reach a steady-state condition within the operating band, starting from a long-term off condition.

STASIS Stagnation or cessation of the life processes within organisms.

STATEMENT OF WORK Narrative description of products or services to be supplied under contract.

STATIC ELECTRICITY Electricity that is at rest, rather than flowing.

STATIC FRICTION COEFFICIENT Index of the force necessary to cause a body to begin to slide over the surface of another body.

STATIC HEAD Vertical distance between the free level of the source of supply and the point of free discharge or level of the free surface.

STATIC PENETRATION TEST Test of soil in which the testing device is pushed into the soil with a measurable force.

STATIC PRESSURE *See* **Pressure.**

STATIC-REPLACEMENT TEST Test method in which the test solution is periodically replaced at specific intervals during the test.

STATIC RESERVE INDEX Time that the known reserves of a resource will last if the rate at which they are being used remains constant.

STATIC SUCTION HEAD Vertical distance from the source of supply, when its level is above the pump, to the centerline of the pump.

STATIC SUCTION LIFT Vertical distance between the center of the suction of a pump and the free surface of the liquid being pumped.

STATIC SYSTEM Test system in which the test solution and test organisms are placed in a test chamber and kept there for the duration of the test, without renewal of the test solution.

STATIC TEST Toxicity test with aquatic organisms in which no flow of test solution occurs. Solutions may remain unchanged throughout the duration of the test.

STATIC WATER LEVEL Level at which water stands in a well when no water is being pumped from, or being added to the aquifer.

STATIC WATER SUPPLY Supply of water at rest that does not supply a pressure head for fire fighting, but which may be employed as a suction source for fire pumps.

STATIONARY COMPACTOR Powered machine designed to compact solid waste or recyclable materials, and which remains stationary when in operation.

STATIONARY DREDGER Bucket-ladder dredge that is not self-propelled and which discharges into a hopper barge or pipeline.

STATIONARY EMISSION SOURCE Stationary facility that releases combustion gases or vapors to the environment.

STATIONARY GAS TURBINE Simple cycle gas turbine, regenerative cycle gas turbine or any gas turbine portion of a combined-cycle steam/electric generating system that is not self propelled, but which may be mounted on a vehicle for portability.

STATIONARY PACKER *See* **Refuse truck.**

STATIONARY SOURCE Building, structure, facility, or installation which emits or may emit an air pollutant.

STATION RATING CURVE Relation between a station gauge height and the discharge of a stream or conduit.

STATISTICAL PROCESS CONTROL CHART Plot of the daily performance of a plant and/or process, such as a trend chart.

STATISTICAL SIGNIFICANCE Statistical measurement obtained using appropriate standard techniques of multivariate or other analysis, with results interpreted at the 95% or greater confidence level and based on data-relating species which are present in sufficient numbers at control areas to permit a valid statistical comparison with the areas being tested.

STATISTICAL SOUND LEVEL Level in decibels that is exceeded in a stated percentage (x) of the duration of the measurement period.

STATOR Portion of a machine that contains the stationary parts which surround the moving parts.

STEADY FLOW Stream line flow; flow in which either the quantity, or the velocity of water passing a given point per unit of time remains constant, but not both.

STEADY NONUNIFORM FLOW Flow in which the quantity of water flowing per unit of time remains constant at every point along a conduit, but where the velocity varies due to changes in hydraulic characteristics.

STEADY STATE Time period during which the amounts of test substance being taken up and depurated by the test organisms are equal, i.e. equilibrium.

STEADY-STATE BIOCONCENTRATION FACTOR Mean concentration of the test chemical in test organisms during steady-state, divided by the mean concentration of the test chemical in the test solution during the same period.

STEADY-STATE OR APPARENT PLATEAU Condition in which the amount of test material being taken up and depurated is equal at a given water concentration.

STEADY UNIFORM FLOW Flow in which the quantity and velocity of water flowing per unit of time remains constant.

STEAM Water in the vapor state.

STEAMER CONNECTION Larger of two or more nozzles on a hydrant, used to deliver water to pumpers.

STEAM FOG Fog formed when cold air overlies a body of warm air.

STEAM GENERATING UNIT Furnace, boiler, or other device used for combusting fuel for the purpose of producing steam.

STEAM STRIPPING Distillation operation in which vaporization of the volatile constituents of a liquid mixture takes place by the introduction of steam directly into the charge.

STEAM TRAP Pipe arrangement allowing the passage of condensate, or air and condensate, and preventing the passage of steam.

STEEL PIPE *See* **Pipe.**

STEENING Lining a well or soakpit with brick, stone, etc.

STEM Shaft of a faucet to which the handle is attached.

STEMFLOW Moisture fed to the ground down the stems of plants.

STEP AERATION Technique of adding increments of settled wastewater at intervals along the line of flow in an aeration tank(s) in the activated sludge treatment process.

STEP IRON Heavy U-shaped metal shape built horizontally into a wall and projecting from it to form the steps of a vertical ladder.

STERADIAN One of the two supplementary units of the SI system: a unit of solid angle equal to the measure of a solid angle with its vertex at the center of a sphere and enclosing an area of the spherical surface equal to that of a square with sides equal in length to the radius. Symbol: sr. *See also the appendix*: **Metric and nonmetric measurement.**

STERE Obsolete metric unit of volume, equal to 1 m^3, that should not be used with the SI system of measurement. Symbol: st. *See also the appendix*: **Metric and nonmetric measurement.**

STERILIZATION Treatment of raw water and new potable water supply pipes to destroy potentially harmful bacteria, by chlorination, bromine or ozone treatment. *See also* **Sewage treatment.**

STERILIZED WASTEWATER Treatment plant effluent in which all microorganisms have been destroyed.

STEVENSON SCREEN Standardized wooden container in which instruments used to measure surface weather are housed. It is of louvered construction through which the air mass may pass freely, while thermometers are shielded from direct sunlight. The white-painted box is placed so that the thermometers are 1.25 m above ground.

STILLING POOL Artificial deepening of a river bed at the foot of a dam spillway that reduces hydraulic velocity and minimizes scour.

STILL WATER Body of water in which there is no apparent flow.

STILL-WATER LEVEL Assumed surface of turbulent water if all wave and wind action were to cease.

STIMULATION Techniques used to artificially increase the output from a well, i.e. application of an acid, use of explosives, injection of water under high pressure, etc.

STIPULATED PRICE CONTRACT Contract for work to be done that does not allow for variation, for whatever reason, from the negotiated price. *Also called* **Fixed price contract** and **Lump sum agreement.** *See also* **Lump sum contract.**

STIPULATION Requirement or term within a written contract.

STOCHASTIC EFFECTS Condition where the probability of an adverse medical effect occurring is proportional to the radiation dose received.

STOCK POND Excavation around a spring that impounds drinking water for livestock.

STOCK SOLUTION Source of a test solution, prepared by dissolving the test substance in dilution water or a carrier which is then added to dilution water at a specified, selected concentration by means of the test substance delivery system.

STOICHIOMETRIC Exact or fixed proportions of elements in a chemical or of reactants to produce a compound.

STOKE Obsolete metric unit of kinematic viscosity that should not be used with the SI system of measurement. Symbol: St. *See also the appendix*: **Metric and nonmetric measurement.**

STOKER Mechanical device used to regulate the feed of a solid fuel into a boiler.

STOKES' LAW Mathematical expression for the drag of a small sphere falling through an infinite fluid.

STOMA One of the very tiny openings in the surface of a leaf, through which water vapor and gases pass in and out.

STOMATAL TRANSPIRATION Outward diffusion of water vapor into the atmosphere through the stomata of plants.

STONE DRAIN Rubble drain; trench filled with stones and aggregate so as to drain groundwater or surface water.

STOP-AND-WASTE COCK *See* **Valve.**

STOPCOCK *See* **Valve.**

STOP LOG Removable log or board in an outlet box, weir or other device, used to control the overflow level.

STOP SWITCH *See* **Switch.**

STOP VALVE *See* **Valve.**

STOP WORK ORDER Written instruction by the owner's representative, or by a regulatory authority, to cease work on a project for specified cause, e.g. failure to meet the terms of the contract documents, failure to comply with building regulations or codes, labor disputes, etc.

STORAGE Known volume of water, either in surface or underground reservoirs, held in reserve against some future requirement.

STORAGE CAPACITY Volume that a vessel or container is designed to hold.

STORAGE COEFFICIENT Volume of water released from an aquifer per unit area per unit decline of head.

STORAGE GALLERY Water-collecting or -transporting formation.

STORAGE PIT Pit in which solid waste is held prior to processing.

STORAGE RATIO Ratio of the net available storage of an impounding reservoir to the annual mean flow of the stream feeding it.

STORAGE RESERVOIR Impounding reservoir in which raw water is retained for a considerable period.

STORAGE TANK Closed container that has a capacity of more than 250 L (66 gal), designed to be installed in a fixed location.

STORM Atmospheric disturbance manifested in strong winds accompanied by rain, snow, or other precipitation, and often by thunder and lightning; wind ranging from 102 to 115 km/h (64 to 72 mph).

STORM CENTER Central area covered by a storm; especially, the point of lowest barometric pressure within a storm.

STORM DRAIN Drain that conveys rain or groundwater, but which excludes sewage and polluted industrial wastes.

STORM FLOW 1. Portion of precipitation that leaves the drainage area in a comparatively short time on or near the surface; **2.** Additional flow within a combined sewer system that is attributable to precipitation that has entered the system.

STORM OVERFLOW Weir that permits discharge from a combined sewer of flows in excess of design capacity.

STORM OVERFLOW SEWER Sewer used to carry excess flows from a combined system.

STORM SEWER Sewer that conveys water collected by storm drains, but which excludes sewage and polluted industrial wastes.

STORM SURGE Unusual variation in the amplitude of the tide, caused by atmospheric conditions of wind and pressure gradients.

STORM TIDE Higher, or lower than predicted tide levels caused by high winds pushing water onto or away from the shore.

STORMWATER Excess rainwater that is not absorbed into the ground.

STORMWATER CHANNEL Gully or channel built to carry storm water flows.

STORMWATER OR WASTEWATER COLLECTION SYSTEM Piping, pumps, conduits, and any other equipment necessary to collect and transport the flow of surface water runoff resulting from precipitation, or domestic, commercial, or industrial wastewater to and from retention areas or any areas where treatment is designated to occur.

STORMWATER OVERFLOW WEIR In a combined or partially combined system of sewers, a weir within the system that allows flows above a certain volume, calculated to occur following a local heavy storm, to overflow into a storm sewer, or be diverted into a tank, temporary holding pond, etc.

STORMWATER POINT SOURCE Conveyance or system (including pipes, conduits, ditches, and channels) primarily used for collecting and conveying stormwater runoff and which is located at an urbanized area or which discharges from lands or facilities used for industrial or commercial activities.

STORMWATER SEWER SYSTEM Drain and collection system designed and operated for the sole purpose of collecting stormwater and which is segregated from the process wastewater collection system.

STORMWATER TANK Sedimentation tank through which storm water flows prior to discharge to a receiving body of water, to permit settlement of solids.

STOSS Direction from which ice has come: stoss-and-lee topography is a landform showing rocks with smoothly abraded slopes on one side and broken, steep slopes on the other.

STRAIGHT TEE Pipe fitting consisting of a tee with all openings of the same size.

STRAINER 1. Wire or other metal guard used to prevent debris from clogging an intake hose or pipe; **2.** Filtering device for the removal of coarse solids from a fluid.

S-TRAP Sanitary fitting that provides a water seal where the inlet and outlet are vertical, offset from each other. *See also* **P-trap.** *Also called* **Siphon trap.**

STRAPPED ELBOW Drop elbow: a plumbing fitting.

STRATIFICATION Separation of a material into layers.

STRATIFIED SOILS Soils or rocks having pronounced horizontal layering indicating that they were laid down by deposition from the waters of rivers, lakes or seas.

STRATIGRAPHY Study of soils and rocks, their properties and geological time sequence.

STRATOCUMULUS Cumulus clouds in a layer generated by convection.

STRATOPAUSE Boundary at a height of approximately 50 km (31 miles) between the stratosphere and the mesosphere.

STRATOSPHERE Region of the Earth's atmosphere just above the troposphere, extending to a height of about 50 km (31 miles) where the mesosphere begins, characterized by a concentration of ozone, chiefly horizontal winds, and temperatures that increases with altitude.

STRATUM Homogeneous soil layer in a stratified soil deposit.

STRAW Stalks or stems of grain after drying and threshing.

STREAK CLOUDS Fibrous patches of cloud drawn out in the direction of the wind shear.

STREAM Body of running water in a narrow, clearly defined natural watercourse or channel. Includes rivers, creeks, and brooks.

STREAM DEGRADATION Gradual lowering of the bed of a stream due to such causes as erosion and storm flows, etc.

STREAM DISCHARGE Rate of flow or volume of water per unit of time, flowing in a stream at a given place.

STREAMFLOW DEPLETION Volume of water flowing into a valley minus the water flowing out of the valley.

STREAMFLOW RECORD Data recording the flow of a stream over time.

STREAMFLOW REGULATION Legislative and physical measures established to control the quantity, or quality, of water in a stream.

STREAMFLOW SOURCE ZONE Upstream headwaters area that drains into a recharge zone.

STREAMFLOW WAVE Travelling wave resulting from a sudden increase of flow.

STREAM GAUGING Measurements taken at a place on a stream which enable calculation of its discharge.

STREAM GRADIENT General slope, or rate of vertical drop per unit of length of a flowing stream.

STREAMLINE Path of water or air that is flowing without turbulence; the device causing such flow.

STREAMLINE FLOW Fluid flow characterized by continuous steady motion in one direction.

STREAMSIDE MANAGEMENT ZONE *See* **Buffer strip.**

STREAM SIZE Width or depth of a channel, and volume of water flowing.

STREET ELBOW Elbow pipe fitting with one male end and one female end. *Also called* **Service L** and **Street L.**

STREET L *See* **Street elbow.**

STREET REFUSE *See* **Solid waste.**

STREET SWEEPER Specialized vehicle equipped to sweep debris from paved streets and collect it into a self-contained vessel prior to discharge. The vehicle also has a sprinkler system to lay dust while sweeping. *Also called* a **Sweeper.**

STREET T (or Tee) Tee fitting with one internal and one external threaded opening, plus an outlet opening with an internal thread. *Also called* **Service T.**

STREET WASH Surface runoff from paved areas that is channelled into sewers or storm drains.

STREET WASTES Materials picked up by manual or mechanical sweeping of alleys, streets, and sidewalks; wastes from public waste receptacles; and material removed from catch basins.

STRIPPED GASES Gases released from a liquid by bubbling air through the liquid or by causing the liquid to be sprayed or tumbled over media.

STRIPPED ODORS Odors released from a liquid by bubbling air through the liquid or by causing the liquid to be sprayed or tumbled over media.

STRIPPING Removal of topsoil form an area.

STRONTIUM-90 Radioactive isotope of strontium produced in nuclear reactions as a fission product and during nuclear power generation. It has a half-life of 29 years. The maximum acceptable concentration for strontium-90 in drinking water is 5 Bq/L.

STRUCTURAL FILL Material that is placed and compacted in layers under carefully controlled conditions to achieve a uniform and dense soil mass that is capable of supporting structural loading. *See also* **Backfill,** and **Fill.**

STRUCTURE CONTOUR Theoretical line that passes through all points on the upper or lower surface of a geologic formation, or aquifer, having the same elevation above a datum.

STRUVITE Deposit or precipitate of magnesium ammonium phosphate hexahydrate found on the rotating components of centrifuges and centrate discharge lines.

STUDY Experiment at one or more test sites, in which a test substance is studied in a test system under laboratory conditions, or in the environment, to define or help predict its effects, metabolism, product performance, environmental and chemical fate, persistence and residue, or other characteristics in humans, other living organisms, or media.

STUFFING BOX Space around a shaft filled with pliable packing to prevent fluids or gases from leaking along it.

SUBACUTE TOXICITY Property of a substance or mixture sufficient to cause adverse effects in an organism upon repeated or continuous exposure within less than one-half the lifetime of that organism.

SUBAQUEOUS Formed or adapted for underwater use or operation; found or occurring under water.

SUBAQUEOUS PIPE Submerged pipe; a pipeline under a body of water.

SUBARTESIAN WELL Nonflowing well in which the static level is above that of the saturation zone, but below ground level.

SUBCHRONIC DERMAL TOXICITY Adverse effects occurring as a result of the repeated daily exposure of experimental animals to a chemical by dermal application for part (approximately 10%) of a life span.

SUBCHRONIC INHALATION TOXIC-ITY Adverse effects occurring as a result of the repeated daily exposure of experimental animals to a chemical by inhalation for part (approximately 10%) of a life span.

SUBDUCTION ZONE Areas on the Earth's crust where lithospheric plates are descending into the mantle.

SUBHARMONIC Harmonic of a frequency that is an integral number of times lower than the fundamental frequency in a periodic wave.

SUBLIMATION Process by which a solid is converted directly to the vapor phase by application of heat.

SUBLITTORAL ZONE 1. That part of a lake which is too deep for rooted plants to grow; **2.** Zone of a sea lying below the intertidal zone and extending to the limit of the continental shelf.

SUBMERGED ORIFICE Orifice discharging wholly under water.

SUBMERGED OUTLET Outlet entirely covered by water.

SUBMERGED PIPE Pipeline laid on the bed of a body of water.

SUBMERSIBLE PUMP *See* **Pump.**

SUBSEQUENT Rivers that follow channels cut by themselves into easily erodible rock.

SUBSERE Series of plant communities making up the stages in a secondary succession, or any one of these stages.

SUBSOIL Weathered portion of the Earth's crust between the topsoil and the unweathered material below.

SUBSOIL DRAIN Drain that receives only groundwater and conveys it to a storm drain.

SUBSOIL DRAINAGE PIPE Perforated pipe that is installed underground to intercept and convey groundwater.

SUBSOIL EXPLORATION Determination of the disposition and characteristics of the materials below the surface of the ground to a specified depth, to a defined degree of detail, and for an express purpose.

SUBSTANTIAL COMPLETION Date the works or any designated part thereof that the owner has agreed to accept separately, is sufficiently complete in accordance with the contract documents for the owner to occupy or utilize it for the purpose for which it is intended, and is so certified by the designing authority. *See also* **Substantial performance.**

SUBSTANTIAL PERFORMANCE Is achieved when:

(a) The work or a substantial part of it is ready for use or is being used for the purpose intended; and

(b) The work to be done under the prime contract is capable of completion or correction at a cost of not more than:

(i) 3% of the first $250,000 of the contract price

(ii) 2% of the next $250,000 of the contract price, and

(iii) 1% of the balance of the contract price and is so certified by the Certificate of Substantial Performance.

SUBSTRATE 1. Material or substance upon which an enzyme acts; a surface on which a plant or animal grows or is attached; **2.** Liquor in which activated sludge or other matter is kept in suspension.

SUBSTRATUM Soil layer beneath the solum.

SUBSURFACE Below the exposed surface of the earth.

SUBSURFACE AIR Gases present in the interstices of an aeration zone that is open to the atmosphere.

SUBSURFACE INVESTIGATION Investigative and analytical techniques, using samples obtained from boring, aimed at revealing the nature and characteristics of the subsurface materials likely to influence the design of a structure.

SUBSURFACE WATER Rainfall that is not evaporated or which does not flow away as surface runoff and which penetrates into the ground.

SUBTERRANEAN STREAM Well defined flow of underground water having a measurable velocity in a definite direction.

SUCTION 1. Inlet, pull side of a pump; **2.** Effect of atmospheric pressure that causes objects to resist being lifted from or pulled out of soft soil or mud; **3.** Atmospheric pressure against a partial vacuum; **4.** Partial vacuum on the downwind face of a structure.

SUCTION BOOSTER Type of jet siphon device used to bring water to a pumper from greater distances and to higher elevations than is possible with suction (depending on atmospheric pressure).

SUCTION-CUTTER DREDGE Suction dredge having a rotating cutter at the working end of its suction pipe.

SUCTION DREDGE Dredge without digging buckets that digs by the use of pumps to suck a mud-and-water mixture from the bottom and pumping it through pipes to land or to a hopper barge.

SUCTION HEAD *See* **Suction lift.**

SUCTION HOSE Hose reinforced against collapse due to atmospheric pressure and used for drafting water into fire pumps where a partial vacuum is created in the pump, causing atmospheric pressure to push water through the hose upward into the pump.

SUCTION LINE 1. Tubular connection between a reservoir or tank and the inlet of a hydraulic pump; **2.** Pump intake line in which the fluid is below atmospheric pressure.

SUCTION LIFT Distance from the surface of the water supply to the center of the pump impeller. *Also called* **Suction head.**

SUCTION PIPE Pipe connecting the suction inlet of a pump with its source of supply, which is commonly at a lower elevation than that of the pump.

SUCTION PIT Lined sump dug into the bed of a stream or river and into which the suction pipe of a pump is placed.

SUCTION PUMP Pump positioned above the surface of the body of water into which it's suction pipe is placed.

SUCTION VALVE *See* **Valve.**

SUDDEN ACCIDENTAL OCCURRENCE Occurrence that is not continuous or repeated in nature.

SUDDEN DRAWDOWN Rapid drop in water level behind a dam or alongside a quay or earth embankment that may result in an unstable condition.

SUGAR Any of a class of carbohydrates to which this substance belongs, i.e. compounds of carbon, hydrogen and oxygen, which dissolve in water to give a sweet-tasting solution, and which may be classified by molecular structure, e.g. as mono-, di-, tri-, or polysaccharides.

SULFATE Salt or ester of sulfuric acid that occurs naturally in many minerals, and which is discharged into the aquatic environment in wastes from industries. A wide range of sulfate concentrations may be found in private and public water supplies. The aesthetic objective for sulfate in drinking water is <500 mg/L.

SULFATE-REDUCING BACTERIA *See* **Bacteria.**

SULFITE Salt or ester of sulfurous acid.

SULFONAMIDE Any of a group of sulfa drugs derived from sulfanilamide, that check bacterial infection.

SULFUR Light-yellow, nonmetallic chemical element, S, that burns with a blue flame and stifling odor, having an atomic number of 16 and an atomic weight of 32.064.

SULFUR BACTERIA *See* **Bacteria.**

SULFUR DIOXIDE Heavy, colorless gas (SO_2) that has a sharp odor; the products of combustion of a wide range of fuels.

SULFUR RECOVERY UNIT Process device that recovers elemental sulfur from acid gas.

SULLAGE Domestic wastewater from basins, baths, showers, etc., but not sewage.

SUMP Tank or pit that receives and holds the discharge from a drainage pipe.

SUMP FILTER *See* **Reservoir filter.**

SUMP PUMP *See* **Pump.**

SUMP TANK *See* **Reservoir.**

SUNLIGHT DIRECT AQUEOUS PHOTOLYSIS RATE CONSTANT First-order rate constant in the units of day^{-1} and is the rate of disappearance of a chemical dissolved in a water body in sunlight.

SUPERCHLORINATION Application of sufficient chlorine as to produce free or combined residuals in large enough quantities to require subsequent dechlorination.

SUPERCOOLING Metastable state of a liquid in which it remains in the liquid phase although at a lower temperature than the freezing point.

SUPERCRITICAL OXIDATION Water heated beyond 374°C to pressures greater than 217.7 atmospheres, enabling it to break down complex organic compounds into simpler chemical building blocks.

SUPERNATANT Liquid standing above a sediment or precipitate.

SUPERNATANT LIQUOR Liquid between the bottom sludges and surface scum in a sludge digestion tank.

SUPERSATURATION Vapor having a density greater than that normal for equilibrium under a given circumstance.

SUPERVISORY INDICATOR Visual or audible signalling indicator that advises of a condition.

SUPERVISORY SWITCH *See* **Switch.**

SUPPLEMENTARY CONTROL SYSTEM Any technique for limiting the concentration of a pollutant in the ambient air by varying its emission.

SUPPLIER OF WATER Any person who owns or operates a public water system.

SUPPLY LINES Conduits connecting a water source to a distribution system.

SUPPLY MAIN Primary piping system bringing water, gas, etc. into a facility.

SUPPLY PIPE Extension of a service pipe that is the responsibility of the owner to maintain.

SUPPLY PUMP *See* **Pump.**

SUPPRESSED WEIR Measuring weir notch whose sides are flush with the channel, eliminating or suppressing end contractions of the overflowing water.

SUPRAPERMAFROST WATER Groundwater lying above the permafrost.

SURCHARGE 1. Fill, temporarily placed on a site soon to be developed, in sufficient quantities to impose a load calculated to compact the natural soils to a calculated degree; **2.** Static or live elevated load above the top of a retaining wall; **3.** Charge above the customary cost.

SURFACE ACTIVE Having the ability to modify surface energy and to facilitate wetting, penetrating, emulsifying, dispersing, solubilizing, foaming, frothing, etc., of other substances.

SURFACE-ACTIVE AGENT Substance that affects markedly the interfacial or surface tension of solutions even when present in very low concentrations. *See also* **Surfactant.**

SURFACE AERATION 1. Air absorbed through the surface of a liquid; **2.** Any of several devices that disturb the surface of a pond or basin to the degree that the disturbed water will absorb oxygen.

SURFACE CASING First string of well casing to be installed in a well.

SURFACE CURVE Longitudinal profile of the surface of water flowing in a channel.

SURFACE DRAIN Surface channel that primarily removes surface water.

SURFACE EVAPORATION Evaporation from the surface of a body of water, moist soil, snow, or ice.

SURFACE FILTRATION Filtration which primarily retains a contaminant on the influent face.

SURFACE IMPOUNDMENT 1. Natural topographic depression, man-made excavation, or diked area formed primarily of earthen materials (although it may be lined with man-made materials) that is not an injection well; **2.** Waste management unit that is a natural topographic depression; a man-made excavation, or diked area formed primarily of earthen materials (although it may be lined with man-made materials), designed to hold an accumulation of liquid wastes or containing free liquids. Examples are holding, storage, settling, and aeration pits, ponds, and lagoons.

SURFACE INVERSION Atmospheric surface or ground temperature inversion based at the Earth's surface.

SURFACE LOADING Operation parameter for settling tanks and clarifiers determined by dividing the flow rate by the surface area of the vessel to produce the flow per surface area.

SURFACE MOISTURE Water that is not chemically bound to a metallic mineral or concentrate.

SURFACE PRESSURE CHART Chart of the surface atmospheric pressure, plotted as isobars over a geographical area.

SURFACE RECYCLING Process in which an asphalt pavement surface is heated in place, scarified, remixed, relaid, and compacted.

SURFACE TENSION Property, due to molecular forces, that exists in the surface film of all liquids and which tends to prevent the liquid from spreading.

SURFACE WATER Water lying on or flowing over the surface of the ground.

SURFACE-WATER DRAIN Pipe to convey runoff and rainwater.

SURFACE WIND Wind at an elevation of 10 m (33 ft) above a flat, smooth, unobstructed area of ground.

SURFACTANT Any substance that alters the energy relationship at interfaces; organic compounds displaying surface activity, such as detergents, wetting agents, dispersing agents and emulsifiers. *Also called* **Emulsifier** and **Surface-active agent.**

SURGE Sudden and temporary increase in a state or condition.

SURGE CONTROL TANK Large-sized pipe or storage reservoir sufficient to contain the surging liquid discharge of the process tank to which it is connected.

SURGE PIPE Open-top standpipe, used to release surge pressure.

SURGE SUPPRESSOR Device associated with automatic pump controls that minimizes surges in a pipeline.

SURGE TANK Chamber or tank that acts to cushion the effects of a sudden change in the rate of flow and/or pressure in an otherwise closed hydraulic system.

SUSCEPTIBILITY Degree to which an organism is affected by a pesticide at a particular level of exposure.

SUSPENDED GROWTH PROCESS Wastewater treatment process in which the microbiological organisms effecting waste reduction are suspended in the flows being treated to effect BOD removal, nitrification, and denitrification.

SUSPENDED LOAD Sediment and other particulate matter in suspension in a stream, and which is transported at essentially the velocity of the water.

SUSPENDED SEDIMENT LOAD Volume of the particles small enough to be carried in suspension by moving water.

SUSPENDED SOLIDS Solids and particulate matter of a buoyancy and/or specific gravity that prohibits their settling.

SUSPENSION Solid particles distributed through a liquid.

SUSPENSION OF WORKS Temporarily stopping work on site.

SUSTAINABLE YIELD Maximum extent to which a renewable resource may be exploited without depletion.

SWABBING Removal of attached deposits from the inside of pipes using a plastic foam swab driven through the bore by water pressure.

SWALE Shallow dip made to allow the passage of surface water.

SWAMP Area saturated with water throughout much of the year, but with the surface of the soil usually not deeply submerged. Usually characterized by specific types of tree or shrub vegetation.

SWAMP BOAT Shallow-draft, boat-shaped vehicle powered by an aircraft-type, propeller-equipped engine mounted at the rear, and steered by vanes mounted to deflect the air flow produced by the propeller, used to navigate swamp and marsh.

SWEATING Formation of water droplets on the outside of a pipe or tank containing still or flowing water at a significantly lower temperature than the surrounding atmosphere, due to condensation.

SWEEP T Plumbing fitting in which the T branch leaves via a curve.

SWEET GAS Natural gas containing little hydrogen sulfide.

SWEET WATER Solution of 8% to 10% crude glycerine and 90% to 22% water that is a by-product of saponification or fat splitting.

SWEETENING UNIT Process device that separates the H_2S and CO_2 contents from a sour natural gas stream.

SWING CHECK VALVE *See* **Valve.**

SWITCH Device to open and close an electrical circuit, including the following:

Access: Keyed switch by which a circuit may be activated.

Alarm: 1. Switch that, when activated by a condition or signal, causes an alarm to sound or show; **2.** Automatic sprinkler system switch used to open or close (generally close) an electric circuit to sound an electric alarm.

Bypass: Specific device or combination of devices designed to bypass a regulator or an automatic transfer switch.

Direction: Contactor that determines the direction of rotation of an electric motor when power is applied.

Direction limit: Mechanical switches that are activated when a moving part reaches the design limit of travel, cutting off power to the part.

Disconnecting: Mechanical switching device used for isolating a circuit or equipment from a source of electric power. *Also called* an **Isolating switch.**

Double-pole: Switch that opens or closes two isolated circuits simultaneously.

Double-throw: Switch that connects one circuit to either of two other isolated circuits.

Emergency stop: Switch that, when activated, interrupts the circuit of, or to a motor or engine.

Float: Electric switch that is responsive to liquid level.

Flow: Electric switch operated by a liquid flow.

Four-way: *See* **Three-way switch,** below.

Fused: Switch containing a fuse.

General-use: Switch intended for use in general distribution and branch circuits.

General-use snap: Form of general-use

switch so constructed that it can be installed in flush device boxes or on outlet box covers, or otherwise used in conjunction with wiring systems.

Governor: 1. Mechanically operated switch mounted on a governor that removes power from a motor when the object the motor is driving or powering overspeeds; **2.** Mechanically operated switch mounted on a governor that actuates a control circuit to reduce speed.

Ignition: Keyed switch, part of the ignition circuit of an engine, that commonly also serves to control other functions, such as initiation of the starter motor, lights, auxiliary equipment, etc. *Also called* **Ignition key.**

Isolating switch: Switch intended for isolating an electric circuit from the source of power.

Key: Electric switch operated by a removable key.

Key operated: Switch that incorporates a keyed lock as part of its mechanical operation.

Knife: Electrical switch consisting of a thin blade that makes contact between two flat surfaces.

Knife-blade: Electrical switch where the moving contact is sandwiched between the arms of a fixed contact.

Limit: Electric switch that restricts the mechanical travel of an electrically operated or electrically controlled device.

Linked: Two or more switches physically joined so that they open and close simultaneously.

Magnetic: Electric switch whose switch contacts are controlled by an electromagnet.

Master: Switch controlling two or more electrical circuits.

Mercury contact: Switch containing a sealed, pivoted vial containing mercury which, when moved or tilted, causes an electrical circuit to be closed or opened.

Momentary-contact: Electrical switch whose pole will always return to its default position when operating pressure is removed.

Motor-circuit: Switch rated in horsepower, capable of interrupting the maximum operating overload current of a motor of the same horsepower rating as the switch at the rated voltage.

Operation selector: Multiposition switch that can be set to the selected mode of operation.

Override: Switch to allow manual operation of an automatic control.

Over-speed: Switch that is actuated to remove power from an electric motor when the device the motor is powering exceeds a preset speed.

Pendant: Electrical toggle switch, usually fitted to a lighting fixture, operated by a hanging cord.

Potential: Contactor whose coil is in series with a safety circuit: the contacts are in the power line so that if the safety circuit is not closed, the power supply to the principal device cannot be completed.

Pressure: Electric switch operated by a fluid or physical pressure.

Pressure-differential: Electric switch operated by a difference in pressure.

Proximity: Limit switch operated electronically, usually by capacitance or inductance, commonly used in a dirty environment that could jam a mechanical switch.

Pull: Switch fixed to a ceiling box and operated by a hanging cord.

Rotary: Electrical switch where a rotating central lever causes alternating make and break of contacts.

Safety: Switch used in interior electric wiring and which is mounted inside a metal box and operated from outside the housing by means of a handle or lever connected to the switch mechanism. The box can only be opened when the switch is off, isolating the source of energy.

Safety-operated: Mechanically operated switch that removes power from an engine or motor when a safety mechanism is actuated.

Sensitive: Switch actuated by a spring mechanism that operates independently of the amount of force applied to it.

Service: Electric switch that controls all the energy registered by a meter.

Service entrance: Switched main panel through which service conductors are brought into a building and fed to a fused distribution box.

Single-pole: Switch that makes or breaks one side of an electric circuit.

Stop: Switch that turns off the engine's ignition system and stops the engine, typically on a chainsaw.

Summer: Switch on the electrical circuit of a forced-air furnace permitting manual operation of the fan alone; used to circulate air without the furnace operating.

Supervisory: Monitoring switch activated by the closing of a valve, power failure, abnormal air pressure drop, etc.

Three-way: Switch used to make or break a circuit at more than one location.

Toggle: Electrical switch in which contacts are made and opened through the action of a toggle.

Transfer: Switch designed so that it will disconnect the load from one power source and reconnect it to another source.

Tumbler: Lever-actuated switch that causes a connected part to make, or break an electrical contact.

Two-way: One of two switches within a circuit, wired for making or breaking the circuit from two separated position.

Water flow: Switch, activated by either pressure drop or water flow within an automatic sprinkler system that causes a local or remote alarm to indicate system operation.

SWITCHBOARD Large single panel, frame, or assembly of panels that have switches, buses, instruments, overcurrent, and other protective devices mounted on the surface, or the back, or both.

SWITCHBOX Accessory box built into a wall for a flush-mounted switch.

SWITCHFUSE Switch built into an enclosure that also contains a fuse(s).

SWITCHGEAR Devices, including switches and circuit breakers, protecting and/or controlling a power circuit.

SWITCHING DEVICE Device designed to close and/or open one or more electric circuits.

SWITCHING SPEED Time required to turn a device on or off.

SYMBIOSIS Relationship of two or more different organisms in a close association that may be, but are not necessarily, of benefit to each other.

SYMMETRICAL DISTRIBUTION Set of values or observations that are distributed evenly about the mean; often graphically represented as a bell-shaped curve.

SYNCHRONOUS SATELLITE Satellite in an orbit that maintains its position relative to a particular point on the Earth's surface.

SYNERGISM Action of two or more substances, organs, or organisms to achieve an effect of which each is individually incapable.

SYNGAS Synthetic gas resulting from pyrolysis of organic material produced by incomplete combustion of organic matter. The combustible components are primarily carbon monoxide and hydrogen (usually about 300 Btu/scf, but less than 900 Btu/scf).

SYNTHETIC Produced by synthesis; especially, not of natural origin.

SYNTHETIC DETERGENT Anionic, cationic, or nonionic synthetic cleaning agent containing surface-active agents.

SYNTHETIC FIBER Fiber composed partially or entirely of materials made by chemical synthesis, or made partially or entirely from chemically-modified naturally-occurring materials.

SYNTHETIC SLUDGE Sludge manufactured by aerating any of a range of organic materials and seeded with wastewater organisms, used for experimental purposes.

SYSTEM Set of things or parts forming a whole.

SYSTEM DESIGN CAPACITY Capacity of a collection or treatment system, established at the design stage and in accordance with engineering standards.

SYSTEM DISPLAY MONITOR Cathode ray tube (CRT), mounted in proximity to and visible from the operator's position, that displays information relating to machine operating state, conditions, selection, etc.

SYSTEME INTERNATIONAL Metric system of measurement. *See the appendix* **Metric and nonmetric measurement.**

SYSTEMIC AGENT Agent that effects the body as a whole and not a particular part of organ.

SYSTEM PRESSURE *See* **Pressure.**

SYSTEMS ENGINEERING Design and specification of the means to achieve an objective part of a whole: heating/ventilating/cooling of a building, for instance, production of a given quantity per hour of concrete,

warehousing of a type of goods, etc.

T

TABULATED DATA Tables and charts approved by a registered professional engineer and used to design and construct a protective system.

TAG 1. System or method of identifying circuits, systems, or equipment for the purpose of alerting persons that the circuit, system, or equipment is being worked on; **2.** Temporary sign, usually attached to a piece of equipment or part of a structure to warn of existing or immediate hazard.

TAIL Short length of pipe or cable, used to make a connection.

TAIL BAY Section of a canal downstream of the tailgate of a lock.

TAIL GATE Gate at the low-level end of a lock.

TAILINGS Waste material separated from pay material during processing or screening.

TAILINGS DAM Embankment of fine refuse from mineral benefaction built up to form a silt pond.

TAIL PIPE Suction line of a pump.

TAILRACE 1. Channel for floating away mine tailings and refuse; **2.** Part of a millrace below the water wheel through which spent water flows. *Also called* **Millrace** and **Mill tail.**

TAIL WATER Water immediately downstream from a structure, having been spilled from it or lost from a system.

TAINTER GATE *See* **Canal lock.**

TAKEOFF List of materials prepared from working drawings and specifications giving the type of material, number, weight, volume, size, etc.

TAKEOFF LINE Pipeline leading from the high pressure side of a main circulating oil loop to the branch circuits.

TALBOT PROCESS Protective coating of sand and bitumen inside cast iron pipe.

TALUS Slope formed by the accumulation of debris, especially at the base of a cliff.

TALUS SPRING Spring at the foot of a talus slope.

TAMPING FOOT Specially shaped and arranged projections on a compactor drum, used to achieve compaction by a series of blows. *See also* **Sheepsfoot pad** and **Pad foot.**

TAMPING ROLLER Steel drum fitted with projecting feet, used singly or in a group in a machine, self propelled, or to be towed: a sheepsfoot roller.

TANDEM-CENTER VALVE *See* **Valve.**

TANGENTAL VELOCITY Component of velocity along the tangent in a curved flow.

TANK Pool, pond, cistern, reservoir, etc., through which liquids pass or in which they are held for a period.

TANK BODY Fully enclosed truck, tractor, or trailer body designed to transport fluid.

TANKER Vehicle fitted with a storage tank capable of storing, and often of distributing, liquids.

TANK SYSTEM Hazardous waste storage or treatment tank and its associated ancillary equipment and containment system.

TANK VEHICLE Any vehicle, other than a railroad tank car, or a boat, with a cargo tank having a capacity of more than 450 L (118 gal), mounted or built as an integral part of the vehicle.

TAP 1. Building connection to a gas or water service main; **2.** *See* **Faucet.**

TAP BOLT Machine bolt, threaded close to the head.

TAPER BEND Taper pipe formed as a bend.

TAPERED AERATION Technique of introducing progressively smaller amounts of air into different parts of an aeration tank, from the inlet to the outlet, in the activated sludge sewage treatment process.

TAPERED PIPE THREAD Pipe threads in which the pitch diameter follows a helical cone to provide interference in tightening. *See also* **Dry-seal pipe thread** and **Pipe thread.**

TAPERED REAMER Tool for deburring and cleaning the inside ends of pipes.

TAPERED THREAD Screw thread used on the outside of screwed pipe ends to ensure a pressure-tight joint when screwed tight over jointing tape. The amount of taper is 1:16.

TAPPED T Cast iron T with at least one branch tapped to receive a threaded pipe or fitting.

TAPPING MACHINE Machine used to cut and tap (internally thread) a hole in metal.

TAPPING SLEEVE Split sleeve used to make a wet connection where a single branch line is to be tapped into a water main under pressure.

TARE WEIGHT Weight of an empty weighing dish or vessel.

TARGET COMPLETION DATE Imposed date which constrains or otherwise modifies the network analysis.

TARGET FINISH DATE Date work is planned (targeted) to finish on an activity.

TARGET START DATE Date work is planned (targeted) to start on an activity.

TAR SANDS Deposit of bitumen mixed with sand, clay, and various minerals, having the appearance and texture of asphalt paving and found near the surface or hundreds of meters deep. The bitumen in the tar sands can be processed into a lighter liquid form which is called synthetic crude oil to distinguish it from the traditional petroleum occurring naturally in liquid state. The Alberta tar sands, for instance, consisting of approximately 900 000 million barrels of in-place heavy oil, constitute the largest known such reserve in the world

TASK Item of work assigned or required. *See also* **Activity.**

TASTE There is no method for the objective measurement of taste in drinking water, and there is considerable variation among consumers as to which tastes are acceptable. Also, consumers frequently mistake odors for taste.

TAXABLE HORSEPOWER *See* **Horsepower.**

TECHNICAL LIFE LENGTH Time from when a machine goes into operation until it is no longer used in any operation; machine productive time, expressed in hours.

TECHNOSPHERE That part of the physical environment built or modified by humans.

TECTONIC Of or having to do with structures that build up on the Earth's crust.

TECTONIC CREEP Slight, and apparently continuous, movement along a fault.

TECTONICS Geological study of the Earth's crust.

TECTONIC STRESS Stress caused by deformation of the Earth's crust; may occur near the surface and may greatly exceed the stress in the rock due to gravity.

TEE Three-way pipe fitting shaped like the letter T.

TELEMETRY Science concerned with measurement, transmitting the results to a distant station, and interpreting, indicating, or recording the transmitted data.

TELLTALE Any of various devices that indicate or register information; typically, a water level indicator installed in a reservoir; an overflow pipe that indicates, by dripping or running, when a cistern is full.

TEMPERATURE 1. Degree of hotness or coldness of something; **2.** An aesthetic objective of <15°C (59°F) has been established for the temperature of drinking water.

TEMPERATURE APPROACH Least temperature difference between the two fluids within a discrete-surface heat exchanger.

TEMPERATURE INVERSION Condition where a layer of warmer air overlies that of cooler air, which is then trapped at the Earth's surface.

TEMPERATURE SCALES Arbitrary thermometric calibrations that serve as convenient references for temperature determination. There are two widely used thermometric scales based on freezing and boiling point of water at a pressure of one atmosphere: the Fahrenheit (F) scale (32° = freezing, 212° = boiling) and the Celsius (C), or Centigrade, scale (0° = freezing, 100° = boiling). Additionally, there are two scales in which 0° = absolute zero, the temperature at which all molecular movement theoretically ceases: the Kelvin (K), or Absolute (°A), scale and the Rankine (°R) scale, which are related to the Celsius and the Fahrenheit scales, respectively (0°K = 273.16°C; 0°R = 459.69°F). The four scales can be related to each other by the following formulas:

$$°C = 5/9 \ (°F - 32)$$

$$°K = °C + 273.16$$

$$°F = 9/5 \ °C + 32$$

$$°R = °F + 459.69$$

Another scale based on the thermometric properties of water is the Reaumur scale, in which the freezing point is set at zero degrees and the boiling point at 80 degrees. This scale has only limited application.

TEMPERATURE SENSOR Device capable of measuring temperature and either indicating it, recording it or transmitting it.

TEMPORARILY ABSORBED WATER Water within the soil water belt or intermediate saturation zone, from which it is removed by evaporation or root absorption and vegetable transpiration, and which never reaches the saturation zone.

TEMPORARY SPRING An intermittent spring.

TENDER 1. Offer or bid to complete work or supply materials; **2.** Laborer who tends masons. A general name covering hod and pack carriers and wheelbarrow handlers.

TENSIOMETER Device that uses porous clay filled with water and connected to a manometer to measure the potential of soils or their moisture content.

TEN-YEAR PRECIPITATION EVENT Maximum 24-hour precipitation event with a probable recurrence interval of once in 10 years.

TEPHIGRAM Diagram showing the temperature and entropy at different elevations in the atmosphere.

TEPHRA Pyroclastic material of all types ejected from a volcanic vent into the atmosphere.

TERA Prefix representing 10^{12}. Symbol: T. Used in the SI system of measurement. *See also the appendix*: **Metric and nonmetric measurement.**

TERATOGENIC Property of a substance or mixture to produce or induce functional deviations or developmental anomalies, not heritable, in or on an animal embryo or fetus.

TERBUFOS Organophosphorus insecticide for which a maximum acceptable concentration of 0.001 mg/L has been established for drinking water.

TERMINAL MORAINE Gravel bed at the end of a glacier or on the edge of an ice sheet.

TERMINAL TEMPERATURE Maximum temperature attained by feedwater in a still.

TERMINAL VELOCITY Maximum velocity at which a particle settles out in air or water.

TEROTECHNOLOGY Design, installation, commissioning, operation, maintenance, and replacement of machinery, plant and equipment.

TERRACE Raised bank of earth having vertical or sloping sides and a flat top.

TERRAIN Ground, considered for its fitness or desirability for some purpose.

TERRESTRIAL Of or pertaining to the Earth or its inhabitants.

TERRIGENOUS Sediments derived from the land.

TERRITORIAL SEAS Belt of the seas measured from the line of ordinary low water along that portion of the coast that is in direct contact with the open sea and the line marking the seaward limit of inland waters, and extending seaward a promulgated and internationally recognized distance.

TERTIARY (SEWAGE) TREATMENT *See* **Sewage treatment.**

TESLA SI unit of magnetic induction equal to 1 Wb of magnetic flux per m². (1 T=1 Wb/m²). Symbol: T. *See also the appendix*: **Metric and nonmetric measurement.**

TEST Means of examination, trial, or proof.

TEST CERTIFICATE Document certifying that the product it refers to meets the requirements of an established standard or performance criteria.

TEST CHAMBER Individual container in which a test organism is maintained during exposure to a test solution.

TEST DATA Data from a formal or informal study, test, experiment, recorded observation, monitoring, or experiment, or information concerning the objectives, experimental methods and materials, protocols, results, data analyses (including risk assessments), and conclusions from a study, test, experiment, recorded observation, monitoring, or measurement.

TESTED WELL CAPACITY Maximum

rate at which a well is demonstrated to yield water without significant increase in drawdown.

TEST PIT Hole dug into the ground for the purposes of evaluation, typically to ascertain the permeability of the ground and its ability to absorb an effluent.

TEST PRESSURE *See* **Pressure.**

TEST SOLUTION Test chemical and the dilution water in which it is dissolved or suspended.

TEST SUBSTANCE Substance or mixture administered or added to a test system in a study.

TEST WELL Well installed to assess aquifer conditions.

TETRAETHYL LEAD Tetraethyl lead and tetramethyl lead are the principal additives to gasoline, used to raise the octane rating.

TETRAPOD Four-legged equiangular block of reinforced concrete weighing up to 20 t, used in the construction of breakwaters.

TEXTURE Composition or structure of a substance; grain.

THALWEG Line that follows the deepest part of a stream.

T-HEAD Top of a shore formed with a braced horizontal member projecting on two sides forming a T-shaped assembly.

THEORETICAL DISCHARGE Calculated discharge over a weir or through an orifice, tube or pipeline, in contrast to the actual discharge.

THEORETICAL PUMP DISPLACEMENT Manufacturer's published rating.

THERM Nonmetric unit of heat, equal to 100,000 British thermal units. Symbol: th. Multiply by 105.506 to obtain megajoules, symbol: MJ. *See also the appendix*: **Metric and nonmetric measurement.**

THERMAL COLUMN Column of smoke and gases given off by forest fires, moving upward because heated gases expand and become lighter and rise, while cooler air, bringing additional oxygen, is drawn in toward the base of a fire.

THERMAL CONDUCTIVITY Quantity of heat that flows in one second across the unit area of a slab of a substance of unit thickness when the temperatures of the faces differ by one degree.

THERMAL CONVECTION STORM Result of local temperature inequalities in which rainfall is intense, of short duration, and local.

THERMAL CONVERSION Conversion of organic waste into energy through combustion.

THERMAL EFFICIENCY Measured efficiency of a thermal energy conversion device.

THERMAL POLLUTION Uncontrolled release of heat from combustion into the atmosphere without regard for the potential effect on the environment.

THERMAL SPRING Spring issuing water having a temperature appreciably above the mean annual temperature of the local atmosphere.

THERMAL STRATIFICATION Formation of layers of different temperature in a body of water.

THERMIONIC CONVERTER Device that produces electricity by direct energy conversion resulting from the heating of a suitable metal and the electrons it produces.

THERMISTOR Resistor made of semiconductors having a resistance that varies rapidly and predictably with temperature.

THERMOCLINE Central of three horizontal strata of water in a body of water in which the temperature declines rapidly between the temperature of the top stratum and that of the bottom stratum.

THERMOCOUPLE Heat-sensing device consisting of two conductors of different metals joined at their ends.

THERMODYNAMICS Branch of physics

that deals with the relationships between heat and mechanical energy or work. The laws of thermodynamics comprise:

Zeroth law — Two objects are in thermal equilibrium when no heat passes between them when they are placed in contact with one another.

First law (of thermodynamics) — Energy can be neither created nor destroyed, but it can be changed from one form to another (the conservation of energy).

Second law (of thermodynamics) — When two bodies at different temperatures are placed in contact, heat will flow from the warmer to the cooler, and there is no continuous, self-sustaining process by which the heat can be transferred from the cooler to the warmer body. Thus energy tends to become distributed evenly throughout a closed system (i.e. the entropy of a closed system increases with time).

Third law (of thermodynamics) — Absolute zero ($0°K$) can never be attained.

THERMOGRAPH Thermometer that records the temperature it indicates.

THERMOPHILIC Requiring high temperatures for normal development, as certain bacteria.

THERMOPHILIC DIGESTION Sludge digestion carried on between 45°C and 63°C (113°F and 145°F).

THERMOPHILIC RANGE Temperature most conducive to thermophilic bacteria, and ranging between 49°C and 57°C (120°F and 135°F).

THERMOPHONE Instrument used to determine temperature at different depths of water.

THERMOPLASTIC Plastic that deforms on heating and which can be can be heated and reformed repeatedly.

THERMOREGULATING VALVE *See*

474

Valve.

THERMOSIPHON Circulation by gravity: hot substances tend to rise on becoming less dense, or lighter; cold substances tend to sink on becoming more dense, or heavier.

THERMOSPHERE Layer of the Earth's atmosphere lying above the mesosphere, extending from about 85 km (52.8 miles) to about 450 km (280 miles) above the Earth's surface, where the exosphere begins, and characterized by an increase in temperature with increasing altitude.

THIEF HOLE Well that allows sampling of a digester contents without venting of gas.

THIRTIETH HIGHEST HOURLY VOLUME *See* **Volume.**

THIRTY-DAY LIMITATION Value that should not be exceeded by the average of daily measurements taken during any 30-day period.

THIXOTROPY Infinitely reversible property of certain clay minerals in that, when mixed with water, the mineral forms a gel and on agitation becomes a highly viscous fluid.

THREADED JOINT Joint that uses mating threads formed on the ends of the two pieces to be connected, typically pipe.

THREADER Device or tool used to cut threads on the end of a pipe.

THREE-MINUTE MEAN CONCENTRATION Maximum permissible concentration of a pollutant at ground level from a stationary source.

THREE-POSITION VALVE *See* **Valve.**

THREE-QUARTER S-TRAP Sanitary trap that provides a water seal and that has a vertical inlet and an outlet that is 45° to the horizontal.

THREE-WAY VALVE *See* **Valve.**

THRESHOLD Point, value, line or other demarcation that denotes a difference, one side of which may constitute acceptance, the other rejection, or some other distinction.

THRESHOLD LIMITING VALUE Value of airborne toxic materials that is to be used as a guide in control of health hazards and which presents time-weighed concentrations to which all workers may be exposed eight hours per day over extended periods without adverse effects.

THRESHOLD ODOR Minimum odor emanating from a water sample that can be detected.

THRESHOLD OF AUDIBILITY Minimum sound pressure level at which a person can hear a sound of a given frequency.

THRESHOLD TREATMENT Addition of a small quantity of sodium hexametaphosphate to softened water to prevent the deposition of calcium carbonate.

THRUST BLOCK Concrete poured behind a bend or angle of a pipe to support the pipe against the thrust of fluids being transported through the pipe. *Also called* **Kicker block.**

THRUST BORER Equipment that drills an underground hole, primarily to insert pipes or cables.

THUNDERSTORM Electrical storm accompanied by heavy rain.

TIDAL BASIN Relatively enclosed body of water connected with an ocean or tidal estuary, in which the water level changes with the tides.

TIDAL BORE Wave with a nearly vertical front, promoted by a rapidly flowing tide, that advances upstream against a vigorously flowing stream.

TIDAL CURRENT Current originated by tidal forces.

TIDAL FLAT Land that is successively inundated and exposed by the rise and fall of the tide.

TIDAL LAG Delay between tidal extremes at an estuary and the level of adjacent groundwater.

TIDAL MARSH Low flat marshlands traversed by interlaced channels and tidal sloughs and subject to tidal inundation; normally, the only vegetation present is salt-tolerant bushes and grasses.

TIDAL OUTLET Channel(s) from a drainage system discharging to a tidal waterway that will only discharge when there is sufficient hydraulic head in the conduit(s) to overcome external hydraulic pressure.

TIDAL POWER Electrical generation using low-head turbines powered by tidal water held in a head pond.

TIDAL RANGE Maximum difference between the levels of high and low tides at a location measured at the spring equinox.

TIDAL RIVER River in which the direction and speed of flow, and the surface elevation, are affected by the tides.

TIDAL VALVE *See* **Valve.**

TIDE GATE *See* **Backwater gate.**

TIDE GAUGE Instrument that measures the height of a tide.

TIDE TABLE Data that give daily predictions of the times and heights of the tide for a specific location.

TIGHT COVER Manhole cover without openings.

TILE BED *See* **Leach field.**

TILE DRAIN Length of unglazed clay pipe, sometimes perforated, laid with open joints in a gravel trench to remove surplus groundwater.

TILE SHOE Box towed behind a ditching machine through which land tiles can be laid on a ditch bottom. *See also* **Shoe.**

TILL Dense glacially deposited soil formations consisting of a heterogeneous mixture of fine-grained and coarse-grained material, often including significant quantities of boulders and cobbles.

TILTH Cultivation of land.

TILTING GATE Crest gate of a dam hinged so that it will tilt open when the water level has risen to create a specified pressure, and close once the level has dropped to normal.

TILTMETER Instrument used to measure changes in slope of the ground surface.

TIME CYCLE 1. Time required for one complete sequence of events; **2.** *See* **Traffic signal.**

TIME FOR COMPLETION *See* **Contract time.**

TIME FRAME Time necessary to do something, or the time that something is due to happen, put in perspective of the time of related circumstances.

TIME LAG Time interval between two closely related events or phenomena.

TIME OF COMPLETION Date specified in the contract documents for substantial completion of the works.

TIME OF CONCENTRATION Time required for water contained within a storm water system to travel from the most remote section of the tributary area to an inlet or drain.

TIME OF FLOW Time required for water or wastewater to enter a sewer and move to its outlet, or any other point.

TIMER Device that monitors the passage of time and which is capable of initiating action at preset points: typically, opening or closing an electrical circuit; turning on or turning off lights.

TIME-RESPONSE CURVE Curve relating cumulative percentage response of a test batch of organisms, exposed to a single dose or single concentration of a chemical, to a period of exposure.

TIME-WEIGHTED AVERAGE Average concentration of a substance based on the times and levels of its concentrations.

TINKER'S DAM In plumbing, a small dam made to enclose a work area that is to be flooded with solder.

TIPPING BAY Opening, typically 4 to 5 m (12 to 15 ft) wide, that allows vehicles carrying wastes to discharge their load into a storage pit or transfer station hopper.

TIPPING FEE Charge to unload waste materials at a transfer station, processing plant, landfill, or other disposal site.

TIPPING FLOOR Unloading area for vehicles that are delivering waste materials to a transfer station, incinerator, or other waste processing plant.

TITANIUM Light, strong, silvery or grey metallic element, Ti, that occurs in ilmenite and rutile, the ninth most common element in the Earth's crust.

TITER Concentration of a substance in solution or the strength of such a substance determined by titration.

TITRATION Process or method of determining the concentration of a substance in solution by adding to it a standard reagent of know concentration in carefully measured amounts until a reaction of definite and known proportion is completed, as shown by a color change or electrical measurement, and then calculating the unknown concentration.

TOE 1. Free side of the base of a dam or retaining wall; **2.** Bottom of the working face at a sanitary landfill.

TOEBOARD Vertical barrier at floor level erected along exposed edges of a floor opening, wall opening, platform, runway, or ramp to prevent material from falling.

TOE FILTER Graded filter on the lower end of the free side of an earth dam.

TOE OF DAM Junction of the downstream face of a dam with the ground surface.

TOILET SEAL Wax or putty ring used to seal the joint between a water closet and the pipe on which it sits.

TOLERABLE DAILY INTAKE 1. Threshold value for individual additives or trace chemicals in food, below which no adverse health effects may be expected; **2.** Amount of a substance that can be consumed from all sources each day by an adult, even for a lifetime, without any significant increased risk to health, based on current knowledge.

TOLERANCE 1. Permitted variation from a given dimension, quality, or quantity; **2.** Range of variation permitted in maintaining a specified dimension; **3.** Permitted variation from location or alignment. *See also* **Float.**

TON (LONG, 2240 lb) Nonmetric unit of mass. Symbol: ton. Multiply by 1016.047 to obtain kilograms, symbol: kg; by 1.016 047 to obtain tonnes, symbol: t. *See also the appendix*: **Metric and nonmetric measurement.**

TON (SHORT, 2000 lb) Nonmetric unit of mass. Symbol: t(s). Multiply by 907.188 to obtain kilograms, symbol: kg; by 0.908 to obtain tonnes, symbol: t. *See also the appendix*: **Metric and nonmetric measurement.**

TON-MILE Movement of a ton of freight or cargo a distance of one mile.

TONNAGE 1. Total weight expressed in tons; **2.** Charge per unit of weight on cargo.

TONNE Non-SI unit of mass, equal to 1000 kg (1 Mg), permitted for use with the SI system of measurement. Symbol: t. Multiply by 2205 to obtain pounds, symbol: lb; by 0.984 to obtain long tons (2240 lb), symbol: ton; and by 1.102 to obtain short tons (2000 lb), symbol: t(s). *See also the appendix*: **Metric and nonmetric measurement.**

TONNE-PER-DAY Annual metric tonnage divided by the number of days worked.

TOO NUMEROUS TO COUNT Means that the total number of bacterial colonies exceeds 200 on a 47-mm-diameter membrane filter used for coliform detection.

TOPOGRAPHIC DIVIDE Line following the height of land elevation forming the boundary of, or between, one or more drainage basins.

TOPOGRAPHIC MAP Map showing such features as hills and valleys, mountain ranges, contours, towns and cities, etc.

TOPOGRAPHY Detailed and accurate description of a place or region.

TOPOTYPE Population of one geographical region whose characteristics differ from those of another region.

TOPSOIL *See* **Loam,** and **Soil types.**

TORICELLI'S THEOREM Statement that the liquid velocity at an outlet discharging into the free atmosphere is proportional to the square root of the head.

TORNADO Rotating column of air, usually accompanied by a funnel-shaped downward extension of cumulonimbus cloud and having a vortex several hundred meters in diameter whirling destructively at speeds up to 500 km/h (300 mph).

TOTAL CAPACITY 1. Maximum rate at which a well or spring will yield water; **2.** Maximum hydraulic flow that can be passed; **3.** Largest amount that can be carried, stored, lifted, pumped, etc.

TOTAL DISSOLVED SOLIDS Solids residue after evaporating a sample of water or liquid effluent.

TOTAL DYNAMIC HEAD Same as total head and is the dynamic suction head plus the dynamic discharge head.

TOTAL ENERGY Integrated use of all or most of the heat generated by the combustion of fossil or nuclear fuels.

TOTAL HEAD *See* **Head.**

TOTALIZER Device or meter that continuously measures and calculates a process rate variable in cumulative fashion.

TOTALLY ENCLOSED TREATMENT FACILITY Facility for the treatment of hazardous waste which is directly connected to an industrial production process and which is constructed and operated in a manner which prevents the release of any hazardous waste into the environment during treatment.

TOTAL METAL Sum of the concentration or mass of Copper (Cu), Nickel (Ni), Chromium (Cr) (total) and Zinc (Zn).

TOTAL ORGANIC CARBON Total of all organic compounds in a wastewater sample as determined by the combustion–infrared method prescribed by approved laboratory procedures.

TOTAL POPULATION Sum of the population density of individuals of all stages plotted against time.

TOTAL PRESSURE *See* **Pressure.**

TOTAL PUMPING HEAD Amount of energy increase per unit measure of the material being pumped as imparted to it by the pump: the difference between the total discharge head and the total suction head.

TOTAL RESIDUAL CHLORINE Amount of available chlorine remaining after a given contact time.

TOTAL SOLIDS Sum of dissolved and undissolved constituents in water.

TOTAL TOXIC ORGANICS Sum of the mass of each of the toxic organic compounds found at a concentration greater than 0.010 mg/L.

TOTAL WATER DEMAND Volume of water required by a complete plumbing system, i.e. volume required to fill all the piping beyond the water meter and to operate all of the attached fixtures.

TOTAL WELL CAPACITY Maximum rate at which water can be pumped from a well after any stored water has been removed.

TOTE BOX Small- to medium-sized rectangular box used for the collection of solid waste recyclables.

TOUCHLESS CONTROLS Electronic automatic controls operated by sensors that react to a defined situation or stimulus. Typically, the opening of doors from detection of the presence of people; the operation of a faucet from the sensing of hands, etc.

TOXIC Substance poisonous to a living organism.

TOXIC ACTION OF POLLUTANTS Of the various manifestations of toxic effects upon the human organism, the following three are considered to be most critical:

a) Enzymatic action by, for instance, combining with the enzyme so that it cannot function.

b) Chemical combination with the

constituents of cells as, for example, carbon monoxide combining with blood.

c) Secondary action due to their presence. i.e., hay fever is brought about by pollen and the system reacts to produce histamine.

TOXICITY Relative degree of being poisonous or toxic.

TRACE ELEMENTS Elements that occur in minute quantities and which are natural essential constituents of living organisms and tissues.

TRACE OF PRECIPITATION Rainfall in amounts insufficient to be measured in a standard gauge.

TRACER Object or substance that can be found or tracked when placed in, or mixed with something else.

TRAILER PUMP Mobile water pumping unit, typically mounted on a two-wheel trailer.

TRAINING WALL Structure constructed along a river consisting of loose mounds or heaps of rubble, with or without a surmounting wall, timber, close timber piling, wood sheet piling, steel sheet piling or reinforced concrete, used to direct the flow of the river into a more favorable, fixed channel.

TRAMP OIL Oil present on the surface of a body of water due to natural flotation.

TRANSDUCER Device that can change energy from one form to another, i.e. electrical energy into mechanical energy. *See also* **Load sensor.**

TRANSECT Cross-section of an area, used as a sampling line for recording vegetation.

TRANSFER STATION Supplemental transportation system, an adjunct to route collection vehicles to reduce haul costs or add flexibility to a waste collection and disposal operation. Typically, route vehicles empty into a large hopper from which large semitrailers, railroad gondolas, or barges are filled. There may be some compaction of refuse. Transfer stations may be fixed or mobile, since the larger compacting collection vehicles can serve this function.

TRANSITION Fitting that converts one shape or size to another.

TRANSITION BELT Short belt carrying material from a loading point to a main conveyor belt.

TRANSITION ZONE Relatively thin layer between the upper and lower layers of impounded water and characterized by sudden and significant temperature and other changes.

TRANSMISSION CONSTANT Quantity used to express the ability of a granular material to transmit water and equal to the volume of discharge per unit of time through each unit measure of the cross-sectional area when the hydraulic gradient is 100%.

TRANSMISSION LOSS Measure of the sound insulation qualities of a barrier or obstruction.

TRANSPIRATION Act or process of loss of liquid, especially through the stomata of plant tissue or the pores of the skin.

TRANSPIRATION RATIO Ratio of the weight of water transpired by a plant to the weight of dry plant substance produced.

TRANSPORT CAPACITY Ability of flowing water to carry suspended particles.

TRAP 1. Device that prevents a material flowing or being carried through a conduit from reversing its direction of flow or passing a given point; **2.** U- P- or S-shaped pipe filled with water and located beneath plumbing fixtures to form a seal against the passage of foul odors or gases. *See also* **P-trap, Running trap, S-trap, Sewer trap,** and **Three-quarter S-trap.**

TRAPPED WASTE Waste outlet from a sanitary fitting with a built-in trap.

TRAP SEAL Vertical distance between a crown weir and the dip of a trap.

TRASH *See* **Solid waste.**

TRASH GATE Opening through or over

which floating trash may be removed or discharged.

TRASH-HANDLING CENTRIFUGAL PUMP *See* **Pump.**

TRASH RACK Grid or screen across a stream or channel, designed to catch floating debris.

TRASH SCREEN Screen positioned to prevent the passage of trash.

TRAVELLING SCREEN Rotating trash screen.

TRAY AERATOR Type of aerator in which water is successively discharged through a stack of perforated trays.

TREATABILITY STUDY Study in which a hazardous waste is subjected to a treatment process to determine:

(a) Whether the waste is amenable to the treatment process;

(b) What pretreatment (if any) is required;

(c) The optimal process conditions needed to achieve the desired treatment;

(d) The efficiency of a treatment process for a specific waste or wastes; or

(e) The characteristics and volumes of residuals from a particular treatment process.

TREATED SEWAGE Wastewater that has received partial or complete treatment.

TREATED WATER Water that has been subjected to treatment.

TREATMENT Method, technique, or process, including neutralization, designed to change the physical, chemical, or biological character or composition of any hazardous or nonhazardous waste so as to neutralize it, or so as to recover energy or material resources, or to render it nonhazardous, or less hazardous; safer to transport, store, or dispose of.

TREATMENT PARAMETER Fundamental characteristic of sewage around which treatment is designed, such as, but not limited to, flow, BOD, suspended solids, pH, etc.

TREATMENT WORKS Devices and systems for the storage, treatment, recycling, and reclamation of municipal sewage, domestic sewage, or liquid industrial wastes, or necessary to recycle or reuse water at the most economical cost.

TREE Large perennial plant having a woody trunk, branches, and leaves.

TREMIE Pipe or tube through which concrete is deposited under water, having at its upper end a hopper for filling and a bail for moving the assemblage.

TREMIE CONCRETE Subaqueous concrete placed by means of a tremie.

TREMIE SEAL 1. Depth to which the discharge end of the tremie pipe is kept embedded in the fresh concrete that is being placed; 2. Layer of tremie concrete placed in a cofferdam for the purpose of preventing the intrusion of water when the cofferdam is dewatered.

TRENCH 1. Groove across a member; 2. Long narrow ditch. *Also called* **Ditch.**

TRENCH BOX *See* **Shield.**

TRENCH BRACE Horizontal member of a trench shoring system whose ends bear against the uprights or stringers.

TRENCHER Vehicle or attachment that cuts a ditch. Depending on capacity, may consist of excavating cups attached to a chain rotating about a bar, or buckets or skips mounted to the ladder of an excavator. *Also called* **Bucket ladder excavator.**

TRENCH FORM Vertical sides and semi-circular bottom of a trench shaped to provide full, firm and uniform support for the lower 210° of a cast-in-place concrete pipe.

TRENCHING BUCKET *See* **Bucket.**

TRENCHING OR BURIAL Placement of sewage sludge or septic tank pumpings in a trench, or other natural or man-made depression, and subsequent covering with soil or other suitable material at the end of each operating day such that the wastes do not migrate to the surface.

TRENCH JACK Screw or hydraulic jack used as crossbracing in a trench shoring system.

TRENCHLESS TECHNOLOGY Techniques for conduit installation, replacement, or renovation that minimize excavation from the surface.

TRENCH SHIELD Trench shoring system that can be moved horizontally as work progresses.

TRENCH SHORING Wood, steel, or aluminum shapes used to support and restrain the walls of a trench. May consist of prefabricated shapes making up a demountable and movable shoring system.

TRENCH SOIL CLASSIFICATION SYSTEM Series of classifications that define soil classes, used to determine the most suitable system of shoring necessary to support and restrain the walls of an excavated trench. The system comprises the following classifications:

Type A: Stiff cohesive soil, 11.33 kg/ 0.092 m^2 (25 lb/ft^2) per 300 mm (1 ft) of depth. Clay, silty clay, sandy clay, clay loam with an unconfined compressive strength of 1.36 t/0.092 m^2 (1.5 tons/ft^2) or greater. Not Type A if fissured, subject to vibration, previously disturbed, or part of a sloped layered system where the layers dip into the excavation of a slope of four horizontal to one vertical or greater.

Type B: Medium cohesive to granular soil, 20.4 kg/0.092 m^2 (45 lb/ft^2) per 300 mm (1 ft) of depth. Clay with an unconfined compressive strength between 0.45 and 1.36 t/0.092 m^2 (0.5 and 1.5 tons/ft^2). Cohesionless gravel, silt, silt loam, or sandy loam. Previously disturbed soils may be Type B unless they would be classified as Type C. Soil that meets the requirements of type A but that is subject to vibration or is fissured may be Type

B. Dry rock that is not stable or soil that is part of a sloped, layered system where the layers dip into the excavation on a slope less steep than four horizontal to one vertical are Type B if the material would otherwise be classified as Type B.

Type C: Soft, cohesive to saturated soil, 27 kg/ 0.092 m^2 (60 lb/ft^2) per 300 mm (1 ft) of depth. Clay with an unconfined compressive strength less than 0.36 t/0.092 m^2 (0.5 tons/ft^2), saturated sand, clay, or fractured rock that is not stable. Soil in a sloped, layered system where the layers dip into the excavation on a slope of four horizontal to one vertical or steeper may be Type C. Saturated soils or soils from which water is freely seeping, but is not standing in the trench.

Conditions more severe that Type C would require dewatering or sealing on four sides of the excavation and pumping the trench.

TRENCH TECHNIQUE *See* **Sanitary landfilling.**

TRIANGULAR NOTCH WEIR Measuring weir with a V-shaped notch, used for measuring small discharges.

TRIBUTARY Stream or river flowing into a larger stream or river.

Trichoderma viride Fungus capable of breaking down cellulose by the production of an enzyme of the cellulase group which can accomplish decomposition in a few days.

TRICKLE DRAIN Pond overflow pipe set vertically with its open top at the same elevation as the water surface.

TRICKLING FILTER Biological filter for clarified wastewater in which settling tank effluent is discharged from a rotating distributor over a bed of suitably sized aggregate which promotes the growth of bacteria that oxidize the organic fraction. *Also called* **Biological filter.**

TRIPLE HYDRANT Fire hydrant having three outlets, usually two 65 mm (2.5 in.) outlets and one 115 mm (4.5 in.) outlet.

TRIPLE-TIPPER BODY Dump truck body hinged so that it can be raised by hydraulic lift cylinders to the rear, or to either side, with hinged gates on all three vertical faces.

TRIPOLYPHOSPHATE Water-soluble chelating agent for some metals in solution.

TRITIUM Isotope of hydrogen, H_3, having an atomic number of 3, used as a fuel in fusion power research and having a half-life of 12.3 years and which exists in the environment mainly as water, from which it enters the hydrological cycle and all components of the biosphere. It is produced naturally in the upper atmosphere and artificially in nuclear detonations and during nuclear reactor operations. Typical tritium concentrations in surface waters are on the order of 5 to 10 Bq/L. The maximum acceptable concentration for tritium in drinking water is 7000 Bq/L.

TRIPTON Nonliving matter suspended in a body of water.

TROPHIC LEVEL Nutrient status of a body of water.

TROPHOGENIC REGION Region in a body of water where organic material is produced by photosynthesis.

TROPOPAUSE Upper limit of the troposphere.

TROPOPHYTE Plant that lives under moist conditions for part of the year and under dry condition for the remainder of the time.

TROPOSPHERE Lower portion of the Earth's atmosphere.

TRUCK 1. Vehicle designed for carrying an entire load; **2.** Motorized or towed vehicle for hauling. There are several types, including:

> **Bottom-dump:** Trailer or semitrailer that discharges bulk material by opening doors in the floor of the body.

> **Closed top van:** Van with a sealed top that must be rear loaded.

> **Dump:** *See* **Dump body** (below).

Dump body: Storage and discharging component of a dump truck. Following are some common types:

> **Heavy-duty and severe-duty:** Designed for damaging payloads and bucket loading. They feature thicker steel floors and sides, extra bracing in sides and tailgate, thicker and stronger corner posts, and boxed side rails to withstand shovel impact.

> **Hopper/conveyor spreader:** Can be used for patching jobs, windrowing material, feeding a paver, and for broadcast-spreading of abrasives or chemical compounds for ice control, dust control, or seal coating.

> **Light-duty:** 2.4 to 3 m (8 to 10 ft) long, 0.76 to 2.2 m³ (1 to 3 yd³) capacity, used for light hauling and cleanup. Drop sides are available; a direct-lift hoist is common.

> **Side-dump:** Having a body with a fixed rear wall and with sides that can hinge open and means for the body to be raised on hydraulic rams to either the left or right for discharge of a load such as soil, sand, gravel, etc.

> **Standard-duty:** Designed to handle routine dirt/aggregate hauling, debris removal, and municipal and highway maintenance applications.

> **Two-way dump-spread:** Can discharge at the rear as a normal dump body, or can be raised at the rear for front discharge through an integral spreader system.

Flatbed: Truck fitted with a bed that is completely flat. The bed may contain stake pockets, and can be equipped with a range of accessories at its rear edge, such as a roller to assist in the loading of pipe or similar objects.

Highway: Truck designed to haul a load

that, combined with the net weight of the vehicle, does not exceed legal highway limits.

Off-highway: Truck designed for operations that would not be permitted on facilities with posted weight limits.

Oilfield body: Heavily constructed platform-type truck body equipped with a rear-end roller or bullnose adapted for winch loading.

On-highway: Vehicle designed for operations on paved or improved surfaces only.

On/off-highway: Vehicle designed for operation mostly on hard-surfaced or graded roads with some work over unprepared surfaces.

Open top: Vehicle body without a permanent top assembly.

Platform: Truck with a flat, open body.

Rear-dump: Truck or semitrailer with a box body that can be raised at the front so the load will slide out the rear.

Stake body: Flat platform-type vehicle body with removable stakes around its perimeter.

Utility/Service: Specialized truck body customarily consisting of a narrow platform bed between lockable 'saddlebag' storage compartments (for tools, service parts, special equipment) that open from the exterior side for easy access. Standard bodies are manufactured in a wide range of sizes and configurations for general purpose, electrical, and gas/water plumbing applications. Many options, including aerial devices for overhead work, are available.

TRUE GROUNDWATER VELOCITY *See* **Actual groundwater velocity.**

TRUNK, TRUNKLINE Main pipe from which building drains or water supply branch.

TRUNK SEWER Sewer that discharges to a pumping station, treatment works, or outfall, and to which branches are connected. *Also called* **Main drain.**

TSUNAMI Gravity wave caused by an underwater seismic disturbance.

TUBE Hollow cylinder that conveys a fluid or functions as a passage.

TUBE-AND-COUPLING SHORING Load-carrying assembly of tubing or pipe that serves as posts, braces, and ties, a base supporting the posts, and special couplers that connect the uprights and join various members.

TUBERCULATION Rust blisters that develop on the inner face of cast-iron pipe used for water supply and sewage collection.

TUBERCLE Small, rounded prominence of rust or deposited metallic salts formed on the inside of an iron pipe.

TUBING Lightweight pipe made of materials such as copper, brass, rubber, or plastic.

TUBULAR WELL Well in which a circular section screen or perforated casing is in direct contact with the water-bearing formation.

TUNDRA Vast, level, treeless plain in the Arctic regions.

TUNDRA SOIL Shallow soil above permafrost.

TURBIDIMETER Device for measuring particle-size distribution of a finely divided material by taking successive measurements of the turbidity of a suspension in a fluid.

TURBIDITE Deposits from a moving slurry of sediment and water.

TURBIDITY Lack of clarity of water caused by suspended solids. The principal method of measuring turbidity is the nephelometric turbidity unit (NTU), and an aesthetic objective of <5 NTU has been established for water at the point of consumption.

TURBINE Rotary engine or motor driven by a current of water, steam, or air that pushes against the blades of a wheel or system of

483

wheels attached to a drive shaft, causing the wheel and drive shaft to turn.

TURBULENT Violently agitated or disturbed.

TURBULENT FLOW Flow situation in which the fluid or gas particles move in a random manner. *See also* **Flow** and **Reynold's number.**

TURBULENT MIXERS Devices that mix air bubbles and water and cause turbulence sufficient to dissolve oxygen in water.

TURBULENT VELOCITY Velocity of water flowing in a conduit above which the flow will always be turbulent, and below which the flow may be turbulent or laminar.

TURNDOWN RATIO Ratio of maximum output to minimum output of a plant, boiler, etc., based on continuous operation.

TURNOUT Junction between an irrigation canal and a subsidiary canal to which it supplies water.

TURNOVER Phenomenon occurring in impounded water, usually in spring and fall, in which the surface water replaces bottom water: in the spring, as the surface water warms above the freezing point, it becomes more dense and sinks, producing vertical currents; in the fall as the surface water becomes colder it also tends to sink.

TWO- OR THREE-POSITION VALVE *See* **Valve.**

TWO-PROPENYL Colorless liquid aldehyde with a choking odor, a component of photochemical smog.

TWO-STAGE ELEMENT Filter element assembly composed of two filter media in series.

TWO-STAGE FILTER Filter element assembly composed of two filter elements or media in series.

TWO-, THREE-, OR FOUR-WAY VALVE *See* **Valve.**

TWO-WAY DUMP-SPREAD BODY *See* **Truck.**

TWO-WAY VALVE *See* **Valve.**

TYPE A SOIL *See* **Soil classification system** and **Trench soil classification system.**

TYPE B SOIL *See* **Soil classification system** and **Trench soil classification system.**

TYPE C SOIL *See* **Soil classification system** and **Trench soil classification system**

TYPE K COPPER TUBING *See* **Pipe.**

TYPE L COPPER TUBING *See* **Pipe.**

TYPE M COPPER TUBING *See* **Pipe.**

TYPHOID Waterborne disease caused by drinking sewage-contaminated water.

TYPHOON Tropical cyclone of southeastern Asia.

specific locations throughout a water supply area.

ULTRAFILTRATION *See* **Membrane processes.**

ULTRASONIC TESTING Nondestructive test which applies sound, at a frequency above about 20 HJz, to objects which have been immersed in liquid (usually water) to locate inhomogeneities or structural discontinuities.

ULTRAVIOLET Zone of invisible radiations beyond the violet end of the spectrum and within the wavelength range 310 to 380 nm. Their short wavelengths have enough energy to initiate chemical reactions and degrade some materials and colors.

UNACCEPTABLE ADVERSE EFFECT Impact on an aquatic or wetland ecosystem that is likely to result in significant degradation of municipal water supplies (including surface or groundwater) or significant loss of or damage to fisheries, shellfishing, or wildlife habitat or recreation areas.

UNACCOUNTED-FOR WATER Water delivered into a distribution that is not withdrawn or consumed.

UNAMORTIZED CREDIT OR DEBIT Credit or debit entirely accounted for in the current account period.

UNBURNED HYDROCARBONS Airborne particles of hydrocarbon fuels not consumed in combustion and emitted in exhaust of flue gases.

UNCONFINED WATER *See* **Groundwater.**

UNCONFORMITY In geology: an eroded space, or space caused by lack of deposit that separates younger strata from older rock.

UNCONTROLLED FILL Fill not placed under supervision and not compacted.

UNCONVENTIONAL WASTE *See* **Solid waste.**

UNDERCONSOLIDATION Condition of a soil deposit that is not fully consolidated under the existing overburden pressure, and where excess hydrostatic pore pressures ex-

ULTIMATE ANALYSIS Proportion by weight of a chemical and the inert constituents of a fuel.

ULTIMATE BEARING CAPACITY Average load per unit of area required to produce failure by rupture of a supporting soil mass.

ULTIMATE BEARING PRESSURE Pressure at which shear failure occurs in soil.

ULTIMATE BIOCHEMICAL OXYGEN DEMAND Quantity of oxygen required to satisfy completely the biochemical oxygen demand of a wastewater.

ULTIMATE BIODEGRADABILITY Breakdown of an organic compound to CO_2, water, the oxides or mineral salts of other elements and/or to products associated with normal metabolic processes of microorganisms.

ULTIMATE CARBON DIOXIDE Percentage of carbon dioxide that appears in the dry flue gases when a fuel is burned with its chemically correct air–fuel ratio.

ULTIMATE COMPRESSIVE STRENGTH Maximum compressive strength a material can stand under a gradual and evenly applied load.

ULTIMATE WATER DEMAND Specific requirements as to quantity, pressure, and availability of water for defined purposes at

isting within the material.

UNDERCUT BANK Bank that is undermined by the flow of a stream.

UNDERCUT BEDDING Plane along which failure and sliding may occur because the slope or free face is greater than that of the bedding. *Also called* **Undercut joint.**

UNDERCUT JOINT *See* **Undercut bedding.**

UNDERDRAIN Perforated pipe drain laid in a wedge of stone to intercept groundwater.

UNDERFLOW Water movement in the soil, under ice or under a structure.

UNDERFLOW CONDUIT Permeable layer under the bed of a stream, containing groundwater that percolates downstream.

UNDERFLOW TOW Movement of water along or near the bottom of an impoundment that is of different density to the water immediately above it.

UNDERFOOT CONDITIONS The surface material a machine is working in.

UNDERGROUND DRINKING WATER SOURCE Aquifer supplying drinking water for human consumption in which the ground water contains less than 10,000 mg/L total dissolved solids.

UNDERGROUND INJECTION Subsurface emplacement of fluids through a bored, drilled or driven well; or through a dug well where its depth is greater than the largest surface dimension.

UNDERGROUND SERVICES Buried services.

UNDERGROUND STORAGE TANK One or combination of tanks (including connected underground pipes) used to contain an accumulation of regulated substances.

UNDERGROUND WATER Water within the lithosphere.

UNDERGROWTH Shrubs, saplings and herbaceous plants overstoried by other plant growth.

UNDERSTORY Lower layer of trees in a two-layered woodland.

UNDERTOW Seaward pull of receding waves breaking on the shore.

UNDIGESTED SLUDGE Settled sludge promptly withdrawn from a sedimentation tank before exhaustion of all dissolved oxygen.

UNDULATION Regular rising and falling or movement from side to side.

UNFIT-FOR-USE TANK SYSTEM Tank system that has been determined through an integrity assessment or other inspection to be no longer capable of storing or treating hazardous waste without posing a threat of release of hazardous substances to the environment.

UNIFIED SOIL CLASSIFICATION SYSTEM *See* **Soil classification systems.**

UNIFORM BUILDING CODE *See* **Building code.**

UNIFORM FLOW Flow in which velocities are the same both in magnitude and direction.

UNION Type of fitting used to join lengths of pipe for easy opening of a pipe line.

UNION BEND Bend or other fitting with a union at one end.

UNIT COST Cost of producing a unit of work, measured by time, volume, or pieces completed.

UNIT HYDROGRAPH Method for calculating potential storm flows for a specific basin by plotting the time-varying flows from 25 mm (1 in.) that pass over it. *See also* **Distribution graph.**

UNIT PRICE Price established for a given quantity of material or labor.

UNIT PRICE CONTRACT Contract based on the unit price established for materials and labor, irrespective of the quantity.

UNIT VENT Vent pipe that serves two or more traps.

UNIT WEIGHT Weight per unit volume of a material, expressed as gram/cm^3, kg/m^3, or lb/ft^3.

UNIVERSAL OIL PRODUCT Process for removing sulfur dioxide from flue gases and involving a wet scrubber followed by extraction of the sulfur and regeneration of the reagent.

UNLOADING Release of a contaminant that was initially captured by a filter medium.

UNLOADING BULKHEAD Steel plate that ejects waste out of the rear doors of an enclosed transfer trailer. The movable wall is propelled by a telescoping, hydraulically powered cylinder that traverses the length of the trailer.

UNLOADING VALVE *See* **Valve.**

UNRECLAIMABLE RESIDUES Residual materials of little or no value remaining after incineration.

UNSAFE CONDITION Any condition that could cause undue hazard to life, limb or health of any person authorized or expected to be on or about the premises.

UNSATURATED Capable of dissolving more of a solute at a given temperature.

UNSATURATED ZONE Portion of the lithosphere capable of absorbing or storing a greater volume of water.

UNSATURATED ZONE OF AERATION Zone between the land surface and the watertable.

UNSEALING OF TRAPS Loss of water seal in a trap, permitting escape of sewer gas.

UNSOCKETED PIPE Drain pipe with plain ends, joined by a sleeve coupling.

UNSOUND 1. Not firmly made, placed, or fixed; **2.** Subject to deterioration or disintegration during service exposure.

UNSTABLE SOIL Earth material, other than running, that because of its nature or the influence of related conditions, cannot be depended upon to remain in place without extra support, such as would be furnished by a system of shoring.

UNSTEADY NONUNIFORM FLOW Flow in which the velocity and quantity of water flowing per unit of time at every point along a conduit vary according to time and position.

UPDRAUGHT Air rising in a convection current.

UPFEED DISTRIBUTION Water system that requires pumps to develop pressure.

UPFLOW COAGULATION Technique of passing a coagulating mixture upward through a blanket of settling sludge.

UPFLOW CONTACT CLARIFIER Unit in which water enters the bottom and is discharged at or near the surface.

UPLIFT Upward pressure against the base of a structure, the bottom elevation of which is below the upper surface of the watertable.

UPPER LIMIT Emission level for a specific pollutant above which a certificate of conformity may not be issued or may be suspended or revoked.

UPPERMOST AQUIFER Geologic formation nearest the natural ground surface that is an aquifer, as well as any lower aquifers that are hydraulically interconnected with this aquifer within the area under consideration.

UPRIGHT Vertical member of a trench shoring system placed in contact with the earth and usually positioned so that it and other individual members do not contact each other.

UPRUSH Rapid surge of water up the foreshore following the breaking of a wave.

UPSET Condition where a biological process does not perform as anticipated; typically, in a sewage sludge digester where organic matter will not decompose properly, gas production is reduced, there is a high volatile acid/alkalinity relationship, and poor liquid–solid separation.

UPSURGE Rapid upward swell or rise.

UPTAKE Sorption of a test substance into and onto aquatic organisms during exposure.

UPTAKE PHASE Time during a test when test organisms are being exposed to the test material.

UPTIME Period during which a machine or equipment is available.

UPWARD-FLOW FILTER Water filter through which the water flows from bottom to top.

UPWELLING Sudden upward flow of water from a subsurface current.

URANIUM Principal radioactive element, U, used in the production of nuclear energy.

URANIUM ENRICHMENT Process for the improvement of the properties of natural uranium involving the removal of the diluent uranium-238, which comprises 99.28% of the natural material, increasing the concentration of uranium-235.

URBAN Of, having to do with, or in cities or towns.

URBANIZATION Movement of large numbers of people from rural to urban areas.

URBAN RENEWAL Program, policy, or the process of rehabilitating or replacing rundown or substandard buildings in a city, especially in the downtown core.

URBAN SPRAWL Uncontrolled spread of urban development into rural areas.

URINAL Plumbing fixture used for urinating.

USABLES Secondary materials recovered from discards or waste streams that are salable in their existing form.

USEFUL LIFE Period over which a structure or machine is expected to remain viable.

USEFUL STORAGE Water that can be withdrawn from a storage facility; the volume available between the top of gates and minimum drawdown level.

USER Person, lot, parcel of land, building, premises, municipal corporation or other political subdivision that discharges, causes or permits the discharge of wastewater into a sewage system.

UTILITY Service provided by a public agency, which can include electrical energy, potable water, sewerage, gas, telephone, or cablevision.

UTILIZATION FACTOR Ratio of the total availability of something to the portion that is used.

UTILITY-CONNECTED SYSTEM Natural energy system (wind power, photovoltaic cells) that is connected to the utility grid, enabling the system to draw power from a public system and, if permitted, to pass excess electricity into the public system.

V

VACUUM Air pressure below atmospheric pressure.

VACUUM BREAKER Device that prevents development of a vacuum in a pipe containing fluids and thus prevents backflow.

VACUUM DEAERATION Equipment operating under a partial vacuum and used to remove dissolved gases from liquid.

VACUUM FILTER Cylindrical drum rotating on a horizontal axis, covered with a filter cloth behind which a partial vacuum is maintained, and partially submerged in digested sludge or a slurry. Moisture is drawn through the filter cloth and the dewatered sludge cake is scraped off continuously.

VACUUM GAUGE *See* **Gauge.**

VACUUM LOADER Specialized vacuum truck body used for street sweeping, sewer cleaning, leaf pickup, and other municipal and industrial applications.

VACUUM PUMP *See* **Pump.**

VACUUM RELIEF VALVE *See* **Valve.**

VADOSE ZONE Subsurface zone between the surface and the watertable that contains water under pressure less than that of the atmosphere.

VADOSE WATER Water moving in aerated ground above the watertable.

VALLEY Low land between hills or mountains, usually having a stream or river flowing through it.

VALLEY DRAINAGE Where water collects in a natural or man-made depression, typically along a gutter. *See also* **Crowned drainage.**

VALLEY FOG Fog that forms in air which has cooled on hillsides by radiation at night.

VALUATION Estimation of worth.

VALUE ANALYSIS *See* **Value Engineering.**

VALUE ENGINEERING Analysis of materials, processes, and products in which functions are related to cost and from which a selection may be made for the purpose of achieving the required function at the lowest overall cost consistent with the requirements for performance, reliability, and maintainability. *Also called* **Value Analysis.**

VALVE 1. Device that directs the flow and direction of hydraulic fluid to a specific actuator; **2.** Mechanism that directs the flow rate and flow direction of a fluid contained within a system. The following are some of the many types of valve:

> **Air-admittance:** Device to let air into a soil pipe system to relieve minor difference in air pressure.
>
> **Air purge:** Device that eliminates trapped air from a piping system.
>
> **Air release:** Small valve used to bleed air from pipework, casings and other vessels under pressure. *Also called* **Petcock.**
>
> **Air relief:** Air valve placed at the apex of a pipeline to automatically vent air and prevent the pipeline from becoming air-bound.
>
> **Air valve:** *See* **Air release,** above.
>
> **Altitude control:** Valve that automatically shuts off the flow when the water level in a storage tank reaches a preset elevation and opens when the pressure on the system side is

489

less than that on the tank side.

Angle: Similar to a globe valve but with pipe connections at right angles to each other.

Angle gate: Gate valve with an elbow cast on one end integral with the body.

Angle globe: Valve whose inlet and outlet are at 90° to each other and which can be used where piping changes direction, saving an extra fitting.

Angle stop: Type of shutoff valve, often mounted beneath sinks, lavatories, and toilets.

Antiflood: Check valve in a drain which is at flood level or high-tide level. *Also called* **Tidal valve.**

Antisiphon: Type of check valve that prevents siphoning of potentially contaminated water back into a potable supply system.

Automatic: Valve that opens or closes without human assistance when prescribed conditions are reached.

Automatic control: Valve to control the flow of steam, water, gas, or other fluids, by means of a variable orifice which is positioned by an operator in response to signals from a sensor or controller.

Backflow preventer: Device used to protect potable water supplies against potential contamination due to cross-connections with nonpotable water supplies.

Back pressure: Valve provided with a disk hinged on the upper edge so that it opens in the direction of normal flow and closes with reversal of flow.

Backwater: Flap valve that prevents the reverse flow of water or sewage in a pipe or channel.

Balanced: Valve that has both sides of its closing mechanism in contact with water and which opens and closes as a result of the difference in water

pressure, or water pressure plus other pressures, exerted on the sides of the mechanism.

Balancing: Discharge valve with no handle on its spindle.

Ball: Valve in which the rate of flow is regulated by a drilled ball that rotates against a flexible seat.

Bleed: Valve for removing unwanted, or small quantities from a system.

Blending: Three-way valve that permits entering liquid to be mixed with liquid that circulates through the valve; commonly used to blend hot and cold water to obtain a desired temperature.

Blowoff: Valve positioned in a low point or depression on a pipeline to allow its drainage.

Butterfly: Damper or throttle valve in a pipe, consisting of a disk that rotates about its diameter as an axis.

Bypass: Device which allows flow to divert around a filter when differential pressure becomes too great. *Also called* **Relief valve.**

Check: Valve that permits the flow of fluid in only one direction. *Also called* **Nonreturn valve.**

Ceramic-disc Faucet with one fixed and one movable hard, smooth ceramic disc, with matching holes. The tap is opened by turning one disc until the holes are opposite those of the other.

Clack: Type of check valve consisting of a controlling element hinged on one edge that opens for flow in one direction and closes when the flow is reversed.

Closed center: Valve in which all ports are closed in the center position.

Compression: Faucet or valve designed to stop the flow of water by the action of a washer closing against a seat.

Control: Device that controls the flow of oil within a hydraulic system.

Corporation cock: Valve joining a service pipe to a street water main.

Counterbalance: Valve that regulates fluid flows by maintaining resistance in one direction, but allows free flow in the other. *Also called* **Holding valve.**

Cone: Valve in which the moving plug is conical; the valve is opened by unscrewing the plug from the seat and turning it through an angle of 90°.

Dead weight relief: Valve in which the unrelieved weight of the plug is the force that tends to keep the valve closed. The valve opens when the pressure increases sufficiently to lift the plug against the force of gravity.

Detector check: Swing check valve, typically used in an automatic sprinkler system, that has its clapper weighted to divert small flows away from the main line through a bypass, where it will be measured by a meter.

Differential pressure: Valve operated by opposing pressures of different values.

Directional control: Valve whose primary function is to direct flow through selected passages.

Discharge: Control valve for reducing and increasing flow in a pipe, but not stopping the flow.

Diversion: Valve that permits flow to be directed into any one of several pipes.

Diverter: Hydraulic valve that permits a change in the direction of flow of a fluid.

Dry: *See* **Dry-pipe valve,** below.

Dry-pipe: Valve used on an automatic sprinkler system to control water supply to a dry-pipe system and, under defined conditions, to cause an alarm

to sound. *Also called* **Dry valve.**

Expansion: Valve in a refrigeration system that controls the flow of refrigerant to the cooling element.

Faucet: Valve that will control the flow of water.

Flap: Check valve with a hinged disk that opens when the flow is normal and closes by gravity, or by the flow when the flow tends to reverse.

Flapper action: Valve design in which output control pressure is regulated by a pivoted flapper in relation to one or two orifices.

Float: Valve actuated by a floating ball on a lever; a ball cock.

Flow control: Valve whose primary function is to control flow rate.

Flow control (deceleration): Flow control valve that gradually reduces flow rate to provide deceleration.

Flow divider: Valve that divides a flow into two streams.

Flush: 1. Valve in the bottom of a water closet tank that controls the flow of water into the toilet bowl; **2.** Pressure-controlled valve that controls the flow, and duration of flow, of water into the bowl of a water closet or to a urinal.

Foot: Check valve fitted to the inlet of a pump suction hose.

Footbrake: Driver-controlled valve that controls the air pressure delivered to or released from the service brake chambers.

Fourway: Valve that can be set to direct flow to all or any of four distinct settings.

Full open: Shutoff valve whose cross-section, in the open position, equals 85% of the cross-sectional area of the connecting pipes.

Gas cock: Shutoff valve on a gas line.

Gate: Valve that regulates flow within a pipe. *Also called* **On-off valve.**

Globe: Valve that allows for throttling the flow of water in a pipe.

Holding: Device placed in a hydraulic circuit to prevent loss of fluid from a load-bearing cylinder. This device also has thermal relief properties that prevent temperature increases from causing cylinder damage.

Holding (integral): Device attached directly to a cylinder to protect against fluid loss from the cylinder.

Hydraulic control: Mechanical device to divert or control the flow of fluid in a hydraulic circuit.

Hydraulic relief: Mechanical device used to limit the pressure in a hydraulic circuit.

Indicator: Valve that indicates by a sign, an open or shut position.

Integral holding: *See* **Holding valve (integral),** above.

Isolating: Stopvalve to close off the water supply to part of an installation but not to regulate flow.

Key: Valve operated by a removable key.

Lowering: Valve that allows a supported load to travel downward at a rated speed.

Mixing: Valve that mixes flows from two or more inlets, to a single outlet, adjusting the proportions by either manual or automatic control.

Mixing faucet: Faucet for hot and cold water supply that discharges through a common spout.

Moisture ejection: Valve that automatically expels moisture accumulation from the compressed air tanks of air-actuated brake systems or air compressors.

Motorized: Valve operated by an electric motor.

Needle: Valve with an externally adjustable tapered closure that regulates the flow passage.

Nonreturn: *See* **Check valve,** above.

On/off: *See* **Gate valve,** above.

Open-center: Valve in which all ports are interconnected in the center position.

Open spool: Valve that permits the flow of hydraulic oil in an actuator when that particular function's control lever is placed in neutral; commonly employed in circuits using hydraulic motors in order to provide the components with enough flow to allow them to glide to a stop once the valve is returned to neutral.

OS&Y: Type of outside screw-and-yoke valve used on piping in sprinkler systems. The position of the stem indicates whether the valve is open or closed.

Petcock: *See* **Air-release valve,** above.

Pilot: Valve applied to operate another valve or control.

Pilot-operated: Valve in which operating parts are actuated by pilot pressure.

Pilot-operated check: 1. Used in conjunction with outrigger jacks, this device is mounted on the side of the vertical jack to prevent the loss of hydraulic fluid from the jack cylinder. Due to the shock loading this cylinder must endure, even slight cylinder retraction cannot be tolerated. Therefore, this cylinder is not equipped with a holding valve that could allow fluid to escape; **2.** One way-flow valve whose operating parts are actuated by pilot pressure.

Plug cock: Valve where the fluid passes through a hole in a tapered plug, and that is closed by turning the plug through 90°.

Pneumatic: Valve for controlling gas flow or pressure.

Pop-off: Safety valve that opens automatically when pressure or some other factor exceeds a predetermined value.

Poppet: Mushroom-shaped valve that rests on a circular seat and that is opened by pressure on the stem (usually by the cam of a rotating shaft) and closed by the action of a confined spring mounted round the stem.

Post indicating: Valve that provides a visual means of indicating 'open' or 'shut' positions; found on the supply main of installed fire protection systems.

Prefill: Valve that permits full flow from a tank to a 'working' cylinder during the advance portion of a cycle, which permits the operating pressure to be applied to the cylinder during the working portion of the cycle, and permits free flow from the cylinder to the tank during the return portion of the cycle.

Pressure control: Valve whose primary function is to control pressure. There are several types:

Counterbalance: Pressure control valve that maintains back pressure to prevent a load from falling.

Decompression: Pressure control valve that controls the rate at which the contained energy of compressed fluid is released.

Differential dry-pipe: Valve in a dry-pipe sprinkler system in which air-pressure is used to hold the valve closed and thus hold the water back.

Load-dividing: Pressure control valve used to proportion pressure between two pumps in series.

Pressure-reducing: Pressure control valve whose primary function is to limit outlet pressure.

Pressure-reducing (gas): Valve used to reduce gas line pressure to usable limits of a gas carburettor.

Pressure-reducing (water): Valve used to reduce water pressure, typically between the main and an engine cooling system.

Pressure-relief: Pressure control valve whose primary function is to limit system pressure.

Pressure-unloading: Pressure control valve whose primary function is to permit a pump or compressor to operate at minimum load.

Pressure regulator: Valve used to automatically reduce and maintain pressure.

Pressure sensing: Device similar to an electrical pressure switch in which a signal to be sensed enters a control point and actuates a mechanism that, at the proper pressure level, causes one or more flow passages to change condition. Removal of the signal allows the pressure sensing valve to reset.

Relay: Valve that is actuated upon receipt of a signal or because of an activity of another valve or device.

Relay emergency: Combination relay valve with provision for the automatic application of trailer brakes in the event pressure is lost in the trailer supply air line.

Relief: Valve whose primary function is to limit system pressure by opening at a preset pressure and/or temperature. *See also* **Bypass valve.**

Reverse-acting diaphragm: Valve that opens when pressure is applied on a diaphragm and closes when pressure is released.

Rotary plug: Type of valve in which a ported sleeve or plug is rotated past an opening in the valve body.

Safety: Combination temperature and pressure relief valve required on hot water tanks.

Safety shutoff: Valve that automatically and completely shuts off a process or machine upon detection of an out-of-limit condition.

Scouring sluice *See* **Washout valve,** below.

Screw-down: Valve with a plate or disc across its waterway that is moved straight up or down by a spindle to open or close it.

Sequence: Valve whose primary function is to direct flow in a predetermined sequence.

Servicing: Stopvalve on the intake to a cistern or storage tank, permitting it to be isolated for repair or maintenance.

Servo-: Valve that modulates output as a function of an input command.

Shutoff: 1. Valve that operates fully open or fully closed; **2.** Valve installed in a water line whenever a cutoff is required.

Shuttle: Connective valve that selects one of two or more circuits because of flow or pressure changes between the circuits.

Slave: Valve that helps protect gears and components in a transmission's auxiliary section by permitting range shifts to occur only when the transmission's main gearbox is in neutral. Air pressure from a regulator signals the slave valve into operation.

Stop: Valve used to turn on or close a supply.

Stop-and-waste cock: Valve used to stop the flow of water in a pipe and permit the water downstream of the valve to be drained by allowing air into the line.

Stopcock: Small valve that turns off or on a water or gas supply, sometimes operated by a turn cock or special key.

Suction: Check valve on a suction pipe.

Swing, check: Check valve having a hinged gate that permits fluid to pass through the valve in one direction only.

Tandem-center: Valve in which advance and return ports are closed in the center position and the pump and tank ports are open.

Thermoregulating: Heat-actuated valve that limits the amount of municipal or raw cooling water into a system to conserve water and regulate cooling.

Tidal: *See* **Antiflood valve,** above.

Transfer: Valve used for placing multi-stage centrifugal pumps in volume or pressure operation.

Two- or three-way: Valve having two or three positions to give various selections of flow.

Two-, three-, or fourway: Directional control valve having 2, 3 or 4 ports for direction of flow.

Unloading: Valve that bypasses flow to a tank when a set pressure is maintained on its pilot port.

Vacuum relief: Valve that automatically opens and closes a vent to relieve a vacuum within a hot water supply system.

Variable air: In a heating/ventilating/air-conditioning system, a control unit consisting of a enclosure containing damper-position control equipment, a controller, and a sensor. The box is provided with primary air via a duct from the main distribution system; its output delivers air to diffusers located in the space being served.

Wet alarm: Valve that a) permits the flow of water into a wet-pipe sprinkler system, b) prevents the reverse flow of water, and c) incorporates a mechanism that actuates an alarm under specified flow conditions.

Washout: Gated opening in the lower part of a dam through which accumulated sand, gravel and other detritus can be expelled. *Also called* **Scouring sluice.**

VALVE ACTUATOR Valve part(s) through which force is applied to move or position flow-directing elements.

VALVE BANK Enclosed grouping of valves fed by its own hydraulic pump.

VALVE BEAT-IN Wear on the valve face or valve seat in internal combustion engines resulting from the pounding of the valve on the seat.

VALVE BODY Main part of a valve into which the stem and other parts are installed.

VALVE BOSS Flat area on the circumference of a storage cylinder.

VALVE BOX Covered concrete or metal box set over a valve stem and rising to ground level that permits operation of the valve from that level.

VALVE BRIDGE Part that allows two valves to be operated by a single rocker arm. *Also called* **Crosshead**

VALVE KEY Socket mounted on the end of a long rod for operating a valve that is installed several feet below ground level.

VALVE MOUNTING Mounting characteristics of a valve.

VALVE STEM Rod that actuates a valve mechanism, and by means of which it is opened of closed.

VALVE TOWER Tower built up from the bed of a reservoir to above the highest water level that houses the controls for valves, etc.

VANE Any of several flat or curved pieces set around an axle, moved by the passage of air or water.

VANE HYDRAULIC PUMP *See* **Pump, hydraulic.**

VANE PUMP *See* **Pump.**

VANE-TYPE HYDRAULIC PUMP *See* **Pump.**

VAPOR Gaseous phase of a substance normally in a liquid or solid state.

VAPOR BLANKET Layer of air above a body of water that has a higher moisture content than the surrounding atmosphere.

VAPOR CAPTURE SYSTEM Device or combination of devices designed to contain, collect, and route organic solvent vapors.

VAPOR DENSITY Weight of a vapor or gas compared to the weight of an equal volume of air.

VAPOR DIFFUSION Transfer of water in a partially dry solid from regions of high concentration to those of low.

VAPOR INCINERATOR Enclosed combustion device used for destroying organic compounds and which does not extract energy in the form of steam or process heat.

VAPORIZATION Passage from a liquid to a gaseous state.

VAPOR LOCK Vapor trapped in a service line that prevents the normal flow of a fluid; typically, the boiling of gasoline in a fuel line that prevents liquid fuel from reaching the carburettor.

VAPOR LOCK DEVICE Any device, such as an orifice or capillary tube, that eliminates, minimizes or prevents the collection of vapor in a pipe.

VAPOR PLUME Stack effluent consisting of flue gas made visible by condensed water droplets or mist.

VAPOR PRESSURE *See* **Pressure.**

VAPOR PROCESSING SYSTEM Equipment used for recovering or oxidizing total organic compound vapors displaced from the affected facility.

VAPOR RECOVERY SYSTEM Vapor gathering system capable of collecting all hydrocarbon vapors and gases discharged from a storage vessel, plus a vapor disposal system capable of processing such hydrocarbon vapors and gases so as to prevent their emission to the atmosphere.

VARIABLE Factor (flow, temperature, etc.) that can be sensed and quantified.

VARIABLE AIR VALVE *See* **Valve.**

VARIABLE CAN RATE Charge made for solid waste services based on the volume of waste generated, and as measured by the number of containers set out for collection.

VARIABLE COST Operational cost that results from running a machine, calculated on an hourly basis; includes the cost of labor and items such as fuel, oil, wire rope, and other replacement parts. *Also called* **Operating cost.**

VARIABLE-SIZE CREW *See* **Collection method.**

VARIABLE TIME Those times (i.e. haul, return) that vary with distance and speed.

VARIANCE 1. Permission granted by a zoning authority for a specified difference to established zoning requirements; **2.** Statistically, the square of the standard deviation.

VARIATE Quantity that can have one of a range of specific values, each with a specific probability.

VARIATION Act of changing in condition or degree.

VARVED SILT or CLAY Fine-grained, glacial lake deposit with alternating layers of silt or fine-grained sand and clay, formed by variations in sedimentation from winter to summer during the year.

VECTOR 1. Physical quantity that cannot be described completely without reference to a direction; **2.** Carrier that is capable of transmitting a pathogen from one organism to another.

VEGETATION Plant life; growing plants.

VEHICLE Agent which facilitates the mixture, dispersion, or solubilization of a test substance with a carrier.

VEIL OF CLOUD Layer of almost featureless cloud that characteristically covers the sky ahead of a warm front.

VELOCITY Distance travelled in a specified amount of time.

VELOCITY–AREA METHOD Determination of the discharge of a stream or open channel based on the sum of the velocity of the flowing water measured at several points within the cross section, multiplied by the respective fraction of the total area that each point represents.

VELOCITY COEFFICIENT Ratio between the actual velocity issuing from an orifice or hydraulic structure and the theoretical velocity.

VELOCITY HEAD Theoretical vertical height to which a liquid may be raised by its kinetic energy; equal to the square of the velocity divided by two times the acceleration due to gravity.

VELOCITY OF APPROACH Mean velocity in the channel of a measuring weir at the point where the depth over the weir is measured.

VELOCITY OF RETREAT Mean velocity of a flow immediately downstream of a measuring weir.

VELOCITY PRESSURE *See* **Pressure.**

VENT Opening permitting the escape or passage of liquids, fumes, steam, and the like.

VENTED Discharged through an opening, typically an open-ended pipe or stack, allowing the passage of a stream of liquids, gases, or fumes into the atmosphere.

VENTILATION Movement of air into, out of, or through a space, naturally, or as induced and controlled by equipment.

VENTILATION RATE Number of complete air changes within an area over a period of time.

VENTING SYSTEM Assembly of pipes and fittings that connects a drainage system with outside air to assure circulation of air and the protection of trap seals within the system.

VENT PIPE 1. Pipe or flue connecting any interior space to the outside atmosphere for the purpose of ventilation; **2.** Pipe connecting a plumbing fixture or its drain to the vent stack.

VENT STACK Vertical pipe connecting all the individual vent pipes to carry off foul air and gases from a building, and in particular from the drainage system with which a building is equipped.

VENTURI Pressure jet that draws in and mixes two or more gases or fluids.

VENTURI EFFECT Acceleration of a stream (gaseous or fluid) as it passes through a constriction in a channel or tube, which can be correlated with the flow rate ahead of the restriction to determine flow volume.

VENTURI FLUME Tube or channel having a constricted section to smooth pressure surges in materials being conveyed or to increase pressure or flow through that section.

VENTURI METER Meter for measuring the flow within a closed pipe having a throat followed by an expansion to normal diameter.

VENTURI TUBE Short tube with a constricted throat that is used to determine fluid pressures and velocities by measurement of the differential pressures generated at the throat as a fluid traverses the tube.

VERMICULITE Any of various silicate minerals that occur naturally as a magnesium hydroaluminosilicate in the form of small, lightweight granules or flakes that readily absorb water.

VERTICAL DRAIN Column of sand or other porous material used to vent water squeezed out of waterlogged or saturated soil.

VERTICAL PUMP *See* **Pump.**

VERTICAL SERVICE SPACE Shaft oriented essentially vertically that is provided in a building to facilitate the installation of building services.

VIABLE Capable of living, developing, or germinating under favorable conditions.

VIBRATORY PLOW Blade that is towed behind a tractor, and to which a vibratory force is imparted, permitting it to slice a trench through even densely packed material. The blade is frequently shaped to permit it to bury small-diameter cable and pipe as it progresses.

VICTAULIC COUPLING Rubber-gasketed pipe-coupling fitting that permits limited movement after fixing.

VICTAULIC PIPE Cast iron water pipe with ends formed to accept victaulic couplings.

VIOLATION Anything contrary to law regarding the use of property.

VIRTUAL SLOPE Hydraulic gradient, showing the loss from friction in pressure per unit length.

VIRUS Any of various submicroscopic pathogens consisting essentially of a core of nucleic acid surrounded by a protein coat, having the ability to replicate only inside a living cell.

VISCOSIMETER Instrument for determining the viscosity of liquids and slurries.

VISCOSITY *Also called* **Body.** Measure of the internal friction or the resistance of a fluid to flow. There are several measures, including:

> **Absolute:** Product of a fluid's kinematic viscosity times its density; a measure of a fluid's tendency to resist flow without regard to its density.

> **Apparent:** Ratio of shear stress to rate of shear of a non-Newtonian fluid, calculated from Poiseuille's equation

497

and measured in poises. The apparent viscosity of most greases varies with changing rates of shear and temperature, and must therefore be reported as the value at a given shear rate and temperature.

Kinematic: Flow property of liquid.

VISCOSITY INDEX Measure of the viscosity–temperature characteristics of a fluid as referred to that of two arbitrary reference fluids; a number derived from viscosity measurements at 40°C and 100°C that indicates the extent to which the viscosity decreases with increasing temperature.

VISCOUS FLOW Fluid condition in which there is a continuous steady motion in the direction of flow.

VISIBLE AREA Ground, or vegetation, that can be directly seen from a given lookout point under favorable atmospheric conditions.

VISIBLE EMISSIONS Emissions that are visually detectable without the aid of instruments.

VISIBILITY Greatest distance under given weather conditions to which it is possible to see with the aid of an instrument.

VISUAL ALARM Device that produces a visual indication (light, flag, etc.) of a specific operational condition or mode of operation.

VISUAL IMPACT ASSESSMENT Evaluation of the effect likely to be created on an existing landscape by development of new structures and an appraisal of the aesthetic impact.

VITREOUS Material that resembles glass, frequently used to surface sanitary fittings.

VITRIFIED CLAY PIPE Fired clay pipe, generally used for sewers.

VOID Open space between solid material.

VOID RATIO Ratio of the void space volume to the volume of solids in such as soils, or filter media, or other porous materials.

VOLATILE Readily vaporized organic material that, when mixed with oxygen, is easily ignited.

VOLATILE ACIDS Mainly acetic, propionic and butyric acids that are produced during anaerobic decomposition of organic wastes.

VOLATILE FLAMMABLE LIQUID Flammable liquid having a flash point below 38°C (100°F).

VOLATILE MATERIAL 1. Material that is subject to release as a gas or vapor; **2.** Liquid that evaporates readily.

VOLATILE MATTER Material weight lost from a dry powdered fuel sample that is heated at 950°C (1742°F) for seven minutes in a closed crucible.

VOLATILE ORGANIC COMPOUND Any organic compound that reacts with nitrogen oxides in the presence of sunlight to form ozone, a major component of photochemical smog.

VOLATILE ORGANIC LIQUID Organic liquid that can emit volatile organic compounds into the atmosphere.

VOLATILE SOLID Sum of the volatile matter and fixed carbon of a sample, as determined by allowing a dried sample to burn to ash in a heated and ventilated furnace.

VOLATILITY Expression of evaporation tendency.

VOLCANIC Of or resembling an erupting volcano.

VOLCANIC DUST Finest pyroclastic ash ejected by an explosive volcanic eruption.

VOLCANIC GLASS Volcanic igneous rock of vitreous or glassy texture.

VOLCANIC ROCK Rock formed by volcanic action at the Earth's surface.

VOLCANIC SPRING Water brought to the surface from considerable depths by volcanic forces.

VOLCANISM Volcanic force or activity.

VOLCANO Vent in the Earth's crust through which molten lava and gases are ejected.

VOLCANOLOGY Science concerned with volcanic phenomena.

VOLT SI unit of potential difference between two points of a conducting wire carrying a constant current of 1 A, when the power dissipated between these points is equal to 1 W. (1 V=1 W/A). Symbol: V. *See also the appendix*: **Metric and nonmetric measurement.**

VOLTAGE Measure of electric pressure between any two wires of an electric circuit. *See also* **Nominal voltage.**

VOLTAGE AMPLIFIER Amplifier that increases a signal voltage.

VOLTAGE DIP Reduction in voltage resulting from a sudden application of load, usually expressed in percentage.

VOLTAGE DRIFT Gradual deviation of the mean regulated voltage above or below the desired voltage under constant operating conditions.

VOLTAGE DROP Difference in voltage measured across a current-carrying device.

VOLTAGE GAIN Ratio of input voltage to output voltage.

VOLTAGE MULTIPLIER Circuit that produces high-voltage DC from low-voltage AC current.

VOLTAGE OPERATING BAND Span of voltage through which a generator can be adjusted and operated.

VOLTAGE REGULATION Difference between the steady-state no load and steady-state full load output, expressed as a percentage of the full-load voltage.

VOLTAGE REGULATOR Device that automatically controls the voltage output of a generator at its specific value.

VOLTAGE-TO-GROUND For grounded circuits, the voltage between the given conductor and that point or conductor of the circuit that is grounded; for ungrounded circuits, the greatest voltage between the given conductor and any other conductor of the circuit.

VOLT–AMP CURVE Graph that shows the output characteristics of a welding power source.

VOLT–AMPERE Product of volts and amperes flowing in a circuit: in resistive circuits, the product of volts and amperes equals watts.

VOLTMETER Meter used to measure voltage in an electrical system.

VOLTOHMMETER Multipurpose electrical testing instrument.

VOLT PER METER A derived unit of electric field strength in the SI system of measurement. Symbol: V/m. *See also the appendix*: **Metric and nonmetric measurement.**

VOLUMETRIC Of or pertaining to measurement of volume

VOLUNTARY SEPARATION Separation of various fractions (glass bottles, plastics, newspaper, etc.) from domestic refuse, by the originator, prior to collection.

VOLUNTARY STANDARD Standard to which many adhere but which is not enforced by any regulatory or recognized authority.

VOLZ PHOTOMETER Instrument capable of low-precision measurements of the intensity of direct sunlight at wavelengths defined by narrow-band interference filters.

VORTEX Fluid flow involving rotation about an axis.

VORTEX SEPARATOR Vertical cylindrical tank into which air- or liquid-borne particulate matter is introduced tangentially, separation of the heavier particulates and their ultimate deposition in the conical base for removal being effected by centrifugal force.

V-PLOW Forest plow with a V-shaped blade, used to prepare strips for hand planting by removing surface debris and competing vegetation.

WAITING REPAIR TIME *See* **Machine time.**

WAIVER Intentional relinquishing of a right, claim, or privilege.

WAKE Recurring eddies downstream of an obstacle.

WALE 1. Long formwork member (usually double), used to gather loads from several studs (or similar members) to allow wider spacing of the restraining ties; **2.** When used with prefabricated panel forms, this member is used to maintain alignment. *Also called* **Ranger** and **Waler.**

WALER Horizontal brace, used to hold timbers in place against the sides of an excavation. *Also called* **Wale** and **Whaler.**

WALL-HUNG BASIN Wash basin fixed to the wall, usually on brackets.

WALL HYDRANT Hydrant that protrudes through the wall of a building.

WARM FRONT Front along which an advancing mass of warm air rises over a mass of cold air.

WARM SPRING Terminal spring with water having a temperature lower than 37°C (98°F).

WARNING DEVICE Device that warns an operator when a predetermined operating parameter has been reached.

WARNING LIGHT Illuminated indicator that warns of a situation or condition.

WARRANTY Guarantee or assurance that work or a product shall be or will perform as represented. In addition to a warranty used typically within the construction industry, there are two forms:

> **Construction:** Guarantee that the entity has been constructed in accordance with applicable codes and regulations, with the materials specified, in the manner described in the contract documents, and that installed equipment will operate or perform as described by the manufacturer or supplier.

> **Expressed:** One that is defined and particular to a product.

> **Implied:** One that is not written but exists under law.

WARRANTY DEED Deed containing a covenant by the grantor to the grantee to warrant and defend the title of the estate conveyed. *See also* **Deed** and **Quitclaim deed.**

WASH 1. Passing water over and through earth, gravel, and other material to cause separation and segregation; **2.** Weathered slope subject to occasional flooding.

WASHBASIN Sanitary fixture designed to facilitate personal washing.

WASH BORING Method of examining soil, usually in soft soil or clay, by driving a pipe into the ground and then inserting a small pipe inside of it, through which water is forced to wash out soil particles in water suspension for examination. *Also called* **Jetting** and **Water jet.** *See also* **Air lift.**

WASH DOWN Flushing spilled combustible or flammable liquids.

WASH FILTER Filter element in which a larger unfiltered portion of the fluid flowing parallel to a filter element axis is utilized to continuously clean the influent surface which filters the lesser flow.

WASHLAND Land between a river and a flood embankment.

WASHOUT VALVE *See* **Valve.**

WASHROOM Toilet facility; also a laundry room.

WASH WATER Water used to clean the filter beds of a rapid sand filter.

WASHWATER GUTTER Trough through which used wash water is carried away from a rapid sand filter.

WASHWATER RATE Rate at which wash water is applied to a rapid sand filter to clean it.

WASHWATER TANK Tank in which water to be used to clean the filter beds of a rapid sand filter is stored at an elevation above the beds sufficient to ensure an adequate pressure.

WASTE 1. Material of no value. *See also* **Residual**; **2.** Solid, liquid or gaseous substance that remains as a residue, or is a by-product, of a process and for which no use can be found; **3.** Drain line in plumbing; **4.** Overflow line in piping.

WASTE BRANCH Pipe that carries waste water away from a sanitary fitting and which slopes or falls to a connection with a waste stack.

WASTE CONTAINER There are several types, including:

Carrying: Receptacle of 9.2- to 13.2-L (35- to 50-gal) capacity, usually constructed of plastic or aluminum, that is carried by a collector in a back yard carry out service. *Also called* **Tote barrel.**

Disposable: Plastic or paper sacks designed for storing solid waste.

Dumping: Metal container, 2.3- to 4.6-m³ (3- to 6-yd³) capacity, with or without wheels and having one or more hinged lids, fitted with a horizontal slot on each side face into which the lifting arms of a specially-equipped packer can be inserted and the bin raised and inverted above the body of the truck where its contents is discharged.

Lift-and-carry: A large container that can be lifted onto a service vehicle and transported to a disposal site for emptying. *Also called* **Detachable container** and **Drop-off box.**

Roll-on/roll-off: Large container of 15- to 30-m³ (20- to 40-yd³) capacity that can be pulled onto a service vehicle mechanically and carried to a disposal site for emptying.

WASTE-DERIVED FUEL Combustible fuel, which may be gaseous, liquid, or solid, in a form suitable for its intended use (i.e. pelletized, liquefied, etc.) generated from the conversion and/or management of wastes.

WASTE DISPOSAL UNIT Electrically driven grinder designed to fit under the waste outlet of a kitchen sink and connected to the drain. *Also called* a **Garburator.**

WASTE EXCHANGE Use by one company of an industrial waste generated by another firm.

WASTE FACTOR Departure from the calculated ideal energy necessary to effect a change.

WASTE GAS Surplus gas vented from a sludge digestion tank.

WASTE-GAS BURNER Device used to flare off surplus gas vented from a sludge digestion tank.

WASTE GATE Gate in a canal wall, used to waste surplus water.

WASTE HEAT Heat that has been generated but not used prior to rejection to the environment.

WASTE LOAD ALLOCATION Portion of a receiving water's loading capacity that is allocated to one of its existing or future point sources of pollution.

WASTE MANAGEMENT Management of wastes at all stages from generation to ultimate disposal, including management of reclaimed or recycled fractions out of the waste stream..

WASTE MANAGEMENT UNIT Piece of equipment, structure, or transport mechanism used in handling, storage, treatment, or disposal of waste.

WASTE MANAGEMENT UNIT BOUNDARY Vertical surface located at the hydraulically downgradient limit of the unit and extending down into the uppermost aquifer.

WASTE MINIMIZATION Activity that eliminates or reduces the amount of any pollutant from entering a waste stream or the environment.

WASTE PILE Noncontainerized accumulation of solid, nonflowing waste.

WASTE PIPE Pipe for carrying off waste flows.

WASTE PROCESSING Operation such as shredding, compaction, composting, or incineration, in which the physical or chemical properties of wastes are changed.

WASTE RECEPTACLE Container for holding or facilitating the removal of refuse.

WASTE REDUCTION Practice of producing smaller quantities of disposable waste.

WASTE STACK Vertical pipe that collects waste flows from appliances and their individual waste pipes.

WASTE STREAM Waste generated by a particular process unit, product tank, or waste management unit.

WASTE TAIL Pipe tail on the outlet from a basin, bath, sink, etc.

WASTE TREATMENT Process by which the waste component of wastewater is removed or treated to render it harmless.

WASTE WATER Water that is not, or is no longer required.

WASTEWATER Water containing biological, organic and/or chemical contaminants, primarily as a result of human activities.

WASTEWATER ANALYSIS Determination of the organic and mineral content, chemical composition, and biological condition of a waste or effluent stream.

WASTEWATER CHARGE Levy made for the provision of wastewater collection and treatment.

WASTEWATERS Wastes that contain less than 1% by weight total organic carbon (TOC) and less than 1% by weight total suspended solids (TSS).

WASTEWATER SURVEY Investigation of the sources and characteristics of wastewater discharges to a collecting sewer, and/or at the point of entry to a wastewater treatment plant.

WASTEWATER TREATMENT PLANT Water pollution control plant.

WASTEWATER TREATMENT PROCESS Process that modifies characteristics such as BOD, COD, TSS, and pH, usually for the purpose of meeting effluent guidelines and standards.

WASTEWATER TREATMENT TANK Tank designed to receive and treat an influent wastewater through physical, chemical, or biological methods.

WASTEWAY A spillway.

WATER Clear, colorless, nearly odorless and tasteless liquid, H_2O, essential for most plant and animal life, and the most widely used of all solvents. Melting point 0°C (32°F), boiling point at sea level 100°C (212°F), specific gravity (4°C) 1.000.

WATER ABSORPTION Act of a material, particularly one that has been manufactured, picking up water when wet or exposed to rain.

WATER ABSORPTION CHARACTERISTIC *See* **Water resistance.**

WATER ANALYSIS Chemical analysis of the dissolved materials in a sample of water, including the amount and type of suspended solids, and its pH value.

WATER BALANCE Method of determining the gross loss or gain of water based on precipitation, evaporation, transpiration, and runoff.

WATER BALLOON TEST METHOD *See* **Soil compaction tests.**

WATER BAR Shallow trench cut into the surface of a forest road or created by an embankment (e.g. log and soil) to collect and channel water off the surface to avoid erosion. *See also* **Cross-ditch.**

WATER-BEARING GROUND Permeable ground below the standing-water level.

WATER BLAST System of cutting or abrading a surface by a stream of water ejected from a nozzle at high velocity.

WATERBORNE DISEASE Disease caused by organisms or toxic substances carried by water.

WATERBORNE DISEASE OUTBREAK Significant occurrence of acute infectious illness, epidemiologically associated with the ingestion of water from a public water system which is deficient in treatment.

WATER BREAK Barrier in the bed of a ditch or channel that reduces flow velocity and minimizes scour.

WATER BUDGET Measure of the incoming and outgoing water of a region by all means.

WATER CHANNEL Groove to allow water or condensation to drain away.

WATER CLOSET A toilet.

WATER COLUMN Water held above a valve; a measure of the head or pressure in a closed pipe or conduit.

WATER COMPENSATION Volume of water that must escape a diversion works so as to satisfy the needs of holders of prior rights on the stream.

WATER CONDITIONER Device that changes the condition of water passed through it: by removing dissolved minerals, etc.

WATER CONSUMPTION Per capita vol-

ume of water supplied to, or used in a water management area.

WATER CONTENT Ratio of the quantity (by weight) of water in a given volume of soil mass to the weight of the soil solids, typically expressed as a percentage.

WATER COOLING TOWER Structure in which water is cascaded over a series of perforated plates, or stages, against the countercurrent of air blown in the opposite direction.

WATER CORROSION RESISTANCE *See* **Water resistance.**

WATERCOURSE Channel in which a flow of water occurs, either continuously or intermittently.

WATER CYCLE Various stages taken by water in moving from the surface of the earth, through the atmosphere, and back to the surface, including evaporation, precipitation, runoff, infiltration, percolation, transpiration, etc.

WATER DEMAND Volume of water required by an individual plumbing fixture, or group of fixtures. *See also* **Total water demand.**

WATER DISTRIBUTION PIPE Internal pipe that conveys water from the water service pipe to plumbing fixtures or other water outlets.

WATER DISTRICT Administrative area established for the provision of potable and irrigation water.

WATER EQUIVALENT OF SNOW Depth of water that would result at a point or over an area if an accumulation of snow were reduced to water.

WATER FILTER Device for the removal or reduction of suspended solids from a flow of water.

WATER-FLOW FORMULA Any of a number of mathematical means to determine the velocity of discharge of water in or from a conduit, channel, stream, etc.

WATER-FLOW SWITCH *See* **Switch.**

WATER GAUGE *See* **Gauge.**

WATER HAMMER Condition occasioned by the sudden stopping of water flow in a pipe resulting in a pressure wave that impacts upon closed valves and pipe walling at extreme direction changes.

WATER-HAMMER ARRESTER Device installed in a piping system to absorb hydraulic shock waves and eliminate water hammer.

WATER HOLE Small natural depression in which water collects.

WATER INJECTION 1. Technique for reducing the formation, and subsequent emission, of pollutants by internal combustion; **2.** Technique of injecting water into oil-bearing strata to increase the availability and flow of the oil; **3.** Process of recharging a depleted water table by injecting water.

WATER INTAKE Hydraulic structure through which water is obtained from a source.

WATER LANCE Pipe on the end of a hose, used to hydraulically blast out solids.

WATER LEVEL Elevation above a known datum of the surface of a standing body of water.

WATER-LEVEL GAUGE Device that shows and/or records the elevation of a water surface.

WATER LINE Highest level to which a ball valve should permit refill water to rise in a cistern.

WATERLOGGED Soaked and saturated with water.

WATERLOGGED SOIL Soil so continuously saturated as to drive out all gases, and in which upland plants cease to grow.

WATER LOSS Water that has once been accounted for, and which can no longer be found.

WATER MAIN Large water supply pipe to which branches are connected. *Also called* **Main water line.**

WATER MARK Water surface elevation of the design peak flow. *See also* **High water mark** and **Low water mark.**

WATER METER Device for recording the quantity of water flowing through a pipe.

WATER OF ADHESION *See* **Adhesive water.**

WATER OF CAPILLARITY *See* **Held water.**

WATER PHONE Device used to locate leaks in a buried water pipe under pressure by amplification of the sound made by escaping water.

WATER PLANT Water treatment plant.

WATER POLLUTION Introduction to water of substances that diminish its quality.

WATER POLLUTION CONTROL PLANT Facility designed and equipped for the control of waterborne pollution and the removal of substances likely to degrade the receiving stream.

WATER POWER Energy developed through the use of moving and/or falling water.

WATERPROOF Impervious to water in either liquid or vapor state.

WATERPROOFING Sealing or coating with a material or substance that will prevent the passage of water. *See also* **Dampproofing** and **Vapor barrier.**

WATERPROOFING COMPOUND Material used to impart water repellency to a structure or a constructional unit.

WATERPROOF MEMBRANE Sheet materials applied to a roof or wall surface to prevent the penetration of water, often in several layers or 'plies.'

WATER QUALITY Measure of the chemical, physical, and biological characteristics of water.

WATER QUALITY OBJECTIVES Minimum quality objectives for water in its natural state, under storage, or intended for specific uses.

WATER QUALITY PROGRAMS Programs required by the national pollutant discharge elimination system permit for industrial pretreatment enforcement and monitoring, customer education, and plant laboratory analysis and monitoring.

WATER QUALITY STANDARDS Minimum quality standards for water in its natural state, under storage, or intended for specific uses.

WATER RAM Device that periodically checks the flow of water in a supply pipeline, then releases it in a manner designed to produce an impulse, or water hammer, the force of which is sufficient to lift the exiting water to a new elevation.

WATER RATE Amount charged per given volume or per time period for the supply of water. *Also called* **Water tax.**

WATER RECLAMATION Process by which suspended and dissolved solids remaining in a waste stream following secondary or advanced treatment are removed to a level that allows the sewage treatment plant effluent to be used beneficially. This may include flocculation, coagulation and filtration.

WATER REPELLENT Property of a surface that resists wetting but which permits the passage of water when hydrostatic pressure occurs.

WATER-REPELLENT PRESERVA-TIVE Liquid formulated to penetrate into the pores of wood, impregnating the surface fibers causing them to repel moisture.

WATER RESISTANT 1. Having the ability to withstand the passage, or deteriorating effect of water; **2.** *See* **Water resistance.**

WATER RIGHTS Powers or privileges concerning water, recognized by a legal authority or precedent.

WATER SEAL Depth of water trapped in a sanitary fitting due to the shaping of the pipe in order to prevent the passage of vapor or gases from the sewer through the fixture.

WATER SERVICE Provision to the lot line or water meter by a water authority of the supply of potable water and/or water for fire protection.

WATER SERVICE PIPE Pipe that is part of a water system and that conveys water from a public water main or a private water source to the inner side of the wall or floor through which the system enters the building.

WATERSHED Area that drains into a stream or other water passage.

WATERSHED DIVIDE Theoretical line that follows ridges or summits and which separates watersheds.

WATERSHED PLANNING Detailed evaluation of the potential uses to which all of the known water present in a watershed, and that arriving through precipitation and other means, could be put without harming the resource.

WATER SOFTENER Attachment to a water distribution system that receives water from the potable supply, passes it through a bed composed of a calcium- and magnesium-absorbing medium (usually zeolite), and sometimes through an additional sand-and-gravel filter, before passing the softened and filtered water to the appliances and faucets.

WATERSPOUT 1. Cyclonic storm similar to a tornado that occurs over water and which forms a dense funnel-shaped cloud by entraining water droplets from the surface. **2.** Pipe or other orifice through which water is spouted or conveyed.

WATER STANDARDS Definitions of water quality and requirements for supply.

WATER SUPPLY 1. Potential sources of supply within a watershed, or administrative area such as a water management district; **2.** Provision of water of stated quality in sufficient quantity at a required pressure to meet a defined need.

WATER SUPPLY FACILITIES Works, structures, equipment and processes necessary to obtain, treat, and deliver water for domestic, industrial, and fire use.

WATER SUPPLY MAP Map showing the location of supplies of water readily available for pumps, tanks, trucks, camp use, etc.

WATER SUPPLY SOURCE Location where raw water may be obtained in sufficient quantities to meet a predicted need.

WATER SUPPLY SYSTEM Array of piping, valves, and fittings from the source of water to its point of use.

WATERTABLE 1. Level in the ground above which water will not naturally rise (not a constant measure, influenced by many natural phenomena); **2.** Projection of masonry on the outside of a wall, slightly above ground level. Often a damp course is placed at the level of the watertable to prevent upward penetration of groundwater.

WATER TAX *See* **Water rate.**

WATER TEST Pressure test applied to pipes and drains.

WATER THIEF Variation of a wye adapter having three gated outlets, usually two 38 mm (1.5 in.) and one 65 mm (2.5 in.), with a single inlet, 65 mm (2.5 in.) or larger.

WATERTIGHT Impermeable to water except when under hydrostatic pressure sufficient to produce structural discontinuity by rupture.

WATER TOWER Tank mounted above ground level for the storage of water.

WATER TREATMENT Processes that convert raw water into potable water conforming to a standard.

WATER TREATMENT PLANT Facilities provided for the treatment of raw water to improve its quality and ensure its purity.

WATER TREATMENT SYSTEM Component, piece of equipment, or installation that receives, manages, or treats process wastewater, product tank drawdown, or landfill leachate prior to direct or indirect discharge.

WATER TURBINE Wheel turned by the force of flowing water.

WATER USE Classification of a body of water according its actual or potential uses, i.e. potable supply, transportation, recreation, irrigation, fish culture, power production, etc.

WATER VAPOR Water in a vaporous form and diffused in the atmosphere.

WATER VAPOR TRANSMISSION Rate of water vapor flow, under defined conditions, through a unit area of a material.

WATER VELOCITY Speed of a stream. Usually refers to the average water speed through a stream's cross section.

WATER WASTE SURVEY Location and evaluation of actual and potential loss through leakage from a water supply system.

WATERWAY River, channel, canal, or other navigable body of water used for travel or transport.

WATER WELL Artificial excavation into the water table for the purpose of obtaining water.

WATER WHEEL Wheel made to rotate by falling water; a wheel designed to lift water to a higher level.

WATERWORKS System by which a public water supply is conditioned to an acceptable level of quality and pumped into the distribution system.

WATER YEAR Continuous 12-month period, not necessarily a calendar year, embracing a complete annual cycle of hydraulic events and conditions.

WATT SI unit of power available when energy of 1 J is expended in 1 s. (1 W=1 J/s). Symbol: W. *See also the appendix:* **Metric and nonmetric measurement.**

WATT HOUR Non-SI unit of energy. Symbol: W/h. Multiply by 3.600 to obtain kilojoules, symbol: kJ. *See also the appendix:* **Metric and nonmetric measurement.**

WATT-HOUR METER Indicating instrument that displays the kilowatt-hour output continuously for record purposes.

WATTMETER Instrument that measures

the real power of the circuit in which it is connected.

WATT PER METER KELVIN A derived unit of heat in the SI system of measurement. Symbol: W/(m°K). *See also the appendix*: **Metric and nonmetric measurement.**

WATT PER SQUARE METER A derived unit of heat flux density in the SI system of measurement. Symbol: W/m². *See also the appendix*: **Metric and nonmetric measurement.**

WATT PER STERIDIAN A derived unit of radiant intensity in the SI system of measurement. Symbol: W/sr. *See also the appendix*: **Metric and nonmetric measurement.**

WAVE Motion characterized by a regular progression of equally spaced peaks separated by valleys of uniform depth and form.

WAVEBAND Segment of the spectrum of wave frequencies.

WAVE-BUILT TERRACE Upper elevations of a beach, characterized by layers of coarse shingle, parallel to the shoreline.

WAVE-CUT TERRACE Flat or gently sloping surface cut by wave action at the base of a cliff adjacent to the shoreline.

WAVE EQUATION Partial differential equation in one, two, or three dimensions, the solution of which represents the propagation of a wave with constant velocity.

WAVE FRONT Surface of a propagating wave that is the locus of all points having identical phase, the surface being usually, but not always, perpendicular to the direction of propagation.

WAVE HEIGHT Vertical distance between the bottom of the trough and the top of the crest of a wave.

WAVE LENGTH Horizontal distance between corresponding parts of succeeding waves.

WAVELENGTH Distance between the crests of a sine wave.

WAVELET Small wave or ripple.

WAVE PERIOD Time required for two successive wave crests to pass a fixed point.

WAVE POWER Electricity developed by harnessing the energy present in waves.

WAVE PRESSURE Pressure on breakwaters and other marine structures caused by the wave action of the adjacent water body.

WAVE STEEPNESS Ratio of a wave's length to its height.

WAVE VELOCITY Speed at which the front of a wave advances.

WEAR Removal of materials from surfaces in relative motion. Three types of wear are:

    Abrasive wear: Removal of materials from surfaces in relative motion by a cutting or abrasive action of a hard particle (usually a contaminant).

    Adhesive wear: Removal of materials from surfaces in relative motion as a result of surface contact. Galling and scuffing are extreme cases.

    Corrosive wear: Removal of materials by chemical action.

WEAR AND TEAR Physical deterioration of a capital asset through use or exposure to the elements.

WEAR CONTROL FILTER Filter capable of removing all particles contributing to component wear from a system operating fluid.

WEAR PLATE Metal accessory used to protect and extend the life of a dump truck floor used to haul chunked concrete and other damaging materials. *Also called* **Wood cushion.**

WEATHER State of the atmosphere at a given time and place, described by specification of variables such as temperature, moisture, wind velocity and direction, and pressure.

WEATHER BALLOON Device used to carry instruments into the atmosphere to gather meteorological data.

WEATHERING Any process of decay brought about by the effect of weather conditions. *See also* **Deterioration** and **Disintegration.**

WEATHERING ZONE Top layer of the lithosphere in the katamorphic zone, between the surface of the ground and the groundwater table.

WEATHERING Any of the chemical or mechanical processes by which natural and artificial materials exposed to weather degrade and decay.

WEATHER MAP Map or chart depicting or predicting the meteorological conditions over a specific geographical area at a specific time.

WEATHERPROOF Able to withstand exposure to the weather without damage.

WEBER SI unit of magnetic flux which, linking a circuit of one turn, produces in it an electromotive force of 1 V as it is reduced to zero at a uniform rate of 1 s. (1 Wb=1 V·s). Symbol: Wb. *See also the appendix*: **Metric and nonmetric measurement.**

WEIGHBRIDGE Steel weighing platform large enough to accommodate all the wheels of the largest highway-legal truck and coupled to a device that will calculate the weight of the vehicle.

WEIGH SCALE In ground device for weighing vehicles, often up to 21 m (70 ft) long.

WEIGH STATION Facility for weighing vehicles and recording the results.

WEIR Structure across a waterway, used for measuring flow, stream diversion, or for catching fish or sediment.

WEIR HEAD Depth of water measured from the bottom of the notch in a weir plate to the upstream water surface.

WEIR LOADING Rate at which the contents of a tank or vessel leaves the unit by overflowing a weir.

WEIR PLATE Metal plate set vertically into the top face of a weir, often containing one or more V-shaped notches either to regulate the overflow or for measurement purposes.

WELDED STEEL PIPE Pipe fashioned from steel sheet curved into a cylinder with the edges welded together, the resulting seam being longitudinal or helical.

WELL Hole dug or drilled into the ground and penetrating the watertable, from which water may be drawn or pumped. *See also* **Artesian well** and **Dry well**.

WELL CAPACITY Maximum rate at which a well will yield water under a stipulated set of conditions.

WELL CASING Solid or perforated metal pipe used to line the borehole of a well.

WELL CONE OF INFLUENCE Approximately conical depression produced in the watertable as result of extraction of water from a well.

WELL CURB Parapet built around a well at ground level to prevent drainage of surface water into the borehole.

WELL DRAIN Well to drain surface and groundwater into a lower strata. *Also called* **Absorbing drain.**

WELL FIELD Number of wells in relative proximity and drawing from the same watershed.

WELL FLOODING Diversion of water from streams into the borehole of wells so as to recharge or increase the availability of groundwater.

WELL HYDROGRAPH Graphical representation of the fluctuations over time of the water surface in a well.

WELL INFILTRATION AREA Area of water-bearing formation contributory to a well.

WELL INTAKE Well screen, perforated casing, or other means by which water is admitted from a water-bearing strata to a well.

WELL INTERFERENCE Result of two or

more wells in relative proximity and drawing from the same watertable at similar depths at rates above the capacity of the available supply.

WELL LOG Detailed chronological record of the sinking and development of a well.

WELL LOSS Head loss associated with friction and turbulence as water enters the wellbore.

WELLPOINT Pipe, 38 to 65 mm (1.5 to 2.5 in.) in diameter, fitted with a driving point and a fine mesh screen, used to remove underground water.

WELLPOINT DRILL Centrifugal pump that can accommodate large quantities of air, used to remove underground water from the vicinity of a planned or completed excavation.

WELLPOINT PUMP *See* **Pump.**

WELLPOINT SYSTEM Machinery and equipment consisting principally of a pump, header pipe, lateral pipes, riser pipe, valves, and a number of wellpoints, laid out in a systematic manner to remove underground water.

WELL RECORD Data recording the hydraulic performance of a well.

WELL SCREEN Filtering device used to keep sediment from entering a water well.

WELL SEEPAGE AREA Area of water-bearing formation tributary to a well.

WELL SHOOTING Detonation of an explosive in the bottom of a well so as to increase the flow of water.

WELL STIMULATION Any of several processes used to clean the well bore, enlarge channels, and increase pore space.

WELL STRAINER Basket screen through which water is admitted to the suction of a well pump.

WELL WATER Raw groundwater obtained from a well dug or drilled into the zone of saturation.

WELL WORKOVER Reentry of an injection well; including, but not limited to, the pulling of tubular goods, cementing or casing repairs; and excluding any routine maintenance (e.g. resealing the packer at the same depth, or repairs to surface equipment).

WET ADIABATIC Process in which air saturated with water is expanded or compressed without external heat addition.

WET ADIABATIC INSTABILITY Atmospheric condition where vertical movements of moist air tend to increase.

WET ADIABATIC LAPSE RATE Drop in temperature of a moving parcel of saturated air per unit increase in height in adiabatic ascent.

WET AIR POLLUTION CONTROL SCRUBBERS Air pollution control devices used to remove particulates and fumes from air by entraining the pollutants in a water spray.

WET ANALYSIS Mechanical analysis of soil samples smaller than 0.06 mm (0.0024 in.) made by mixing the sample in a measured volume of water and checking its density at intervals using a hydrometer.

WET-BULB DEPRESSION Difference between dry-bulb and wet-bulb temperatures.

WET-BULB TEMPERATURE Temperature of the wet-bulb thermometer in a relative humidity measurement. It is compared with the dry-bulb temperature to determine relative humidity (RH).

WET-BULB THERMOMETER Thermometer whose bulb is kept moist.

WET CONNECTION Connection to a water main completed while the main is under service pressure.

WET DIGESTION Solid waste stabilization process in which solid organic wastes are placed in an open digestion pond to decompose anaerobically. The carbonaceous matter is converted into carbon dioxide and methane; the soluble and suspended fraction is converted aerobically by algae in a bio-oxidation pond.

WET-HEAD HYDRANT Fire hydrant with the valve in the top.

WETLAND 1. Transitional area between dry land and aquatic areas, having a high watertable or shallow water; **2.** Land with one of the following attributes:

(a) Periodically supports hydrophytes;

(b) Substrate is predominately undrained hydric soil; or

(c) Substrate is not soil and is saturated or covered with water during part of the growing season each year.

WET OR LOW-LYING AREA TECHNIQUES *See* **Sanitary landfilling.**

WET OXIDATION PROCESS Sludge disposal method in which the oxidation of a suspension of sludge solids is conducted under increasing pressure and temperature.

WET ROT Fungus decay of wood promoted by warm, moist, unventilated conditions.

WET SCRUBBER UNIT Emission control device that mixes an aqueous stream of slurry gases from a steam generating plant with the exhaust unit to control emissions of particulate matter or sulfer dioxide.

WET SILT Silt resting below water where it occupies about five times its dry volume.

WET SOIL Soil that contains significantly more moisture than moist soil, but in such a range of values that cohesive material will slump or begin to flow when vibrated. Granular material that would exhibit cohesive properties when moist will lose those properties when wet.

WETTED PERIMETER Length of the surface in contact with water in a channel.

WETTING Process in which a liquid is absorbed by a solid surface and forms a liquid film.

WETTING AGENT Any of several compounds that cause a liquid to spread more easily across or penetrate into the surface of a solid by reducing the surface tension of the liquid.

WET VENT Soil or waste pipe serving also as a vent.

WET WELL Sump of a pumping station.

WHALER *See* **Waler.**

WHEEL DITCHER Machine that digs trenches by rotation of a wheel fitted with toothed buckets. *See also* **Ditcher** and **Ladder ditcher.**

WHEEL HORSEPOWER *See* **Horsepower.**

WHIRLING PSYCHROMETER Psychrometer in which wet- and dry-bulb thermometers are mounted on a pivot about which they can be swung to provide an air flow past the wet bulb so that the humidity of the air next to the bulb is always close to the ambient value.

WHIRLWIND Small, near-vertical vortex that forms in conditions of exceptionally strong convection.

WHITE NOISE Random noise that has equal energy at every frequency in a particular waveband.

WHMIS Workplace Hazardous Materials Information System, a system which requires evaluation of the potentially harmful effects of materials used in the workplace and includes requirements for special labelling, Material Safety Data Sheets (MSDS) and employee training in handling hazardous materials.

WHMIS CLASSIFICATION Classification of products that are controlled under WHMIS according to the nature of the hazard involved in their handling (e.g. flammability, toxicity).

WHOLE EFFLUENT TOXICITY Aggregate toxic effect of an effluent measured directly by a toxicity test.

WILDERNESS Wild or desolate region with few or no people living in it.

WIDE-CRESTED WEIR *See* **Broadcrested weir.**

WIND Moving air; especially, a natural and

perceptible movement of air parallel to or along the ground; a movement or current of air.

WINDBLOWN Blown or dispersed by the wind; growing or shaped in a manner governed by the prevailing winds.

WINDBREAK Hedge, row of trees, or fence serving to lessen or break the force of the wind.

WIND-BORNE Carried by the wind.

WIND CONE Wind indicator.

WIND DEPOSIT Soil and granular material lifted, carried, and deposited in a new location by the action of the wind.

WIND DEW Moisture deposited as dew from tropical maritime air blowing against a cold surface.

WIND DIRECTION Point of the compass from which the wind blows.

WIND ENERGY Electricity generated by propeller-equipped turbines driven by the wind.

WIND FARM Area where, due to the prevailing conditions, a large number of wind turbines are installed for the purpose of generating electricity.

WINDFLOW Sudden gust or blast of wind.

WIND GAP Shallow notch or ravine on the side of a steep mountain ridge.

WIND GAUGE Instrument that measures the force of wind; an anemometer.

WIND MEASUREMENT Techniques used to measure the direction and strength of the wind.

WINDMILL Machine with a rotor which is fitted with vanes that are oriented so as to cause wind to effect their rotation: the resulting energy generated by a revolving rotor is used to perform work.

WINDOW PIPE Dredge discharge pipe with one or more openings in the bottom.

WIND POWER Mechanical or electrical power generated by a windmill using the kinetic energy of the wind as the prime energy source.

WIND PROFILE Graphical representation of variations in wind speed as a function of height or distance.

WIND PUMP Pump operated by the force of the wind rotating a propeller connected to a crank which, in turn, actuates a rod connected to a well pump.

WINDROW 1. Ridge of loose material thrown up by a machine; **2.** Composting material stacked in a pyramid shape.

WINDROWING Concentration of slash, branchwood, and debris into rows to clear the ground for regeneration. Windrows are often burned.

WIND SHEAR 1. Effect of wind in raising the level of water on a lee shore; **2.** Line of extreme turbulence between air masses of widely different characteristics.

WIND SPEED Wind velocity, measured at 6 m (20 ft) above the ground, or the average height of vegetation cover, and averaged over at least a 10-minute period.

WING DAM Breakwater.

WING WALL Short retaining wall at the end of a bridge or culvert to retain the earth; wall that guides a stream into a bridge opening or culvert barrel.

WINTER DESIGN TEMPERATURE Lowest ambient temperature anticipated for a location and for which adequate insulation should be provided.

WOODLAND Land covered by trees.

WOOD PRESERVATIVE Chemical product used to prevent or halt decay in exterior wood, applied by pressure treatment, soaking, or brush.

WORK 1. The entire project (or separate parts) that is required to be completed under the Contract Documents. Work is the result of performing services, furnishing labor, and supplying and incorporating materials and equipment into the construction, as required by the contract documents; **2.** Unit or piece being handled or worked on; **3.** Product of force and the distance through which it moves; **4.** *See* **Tire types.**

WORKABLE SLUDGE Digested sewage sludge, generally having 75% or better moisture content, that can be readily transported from a sludge drying bed.

WORKAROUND Response to a negative risk event, distinguished from contingency plan in that it is not planned in advance of the occurrence of the risk event.

WORK DAY Portion of a 24-hour day during which work is contracted for, and/or paid for. *See also* **Day** and **Calendar day.**

WORKING CYCLE Complete set of repetitive and productive operations of a machine.

WORKING DAY Calendar day during which normal construction operations could proceed for a major part of a shift; normally excludes Saturdays, Sundays, and holidays. *See also* **Calendar day.**

WORKING DRAWING Drawing that shows sufficient detailed information, including sizes and shapes, from which to properly build the object shown and described. *Also called* **Plans, Shop drawing** and **Standard plans.**

WORKING HORSEPOWER *See* **Horsepower.**

WORKING PRESSURE *See* **Pressure.**

WORKING SHAFT Shaft sunk to excavate a tunnel or sewer and filled in on completion of the work.

WORKING SPACE Clear space around equipment to allow access and maneuverability.

WORK IN PROGRESS Construction work that has commenced but is not yet complete.

WORKSITE Jobsite.

WOVEN MEDIUM Filter medium made from strands of fiber, thread, or wire interlaced into a cloth on a loom.

WROUGHT Shaped by hammering.

WROUGHT PIPE Pipe made by hammering steel plate to shape and welding the seams.

WYE Three-ended plumbing fitting, usually consisting of two pipes joined to a third at 45° angles.

X-CRACK Tension crack characteristic of material failure due to vibration.

XENON Rare, heavy, colorless, gaseous chemical element, Xe, that occurs in the atmosphere in minute quantities and which is chemically inactive.

XENON EFFECT Rapid but temporary poisoning of a nuclear reactor from the buildup of xenon-135 caused by the radioactive decay of the fission product iodine-135.

XEROPHREATOPHYTE Phreatophyte able to resist drought.

XEROPHYTE Plant capable of living in areas of limited water supply.

X-RAY Ionizing or electromagnetic radiation having an extremely short wavelength that will penetrate substances that ordinary light cannot go through. When directed at a photographic film, the rays are capable of producing an image of objects and matter wholly or partially obstructing their passage.

X-RAY FLUORESCENCE Characteristic secondary radiation emitted by an element as a result of excitation by X-rays, used to yield chemical analysis of a sample.

XYLEM Principal strengthening and water-conducting tissue of stems, leaves, and roots; a component of wood.

YELLOWCAKE Concentrated crude uranium oxide.

Y GROWTH RATE Experimentally-determined constant, used to estimate the unit growth rate of bacteria while degrading organic wastes.

YIELD Amount produced.

YOUNG RIVER Stream that is actively eroding its channel.

YOUNG VALLEY Valley in its early stages of development and characterized by a narrow bottom bounded by steep slopes.

YARD 1. Nonmetric unit of length, equal to 3 feet. Symbol: yd. Multiply by 0.9144 to obtain meters, symbol: m. *See also the appendix*: **Metric and nonmetric measurement**; **2.** Area for the storage and maintenance of equipment, storage of materials, site office, etc.; **3.** Place where logs are accumulated. *See also* **Harvest functions**; **4.** Open space on a building lot, usually to the rear of the building.

YARDAGE Volume of material filled or excavated, measured in cubic yards.

YARDANGS Ridges formed by wind carrying sand at low level.

YARD DRAIN Surface drain, used to drain an area of surface water.

YARD HYDRANT Private hydrant, similar to other fire hydrants, located on private property and installed for its protection.

YARD RUBBISH Prunings, grass clippings, weeds, leaves, and general yard and garden wastes.

YARD WASTE *See* **Solid waste.**

Y-BRANCH Plumbing fitting joining a single pipe to two others at 45° angles.

YEAST Any of various unicellular fungi of the genus *Saccharomyces* and related genera, reproducing by budding and capable of fermenting carbohydrates.

Z

ZEOLITE Any of a group of approximately 30 hydrous aluminum silicate minerals or their corresponding synthetic compounds, used chiefly as molecular filters and ion-exchange agents.

ZEOLITE FILTER Filter capable of removing certain chemical constituents from water by base exchange.

ZEOLITE SOFTENING Water softening process employing the mineral zeolite that functions by base exchange.

ZERO MOISTURE INDEX Condition when the annual amount of precipitation is just adequate to supply all the water needed for maximum evaporation and transpiration.

ZERO POPULATION GROWTH Concept in which the rate of human births is equal to the rate of human deaths.

ZINC Bluish-white metallic element, Zn, having an atomic number of 30 and atomic mass of 65.38. Although surface waters seldom contain zinc at concentrations above 0.1 mg/L, levels in tap water can be considerably higher due to the use of zinc in plumbing materials. The aesthetic objective for zinc in drinking water is <5.0 mg/l.

ZINC EQUIVALENT Factor used to assess the potential effect on crop growth from the addition of waste sludge containing cop-

per, zinc, and nickel, to the soil.

ZIRCALOY Alloy of zirconium and tin, used for fuel cladding in water reactors.

ZONAL INDEX Measure of the strength of the atmospheric circulation in a specified large area of the globe.

ZONE Area defined by a local authority for specific use and subject to defined conditions and/or restrictions. *Also called* **District**.

ZONE OF AERATION Area above a watertable where the interstices are not completely filled with water.

ZONE OF CAPILLARITY Area above a watertable where some or all of the interstices are filled with water that is held by capillarity.

ZONE OF SATURATION Ground below the watertable.

ZONE SWITCH *See* **Switch.**

ZONE TIME Local time system in which 24 time zones, each covering 15° of longitude, are designated by letters of the alphabet (Greenwich Mean Time, a commonly-used base reference, is designated Z, or Zulu time).

ZONING Division of an area into districts and the public regulation of the character and intensity of use of land and improvements thereon. *See also* **Roadside zoning.**

ZONING BYLAW *See* **Zoning ordinance.**

ZONING ORDINANCE Regulation of property by local government. *Also called* **Zoning bylaw.**

ZONING PERMIT Permit issued by a regulatory agency authorizing land to be used or developed for a specific use.

ZOOGLEA Any of various bacteria of the genus *Zoogloea,* forming colonies in a jelly-like secretion.

ZOOGLEAL MATRIX Floc formed primarily by slime-producing bacteria in the activated sludge treatment process or in biological beds.

ZOONOSES Infections that can be transmitted from animals to humans.

ZOOPLANKTON Floating and drifting aquatic animal life.

METRIC AND NONMETRIC MEASUREMENTS:

The SI, U.S. Customary, and Imperial Systems

America still going metric

Once, the United States almost went metric: in 1975, Congress passed the *Metric Conversion Act* "...to coordinate and plan the increasing use of the metric system in the United States". However, it never happened; the US Metric Board that was established to do the job found significant disinterest in the subject by the general public and, seven years later, in 1982, was 'disestablished'.

In reality, America's involvement with metric measurement is both long and considerable, dating back to the early 1800s when the US Coast and Geodetic Survey used the meter and kilogram. In 1866, Congress authorized the use of the metric system and supplied each state with a set of standard metric weights and measures. In 1875, the United States was one of the original seventeen signatory nations to the *Treaty of the Meter*, and in 1893, the metric measurement standards established by the International Bureau of Weights and Measures, established by the treaty, were adopted as the fundamental standards for length and mass in the US.

A century later, not much has happened. According to the US Department of Commerce (which is charged with encouraging the use of metric), 'The United States is the only industrialized country in the world not officially using the metric system'. While this is true of virtually all products used domestically, there is a growing trend among those companies doing business overseas — including many of the country's largest — to wholly convert to metric measurements both for the home market and for exports. Also, most NASA projects are now designed and built to metric specifications; the General Services Administration is establishing metric specifications for products it buys for government — even the Congressional Record and Federal Register will be produced to metric sizes by the Government Printing Office.

There is no law that requires US companies to convert to metric measurements. Increasing global trade, NAFTA, and the other economic trading blocks, and the increasing isolation of American industry in a metric world may at last hasten its move to the International System of Units (SI), the name adopted under the *Treaty of the Meter* in 1960 for the metric system.

Is Canada metric?

Canadians buy their gasoline and milk by the litre (note the spelling), and drive on roads that are measured in kilometers at speeds registered in km/h. In a bilingual country where all consumer products sold in other than their province of manufacture must, by law, be labeled in English and French, so too must all products be measured in metric terms (although most also carry nonmetric equivalents, which the majority of purchasers continue to prefer). At one time, about a decade ago, merchants in several 'test' communities where full conversion to metric was staged were prosecuted for displaying price tickets in other than metric. But those days have gone, and metric conversion in Canada has gone far enough to make a dent in the old 'Imperial' system, but not far enough to encourage its population to make the full conversion — in their heads or in the marketplace.

The Canadian construction industry was scheduled to go metric on 'M-Day', January 1, 1978. After the expenditure of millions of dollars in committees, special research, seminars, tons upon tons (or, rather, tonnes upon tonnes) of literature aimed at all those involved in building and the supply of construction materials and equipment, it didn't happen. The country is left with a hodgepodge of often conflicting requirements and standards. Some public authorities insist on all quotations, contracts, and working drawings being prepared in SI — the International System of Units, as the metric system is known. However, only a diminutive number of products are yet produced to 'hard' metric dimensions, most documents are cluttered with a confusion of 'soft' converted numbers, interspersed by the dimensions of the relatively few truly metric products yet available.

A word of explanation. 'Hard' conversion means that a product traditionally manufactured to a rational imperial set of dimensions would be 'resized' to a set of rational metric dimensions. Plywood, for instance, which is produced in 4 x 8 ft (1219.2 x 2438.4 mm) sheets, in thicknesses like 1/2 in. (12.7 mm) and 3/4 in. (19.05 mm), among others, would in future be manufactured in 1200 x 2400 mm sheets in equivalent thicknesses of 12.5 mm, 18.5 mm, etc. 'Soft' conversion means that the Imperial dimensions of the product remain the same. However, those Imperial dimensions are converted to metric. Most plywood sheets still typically measure 1219.2 x 2438.4 mm.

Since Canada's heady days of the 'metric police', some change toward metrification has taken place. Children in school are mostly taught measurement using the metric system, with acknowledgment for the historic (and still used) nonmetric equivalents. A (very) slowly growing range of products is made to fully metric measurements, and there is complete conversion by public agencies to metric. Canada will become a fully metric country — likely later rather than sooner and, as always, subject to the pleasure of US practice.

The metric, or SI system of measurement

The meter was introduced as a unit of linear measurement in 1791, being defined as 10^{-7} of the Earth's quadrant. This was changed in 1799 when the meter was defined by the length of a standard meter bar. In 1960 it was again redefined, by the Eleventh General

Conference of Weights and Measures, in terms of the wavelength of the radiation produced by a specified quantum transition of the Krypton-86 isotope. Also in 1960, the metric system was renamed at the General Conference of Weights and Measures as *Le Système International d'Unités* or the International System of Units, abbreviated SI.

SI has been described as 'a coherent set of units', so that when SI units are used throughout the solution of an equation, the answer is automatically equal to the value of the quantity.

Well, it's not quite that simple. The above description works fine so long as only the base, supplementary, and derived units are used, but not their multiples. And what's this about 'supplementary' units? Isn't SI supposed to be a logical and **complete** measuring system. Nonmetric is being thrown out because of such supplementary units as 'peck' 'bushel' and 'skip.' Now we have base units, supplementary units, derived units, derived units with special names, derived units expressed by means of special names, derived units formed by using supplementary units, and units in use temporarily with SI!

Let's put all this in order.

SI is constructed from seven base units for independent quantities, plus two supplementary derived dimensionless units:

SI BASE UNITS

Quantity	Name	Symbol
length	meter	m
mass	kilogram	kg
time	second	s
electric current	ampere	A
thermodynamic temperature	kelvin	K
amount of substance	mole	mol
luminous intensity	candela	cd

SUPPLEMENTARY UNITS

Quantity	Name	Symbol
plane angle	radian	rad
solid angle	steradian	sr

Units for all other quantities are derived from these nine units and are expressed as products and quotients without numerical factors:

EXAMPLES OF DERIVED UNITS EXPRESSED IN TERMS OF BASE UNITS

Quantity	Name	Symbol
area	square meter	m^2
volume	cubic meter	m^3

speed, velocity	meter per second	m/s
acceleration	meter per second squared	m/s^2
wave number	reciprocal meter	m^{-1}

EXAMPLES OF DERIVED UNITS EXPRESSED IN TERMS OF BASE UNITS

Quantity	Name	Symbol
density, mass density	kilogram per cubic meter	kg/m^3
specific volume	cubic meter per kilogram	m^3/kg
current density	ampere per square meter	A/m^2
magnetic field strength	ampere per meter	A/m
concentration (of amount of substance)	mole per cubic meter	mol/m^3
luminance	candela per square meter	cd/m^2

Certain derived units have been given special names, and may themselves be used to express other derived units:

SI DERIVED UNITS WITH SPECIAL NAMES

Quantity	Name	Symbol	Expression in terms of other units
frequency	hertz	Hz	s^{-1}
force	newton	N	$(m \bullet kg)/s^2$
pressure, stress	pascal	Pa	N/m^2
energy, work, quantity of heat	joule	J	$N \bullet m$
power, radiant flux	watt	W	J/s
electric charge, quantity of electricity	coulomb	C	$s \bullet A$
electric potential, potential difference, electromotive force	volt	V	W/A
capacitance	farad	F	C/V
electric resistance	ohm	Ω	V/A
electric conductance	siemen	S	A/V
magnetic flux	weber	Wb	$V \bullet s$
magnetic flux density	tesla	T	Wb/m^2
inductance	henry	H	Wb/A
luminous flux	lumen	lm	$cd \bullet sr$
illuminance	lux	lx	lm/m^2
activity (of a radionuclide)	becquerel	Bq	s^{-1}
absorbed dose, specific ionising energy imparted, kerma, absorbed dose index	gray	Gy	J/kg
dose equivalent, dose equivalent index	sievert	Sv	J/kg

SOME SI-DERIVED UNITS EXPRESSED USING SPECIAL NAMES

Quantity	Name	Symbol
absorbed dose rate	gray per second	Gy/s
dynamic viscosity	pascal second	Pa•s
electric charge density	coulomb per cubic meter	C/m^3
electric field strength	volt per meter	V/m
electric flux density	coulomb per square meter	C/m^2
energy density	joule per cubic meter	J/m^3
exposure (X and gamma rays)	coulomb per kilogram	C/kg
heat capacity, entropy	joules per kelvin	J/K
heat flux density, irradiance	watt per square meter	W/m^2
kinematic viscosity	square meter per second	m^2/s
molar energy	joule per mole	J/mol
molar energy, molar heat capacity	joule per mole kelvin	J/(mol•K)
moment of force	newton meter	N•m
permeability	henry per meter	H/m
permittivity	farad per meter	F/m
radiant intensity	watt per steradian	W/sr
specific energy	joule per kilogram	J/kg
specific heat capacity, specific entropy	joules per kilogram kelvin	J(kg•K)
surface tension	newton per meter	N/m
thermal conductivity	watt per meter kelvin	W(m•K)

Supplementary units may be used in the expression of derived units:

EXAMPLES OF SI-DERIVED UNITS FORMED BY USING SUPPLEMENTARY UNITS

Quantity	SI unit	
	Name	Symbol
angular velocity	radian per second	rad/s
angular acceleration	radian per second squared	rad/s^2
radiant intensity	watt per steradian	W/sr
radiance	watt per square meter steradian	$W/(m^2•sr)$

In fields where their usage is already well established, the use of the following units is accepted, but subject to future review:

UNITS PERMITTED FOR USE WITH SI

Quantity	Name	Symbol
area	hectare	ha
energy	electronvolt	eV

length	parsec (astronomical unit)	pc
linear density	tex	tex
mass	tonne (metric ton)	t
mass of an atom	unit of unified atomic mass	u
plane angle	degree	°
	minute	'
	second	"
	revolution	r
time	minute	min
	hour	h
	day	d
	year	a
volume	liter	L or l

UNITS IN USE TEMPORARILY WITH SI

Name	Symbol	Definition
barn	b	$1\ b = 100\ fm^2$
knot	kn	1 nautical mile per hour
		$1\ kn = (1852/3600)\ m/s$
millibar	mbar	$1\ mbar = 100\ Pa$
nautical mile	M	1 nautical mile = 1852 m

UNITS NOT ACCEPTABLE FOR USE WITH SI

Quantity	Name	Symbol
absorbed dose of ionizing radiation	rad	rad
acceleration	gal	Gal
activity (radioactive)	curie	Ci
amount of substance	equivalent	eq
area	are	a
conductance	mho	mho
dose equivalent	rem	rem
dynamic viscosity	poise	P
energy	calorie (IT)	cal
	erg	erg
force	kilogram-force	kgf
	kilopoid	kp
	dyne	dyn

Quantity	Name	Symbol
illuminance	phot	ph
illuminance	stilb	sb
kinematic viscosity	stoke	St
length	angström	Å
	micron	μ
	fermi	fm
	X unit	-

magnetic field strength	oersted	Oe
magnetic flux	maxwell	Mx
magnetic flux density	gauss	Gs, G
magnetic induction	gamma	g
mass	metric carat	-
pressure	torr	Torr
	millimeter of mercury	mm Hg
	bar	bar
	standard atmosphere	atm
radiation exposure	röntgen	R
volume	stere	st
	lambda	λ

There are 16 prefixes used in SI, ranging from 10^{18} to 10^{-18}. In practical use, in construction-related activities, the following six prefixes are generally used:

PREFIXES

factor	multiple	prefix	symbol
10^{24}	1 000 000 000 000 000 000 000 000	yotta	Y
10^{21}	1 000 000 000 000 000 000 000	zetta	Z
10^{18}	1 000 000 000 000 000 000	exa	E
10^{15}	1 000 000 000 000 000	peta	P
10^{12}	1 000 000 000 000	tera	T
10^{9}	1 000 000 000	giga	G
10^{6}	1 000 000	mega	M
10^{3}	1000	kilo	k
10^{2}	100	hecto	h
	10	deca	da
	1 unit	—	—
10^{-1}	0.1	deci	d
10^{-2}	0.01	centi	c
10^{-3}	0.001	milli	m
10^{-6}	0.000 001	micro	μ
10^{-9}	0.000 000 001	nano	n
10^{-12}	0.000 000 000 001	pico	p
10^{-15}	0.000 000 000 000 001	emto	f
10^{-18}	0.000 000 000 000 000 001	atto	a
10^{-21}	0.000 000 000 000 000 000 001	zepto	z
10^{-24}	0.000 000 000 000 000 000 000 001	yocto	y

How to write SI

There are some rules. For instance, each unit of measurement has an individual name: mass is expressed in kilograms; force in newtons (both are expressed in pounds in the nonmetric system of measurement). Likewise, each quantity has only one unit: power, be it electrical, thermal, or mechanical, is expressed only in watts.

Symbols are written in lower case (m for meter, g for gram) except for those derived from proper names (C for Celsius, N for Newton), and for liter (L or l). When written out in full, the names of all units, excepting Celsius are written in lower case (newton, pascal, liter, etc.). The plural form for symbols is identical to the singular (5 m, not 5 ms); a space is left between a numeral and the character denoting its symbol (5 m not 5m) except where the numeral is followed by characters other than letters (18°C or 18 °C). Prefixes are used with SI names or symbols to signify multiples or submultiples. The prefixes most commonly used in the construction/contracting industry are: 10^6 mega (M), 10^3 kilo (k), 10^{-2} centi (c), and 10^{-3} milli (m).

When a triad separator is required to facilitate the reading of long numbers, the separator is a space (e.g. 12 345). However, a space is not necessary with a four-digit group (e.g. 1234) except when required for consistency, eg, when it is in a column with other numbers having five or more digits. If a numerical value is less than one written in decimal form, a zero precedes the decimal marker (e.g., 0.10, not .10).

This dictionary has been written using metric terminology (in American-English), with nonmetric equivalents in parentheses. The following conversion table includes terms commonly used in the fields covered by the dictionary (bold face are exact).

SOME CONVERSION FACTORS

On the following pages, factors are given that permit conversion of nonmetric values to metric values, and metric values to nonmetric.

MULTIPLY THIS	UNIT	BY THIS	TO OBTAIN	SYMBOL
acre	area	0.404 685 6	hectare	ha
acre	area	4 046.856	square meter	m²
acre foot	volume	1 233.482	cubic meter	m³
ampere hour	electric charge	**3.6**	kilocoulomb	kC
ampere turn	electric charge	1.0	ampere	A
ångström	length	**0.1**	nanometer	nm
arpent (French measure)	area	0.341 889 4	hectare	ha
arpent (French measure)	length	58.471 31	meter	m
astronomical unit	length	149.597 870	gigameter	Gm
atmosphere, standard (760 torr)	pressure	101.325	kilopascal	kPa
atmosphere, technical (1 kgf/cm²)	pressure	98.066 5	kilopascal	kPa
bar	pressure	**100**	kilopascal	kPa
barrel (oil, 42 US gal)	volume	0.158 987 3	cubic meter	m³
barrel (36 UK gal)	volume	0.163 659 2	cubic meter	m³
barrel (US dry, 7056 in.³)	volume	0.115 627 1	cubic meter	m³
biot	electric charge	10	ampere	A
board foot	volume	2.359 737	cubic decimeter	dm³
British thermal unit (Btu) (Int. table)	energy	1.055 06	kilojoule	kJ
Btu foot per (square foot hour °F)	heat flow	1.730 735	watt per (meter kelvin)	W/(m•K)
Btu inch per (square foot hour °F)	heat flow	0.144 227 9	watt per (meter kelvin)	W/(m•K)
Btu inch per (square foot second °F)	heat flow	519.220 4	watt per (meter kelvin)	W/(m•K)
Btu (59°F, 15°C)	energy	1.054 80	kilojoule	kJ
Btu (60.5°F)	energy	1.054 615	kilojoule	kJ
Btu (390°F)	energy	1.059 67	kilojoule	kJ
Btu (mean)	energy	1.055 87	kilojoule	kJ
Btu per cubic foot	heat conductivity	37.258 95	kilojoule per cu m	kJ/m³
Btu per (cubic foot °F)	heat conductivity	67.066 11	kilojoule per (cu m kelvin)	kJ/(m³•K)
Btu per gallon	heat	232.08	kilojoule/cubic meter Celsius	kJ/(m.³°C)

MULTIPLY THIS	BY THIS	TO OBTAIN	SYMBOL
Btu per gallon (U.S.)	278.717	kilojoule/cubic meter Celsius	kJ/(m.³°C)
Btu per hour	0.293 071 1	watt	W
Btu per hour (thermochemical)	0.292 875 1	watt	W
Btu per minute (thermochemical)	17.572 50	watt	W
Btu per pound	2.326	kilojoule per kilogram	kJ/kg
Btu per pound (°F)	**4.186 8**	kilojoule per (kilogram kelvin)	kJ/(kg•K)
Btu per second (thermochemical)	1.054 350	kilowatt	kW
Btu per (square foot hour °F)	5.678 263	watt per (square meter kelvin)	W/(m²•K)
Btu per (square foot hour)	3.154 60	watt per square meter	W/m²
Btu per ton (2000 lb)	1.163	kilojoules per tonne	kJ/t
Btu (thermochemical)	1.054 35	kilojoule	kJ
Btu (thermochemical) per (square foot hour)	3.152 481	watt per square meter	W/m²
Btu (thermochemical) per (sq ft minute)	189.148 8	watt per square meter	W/m²
Btu (thermochemical) per (sq ft second)	11.348 93	kilowatt per square meter	kW/m²
Btu (thermochemical) per (sq in. second)	1 634 246	megawatt per square meter	MW/m²
calorie (international)	**4.186 8**	joule	J
calorie (thermochemical)	4.184	joule	J
candela per square foot	10.763 91	candela per square meter	cd/m²
candela per square inch	1.550 003	kilocandela per square meter	kcd/m²
carat	**200**	milligram	mg
centimeter	0.393 70	inch	in.
centistoke	**1.0**	square millimeter per second	mm²/s
cental (100 lb)	45.359 237	kilogram	kg
chain (66 ft)	**20.116 8**	meter	m
circular mil	506.707 5	square micrometer	μm²
cord (stacked wood 128 ft³)	3.624 6	cubic meter	m³
cubic centimeter	0.061 02	cubic inch	in.³
cubic centimeter	0.035 195	ounce (fluid)	fl oz
cubic centimeter	**1.0**	milliliter	mL
cubic centimeter	**0.001**	liter	L
cubic foot	28.316 85	cubic decimeter	dm³
cubic foot	0.028 316 85	cubic meter	m³
cubic foot	28.316 85	liter	L
cubic foot per hour	28.316 85	liter per hour	L/h

527

MULTIPLY THIS	UNIT	BY THIS	TO OBTAIN	SYMBOL
cubic foot per hour	volume rate of flow	0.007 865 79	liter per second	L/s
cubic foot per minute	volume rate of flow	0.000 471 947 4	cubic meter per second	m³/s
cubic foot per minute	volume rate of flow	0.471 947 4	liter per second	L/s
cubic foot per second	volume rate of flow	0.028 316 85	cubic meter per second	m³/s
cubic foot per second	volume rate of flow	28.316 85	liter per second	L/s
cubic inch	volume	**16.387 064**	cubic centimeter	cm³
cubic inch	volume	**0.016 387 064**	liter	L
cubic inch per minute	volume rate of flow	0.273 117 7	cubic centimeter per second	cm³/s
cubic inch per second	volume rate of flow	16.387 06	cubic centimeter per second	cm³/s
cubic meter	volume	0.275 9	cord	cd
cubic meter	volume	1.308	cubic yard	yd³
cubic meter	volume	35.314 7	cubic foot	ft³
cubic meter	volume	219.97	gallon (Imperial)	gal(Imperial)
cubic meter	volume	264.17	gallon (US)	gal(US)
cubic millimeter	volume	**0.001**	cubic centimeter	cm³
cubic millimeter	volume	61.023 x 10⁶	cubic inch	in.³
cubic millimeter	volume	1.0 x 10⁻⁹	cubic meter	m³
cubic yard	volume	0.764 555	cubic meter	m³
cubic yard per minute	volume rate of flow	12.742 58	cubic decimeter per second	dm³/s
cunit (100 ft³ solid wood)	volume	2.831 685	cubic meter	m³
darcy	permeability	0.986 923 3	square micrometer	µm²
day (mean solar)	time	**86 400**	kilosecond	ks
day (sidereal)	time	86.164 09	kilosecond	ks
decibel	sound power level	**1.0**	decibel	dB
degree (angle)	plane angle	0.017 453	radian	rad
degree (Celsius)	temperature	(°C x 9/5) + 32	degree Fahrenheit	°F
degree Celsius	interval	1.0	kelvin	K
degree Fahrenheit	temperature	(°F - 32) x 5/9	degree Celsius	°C
degree Fahrenheit	interval	5/9	kelvin	K
degree Rankine	interval	5/9	kelvin	K
dyne	force	**10**	micronewton	µN
electronvolt	energy	0.160 217 7	attojoule	aJ
ell (45 in.)	length	1.143	meter	m
erg	energy	**0.1**	microjoule	µJ

MULTIPLY THIS	UNIT	BY THIS	TO OBTAIN	SYMBOL
fathom	length	1.828 8	meter	m
foot	length	**0.304 8**	meter	m
foot	length	**304.8**	millimeter	mm
foot candle	illuminance	10.763 91	lux	lx
foot lambert	luminance	3.426 259	candela per square meter	cd/m^2
foot of water (39.2°F, 4°C)	pressure	2.988 98	kilopascal	kPa
foot per hour	velocity	0.084 666 7	millimeter per second	mm/s
foot per hour	velocity	**304.8**	millimeter per hour	mm/h
foot per minute	velocity	**5.08**	millimeter per second	mm/s
foot per minute	velocity	304.8	millimeter per minute	mm/m
foot per second	velocity	**0.304 8**	meter per second	m/s
foot per second squared	acceleration	**0.304 8**	meter per second squared	m/s^2
foot poundal	energy	42.140 11	millijoule	mJ
foot pound-force	energy	1.355 818	joule	J
foot pound-force per hour	energy	0.376 616 1	milliwatt	mW
foot pound-force per minute	energy	22.596 97	milliwatt	mW
foot pound-force per second	energy	1.355 818	watt	W
foot (US survey)	length	0.304 800 6	meter	m
furlong	length	0.201 168	kilometer	km
gal	acceleration	1.0	centimeter per second squared	cm/s^2
gallon (Imperial)	volume	**0.004 546 09**	cubic meter	m^3
gallon (Imperial)	volume	**4.546 09**	liter	L
gallon (Imperial) per day	volume rate of flow	**0.004 546 09**	cubic meter per day	m^3/d
gallon (Imperial) per day	volume rate of flow	0.000 052 616 8	liter per second	L/s
gallon (Imperial) per minute	volume rate of flow	75.768 17	cubic centimeter per second	cm^3/s
gallon (Imperial) per minute	volume rate of flow	0.075 768	liter per second	L/s
gallon (US)	volume	0.003 785 412	cubic meter	m^3
gallon (US)	volume	3.785 412	liter	L
gallon (US) per minute	volume rate of flow	63.090 20	cubic centimeter per second	cm^3/s
gallon (US) per minute	volume rate of flow	0.063 090 2	liter per second	L/s
gamma	mass	1.0	microgram	µg
grain	mass	64.798 91	milligram	mg
gram	mass	0.035 273 96	ounce	oz
gram	mass	0.002 204 62	pound	lb

MULTIPLY THIS	UNIT	BY THIS	TO OBTAIN	SYMBOL
gram per cubic centimeter	density	1.0	kilogram per liter	kg/L
gram per cubic centimeter	density	62.428 7	pound per cubic foot	lb/ft³
gram per cubic centimeter	density	10.022 41	pound per gallon (Imperial)	lb/gal
gram per cubic meter	density	0.436 995 7	grain per cubic foot	gr/ft³
gram per cubic meter	density	1.0	milligram per liter	mg/L
gram per cubic meter	density	0.000 062 428	pound per cubic foot	lb/ft³
gram per liter	density	0.062 428 7	pound per cubic foot	lb/ft³
gram per liter	density	0.010 022 41	pound per gallon (Imperial)	lb/gal
grain per cubic foot	mass density	2.288 352	gram per cubic meter	g/m³
grain per gallon (Imperialerial)	mass density	14.253 77	gram per cubic meter	g/m³
grain per gallon (US)	mass density	17.118 06	gram per cubic meter	g/m³
hectare	area	2.471 054	acre	ac
hectare	area	10 000	square meter	m²
horsepower (boiler)	power	9.809 50	kilowatt	kW
horsepower (electric)	power	746	watt	W
horsepower (water)	power	746.043	watt	W
horsepower (550 ft/lbf/s)	power	745.699 9	watt	W
horsepower hour	energy	2.684 52	megajoule	MJ
hour (mean solar)	time	3.6	kilosecond	ks
hour (sidereal)	time	3.590 17	kilosecond	ks
hundredweight (100 lb)	mass	45.359 237	kilogram	kg
hundredweight (112 lb)	mass	50.802 345	kilogram	kg
inch	length	2.54	centimeter	cm
inch	length	0.025 4	meter	m
inch	length	25.4	millimeter	mm
inch³	section modulus	16.387 064	cubic centimeter	cm³
inch of mercury (conventional 0°C)	pressure	3.386 39	kilopascal	kPa
inch of mercury (60°F)	pressure	3.376 85	kilopascal	kPa
inch of mercury (68°F, 20°C)	pressure	3.374 11	kilopascal	kPa
inch of mercury (0°C)	pressure	25.4	millimeter of mercury	mm
inch of water (conventional)	pressure	249.082 89	pascal	Pa
inch of water (39.2°F, 4°C)	pressure	249.082	pascal	Pa
inch of water (60°F)	pressure	248.843	pascal	Pa

MULTIPLY THIS	UNIT	BY THIS	TO OBTAIN	SYMBOL
inch of water (68°F, 20°C)	pressure	248.641	pascal	Pa
inch per minute	velocity	25.4	millimeter per minute	mm/min
inch per second	velocity	25.4	millimeter per second	mm/s
inch per second squared	acceleration	25.4	millimeter per second squared	mm/s²
inch pound-force	work	0.112 985	newton meter	N•m
joule	heat	0.000 947 8	Btu (international)	Btu
joule	work	0.737 562	foot pound-force	ft/lb
joule	work	0.3725 x 10⁶	horsepower hour	hp•h
joule	power	0.2778 x 10⁶	kilowatt hour	kW•h
joule	work	1.0	newton meter	N•m
joule per liter	heat	0.026 839	Btu per cubic foot	Btu/ft³
joule per liter	heat	1.0	kilojoule per cubic meter	kJ/m³
kilogram	mass	35.273 96	ounce	oz
kilogram	mass	2.204 62	pound	lb
kilogram	mass	0.000 984 21	ton (long)	t(l)
kilogram	mass	0.001 102 3	ton (short)	t(s)
kilogram calorie (international)	heat	3.968 3	Btu (international)	Btu
kilogram calorie (international)	heat	4 186.8	joule	J
kilogram-force	force	9.806 65	newton	N
kilogram per cubic centimeter	mass	32.127 292	pound per cubic inch	lb/in.³
kilogram per cubic meter	mass	1.0	gram per liter	g/L
kilogram per cubic meter	mass	1.685 556	pound per cubic yard	lb/yd³
kilogram per cubic meter	mass	0.010 022	pound per gallon (Imperial)	lb/gal(I)
kilogram per kilometer	mass	1.0	gram per meter	g/m
kilogram per kilometer	mass	0.6719 x 10³	pound per foot	lb/ft
kilogram per kilometer	mass	3.458	pound per mile	lb/mi
kilogram per liter	mass	1.0	gram per milliliter	g/mL
kilogram per meter	mass	0.671 97	pound per foot	lb/ft
kilogram per meter	mass	2.015 91	pound per yard	lb/yd
kilogram per square centimeter	mass	2 048.16	pound per square foot	lb/ft²
kilogram per square centimeter	mass	14.223	pound per square inch	lb/in.²
kilogram per square meter	mass	0.204 816	pound per square foot	lb/ft²
kilojoule per cubic meter	heat	0.026 839	Btu per cubic ft	Btu/ft³

MULTIPLY THIS	BY THIS	TO OBTAIN	SYMBOL
kilojoule per cubic meter	0.004 309	Btu per gallon (Imperial)	Btu/gal
kilojoule per cubic meter	1.0	joule per liter	J/L
kilojoule per kilogram	0.429 923	Btu per pound	Btu/lb
kiloliter	35.315	cubic foot	ft³
kiloliter	219.969	gallon (Imperial)	gal
kiloliter	264.172	gallon (US)	gal
kilometer	0.621 371	mile	mi
kilometer per hour	0.539 96	knot	kn
kilometer per hour	0.277 778	meter per second	m/s
kilometer per hour	0.621 371	mile per hour	mph
kilonewton	0.112 40	ton-force (short ton)	
kilopascal	0.295 3	inch of mercury (0°C)	
kilopascal	4.014 74	inch of water (4°C)	
kilopascal	1 000	newton per square meter	N/m²
kilopascal	20.885 43	pound-force per square foot	psf
kilopascal	0.145 037 7	pound-force per square inch	psi
kilopoid	9.806 65	newton	N
kilowatt	0.947 81	Btu (international) per second	Btu/sec
kilowatt	1.340 48	horsepower (electric)	hp(e)
kilowatt hour	3 412	Btu (international)	Btu
kilowatt hour	1.340 5	horsepower hour	hp/h
kilowatt hour	3.6	megajoule	MJ
kip (thousand pound-force)	4.448 222	kilonewton	kN
knot (international)	1.852	kilometer per hour	km/h
knot (international)	0.514 444 4	meter per second	m/s
knot (UK)	1.853 184	kilometer per hour	km/h
ksi (kip per square inch)	6.894 757	megapascal	MPa
lambert	3 183.099	candela per square meter	cd/m²
langley per minute	0.697 8	kilowatt per square meter	kW/m²
league (International nautical)	5.556	kilometer	km
league (UK nautical)	5.559 552	kilometer	km
league (US)	4.828 032	kilometer	km

MULTIPLY THIS	UNIT	BY THIS	TO OBTAIN	SYMBOL
legal subdivision (40 acres)	area	0.161 874 2	square kilometer	km²
link (1/100 chain)	length	0.201 168	meter	m
liter	volume	**1.0**	cubic decimeter	dm³
liter	volume	0.035 315	cubic foot	ft³
liter	volume	61.023 744	cubic inch	in.³
liter	volume	**0.001**	cubic meter	m³
liter	volume	0.219 969	gallon (Imperial)	(I)gal
liter	volume	0.264 172	gallon (US)	(US)gal
iter	volume	35.195 1	ounce (fluid)	fl oz
liter	volume	0.879 877	quart (Imperial)	qt
liter per second	velocity	2.118 88	cubic foot per minute	ft³/min
liter per second	velocity	13.198 2	gallon (Imperial) per minute	gpm
liter per second	velocity	15.850 3	gallon (US) per minute	gpm
lumen per square foot	illuminance	10.763 91	lux	lx
meter	length	39.370	inch	in.
meter	length	3.280 84	foot	ft
meter	length	1.093 6	yard	yd
meter	volume	3.280 84	foot of water	
meter of water (4°C)	pressure	9.806 378	kilopascal	kPa
meter of water (4°C)	pressure	1.422 29	pound per square inch	psi
meter of water (4°C)	velocity	3.280 8	foot per minute	fpm
meter per minute	velocity	0.054 68	foot per second	fps
meter per minute	velocity	0.037 28	mile per hour	mph
meter per minute	velocity	2.236 9	mile per hour	mph
meter per second	velocity	196.85	foot per minute	fpm
meter per second	velocity	3.280 8	foot per second	fps
meter per second squared	velocity	3.280 8	foot per second squared	fps²
metric carat	mass	200	milligram	mg
microinch	length	25.4	nanometer	nm
micrometer	length	0.039 370	mil	mil
micron	length	**1.0**	micrometer	μm
mil (0.001 in.)	length	**25.4**	micrometer	μm
mile (international nautical)	length	**1.852**	kilometer	km
mile	length	**1.609 344**	kilometer	km

533

MULTIPLY THIS	UNIT	BY THIS	TO OBTAIN	SYMBOL
mile	length	1 609.344	meter	m
mile (UK nautical)	length	1.853 184	kilometer	km
mile per hour	velocity	1.609 344	kilometer per hour	km/h
mile per hour	velocity	0.447 04	meter per second	m/s
mile per minute	velocity	26.822 4	meter per second	m/s
mile per second squared	acceleration	1.609 344	kilometer per second squared	km/s²
millibar	pressure	0.1	kilopascal	kPa
milligram	mass	0.015 432	grain	gr
milligram	mass	35.274 x 10⁶	ounce	oz
milligram	mass	2.204 62 x 10⁶	pound	lb
milliliter	mass	0.061 02	cubic inch	in.³
milliliter	mass	0.035 195	ounce (fluid)	fl oz
millimeter	length	0.039 37	inch	in.
millimeter	length	39.37	mil	mil
millimeter	length	1000	micrometer	μm
millimeter mercury (0°C)	pressure	133.322 4	pascal	Pa
millimeter water (4°F)	pressure	9.806 378	pascal	Pa
million Btu per ton	heat	1.163	gigajoule per tonne	GJ/t
million gallons (Imperial) per day	volume rate of flow	52.616 78	cubic decimeter per second	dm³/s
million gallons (US) per day	volume rate of flow	4 546.09	cubic meter per day	m³/d
million gallons (US) per day	volume rate of flow	0.052 616 8	cubic meter per second	m³/s
minute (mean solar)	time	60	second	s
minute (sidereal)	time	59.836 17	second	s
newton	force	0.224 808 9	pound-force	lb
newton meter	force	0.737 562	foot pound-force	ft/lb
ounce	mass	28.349 523	gram	g
ounce (avoirdupois)	mass	28.35	gram	g
ounce (fluid)	mass	28 413 062	milliliter	mL
ounce inch squared	moment of inertia/mass	0.182 899 8	kilogram square centimeter	kg•cm²
ounce-force	force	0.278 013 9	newton	N
ounce-force inch	torque	7.061 552	millinewton meter	mN•m
ounce-force per square inch	pressure	0.430 922 3	kilopascal	kPa
ounce (mass) per square foot	area density	305.151 7	gram per square meter	g/m²

534

MULTIPLY THIS	UNIT	BY THIS	TO OBTAIN	SYMBOL
ounce (mass) per square yard	area density	33.905 75	gram per square meter	g/m²
ounce per cubic foot	mass density	1.001 154	kilogram per cubic meter	kg/m³
ounce per gallon (Imperial)	mass density	6.236 023	kilogram per cubic meter	kg/m³
ounce per gallon (US)	mass density	7.489 152	kilogram per cubic meter	kg/m³
ounce per inch	linear density	1.116 123	kilogram per meter	kg/m
parsec	length	30.856 78	petameter	Pm
perch (French measure)	area	34.188 94	square meter	m²
perch (French measure)	length	5.847 130 8	meter	m
perm (23°C)	permeability	57.452 5	nanogram/(pascal second square meter)	ng/(Pa•s•m²)
perm inch (23°C)	permeability	1.459 29	nanogram per (pascal second meter)	ng/(Pa•s•m)
Petrograd standard (165 ft³ sawn timber)	volume	4.672 280	cubic meter	m³
phot	illuminance	10	kilolux	klx
pennyweight	mass	1.555 174	gram	g
pica (printer's)	length	4.217 518	millimeter	mm
pint (Imperial)	volume	0.568 261 2	cubic decimeter	dm³
pint (Imperial)	volume	0.568 261	liter	L
pint (US)	volume	0.473 176 5	cubic decimeter	dm³
pint (US)	volume	0.473 176	liter	L
point (Didot)	length	0.375 972 9	millimeter	mm
point (pica)	length	0.351 459 8	millimeter	mm
poise	dynamic viscosity	0.1	pascal second	Pa•s
pound	mass	**453.592 37**	gram	g
pound	mass	**0.453 592 37**	kilogram	kg
poundal	force	0.138 255 0	newton	N
poundal per square foot	force per area	1.488 164	pascal	Pa
poundal second per square foot	dynamic viscosity	1.488 164	pascal second	Pa•s
pound (avoirdupois)	mass	453.6	gram	g
pound foot per second	momentum	0.138 255	kilogram meter per second	kg•m/s
pound foot squared	moment of inertia of mass	42.140 11	gram square meter	g•m²
pound foot squared per second	angular momentum	42.140 11	gram square meter per second	g•m²/s
pound force	force	4.448 222	newton	N
pound-force foot	torque	1.355 818	newton meter	N•m
pound-force inch	torque	0.112 985	newton meter	N•m
pound-force per square foot	force per area	47.888 26	pascal	Pa

MULTIPLY THIS	UNIT	BY THIS	TO OBTAIN	SYMBOL
pound-force per square foot	pressure	0.047 888 26	kilopascal	kPa
pound-force per square inch	pressure	0.703 1	meter of water (4°C)	m
pound-force per square inch (psi)	force per area	6.894 757	kilopascal	kPa
pound-force second per square foot	dynamic viscosity	47.880 26	pascal second	Pa•s
pound inch squared	moment of inertia of mass	2.926 397	kilogram square centimeter	kg•cm²
pound (mass) per square foot	area density	4.882 428	kilgram per square meter	kg/m²
pound (mass) per square inch	area density	703.069 6	kilgram per square meter	kg/m²
pound per cubic foot	mass	16.018 46	gram per liter	g/L
pound per cubic foot	mass density	16.018 46	kilogram per cubic meter	kg/m³
pound per cubic inch	mass density	27.679 90	megagram per cubic meter	Mg/cm³
pound per cubic inch	mass	27.679 90	kilogram per liter	kg/L
pound per cubic yard	mass	0.593 276	kilogram per cubic meter	kg/m³
pound per foot	linear density	1.488 164	kilogram per meter	kg/m
pound per (foot second)	dynamic viscosity	1.488 164	pascal second	Pa•s
pound per gallon (Imperial)	mass density	99.776 37	kilogram per cubic meter	kg/m³
pound per gallon (U.S.)	mass density	119.826 4	kilogram per cubic meter	kg/m³
pound per hour	force	0.125 997 9	gram per second	g/s
pound per hour	force	**0.453 592 37**	kilogram per hour	kg/h
pound per inch	linear density	17.857 97	kilogram per meter	kg/m
pound per mile	force	0.281 849	kilogram per kilometer	kg/km
pound per minute	force	7.559 87	gram per second	g/s
pound per minute	force	0.007 559 87	kilogram per second	kg/s
pound per second	force	**0.453 592 37**	kilogram per second	kg/s
pound per ton (short)	mass	**0.50**	kilogram per tonne	kg/t
pound per yard	linear density	0.496 055	kilogram per meter	kg/m
pound (2000) per acre	area density	0.224 170 2	kilogram per square meter	kg/m²
pounds (2000) per square mile	area density	350.266	milligram per square meter	mg/m²
quart (Imperial)	volume	1.136 522	cubic decimeter	dm³
quart (Imperial)	volume	1.136 522	liter	L
quart (US)	volume (liquid)	0.946 352 9	cubic decimeter	dm³
quart (US)	volume (liquid)	0.946 353	liter	L
quart (US)	volume (dry)	1102.2	cubic centimeter	cm³
quarter (28 lb UK)	mass	12.700 58	kilogram	kg

MULTIPLY THIS	UNIT	BY THIS	TO OBTAIN	SYMBOL
radian	plane angle	180/π	degreee (angle)	°
radian per second	velocity	30/π	revolution per minute	rpm
radian per second	velocity	1/(2 π)	revolution per second	rps
revolution per minute	velocity	π/30	radian per second	rad/s
revolution per second	velocity	2π	radian per second	rad/s
second (sidereal)	time	0.997 269 6	second	s
section (1 mi², 640 acres)	area	2.589 988	square kilometers	km²
square centimeter	area	0.155	square inch	in.² or sq in.
square centimeter	area	0.0001	square meter	m²
square centimeter	area	100	square millimeter	mm²
square foot	area	929.030 4	square centimeter	cm²
square foot (French measure)	area	1 055.214	square centimeter	cm²
square foot hour °F per Btu	heat	0.176 110 1	square meter kelvin per watt	m²•°K/W
square foot per second	kinematic viscosity	92 903.04	millimeters squared per second	mm²/s
square inch	area	6.4516	square centimeter	cm²
square inch	area	645.16	square millimeter	mm²
square inch per second	kinematic viscosity	645.16	millimeter squared per second	mm²/s
square kilometer	area	247.1	acre	ac
square kilometer	area	100	hectare	ha
square kilometer	area	0.386 10	square mile	mi²
square meter	area	0.000 1	hectare	ha
square meter	area	10 000	square centimeter	cm²
square meter	area	10.763 9	square foot	ft² or sq ft
square meter	area	1.195 99	square yard	yd² or sq yd
square mile	area	258.998 8	hectare	ha
square mile	area	2.589 988	square kilometer	km²
square millimeter	area	0.01	square centimeter	cm²
square millimeter	area	0.001 550	square inch	in.² or sq in.
square yard	area	0.836 127 4	square meter	m²
stilb	luminance	1.0	candela per square meter	cd/m²
stoke	kinematic viscosity	100	millimeter squared per second	mm²/s
tex	linear density	1.0	milligram per meter	mg/m
therm	energy	105.506	megajoule	MJ